Grundlehren der mathematischen Wissenschaften 159

A Series of Comprehensive Studies in Mathematics

Gottfried Köthe

Topological Vector Spaces I

Translated by D. J. H. Garling

Second printing, revised

Springer-Verlag Berlin · Heidelberg · New York 1983

Prof. Dr. Dr. h.c. Gottfried Köthe

Institut für angewandte Mathematik
der Johann-Wolfgang-Goethe-Universität, Frankfurt am Main

Translation of
Topologische Lineare Räume I, 1966
(Grundlehren der mathematischen Wissenschaften, Vol. 107)

ISBN-13: 978-3-642-64990-5 e-ISBN-13: 978-3-642-64988-2
DOI: 10.1007/978-3-642-64988-2

Softcover reprint of the hardcover 1st edition 1983

Offsetprinting and Bookbinding: Zechnersche Buchdruckerei Speyer
2141/3020-543210

Preface to the First Edition

It is the author's aim to give a systematic account of the most important ideas, methods and results of the theory of topological vector spaces. After a rapid development during the last 15 years, this theory has now achieved a form which makes such an account seem both possible and desirable.

This present first volume begins with the fundamental ideas of general topology. These are of crucial importance for the theory that follows, and so it seems necessary to give a concise account, giving complete proofs. This also has the advantage that the only preliminary knowledge required for reading this book is of classical analysis and set theory. In the second chapter, infinite dimensional linear algebra is considered in comparative detail. As a result, the concept of dual pair and linear topologies on vector spaces over arbitrary fields are introduced in a natural way. It appears to the author to be of interest to follow the theory of these linearly topologised spaces quite far, since this theory can be developed in a way which closely resembles the theory of locally convex spaces. It should however be stressed that this part of chapter two is not needed for the comprehension of the later chapters.

Chapter three is concerned with real and complex topological vector spaces. The classical results of Banach's theory are given here, as are fundamental results about convex sets in infinite dimensional spaces. The subsequent chapters contain a full account of the properties of locally convex spaces. This account is concerned above all with the general theory, but some important classes of spaces, such as for example (F)-spaces, barrelled spaces and bornological spaces, are considered in greater detail. A large number of examples and counterexamples are intended to enable both the scope and the limits of the theory to be seen.

The second volume will contain the theory of linear mappings and the special spaces and classes of spaces which are important in analysis. The theory of Hilbert space will not be dealt with, since there are plenty of excellent textbooks on this topic.

Information about the contents of the book is given in the detailed table of contents at the beginning of the book, and in the short summaries at the beginning of each chapter. No claim for completeness is made for the bibliography at the end of the book, but it should nevertheless be detailed enough to enable further independent work to be done.

My teacher O. Toeplitz provided the first impulse for work on the theme of this book. In § 30, I have endeavoured to give an account of the theory of perfect spaces, which was developed by us together, I have to thank repeated personal contact with my French colleagues J. Dieudonné, A. Grothendieck and L. Schwartz since the war, for detailed knowledge of the theory developed by them; this forms the main subject-matter of this book. The present account is frequently based on the two volumes of Bourbaki (Bourbaki [6] in the bibliography) and on the lectures of Grothendieck [11].

I am particularly indebted to W. Neumer and H. G. Tillmann who have respectively read through the first half, and the whole of the manuscript, carefully and critically. M. Landsberg, H. Schaefer and J. Wloka have made important suggestions and observations.

Finally I thank the publishers for their speedy and excellent printing.

Heidelberg, August 1960 G. Köthe

Preface to the Second Edition

The second edition contains a number of corrections, the need for which was kindly pointed out to me by various readers, together with reference to recent articles in which some of the open problems mentioned in the first edition are solved. Apart from this, the text remains unaltered.

Frankfurt, October 1965 G. KÖTHE

Preface to the English Edition

This English edition is a translation of the second German edition. It differs from the German edition only in several corrections, mainly due to Dr. D. J. H. Garling.

I wish to express my sincere gratitude to Dr. Garling for the excellent and careful translation. I am also indebted to Dr. D. Findley for preparing the index.

Frankfurt, July 1969 G. KÖTHE

Contents

Chapter One

Fundamentals of General Topology

Chapter Two

Vector Spaces over General Fields

Chapter Three

Topological Vector Spaces

Chapter Four

Locally Convex Spaces. Fundamentals

Chapter Five

Topological and Geometrical Properties of Locally Convex Spaces

Chapter Six

Some Special Classes of Locally Convex Spaces

CHAPTER ONE

Fundamentals of General Topology

In this preliminary chapter we gather together those ideas and theorems of general topology which we shall need later. We have also given proofs of the theorems, since an understanding of the methods of topology is essential for the study of vector spaces.

For detailied information one must of course refer to texts on general topology; we mention BOURBAKI [5], KELLEY [2], LEFSCHETZ [1] and SCHUBERT [1]. The account given here follows BOURBAKI closely.

§ 1. Topological spaces

1. The notion of a topological space. A topology $\mathfrak{T}$ is defined on a set R when a class $\mathfrak{O}$ of subsets of R is given, which satisfies the conditions:

(O 1) *R and the empty set are in $\mathfrak{O}$;*

(O 2) *$\mathfrak{O}$ contains with every finite collection of sets their intersection, and with every arbitrary collection of sets their union.*

The sets of $\mathfrak{O}$ are called the open sets of R. A set R with a topology $\mathfrak{T}$ defined on it is called a topological space. The elements of R are called the points of the space.

A subclass $\mathfrak{B}$ of $\mathfrak{O}$ is called a basis of open sets of R if every open set is a union of sets of $\mathfrak{B}$. A subclass of $\mathfrak{O}$ is called a sub-basis when the finite intersections of its sets form a basis.

The topology $\mathfrak{T}$ is determined by a basis or sub-basis of open sets.

The complement $R \sim O$ of an open set O is called a closed set of R. The class $\mathfrak{A}$ of all closed sets of R clearly has the properties:

(A 1) *R and the empty set are in $\mathfrak{A}$;*

(A 2) *$\mathfrak{A}$ contains with every finite collection of sets their union, and with every arbitrary collection of sets their intersection.*

As a result, a topology on a set R can also be defined by giving a class $\mathfrak{A}$ with the properties (A 1) and (A 2). In this case the open sets are the complements of the closed sets. Bases and sub-bases of $\mathfrak{A}$ are defined as above, exchanging the notions of "union" and "intersection". When we speak of a basis of R, however, we shall always mean a basis of open sets of R.

2. Neighbourhoods. A third way of introducing a topology is by giving the collection of all neighbourhoods.

A subset of the topological space R which contains an open set containing the point x is called a neighbourhood of x. Let $\mathfrak{N}(x)$ be the class of all neighbourhoods of x. It is easy to confirm the following properties of $\mathfrak{N}(x)$:

(N 1) *$\mathfrak{N}(x)$ is non-empty and x belongs to each set of $\mathfrak{N}(x)$;*
(N 2) *If a set belongs to $\mathfrak{N}(x)$ then so does every larger subset of R;*
(N 3) *The intersection of a finite collection of sets of $\mathfrak{N}(x)$ lies in $\mathfrak{N}(x)$;*
(N 4) *For every U in $\mathfrak{N}(x)$ there is a V in $\mathfrak{N}(x)$ such that $U \in \mathfrak{N}(y)$ for each y in V.*

For (N 4) we observe that every open neighbourhood V of x contained in U has the required property.

Conversely, suppose that for each x in a set R a non-empty class $\mathfrak{N}(x)$ of subsets of R is given, and that (N 1) to (N 4) are satisfied. If $\mathfrak{N}(x)$ is to be, for each x, the class of all neighbourhoods of x in some topology $\mathfrak{T}$ on R, then the non-empty open sets must be identical with those subsets O of R for which $O \in \mathfrak{N}(x)$ whenever $x \in O$.

The class $\mathfrak{O}$ of all these sets O together with the empty set satisfies (O 1) and (O 2). For the empty set is in $\mathfrak{O}$, and by (N 2) so is R; by (N 3) the intersection of finitely many sets O is again a set of $\mathfrak{O}$, and by (N 2) so is an arbitrary union of sets O. Thus $\mathfrak{O}$ defines a topology $\mathfrak{T}$.

We still have to show that the $\mathfrak{T}$-neighbourhoods of x coincide with the sets of $\mathfrak{N}(x)$. Every $\mathfrak{T}$-neighbourhood of x contains a set O which belongs to $\mathfrak{N}(x)$, so that by (N 2) every $\mathfrak{T}$-neighbourhood of x is in $\mathfrak{N}(x)$. Conversely, suppose that U belongs to $\mathfrak{N}(x)$. We consider the subset U_1 of all y in U for which $U \in \mathfrak{N}(y)$. Since x is in U_1 it is enough to show that U_1 belongs to $\mathfrak{O}$. By (N 4) there is for each y in U a V in $\mathfrak{N}(y)$ for which $U \in \mathfrak{N}(z)$ for each z in V. From the definition of U_1, z lies in U_1, so that $V \subset U_1$ and so, by (N 2), $U_1 \in \mathfrak{N}(y)$. Thus we have shown

(1) *Suppose that for each x in a set R a class $\mathfrak{N}(x)$ of subsets is given, and that* (N 1) *to* (N 4) *are satisfied. Then there is a unique topology on R for which $\mathfrak{N}(x)$ is the class of all neighbourhoods of x, for each x in R.*

Two topologies $\mathfrak{T}$ and $\mathfrak{T}'$ on a set R thus give rise to the same topological space when they give either the same open sets or the same closed sets or the same neighbourhoods of each point.

Two topological spaces R_1 and R_2 are homeomorphic when there is a one-one mapping of the points of R_1 onto the points of R_2 which sends every open set of R_1 into an open set of R_2, and conversely. Such a mapping is called a homeomorphism. Closed sets or the classes $\mathfrak{N}(x)$ can be used instead of open sets in the definition of homeomorphism.

3. Bases of neighbourhoods. If $\mathfrak{N}(x)$ is the class of all neighbourhoods of a point x of the topological space R, a subclass $\mathfrak{B}(x)$ of $\mathfrak{N}(x)$ is called a base of neighbourhoods of x (or fundamental system of neighbourhoods of x) if every neighbourhood in $\mathfrak{N}(x)$ contains one in $\mathfrak{B}(x)$; to put it another way, $\mathfrak{N}(x)$ is obtained from $\mathfrak{B}(x)$ by taking all those subsets of R which contain some set in $\mathfrak{B}(x)$. If a base of neighbourhoods $\mathfrak{B}(x)$ is given for each x, we speak of a base $\mathfrak{B}$ of neighbourhoods in R.

For a base $\mathfrak{B}$ of neighbourhoods consisting solely of open sets HAUSDORFF's three axioms follows easily from (N 1) to (N 4), together with the characterization of open sets given in 2.:

(H 1) *Every point x has at least one neighbourhood in $\mathfrak{B}(x)$, and lies in each of its neighbourhoods;*

(H 2) *The intersection of two neighbourhoods in $\mathfrak{B}(x)$ contains a neighbourhood in $\mathfrak{B}(x)$;*

(H 3) *If y lies in $V\in\mathfrak{B}(x)$, there is a $W\in\mathfrak{B}(y)$ with $W\subset V$.*

Here also the converse holds, that a unique topology is defined by a base $\mathfrak{B}$ of neighbourhoods in R which satisfies (H 1) to (H 3). This follows, since we obtain a class of neighbourhoods satisfying (N 1) to (N 4) by taking as neighbourhoods all subsets of R larger than the given neighbourhoods.

Thus we have a fourth method of introducing a topology on a set R.

Starting from a base of neighbourhoods, a set is open if and only if whenever it contains a point it contains a basic neighbourhood of the point. The open sets of a topological space other than the empty set always form a base of neighbourhoods.

Two base of neighbourhoods $\mathfrak{B}$ and $\mathfrak{B}'$ on the same set R are called equivalent when they define the same topology. This is obviously the case if and only if the classes of all neighbourhoods determined by them are the same. From this, Hausdorff's criterion, which will often be used later, follows immediately:

(1) *Two bases of neighbourhoods $\mathfrak{B}$ and $\mathfrak{B}'$ on the same set R are equivalent if and only if for each x in R every neighbourhood of x in either base always contains a neighbourhood of x in the other base.*

4. Hausdorff spaces. A topological space R is said to be a Hausdorff space, or separated, if it satisfies the fourth of HAUSDORFF's axioms:

(H 4) *Any two distinct points of R possess neighbourhoods in $\mathfrak{B}$ without common points.*

This can also be expressed in the following way

(T_2) *Two distinct points of R always lie in disjoint open sets.*

T_2 is often called the second, or Hausdorff, separation axiom, and Hausdorff spaces are called T_2-spaces (cf. LEFSCHETZ [13]).

If R is a general set, and *all* the subsets of R containing x are taken as neighbourhoods of x, for each x in R, then R, with the discrete topology defined in this way, is a Hausdorff space.

5. Some simple topological ideas. A point x is called an interior point of a subset M of a topological space, if a whole neighbourhood of x lies in M. The collection of all interior points of M forms an open set, the interior of M. A point x is called an exterior point of M if it is an interior point of the complement $R \sim M$. A set $U \supset N$, where N is an open set containing M, is called a neighbourhood of M.

A point x is called a closure point of the set M if every neighbourhood of x contains at least one point of M. The set of all closure points of a set M is called the closure $\bar{M}$ of M. Since the complement of $\bar{M}$ is an open set, $\bar{M}$ is a closed set, and indeed is the intersection of all the closed subsets of R which contain M. Thus a set is closed if and only if it coincides with its closure. In particular $\bar{\bar{M}} = \bar{M}$, for any set M.

A point x is called an accumulation point of the set M if every neighbourhood of x contains at least one point of M distinct from x. A closure point of M fails to be an accumulation point of M if and only if it is an isolated point of M—i.e. a point which has a neighbourhood containing no other point of M. Clearly M is closed if and only if it contains all its accumulation points.

The boundary of a set M is the intersection of the closures of M and $R \sim M$. A boundary point of M is thus a closure point of M and $R \sim M$. Every closed set contains its boundary and every open set is disjoint from its boundary.

The set N is said to be dense in M when $M \subset \bar{N}$, everywhere dense when $\bar{N} = R$, and nowhere dense when $\bar{N}$ has no interior points. The boundary of any open or closed set is nowhere dense.

As an application of these ideas we show

(1) *In a Hausdorff space the intersection of the closed neighbourhoods of a point contains the point alone.*

If x_0 is the given point and y is a point different from x_0 then by (H4) there exist a neighbourhood $U(x_0)$ and a neighbourhood $V(y)$ with $U \cap V$ empty. But then y is an interior point of $R \sim U$, and so an exterior point of U, and thus y does not lie in the closure of U.

From (1) there follows immediately

(2) *The only Hausdorff topology on a finite set is the discrete topology.*

6. Induced topologies and comparison of topologies. Connectedness. If S is a subset of the topological space R, the topology $\mathfrak{T}$ of R induces a topology on S when the sets $S \cap O$, O open in R, are taken as open sets in S. The induced topology is also obtained by considering the intersections of the closed sets with S or the intersections with S of neighbourhoods of the points of S. Notice that a set which is open or closed in S need not be so in R.

If R is Hausdorff, then so is S in the induced topology, which in general we shall denote by $\mathfrak{T}$ again.

If two topologies $\mathfrak{T}_1$ and $\mathfrak{T}_2$ are defined on a set R, $\mathfrak{T}_1$ is said to be finer (or stronger) than $\mathfrak{T}_2$ if every $\mathfrak{T}_2$-neighbourhood is also a $\mathfrak{T}_1$-neighbourhood; $\mathfrak{T}_2$ is said to be coarser (or weaker) than $\mathfrak{T}_1$. That $\mathfrak{T}_1$ is finer than $\mathfrak{T}_2$ can also be expressed by saying that the class $\mathfrak{O}_1$ of $\mathfrak{T}_1$-open sets includes the class $\mathfrak{O}_2$ of $\mathfrak{T}_2$ open sets. The same holds for the classes of closed sets; for this reason we write $\mathfrak{T}_1 \supset \mathfrak{T}_2$. If $\mathfrak{T}_1$ is finer than $\mathfrak{T}_2$, every $\mathfrak{T}_1$-closure point of a set M is also a $\mathfrak{T}_2$-closure point, but not conversely; in general a smaller set is obtained by forming the $\mathfrak{T}_1$-closure than by forming the $\mathfrak{T}_2$-closure.

If finitely or infinitely many topologies $\mathfrak{T}_\alpha$ are defined on a set R, there is a finest topology $\mathfrak{T}$ among the topologies on R which are coarser than every $\mathfrak{T}_\alpha$: the $\mathfrak{T}$-neighbourhoods of a point x are those sets which are $\mathfrak{T}_\alpha$-neighbourhoods of x for each α. If $\mathfrak{O}_\alpha$ are the classes of $\mathfrak{T}_\alpha$-open-sets, then the class $\mathfrak{O}$ of $\mathfrak{T}$-open sets satisfies $\mathfrak{O} = \bigcap_\alpha \mathfrak{O}_\alpha$, since $\mathfrak{O} \subseteq \mathfrak{O}_\alpha$ for each α. $\mathfrak{T}$ is called the intersection of the $\mathfrak{T}_\alpha$.

$\mathfrak{T}$ need not be Hausdorff, even if the $\mathfrak{T}_\alpha$ are Hausdorff. If $\bigcap_\alpha \mathfrak{O}_\alpha$ consists only of the empty set and R, then $\mathfrak{T}$ is the trivial topology, in which every point has the single neighbourhood R.

Similarly, given $\mathfrak{T}_\alpha$, there is a coarsest topology $\mathfrak{T}$ among the topologies which are finer than every $\mathfrak{T}_\alpha$. This is called the union of the $\mathfrak{T}_\alpha$. A base of $\mathfrak{T}$-neighbourhoods of x in R is formed by the $\mathfrak{T}_\alpha$-neighbourhoods of x, for all α, together with their finite intersections. The discrete topology (4.) can occur here as an extreme case. The union of Hausdorff $\mathfrak{T}_\alpha$ is again Hausdorff.

A topological space R is called connected when it is not the union of two non-empty disjoint open sets. This is equivalent to saying that R is not the union of two non-empty disjoint closed sets, or that R contains no proper non-empty sets which are both open and closed.

A subset S of R is said to be connected, when S is connected as a topological space with the induced topology.

R is connected whenever every two points of R lie in a connected subset. If R could be divided into two non-empty open sets R_1 and R_2, than any subset S containing two points from R_1 and R_2 would divide into two non-empty open subsets $R_1 \cap S$ and $R_2 \cap S$ of S.

We denote by P the field of real numbers, and by P^n real n-dimensional space in the natural (Euclidean) topology.

Since the straight line joining two points of P^n is connected, n-dimensional space P^n is also connected.

A topological space is called totally disconnected when it has no connected subsets other than the one-point sets. Every discrete space is totally disconnected, but not conversely; for example the rational numbers form a totally disconnected space in the topology induced by the natural topology of P.

7. Continuous mappings. Let A be a mapping from the topological space R_1 into the topological space R_2—that is, an assignment which sends each $x \in R_1$ to $Ax \in R_2$. We also speak of a function on R_1 with values in R_2, although this term will generally only be used when R_2 is a space of numbers.

Every such point-mapping gives rise to a mapping from the class of subsets of R_1 into the class of subsets of R_2, which will again be denoted by A. In detail, if M is a subset of R_1, the set of all Ax, $x \in M$, forms a subset $A(M)$ of R_2, which is called the image set or image of M. In particular $A(R_1)$ is called the image space of A.

If A is one-one and $y = Ax$, the correspondence $A^{(-1)}y = x$ defines a one-one mapping $A^{(-1)}$ from $A(R_1)$ onto R_1. We call $A^{(-1)}$ the inverse of A. This also gives rise to a mapping $A^{(-1)}$ from the class of subsets of $A(R_1)$ onto the class of subsets of R_1.

If A is not one-one, the point-mapping has no inverse. To every subset N of $A(R_1)$, however, we can make correspond its inverse image $A^{(-1)}(N)$, the set of all x in R_1 with $Ax \in N$. In this way, for every mapping A the inverse $A^{(-1)}$ is defined as a mapping of the class of subsets of $A(R_1)$ into the class of subsets of R_1.

If M is a general subset of R_2, by $A^{(-1)}(M)$ we shall always mean $A^{(-1)}(M \cap A(R_1))$.

If a mapping A from R_1 into R_2 maps every open (respectively closed) subset of R_1 into a set which is open (respectively closed) in $A(R_1)$ (but not necessarily open or closed in R_2!), then A is called an open (respectively closed) mapping. In the same way the inverse $A^{(-1)}$ is said to be open (closed) when every open (closed) subset of $A(R_1)$ has an open (closed) inverse image.

A mapping A from R_1 into R_2 is said to be continuous at x_0, when for each neighbourhood V of Ax_0 in R_2 there exists a neighbourhood U of x_0 in R_1 whose image lies entirely in V. Clearly we can restrict attention to neighbourhoods lying in a given base of neighbourhoods. If A is continuous at every point x of R_1, A is said to be continuous (on R_1).

(1) *The following properties of A are equivalent: a) A is continuous, b) $A^{(-1)}$ is open, c) $A^{(-1)}$ is closed.*

Proof. If $A^{(-1)}$ is open, then for a given open neighbourhood V of Ax_0 the set $A^{(-1)}(V)$ is open, and so is a neighbourhood of x_0, whose image under the mapping A is contained in V.

If conversely A is continuous and M is a subset of $A(R_1)$ which is open in $A(R_1)$, then if $Ax_0 \in M$ there is a whole neighbourhood contained in M. The image of this neighbourhood under the mapping $A^{(-1)}$ contains a neighbourhood of x_0, so that $A^{(-1)}(M)$ is open in R_1.

Since $A^{(-1)}$ maps $A(R_1) \sim M$ into $R_1 \sim A^{(-1)}(M)$, $A^{(-1)}$ is open if and only if it is closed.

If the continuous mapping A is one-one, $A^{(-1)}$ need not be continuous.

(2) *A one-one mapping A from R_1 onto R_2 is a homeomorphism if and only if A and $A^{(-1)}$ are continuous, and if and only if A and $A^{(-1)}$ are open (closed).*

Under a homeomorphism there is a one-one correspondence between the neighbourhoods, the open sets and the closed sets of R_1 and those of R_2. On the other hand, it follows by (1) from the continuity of A and $A^{(-1)}$ that $A^{(-1)}$ and A are open—i.e. the open sets of R_1 and R_2 are in one-one correspondence, and A is a homeomorphism.

(3) *A mapping A is continuous if and only if it sends every closure point of a set into a closure point of the image set.*

Proof. It follows directly from the definition of continuity that a closure point is sent into a closure point of the image set. We remark that the image of an accumulation point need not be an accumulation point.

The other part of (3) does however hold for accumulation points: if A is not continuous, there exists at least one neighbourhood V of a point Ax_0 for which points x_U can be found in every neighbourhood U of x_0, whose images do not belong to V. The set of these x_U has x_0 as accumulation point, and Ax_0 is not a accumulation point of the set of images Ax_U.

The composition of finitely many continuous mappings is always a continuous mapping.

8. Topological products. If $R_1, \ldots, R_n$ are given sets, the set R of all n-tuples $x = (x_1, \ldots, x_n)$, $x_i \in R_i$, is denoted by $R_1 \times \cdots \times R_n = \prod_{i=1}^{n} R_i$. If the R_i are topological spaces, R becomes the topological product of the R_i when the class of all sets $U = \prod_{i=1}^{n} U_i(x_i)$ is taken as base of neighbourhoods of x, where $U_i(x_i)$ is a general neighbourhood of x_i in R_i. We use the expression $\prod_{i=1}^{n} R_i$ for these topological spaces as well.

These definitions can be extended to arbitrarily many factors. If R_α, $\alpha \in \mathsf{A}$, are given, $R = \prod_{\alpha\in\mathsf{A}} R_\alpha$ denotes the set of all functions $x(\alpha) = x_\alpha \in R_\alpha$. R is called the set-theoretic product or Cartesian product of the R_α. If the R_α are topological spaces, R becomes the topological product of the R_α when all subsets $U = \prod_{\alpha\in\mathsf{A}} W_\alpha$ are taken as base of neighbourhoods of the point x, where $W_\alpha = R_\alpha$ for all but finitely many α, and $W_\beta = U_\beta(x_\beta)$ for the others, where $U_\beta(x_\beta)$ is a general neighbourhood of x_β in R_β.

If $R_\alpha = S$ for all α, we write S^{A} for the topological product; in particular we write S^n when there are n equal factors and S^ω when there are countably many. Thus if P is the set of real numbers with its natural topology, P^n is n-dimensional space. P^ω is the space of all sequences, with the topology which has just been defined.

By the parallelotope $\mathfrak{P}^{\mathsf{A}}$ we mean the topological product S^{A}, where S is the closed interval $[0,1]$.

If the R_α are Hausdorff, $\prod_{\alpha\in\mathsf{A}} R_\alpha$ is also Hausdorff.

The mapping P_α, which sends each $x \in R$ to its component $x_\alpha \in R_\alpha$, is called the projection of R onto R_α. It is a continuous mapping from R onto R_α, and *the topology of R is the coarsest topology for which all the projections P_α are continuous.* For if P_α is continuous, the inverse image of $U_\alpha(x_\alpha)$ must be a neighbourhood of x in R, by 7.(1). By taking finite intersections of these inverse images we obtain all the neighbourhoods in the given base of neighbourhoods.

The projection P_α is open, since an open set contains a neighbourhood $\prod W_\alpha$ of each point, the projection contains the neighbourhood W_α of the image, and so the α-components of the points of an open set form an open set. P_α need not be closed, as is shown for example by considering the closed set of all $(n, 1/n)$, $n = 1, 2, \ldots$, in P^2.

(1) *If M_α are subsets of R_α, the closure of the set $\prod_{\alpha\in\mathsf{A}} M_\alpha$ is $\prod_{\alpha\in\mathsf{A}} \bar{M}_\alpha$, where $\bar{M}_\alpha$ is the closure of M_α in R_α.*

Proof. It is immediately clear that at least one element of $\prod M_\alpha$ lies in each of the neighbourhoods belonging to the given base of neighbourhoods of $x \in \prod \bar{M}_\alpha$, and conversely only elements of $\prod \bar{M}_\alpha$ can be closure points.

If the M_α are all closed, so is $\prod M_\alpha$. If the M_α are open, $\prod M_\alpha$ need not be open, when A has infinitely many elements.

If A is a mapping from the topological product $R_1 \times R_2$ into the topological space S which is continuous at the point $(x_1^{(0)}, x_2^{(0)})$, the mapping $x_1 \to A(x_1, x_2^{(0)})$, from R_1 into S is continuous at the point $x_1^{(0)}$ (partial continuity).

§ 2. Nets and filters

1. Partially ordered and directed sets. A set H is said to be partially ordered or semi-ordered if a relation $x \leqq y$ (x less than or equal to y) is defined for certain pairs of its elements, which is reflexive ($x \leqq x$), transitive ($x \leqq y$ and $y \leqq z$ imply $x \leqq z$) and antisymmetric ($x \leqq y$ and $y \leqq x$ imply $x = y$).

For $x \leqq y$ we also write $y \geqq x$; $x < y$ means $x \leqq y$ and $x \neq y$.

A partially ordered set is called totally ordered or simply ordered if one of the relations $x \leqq y$ or $y \leqq x$ always holds for any two of its elements x and y.

A partially ordered set H is called a directed set when for any two elements x and y there always exists $z \in H$ for which $x \leqq z$ and $y \leqq z$. H is said to be inversely directed when for any two elements x and y there always exists $z \in H$ for which $z \leqq x$ and $z \leqq y$.

Every totally ordered set is both directed and inversely directed.

If x is a point of a topological space, the neighbourhoods of a base of neighbourhoods of x form a directed set under the set-theoretic relation $\supset$.

Let M be a subset of the partially ordered set H. It is partially ordered under $\leqq$. M is said to be bounded above (bounded below) if there exists $y \in H$ for which $x \leqq y$ ($y \leqq x$) for all $x \in M$. y is called an upper (lower) bound of M. Every finite subset of a directed set is bounded above.

If the set of upper (lower) bounds of M has a least (greatest) element y_0, y_0 is called the least upper bound (greatest lower bound) of M.

M is called a maximal (minimal) element of M if there exists no x in M with $z < x$ ($x < z$). A least element of M is always minimal, but not conversely.

2. ZORN's Lemma. A totally ordered set is said to be well-ordered if each of its non-empty sets has a least element.

We shall take the results of classical set theory for granted (cf. HAUSDORFF [2] and KAMKE [1], for example). In particular, we shall assume the validity of the axiom of choice. Every set can then be well-ordered, using the ordinals as index-set. We shall also make occasional use of transfinite induction and the theory of cardinal numbers.

As an example of transfinite induction, we mention ZORN's lemma

(1) *If every totally ordered subset of a partially ordered set H has an upper bound, H has at least one maximal element.*

Proof. Let x_α, $\alpha = 0, 1, \ldots$, be a well-ordering of the elements of H. We determine a totally ordered subset G of H by transfinite induction: x_0 belongs to G; if it has been determined which x_β belongs to G, for all $\beta<\gamma$, then x_γ belongs to G if and only if $x_\beta<x_\gamma$ for all $x_\beta\in G$. By hypothesis G has an upper bound z. Since z is an x_α, z must lie in G and so must be the largest element of G and a maximal element of H.

ZORN's lemma is so general that most applications of the well-ordering principle are special cases of this result, so that we shall not need to make repeated use of the well-ordering principle.

As a special case, for subsets of a set M partially ordered by $\subset$ we have

(2) *If H is a class of subsets of a set M with the property that if a collection of subsets in H is totally ordered by $\subset$, then its union also belongs to H, then H has at least one maximal subset.*

Let us remark that ZORN's lemma can be deduced diretly from the axiom of choice, without the aid of the well-ordering principle (cf. KAMKE [1], for example). In fact the axiom of choice, the well-ordering principles and ZORN's lemma are equivalent assumptions (cf. BIRKHOFF [3] or HERMES [1] as well).

3. Nets in topological spaces. If x_n is a sequence of points in a topological space R, x_n is said to be convergent to $x_0\in R$ if for each neighbourhood U of x_0 there exists $n_0(U)$ such that $x_n\in U$ for all $n\geqq n_0(U)$. x_0 is called limit of the sequence x_n, and we write $x_n\to x_0$. If R is Hausdorff, a sequence can only have one limit. In this case we write $x_0=\lim_n x_n$.

In a general topological space R an accumulation point of a set M need not be limit of a sequence of points of M. The parallelotope $\mathfrak{P}^{\mathsf{A}}$, with uncountable A, gives an example of this (cf. § 3, 4.). By using directed sets, however, the concept of limit is generalized in such a way that every accumulation point becomes a limit.

Let A be a directed set, and let M be a general set. If for each $\alpha\in\mathsf{A}$ an $x_\alpha\in M$ is given, the x_α form a net in M. When $\mathsf{A}=1,2,\dots$ we obtain sequences as special cases of nets.

The net x_α, $\alpha\in\mathsf{A}$, is said to be convergent to $x_0\in R$ if for each neighbourhood U of x_0 there exists $\beta(U)\in\mathsf{A}$ such that $x_\gamma\in U$ for all $\gamma\geqq\beta(U)$. x_0 is called limit of x_α, and we write $x_\alpha\to x_0$.

Again, the limit of a net in a Hausdorff space must be unique, and this property characterizes Hausdorff spaces (proof!). In this case we also write $x_0=\lim_\alpha x_\alpha$.

The concept of subsequence can also be generalized: a subset B of a directed set A is cofinal if for each $\alpha\in\mathsf{A}$ there exists $\beta\in\mathsf{B}$ with $\beta\geqq\alpha$. If a subset of a directed set is not cofinal, then its complement is. If x_α, $\alpha\in\mathsf{A}$, is a net, the x_β form a cofinal subnet if the β form a cofinal subset of A.

If x_0 is limit of the net x_α, x_0 is also limit of every cofinal subnet. A point y_0 is called an adherent point of the net x_α if every neighbourhood of y_0 contains a cofinal subnet. Every adherent point of the net x_α is a closure point of the subset of R consisting of the distinct x_α, but not conversely. Every limit of x_α is an adherent point. From this follows one half of

(1) *A subset M of a topological space R is closed if and only if it contains the limits of all the convergent nets of elements of M.*

On the other hand if a is a closure point of M, an $x_U \in M$ can be picked out of every neighbourhood U of a base of neighbourhoods of a. The sets U form a directed system under $\supset$, by 1. The net x_U clearly has limit a.

(2) *A mapping A is continuous at x_0 if and only if $x_\alpha \to x_0$ always implies $A x_\alpha \to A x_0$.*

Necessity follows immediately from the definition of continuity of A and the definition of convergence of a net.

Conversely, suppose that A is not continuous at x_0, and that $\mathfrak{U} = \{U\}$ is a base of neighbourhoods of x_0. Then there exists a neighbourhood V of $A x_0$ and an x_U in each $U \in \mathfrak{U}$ such that $A x_U$ does not lie in V. But then the x_U form a net with $x_U \to x_0$, and the image net $A x_U$ does not converge to $A x_0$.

We remark that we obtain a proof of continuity at x_0 using nets x_U, where U runs through a fixed base of neighbourhoods of x_0. An analogous remark applies to (1).

If $A x_n \to A x_0$ for all sequences $x_n \to x_0$, we call A sequentially continuous at x_0. Sequential continuity of A does not in general imply continuity of A.

4. Filters. Closely related to the concept of net is the concept of filter.

A non-empty class $\mathfrak{F} = \{F_\alpha\}$ of subsets of a set M is called a filter on M if

(F 1) *Every subset of M containing an F_α belongs to $\mathfrak{F}$;*
(F 2) *The intersection of finitely many F_α belongs to $\mathfrak{F}$;*
(F 3) *The empty set does not belong to $\mathfrak{F}$.*

The class of all subsets of the natural numbers with finite complements forms a filter.

More generally a filter on M is obtained from a net x_α, $\alpha \in \mathsf{A}$, $x_\alpha \in M$, by forming the set F_α of all distinct x_β with $\beta \geqq \alpha$, for each α, and collecting these sets and the subsets of M containing them into a class $\mathfrak{F}$.

We call this filter the filter corresponding to the net x_α. Conversely the sets F_α of a filter $\mathfrak{F}$ form a directed set under $\supset$; moreover we obtain a net x_α if we choose an x_α from each F_α, and order the α by the partial order of the F_α which has just been described. Nets formed in this way are called the nets corresponding to a filter.

An especially important example is given by the filter which consists of all the neighbourhoods of a point x_0 of a topological space, the neighbourhood filter of x_0.

A non-empty subclass $\mathfrak{B}$ of a filter $\mathfrak{F}$ on M is called a filter-base of $\mathfrak{F}$ if it satisfies the conditions

(B 1) *The intersection of two sets of* $\mathfrak{B}$ *contains a set of* $\mathfrak{B}$;

(B 2) *The empty set does not belong to* $\mathfrak{B}$,

and if $\mathfrak{F}$ consists of all those subsets of M which contain a set of $\mathfrak{B}$.

Conversely, given a collection $\mathfrak{B}$ of subsets of a set M which satisfies (B 1) and (B 2), a filter is obtained by taking all the larger sets. Because of this, such a collection is called a filter-base.

In this context a base of neighbourhoods of a point is nothing else than a filter-base of the filter of all neighbourhoods of the point.

Hausdorff's criterion in § 1, 3. generalizes to

(1) *Two filter-bases* $\mathfrak{B}$ *and* $\mathfrak{B}'$ *define the same filter if and only if every set of either base contains a set of the other base.*

Such bases are called equivalent.

A collection $\mathfrak{S}$ of subsets of a set M, for which every finite collection of sets has a non-empty intersection, gives rise to a filter-base, by taking all these finite intersections. Such a collection is called a sub-base of the filter which it determines.

If $\mathfrak{F}$ and $\mathfrak{F}'$ are two filters on the same set M, and if $\mathfrak{F} \subset \mathfrak{F}'$, that is if $\mathfrak{F}$ is a subclass of $\mathfrak{F}'$, then $\mathfrak{F}$ is said to be coarser than $\mathfrak{F}'$, and $\mathfrak{F}'$ finer than $\mathfrak{F}$. If $\mathfrak{T}'$ is a finer topology on M than $\mathfrak{T}$, then the neighbourhood filter of a point x_0 relative to $\mathfrak{T}'$ is finer than that relative to $\mathfrak{T}$.

If $M \subset N$ and $\mathfrak{F} = \{F_\alpha\}$ is a filter on M, then the F_α form the base of a filter on N which we shall in general denote by $\mathfrak{F}$ again. Conversely if $\mathfrak{F} = \{F_\alpha\}$ is a filter on N, then, provided that all of them are non-empty, the sets $F_\alpha \cap M$ form a filter on M, the restriction of $\mathfrak{F}$ to M.

5. Filters in topological spaces. Following the pattern of 3. we make the following definition. A filter $\mathfrak{F} = \{F_\alpha\}$ on a topological space R converges to x_0 if there exists an $F_\beta \subset U$ for every neighbourhood U of x_0. x_0 is called limit of the filter, and we write $F_\alpha \to x_0$. If R is Hausdorff, there is at most one limit, and then we write $\lim \mathfrak{F} = x_0$ or $\lim_\alpha F_\alpha = x_0$. If $\mathfrak{F}$ converges to x_0, so does every finer filter $\mathfrak{F}'$.

If the filter is given by a base $\{B_\alpha\}$, the condition for convergence reads: every $U(x_0)$ must contain a B_α.

A point x_0 which is a closure point of all the F_α is called an adherent point of the filter. For this it is sufficient for x_0 to be a closure point of all the sets of a base of the filter. If $\mathfrak{F}$ has limit x_0 in a Hausdorff space, x_0 is the unique adherent point of $\mathfrak{F}$. The converse is not true, as the filter on P with base $F_n=\{0\}\cup[n,\infty)$, $n=1,2,\ldots$, shows.

(1) *If x_0 is an adherent point of $\mathfrak{F}=\{F_\alpha\}$ and $\{U_\beta\}$ is the neighbourhood filter of x_0, then all the $F_\alpha\cap U_\beta$ form a filter convergent to x_0, which is finer than $\mathfrak{F}$.*

The following relations between these ideas and the corresponding ones for nets are easily established:

(2) *The filter $\{F_\alpha\}$ has x_0 as limit if and only if $x_\alpha\to x_0$ for every net corresponding to it. The net x_α has limit x_0 if and only if the corresponding filter has limit x_0.*

(3) *If x_α is a net corresponding to $\{F_\alpha\}$, the filter corresponding to the net x_α is finer than $\{F_\alpha\}$, and has exactly the same adherent points as x_α. The adherent points of the nets corresponding to $\{F_\alpha\}$ are thus adherent points of $\{F_\alpha\}$.*

We remark that conversely an adherent point of $\{F_\alpha\}$ need not always be an adherent point of every corresponding net.

Theorems (2) and (3) also hold for filter-bases $\{B_\beta\}$ and the corresponding nets $x_\beta\in B_\beta$.

If M is a subset of R and x_0 is a closure point of M, consideration of the filter $\{M\cap U_\beta\}$, where $\{U_\beta\}$ is the neighbourhood filter of x_0 in R, shows the following, the direct analogue of 3.(1):

(4) *A subset M of a topological space R is closed if and only if it contains all the limits in R of filters on M.*

If A is a mapping of the set M_1 into the set M_2 and $\mathfrak{F}=\{F_\alpha\}$ is a filter on M_1, then since $A(F_\alpha)$ is non-empty and $A(F_\alpha\cap F_\beta)\subset A(F_\alpha)\cap A(F_\beta)$ the image sets $A(F_\alpha)$ form the base of a filter on $A(M_1)$, and also on M_2, which we call the image-filter $A(\mathfrak{F})$. The images of a base also form a base of the image-filter. If $\mathfrak{G}=\{G_\beta\}$ is a filter on $A(M_1)$, the sets $A^{(-1)}(G_\beta)$ generate a filter on M_1, which we call the inverse-image filter $A^{(-1)}(\mathfrak{G})$.

From 3.(2) and (2), or directly, we have

(5) *A mapping A from a topological space R_1 into a topological space R_2 is continuous at x_0 if and only if $F_\alpha\to x_0$ always implies that $AF_\alpha\to Ax_0$.*

6. Nets and filters in topological products. Let $R=\prod_{\beta\in B} R_\beta$ be a topological product.

(1) *A net $x^{(\alpha)}\in R$ has $x^{(0)}$ as limit if and only if $x_\beta^{(\alpha)}\to x_\beta^{(0)}$, for each β.*

Necessity follows from the continuity of the projections P_β of R onto R_β (§ 1, 8.) and 3.(2); sufficiency results from the fact that a neighbourhood $\prod_\beta W_\beta$ of $x^{(0)}$ has only finitely many $W_\beta \neq R_\beta$, so that an index γ exists for which $x_\beta^{(\delta)} \in W_\beta(x_\beta^{(0)})$, for all β and all $\delta \geqq \gamma$; i.e. $x^{(\delta)} \in \prod_\beta W_\beta$.

If A is a mapping of the topological space S into the product R, then $P_\beta A x = A_\beta x$ is a mapping of S into R_β. From (1), we have

(2) *The mapping A of a topological space S into $R = \prod R_\beta$ is continuous at x_0 if and only if all the A_β are continuous at x_0.*

Let $\mathfrak{F} = \{F^\alpha\}$ be a filter on $R = \prod R_\beta$. The projections of the elements of F^α on R_β form a set F_β^α. The filter $P_\beta(\mathfrak{F}) = \mathfrak{F}_\beta$ with sets F_β^α is called the projection of the filter $\mathfrak{F}$ on R_β.

(3) *The filter $\mathfrak{F}$ converges to $x^{(0)}$ on R if and only if every projection $\mathfrak{F}_\beta$ converges to $x_\beta^{(0)}$.*

By 5.(2), this is just another wording of (1).

If a filter $\mathfrak{F}_\beta$ is given on each R_β, the product-filter $\prod \mathfrak{F}_\beta$ is defined as the filter $\mathfrak{F}$ on $\prod R_\beta$ which has as base all sets $\prod A_\beta$, where $A_\beta = R_\beta$ for all but finitely many β, and A_β is an arbitrary set of $\mathfrak{F}_\beta$ for finitely many β.

The product $\prod \mathfrak{B}_\beta$ of filter-bases of F_β is likewise a base for $\prod \mathfrak{F}_\beta$. In this setting, the neighbourhood filter of a point x of $\prod R_\beta$ is the product of the neighbourhood filters of its components.

7. Ultrafilters. The filters on M form a partially ordered set under $\subset$. If a collection $\mathfrak{F}_\alpha$ of filters is given, $\bigcap_\alpha \mathfrak{F}_\alpha$ is again a filter, since (F_1), (F_2) and (F_3) are satisfied and M always belongs to it. $\bigcap_\alpha \mathfrak{F}_\alpha$ is the greatest lower bound of the filters $\mathfrak{F}_\alpha$. The union $\bigcup_\alpha \mathfrak{F}_\alpha$ forms a filter (the least upper bound of the $\mathfrak{F}_\alpha$) if and only if the intersection of finitely many sets from distinct $\mathfrak{F}_\alpha$ is never empty.

If $\{\mathfrak{F}_\alpha\}$ is a totally ordered collection of filters on M, the least upper bound $\bigcup_\alpha \mathfrak{F}_\alpha$ exists. Using ZORN's lemma [2.(1)] we consequently have

(1) *For every filter $\mathfrak{F}$ on M there exists a finer maximal filter, a so-called ultrafilter.*

(2) *A filter $\mathfrak{F}$ is an ultrafilter on M if and only if the following holds for any two subsets A and B of M: if $A \cup B \in \mathfrak{F}$, then $\mathfrak{F}$ contains at least one of the two sets A and B.*

Proof. If $\mathfrak{F}$ does not contain A and B, all the subsets N of M with $N \cup A \in \mathfrak{F}$ form a filter, since $(N_1 \cap N_2) \cup A = (N_1 \cup A) \cap (N_2 \cup A) \neq A$, so that $N_1 \cap N_2$ is non-empty. This filter is then finer than $\mathfrak{F}$, since the set B has been added to the sets of $\mathfrak{F}$. On the other hand, if the condition is satisfied $\mathfrak{F}$ contains at least one of every pair of sets A and $M \sim A$.

If there were a finer filter, then it would have to contain some A and its complement $M \sim A$, and so would also contain the empty set $A \cap (M \sim A)$.

(3) *The image of an ultrafilter is again an ultrafilter.*

Let $\mathfrak{F} = \{F^\beta\}$ be an ultrafilter on M, where A maps M into N. If $A(F^\beta)$ were not the base of an ultrafilter on N, then by 4.(1) there would be a finer filter $\mathfrak{G} = \{G^\gamma\}$ on N, with at least one $G^{\gamma_0} \subset A(M)$ containing no $A(F^\beta)$. The filter defined by $A^{(-1)}(G^{\gamma_0})$ and $\mathfrak{F}$ on M would then be finer than $\mathfrak{F}$, since $A^{(-1)}(G^{\gamma_0})$ could contain no F^β.

If $\mathfrak{F}$ is an ultrafilter on $M \subset N$, then by (3) $\mathfrak{F}$ defines an ultrafilter on N. Equally, by (2) the restriction to M of an ultrafilter on N is again an ultrafilter; the restriction exists if and only if M belongs to the ultrafilter.

(4) *In a topological space an ultrafilter $\mathfrak{F}$ is either convergent or has no adherent points.*

If $\mathfrak{F}$ has an adherent point x_0, then by 5.(1) there is a finer filter convergent to x_0. Since $\mathfrak{F}$ is maximal, this filter coincides with $\mathfrak{F}$.

(5) *A mapping A from a topological space R_1 into a topological space R_2 is continuous at x_0 if and only if for every ultrafilter $\mathfrak{F} = \{F^\beta\}$, $F^\beta \to x_0$ always implies that $A(F^\beta) \to A(x_0)$.*

Proof. By 5.(5) it is enough to show that, given a filter $\mathfrak{G} = \{G^\beta\}$ convergent to x_0 but with an image-filter which is not convergent to $A x_0$, there exists an ultrafilter with the same property.

Since $A(G^\beta)$ does not converge to $A x_0$, there is a neighbourhood V of $A x_0$ for which all the sets $H^\beta = (R_2 \sim V) \cap A(G^\beta)$ are non-empty. But then the sets $M^\beta = A^{(-1)}(H^\beta) \cap G^\beta$ form the base of a filter $\mathfrak{M}$ which is finer than $\mathfrak{G}$. Any ultrafilter $\mathfrak{F}$ which is even finer is convergent to x_0, and its image $A(\mathfrak{F})$ is an ultrafilter which does not converge to $A(x_0)$, since the sets $A(M^\beta) = H^\beta$ of the filter are disjoint from V.

8. Regular spaces. A Hausdorff space R is said to be regular if it satisfies the condition

(R) *The closed neighbourhoods of each point form a base for the neighbourhood filter.*

If $\mathfrak{B}(x)$ is a base of neighbourhoods of x, and U is an arbitrary closed neighbourhood of x, then there exists $V \in \mathfrak{B}(x)$ with $V \subset U$, and so also $\bar{V} \subset U$: *The closures of the neighbourhoods of a base of neighbourhoods of a regular space form a base of neighbourhoods for R.*

Condition (R) is equivalent to the condition

(R′) *If M is closed in R and x does not belong to M, then disjoint neighbourhoods of M and x can always be found.*

The simple proof is left to the reader.

In regular spaces $F_\alpha \to x_0$ always implies that $\bar{F}_\alpha \to x_0$.

(1) *Every subspace S of a regular space R is regular.*

For S is Hausdorff and the intersections with S of the closed neighbourhoods in R of a point x of S are closed in S, and form a base of neighbourhoods in S.

(2) *The topological product of regular spaces is regular.*

In certain circumstances continuous mappings into regular spaces can be extended.

Let M be dense in the topological space R_1. Let A be a continuous mapping from M into the regular space R_2. If x_0 is a point of R_1 which is not in M and $U'_\alpha = U_\alpha \cap M$, where $\{U_\alpha\}$ is the neighbourhood filter of x_0 in R_1, then A can be extended continuously to x_0 only if $A(U'_\alpha)$ converges to a point of R_2, which we define to be the image $A x_0$.

This requirement is stronger than that of continuity of A on M. It can also be expressed in terms of nets: if $x_0 \in R_1$, then for all nets $x_\alpha \to x_0$, with $x_\alpha \in M$, the image nets $A x_\alpha$ must all converge to one and the same element of R_2.

We still have to show that the mapping which has now been defined on the whole of R_1 is continuous at each point. Let V be a neighbourhood of $A x_0$. Because R_2 is assumed to be regular we can suppose that V is closed. There exists an open neighbourhood U of x_0 with $A(U \cap M) \subset V$. If y is an arbitrary point of U, there is a net y_α in $U \cap M$ with $y_\alpha \to y$, so that $A y$ is limit of the net $A y_\alpha$, which lies in $A(U \cap M) \subset V$. Thus $A y$ is in $\bar{V} = V$, and $A(U) \subset V$. Thus we have established

(3) *A continuous mapping A from a dense subset M of a topological space R_1 into a regular space R_2 can be continuously extended to the whole of R_1 provided that for each point $x \in R_1$ with neighbourhood filter $\{U_\alpha\}$ $A(U_\alpha \cap M)$ always converges.*

The extension is clearly unique.

§ 3. Compact spaces and sets

1. Definition of compact spaces and sets. A Hausdorff space R is said to be compact if every filter on R has at least one adherent point. Since every filter is contained in an ultrafilter, by § 2, 7.(1), we can also say

(1) *R is compact if and only if every ultrafilter on R is convergent.*

(2) *A Hausdorff space R is compact if and only if every collection $\{A_\alpha\}$ of closed subsets A_α with empty intersection $\bigcap_\alpha A_\alpha$ contains a finite collection of subsets with an empty intersection.*

Proof. Suppose that R is compact, that $\bigcap_\alpha A_\alpha$ is empty and that the intersection of finitely many A_α is always non-empty. Then the A_α would form a sub-base of a filter on R which could have no adherent point, as this would belong to all the A_α, since the A_α are closed.

If R is not compact, there is a filter $\{F_\alpha\}$ with no adherent points. The sets $\bar{F}_\alpha$ form a collection of closed subsets with $\bigcap_\alpha \bar{F}_\alpha$ empty, although the intersection of finitely many $\bar{F}_\alpha$ is always non-empty.

Taking complements, we immediately get

(3) *A space R is compact if and only if every cover of R by open sets contains a finite sub-cover.*

A subset M of a Hausdorff space R is said to be compact if it is a compact space in the induced topology. M is thus compact if every filter on M has at least one adherent point in M. Since the closed (respectively open) sets of M are intersections of closed (respectively open) sets of R with M, we also have

(4) *A subset M of a Hausdorff space R is compact if and only if*

a) *every collection $\{A_\alpha\}$ of closed subsets of R with $\bigcap_\alpha (A_\alpha \cap M)$ empty contains a finite collection $\{A_{\alpha_i}\}$ with $\bigcap_\alpha (A_{\alpha_i} \cap M)$ empty,*
or if and only if

b) *every cover of M by open subsets of R contains a finite sub-cover.*

From § 2, 5.(3) there follows immediately

(5) *M is compact if and only if every net on M has an adherent point in M.*

2. Properties of compact sets. (1) *Every compact set is closed.*

If x_0 is a closure point of the compact set $M \subset R$ and if $\{U_\alpha\}$ is the neighbourhood filter of x_0, then $\{U_\alpha \cap M\}$ is a filter on M, which can only have x_0 as adherent point, since R is Hausdorff. By the compactness condition, x_0 must thus lie in M.

A set $M \subset R$ is called relatively compact if its closure $\bar{M}$ is compact. Every filter on M then has an adherent point in $\bar{M} \subset R$.

(2) *Every subset of a compact set M is relatively compact, and every closed subset of M is compact.*

(3) *Every compact space R is regular.*

Were this not the case, there would be a point x_0 in R with an open neighbourhood U, for which $A_\alpha \cap (R \sim U)$ would be non-empty, for every closed neighbourhood A_α of x_0. By § 1, 5.(1), $\bigcap_\alpha A_\alpha = \{x_0\}$, so that, as $x_0 \notin R \sim U$, the closed sets $A_\alpha \cap (R \sim U)$ have an empty intersection.

Since every finite collection of them has a non-empty intersection, we obtain a contradiction to 1.(2).

(4) *The union of finitely many compact sets is compact.*

This follows directly from the covering property 1.(3); if $M=\bigcup_{i=1}^{n} M_i$ is covered by open sets, each M_i is covered by finitely many open sets, and so therefore is M.

(5) *The continuous image $A(R)$ of a compact space R in a Hausdorff space is again compact, so that the mapping A is closed.*

Let A be a continuous mapping of a Hausdorff space R_1 into a Hausdorff space R_2. The image $A(M)$ of a compact (respectively relatively compact) set $M\subset R_1$ is again compact (respectively relatively compact).

If $\mathfrak{G}=\{G^\beta\}$ is a filter on $A(M)$, the sets $A^{(-1)}(G^\beta)$ generate a filter $A^{(-1)}(\mathfrak{G})$ on M. The compactness condition implies the existence of an adherent point x_0 of $A^{(-1)}(\mathfrak{G})$. $A(x_0)$ is then an adherent point of $\mathfrak{G}$, so that $A(M)$ is compact. (1) and (2) establish the other assertions.

As a special case we obtain from § 1, 7.(2)

(6) *A continuous one-one mapping of a compact space onto a Hausdorff space is a homeomorphism.*

If R is compact under the topology $\mathfrak{T}$, every coarser Hausdorff topology coincides with $\mathfrak{T}$.

The second assertion follows from the first by considering the identity mapping of R onto itself.

3. Tychonoff's Theorem. This says

(1) *The topological product $R=\prod_\alpha R_\alpha$ of arbitrarily many compact spaces is again compact.*

Proof. By 1.(1) we must show that every ultrafilter $\mathfrak{F}=\{F^\beta\}$ on R converges. We form the projections $\mathfrak{F}_\alpha$ in each R_α (cf. § 2, 6.). These are again ultrafilters. By hypothesis, each $\mathfrak{F}_\alpha$ converges to an $x_\alpha\in R_\alpha$. By § 2, 6.(3), $\mathfrak{F}$ converges to the element x of R whose components are the x_α.

In particular, by 1.8. every parallelotope $\mathfrak{P}^A$ is compact.

4. Other concepts of compactness. The concept "compact" can be weakened in various ways. If $\aleph_\alpha$ is some cardinal number, a subset M of the Hausdorff space R is called $\aleph_\alpha$-compact if every filter with a base of at most $\aleph_\alpha$ sets of M has an adherent point in M. This is the case if and only if every net x_β of at most $\aleph_\alpha$ elements of M always has an adherent point in M. The proof of this is given by the generalization of § 2, 5.(3) to filter-bases.

The criteria 1.(2) and 1.(3) are also carried over, if the cardinal of the collection of closed sets (respectively of the collection of open sets of the cover) is at most $\aleph_\alpha$.

For $\aleph_0$-compact, we also say countably compact. If M is countably compact, every sequence of points of M has at least one adherent point. On the other hand if this is the case M is countably compact: indeed if $\mathfrak{F}$ is a filter on M with countable base F_i, $i=1,2,\ldots$, then a sequence $x_n \in \bigcap_{i=1}^{n} F_i$ has an adherent point in M, which is also an adherent point of all the F_i.

A set M is called sequentially compact if every sequence of points of M contains a subsequence which is convergent to a point of M. A sequentially compact set is always countably compact. The converse is not true, for there even exist compact sets which are not sequentially compact, and equally there exist sequentially compact sets which are not compact.

In the parallelotope $\mathfrak{P}^{\mathrm{A}}$, A uncountable, the set M of all $x=\{\xi_\alpha\}$ with only countably many coordinates different from zero is sequentially compact, but is not compact, since $\bar{M}=\mathfrak{P}^{\mathrm{A}}$.

If A is the interval $[0,1]$ of the real line, we can consider $\mathfrak{P}^{\mathrm{A}}$ as the set of all functions on $[0,1]$ with values in $[0,1]$. If $f_n(x)$ is the function which goes linearly from 0 to 1 in every sub-interval $[k\cdot 10^{-n},(k+1)10^{-n})$ of $[0,1]$, the sequence $f_n(x)$ has no convergent subsequence, so that $\mathfrak{P}^{\mathrm{A}}$ is compact, but not sequentially compact.

In P^n all these concepts of compactness coincide. The corresponding "relative" concepts are obtained if the adherent points (respectively limits) are only required to lie in R.

We shall later be concerned with the question of finding conditions which ensure that compactness or relative compactness follows from one of these weaker concepts.

We mention one of the properties resulting from these ideas, which follows directly from the definitions

(1) *If $N_1 \supset N_2 \supset \cdots$ is a decreasing sequence of closed non-empty subsets of a countably compact or sequentially compact set M, then $\bigcap_i N_i$ is non-empty.*

5. Axioms of countability. In the following we shall again suppose that R is Hausdorff.

First axiom of countability. *Every neighbourhood filter of a point of R has a countable base.*

The base can then be chosen as a decreasing sequence $U_1 \supset U_2 \supset \cdots$. Every subspace also satisfies the first axiom of countability.

(1) *If R satisfies the first axiom of countability, every closure point of a subset M of R is limit of a convergent sequence of points of M.*

To prove this, form the sets $M \cap U_i$ from the basic neighbourhoods U_i of a closure point x_0, and choose a point x_i from each. A sequence convergent to x_0 is obviously obtained.

In such spaces countably compact sets are sequentially compact. Furthermore it follows from § 1, 7.(3) that sequentially continuous mappings are continuous.

Second axiom of countability. *R has a countable basis.*

For the definition of basis, see § 1, 1. Every space which satisfies the second axiom also satisfies the first. Every subspace again has a countable basis.

(2) *Every open cover of a Hausdorff space with a countable basis contains a countable subcover.*

Proof. Let R be covered by the open sets Q_α and let O_i, $i=1,2,\ldots$, be a basis of the open sets. Let O_{i_n}, $n=1,2,\ldots$, be the collection of those O_i from which all the Q_α can be formed by taking suitable unions. For each O_{i_n} there is thus at least one Q_{α_n} with $O_{i_n} \subset Q_{\alpha_n}$. But then $R \subset \bigcup_\alpha Q_\alpha = \bigcup_n O_{i_n} \subset \bigcup_n Q_{\alpha_n}$.

To know whether a set is compact it is therefore only necessary to examine countable covers by open sets, i.e.

(3) *In a Hausdorff space with a countable basis every countably compact set is compact.*

In such a space the compact, sequentially compact and countably compact sets thus coincide.

In these spaces, moreover, filters and nets are superfluous, and everything can be analysed using sequences of points.

We remark that (3) is not always true for Hausdorff spaces satisfying the first axiom of countability (cf. Bourbaki [5], Vol. 4, p. 32, example 21).

(4) *If the Hausdorff space R has a countable basis $\{U_i\}$, every basis $\{V_\alpha\}$ has a countable subsystem which is again a basis.*

For every U_i is, by (2), union of countably many $V_{\alpha_{ik}}$, $k=1,2,\ldots$, and these, taken together, also form a countable basis.

6. Locally compact spaces. A Hausdorff space is called locally compact if every point has a neighbourhood whose closure is compact. Every compact space is locally compact.

P^n is locally compact but not compact, and every discrete topological space is locally compact.

Every closed subspace of a locally compact space is, by 2.(2), again locally compact.

(1) *Every locally compact space is regular, and the compact neighbourhoods form a base of neighbourhoods.*

Every point x has a compact neighbourhood U. If V is any neighbourhood, $V \cap U$ is a neighbourhood of x in the compact space U, which is regular, by 2.(3). Thus $V \cap U$ contains a closed neighbourhood W of x for the topology of U. W, being the intersection of a neighbourhood in R with U, is itself a neighbourhood of x in R. W is compact, by 2.(2), and so is closed in R. Thus the compact neighbourhoods of x form a base of all the neighbourhoods.

It follows directly from (1) that *every open subspace of a locally compact space is again locally compact.*

The topological product of finitely many locally compact and arbitrarily many compact spaces is again locally compact, by TYCHONOFF's theorem. If x_0 is a point of a compact space R, $R \sim x_0$ is clearly locally compact in the topology induced by R. Conversely

(2) ALEXANDROFF's theorem. *Every space R which is locally compact and not compact can be enlarged by the addition of one point to give a compact space, the one-point compactification of R.*

Proof. Let R' be the space consisting of the points of R and one further point z. We define the closed sets of R' to be all the compact sets K of R together with the sets $A \cup z$, A closed in R. The axioms (A 1) and (A 2) are clearly satisfied, so that R' is a topological space. The subspace topology induced on R is the original one, since the intersections with R of the closed sets of R' which have just been defined are exactly the closed sets of R.

R' is Hausdorff: it is only necessary to show that there are disjoint neighbourhoods of $x \in R$ and z. If U is a compact neighbourhood of x in R, U is closed in R', so that $R' \sim U$ is an open neighbourhood of z which has no point in common with U.

R' is compact: given closed sets B_α with an empty intersection, they cannot all be of the form $A \cup z$, for otherwise z would lie in their intersection. If $B_{\alpha_0} = K_0 \subset R$, then $\bigcap_\alpha (K_0 \cap B_\alpha)$ is also empty. The sets $K_0 \cap B_\alpha$ are closed sets of the compact set K_0, however, so that finitely many of them have an empty intersection.

(3) *The compactification of* (2) *is unique up to homeomorphism.*

It is enough to show that the closed sets of a compactification $\check{R} = R \cup z'$ must coincide with the sets K and $A \cup z'$, K compact in R and A closed in R. $\check{R}$, being compact, is Hausdorff, so that z' is a closed set and the sets K and $A \cup z'$ are closed in $\check{R}$. If conversely $B \supset z'$ is closed, then $B = \overline{(B \sim z')} \cup z'$. Thus $B \sim z'$ has as closure points only elements of $B \sim z'$, and possibly z', and is therefore closed in R.

Since every subspace of a regular space is regular, (1) is also a consequence of (2).

The point z adjoined to R is called the point at infinity.

A space which is locally compact but not compact is said to be countable at infinity if it is the union of countably many compact sets.

(4) *A space which is locally compact but not compact is countable at infinity if and only if the point z at infinity in the compactification R' has a countable base of neighbourhoods.*

Proof. The condition is sufficient, for if V_n is a countable open base of neighbourhoods of z, the countable collection of compact sets $R' \sim V_n$ cover the space R.

Conversely, let $K_1 \subset K_2 \subset \cdots$ be a covering of R by countably many compact sets (by forming finite unions an increasing sequence can always be obtained). Every point of K_1 has a relatively compact open neighbourhood. The union U_1 of finitely many of these neighbourhoods cover K_1. In the same way we find an open relatively compact set U_2 which covers $\bar{U}_1 \cup K_2$, and so on. In this way we obtain a sequence U_n, with $\bar{U}_{n-1} \subset U_n$, which covers R. We now show that the $V_n = R' \sim \bar{U}_n$ form a base of neighbourhoods of z. From the definition of the closed sets of R' it follows that the sets $R' \sim K$, K compact in R, form a base of neighbourhoods of z in R'. It is therefore enough to show that for every compact subset K there exists a U_n with $K \subset U_n$. K, being a compact set, is covered by finitely many U_k, and so by one U_n, for sufficiently large n.

7. Normal spaces. The properties of being a Hausdorff or regular topological space are not sharp enough for many purposes.

A Hausdorff space is said to be normal if it satisfies the condition

(N) *If A_1 and A_2 are two disjoint closed subsets of R, there always exist two disjoint open subsets $U_1 \supset A_1$ and $U_2 \supset A_2$.*

An equivalent condition to (N) is

(N′) *If A is a closed subset of R, and if U is open and $U \supset A$, then there is an open neighbourhood V of A with $\bar{V} \subset U$.*

Proof. Suppose (N′) holds. If A and B are disjoint and closed, $R \sim B$ is an open neighbourhood of A, and so by (N′) there exists $U_1 \supset A$ with $\bar{U}_1 \cap B$ empty. U_1 and $R \sim \bar{U}_1$ are disjoint open neighbourhoods of A and B.

Conversely suppose that (N) holds. Applying (N) to the closed sets A and $R \sim U$, where U is an open neighbourhood of A, open sets $U_1 \supset A$ and $U_2 \supset R \sim U$ are obtained, with $U_1 \cap U_2$ empty. But then $\bar{U}_1 \cap (R \sim U)$ is empty, so that $\bar{U}_1 \subset U$ and (N′) is satisfied.

Taking A to be one point, (N′) gives

(1) *Every normal space is regular.*

Subspaces of normal spaces are not always normal, and there are locally compact spaces which are not normal. On the other hand

(2) *Every compact space is normal.*

The closed subsets of a compact space R are again compact. Let A and B be two disjoint compact subsets of R. Since R is regular, by 2.(3), for each $x \in A$ there exists an open neighbourhood $U(x)$ with $\bar{U}(x) \cap B$ empty. As x runs through the whole of A, the $U(x)$ form an open cover of A, and, by 1.(4), $A \subset \bigcup_{i=1}^{n} U(x_i)$. $A' = \bigcup_{i=1}^{n} \bar{U}(x_i)$ is a closed set disjoint from B. Thus $U_1 = \bigcup_{i=1}^{n} U(x_i)$ and $U_2 = R \sim A'$ are disjoint open sets with $U_1 \supset A$ and $U_2 \supset B$.

§ 4. Metric spaces

1. Definition. A set R is called a metric space if a real number $|x,y|$, the distance between x and y, is defined for every pair x,y of elements of R, with the following properties:

(D1) $|x,y| \geqq 0$,
(D2) $|x,y| = 0$ if and only if $x = y$,
(D3) $|x,y| = |y,x|$,
(D4) $|x,z| \leqq |x,y| + |y,z|$ (triangle inequality).

We also say that a metric is defined on R by the function $|x,y|$.

Every subspace of a metric space is again a metric space, using the same definition of distance.

A one-one mapping $x \to x'$ of a metric space R onto a metric space R' is called isometric if $|x,y| = |x',y'|$ holds for all pairs x,y.

The simplest example is P^n with distance defined by $|x,y| = \sqrt{\sum_{i=1}^{n} |\eta_i - \xi_i|^2}$. This is called n-dimensional Euclidean space.

The set of all real sequences $x = (\xi_1, \xi_2, \ldots)$ with $\sum_{i=1}^{\infty} |\xi_i|^2 < \infty$ forms a metric space with distance defined by $|x,y| = \sqrt{\sum_{i=1}^{\infty} |\eta_i - \xi_i|^2}$. This space is called Hilbert space.

2. Metric space as a topological space. If x_0 is a point of the metric space R, the set of all $y \in R$ with $|x_0, y| < r$ (respectively $|x_0, y| \leqq r$) is called the open (respectively closed) ball of radius r and centre x_0. The set of y with $|x_0, y| = r$ is called the sphere of radius r about x_0.

If A, B are two subsets of R the greatest lower bound $\inf_{x \in A, y \in B} |x, y|$ is called the distance between the two sets. The number $\sup_{x, y \in A} |x, y|$ is called the diameter of the set A. A set A is said to be bounded (with respect to the metric) if its diameter is finite.

(1) *Every metric space becomes a Hausdorff topological space when the open balls with centre x are taken as a base of neighbourhoods of the point x.*

Every metric space is a normal space with a countable base of neighbourhoods at each point.

Proof. Axioms (H1) and (H2) of § 1.3 are clearly satisfied. (H3) results from the following: if y lies in the open ball $K_r(x)$ of radius r about x, then $d = |x, y| < r$, so that there exists $\varepsilon > 0$ with $d + \varepsilon < r$. But then $K_\varepsilon(y)$ lies inside $K_r(x)$, since if $z \in K_\varepsilon(y)$, $|x, z| \leqq |x, y| + |y, z| < d + \varepsilon < r$. Further if $x \neq y$, $d = |x, y| \neq 0$. The open balls $K_{d/2}(x)$ and $K_{d/2}(y)$ have no common point z, for otherwise $|x, y| \leqq |x, z| + |z, y|$ would be less than d. Thus (H4) is satisfied. Axiom (N) for a normal space is satisfied: if A and B are disjoint closed sets, the set U of all x with $|x, A| < \frac{1}{2}|x, B|$ and the set V of all y with $|y, B| < \frac{1}{2}|y, A|$ are disjoint open neighbourhoods of A and B. Finally the balls about x with radii $1/n$, $n = 1, 2, \ldots$, form a countable base of neighbourhoods of x.

Two isometric spaces are homeomorphic, but the converse does not always hold.

3. Continuity in metric spaces. Since a metric space satisfies the first axiom of countability, the results of § 3, 5 apply. A mapping A from a metric space R_1 into a metric space R_2 is therefore continuous if $x^{(n)} \to x_0$, i.e. $|x^{(n)}, x_0| \to 0$, always implies that $A x^{(n)} \to A x_0$. This can clearly also be expressed in terms of ε and δ: A is continuous if for each $\varepsilon > 0$ there exists $\delta > 0$ for which $|x, x_0| < \delta$ always implies that $|A x, A x_0| < \varepsilon$.

Closure points and accumulation points of a set are always limits of convergent sequences of elements of the set, and so a limit of limits of a set is again limit of a sequence of elements of the set.

The metric allows what has not been possible in the topological spaces considered up to now, namely the comparison of the size of neighbourhoods of different points. This produces a series of further concepts.

For example, uniform continuity of a mapping A can be defined as follows: A is uniformly continuous if for every $\varepsilon > 0$ there exists $\delta > 0$ such that $|A x, A y| < \varepsilon$ whenever $|x, y| < \delta$.

The concept of a Cauchy sequence can also be defined: $x^{(n)}$, $n=1,2,\ldots$, is a Cauchy sequence if for each $\varepsilon>0$ there exists $n_0(\varepsilon)$ such that $|x^{(m)},x^{(n)}|<\varepsilon$ whenever $m,n\geqq n_0$.

A metric space is called complete if every Cauchy sequence has a limit in R.

4. Completion of a metric space. Following the method of establishing the real numbers of CANTOR and MÉRAY, we prove the following theorem

(1) *Every metric space can be embedded in a smallest complete space $\tilde{R}$, which is unique up to isometry, and is called the completion of R.*

For every two Cauchy sequences $x=(x^{(n)})$ and $y=(y^{(n)})$, $\lim|x^{(n)},y^{(n)}| = |x,y|$ always exists, for since

$$|x^{(m)},y^{(m)}|\leqq|x^{(m)},x^{(n)}|+|x^{(n)},y^{(n)}|+|y^{(n)},y^{(m)}|$$

we have

$$\big||x^{(m)},y^{(m)}|-|x^{(n)},y^{(n)}|\big|\leqq|x^{(m)},x^{(n)}|+|y^{(n)},y^{(m)}|.$$

If $|x,\hat{x}|=0$ and $|y,\hat{y}|=0$, then $|x,y|=|\hat{x},\hat{y}|$: for from

$$|\hat{x}^{(n)},\hat{y}^{(n)}|\leqq|\hat{x}^{(n)},x^{(n)}|+|x^{(n)},y^{(n)}|+|y^{(n)},\hat{y}^{(n)}|$$

$\lim|\hat{x}^{(n)},\hat{y}^{(n)}|\leqq\lim|x^{(n)},y^{(n)}|$ follows directly, and the interchange of x, y with $\hat{x}$, $\hat{y}$ establishes the assertion.

The relation $|x,y|=0$ defines an equivalence relation on the set of all Cauchy sequences. We denote the class to which x belongs by $\tilde{x}$. By what we have just shown, a unique real number $|\tilde{x},\tilde{y}|=\lim\limits_{n\to\infty}|x^{(n)},y^{(n)}|$ is associated with each pair of classes. Under this the set $\tilde{R}$ of all classes becomes a metric space: (D1), (D2) and (D3) are trivially satisfied, and (D4) follows from $|x^{(n)},z^{(n)}|\leqq|x^{(n)},y^{(n)}|+|y^{(n)},z^{(n)}|$, by taking limits.

If we assign the element $a\in R$ to the class $\tilde{a}$ determined by the Cauchy sequence $(x^{(n)})$ with $x^{(n)}=a$, an isometry is obtained, and we can identify R with a subspace of $\tilde{R}$. R is dense in $\tilde{R}$, for if x is the Cauchy sequence $(x^{(n)})$, $\tilde{x}=\lim\tilde{x}^{(n)}$: indeed $|\tilde{x},\tilde{x}^{(n)}|=\lim\limits_{m\to\infty}|x^{(m)},x^{(n)}|\leqq\varepsilon$ for $n\geqq n_0(\varepsilon)$.

$\tilde{R}$ is complete: let $(\tilde{x}_n)$ be a Cauchy sequence in $\tilde{R}$. For each $\tilde{x}_n$ there is an $a^{(n)}\in R$ with $|\tilde{x}_n,\tilde{a}^{(n)}|<1/n$. Since

$$|a^{(m)},a^{(n)}|\leqq|\tilde{a}^{(m)},\tilde{x}_m|+|\tilde{x}_m,\tilde{x}_n|+|\tilde{x}_n,\tilde{a}^{(n)}|$$

the $a^{(n)}$ form a Cauchy sequence $\tilde{a}$ in R, and from $|\tilde{a},\tilde{x}_n|\leqq|\tilde{a},\tilde{a}^{(n)}|+|\tilde{a}^{(n)},\tilde{x}_n|$ it follows that $\tilde{a}=\lim\tilde{x}_n$.

Since R is dense in $\tilde{R}$ and the distances $|\tilde{x},\tilde{y}|$ are uniquely defined as $\lim|x^{(n)},y^{(n)}|$, they are the same in each extension of R, and $\tilde{R}$ is uniquely determined up to isometry as the completion of R.

5. Separable and compact metric spaces. A topological space is called separable if there is a countable set of points which is dense in it.

(1) *A metric space R is separable if and only if it has a countable basis.*

In one direction the result holds for general topological spaces: if O_i, $i=1,2,\ldots$, is a countable basis and $x_i \in O_i$, the x_i are clearly dense in R. If conversely x_i is a sequence which is dense in the metric space R, we shall show that the open balls $K_\rho(x_i)$ with rational ρ form a basis: it is enough to show that each $x \in R$ has a base of neighbourhoods formed of suitable $K_\rho(x_i)$. But given $K_\rho(x)$ and $|x, x_i| < \rho/2$, $K_{\rho/2}(x_i)$ is a neighbourhood of x contained in $K_\rho(x)$.

A compact metric space is called a compactum. Every compactum is complete, as § 3, 1. (3) shows immediately.

(2) *Every compactum, and equally every relatively compact subset M of a metric space, is separable and has a finite diameter.*

For every n there are finitely many open balls $K_{1/n}(x_i^{(n)})$, $i=1,\ldots,N(n)$, which cover the whole of M. The second part of the proposition follows from this. Further the countably many $x_i^{(n)}$ taken together clearly form a dense set in M.

(3) *Every sequentially compact metric space is a compactum.*

By § 3, 5. (3) it is enough to show that R has a countable basis, and thus that it is separable. For each n there exist finitely many $x_i^{(n)}$ with the property that each x is at a distance of less than $1/n$ from a suitable $x_i^{(n)}$. For otherwise there would be for some $n_0 > 0$ a sequence $x^{(n)}$ with $|x^{(n)}, x^{(m)}| \geqq 1/n_0$ for all n, m, and this sequence would have no convergent subsequence. Again the countable set of all $x_i^{(n)}$ is dense in R.

Applying § 3, 5. (3) again, we see that

(4) *In a metric space the countably compact, sequentially compact and compact sets coincide.*

A metric space R is called precompact if its completion $\tilde{R}$ is compact. This is the case if and only if every infinite subset of R contains a Cauchy sequence; it is easy to show that $\tilde{R}$ has this property whenever R does, and thus is sequentially compact.

A subset M of a metric space is called totally bounded if for every $\varepsilon > 0$ there is a finite cover of M by sets of diameter $\leqq \varepsilon$.

(5) *A subset of a metric space is precompact if and only if it is totally bounded.*

Proof. Every precompact set is totally bounded, since $\tilde{M}$, and so also M, is covered by finitely many sets N_i of diameter $\leqq \varepsilon$.

If conversely M is totally bounded and infinite it is covered by finitely many sets of diameter $\leqq 1$, and so in one of these sets there exist infinitely many $x_i^{(1)} \in M$ with $|x_i^{(1)}, x_k^{(1)}| \leqq 1$ for all i, k. From these in turn we can find infinitely many $x_i^{(2)}$ with $|x_i^{(2)}, x_k^{(2)}| \leqq \frac{1}{2}$, and so on. The diagonal sequence $x_1^{(1)}, x_2^{(2)}, \ldots$ clearly forms a Cauchy sequence.

(6) *If* $N_1 \supset N_2 \supset \cdots$ *is a decreasing sequence of closed sets in a complete metric space, and if their diameters tend to zero, then* $\bigcap_{i=1}^{\infty} N_i$ *consists of one point.*

For every sequence $x_i \in N_i$ is a Cauchy sequence, and all such sequences converge to one and the same point.

6. BAIRE's theorem. We give this in several versions.

(1) *If the complete metric space* R *is the union* $\bigcup_{i=1}^{\infty} M_i$ *of countably many subsets, then at least one* $\bar{M}_i$ *contains an entire ball.*

Were this not the case, each M_i would be nowhere dense in R (cf. § 1, 5), and there would then be a closed ball K_1 of diameter $\leqq 1$ with $K_1 \cap M_1$ empty, a closed ball K_2 in K_1 of diameter $\leqq \frac{1}{2}$ with $K_2 \cap M_2$ empty, and so on. The decreasing sequence $K_1 \supset K_2 \supset \cdots$ would by 5.(6) have one common point, which could not belong to any M_i and so could not belong to R.

This theorem of BAIRE's also holds for locally compact spaces.

(2) *If the locally compact space* $R = \bigcup_{i=1}^{\infty} M_i$, *then at least one* $\bar{M}_i$ *contains a non-empty open set.*

The proof proceeds analogously: were all the M_i nowhere dense in R, then there would be a non-empty open set O_1 with $O_1 \cap M_1$ empty. By § 3, 6.(1) R is regular, so that O_1 would contain a closed subset K_1 with interior points and $K_1 \cap M_1$ would be empty. We could take K_1 to be compact. In K_1 there would be a compact subset K_2 with $K_2 \cap M_2$ empty, and so on. By § 3, 4.(1) the sequence $K_1 \supset K_2 \supset \cdots$ would have a non-empty intersection which again would not lie in $R = \bigcup_{i=1}^{\infty} M_i$.

(1) and (2) can be extended to

(3) *If an open subset* O *of a complete metric or locally compact space* R *is the union of countably many subsets* M_i, *then at least one* $\bar{M}_i \cap O$ *contains a non-empty open subset of* R.

If R is locally compact, O is also locally compact in the induced topology, by § 3,6. Hence the assertion follows directly from (2).

Suppose that R is a complete metric space. Since R is regular the open set O contains a closed ball K. Then $K = \bigcup_{i=1}^{\infty} N_i$, where $N_i = M_i \cap K$. Applying (1), we find an $\bar{N}_i$ which contains a closed ball, and *a fortiori* an open ball K_0. Consequently $K_0 \subset \bar{M}_i \cap K \subset \bar{M}_i \cap O$.

If the M_i are all nowhere dense in R, there is no non-empty open set contained in $\bigcup_{i=1}^{\infty} M_i$, so that

(4) *In a complete metric or locally compact space R the complement of a countable union of nowhere dense sets is dense in R.*

A countable union of nowhere dense subsets of R is also called a set of the first category (BAIRE) or meagre (BOURBAKI). If a set is not of the first category in R, it is said to be of the second category. In this case BAIRE's theorem takes the following form:

(5) *Every open subset of a complete metric or locally compact space R is of the second category in R.*

7. The topological product of metric spaces. If $R_1,\dots,R_n$ are finitely many metric spaces, we define, for elements $x=(x_1,\dots,x_n)$, $y=(y_1,\dots,y_n)$ of $R_1\times\cdots\times R_n$,

$$|x,y|=\sum_{i=1}^{n}|x_i,y_i|;$$

if $R_1,R_2,\dots$ is a sequence of metric spaces, we define in a corresponding way, for $x,y\in\prod_{i=1}^{\infty}R_i$,

$$|x,y|=\sum_{i=1}^{\infty}\frac{1}{2^i}\,\frac{|x_i,y_i|}{1+|x_i,y_i|}.$$

In either case it is easily confirmed that in this way a metric space is produced. Instead of $1/2^i$ one could take any $c_i>0$ for which $\sum_{i=1}^{\infty}c_i<\infty$. To be sure, by doing this, different metrics are given, but we have

(1) *Every such metric defines the product topology on $\prod_{i=1}^{\infty}R_i$.*

Proof. An elementary argument shows that metric convergence $x^{(n)}\to x^{(0)}$ in $\prod R_i$ is equivalent to the metric convergence $x_i^{(n)}\to x_i^{(0)}$ of the components in each R_i. By § 2, 6.(1), convergence in the topological product sense also has this property, so that the two concepts of convergence are identical. Since the first axiom of countability holds in the R_i, it also holds in the topological product, and thus, by § 3, 5.(1), every closure point is limit of a convergent sequence. It follows that the closed sets of R in the metric sense and in the product topology sense coincide, and so therefore do the topologies.

(2) *$\prod_{i=1}^{\infty}R_i$ is complete if and only if the metric spaces R_i are complete.*

The parallelotope $\mathfrak{P}^\omega$ (cf. § 1, 8.) is also called the Hilbert parallelotope, since it is homeomorphic with the product $R=\prod_{i=1}^{\infty} R_i$, where R_i is the interval $[0,1/i]$, for which we have

(3) *R is a closed subspace of Hilbert space, and its topology is induced by the metric on Hilbert space.*

Proof. The first part of the assertion is trivial. If $|x,y^{(n)}|=\sqrt{\sum_{i=1}^{\infty}|\xi_i-\eta_i^{(n)}|^2}\to 0$, for $x,y^{(n)}\in R$, then $|\xi_i-\eta_i^{(n)}|\to 0$ for each i. Conversely, if this holds, then given $\varepsilon>0$, $\sum_{i=m}^{\infty}|\xi_i-\eta_i^{(n)}|^2<\varepsilon^2/2$ for a sufficiently large m and for each n, and $\sum_{i=1}^{m-1}|\xi_i-\eta_i^{(n)}|^2<\varepsilon^2/2$ for sufficiently large n, so that $|x,y^{(n)}|<\varepsilon$. The assertion now follows from the fact that the two concepts of convergence agree, as in the proof of (1).

R is called the Hilbert cube.

§ 5. Uniform spaces

1. Definition. The topological spaces in which we shall be interested later are by no means all metric spaces, but ideas can be used for them which are similar to those used for metric spaces in the preceding paragraph. These spaces, which, like metric spaces, have an additional structure defining the topology, were first considered by A. WEIL [1].

Let R be a set, and $R\times R$ the set of ordered pairs (x,y) of elements of R. If N is a subset of $R\times R$, with elements (x,y), we denote the set of all transposed pairs (y,x) by N^{-1}. The product MN of two such sets consists of all (x,z) for which there exists a y with (x,y) in M and (y,z) in N. NN is also written as N^2. Using this terminology, we make the following definition:

R is a uniform space when a filter $\mathfrak{N}$ is given on $R\times R$ with the following properties:

(V 1) *Every $N\in\mathfrak{N}$ contains the diagonal, that is, the set of all (x,x) with $x\in R$,*

(V 2) *If $N\in\mathfrak{N}$, so does N^{-1},*

(V 3) *For each $N\in\mathfrak{N}$ there exists $M\in\mathfrak{N}$ with $M^2\subset N$.*

The sets of the filter are called the vicinities of the uniform space; if $(x,y)\in N$, x and y are called close, of order N. $\mathfrak{N}$ defines a uniform structure or uniformity on R.

A base of the filter $\mathfrak{N}$ is called a base of the uniform space or of the uniformity. A filter base $\mathfrak{B}$ of subsets of $R\times R$ forms the base of a uniformity if (V 1) und (V 3) are satisfied for $\mathfrak{B}$ instead of $\mathfrak{N}$, together with

(V 2′) *If* $N\in\mathfrak{B}$, *there exists* $N'\in\mathfrak{B}$ *with* $N'\subset N^{-1}$.

There is always a base of symmetric vicinities, i.e. those with $N=N^{-1}$, since the symmetric vicinity $M\cap M^{-1}$ is contained in the vicinity M.

Every metric space is a uniform space, taking as base the countably many sets $N_{1/n}$ of all (x,y) with $|x,y|<1/n$.

Every subspace S of a uniform space R becomes a uniform space when the intersections with $S\times S$ of the sets of a base for R are taken as base. This is the uniformity induced on S by R.

Two uniform spaces R and R' are called isomorphic or equivalent if there is a one-one mapping of R onto R', and so of $R\times R$ onto $R'\times R'$, under which the filters defining the uniformities are transformed into each other.

Two metric spaces can be equivalent without being isometric. For example, one need only multiply all the distances in a metric space by a fixed positive factor to obtain a new metric equivalent to the original one.

If two uniformities on a set R are given by filters $\mathfrak{N}_1$ and $\mathfrak{N}_2$, the uniformity given by $\mathfrak{N}_1$ is called finer (respectively coarser) than that given by $\mathfrak{N}_2$, if $\mathfrak{N}_1\supset\mathfrak{N}_2$ (respectively $\mathfrak{N}_2\supset\mathfrak{N}_1$).

2. The topology of a uniform space. If x is a point of the uniform space R and if N is a vicinity, the set $U_N(x)$ of all y with $(x,y)\in N$ is defined to be a neighbourhood of x. The neighbourhoods defined in this way for all $N\in\mathfrak{N}$ and for all $x\in R$ satisfy axioms (N 1) to (N 4) of § 1, 2: (N 1) follows from (V 1), and (N 2) and (N 3) follow from the filter properties of $\mathfrak{N}$. To show (V 4) we choose an $M^2\subset N$, using (V 3). Then if $y\in U_M(x_0)$, every point z of $U_M(y)$ is in $U_N(x_0)$, since from $(x_0,y)\in M$ and $(y,z)\in M$ it follows that $(x_0,z)\in M^2\subset N$. Thus $U_N(x_0)$ is also a neighbourhood of y. The topology defined in this way is called the topology of the uniform space R.

In the case of a metric space we obtain the topology introduced in § 4, 2. once again.

A topology is defined on $R\times R$ as well, namely the product topology.

(1) *Every uniformity has a base of symmetric vicinities which are closed in the topology of* $R\times R$.

If N is a vicinity and M is a symmetric vicinity with $M^3\subset N$, then $\bar{M}\subset N$: if (x_0,y_0) is a closure point of M there is an $(x,y)\in M$ with $(x,x_0)\in M$ and $(y,y_0)\in M$, so that $(x_0,y_0)\in M^3\subset N$.

(2) *The topology of a uniform space is Hausdorff if and only if the further axiom*

(V 4) *The intersection of all $N \in \mathfrak{N}$ is the diagonal, is satisfied.*

If the intersection is the diagonal, the diagonal is, by (1), also the intersection of all the closed vicinities, and is itself closed. If $x \neq y$, there is a neighbourhood $U \times V$ of (x, y) which has no point in common with the diagonal, so that the neighbourhood U of x and the neighbourhood V of y have no common point.

On the other hand if R is Hausdorff and $x \neq y$, then there is a $U_N(x)$ which does not contain y. This means that (x, y) does not lie in N, and so the intersection of all vicinities is the diagonal.

If the diagonal is itself a vicinity, the uniformity is called the discrete uniformity; we obtain the discrete topology on R from it.

If we take the sequence $1/n$, $n=1,2,\ldots$, as a subset S of the metric space P, the induced uniformity on S is not the discrete one, although S is discrete as a topological space.

If $\bar{M}$ is a closed vicinity of the Hausdorff uniform space R, the set of all (x_0, y) in $\bar{M}$ is closed in $R \times R$, being the intersection of the closed sets $\{x_0\} \times R$ and $\bar{M}$, so that $U_{\bar{M}}(x_0)$ is closed in R; (1) gives

(3) *Every Hausdorff uniform space is regular.*

(4) *If A is any subset of $R \times R$, $\bar{A} = \bigcap\limits_N N A N$, as N runs through all the symmetric vicinities of R.*

If N_1 and N_2 are two symmetric vicinities and if $N = N_1 \cap N_2$, then the neighbourhood $U_{N_1}(x) \times U_{N_2}(y)$ of (x, y) contains the neighbourhood $U_N(x) \times U_N(y)$; hence these neighbourhoods form a base of neighbourhoods of (x, y), as N varies. If (x_0, y_0) is a closure point of A, there is, for each N, an $(x, y) \in A$ with $(x, y) \in U_N(x_0) \times U_N(y_0)$, and by the symmetry of N we also have that $(x_0, y_0) \in U_N(x) \times U_N(y)$, so that $(x_0, y_0) \in N A N$ and $\bar{A} \subset \bigcap\limits_N N A N$. If conversely $(x_0, y_0) \in \bigcap\limits_N N A N$, there is, for each N, an $(x, y) \in A$ with $(x_0, y_0) \in U_N(x) \times U_N(y)$, so that $(x, y) \in U_N(x_0) \times U_N(y_0)$, i.e. $(x_0, y_0) \in \bar{A}$.

Corresponding to (1) we have

(5) *Every uniformity has a base of symmetric vicinities which are open in the topology of $R \times R$.*

It is enough to show that the interior of a vicinity contains a vicinity. If M is a symmetric vicinity with $M^3 \subset N$, and if $(x_0\, y_0) \in M$, then all the points (x, y) of the neighbourhood $U_M(x_0) \times U_M(y_0)$ lie in N, so that (x_0, y_0) is an interior point of N.

3. Uniform continuity. As can be seen in the proofs which have just been given, a set whose pairs of points belong to a sufficiently small symmetric vicinity corresponds to a set of diameter $<\varepsilon$ in a metric space; such a set is called small of order N. A symmetric vicinity M with $M^2 \subset N$ then corresponds to a diameter $<\varepsilon/2$.

A mapping $Ax = x'$ of a uniform space R into a uniform space R' is called uniformly continuous, if to every vicinity N' in R' there corresponds a vicinity N in R with the property that $(x', y') \in N'$ whenever x, y are two points with $(x, y) \in N$. It is clear that every uniformly continuous mapping is continuous.

Every isomorphism of two uniform spaces is uniformly continuous in both directions.

If real functions $f_\beta(x), \beta \in \mathsf{B}$ are defined on a set R, they can be used to induce a uniformity on R. For every finitely many f_{β_i}, $i = 1, \ldots, n$, and $\varepsilon > 0$, we form the set N of all pairs $(x, y) \in R \times R$ with $|f_{\beta_i}(x) - f_{\beta_i}(y)| < \varepsilon$ for $i = 1, \ldots, n$. These sets N form the base of a filter of vicinities on $R \times R$. The f_β are clearly uniformly continuous with respect to this uniformity on R, and it is the coarsest uniformity on R with this property.

The uniformity is Hausdorff if and only if every two points x, y of R are separated by at least one f_β, i.e. $f_\beta(x) \neq f_\beta(y)$.

This idea can be generalized directly to the case where a collection of maps A_β from R into uniform spaces S_β is given. The sets of all (x, y) with $(A_{\beta_i}(x), A_{\beta_i}(y)) \in N_{\beta_i, \gamma_i}$ for $i = 1, \ldots, n$, where N_{β_i, γ_i} is an arbitrary vicinity in S_{β_i}, are then taken as vicinities.

4. Cauchy filters. Cauchy filters and Cauchy nets in uniform spaces correspond to Cauchy sequences in metric spaces: a filter $\{F_\alpha\}$ on R is called a Cauchy filter if to each vicinity N there corresponds at least one F_α which is small of order N, i.e. $(x, y) \in N$ for all $x, y \in F_\alpha$. An adherent point of a Cauchy filter is a limit of the filter, and in a Hausdorff uniform space a Cauchy filter has at most one adherent point.

Every convergent filter is a Cauchy filter. If every Cauchy filter on R is convergent, R is said to be complete.

In an analogous way, a net x_α, $\alpha \in \mathsf{A}$ is called a Cauchy net if to each vicinity N there corresponds a $\beta \in \mathsf{A}$ for which $(x_\gamma, x_\delta) \in N$ whenever $\gamma, \delta \geqq \beta$.

From the connection between nets and filters (§ 2, 4. and § 2, 5.) there follows directly

(1) *R is complete if and only if every Cauchy net has a limit.*

For metric spaces we have

(2) *A complete metric space is complete as a uniform space.*

If $\{F_\alpha\}$ is a Cauchy filter on R, there is a decreasing sequence F_{α_n} with $|x, y| \leqq 1/n$ for all $x, y \in F_{\alpha_n}$. A sequence $x_n \in F_{\alpha_n}$ is then a Cauchy sequence in R. Its limit x_0 is an adherent point of each F_α, since $F_\alpha \cap F_{\alpha_n}$ is non-empty.

In general a uniform space can very well contain limits of all the Cauchy sequences in it, without being complete, for the limits of certain Cauchy filters can be missing (cf. the example $\mathfrak{P}^{\mathsf{A}}$ in § 3, 4.).

(3) *Every discrete uniform space is complete.*

Every uniformly continuous mapping sends every Cauchy filter into a Cauchy filter, and equally sends every Cauchy net into a Cauchy net.

Theorem § 2, 8.(3) on the extension of continuous mappings leads to

(4) *A uniformly continuous mapping from a dense subspace S of a uniform space R into a complete Hausdorff uniform space R′ can be extended in a unique way to a uniformly continuous mapping from the whole of R.*

The condition of § 2, 8.(3) is satisfied. For every point $x \in R$, the sets $U_\alpha \cap S$ form a Cauchy filter $\mathfrak{F}$ on S; its image $A(\mathfrak{F})$ is a Cauchy filter on R', which defines the image Ax. Moreover R' is regular, by 2.(3). Thus A is continuously defined on the whole of R. A is uniformly continuous on R: let N' be a closed vicinity in R', and let N be a vicinity in R with the property that $(Ax, A, y) \in N'$ whenever $(x,y) \in N \cap (S \times S)$. Let M be a symmetric vicinity with $M^3 \subset N$. If x_0, y_0 are points of R with $(x_0, y_0) \in M$, then (x_0, y_0) is a closure point in $R \times R$ of $S \times S$, and in particular of the set of all (x,y) in $S \times S$ with $(x, x_0) \in M$ and $(y, y_0) \in M$. If (x,y) belongs to this set, $(x,y) \in M^3 \subset N$, so that $(Ax, Ay) \in N'$. Since (Ax_0, Ay_0) is a closure point of the (Ax, Ay), and since N' is closed, (Ax_0, Ay_0) lies in N'. Thus (Ax_0, Ay_0) belongs to N' for all (x_0, y_0) in M.

5. The completion of a Hausdorff uniform space. Let $\mathfrak{B} = \{N_\alpha\}$, $\alpha \in \mathsf{A}$, be a base of symmetric vicinities for the uniformity on R [cf. 2.(1)]. If we put $\alpha \leqq \beta$ when $N_\alpha \supset N_\beta$, A becomes a directed set. 4.(1) can be sharpened to give

(1) *The uniform space R is complete if and only if every net with index set* A *has a limit in R.*

It is enough to show that every Cauchy filter $\mathfrak{F} = \{F_\alpha\}$ has a limit in R. For every $\alpha \in \mathsf{A}$ we choose an $F_\alpha \in \mathfrak{F}$ which is small of order N_α, and an $x_\alpha \in F_\alpha$. If $\beta \geqq \alpha$, $\gamma \geqq \alpha$ and $z \in F_\beta \cap F_\gamma$, then (x_β, z) lies in $N_\beta \subset N_\alpha$ and (x_γ, z) lies in $N_\gamma \subset N_\alpha$, so that (x_β, x_γ) lies in N_α^2. The net (x_α) is thus a Cauchy net and by hypothesis has a limit x_0 in R.

Now let $N_\delta \in \mathfrak{B}$ be given and let $N_\rho \in \mathfrak{B}$, with $N_\rho^3 \subset N_\delta$. There exists an x_σ with $\sigma \geqq \rho$ and $(x_0, x_\sigma) \in N_\rho$. If $z \in F_\sigma \cap F_\rho$, $(x_\sigma, z) \in N_\rho$. If $y \in F_\rho$, then $(z, y) \in N_\rho$. Hence $(x_0, y) \in N_\rho^3 \subset N_\delta$, for all $y \in F_\rho$. But this means that $\mathfrak{F}$ has x_0 as limit.

As analogue of § 4, 4. we now show

(2) *Every Hausdorff uniform space R may be embedded in a smallest complete Hausdorff uniform space* $\tilde{R}$, *which is unique up to isomorphism.*

$\tilde{R}$ is called the completion of R.

Proof. a) *Definition of* $\tilde{R}$. We consider the Cauchy nets $(x_\alpha), \alpha \in \mathsf{A}$, on R, where A is the index set of the base fixed above. Two such nets (x_α) and (y_α) are said to be equivalent if for each symmetric vicinity N of the uniformity on R there exists an index $\beta(N)$ for which $(x_\gamma, y_\gamma) \in N$ for all $\gamma \geqq \beta$. If (x_α) is equivalent to (y_α), and (y_α) to (z_α), then (x_α) is also equivalent to (z_α): for if M is a symmetric vicinity with $M^2 \subset N$, it follows from the fact that $(x_\gamma, y_\gamma) \in M$ and $(y_\gamma, z_\gamma) \in M$ for sufficiently large γ that $(x_\gamma, z_\gamma) \in M^2 \subset N$. We collect equivalent Cauchy nets together into equivalence classes, which we define to be the points of $\tilde{R}$. The class in which the Cauchy net (x_α) lies will be denoted by $\tilde{x}$.

b) *The uniform structure of* $\tilde{R}$. If M is a symmetric vicinity of R, let $\tilde{M}$ denote the set of all pairs $(\tilde{x}, \tilde{y})$ with the property that for each symmetric vicinity N there exists an index $\beta(N)$ such that $(x_\gamma, y_\gamma) \in NMN$, for $\gamma \geqq \beta(N)$. This definition is independent of the representatives of the classes: suppose that (z_α) is equivalent to (y_α). If $N_1^2 \subset N$, $(x_\gamma, y_\gamma) \in N_1 M N_1$ for $\gamma \geqq \beta(N_1)$. If, further, γ is chosen large enough for (y_γ, z_γ) to belong to N_1, then $(x_\gamma, z_\gamma) \in N_1 M N_1^2 \subset NMN$. Thus $(\tilde{x}, \tilde{z})$ belongs to $\tilde{M}$.

The $\tilde{M}$ form the base of a uniformity on R: (V 1) is clearly satisfied, and so is (V 2), as $\tilde{M}^{-1} = \tilde{M}$, because of the symmetry of M. $M \subset N_1 \cap N_2$ implies that $\tilde{M} \subset \tilde{N}_1 \cap \tilde{N}_2$, so that the $\tilde{M}$ form a filter base on $\tilde{R} \times \tilde{R}$. Finally, if $M^3 \subset N$, it is easily confirmed that $\tilde{M}^2 \subset \tilde{N}$, so that (V 3) also holds.

c) $\tilde{R}$ *is Hausdorff*. If $(\tilde{x}, \tilde{y})$ lies in every $\tilde{M}$, $(x_\gamma, y_\gamma) \in M^3$ for $\gamma \geqq \beta(M)$, for all $\tilde{M}$; (x_α) and (y_α) are thus equivalent. The intersection of all the $\tilde{M}$ is therefore the diagonal.

d) R *is a subspace of* $\tilde{R}$. We denote the equivalence class containing the Cauchy net $x_\alpha = a$, $a \in R$, by $\tilde{a}$. $(\tilde{a}, \tilde{b})$ lies in $\tilde{M}$ if and only if $(a,b) \in \bigcap_N NMN$. By 2.(4) this means that $(a,b) \in \bar{M}$. If we identify a and $\tilde{a}$, then, from what we have just shown, $\tilde{M} \cap (R \times R) = \bar{M}$, so that by 2.(1) the uniformity induced on R by $\tilde{R}$ coincides with the original structure.

e) R *is dense in* $\tilde{R}$. If $\tilde{x}$ is a general element of $\tilde{R}$ and (x_α) is a net defining $\tilde{x}$, there is for each M a $\beta(M)$ for which $(x_{\gamma'}, x_\gamma) \in M$ for all $\gamma, \gamma' \geqq \beta$. But this means that $(\tilde{x}, x_\gamma) \in \tilde{M}$, when x_γ is considered as an element of $\tilde{R}$.

f) $\tilde{R}$ *is complete*. If $\mathfrak{B} = \{N_\alpha\}$ is the base defined at the beginning, then, by b), $\tilde{\mathfrak{B}} = \{\tilde{N}_\alpha\}$ is a base for the uniformity on $\tilde{R}$. By (1) it is enough to show that every Cauchy net $(\tilde{x}_\alpha)$, $\alpha \in \mathsf{A}$, has a limit in $\tilde{R}$. By e) we choose for each $\tilde{x}_\alpha$ a point z_α in R with $(\tilde{x}_\alpha, z_\alpha) \in \tilde{N}_\alpha$. (z_α) is then a Cauchy net in R: let $\rho \in \mathsf{A}$ be specified so that $\tilde{N}_\rho^3 \subset \tilde{N}_\alpha$. Then $(\tilde{x}_\gamma, \tilde{x}_{\gamma'}) \in \tilde{N}_\rho$ whenever $\gamma, \gamma' \geqq \sup(\beta(\rho), \rho)$. Since, moreover, $(\tilde{x}_\gamma, z_\gamma) \in \tilde{N}_\rho$ and $(\tilde{x}_{\gamma'}, z_{\gamma'}) \in \tilde{N}_\rho$, it follows that $(z_\gamma, z_{\gamma'}) \in \tilde{N}_\rho^3 \subset \tilde{N}_\alpha$, i.e. (z_α) is a Cauchy net $\tilde{z}$. Since $(\tilde{z}, z_\gamma) \in \tilde{N}_\rho$

for sufficiently large γ, and since $(z_\gamma, \tilde{x}_\gamma)\in\tilde{N}_\rho$ and $(\tilde{x}_\gamma, \tilde{x}_{\gamma'})\in\tilde{N}_\rho$, it follows that $(\tilde{z}, \tilde{x}_{\gamma'})\in\tilde{N}_\alpha$ for all sufficiently large γ', i.e. the Cauchy net $(\tilde{x}_\alpha)$ has $\tilde{z}$ as limit.

$\tilde{R}$ is thus a completion of R.

Finally, the fact that $\tilde{R}$ is uniquely determined up to isomorphism is a direct consequence of

(3) *If R_1 and R_2 are isomorphic uniform spaces which are dense in the complete Hausdorff uniform spaces S_1 and S_2 respectively, then any isomorphism between R_1 and R_2 can be extended uniquely to an isomorphism between S_1 and S_2.*

Let $A(R_1)=R_2$, $B(R_2)=R_1$ be the mappings given by the isomorphism. Since an isomorphism is uniformly continuous in both directions, the existence of two uniformly continuous mappings $\tilde{A}$ and $\tilde{B}$ which map S_1 into S_2 and S_2 into S_1 respectively follows, by extension, from 4.(4). The class of equivalent Cauchy nets in R_1 corresponds to the class of equivalent Cauchy nets in R_2 under the isomorphism of R_1 and R_2, so that $\tilde{A}$ and $\tilde{B}$ are uniformly continuous mappings of S_1 onto S_2 and S_2 onto S_1 respectively, and are each the inverse of the other. Consequently the vicinity filters of S_1 and S_2 are also sent into each other, and $\tilde{A}$ and $\tilde{B}$ define an isomorphism of S_1 and S_2.

(4) *If M is a vicinity in $R\times R$, the vicinity $\tilde{M}$ in $\tilde{R}\times\tilde{R}$ is the closure of M in $\tilde{R}\times\tilde{R}$.*

The closure in $\tilde{R}\times\tilde{R}$ of the vicinities of a base in $R\times R$ form a base in $\tilde{R}\times\tilde{R}$.

If $(\tilde{x}^{(0)}, \tilde{y}^{(0)})$ is a closure point of $\tilde{M}$, then for each symmetric vicinity N there exists an $(\tilde{x}, \tilde{y})\in\tilde{M}$ for which $(\tilde{x}^{(0)}, \tilde{x})\in\tilde{N}$ and $(\tilde{y}, \tilde{y}^{(0)})\in\tilde{N}$, so that, for sufficiently large γ, $(x_\gamma^{(0)}, x_\gamma)\in N^3$, $(x_\gamma, y_\gamma)\in NMN$ and $(y_\gamma, y_\gamma^{(0)})\in N^3$; thus $(x_\gamma^{(0)}, y_\gamma^{(0)})\in N^4MN^4$. But this means that $(\tilde{x}^{(0)}, \tilde{y}^{(0)})\in\tilde{M}$, and that $\tilde{M}$ is closed.

On the other hand if $(\tilde{x}, \tilde{y})\in\tilde{M}$, then for each symmetric vicinity N there exists elements x_γ, y_γ in R such that $(\tilde{x}, x_\gamma)\in\tilde{N}$, $(\tilde{y}, y_\gamma)\in\tilde{N}$ and $(x_\gamma, y_\gamma)\in NMN$. Thus there exist $z, t\in R$ with $(z,t)\in M$, $(z, x_\gamma)\in N$ and $(t, y_\gamma)\in N$, so that $(\tilde{z}, \tilde{t})\in U_{N^2}(\tilde{x})\times U_{N^2}(\tilde{t})$, and M is dense in $\tilde{M}$.

For the proof of the second assertion we observe that, according to the definition of the uniform structure on $\tilde{R}$, the $\tilde{M}$ with symmetric M form a base in $\tilde{R}\times\tilde{R}$. Every M contains a vicinity N of the given base in $R\times R$, so that $\tilde{M}\supset\tilde{N}$ as well, and the $\tilde{N}$ form a base in $\tilde{R}\times\tilde{R}$.

6. Compact uniform spaces. (1) *In a compact uniform space every ultrafilter is a Cauchy filter.*

If M is a vicinity of R, R is covered by finitely many open sets M_i, $i=1,\dots,n$, which are small of order M. By § 2, 7.(2) one of the M_i belongs to the ultrafilter, which thus contains arbitrarily small sets.

(1) also follows directly from § 3, 1.(1).

Every compact uniform space is complete, since every Cauchy filter has an adherent point, and thus has a limit.

As in § 4, 5. we call a Hausdorff uniform space R precompact if its completion is compact.

A Hausdorff uniform space is said to be totally bounded if for every vicinity N of R there exists a covering of R by finitely many sets which are all small of order N.

Analogous to § 4, 5. we have

(2) a) *A Hausdorff uniform space R is precompact if and only if it is totally bounded.*

b) *A Hausdorff uniform space is precompact if and only if every ultrafilter is a Cauchy filter.*

Proof. a) If R is precompact, the completion $\tilde{R}$, and thus also R, is covered by finitely many sets M_i which are small of order $\tilde{N}$. R is covered by the sets $M_i \cap R$, which are small of order N since $\tilde{N}$ is obtained from the closed vicinity N in $R \times R$ by forming its closure in $\tilde{R} \times \tilde{R}$.

If conversely R is totally bounded and finitely many M_i of order N cover the whole of R, then the closures $\tilde{M}_i$ in $\tilde{R}$ cover the whole of $\tilde{R}$ and are small of order $\tilde{N}$. By 5.(4) the $\tilde{N}$ form a base for the uniformity of $\tilde{R}$. If now $\mathfrak{F}$ is an ultrafilter on $\tilde{R}$, then, as in (1), by § 2, 7.(2) at least one $\tilde{M}_i$ is in $\mathfrak{F}$. $\mathfrak{F}$ thus contains arbitrarily small sets and is therefore a Cauchy filter, which converges in $\tilde{R}$, since $\tilde{R}$ is complete. $\tilde{R}$ is therefore compact.

b) A precompact space is totally bounded, by a), and it follows from this, as in (1), that every ultrafilter is a Cauchy filter.

If R is not precompact, $\tilde{R}$ is not compact, and so there is a filter $\mathfrak{F} = \{F^\alpha\}$ on $\tilde{R}$ which has no adherent points. For every F^α and every symmetric vicinity $\tilde{N}$ we form the set $F^\alpha_{\tilde{N}}$ of all $z \in \tilde{R}$ with the property that $(x,z) \in \tilde{N}$ for some $x \in F^\alpha$. The sets $F^\alpha_{\tilde{N}}$ form a filter base on $\tilde{R}$, and, as $F^\alpha_{\tilde{N}} \cap R$ is always non-empty, the sets $F^\alpha_{\tilde{N}} \cap R$ form the base of a filter $\mathfrak{F}'$ on R. $\mathfrak{F}'$ has no adherent points in $\tilde{R}$, since such a point would also be an adherent point of $\mathfrak{F}$. Any ultrafilter on R which is finer than $\mathfrak{F}'$ is therefore not a Cauchy filter.

(3) *A Hausdorff uniform space R is precompact if and only if every sequence in R has an adherent point in $\tilde{R}$.*

The condition is clearly necessary. If conversely R is not precompact, there is, by (2) a), a vicinity N for which there is no finite cover of R by sets of order smaller than N. Let N_1 be a symmetric vicinity with $N_1^4 \subset N$. We choose an x_1 and form the neighbourhood $U_{N_1}(x_1)$. This is of order N_1^2. Now suppose that points $x_1, \ldots, x_k$ have already been chosen so that the $U_{N_1}(x_i)$ are pairwise disjoint. Were there no x_{k+1} for which $U_{N_1}(x_{k+1})$ was disjoint from the $U_{N_1}(x_i)$ already determined, then all the points of R would be contained in the sets $U_{N_1^2}(x_i)$, $i=1,\ldots,k$, so that k sets of order smaller than $N_1^4 \subset N$ would cover the whole of R. For the sequence x_k, $k=1,2,\ldots$, determined in this way, $(x_k, x_m) \notin N_1$ for all k, m, so that the sequence has no adherent point in $\tilde{R}$.

(4) *A complete Hausdorff uniform space is compact if it is countably compact.*

For by (3) a countably compact Hausdorff uniform space is precompact, and compactness follows from completeness.

(5) *In a compact uniform space, the filter $\mathfrak{N}$ of vicinities consists of all the neighbourhoods of the diagonal in $R \times R$.*

Every vicinity N is a neighbourhood of the diagonal in $R \times R$. Were there, on the other hand, an open neighbourhood N_0 of the diagonal which did not belong to $\mathfrak{N}$, then, for each $N \in \mathfrak{N}$, $(R \times R \sim N_0) \cap N$ would be non-empty, since $N \not\subset N_0$; the sets $(R \times R \sim N_0) \cap N$ would thus define a filter $\mathfrak{N}' \supset \mathfrak{N}$, which, because of the compactness of $R \times R$, would have an adherent point belonging to $R \times R \sim N_0$, and so not belonging to the diagonal. But $\mathfrak{N}$, and also $\mathfrak{N}'$, have only points of the diagonal as adherent points, since R is Hausdorff and regular.

(6) *Every continuous mapping A of a compact uniform space R into a uniform space R' is uniformly continuous.*

If A is continuous, by § 2, 6.(2) the mapping $(x, y) \to (Ax, Ay)$ from $R \times R$ into $R' \times R'$ is continuous. If, therefore, N' is an open vicinity in R' [cf. 2.(5)], the set of the inverse images (x, y) of all $(Ax, Ay) \in N'$ is a neighbourhood of the diagonal in $R \times R$, and so, by the preceding theorem, is a vicinity in R.

7. The product of uniform spaces. If R_α, $\alpha \in \mathsf{A}$, are uniform spaces with vicinity filters $\mathfrak{N}_\alpha$, $R = \prod_\alpha R_\alpha$ becomes the uniform product of the R_α when one takes as basic vicinities N in $R \times R = \prod_\alpha (R_\alpha \times R_\alpha)$ all sets $\prod M_\alpha$, where $M_\alpha = R_\alpha \times R_\alpha$ for all but finitely many α and $M_\alpha = N_\alpha \times N_\alpha$ for the rest, where N_α is an arbitrary vicinity in $\mathfrak{N}_\alpha$.

It is easy to see that the topology defined by this uniformity on R is the product topology (§ 1, 8.). We can add the following easily established results to those already obtained in § 2, 6.:

(1) *A filter $\mathfrak{F}$ on R is a Cauchy filter if and only if all its projections $\mathfrak{F}_\alpha$ are Cauchy filters.*

(2) *The uniform product $R = \prod R_\alpha$ of the uniform spaces R_α is complete if and only if all the R_α are complete.*

The completion of R is the uniform product of the completions of the R_α.

This follows from (1) and § 2, 6.(3).

A mapping A of a uniform space into the uniform product $\prod R_\alpha$ is uniformly continuous if and only if all the projections $A_\alpha = P_\alpha A$ are.

§ 6. Real functions on topological spaces

1. Upper and lower limits. If we add $+\infty$ and $-\infty$ to the set P of real numbers, we obtain a set which we denote by $\bar{\mathsf{P}}$. The order relations and algebraic operations in $\bar{\mathsf{P}}$ are defined in the usual way. $\bar{\mathsf{P}}$ becomes a topological space when we take the intervals $[-\infty, a]$, $[b, +\infty]$ as a sub-basis for the collection $\mathfrak{A}$ of closed sets. The original topology of P is induced by this. The mapping $x \to \operatorname{tg} x$ maps $[-\pi/2, \pi/2]$ homeomorphically onto $\bar{\mathsf{P}}$, so that $\bar{\mathsf{P}}$ is compact.

The least upper bound $\sup_{a \in A} a$ and the greatest lower bound $\inf_{a \in A} a$ are defined for any subset A of $\bar{\mathsf{P}}$. They can of course be $+\infty$ or $-\infty$.

If M is any set and if an element $f(x)$ in $\bar{\mathsf{P}}$ is ascribed to each $x \in M$, we speak of a real function on M. If the values lie only in P, $f(x)$ is called a finite-valued real function. Least upper bounds and greatest lower bounds of the values of the function on M are denoted by $\sup_{x \in M} f(x)$ and $\inf_{x \in M} f(x)$ respectively.

Since we allow both $+\infty$ and $-\infty$ as values of functions, the sum $f(x)+g(x)$ need not be defined for all x ($+\infty$ and $-\infty$ cannot be added). But if $f(x)+g(x)$ is defined for at least one $x \in M$, then the following hold, provided that the right-hand sides are meaningful:

$$\sup_{x \in M} (f(x)+g(x)) \leqq \sup_{x \in M} f(x) + \sup_{x \in M} g(x), \tag{1a}$$

$$\inf_{x \in M} (f(x)+g(x)) \geqq \inf_{x \in M} f(x) + \inf_{x \in M} g(x). \tag{1b}$$

The formulation and proof of the corresponding formulae for multiplication and division are left to the reader.

(2) *If $f(x) \leqq g(x)$ for all $x \in M$, then $\sup_{x \in M} f(x) \leqq \sup_{x \in M} g(x)$ and $\inf_{x \in M} f(x) \leqq \inf_{x \in M} g(x)$.*

A net ξ_α on $\bar{\mathsf{P}}$ is called monotonic increasing (monotonic decreasing) if $\alpha < \beta$ implies that $\xi_\alpha \leqq \xi_\beta$ ($\xi_\alpha \geqq \xi_\beta$).

(3) *Every monotonic increasing (decreasing) net ξ_α on $\bar{\mathsf{P}}$ has a limit, and* $\lim \xi_\alpha = \sup \xi_\alpha (= \inf \xi_\alpha)$.

For if $\sup \xi_\alpha = \gamma$, then for each $\delta < \gamma$ there exists a β for which $\delta < \xi_{\beta'} \leqq \gamma$ for all $\beta' \geqq \beta$, i.e. γ is the limit of the set ξ_α.

If ξ_α is an arbitrary net on $\bar{\mathsf{P}}$, then the net $\eta_\alpha = \sup_{\beta \geqq \alpha} \xi_\beta$ derived from it is monotonic decreasing. By (3), $\lim \eta_\alpha = \inf_\alpha \sup_{\beta \geqq \alpha} \xi_\beta$ exists; we call it the upper limit or limes superior of ξ_α and denote it by $\overline{\lim}\, \xi_\alpha$. Similarly the lower limit or limes inferior is defined as $\underline{\lim}\, \xi_\alpha = \sup_\alpha \inf_{\beta \geqq \alpha} \xi_\beta$.

The inequality $\underline{\lim}\, \xi_\alpha \leqq \overline{\lim}\, \xi_\alpha$ always holds.

If $\mathfrak{F} = \{F_\alpha\}$ is a filter on $\bar{\mathsf{P}}$, whose indices are ordered by the rule that $\alpha \leqq \beta$ if $F_\alpha \supset F_\beta$, then we have the analogous definitions

$$\overline{\lim}\, \mathfrak{F} = \overline{\lim}\, F_\alpha = \lim_\alpha \sup_{\xi \in F_\alpha} \xi = \inf_\alpha \sup_{\xi \in F_\alpha} \xi,$$

$$\underline{\lim}\, \mathfrak{F} = \underline{\lim}\, F_\alpha = \lim_\alpha \inf_{\xi \in F_\alpha} \xi = \sup_\alpha \inf_{\xi \in F_\alpha} \xi.$$

We observe that in order to form $\overline{\lim}\, \mathfrak{F}$ it is only necessary to use the F_γ from a base for $\mathfrak{F}$, since the net $\eta_\gamma = \sup_{\xi \in F_\gamma} \xi$ is cofinal with the net of all η_α, and so has the same limit.

(4) *The upper limit of a net or filter on* $\bar{\mathsf{P}}$ *is the greatest adherent point of the net or filter.*

We shall only show this for nets. Let $\eta = \overline{\lim}\, \xi_\alpha$. Since $\eta = \inf \eta_\alpha$, and since η_α is monotonic descreasing, given $\varepsilon > 0$ there is an α_0 for which $|\eta - \eta_\alpha| < \varepsilon/2$ for $\alpha \geqq \alpha_0$. For each η_α there is a ξ_{β_α}, with $\beta_\alpha \geqq \alpha$, for which $|\eta_\alpha - \xi_{\beta_\alpha}| < \varepsilon/2$. The net ξ_{β_α}, $\alpha \geqq \alpha_0$, which is cofinal with ξ_α, therefore lies in the ε-neighbourhood of η, and η is an adherent point of ξ_α.

On the other hand if ξ is an adherent point, $\xi \leqq \eta_\alpha$ for each α, so that $\xi \leqq \inf \eta_\alpha = \eta$, by (2).

Now let $f(x)$ be a real function on the topological space R. If x_α is a net convergent to a then the values $f(x_\alpha)$ of the function form a net whose upper and lower limits we denote by $\overline{\lim}_{x_\alpha \to a} f(x)$ and $\underline{\lim}_{x_\alpha \to a} f(x)$ respectively.

On the other hand if $\mathfrak{U} = \{U_\alpha\}$ is the filter of neighbourhoods of a, the image filter $f(\mathfrak{U})$ is a filter on $\bar{\mathsf{P}}$ with base $f(U_\alpha)$, and we can form the upper and lower limits of $f(\mathfrak{U})$. They are denoted by

$$\bar{f}(a) = \overline{\lim}_{x \to a} f(x) = \inf_{U \in \mathfrak{U}} \sup_{x \in U} f(x) = \lim_{U \in \mathfrak{U}} \sup_{x \in U} f(x),$$

$$\underline{f}(a) = \underline{\lim}_{x \to a} f(x) = \sup_{U \in \mathfrak{U}} \inf_{x \in U} f(x) = \lim_{U \in \mathfrak{U}} \inf_{x \in U} f(x).$$

$\bar{f}(x)$ and $\underline{f}(x)$ are called the upper and lower limits of $f(x)$ respectively.

The two concepts which have just been introduced are generalizations of the two well known concepts $\overline{\lim\limits_{x_n\to a}}\, f(x)$ and $\overline{\lim\limits_{x\to a}}\, f(x)$ of analysis. Since the filter corresponding to a net $x_\alpha\to a$ is finer than the filter of neighbourhoods of a, $\overline{\lim\limits_{x_\alpha\to a}}\, f(x)\leqq \bar{f}(x)$. For a suitable choice of x_α, however, equality holds: let I_n be a sequence of neighbourhoods of $\bar{f}(a)$ with intersection $\bar{f}(a)$. For each neighbourhood U of a we form the set U_n of all $x\in U$ for which $f(x)\in I_n$. The collection of all these U_n defines a filter $\mathfrak{F}$ whose image $f(\mathfrak{F})$ has $\bar{f}(a)$ as limit. If $x_\alpha\to a$ is a corresponding net, then $\lim\limits_{x_\alpha\to a} f(x)=\bar{f}(a)$, by § 2, 5.(2).

(5) *The inequalities* $\underline{f}(a)\leqq f(a)\leqq \bar{f}(a)$ *always hold.*

(6) *If* $f(x)\leqq g(x)$ *in some neighbourhood of* a, *then* $\underline{f}(a)\leqq \underline{g}(a)$ *and* $\bar{f}(a)\leqq \bar{g}(a)$.

At every point a for which $f(x)+g(x)$ is defined, we have, by taking limits in (1 a) and (1 b) respectively,

$$\text{(7)}\qquad \overline{(f+g)}\,(a)\leqq \bar{f}(a)+\bar{g}(a),\qquad \underline{(f+g)}\,(a)\geqq \underline{f}(a)+\underline{g}(a),$$

provided that the right hand sides are meaningful.

2. Semi-continuous functions. A real function $f(x)$ on a topological space R is said to be l o w e r s e m i - c o n t i n u o u s at a if $f(a)=\underline{f}(a)$. It follows from 1.(5) that it is enough to require that $f(a)\leqq \underline{f}(a)$.

A function which is lower semi-continuous at each point is called lower semi-continuous on R. U p p e r s e m i - c o n t i n u i t y is defined similarly.

A function which is both upper and lower semi-continuous at a is continuous at a.

Since $\underline{f}(a)$ is, by the preceding number, the greatest lower bound of the $\underline{\lim\limits_{x_\alpha\to a}}\, f(x)$, we have

(1) $f(x)$ *is lower semi-continuous at* a *if and only if* $f(a)\leqq \underline{\lim\limits_{x_\alpha\to a}}\, f(x)$, *whenever* $x_\alpha\to a$.

A further criterion is

(2) $f(x)$ *is lower semi-continuous at* a *if and only if whenever* $\gamma<f(a)$ *there exists a neighbourhood* $U(a)$ *for which* $\gamma<f(x)$ *for all* $x\in U(a)$.

If the condition is satisfied, we have that $\gamma\leqq \inf\limits_{x\in U} f(x)$ for some suitable U, so that $f(a)\leqq \sup\limits_{U}\inf\limits_{x\in U} f(x)=\underline{f}(a)$. The converse is also immediately obvious.

(3) $f(x)$ *is lower semi-continuous on* R *if and only if for each* $\gamma\in\bar{\mathrm{P}}$ *the set* $[f>\gamma]$ *of those* x *with* $f(x)>\gamma$ *is open, or equivalently, if and only if for each* $\gamma\in\bar{\mathrm{P}}$ *the set* $[f\leqq\gamma]$ *of those* x *with* $f(x)\leqq\gamma$ *is closed.*

For (2) implies that if a satisfies the inequality $f(x)>\gamma$, then so does a whole neighbourhood of a. The second assertion follows by taking complements.

(4) *The sum of finitely many lower semi-continuous functions is lower semi-continuous wherever it is defined.*

If $f(a)+g(a)$ is defined, then by 1.(7)

$$\underline{(f+g)}\,(a) \geqq \underline{f}(a)+\underline{g}(a)=f(a)+g(a).$$

(5) *The lower limit $\underline{f}(x)$ of an arbitrary real function on R is always lower semi-continuous, i.e. $\underline{\underline{f}}(x)=\underline{f}(x)$.*

For each $\delta<\underline{f}(a)$ there is, from the definition of $\underline{f}(a)$, an open neighbourhood $U_0(a)$ for which $\delta<f(x)$ for all $x\in U_0(a)$. If y lies in $U_0(a)$, so does a whole neighbourhood $V_0(y)$, so that $\underline{f}(y)=\underline{\lim_V}\ \inf_{x\in V(y)} f(x)\geqq\delta$ holds throughout U_0, and $\underline{\underline{f}}(a)=\underline{\lim_U}\ \inf_{y\in U} \underline{f}(y)\geqq\delta$. Since this holds for every $\delta<\underline{f}(a)$, it follows that $\underline{\underline{f}}(a)\geqq\underline{f}(a)$.

It is easy to give examples of semi-continuous functions on any topological space. The characteristic function $\varphi(x)$ of a subset M of R is defined by $\varphi(x)=1$ if $x\in M$, $\varphi(x)=0$ if $x\in R\sim M$. By (3), M is open or closed in R if and only if φ is lower or upper semi-continuous respectively. For the sets $[\varphi>\gamma]$ (respectively $[\varphi<\gamma]$) can only be the empty set, M (respectively $R\sim M$) and R.

For any two distinct points x, y in a Hausdorff space R there is therefore always a finite lower semi-continuous function f with $f(x)\neq f(y)$.

(6) *A lower semi-continuous function $f(x)$ on a compact space R takes a minimum value on R. If it only takes finite values, it is bounded below.*

Since the sets $[f\leqq\gamma]$ are closed, by (3), the non-empty ones among them form a filter base on R. If a is an adherent point of this filter, a lies in all the non-empty $[f\leqq\gamma]$, so that $f(a)\leqq f(x)$ for all x in R.

A direct consequence is

(7) *Every continuous function on a compact space takes a maximum and a minimum value on R. If it is finite valued, it is bounded above and below.*

(6) and (7) also hold for sequentially compact spaces. The proof of (6) is simple: there is a convergent sequence $x_n\to a$ in R for which $f(x_n)\to m$, where m is the greatest lower bound of the values of the function. By (1), $f(a)=m$.

3. The least upper bound of a collection of functions. If $\{f_\beta\}$, $\beta\in\mathsf{B}$, is a collection of real functions on the set M, the function which takes the value $\sup_\beta f_\beta(x)$ at each point x is called the least upper bound $\sup f_\beta$ of the f_β. The greatest lower bound $\inf f_\beta$ is defined similarly.

(1) *The least upper bound of a collection f_β of functions which are each lower semi-continuous at a point a of a topological space is itself lower semi-continuous at a. In particular the least upper bound of a collection of continuous functions on R is lower semi-continuous on R.*

We use criterion 2.(2). If $\gamma<\sup f_\beta(a)$, $\gamma<f_\beta(a)$ for some β, so that $\gamma<f_\beta(x)\leqq\sup\limits_\beta f_\beta(x)$, throughout some neighbourhood $U(a)$.

The next result also follows simply from 2.(2)

(2) *The greatest lower bound of finitely many functions which are lower semi-continuous at a is lower semi-continuous at a.*

From BAIRE's theorem we obtain

(3) *If f is a lower semi-continuous finite-valued real function on a locally compact or complete metric space R, then the set of points which have a neighbourhood in which f is bounded above is open, and dense in R.*

For if M_n is the closed set $[f\leqq n]$, every open set O is the union of the sets $M_n\cap O$, and by § 4, 6.(3) at least one $\overline{M_n\cap O}\cap O=M_n\cap O$ contains an interior point. Thus if O is some neighbourhood of a point $x\in R$, there is a point y in it which has a whole neighbourhood on which f is bounded above.

From (1) and (3) follows immediately

(4) *If the least upper bound of the lower semi-continuous functions f_β on a locally compact or complete metric space R is finite valued, then the set of points possessing a neighbourhood throughout which the f_β are uniformly bounded above is dense in R.*

4. Continuous functions on normal spaces. The question of the existence of non-constant continuous real functions on topological spaces is more difficult to answer; fundamental for this is the following lemma of URYSOHN:

(1) *A Hausdorff space R is normal if and only if given two disjoint closed sets A and B there always exists a continuous function on the whole of R with values in* $[0,1]$, *which takes the value* 0 *on the whole of A and the value* 1 *on the whole of B.*

If the condition is satisfied, the sets $[f<\frac{1}{2}]$ and $[f>\frac{1}{2}]$ are disjoint and open, since f is continuous, and contain A and B respectively, so that (N) is satisfied (cf. § 3, 7.).

Suppose conversely that R is normal. We construct open subsets $B(\rho)$ of R for all dyadic fractions $\rho=k/2^n$, $k=0,1,\dots,2^n$, with $0\leqq\rho\leqq1$.

We set $B(0)$ equal to the empty set and $B(1)=R\sim B$. For $n=1$, applying § 3, 7.(N′), let $B(\frac{1}{2})$ be an open set with $B(\frac{1}{2})\supset A$ and $\bar{B}(\frac{1}{2})\subset B(1)$. If the open sets $B(k/2^n)\supset A$ have already been constructed for all $k=1,\dots,2^n$, in such a way that $\bar{B}(k/2^n)\subset B((k+1)/2^n)$ for each $k<2^n$, then by § 3, 7.(N′) there is always an open set $B((2k+1)/2^{n+1}) \supset A$ with $\bar{B}(k/2^n) \subset B((2k+1)/2^{n+1})$ and $\bar{B}((2k+1)/2^{n+1}) \subset B((k+1)/2^n)$.

It follows that $\bar{B}(\rho) \subset B(\rho')$ whenever ρ and ρ' are dyadic fractions with $\rho < \rho'$.

We now set $f(x) = \sup\limits_{x \notin B(\rho)} \rho$. Then $f(x)=1$ on B, since, if $x \in B$, x lies in no $B(\rho)$ with $0 \leqq \rho \leqq 1$. On the other hand, since, $B(\rho) \supset A$ for all $\rho > 0$, $f(x)=0$ on A. The values of $f(x)$ clearly lie in $[0,1]$. The continuity of $f(x)$ is shown in the following way: let $f(x_0)=\gamma$. If $\gamma=0$, let the interval $[\rho,\rho']$ be equal to $[0,\rho']$ with $\rho'>0$, and if $0<\gamma<1$ let $[\rho,\rho']$ be any interval with dyadic end points ρ,ρ' containing γ in its interior. Then $\bar{B}(\rho)$ is contained wholly in $B(\rho')$ and $B(\rho') \sim \bar{B}(\rho)$ is open and contains x_0 as interior point. But $\rho \leqq f(x) \leqq \rho'$ for each x in $B(\rho') \sim \bar{B}(\rho)$, i.e. $f(x)$ is continuous at x_0. If, finally, $\gamma=1$, then x_0 lies in $R \sim \bar{B}(\rho)$ for any $\rho<1$; but $f(x) \geqq \rho$ for $x \in R \sim \bar{B}(\rho)$, so that $f(x)$ is also continuous for these x_0.

If the function $f(x)$ which has just been constructed is replaced by $a+(b-a)f(x)$, $a,b \in \mathrm{P}$, we obtain

(2) *If A and B are disjoint closed subsets of a normal space R, then there exists a continuous function on R which is equal to a on the whole of A, equal to b on the whole of B, and whose values lie in $[a,b]$.*

Remark. For metric spaces, a function with the properties required in (1) can be given easily. It is enough to put $f(x) = \dfrac{|x,A|}{|x,A|+|x,B|}$.

A topological space is u n i f o r m i z a b l e (respectively m e t r i z a b l e) if a uniformity (respectively a metric) can be defined on R, for which the corresponding topology is the given one.

A first important consequence of (1) is

(3) *Every normal space is uniformizable.*

We consider the set $C(R)$ of all finite-valued real continuous functions on R. By § 5, 3. there is a coarsest uniformity $\mathfrak{N}$ on R, with respect to which all the $f \in C(R)$ are uniformly continuous. It is defined by the vicinities $N_{f,\varepsilon}$ consisting of those $(x,y) \in R \times R$ for which $|f(x)-f(y)|<\varepsilon$. We have to show that the topology $\mathfrak{T}_{\mathfrak{N}}$ defined by this uniformity coincides with the original topology $\mathfrak{T}$. Since each f is continuous, each $U_N(x_0)$, which is the set of all y with $|f(x_0)-f(y)|<\varepsilon$, is an open neighbourhood with respect to $\mathfrak{T}$. If conversely U is an open neighbourhood of x_0 with respect to $\mathfrak{T}$, there is, by (1), a continuous function f with $f(x_0)=0$ and $f(y)=1$ for all $y \in R \sim U$. The set $[f<1]$ is then a $\mathfrak{T}_{\mathfrak{N}}$-neighbourhood of x_0 which is contained in U, so that $\mathfrak{T}$ and $\mathfrak{T}_{\mathfrak{N}}$ are identical.

Since every compact space is normal, by § 3, 7.(2), (1) gives information about the existence of continuous functions on compact spaces. It follows from (3) and § 5, 6.(5) that

(4) *Every compact space is uniformizable in one and only one way.*

For locally compact spaces we have

(5) *If A is a compact subset of a locally compact space R, and if U is an open set containing A, then there exists a continuous function on R which takes the value 1 on the whole of A, and takes the value 0 on the whole of $R \sim U$.*

Let R' be the compactification of R, as in § 3, 6.(2). The result follows by applying (1) to A and $R' \sim U$.

5. The extension of continuous functions on normal spaces. A satisfactory account of the existence of continuous real functions on normal spaces is given by URYSOHN's extension theorem.

(1) *Any bounded continuous function on a closed subset M of a normal space R can be extended continuously to the whole of R, preserving the same bounds.*

Proof. Suppose that $f(x)$ is continuous on the closed subset M of R, and that $|f(x)| \leqq c$. The subsets $M_1 = [f \leqq -c/3]$ and $M_2 = [f \geqq c/3]$ of M are closed and disjoint, so that by 4.(2) there exists a continuous function $g_1(x)$ defined on the whole of R, with $g_1(x) = -c/3$ on M_1, $g_1(x) = c/3$ on M_2 and $|g_1(x)| \leqq c/3$ on R. Then if $h_1(x)$ is the continuous function on M defined by $h_1(x) = f(x) - g_1(x)$, $|h_1(x)| \leqq \frac{2}{3}c$ on M. Applying the same procedure to $h_1(x)$, with bound $\frac{2}{3}c$, a continuous function $g_2(x)$ is obtained on the whole of R, with $|g_2(x)| \leqq \frac{1}{3} \cdot \frac{2}{3} c$, and a continuous function $h_2(x) = h_1(x) - g_2(x)$ is defined on M, with $|h_2(x)| \leqq (\frac{2}{3})^2 c$.

In general, we obtain for each n a continuous function $g_n(x)$ on R with $|g_n(x)| \leqq \frac{1}{3}(\frac{2}{3})^{n-1} \cdot c$, and a continuous function $h_n(x) = h_{n-1}(x) - g_n(x)$ with $|h_n(x)| \leqq (\frac{2}{3})^n c$. The infinite series $\sum_{n=1}^{\infty} g_n(x)$ converges uniformly on the whole of R, and so defines a continuous function $F(x)$, with $|F(x)| \leqq \frac{1}{3} \sum_{n=0}^{\infty} (\frac{2}{3})^n c = c$.

On M, $F(x) = f(x) - h_1(x) + \sum_{n=1}^{\infty} (h_n(x) - h_{n+1}(x)) = \lim_{n \to \infty} (f(x) - h_{n+1}(x))$. Since $|h_{n+1}(x)| \leqq (\frac{2}{3})^{n+1} \cdot c$, $F(x) = f(x)$ on M.

The extension theorem only holds for normal spaces: If A and B are disjoint closed subsets of R, and if the continuous function defined on $A \cup B$ by $f(x) = 0$ on A, $f(x) = 1$ on B may be extended by the extension theorem to the whole of R, then R is normal, by 4.(1).

6. Completely regular spaces. Looking more closely at the proof of 4.(3), it can be seen that the result is still valid, with the following hypothesis about the Hausdorff space R:

(V) *If $x_0 \in R$ and U is a neighbourhood of x_0, there exists a continuous function $f(x)$ on R with values in $[0,1]$ and with $f(x_0)=0$ and $f(x)=1$ on $R \sim U$.*

A Hausdorff space which satisfies (V) is called completely regular, or a Tychonoff space. Thus

(1) *Every completely regular space is uniformizable.*

If x and $U(x_0)$ are given, the set $[f \leqq \frac{1}{2}]$, defined for the function $f(x)$ whose existence is assured by (V), is a closed neighbourhood W of x contained in U, so that *every completely regular space is regular.* Every subspace of a completely regular space is completely regular, so that subspaces of normal spaces, and in particular of compact spaces, are completely regular. From § 3, 6.(2) it follows that *every locally compact space is completely regular* [this also follows from 4.(5)]. Since such spaces are not always normal, the hypothesis of complete regularity is weaker than that of normality. On the other hand the following theorem holds (TYCHONOFF)

(2) *Every completely regular space is homeomorphic to a subspace of a suitable parallelotope.*

Proof. Let $\{f_\alpha\}$, $\alpha \in \mathsf{A}$, be the set of all continuous functions on R with values in $[0,1]$. By (V), the set of all $[f_\alpha < 1]$ forms a basis of open sets in R. Let A be the mapping of R into $\mathfrak{P}^{\mathsf{A}}$ which sends each $x \in R$ to the element $y \in \mathfrak{P}^{\mathsf{A}}$ with $y_\alpha = f_\alpha(x)$. Since R is Hausdorff, for any two distinct x, x' there is always an f_α with $f_\alpha(x) \neq f_\alpha(x')$, so that A is one-one. A is continuous, by § 2, 6.(2), and further is open, since the image of each set $[f_\alpha < 1]$ is open in $A(R)$, being the complement of the closed set consisting of those y with $y_\alpha = 1$.

As a special case of (2) we obtain Urysohn's embedding theorem.

(3) *Every completely regular space with a countable basis, and so for example every separable metric space, is homeomorphic to a subspace of the Hilbert cube, and is therefore normal.*

For since, by § 3, 5.(4), there are countably many functions f_i for which the sets $[f_i < 1]$ form a basis of open sets, a homeomorphism is obtained with a subspace of $\mathfrak{P}^\omega$, which by § 4, 7.(3) is homeomorphic with the Hilbert cube. Normality follows from § 4, 2.(1).

7. Metrizable uniform spaces. Our aim is the converse of 6.(1). In this number we obtain a partial result:

(1) *A uniform space R is metrizable (i.e. its uniform structure can be determined by a metric) if and only if it is Hausdorff and the vicinity filter $\mathfrak{N}$ of R has a countable base.*

The conditions are necessary, for every metric space is Hausdorff, and the vicinities $N_{1/n}$ consisting of those (x,y) with $|x,y|<1/n$ form a base of the vicinity filter.

The second part is proved in a more general form: a function $f(x,y)$ which satisfies conditions (D1), (D3) and (D4) of § 4, 1., and satisfies a weaker form of (D2), namely that $f(x,x)=0$ for each x, is called a gauge; we again use the symbol $|x,y|$. We assert that

(2) *The uniformity of a uniform space R whose vicinity filter $\mathfrak{N}$ has a countable base can be defined by a gauge.*

If R is Hausdorff, the gauge is always a metric, and so (2) implies the second half of (1).

Proof of (2). If N_i', $i=1,2,\ldots$, is the countable base of $\mathfrak{N}$,we form a sequence of symmetric vicinities $N_1,N_2,\ldots$ for which $N_1\subset N_1'$ and $N_{k+1}^3\subset N_k'\cap\bigcap_{i=1}^{k} N_i$, for $k\geqq 1$. The N_i again form a base of $\mathfrak{N}$.

We put $f(x,y)=\inf_k(\frac{1}{2})^k$, where k runs through all those indices for which $(x,y)\in N_k$; if $(x,y)\notin N_1$, we put $f(x,y)=1$. Since the N_k are symmetric, $f(x,y)=f(y,x)$; further $f(x,y)\geqq 0$ and $f(x,x)=0$.

If $(x,y)\in N_k$, $(y,y')\in N_k$ and $(y',z)\in N_k$, then $(x,z)\in N_k^3\subset N_{k-1}$; i.e. from $f(x,y)\leqq(\frac{1}{2})^k$, $f(y,y')\leqq(\frac{1}{2})^k$ and $f(y',z)\leqq(\frac{1}{2})^k$ it follows that $f(x,z)\leqq(\frac{1}{2})^{k-1}$. We deduce from this that

(3) *It follows from $f(x,y)\leqq\varepsilon$, $f(y,y')\leqq\varepsilon$ and $f(y',z)\leqq\varepsilon$ that $f(x,z)\leqq 2\varepsilon$, for every $\varepsilon>0$.*

We now define a function $|x,y|$ by

$$|x,y|=\inf\sum_{k=2}^{n} f(x_{k-1},x_k),$$

where the greatest lower bound is taken over all sequences $x_1=x,x_2,\ldots,x_n=y$ of finitely many points of R which begin with x and end with y. $|x,y|$ is a gauge, for the relations $|x,y|\geqq 0$, $|x,x|=0$ and $|x,y|=|y,x|$ result immediately from the corresponding properties of $f(x,y)$, and the triangle inequality follows from the definition of $|x,y|$, since two sequences linking x to z and z to y combine to give a sequence linking x to y.

The sets of those (x,y) with $f(x,y)\leqq 1/2^k$, $k=1,2,\ldots$, form a base for $\mathfrak{N}$, and so the sets of those (x,y) with $|x,y|\leqq 1/2^k$ again form a base for $\mathfrak{N}$, provided that the relation

$$\tfrac{1}{2}f(x,y)\leqq|x,y|\leqq f(x,y) \tag{4}$$

holds.

The second inequality of (4) is trivial. The first is established in the following way: let a sequence $x=x_1,x_2,\dots,x_n=y$ be given, and let us set $f(x_1,x_2)+\cdots+f(x_{n-1},y)=M$. We shall then prove by induction on n that $f(x,y)\leqq 2M$. This inequality is implied for $n\leqq 4$ by (3).

Suppose now that the inequality holds for all $m<n$, $n>4$.

We divide the sequence $x_1,x_2,\dots,x_{n-1},x_n$ by taking out a link x_h,x_{h+1} in such a way that the inequalities

$$\sum_2^h f(x_{k-1},x_k)\leqq \tfrac{1}{2}M, \qquad \sum_{h+2}^n f(x_{k-1},x_k)\leqq \tfrac{1}{2}M$$

hold for the two remaining sequences. By the induction hypothesis we have that $f(x,x_h)\leqq M$ and $f(x_{h+1},y)\leqq M$, and further $f(x_h,x_{h+1})\leqq M$; it follows from (3) that $f(x,y)\leqq 2M$, so that $\frac{1}{2}f(x,y)\leqq M$. Since this holds for all sequences, it also holds for the greatest lower bound of the values of M, i.e. $\frac{1}{2}f(x,y)\leqq |x,y|$.

In the case of a discrete uniform space the construction of (2) gives the metric $|x,y|=1$ for $x\neq y$.

8. The complete regularity of uniform spaces. Using 7.(2) we now prove

(1) *Every Hausdorff uniform space is completely regular.*

If N is a vicinity of R, there is a sequence N_i of symmetric vicinities with $N_1\subset N$, $N_{i+1}^2\subset N_i$. These define a uniformity on R, which is in general no longer Hausdorff. By 7(2), there is a gauge $|x,y|_N$ on R whose vicinities generate the same vicinity filter as the N_i. The vicinity filter N of R is generated by the sets $|x,y|_N<\varepsilon$, where N is arbitrary in $\mathfrak{N}$.

$|x_0,y|_N$ is a continuous function of y on R, for from $|x_0,y|_N\leqq |x_0,z|_N+|y,z|_N$ and $|x_0,z|_N\leqq |x_0,y|_N+|y,z|_N$ it follows immediately that $\big||x_0,y|_N-|x_0,z|_N\big|\leqq |y,z|_N$; thus if $|y,z|_N<\varepsilon$, then $\big||x_0,y|_N-|x_0,z|_N\big|<\varepsilon$.

If U is an arbitrary neighbourhood of $x_0\in R$, there exists an N and an $\varepsilon>0$ such that $|x_0,z|_N\geqq\varepsilon$ for all $z\in R\sim U$. Since $|x_0,z|_N$ is continuous, the function $f(x)=\operatorname{Min}\left(1,\frac{1}{\varepsilon}|x_0,x|_N\right)$ is continuous on R. But $f(x_0)=0$, and $f(z)=1$ for $z\in R\sim U$, i.e. (V) is satisfied.

CHAPTER TWO

Vector Spaces over General Fields

The first three paragraphs are concerned with the elementary and purely algebraic properties of vector spaces E over a general commutative field. In § 7 the lattice $V(E)$ of linear subspaces of E is studied, and § 8 deals with linear mappings from one vector space into another, and their representation by infinite matrices. The problem of the equivalence of these mappings is completely solved. The algebraic dual space E^* of all linear functionals on E is the theme of § 9. The lattice $\bar{V}(E^*)$ of algebraically closed subspaces of E^* turns out to be dually isomorphic with $V(E)$. The end of § 9 is concerned with the most important elementary properties of tensor products of vector spaces.

The attempt to establish a complete symmetry between the properties of E and those of E^* leads in § 10 to the study of linear topologies on vector spaces. The theory of these linearly topologized spaces is developed in §§ 10—13, following DIEUDONNÉ [4], [6], [10], LEFSCHETZ [1] and MACKEY [4]. In § 10 we define the concept of a dual system, and introduce the weak linear topology $\mathfrak{T}_{ls}$ and the topology $\mathfrak{T}_{lk}$ defined by the linearly compact subsets of the dual space, between which the original topology lies. As a first application of this theory, § 11 contains the complete theory of the solution of row- and column-finite systems of equations. A simpler constructive method is given for the countable case, which had previously been developed by TOEPLITZ [1]. § 12 contains the results of LEFSCHETZ about locally linearly compact spaces, and the theory of equations in the countable case, which was developed by TOEPLITZ and the author (cf. KÖTHE and TOEPLITZ [1]), and which is presented in the form given by DIEUDONNÉ [4]. The general theory of linearly topologized spaces is continued in § 13, with the introduction of the concepts of linear boundedness and the strong linear topology, and ends with results about strongly reflexive spaces and spaces of countable degree.

§ 7. Vector spaces

1. Definition of a vector space. Let K be an arbitrary (commutative) field with elements $\alpha, \beta, \ldots, \xi, \eta$, with zero element 0 and identity element 1. A vector space over K (linear space over K, K-module) is a set E with elements (called points or vectors) $a, b, \ldots, x, y, \ldots$ which has the following properties:

(L 1) *For every two elements $x, y \in E$ a sum $x+y$ is defined in E; under this addition, E is an abelian group, i.e. for all $x, y, z \in E$ we have*

(a) $x+y=y+x$,
(b) $x+(y+z)=(x+y)+z$,
(c) *There exists* $\mathrm{o}\in E$ *with* $x+\mathrm{o}=x$ *for all* $x\in E$,
(d) *There exists for each* $x\in E$ *an* $x'\in E$ *with* $x+x'=\mathrm{o}$.

(L2) *For every* $\xi\in\mathsf{K}$ *and every* $x\in E$ *the product* $\xi x=x\xi$ *of* ξ *with* x *is defined as an element of* E, *and for all* $x,y\in E,\xi,\eta\in\mathsf{K}$ *we have*

(e) $x(\xi+\eta)=x\xi+x\eta$,
(f) $(x+y)\xi=x\xi+y\xi$,
(g) $x(\xi\eta)=(x\xi)\eta$,
(h) $x\cdot 1=x$.

We shall establish some simple consequences.

(1) *For arbitrary* a *and* b *in* E, *the equation* $a+y=b$ *has a unique solution.*

By (d) there is an a' with $a+a'=\mathrm{o}$; application of (b), (a) and (c) shows that
$$a+(a'+b)=(a+a')+b=\mathrm{o}+b=b+\mathrm{o}=b,$$
so that $a'+b$ is a solution.

On the other hand if y_1 and y_2 are two elements with $a+y_1=a+y_2$, then by adding a' to both sides one obtains first that
$$a'+(a+y_1)=(a'+a)+y_1=(a+a')+y_1=\mathrm{o}+y_1=y_1+\mathrm{o}=y_1,$$
and secondly that $a'+(a+y_2)=y_2$, so that $y_1=y_2$. It follows from this, by considering (c), that the additive zero element o in E is uniquely determined, and further, from (d), that x' is uniquely determined. x' is written as $-x$, and the solution $b+(-a)$ of $a+y=b$ as $b-a$ (the difference of b and a).

(2) $a\cdot 0=\mathrm{o}, \mathrm{o}\cdot\alpha=\mathrm{o}$ *and* $a(-\alpha)=-(a\alpha)$ *for all* $a\in E, \alpha\in\mathsf{K}$.

By (e), $a\cdot 1=a(1+0)=a\cdot 1+a\cdot 0$. But the equation $a\cdot 1+y=a\cdot 1$ has the unique solution o, so that $a\cdot 0=\mathrm{o}$. Likewise it follows from (f) that $\mathrm{o}\cdot\alpha=(\mathrm{o}+\mathrm{o})\cdot\alpha=\mathrm{o}\cdot\alpha+\mathrm{o}\cdot\alpha$, so that $\mathrm{o}\cdot\alpha=\mathrm{o}$. Finally, $a\cdot\alpha+a\cdot(-\alpha)=a(\alpha-\alpha)$ $=a\cdot 0=\mathrm{o}$, so that $a(-\alpha)=-(a\alpha)$.

(3) *If* $x\neq\mathrm{o}$ *and* $\alpha\neq 0$ *then* $x\alpha\neq\mathrm{o}$.

For if $x\alpha$ were equal to o, then by (2), (g) and (h), $(x\alpha)\alpha^{-1}=x(\alpha\cdot\alpha^{-1})$ $=x\cdot 1=x$ would also be equal to o.

The most important rules of calculation have now been derived; the rules in which sums of n elements are considered, and the associative laws (b) and (g) with more than three elements, are deduced from them in a familiar way, using complete induction.

If K is the field P of real numbers or the field Γ of complex numbers, then E is called a r e a l or c o m p l e x vector space respectively.

2. Linear subspaces and quotient spaces. A subset H of elements of a vector space E is a vector space provided that whenever it contains x and y it also contains $x\alpha+y\beta$, for arbitrary α, β in K. H is then called a linear subspace of E. For simplicity we shall usually write $\mathfrak{o}$ for the subspace consisting only of the zero element $\mathfrak{o}$.

By a linear manifold in E we mean a subset of E consisting of all elements of the form x_0+y, where y belongs to a linear subspace H. We denote this by x_0+H. We also speak of the manifold x_0+H through x_0 parallel to H. $x_1+H=x_0+H$ for each $x_1\in x_0+H$.

If $x_1,\ldots,x_n$ are finitely many elements of E, every element of the form $x_1\alpha_1+\cdots+x_n\alpha_n$, $\alpha_i\in\mathsf{K}$, is called a linear combination of $x_1,\ldots,x_n$. If M is a finite or infinite subset of E, the collection of all linear combinations of finitely many elements of M forms a linear subspace, the linear span of M. This can also be defined as the intersection of all the linear subspaces containing M. The linear span of the set $\{x_1,\ldots,x_n\}$ is written as $[x_1,\ldots,x_n]$.

If H is a linear subspace of E and x_0 is an arbitrary element of E, the linear manifold x_0+H is also called the H-coset of x_0, and is denoted by $\hat{x}_0$. If we define the sum of two cosets $\hat{x}_0+\hat{y}_0$ to be the coset $\widehat{x_0+y_0}$, and the scalar multiple $\hat{x}_0\alpha$ to be $\widehat{x_0\alpha}$, the collection of cosets becomes a vector space over K, the quotient space E/H of E by H. The operations which have just been introduced in E/H are defined using particular representatives from the cosets. But if, instead of x_0, another element x_0+z_0 is taken from the coset $\hat{x}_0=x_0+H$, then $z_0+H=H$, so that the cosets $(x_0+y_0+z_0)+H$ and $(x_0+y_0)+H$ contain the same elements. The formation of the sum is therefore independent of the choice of representatives. The same is true for multiplication by elements of K, as is immediately verified. Since, further, both operations in E/H are defined in terms of the corresponding operations on the representatives, rules (L 1) and (L 2) carry over directly to calculations with cosets, and so E/H is a vector space over K. The zero element in E/H is clearly $\hat{\mathfrak{o}}=H$.

3. Bases and complements. A finite collection $x_1,\ldots,x_n$ of elements of E is called linearly dependent if there is a linear combination $x_1\alpha_1+\cdots+x_n\alpha_n$, with not all $\alpha_i=0$, which is equal to $\mathfrak{o}$. In this case at least one of the x_i can be written as a linear combination of the others. The elements $x_1,\ldots,x_n$ are called linearly independent if they are not linearly dependent; it then follows from a relation $x_1\alpha_1+\cdots+x_n\alpha_n=\mathfrak{o}$ that all the $\alpha_i=0$.

An infinite collection x_α, $\alpha\in\mathsf{A}$, is said to be linearly independent if each finite collection of them is linearly independent, in the sense which has just been defined.

A set $\{x_\alpha\}$, $\alpha \in \mathsf{A}$, of elements of E is called an algebraic basis of E if the x_α are linearly independent and if each element of E can be expressed as a linear combination of finitely many x_α. It follows directly from the independence of the x_α that this representation is only possible in one way, i.e. each $x \in E$ can be written as $\sum_{\alpha \in \mathsf{A}} x_\alpha \xi_\alpha$, where only finitely many ξ_α are different from 0, and the ξ_α are uniquely determined by x. We shall leave out the prefix "algebraic" for the time being, provided that there is no possible confusion with other concepts of basis.

(1) *Every vector space has a basis.*

Proof. The subsets of E which consist of linearly independent elements clearly satisfy the hypotheses of ZORN's lemma [§ 2, 2.(2)], so that there is a maximal subset $\{x_\alpha\}$ of linearly independent elements of E. Thus if $x \neq 0$ is an arbitrary element of E, the set consisting of x and the x_α cannot consist of linearly independent elements. There is therefore a linear combination of x and of finitely many of the x_α which vanishes, without all the coefficients vanishing. Because of the independence of the x_α, the coefficient of x in the linear combination must be different from zero, and so x can be written as a linear combination of the x_α.

Since, more generally, the subsets of E which consist of independent elements and which contain a fixed independent set $\{y_\beta\}$ also satisfy the hypotheses of ZORN's lemma, we have

(2) *Every system of linearly independent elements of E can be extended to a basis of E.*

Two linear subspaces G and H of E are called algebraically complementary to each other if each $x \in E$ can be represented in one and only one way as a sum $x = y + z$, with $y \in G$ and $z \in H$.

If elements y_β form a basis for G and elements z_γ form a basis for H, then the set consisting of the y_β and the z_γ is clearly a basis for E.

(3) *Every linear subspace G of a vector space E has a complement.*

Proof. Let $\{y_\beta\}$ be a basis for G and let $\{\hat{x}_\gamma\}$ be a basis for E/G. If an element z_γ is picked out of each of the cosets $\hat{x}_\gamma$, the y_β and z_γ together form a basis of E: if $x \in E$, then $\hat{x} = \sum \hat{x}_\gamma \xi_\gamma$, so that $x = \sum z_\gamma \xi_\gamma + y$, $y \in G$. The expression $x = \sum z_\gamma \xi_\gamma + \sum y_\beta \eta_\beta$ then follows from $y = \sum y_\beta \eta_\beta$. On the other hand the y_β and z_γ are linearly independent, for from $\sum y_\beta \eta_\beta + \sum z_\gamma \xi_\gamma = o$ it follows, by going over to cosets, that $\sum \hat{x}_\gamma \xi_\gamma = \hat{o}$. Hence $\xi_\gamma = 0$ for all γ, and so $\eta_\beta = 0$ for all β. The linear span H of the z_γ is clearly a complement for G.

(3) can also be proved by applying (2) to a basis $\{y_\gamma\}$ of G.

We observe that the complement to G is not in general unique, since the z_γ can be chosen arbitrarily in the $\hat{x}_\gamma$. Further let it be stressed that we have only given existence proofs, depending upon the axiom of choice, for (1) and (2). We return to constructive methods in the case of a countable basis in § 11.

4. The dimension of a linear space. A basis $\{x_\alpha\}$, $\alpha \in \mathsf{A}$, of E has a certain magnitude, given by the cardinality of A—the "number" of elements of the basis. We show that this number is the same for different bases of E.

(1) $n+1$ *linear combinations* $y_i = \sum_{k=1}^{n} x_k \alpha_{ki}$, $i=1,\dots,n+1$, *of* n *elements* $x_k \in E$ *are always linearly dependent.*

This is clear for $n=1$. Suppose that the result is true for $n-1$, and that further $\alpha_{11} \neq 0$ (which we can assume without loss of generality). Then by the induction hypothesis, the n elements $y_i' = y_i - y_1 \dfrac{\alpha_{1i}}{\alpha_{11}}$, $i=2,\dots,n+1$, are linearly dependent, being linear combinations of $x_2,\dots,x_n$, and so we can write

$$\sum_{i=2}^{n+1} y_i' \beta_i = \mathrm{o}. \tag{2}$$

But then the y_i are also dependent, for (2) implies the relation

$$\sum_{i=2}^{n+1} y_i \beta_i - y_1 \sum_{i=2}^{n+1} \frac{\alpha_{1i}}{\alpha_{11}} \beta_i = \mathrm{o}$$

and not all the β_i vanish.

If now E has a basis with finitely many elements, let $x_1,\dots,x_d$ be a basis with smallest possible d. If $\{y_\beta\}$ is a second basis with cardinal f then, by (1), f must be less than or equal to d, since on the one hand the y_β are linear combinations of the x_i, and on the other they are linearly independent.

If the bases of E are all infinite, we finish the proof as follows: Let $\{x_\rho\}$ and $\{y_\sigma\}$ be two bases with cardinality d and f respectively. There are equations

(3a) $\quad x_\rho = \sum_\sigma y_\sigma \alpha_{\sigma\rho}, \qquad$ (3b) $\quad y_\sigma = \sum_\rho x_\rho \beta_{\rho\sigma},$

in which for each ρ or σ there are only finitely many non-zero $\alpha_{\sigma\rho}$ or $\beta_{\rho\sigma}$ respectively.

In the equations (3a) each y_σ has at least one coefficient $\alpha_{\sigma\rho} \neq 0$. Indeed if all the $\alpha_{\sigma_0\rho}$ were zero for some σ_0, then we would have

$$y_{\sigma_0} = \sum_\rho x_\rho \beta_{\rho\sigma_0} = \sum_\rho \left(\sum_\sigma x_\sigma \alpha_{\sigma\rho} \right) \beta_{\rho\sigma_0} = \sum_\sigma y_\sigma \sum_\rho \alpha_{\sigma\rho} \beta_{\rho\sigma_0}$$

which expresses y_{σ_0} as a linear combination of finitely many $y_{\sigma'}$, $\sigma' \neq \sigma_0$, contradicting the linear independence of the y_σ.

Using (3a), assign to each x_ρ the set M_ρ of those finitely many y_σ for which $\alpha_{\sigma\rho} \neq 0$. The set M of all the vectors y_σ is the union of these d sets M_ρ, so that the number f of y_σ is less than or equal to $\aleph_0 d$, where $\aleph_0$ is the cardinal of the set of natural numbers. By hypothesis, d is an infinite cardinal; by a well-known theorem of set theory it follows that $\aleph_0 d = d$. $f = d$ then follows from the inequality $f \leqq d$ and the corresponding inequality $d \leqq f$.

We have therefore shown

(4) *Two different bases of a vector space E always have the same number of basic elements.*

This cardinal is called the algebraic dimension $d(E)$ of E.

5. Isomorphism, canonical form. Two vector spaces E_1 and E_2 over the same field K are called algebraically isomorphic if there is a one-one correspondence $x \leftrightarrow x'$ between the elements of E_1 and the elements of E_2, under which $(x\alpha + y\beta)' = x'\alpha + y'\beta$, for all $x, y \in E_1$ and all $\alpha, \beta \in$ K. We shall use the symbol $E_1 \cong E_2$ for this.

The result of the preceding number can now be expressed as follows:

(1) *Two vector spaces E_1 and E_2 over the same field* K *are isomorphic if and only if they have the same dimension.*

Since an isomorphism sends a basis of E_1 into a basis of E_2, the dimensions must coincide. On the other hand if they are equal, then there is a one-one correspondence $x_\alpha \leftrightarrow x'_\alpha$ between a basis of E_1 and a basis of E_2, which defines an isomorphism $x = \sum x_\alpha \xi_\alpha \leftrightarrow \sum x'_\alpha \xi_\alpha = x'$.

To each cardinal number d there is, up to isomorphism, at most one vector space over K of dimension d. On the other hand a vector space of dimension d over K is obtained in the following way:

Let α run through a set A of magnitude d. If a $\xi_\alpha \in$ K is made to correspond to each $\alpha \in$ A, and if only finitely many ξ_α are non-zero, then we call the function defined in this way a finite vector $\mathfrak{x} = \{\xi_\alpha\}$, with d coordinates ξ_α. As usual we define $\mathfrak{x} + \mathfrak{y}$ as the vector with coordinates $\xi_\alpha + \eta_\alpha$, and further define $\mathfrak{x}\rho$, $\rho \in$ K, as the vector with coordinates $\xi_\alpha \rho$; the set E of these vectors is then a vector space over K. The null-vector $\mathfrak{o}$ with all coordinates vanishing is the zero element of E. We denote by $\mathfrak{e}_\alpha$ the α-th unit vector, whose α-coordinate is 1, and whose other coordinates vanish. Clearly $\mathfrak{x} = \sum_\alpha \mathfrak{e}_\alpha \xi_\alpha$, and the $\mathfrak{e}_\alpha$ are linearly independent and therefore form a basis for E.

We call the space E obtained in this way a d-dimensional finite coordinate space φ_d(K) over K. If $d = \aleph_0$, we write simply φ(K).

In this expression the index set A does not appear explicitly, and it can always be replaced by one with the same cardinality. If a uniquely determined canonical form is required, it is natural to take the set Ω_d of those ordinals with smaller cardinality than d. This corresponds to the practice, in the finite and countable cases, of taking natural numbers as indices, and has the advantage that one can fall back upon the order of Ω_d in problems for which an ordering of the coordinates is useful. Let it be stressed, however, that this order is not essentially connected with the concept of basis, but is introduced in addition.

By (1), an isomorphism is obtained between a general d-dimensional vector space E and φ_d, when the elements x_α of a basis of E are made to correspond to the unit vectors $\mathfrak{e}_\alpha$ of φ_d.

6. Sums and intersections of subspaces. A linear subspace of E is always obtained when the set theoretic intersection $\bigcap_\alpha F_\alpha$ of a collection of linear subspaces F_α of E is formed. The intersection of a collection of linear manifolds is either empty or a linear manifold. If in a corresponding way the union $\bigcup_\alpha F_\alpha$ is taken, it need not in general be a linear subspace, although its linear span is. This linear span is called the s u m $\sum_\alpha F_\alpha$ of the F_α. For general subsets M_α of E, $\sum_\alpha M_\alpha$ denotes the set of all finite sums $\sum_{i=1}^{n} x_{\alpha_i} (x_{\alpha_i} \in M_{\alpha_i})$. For linear subspaces M_α, this is again the linear span of the M_α. The plus sign is used as well, when there are finitely many summands, e.g. $M_1 + M_2$.

The sum $F = \sum F_\alpha$ is said to be d i r e c t, if every $x \in F$ can be written in only one way as $\sum x_\alpha$, $x_\alpha \in F_\alpha$ (of course, there are only finitely many $x_\alpha \neq \mathrm{o}$). For direct sums we use the symbols $\bigoplus_\alpha F_\alpha$ and $F_1 \oplus F_2$ respectively. An example of a direct sum is given by two complementary subspaces G and H of E; $E = G \oplus H$.

(1) *The sum* $\sum F_\alpha$ *is direct if and only if* $F_\alpha \cap \sum_{\beta \neq \alpha} F_\beta = \mathrm{o}$, *for each* α.

If the condition is satisfied, it follows from $\sum_\alpha x_\alpha = \mathrm{o}$ that $x_\alpha = \sum_{\beta \neq \alpha} (-x_\beta)$, and so $x_\alpha \in F_\alpha \cap \sum_{\beta \neq \alpha} F_\beta$; this implies that $x_\alpha = \mathrm{o}$ for all α, so that the representation $x = \sum_\alpha x_\alpha$ as a sum is unique.

On the other hand, if this is the case, and if $z \in F_\alpha \cap \sum_{\beta \neq \alpha} F_\beta$, then the representation $\mathrm{o} = z - z$, with $z \in F_\alpha$ and $-z \in \sum_{\beta \neq \alpha} F_\beta$, must coincide with $\mathrm{o} = \mathrm{o} + \mathrm{o}$, so that $z = \mathrm{o}$.

(2) *If F and G are two linear subspaces of E, and if F_1 and G_1 are complementary to $F \cap G$ in F and G respectively, so that*

$$F = F_1 \oplus (F \cap G), \qquad G = G_1 \oplus (F \cap G), \tag{3}$$

then

$$F + G = F_1 \oplus G_1 \oplus (F \cap G). \tag{4}$$

In this way the sum $F+G$ is reduced to a direct sum.

Proof. Clearly $F+G=F_1+G_1+(F\cap G)$. But this sum is direct, for it follows from $x+y+z=o$, $x\in F_1$, $y\in G_1$, $z\in F\cap G$, that $x=-y-z\in G$ and so that $x\in F_1\cap G=F_1\cap(F\cap G)=o$, which shows that $x=o$. But then $y=z=o$ must hold, by (3).

It follows from (2) that

(5) *If F and G are two linear subspaces of E, there is a complement H of F for which $G=(G\cap F)\oplus(G\cap H)$ holds.*

For if G_1 is defined by (3) and if L is a complement of $F+G$ in E, $H=G_1\oplus L$ has the required properties.

The next two isomorphism theorems, well-known in group theory, follow from (2).

(6) *If F and G are two linear subspaces of E, then $(F+G)/G\cong F/(F\cap G)$.*

(7) *If F and G are two linear subspaces of E, with $F\subset G$, then $E/G\cong(E/F)/(G/F)$.*

Proof. By (3), $F/(F\cap G)\cong F_1$, and by (4), $(F+G)/G\cong F_1$, which establishes (6).

If further $F\subset G\subset E$, then we can write $E=F\oplus F_1\oplus H$, where F_1 is a complement of F in G and H is a complement of G in E. $E/G\cong H$ and $E/F\cong F_1\oplus H$; under this isomorphism G/F is mapped onto F_1, so that $(E/F)/(G/F)$ is also isomorphic to H.

7. Dimension and co-dimension of subspaces. Let G be a linear subspace of E, and let H be a complement of G in E. If each $z\in H$ is made to correspond to its coset $\hat{z}$ in the quotient space E/G, an isomorphism is established between H and E/G, so that

(1) *All the complementary subspaces H of a linear subspace G of E have the same dimension, namely the dimension of E/G.*

The dimension of E/G is called the co-dimension, or the defect, $c(G)$ of G in E. We have

$$d(G)+c(G)=d(E), \tag{2}$$

for a basis of G and a basis of one of its complementary spaces together form a basis of E.

More generally, interpreting the sums as sums of cardinal numbers, we have

(3) *If* $E=\bigoplus_{\alpha} F_\alpha$, *then* $d(E)=\sum_{\alpha} d(F_\alpha)$.

(4) *If F and G are two linear subspaces of E, then*

$$d(F+G)+d(F\cap G)=d(F)+d(G), \tag{5}$$

$$c(F+G)+c(F\cap G)=c(F)+c(G). \tag{6}$$

By going over to dimensions, equation (5) can be read off from equations (3) and (4) of the preceding number.

If, further, H is a complementary subspace of $F+G$ in E, then we have by 6.(4)

$$E=H\oplus F_1\oplus G_1\oplus F\cap G.$$

(6) follows by comparing the dimensions of the complements of $F\cap G$, F, G and $F+G$ which appear in this decomposition.

If a linear subspace H has codimension 1 in E, both H and manifolds x_0+H parallel to H are called hyperplanes in E.

(7) *Every linear manifold is the intersection of the hyperplanes containing it.*

It if sufficient to show this for a linear subspace F. Suppose that z does not lie in F. The linear span $[z]$ of z consists of all multiples $z\xi$, $\xi\in\mathsf{K}$. The sum $G=F+[z]$ is direct; thus if H is a complement of G, $E=F\oplus[z]\oplus H$. Clearly $F\oplus H$ has codimension 1 in E, and it is therefore a hyperplane, which does not contain z. Since z is an arbitrary element not lying in F, the intersection of all the hyperplanes containing F is equal to F.

8. Products and direct sums of vector spaces. Given vector spaces E_α, $\alpha\in\mathsf{A}$, the following is a simple method of constructing a new one.

We form the set-theoretic product $E=\prod_{\alpha} E_\alpha$ (cf. § 1, 8.). E becomes a vector space when we define $x+y$ to be the function $x_\alpha+y_\alpha$ on A, and define $x\xi$, $\xi\in\mathsf{K}$, by $(x\xi)_\alpha=x_\alpha\xi$. E is called the product of the E_α. If all the E_α are equal to F, we also write F^{A} for the product, or even F^d, where d is the cardinality of A, when it does not matter going over to an isomorphic space.

In particular if F is equal to K, K^d is the vector space of all vectors $\mathfrak{x}=\{\xi_\alpha\}$, with d arbitrary coordinates in K: this is also called the linear coordinate space $\omega_d(\mathsf{K})$; in the case where $d=\aleph_0$, we simply write $\omega(\mathsf{K})$. As with $\varphi_d(\mathsf{K})$, we may take the ordinals of Ω_d as coordinate indices, if necessary. Let us remark straight away that ω_d need not have dimension d (cf. § 9, 5.).

The collection F of those $x \in E = \prod_\alpha E_\alpha$ with only finitely many non-zero x_α again forms a vector space. This is clearly equal to the direct sum $\bigoplus_\alpha \tilde{E}_\alpha$, where $\tilde{E}_\alpha$ is the vector space, isomorphic to E_α, which consists of those $x \in E$ with $x_\beta = o$ for all $\beta \neq \alpha$. We write $\bigoplus_\alpha E_\alpha$ for F, identifying $\tilde{E}_\alpha$ with E_α, and again call $\bigoplus_\alpha E_\alpha$ the direct sum of the E_α.

If all the E_α, $\alpha \in \mathsf{A}$, are equal to K, then $\bigoplus_\alpha E_\alpha$ is just the space $\varphi_d(\mathsf{K})$ which has already been introduced in 5., where d is the cardinality of A.

$\bigoplus_\alpha E_\alpha$ is a subspace of $\prod_\alpha E_\alpha$; they coincide if and only if the set A of indices α is finite. In this case either expression $E_1 \times \cdots \times E_n$ or $E_1 \oplus \cdots \oplus E_n$ can be used.

A further method of forming products, the tensor product of vector spaces, is dealt with in § 9, 6.

9. Lattices. In 6., we introduced the two operations $+$ and $\cap$ on the linear subspaces of E. We can understand their properties most clearly by using the concept of a lattice. For this we refer to the ideas of § 2.

A partially ordered space V is called a lattice if every set consisting of two points $a, b \in V$ has a least upper bound c and a greatest lower bound d in V.

c is called the union $a \vee b$ of a and b, d the intersection $a \wedge b$ of a and b.

A lattice V is said to be complete if it satisfies the stronger condition that any arbitrary set $\{a_\alpha\}$ of elements of V has a least upper bound and a greatest lower bound. These elements are again called the union $\bigvee_\alpha a_\alpha$ and the intersection $\bigwedge_\alpha a_\alpha$ respectively.

It follows easily from the definition of the operations $\vee$ and $\wedge$ that they are commutative and associative (even for infinitely many terms) and further that they are monotonic with respect to the partial order, i.e.

(1) *If* $a \leqq b$, *then* $a \vee c \leqq b \vee c$ *and* $a \wedge c \leqq b \wedge c$.

As an example, let us mention the topologies on a set, which, by § 1, 6., form a complete lattice with respect to the relation "finer".

Two lattices V_1 and V_2 are said to be isomorphic if there is a one-one correspondence $a_1 \leftrightarrow a_2$ between the elements of V_1 and the elements of V_2, under which $a_1 \leqq b_1$ if and only if $a_2 \leqq b_2$. Unions and intersections of corresponding elements also correspond.

Two lattices V_1 and V_2 are said to be dually isomorphic if there is a one-one correspondence $a_1 \leftrightarrow a_2$ under which $a_1 \leqq b_1$ if and only if $b_2 \leqq a_2$. The union of a collection of elements then corresponds to the intersection of the corresponding elements, and conversely.

To every lattice there corresponds its dual, obtained by interchanging $\leqq$ and $\geqq$.

If there are elements 0 and 1 in a lattice V with $0 \leqq a$ and $a \leqq 1$, for all $a \in V$, then they are called the zero element and unit element of V respectively. Every complete lattice has a zero and a unit element.

A lattice is said to be modular if it satisfies the condition

(2) *If* $a \leqq c$, *then* $a \vee (b \wedge c) = (a \vee b) \wedge c$,

and is said to be distributive if it satisfies the stronger condition

(3) $a \vee (b \wedge c) = (a \vee b) \wedge (a \vee c), \quad a \wedge (b \vee c) = (a \wedge b) \vee (a \wedge c)$

for all a, b and c.

A lattice V with a zero and a unit element is said to be complemented if

(4) *For each* a *there is at least one* a' *with* $a \vee a' = 1$, $a \wedge a' = 0$.

A complemented distributive lattice is called a Boolean algebra.

If the symbols $\vee$ and $\wedge$ are interchanged in (2), (3) and (4) and if $\leqq$ is replaced by $\geqq$ in (2), then the statements remain unaltered, and so these properties are preserved under isomorphisms and dual isomorphisms.

10. The lattice of linear subspaces. We show

(1) *The linear subspaces of a vector space* E *form a complete complemented modular lattice* $V(E)$ *under the relation* $A \subset B$. *The lattice operations* $\bigvee_\alpha$ *and* $\bigwedge_\alpha$ *are just* $\sum_\alpha$ *and* $\bigcap_\alpha$, *respectively.*

Proof. It is immediately clear that $V(E)$ is a complete lattice. To say that it is complemented is to say that each A has a complementary subspace. It remains to show that $V(E)$ is modular.

Suppose that $A \subset C$. If x is an element of $(A+B) \cap C$, then, considered as an element of $A+B$, it can be written as $x = y + z$, $y \in A$, $z \in B$. Since $x \in C$ and $y \in A \subset C$, it follows that $z \in C$, and so $z \in B \cap C$. Thus $x \in A + (B \cap C)$. We have therefore shown that $(A+B) \cap C \subset A + (B \cap C)$.

On the other hand the following holds in any lattice, as is easy to see:

(2) *If* $a \leqq c$, *then* $a \vee (b \wedge c) \leqq (a \vee b) \wedge c$.

From this the proposition follows.

$V(E)$ is not distributive, for although, as in any lattice, one half of the distributive law holds, namely

(3) $\bigcap_{\alpha,\beta} (F_\alpha + G_\beta) \supset \left(\bigcap_\alpha F_\alpha\right) + \left(\bigcap_\beta G_\beta\right), \quad \sum_{\alpha,\beta} (F_\alpha \cap G_\beta) \subset \left(\sum_\alpha F_\alpha\right) \cap \left(\sum_\beta G_\beta\right),$

equality does not always hold in (3), even for the linear subspaces of a finite-dimensional E (example!).

We also remark that, using 6.(1), the idea of direct sum can be defined using $+$ and $\cap$, and so can be defined in terms of the lattice.

§ 8. Linear mappings and matrices

1. Definition and rules of calculation. Let E and F be two vector spaces over K. A correspondence A, which sends each $x \in E$ to an $Ax \in F$, is called a *linear mapping*, a *linear transformation* or a *homomorphism* of E into F if

$$(1) \qquad A(x\alpha + y\beta) = (Aa)\alpha + (Ay)\beta$$

holds, for all α, $\beta \in \mathsf{K}$ and all x, $y \in E$.

The mapping which sends each $x \in E$ to the zero element of F is denoted by 0.

The sum $A+B$ and the product $A\alpha$ of mappings of E into F are defined by

$$(2) \qquad (A+B)x = Ax + Bx, \qquad (A\alpha)x = (Ax)\alpha.$$

Clearly we have

(3) *The set $\mathfrak{S}(E,F)$ of linear mappings of E into F forms a vector space over K.*

If B maps E into F and A maps F into G, then the formula $(AB)x = A(Bx)$ defines a linear mapping of E into G, the *product* AB. This product is associative, the two distributive laws $A(B+C) = AB + AC$, $(A+B)C = AC + BC$ both hold, and by (1) we have $(AB)\alpha = A(B\alpha) = (A\alpha)B$.

A linear mapping of E into itself is also called an *endomorphism*. We write $\mathfrak{S}(E)$ for the set of all endomorphisms of E. The identity endomorphism is denoted by I.

A vector space R over K is called an *algebra* over K if a product ab is defined for any two elements a and b of R, which satisfies the rules

$$(4) \qquad (ab)c = a(bc),$$

$$(5) \qquad a(b+c) = ab + ac, \qquad (a+b)c = ac + bc,$$

$$(6) \qquad (ab)\alpha = a(b\alpha) = (a\alpha)b.$$

With this terminology we have

(7) *The set $\mathfrak{S}(E)$ of endomorphisms of a vector space E over K is an algebra over K, with I as multiplicative identity element.*

2. The four characteristic spaces of a linear mapping. Let A be a linear mapping of E into F. The set $A(H)$ of images of a linear subspace H of E forms a linear subspace of F. In particular, $A(E)$ is called the image space of A.

The set of all $y \in E$ with $Ay = o$ forms a linear subspace of E, the null-space or kernel $N[A]$ of the mapping.

If $N[A] = o$, then A is one-one, and A is called a monomorphism of E into F. In agreement with § 7, 5., A is called an isomorphism of E onto F if $N[A] = o$ and $A(E) = F$. An isomorphism of E onto itself is also called an automorphism. The automorphisms of E form a group, the linear group $\Lambda(E)$ of E.

If H is a linear subspace of E, the monomorphism $J \in \mathfrak{S}(H, E)$ which sends each $y \in H$ to the same element y, now considered as an element of E, is called the embedding (injection) of H in E.

If $A(E) = F$, A is said to be an epimorphism of E onto F. If H is a linear subspace of E, then the mapping K which sends each $x \in E$ to the coset $\hat{x}$ in E/H is an epimorphism of E onto E/H, which we call the canonical mapping of E onto E/H. With this terminology we have

(1) *Every linear mapping A of E into F is the product $J \check{A} K$ of the canonical mapping K of E onto $E/N[A]$, an isomorphism $\check{A}$ of $E/N[A]$ onto $A(E)$, and the embedding J of $A(E)$ into F. The mapping $\hat{A} = J \check{A}$ is a monomorphism of $E/N[A]$ into F.*

We call $E/N[A]$ the inverse-image space of A. By § 7, 7.(1) $N[A]$ has complementary subspaces $U[A]$, which, it is true, are not uniquely determined, but which are all isomorphic to $E/N[A]$. A then maps each such space $U[A]$ isomorphically onto $A[E]$. $U[A]$ will also be called an inverse-image space of A.

The fourth characteristic space for A is $F/A(E)$. The complementary subspaces of $A(E)$ in F are again isomorphic representatives of $F/A(E)$; these we call complements $C[A]$ of the image space, as indeed we call $F/A(E)$.

3. Projections. If $E = E_1 \oplus E_2$ is a decomposition of E into complementary subspaces, then each x in E can be expressed uniquely in the form $x = x_1 + x_2$, with $x_1 \in E_1$ and $x_2 \in E_2$. If we send each x to its component x_1 in E_1, then this mapping is linear, it maps E_1 identically onto itself and it sends E_2 to zero. This endomorphism is called the projection P_{E_1} of E onto E_1, with null-space E_2, or in the direction E_2.

$P_{E_1}^2 = P_{E_1}$, so that the endomorphism is an idempotent element of the algebra $\mathfrak{S}(E)$.

If conversely P is an idempotent endomorphism of E, so that $P^2 = P$, then it follows from $P(Px) = P^2 x = Px$ that the subspace $E_1 = P(E)$ is mapped identically onto itself by P. The set of all $x-Px$ forms a linear subspace E_2 of E, which is sent to zero by P. The decomposition $x=Px+(x-Px)=x_1+x_2$ is unique, for from $x_1+x_2=o$, $x_1\in E_1$, $x_2\in E_2$, it follows, by multiplying by P, that $x_1=o$. E_2 is therefore complementary to E_1. As can be seen immediately, $I-P$ is the projection of E onto E_2, with null-space E_1, and $P(I-P)=(I-P)P=0$. Thus we have established

(1) *To every complementary decomposition* $E=E_1\oplus E_2$ *there correspond two projections* P_{E_1} *and* P_{E_2}, *with* $E_1=P_{E_1}(E)$, $E_2=P_{E_2}(E)$, $P_{E_1}P_{E_2}=P_{E_2}P_{E_1}=0$ *and* $P_{E_1}+P_{E_2}=I$.

Conversely every idempotent mapping $P\in\mathfrak{S}(E)$ *determines a complementary decomposition* $E=P(E)\oplus(I-P)(E)$.

As a first application we show

(2) *If* $E=\bigoplus_\alpha E_\alpha$, *then* $\mathfrak{S}(E,F)$ *is isomorphic, as a vector space over* K, *to* $\prod_\alpha \mathfrak{S}(E_\alpha,F)$.

Let P_α be the projection of E onto E_α with null-space $\bigoplus_{\alpha'\neq\alpha} E_{\alpha'}$. Then for each $x\in E$ we have the representation $x=\sum_\alpha x_\alpha=\sum P_\alpha x$. If $A\in\mathfrak{S}(E,F)$, then we have

$$Ax=\sum_\alpha A(P_\alpha x)=\sum_\alpha (AP_\alpha)(P_\alpha x). \tag{3}$$

AP_α, considered as a mapping from E_α in F, is denoted by A_α. Thus, by (3), to each A there corresponds a vector $\{A_\alpha\}\in\prod_\alpha\mathfrak{S}(E_\alpha,F)$.

Conversely, given such a vector $\{A_\alpha\}$, the formula

$$Ax=\sum_\alpha A_\alpha(P_\alpha x) \tag{4}$$

defines a linear mapping in $\mathfrak{S}(E,F)$.

4. Inverse mappings. If $A\in\mathfrak{S}(E,F)$ and $B\in\mathfrak{S}(F,E)$, and if $BA=I_E$, where I_E is the identity mapping of E, then B is called a left inverse or left reciprocal of A. Similarly $C\in\mathfrak{S}(F,E)$ is called a right inverse of A if $AC=I_F$, where I_F is the identity mapping of F.

If $B\in\mathfrak{S}(F,E)$ is a left inverse of A, and if $C\in\mathfrak{S}(F,E)$ is a right inverse of A, then $B=BI_F=B(AC)=(BA)C=I_EC=C$. In this case we speak of the (two-sided) inverse of A, and denote it by A^{-1}. A mapping A which has an inverse A^{-1} is said to be invertible.

(1) *A mapping* $A\in\mathfrak{S}(E,F)$ *is invertible if and only if it is an isomorphism of E onto F.*

For if A is an isomorphism, the inverse $A^{(-1)}$ is a mapping of F onto E with $A^{(-1)}A=I_E$ and $AA^{(-1)}=I_F$. On the other hand if A is invertible, A is one-one, for it follows from $Ax=\text{o}$ that $A^{-1}(Ax) = (A^{-1}A)x = x = \text{o}$. Further, $A(E)=F$, for given $y \in F$, $y=(AA^{-1})y = A(A^{-1}y)$, so that y is the image of $A^{-1}y \in E$.

If A is not an isomorphism of E onto F, then either $A(E)$ is a proper subspace of F, or A is not one-one, or both, so that the inverse $A^{(-1)}$ is either defined only on a proper subspace of F, or is not a point-mapping of F into E. In neither case, therefore, is it a mapping in $\mathfrak{S}(F,E)$.

We now investigate the extent to which it is generally possible to invert A, using a suitable mapping in $\mathfrak{S}(F,E)$. By 2., we can choose an inverse image space $U[A]$, complementary to the kernel $N[A]$, and a space $C[A]$ complementary to $A(E)$. A maps $U[A]$ isomorphically onto $A(E)$. This correspondence can be inverted in a unique way. If we further stipulate that the elements of $C[A]$ are mapped into the zero element of E, then a linear mapping B is defined on the whole of F, with kernel $C[A]$, inverse image space $A(E)$ and image space $U[A]$. If we combine A and B, BA turns out to be the projection of E onto $U[A]$, with null-space $N[A]$, whereas AB is the projection of F onto $A(E)$, with null-space $C[A]$. Conversely, by 3.(1), every projection $P_{N[A]}$ determines a specific $U[A]$, and every projection $P_{A(E)}$ determines a specific $C[A]$, and so we arrive at the following conclusion:

(2) *If $A \in \mathfrak{S}(E,F)$, if $P_{N[A]}$ is a projection of E onto the null-space of A and if $P_{A(E)}$ is a projection of F onto the image space of A, then there is a mapping $B \in \mathfrak{S}(F,E)$ with*

(3) $$BA = I_E - P_{N[A]}, \qquad AB = P_{A(E)}.$$

Using this, we can answer the question of when A has a left or right inverse.

(4) *Let $A \neq O$ be a linear mapping of E into F. Then one of the four following cases holds:*

1. $N[A]=\text{o}$, $A(E)=F$. *A is invertible.*

2. $N[A]\neq\text{o}$, $A(E)\neq F$. *A has neither a left nor a right inverse.*

3. $N[A]\neq\text{o}$, $A(E)=F$. *A has no left inverse, and has at least two right inverses.*

4. $N[A]=\text{o}$, $A(E)\neq F$. *A has at least two left inverses, but has no right inverse.*

Proof. A can only have a left inverse if $N(A)=\text{o}$, and can only have a right inverse if $A(E)=F$. The conditions are therefore necessary. But if they are satisfied, then the existence of left and right inverses follows from (3). The fact that there is more than one inverse in cases 3.

and 4. follows from the fact that there is more than one projection $P_{N[A]}$ and $P_{A(E)}$ respectively.

Examples. If E is n-dimensional, then cases 3. und 4. cannot arise for endomorphisms of E, as is well-known. If, however, we consider the identity mapping of $\varphi_n(\mathsf{K})$ into itself as a mapping of $\varphi_n(\mathsf{K})$ into $\varphi_m(\mathsf{K})$, with $m>n$, we obtain an example of the fourth case.

If K has only finitely many elements, then in this example there are only finitely many left inverses. If K has infinitely many elements, then there are always infinitely many inverses in cases 3. and 4.

If $x_0, x_1, \ldots$ is a basis of $\varphi(\mathsf{K})$, then the mapping $x_i \to x_{i+1}$, $i=0,1,\ldots$, is an example of case 4., and the mapping $x_0 \to o$, $x_{i+1} \to x_i$, $i=0,1,\ldots$, an example of case 3.

5. Representation by matrices. It is easy to obtain a general picture of all the linear mappings of E into F. To this end, we choose bases $\{x_\nu\}$, $\nu \in \mathsf{N}$, and $\{y_\mu\}$, $\mu \in \mathsf{M}$, in E and F respectively. If we specify an arbitrarily chosen $z_\nu \in F$ as image, for each x_ν, then the correspondence $x = \sum_\nu x_\nu \xi_\nu \to y = \sum_\nu z_\nu \xi_\nu$ defines a linear mapping A, and conversely every linear mapping is determined by the images $A x_\nu = z_\nu$, in this way. Each z_ν can be expressed in terms of the basis y_μ of F

$$z_\nu = \sum_\mu y_\mu \alpha_{\mu\nu}, \qquad \alpha_{\mu\nu} \in \mathsf{K},\ \nu \in \mathsf{N}. \tag{1}$$

If we combine the elements $A x_\nu$, $\nu \in \mathsf{N}$, to form a vector $\{A x_\nu\}$, then A is given by the equation

$$\{A x_\nu\} = \Big\{\sum_\mu y_\mu \alpha_{\mu\nu}\Big\}. \tag{2}$$

The mapping is therefore fully determined by the coefficients $\alpha_{\mu\nu}$. Following normal practice, we call the collection of terms $\alpha_{\mu\nu}$ a m a t r i x $\mathfrak{A} = ((\alpha_{\mu\nu}))$, $\mu \in \mathsf{M}$, $\nu \in \mathsf{N}$, with elements in K, and defined over $\mathsf{M} \times \mathsf{N}$. The terms $\alpha_{\mu_0\nu}$, $\nu \in \mathsf{N}$, form the μ_0-th r o w and the terms $\alpha_{\mu\nu_0}$, $\mu \in \mathsf{M}$, the ν_0-th c o l u m n; the elements of M and N are called row- and column-indices respectively.

It follows from (1) that every column of our matrix $\mathfrak{A}$ has only finitely many non-zero elements, and $\mathfrak{A}$ is therefore said to be c o l u m n - f i n i t e. If d and e are the dimensions of E and F respectively, then $\mathfrak{A}$ has e rows and d columns.

If $\mathfrak{A} = ((\alpha_{\mu\nu}))$, $\mu \in \mathsf{M}$, $\nu \in \mathsf{N}$, is a matrix, and if $\mathfrak{x} = \{\xi_\mu\}$, $\mu \in \mathsf{M}$, is a vector over the set of row indices of $\mathfrak{A}$, then the product $\mathfrak{y} = \mathfrak{x}\mathfrak{A}$ is defined as the vector with components $\eta_\nu = \sum_\mu \xi_\mu \alpha_{\mu\nu}$, provided that each of these sums has only finitely many non-zero summands. In a corresponding way $\mathfrak{A}\mathfrak{z}$ is defined as the vector with components $\sum_\nu \alpha_{\mu\nu} \zeta_\nu$, when $\mathfrak{z} = \{\zeta_\nu\}$ is a vector defined over N.

With this terminology, (2) becomes

$$\text{(3)} \qquad \{A x_\nu\} = \{y_\mu\}\mathfrak{A}.$$

Given $x = \sum x_\nu \xi_\nu \in E$, it follows from (2) that

$$\text{(4)} \qquad A x = \sum_\nu (A x_\nu)\xi_\nu = \sum_\nu \Big(\sum_\mu y_\mu \alpha_{\mu\nu}\Big)\xi_\nu = \sum_\mu y_\mu \sum_\nu \alpha_{\mu\nu}\xi_\nu,$$

or, in terms of matrices and vectors,

$$\text{(5)} \qquad A x = \{A x_\nu\}\mathfrak{x} = (\{y_\mu\}\mathfrak{A})\mathfrak{x} = \{y_\mu\}(\mathfrak{A}\mathfrak{x}).$$

If x is replaced by the vector $\mathfrak{x}$ with respect to the basis x_ν in E, then Ax is represented by $\mathfrak{A}\mathfrak{x}$ with respect to the basis y_μ in F. Thus we have shown

(6) *If bases* $\{x_\nu\}$, $\nu \in \mathsf{N}$, *and* $\{y_\mu\}$, $\mu \in \mathsf{M}$, *are chosen in E and F respectively, then using* (3), *every linear mapping* $y = Ax$ *from E into F corresponds to a column-finite matrix* $\mathfrak{A} = ((\alpha_{\mu\nu}))$, $\mu \in \mathsf{M}$, $\nu \in \mathsf{N}$; *if* $\mathfrak{x}$ *is the vector corresponding to x with respect to the basis* $\{x_\nu\}$, *and* $\mathfrak{y}$ *the vector corresponding to y with respect to the basis* $\{y_\mu\}$, *then the expression* $\mathfrak{y} = \mathfrak{A}\mathfrak{x}$ *is the representation of A.*

Conversely any column-finite matrix with e rows and d columns can be interpreted as representing a linear mapping from a d-dimensional vector space into an e-dimensional vector space.

Let B be a linear mapping of F into G, and suppose that, for a basis $\{z_\lambda\}$, $\lambda \in \Lambda$, we have

$$\text{(7)} \qquad B y_\mu = \sum_\lambda z_\lambda \beta_{\lambda\mu}.$$

Then for the compound mapping BA we have

(8)

$$(BA)x_\nu = B(Ax_\nu) = \sum_\mu (B y_\mu)\alpha_{\mu\nu} = \sum_\mu \Big(\sum_\lambda z_\lambda \beta_{\lambda\mu}\Big)\alpha_{\mu\nu} = \sum_\lambda z_\lambda \sum_\mu \beta_{\lambda\mu}\alpha_{\mu\nu}.$$

The matrix of BA is therefore the matrix $\Big(\Big(\sum_\mu \beta_{\lambda\mu}\alpha_{\mu\nu}\Big)\Big)$, $\lambda \in \Lambda$, $\nu \in \mathsf{N}$, which is written as the m a t r i x p r o d u c t $\mathfrak{B}\mathfrak{A}$; the elements of it are the scalar products of rows of $\mathfrak{B}$ with columns of $\mathfrak{A}$.

The associative law for mappings goes over directly to the representative column-finite matrices, so that we always have $\mathfrak{C}(\mathfrak{B}\mathfrak{A}) = (\mathfrak{C}\mathfrak{B})\mathfrak{A}$. The sum $A+B$ of two mappings in $\mathfrak{S}(E,F)$ corresponds to the sum $\mathfrak{A}+\mathfrak{B} = ((\alpha_{\mu\nu}+\beta_{\mu\nu}))$ of the corresponding matrices with respect to the same bases, and the mapping $A\rho$, $\rho \in \mathsf{K}$, corresponds to the matrix $\mathfrak{A}\rho = ((\alpha_{\mu\nu}\rho))$. The mapping 0 is represented by the zero-matrix $\mathfrak{O}$, all of whose elements are zero, and the identity mapping I by the unit matrix $\mathfrak{E} = ((\varepsilon_{\mu\nu}))$, with $\varepsilon_{\mu\mu} = 1$ and $\varepsilon_{\mu\nu} = 0$ for $\mu \neq \nu$.

6. Rings of matrices. A matrix $\mathfrak{A}=((\alpha_{\mu\nu}))$, $\mu\in\mathsf{M}$, $\nu\in\mathsf{N}$, is said to be square if $\mathsf{M}=\mathsf{N}$. A matrix with the same number of rows and columns need not be square, but it can always be changed into a square one by modifying the indices. The operations $\mathfrak{A}\rho$, $\mathfrak{A}+\mathfrak{B}$ and $\mathfrak{B}\mathfrak{A}$, introduced in 5. for column-finite matrices, may also be defined for general matrices, although it should be observed that the product $\mathfrak{B}\mathfrak{A}$ can only be defined if the set of column-indices of $\mathfrak{B}$ coincides with the set of row-indices of $\mathfrak{A}$, and if each sum $\sum_{\mu}\beta_{\lambda\mu}\alpha_{\mu\nu}$ has only finitely many non-zero summands.

A set $\mathfrak{M}$ of square matrices over the index set N, with elements in K, is called a linear matrix ring over K if $\mathfrak{M}$ forms an algebra over K with respect to the matrix operations $\mathfrak{A}\rho$, $\mathfrak{A}+\mathfrak{B}$ and $\mathfrak{B}\mathfrak{A}$.

Two algebras R and R' over K are said to be isomorphic if there is a one-one correspondence $a\leftrightarrow a'$ between the elements of R and R' for which

$$(1)\qquad (a+b)'=a'+b', \quad (ab)'=a'b', \quad (a\rho)'=a'\rho$$

holds, for all a, $b\in R$ and all $\rho\in\mathsf{K}$. They are said to be anti-isomorphic if the one-one correspondence satisfies the equation $(ab)'=b'a'$ instead of the second equation in (1).

We consider the endomorphisms A of a d-dimensional vector space E. If $\{x_\nu\}$ is a basis of E, we can take the set Ω_d of ordinals of magnitude smaller than d as index set. We take the same basis in $F=E$, so that the matrix corresponding to A, as in 5.(3), is square. The results of 5. then become

(2) *The algebra $\mathfrak{S}(E)$ of endomorphisms of a d-dimensional space E over K is isomorphic to the linear matrix ring $\mathfrak{M}_d^s(\mathsf{K})$ of all $d\times d$ square column-finite matrices over Ω_d, with elements in K.*

A linear matrix ring $\mathfrak{M}$ over K, whose square matrices $\mathfrak{A}$ are defined over the index set N, is said to be maximal, if there is no linear matrix ring $\mathfrak{M}_1$ over K, whose matrices are again defined over N, which contains $\mathfrak{M}$ as a proper subset.

(3) $\mathfrak{M}_d^s(\mathsf{K})$ *is maximal.*

It can clearly be supposed that $d\geqq\aleph_0$. It is not possible to extend $\mathfrak{M}_d^s(\mathsf{K})$ by adding a matrix $\mathfrak{C}$ which is not column-finite: if, for instance, the ν-th column of $\mathfrak{C}$ contains infinitely many non-zero elements, than the $d\times d$ matrix $\mathfrak{A}$ which has the ν-th column of $\mathfrak{C}$ as ν-th row, and otherwise has zero rows, belongs to $\mathfrak{M}_d^s(\mathsf{K})$. The scalar product of the ν-th row of $\mathfrak{A}$ with the ν-th column of $\mathfrak{C}$ cannot be formed, since the sum must be taken over infinitely many non-zero terms.

7. Change of basis. If $\{x'_{\nu'}\}$, $\nu' \in \mathsf{N}'$, is a second basis of E, then it is related to the basis $\{x_\nu\}$, $\nu \in \mathsf{N}$, by equations

$$\text{(1a)} \qquad \{x_\nu\} = \{x'_{\nu'}\}\mathfrak{C}, \qquad \mathfrak{C} = ((\gamma_{\nu'\nu})), \ \nu' \in \mathsf{N}', \ \nu \in \mathsf{N},$$

$$\text{(1b)} \qquad \{x'_{\nu'}\} = \{x_\nu\}\mathfrak{D}, \qquad \mathfrak{D} = ((\delta_{\nu\nu'})), \ \nu \in \mathsf{N}, \ \nu' \in \mathsf{N}'.$$

$\mathfrak{C}$ and $\mathfrak{D}$ are column-finite, and by putting (1a) in (1b) and conversely we obtain $\mathfrak{C}\mathfrak{D} = \mathfrak{E}_{\mathsf{N}'}$, $\mathfrak{D}\mathfrak{C} = \mathfrak{E}_{\mathsf{N}}$, the unit matrices over the sets N' and N respectively.

A column-finite matrix $\mathfrak{X} = ((\xi_{\nu\nu'}))$, $\nu \in \mathsf{N}$, $\nu' \in \mathsf{N}'$, is again called a left inverse of the column-finite matrix $\mathfrak{A} = ((\alpha_{\nu'\nu}))$, if $\mathfrak{X}\mathfrak{A} = \mathfrak{E}_{\mathsf{N}}$, and similarly a right inverse $\mathfrak{Y} = ((\eta_{\nu\nu'}))$ is defined by $\mathfrak{A}\mathfrak{Y} = \mathfrak{E}_{\mathsf{N}'}$. As in 4., it follows from the existence of both a right and a left inverse of $\mathfrak{A}$ that they coincide, and are uniquely determined. In this case $\mathfrak{A}$ is again said to be invertible, and the two-sided inverse is denoted by $\mathfrak{A}^{-1}$.

Thus in the present case $\mathfrak{D} = \mathfrak{C}^{-1}$. Conversely, given a column-finite invertible matrix $\mathfrak{C}$ with two-sided inverse $\mathfrak{D}$, it is easy to see that a change of basis of E is given, using (1a) and (1b).

We remark that $\mathfrak{C}$ is square only if $\mathsf{N} = \mathsf{N}'$.

If x is an element of E, we obtain from (1a) that

$$\text{(2)} \qquad x = \{x_\nu\}\mathfrak{x} = \{x'_{\nu'}\}\mathfrak{C}\mathfrak{x},$$

so that

(3) *Changing the basis $\{x_\nu\}$ to the basis $\{x'_{\nu'}\} = \{x_\nu\}\mathfrak{C}^{-1}$ changes the representative $\mathfrak{x}$ to $\mathfrak{C}\mathfrak{x}$.*

If we also change the basis $\{y_\mu\}$, $\mu \in \mathsf{M}$, of F to a new basis $\{y'_{\mu'}\} = \{y_\mu\}\mathfrak{B}^{-1}$, $\mu' \in \mathsf{M}'$, then, using (1b) and 5.(3), we get the following for the matrix corresponding to A:

$$\text{(4)} \qquad \{Ax'_{\nu'}\} = \{Ax_\nu\}\mathfrak{C}^{-1} = \{y_\mu\}\mathfrak{A}\mathfrak{C}^{-1} = \{y'_{\mu'}\}\mathfrak{B}\mathfrak{A}\mathfrak{C}^{-1},$$

so that

(5) *If the bases $\{x_\nu\}$ and $\{y_\mu\}$ are changed to the bases $\{x'_{\nu'}\} = \{x_\nu\}\mathfrak{C}^{-1}$ and $\{y'_{\mu'}\} = \{y_\mu\}\mathfrak{B}^{-1}$ of E and F respectively, then A is represented by $\mathfrak{B}\mathfrak{A}\mathfrak{C}^{-1}$ instead of by $\mathfrak{A}$.*

In the case where $F = E$, if we take the same change of basis both for the original elements and for the image elements, then from (5) we get

(6) *Changing the basis $\{x_\nu\}$ to the basis $\{x_{\nu'}\} = \{x_\nu\}\mathfrak{C}^{-1}$ changes the representative $\mathfrak{A}$ of an endomorphism $A \in \mathfrak{S}(E)$ to the representative $\mathfrak{C}\mathfrak{A}\mathfrak{C}^{-1}$.*

8. Canonical representation of a linear mapping. We continue with the ideas of 2. We call the dimension of the image space $A(E)$ the rank $r(A)$ of A, the dimension of the kernel $N[A]$ the nullity $s(A)$ and finally the dimension of $F/A(E)$, and so of a complement $C[A]$ of the image space, the defect, or co-nullity, $s'(A)$.

The rank $r(A)$ is clearly also equal to the dimension $E/N[A]$. In order to represent A by a particularly simple matrix, we proceed in the following way. We choose an inverse image space $U[A]$, and then choose a basis $z_{\nu'}$, $\nu' \in \mathsf{N}'$, of this space. By adding a basis $z_{\nu''}$, $\nu'' \in \mathsf{N}''$, of $N[A]$ to this, we obtain a complete basis of E. The elements $w_{\nu'} = A z_{\nu'}$ then form a basis of $A(E)$, and by adding a basis $w_{\mu''}$, $\mu'' \in \mathsf{M}''$, of some $C[A]$ to this we obtain a complete basis of F. A is then given by

(1) $$A z_{\nu'} = w_{\nu'} (\nu' \in \mathsf{N}'), \qquad A z_{\nu''} = o (\nu'' \in \mathsf{N}'').$$

For the given bases of E and F, A is therefore represented by a matrix $\tilde{\mathfrak{D}}$, for which the elements $\tilde{\delta}_{\nu'\nu'}$, $\nu' \in \mathsf{N}'$, are equal to one, and all the others are zero. Thus we have shown

(2) *By choosing bases suitably in E and F, the linear mapping A of E into F is represented by a matrix* $\tilde{\mathfrak{D}}$, *which reduces to an* $r(A) \times r(A)$ *square unit matrix when its* $s'(A)$ *zero rows and* $s(A)$ *zero columns are deleted.*

$\tilde{\mathfrak{D}}$ is called a canonical representation of A, in the wide sense.

If it is required to use given sets N and M as index sets of the bases in E and F respectively, then N' and N'' can be chosen as complementary subsets of N, but for the elements $A z_{\nu'}$ it is only possible to use indices $\mu'(\nu') \in \mathsf{M}' \subset \mathsf{M}$ which are in one-one correspondence with the indices ν'. The canonical representation $\mathfrak{D}$ in the narrow sense which is obtained in this way has $\delta_{\mu'(\nu'),\nu'} = 1$, and so the deletion of the zero rows and zero columns leads to an $r(A) \times r(A)$ matrix, which has exactly one one in each row and column.

In order to obtain a canonical representation of an endomorphism $A \in \mathfrak{S}(E)$, it is necessary not just to take one basis of E, but to take two different bases, one for the elements x, and one for the elements $A x$.

9. Equivalence of mappings and matrices. Two linear mappings A_1 and A_2 in $\mathfrak{S}(E,F)$ are said to be equivalent if there exist an invertible mapping B in $\mathfrak{S}(E)$ and an invertible mapping C in $\mathfrak{S}(F)$ for which $A_2 = C A_1 B$. The equivalence defined in this way is reflexive, symmetric and transitive, so that the mappings in $\mathfrak{S}(E,F)$ fall into classes of equivalent mappings.

If we choose bases x_ν and y_μ in E and F, then for the corresponding row-finite matrices we also have $\mathfrak{A}_2 = \mathfrak{C}\mathfrak{A}_1\mathfrak{B}$, where $\mathfrak{B}$ and $\mathfrak{C}$ are invertible and square. In this case we say that $\mathfrak{A}_1$ and $\mathfrak{A}_2$ are equivalent in the narrow sense, and *the equivalence of the mappings implies the equivalence of the matrices in the narrow sense, and conversely.*

For this matrix equivalence it is necessary for $\mathfrak{A}_1$ and $\mathfrak{A}_2$ to be defined with the same row and column index sets, respectively. More generally, if $\mathfrak{A}_1$ and $\mathfrak{A}_2$ are two column-finite matrices with the same number of rows and the same number of columns, an equivalence in the wide sense can be defined by $\mathfrak{A}_2 = \mathfrak{C}\mathfrak{A}_1\mathfrak{B}$, where now the requirement is that $\mathfrak{B}$ and $\mathfrak{C}$ are invertible. If $\mathfrak{A}_1$ and $\mathfrak{A}_2$ have the same index sets, this is the same concept of equivalence as before, as $\mathfrak{B}$ and $\mathfrak{C}$ must then be square.

Theorem (5) of 7. can now be expressed in the following terms:

(1) *The representative matrix of* $A \in \mathfrak{S}(E,F)$ *obtained by changing the bases in* E *and* F *is equivalent to the original one, in the wide sense. The representatives are equivalent in the narrow sense if the new bases are defined over the same index sets as the old ones.*

From 8. we obtain

(2) *Every column-finite matrix is equivalent in the wide sense to a matrix* $\tilde{\mathfrak{D}}$, *and in the narrow sense to a matrix* $\mathfrak{D}$.

10. The theory of equivalence. It is now easy to give a complete account of the different equivalence classes, both for mappings and for matrices.

From 9.2 and the connection between the equivalence of matrices in the narrow sense and the equivalence of mappings there follows

(1) *Let* A *be a linear mapping of* E *into* F, *let* $\{x_\nu\}$, $\nu \in \mathsf{N}$, *be a basis of* E *and let* $\{y_\mu\}$, $\mu \in \mathsf{M}$, *be a basis of* F. *Let* A *have rank* r, *nullity* s *and defect* s'. *If* $\{x_{\nu'}\}$, $\nu' \in \mathsf{N}'$, *is an arbitrary subset of* r *elements of the basis, whose complement* $\{x_{\nu''}\}$, $\nu'' \in \mathsf{N}'' \subset \mathsf{N}$, *has* s *elements, and if* $\{y_{\mu'(\nu')}\}$, $\mu'(\nu') \in \mathsf{M}' \subset \mathsf{M}$, *is a subset of* $\{y_\mu\}$ *which is mapped in a one-one way, by* $\mu'(\nu')$, *onto* N', *and whose complement* $\{y_{\mu''}\}$, $\mu'' \in \mathsf{M}'' \subset \mathsf{M}$, *has* s' *elements, then* A *is equivalent to the mapping given by*

$$D x_{\nu'} = y_{\mu'(\nu')}, \qquad D x_{\nu''} = \mathrm{o} \qquad (\nu' \in \mathsf{N}', \nu'' \in \mathsf{N}'') \tag{2}$$

$\{x_{\nu'}\}$ is a basis of an inverse-image space of D, $\{y_{\mu'(\nu')}\}$ a basis of the image space, $\{x_{\nu''}\}$ a basis of the null-space, and $\{y_{\mu''}\}$ a basis of a complement of the image space.

(3) *Two linear mappings of* E *into* F *are equivalent if and only if they have equal ranks, equal nullities and equal defects.*

The sufficiency of the condition follows from (1). On the other hand, suppose that $A_2 = CA_1B$. If $A_1 x = \mathrm{o}$, then $CA_1B(B^{-1}x) = \mathrm{o}$ and conversely. The kernel and inverse image space of A_1 are therefore transformed into the kernel and inverse image space of A_2 respectively, under the isomorphism B^{-1}. Thus their dimensions remain unaltered. Finally the image space of A_2 is equal to $C(A_1(E))$, and so has the same dimension and co-dimension as $A_1(E)$.

The three invariants of the equivalence classes cannot be chosen arbitrarily, since they clearly must satisfy the relations

$$r+s=d, \quad r+s'=e, \tag{4}$$

where d and e are the dimensions of E and F respectively.

If d and e are finite, then it is clear that an equivalence class is determined by specifying the rank alone.

For every triple of cardinal numbers r, s, s' satisfying (4), there is a mapping with these as invariants.

If the rank, nullity and defect of a column-finite matrix $\mathfrak{A}$ are defined as the rank, nullity and defect of the corresponding linear mapping (using 5.(3)), then the agreement of r, s and s' is necessary and sufficient for the equivalence of two matrices, in both the narrow and the wide senses.

§ 9. The algebraic dual space. Tensor products

1. The dual space. In § 7, 8. we became acquainted with two methods of obtaining new vector spaces from given ones. The most important construction of this sort however is that of the dual space, and we now turn to the study of this.

Let E be a vector space over the field K. If an element $u(x)=ux$ of K corresponds to each $x\in E$, and if this correspondence is linear, i.e. if $u(x_1\alpha_1+x_2\alpha_2)=u(x_1)\alpha_1+u(x_2)\alpha_2$ for all $x_1, x_2\in E$ and all $\alpha_1,\alpha_2\in$ K, then ux is called a linear functional or linear form on E.

If we set $(u_1+u_2)x=u_1x+u_2x$ and $(\alpha u)x=\alpha(ux)$ for $\alpha\in$ K, then the set of all linear functionals on E clearly forms a vector space over K, the algebraic dual or algebraic conjugate E^* of E.

The prefix "algebraic" will be important later, but we shall leave it out for the time being.

If E_1 and E_2 are isomorphic, with $x_1\leftrightarrow x_2$, then an isomorphism $u_1\leftrightarrow u_2$ between E_1^* and E_2^* is clearly produced by setting $u_2x_2=u_1x_1$.

(1) *Isomorphic vector spaces have isomorphic dual spaces.*

It is easy to give an account of all the linear functionals on E. Let E have dimension d, and let $\{x_\nu\}$, $\nu\in$ N, be a basis of E. ux is known if all the values $ux_\nu=v_\nu$ are known, for then we have

$$ux=u\Big(\sum_\nu x_\nu\xi_\nu\Big)=\sum_\nu(ux_\nu)\xi_\nu=\sum_\nu v_\nu\xi_\nu. \tag{2}$$

On the other hand a linear form (2) is uniquely determined by an arbitrary choice of the values $u x_\nu = v_\nu$, for all $\nu \in \mathsf{N}$.

If to each $u \in E^*$ we make correspond the vector $\mathfrak{u} = \{v_\nu\}$ then E^* is mapped isomorphically onto the space $\omega_d(\mathsf{K})$ which was introduced in § 7, 8., and so onto the product of d copies of K with itself.

If, as in § 7, 5., we replace $x = \sum_\nu x_\nu \xi_\nu$ by the co-ordinate vector $\mathfrak{x} = \{\xi_\nu\}$ which corresponds to it under the isomorphism with $\varphi_d(\mathsf{K})$, then, by (2), $u x$ is equal to the scalar product $\mathfrak{u}\mathfrak{x} = \sum v_\nu \xi_\nu$ of the two vectors $\mathfrak{u}$ and $\mathfrak{x}$; this is meaningful since only finitely many ξ_ν are nonzero. Combining these ideas together,

(3) *If E is a vector space of dimension d over K, then the choice of a basis x_ν of E leads to isomorphisms $x \leftrightarrow \mathfrak{x}$ and $u \leftrightarrow \mathfrak{u}$ of E with $\varphi_d(\mathsf{K})$ and of E^* with $\omega_d(\mathsf{K})$ respectively, under which $u x = \mathfrak{u}\mathfrak{x}$.*

In particular, $(\varphi_d(\mathsf{K}))^* \cong \omega_d(\mathsf{K})$.

It is easily verified that

(4) *If $E = \bigoplus_\alpha E_\alpha$, then $E^* \cong \prod_\alpha E_\alpha^*$.*

The dual space of a finite dimensional space E is isomorphic to E. The unit vectors $\mathfrak{e}_\nu \in \omega_d(\mathsf{K})$ are the images under the isomorphism (3) of the linear functionals $u_\nu \in E^*$ defined by

$$(5) \qquad u_\nu x_\nu = 1, \quad u_\nu x_{\nu'} = 0 \quad \text{for} \quad \nu' \neq \nu, \; \nu, \nu' \in \mathsf{N}.$$

They form the dual system $\{u_\nu\}$ in E^*, dual to the basis $\{x_\nu\}$ of E.

2. Orthogonality. We have just seen that a linear functional u on E is determined by its values on a basis of E. Remembering that a basis of a linear subspace can always be extended to a basis of the whole of E by adding further elements (§ 7, 3.(2)), we immediately obtain the extension theorem

(1) *If a linear functional $l(y)$, $y \in F$, is defined on a linear subspace F of a vector space E, then it can be extended linearly to the whole of E, and so there exists $u_0 \in E^*$ with $u_0 y = l(y)$ for all $y \in F$.*

In a somewhat different version, we have

(1′) *If x_0 is an element of E which does not lie in a linear subspace F of E, then there is a $u_0 \in E^*$ with $u_0 y = 0$ for all $y \in F$, and with $u_0 x_0 = 1$.*

If $u x = 0$ for $x \in E$ and $u \in E^*$, we say that u and x are orthogonal. If M is a subset of E, the set of all $u \in E^*$ which are orthogonal to all $x \in M$ forms a linear subspace of E^*, which we call the orthogonal space $M^\perp$ of M in E^*. If we start from a subset M of E^*, we obtain the orthogonal space $M^\perp$ of M in E in the same way. It follows directly from the definition that

(2) *If* $M_1 \subset M_2$, *then* $M_2^\perp \subset M_1^\perp$, *and so* $M_1^{\perp\perp} \subset M_2^{\perp\perp}$.
(3) M *is always contained in* $M^{\perp\perp}$.

If $M = M^{\perp\perp}$, M is said to be orthogonally closed. An orthogonally closed subset of E or E^* must be a linear subspace. $M^{\perp\perp}$ is called the orthogonal closure of M.

(4) $M^\perp$ *is always orthogonally closed.*

For by (3), $M^\perp \subset M^{\perp\perp\perp}$; on the other hand, applying (2) to (3), we obtain $M^{\perp\perp\perp} \subset M^\perp$, and so $M^\perp = M^{\perp\perp\perp}$.

(5) *Every linear subspace* F *of* E *is orthogonally closed.*

For if x_0 is not in F, then by (1′) there is a $u_0 \in F^\perp$ with $u_0 x_0 = 1$, so that x_0 does not lie in $F^{\perp\perp}$; hence $F = F^{\perp\perp}$.

On the other hand we have

(6) *If* E *is infinite-dimensional, there are always linear subspaces, and indeed hyperplanes in* E^*, *which are not orthogonally closed.*

We can take $E = \varphi_d(\mathsf{K})$, with $d \geqq \aleph_0$. The dual space $\omega_d(\mathsf{K})$ has $F = \varphi_d(\mathsf{K})$ as a proper linear subspace, by § 7, 8. But $F^\perp = \mathfrak{o}$, for if $\mathfrak{x} = \{\xi_\nu\} \in F^\perp$, then $\mathfrak{e}_\nu \mathfrak{x} = \xi_\nu = 0$ for all ν. Since $F^{\perp\perp} = \omega_d(\mathsf{K})$, F is a proper subspace of $F^{\perp\perp}$.

Every proper subspace lying between F and $\omega_d(\mathsf{K})$ is likewise not orthogonally closed, and in particular by § 7, 7.(7) there is a hyperplane which is not orthogonally closed.

Nevertheless, we have

(7) *Every finite-dimensional linear subspace* F *of* E^* *is orthogonally closed.*

This can also be expressed as follows:

(7a) *If* $u_1, \ldots, u_n$ *are* n *linearly independent linear forms on* E, *and if* $F^\perp$ *is the space orthogonal to* $F = [u_1, \ldots, u_n]$, *then each linear form on* E *which vanishes on* $F^\perp$ *is a linear combination of* $u_1, \ldots, u_n$. $F^\perp$ *has codimension* n *in* E.

Proof. There is an x_1 with $u_1 x_1 = 1$. Suppose that it has been shown for $k-1$ that there are elements $x_1, \ldots, x_{k-1}$ in E such that $u_i x_i = 1$ and $u_i x_j = 0$ for $i, j = 1, \ldots, k-1$, and $i \neq j$. Then for each $x \in E$ we have

$$(8) \quad x = \sum_{i=1}^{k-1} x_i(u_i x) + x', \quad \text{with} \quad u_i x' = 0 \quad \text{for} \quad i = 1, \ldots, k-1.$$

If $u_k x' = 0$ for all x', then by (8) $\left[u_k - \sum_{i=1}^{k-1} (u_k x_i) u_i\right] x = 0$, for all $x \in E$, which is impossible, for the linear functional in the square bracket would then vanish identically. Thus there exists an x_k with $u_k x_k = 1$,

$u_i x_k = 0$ for $i=1,\ldots,k-1$. Putting $\tilde{x}_i = x_i - (u_k x_i) x_k$, $i=1,\ldots,k-1$, we have $u_i \tilde{x}_i = 1$, $u_i \tilde{x}_j = 0$ for $j \neq i$ and $i,j = 1,\ldots,k-1$; also $u_k \tilde{x}_i = 0$. If we now write $x_1,\ldots,x_k$ for $\tilde{x}_1,\ldots,\tilde{x}_{k-1}, x_k$, we have established the assertion for k.

We have thus shown, for $k=n$, that $E = F^{\perp} \oplus G$, where G is the n-dimensional subspace of E with basis $x_1,\ldots,x_n$. If now u is a linear functional which vanishes on $F^{\perp}$, and if $u x_i = v_i$, then $u - \sum_{i=1}^{n} v_i u_i$ vanishes on all $x \in E$, and so $u = \sum_{i=1}^{n} v_i u_i$.

3. The lattice of orthogonally closed subspaces of E^*. The orthogonally closed subspaces H of E^* form a lattice $\bar{V}(E^*)$ under the relation $\subset$; the greatest lower bound of a set $\{H_\alpha\}$ exists in $\bar{V}(E^*)$, being the intersection $\bigcap_\alpha H_\alpha$. This is orthogonally closed, for using 2(2) it follows from $\bigcap_\alpha H_\alpha \subset H_\alpha$ that $\left(\bigcap_\alpha H_\alpha\right)^{\perp\perp} \subset H_\alpha^{\perp\perp} = H_\alpha$, so that $\left(\bigcap_\alpha H_\alpha\right)^{\perp\perp} \subset \bigcap_\alpha H_\alpha$, and the assertion follows from 2.(3). Similarly $\bigvee_\alpha H_\alpha$ exists as the intersection of all the orthogonally closed subspaces which contain all H_α; $\bigvee_\alpha H_\alpha$ is the orthogonal closure of the linear span of the H_α.

If to each linear subspace F of E we make correspond its orthogonal space $F^{\perp}$, then by 2.(5) we have a one-one correspondence of $V(E)$ with $\bar{V}(E^*)$; by 2.(2) it transposes the partial order, so that, using § 7, 9. and 10., we have

(1) *The correspondence which sends each linear subspace F of E to its orthogonal space $F^{\perp} \subset E^*$ is a dual isomorphism of the complete lattices $V(E)$ and $\bar{V}(E^*)$. $\bar{V}(E^*)$ is thus also a complemented modular lattice.*

The lattice theoretic intersection in $V(E)$ is the same as in $\bar{V}(E^*)$, namely the set theoretic intersection. The union in $V(E)$ is the linear span, and in $\bar{V}(E^*)$ is the orthogonal closure. If E is infinite-dimensional, with basis x_ν, and if u_ν is the dual system in E^*, the linear span of the one-dimensional and, by 2.(7), orthogonally closed spaces $[u_\nu]$ is different from the orthogonal closure E^* of the $[u_\nu]$, by 2.(6). Since $V(E^*)$ is modular, however, it follows that

(2) *If B_1 and B_2 are orthogonally closed subspaces of E^*, then $B_1 + B_2$ is always orthogonally closed, so that $B_1 \vee B_2 = B_1 + B_2$.*

Suppose first that B_1 is finite-dimensional, that $u_0 \in B_1 \vee B_2$ and that $u_0 \notin B_1 + B_2$. Then the finite dimensional space $C = B_1 \oplus [u_0]$ lies in $\bar{V}(E^*)$. Since $B_1 \subset C$ and $\bar{V}(E^*)$ is modular,

$$(B_1 \vee B_2) \cap C = B_1 \vee (B_2 \cap C). \tag{3}$$

If $v \in B_2 \cap C$, then, being an element of C, it has the form $v = w + u_0 \alpha$, with $w \in B_1$, $\alpha \in \mathsf{K}$. But then $u_0 \alpha = v - w \in B_1 + B_2$, which is only possible if $\alpha = 0$. Thus $v = w \in B_1$, $B_2 \cap C \subset B_1$, and $B_1 \vee (B_2 \cap C) = B_1$, so that, by (3), $C = (B_1 \vee B_2) \cap C = B_1$, which contradicts the assumption about C.

For general B_1 it follows from this that C lies in $\bar{V}(E^*)$; (2) follows by repeating the argument.

In particular it follows from (2) that

(4) *Two orthogonally closed subspaces of E^* which are complements in the lattice sense are algebraic complements, and conversely.*

For, by (2), $B_1 \vee B_2 = E$ means simply that $B_1 + B_2 = E$; together with $B_1 \cap B_2 = 0$, this establishes the assertion.

From (1), (4), and the fact that every linear subspace has a complementary space, there follows

(5) *If $E = F_1 \oplus F_2$ then $E^* = F_1^\perp \oplus F_2^\perp$. If $E^* = G_1 \oplus G_2$ where G_1 and G_2 are orthogonally closed, then $E = G_1^\perp \oplus G_2^\perp$. Every orthogonally closed subspace of E^* has an orthogonally closed complement.*

(6) *If F is a linear subspace of E, then $F^\perp \cong (E/F)^*$.*

For if $u \in F^\perp$, then $u(x+z) = ux$, for each $z \in F$. If we therefore define $u'\hat{x}$, for $\hat{x} \in E/F$, as the common value of ux for all the $x \in \hat{x}$, then each $u \in F^\perp$ corresponds to a $u' \in (E/F)^*$, different u corresponding to different u'. This correspondence is clearly linear. If, conversely, $u' \in (E/F)^*$ is given, the formula $ux = u'\hat{x}$ defines a linear functional on E which lies in $F^\perp$. Since u' is again obtained from this u by the first part of the argument, the correspondence is an isomorphism.

In particular we have

(7) *If F has finite codimension n in E, $F^\perp$ has dimension n.*

The assertion for orthogonally closed subspaces of E^* corresponding to (6) is not in general true; we shall continue the investigation of the relationship between $V(E)$ and $V(E^*)$, using other techniques, in § 10.

4. The adjoint mapping. Suppose that A maps E linearly into F. If v is a linear functional on F, then $v(Ax)$, considered as a function of $x \in E$, is a linear functional u on E, since

$$v[A(x_1\alpha_1 + x_2\alpha_2)] = v[(Ax_1)\alpha_1 + (Ax_2)\alpha_2] = v(Ax_1)\alpha_1 + v(Ax_2)\alpha_2.$$

The mapping $v \to u$ from F^* into E^* which is defined in this way is called the adjoint mapping A' of A. It is therefore defined by

$$(1) \qquad (A'v)x = v(Ax), \quad \text{for all} \quad x \in E,\ v \in F^*.$$

A' is linear, since

$$[A'(\beta_1 v_1+\beta_2 v_2)]\,x=(\beta_1 v_1+\beta_2 v_2)(A\,x)=\beta_1[v_1(A\,x)]+\beta_2[v_2(A\,x)]$$
$$=\beta_1[(A' v_1)\,x]+\beta_2[(A' v_2)\,x]$$

so that we have

(2) *If* $A\in\mathfrak{S}(E,F)$, *then* $A'\in\mathfrak{S}(F^*,E^*)$.

Let x_ν, $\nu\in\mathsf{N}$, and y_μ, $\mu\in\mathsf{M}$, be bases of E and F respectively. A is represented by a matrix $\mathfrak{A}=((\alpha_{\mu\nu}))$ with respect to these bases; with respect to the dual systems of these bases, v and $u=A'v$ are represented by coordinate vectors $\mathfrak{v}=\{\varphi_\mu\}$ and $\mathfrak{u}=\{v_\nu\}$ respectively. By 1. and § 8, 5., we have

(3) $$v_\nu=(A'v)\,x_\nu=v(A\,x_\nu)=\mathfrak{v}(\mathfrak{A}\,\mathfrak{e}_\nu)=\sum_\mu \varphi_\mu\alpha_{\mu\nu}.$$

This implies that $\mathfrak{u}=\mathfrak{A}'\mathfrak{v}$, where $\mathfrak{A}'$ is the transposed matrix of $\mathfrak{A}$; thus we have

(4) *If A is represented by the matrix $\mathfrak{A}$ with respect to certain bases of E and F, then the adjoint mapping A' is represented by the transposed matrix $\mathfrak{A}'$, with respect to the dual systems of the same bases.*

Since $\mathfrak{A}$ is column-finite, $\mathfrak{A}'$ is row-finite.

(5) $$(\alpha A)'=\alpha A,\qquad (A+B)'=A'+B'.$$

for general A, $B=\mathfrak{S}(E,F)$, $\alpha\in\mathsf{K}$. $(B\,A)'=A'B'$, *for general* $A\in\mathfrak{S}(E,F)$, $B\in\mathfrak{S}(F,G)$.

We shall only establish the last relation: It follows from (1) that, for $x\in E$, $v\in G^*$,

$$((B\,A)'\,v)\,x=v((B\,A)\,x)=v(B(A\,x))=(B'\,v)\,(A\,x)=(A'(B'\,v))\,x=((A'\,B')\,v)\,x.$$

The correspondence $A\to A'$ defined by (1) is one-one, by (4), and so (5) follows from § 8, 6.

(6) *The algebra $\mathfrak{S}'(E)$ of mappings A' adjoint to the mappings $A\in\mathfrak{S}(E)$ forms a sub-algebra of $\mathfrak{S}(E^*)$ anti-isomorphic to $\mathfrak{S}(E)$.*

(7) *If E is infinite-dimensional, $\mathfrak{S}'(E)$ is a proper sub-algebra of $\mathfrak{S}(E^*)$.*

Proof. By (3), a mapping $\mathfrak{u}=\mathfrak{A}'\mathfrak{v}$ of $\omega_d(\mathsf{K})$ into itself is identically zero if the individual columns of $\mathfrak{A}'$, which are the images $\mathfrak{A}'\mathfrak{e}_\mu$ of the vectors $\mathfrak{e}_\mu$, are all equal to the zero coordinate vector $\mathfrak{o}$. But if $d\geqq\aleph_0$, there are linear mappings of $\omega_d(\mathsf{K})$ into itself which map all the $\mathfrak{e}_\mu$ into 0, but which map the vector $\mathfrak{e}$, all of whose coordinates are equal to one, and which is linearly independent of the $\mathfrak{e}_\mu$, into a non-zero vector.

5. The dimension of E^*. Because of the equation

(1) $$(\alpha_1 u_1+\alpha_2 u_2)\,x=\alpha_1(u_1\,x)+\alpha_2(u_2\,x)$$

each $x \in E$ defines a linear functional $x(u)$ on E^*, and so the correspondence $x \to x(u)$ is an isomorphism of E with a subspace of E^{**}, which we can identify with E. If E is finite-dimensional, then $E = E^{**}$; on the other hand we have

(2) *If E is infinite-dimensional, then E is a proper subspace of E^{**}, and so every term of the sequence $E \subset E^{**} \subset E^{****} \subset \cdots$ is a proper subspace of its successor.*

Proof. If E is taken in the form $\varphi_d(\mathsf{K})$, then E^* has the form $\omega_d(\mathsf{K})$. Since the vectors of $\varphi_d(\mathsf{K})$ are finite, they only define linear functionals which vanish on all but finitely many $\mathfrak{e}_\nu \in \omega_d(\mathsf{K})$. There exist linear functionals on $\omega_d(\mathsf{K})$, however, which are different from zero on infinitely many $\mathfrak{e}_\nu$ (use a suitable basis of $\omega_d(\mathsf{K})$, and proceed as in 1.(2)).

The following theorem of ERDÖS and KAPLANSKY determines the dimension of E^* exactly.

(3) *The dimension d^* of $\omega_d(\mathsf{K})$ is, for infinite d, equal to the number k^d of elements of $\omega_d(\mathsf{K})$, where k is the number of elements in K.*

Proof (W. NEUMER). Let $\{\mathfrak{u}^\mu\}$ be a basis of $\omega_d(\mathsf{K})$, so that each $\mathfrak{u} \in \omega_d(\mathsf{K})$ is a linear combination of the vectors $\mathfrak{u}^\mu$. There are $(k-1)d^*$ elements in $\omega_d(\mathsf{K})$ of the form $\mathfrak{u}^\mu \alpha$, $\alpha \in \mathsf{K}$, $\alpha \neq 0$, $[(k-1)d^*]^2$ elements of the form $\mathfrak{u}^{\mu_1}\alpha_1 + \mathfrak{u}^{\mu_2}\alpha_2$, with α_1 and α_2 both different from zero, and so on, so that $\omega_d(\mathsf{K})$ contains $\sum_{i=0}^{\infty} [(k-1)d^*]^i$ elements. Since $p^2 = p$ for every infinite cardinal p, we can write the sum as $d^* \sum_{i=0}^{\infty} (k-1)^i$. $\sum_{i=0}^{\infty} (k-1)^i$ is equal to $\aleph_0$, for finite $k \geqq 2$, and to k, for infinite k; in either case we therefore obtain $d^* \sum_{i=0}^{\infty} (k-1)^i = d^* k$. This establishes the equation

$$k^d = d^* k, \tag{4}$$

which we use to determine d^*. If $k \leqq d^*$, then $d^* k \leqq d^{*2} = d^*$, so that $d^* k = d^*$; the assertion $d^* = k^d$ therefore follows from (4).

We now show that $k \leqq d^*$, for all $d \geqq \aleph_0$. It is clearly sufficient to prove this for $d = \aleph_0$. Let us therefore assume that the number d^* of elements $\mathfrak{u}^\mu = (v_1^\mu, v_2^\mu, \ldots)$ in the basis of $\omega(\mathsf{K})$ is less than k. The collection M of all the coordinates v_i^μ of all the vectors $\mathfrak{u}^\mu$ has cardinal at most $\aleph_0 d^* = d^*$. Let K_0 be the subfield of K generated by all the v_i^μ. Since K_0 consists of rational expressions in the v_i^μ with coefficients in the prime field of K, K_0 also has not more than $\aleph_0 d^* = d^*$ elements. Since $d^* < k$, there exists a ξ_1 in $\mathsf{K} \sim \mathsf{K}_0$. We set $\mathsf{K}_1 = \mathsf{K}_0(\xi_1)$, where $\mathsf{K}_0(\xi_1)$ is the field obtained by adjoining ξ_1 to K_0. Again K_1 has at most $\aleph_0 d^* = d^*$ elements. $\xi_2 \in \mathsf{K} \sim \mathsf{K}_1$ can now be chosen, and the procedure repeated.

In this way we obtain a sequence

$$\mathsf{K}_0 \subset \cdots \subset \mathsf{K}_n \subset \cdots, \quad \text{with} \quad \mathsf{K}_n = \mathsf{K}_{n-1}(\xi_n), \qquad \xi_n \in \mathsf{K} \sim \mathsf{K}_{n-1}.$$

Now let $\mathfrak{x} = (\xi_i)$ be the vector in $\omega(\mathsf{K})$ formed by the ξ_i, and let $\mathfrak{x} = \sum_{i=1}^{m} \mathfrak{u}^{\mu_i} \eta_i$ be its representation in terms of the basis. Because of the linear independence of the vectors $\mathfrak{u}^{\mu_i}$, there exists an $m \times m$ sub-matrix $((v_{j_l}^{\mu_i}))$ of the matrix $((v_j^{\mu_i}))$, $i = 1, \ldots, m$, $j = 1, 2, \ldots,$ which has a non-zero determinant. The system of equations

$$\xi_{j_l} = \sum_{i=1}^{m} v_{j_l}^{\mu_i} \eta_i, \qquad l = 1, \ldots, m,$$

can therefore be solved to determine the η_i, and so the η_i lie in K_{n_0}, for some sufficiently large n_0. But it then follows from the fact that $\xi_r = \sum_{i=1}^{m} v_r^{\mu_i} \eta_i$, for $r > n_0$, that all the ξ_r lie in K_{n_0}, which is impossible.

In particular, since $(2^{\aleph_0})^d = 2^d$ for $d \geqq \aleph_0$, we have

(5) *If E is a real or complex vector space of dimension $d \geqq \aleph_0$, then E^* has dimension 2^d.*

A detailed account of the relation between $\varphi_d(\mathsf{K}_1)$ and $\varphi_d(\mathsf{K}_2)$, where K_1 is a sub-field of K_2, can be found in BOURBAKI [4], vol. 2, § 5.

6. The tensor product of vector spaces. Let E and F be two vector spaces over K. We form the set $\Lambda(E \times F)$ of all formal finite linear combinations $\sum_{(x,y) \in E \times F} (x, y) \alpha_{x,y}$ of elements of $E \times F$, with coefficients in K. $\Lambda(E \times F)$ becomes a vector space over K, when we put

$$\left[\sum (x, y) a_{x,y}\right] \beta = \sum (x, y) \alpha_{x,y} \beta$$

and

$$\sum (x, y) \alpha_{x,y} + \sum (x, y) \beta_{x,y} = \sum (x, y) (\alpha_{x,y} + \beta_{x,y}).$$

The zero element is obtained when all the coefficients $\alpha_{x,y}$ are put equal to 0. We write (x, y) for $(x, y) 1$.

We now form the linear span Λ_0 in $\Lambda(E \times F)$ of all elements of the form

$$(1) \qquad \left(\sum_{i=1}^{n} x_i \alpha_i, \sum_{k=1}^{m} y_k \beta_k \right) - \sum_{i=1}^{n} \sum_{k=1}^{m} (x_i, y_k) \alpha_i \beta_k.$$

The quotient space Λ / Λ_0 is called the t e n s o r p r o d u c t, or d i r e c t p r o d u c t, $E \otimes F$ of E and F. The coset to which $(x, y) \alpha$ belongs is denoted by $(x \otimes y) \alpha$. Again, we write $x \otimes y$ instead of $(x \otimes y) 1$.

The following rule of calculation in $E \otimes F$ follows from (1):

$$(2) \qquad \sum_{i=1}^{n} x_i \alpha_i \otimes \sum_{k=1}^{m} y_k \beta_k = \sum_{i=1}^{n} \sum_{k=1}^{m} (x_i \otimes y_k) \alpha_i \beta_k .$$

Accordingly, $\mathrm{o} \otimes y = (\mathrm{o}0) \otimes y = (\mathrm{o} \otimes y)0 = \mathrm{o}$; likewise $x \otimes \mathrm{o} = \mathrm{o}$.

(3) *The tensor product is commutative; that is, $E \otimes F$ is isomorphic to $F \otimes E$ under the correspondence $x \otimes y \leftrightarrow y \otimes x$.*

Since the mapping $(x,y) \to (y,x)$ of $E \times F$ onto $F \times E$ defines an isomorphic mapping of $\Lambda(E \times F)$ onto $\Lambda(F \times E)$ under which the subspaces Λ_0 correspond, the quotient spaces are also isomorphic.

It follows immediately from the equation $(x \otimes y)\alpha = x \otimes (y\alpha)$ that every element of $E \otimes F$ can be written in the form $\sum_{i=1}^{n} (x_i \otimes y_i)$.

(4) *If $y_1, \ldots, y_n$ are linearly independent elements of F, it follows from $\sum_{i=1}^{n} (x_i \otimes y_i) = \mathrm{o}$ that $x_i = \mathrm{o}$ for $i = 1, \ldots, n$.*

Proof. If $u_0 \in E^*$, a linear mapping A is defined from $\Lambda(E \times F)$ into F by putting $A(\sum (x,y)\alpha_{x,y}) = \sum y(u_0 x)\alpha_{x,y}$. Since the elements (1) are mapped into o, A induces a linear mapping A of $E \otimes F$ into F. Under this, $\sum_i (x_i \otimes y_i)$ is mapped to $\sum_i y_i(u_0 x_i)$. If the y_i are linearly independent, it follows that $\sum (x_i \otimes y_i) = \mathrm{o}$ only if $u_0 x_i = 0$ for all $u_0 \in E^*$, i.e. only if $x_i = \mathrm{o}$, $i = 1, \ldots, n$.

(5) *If $\{x_\nu\}$, $\nu \in \mathsf{N}$, is a basis for E and $\{y_\mu\}$, $\mu \in \mathsf{M}$, is a basis for F, then $\{x_\nu \otimes y_\mu\}$, $(\nu, \mu) \in \mathsf{N} \times \mathsf{M}$, is a basis for $E \otimes F$.*

If E has dimension d and F dimension e, then $E \otimes F$ has dimension de.

For if $\sum_{\nu,\mu} (x_\nu \otimes y_\mu)\alpha_{\nu\mu} = \mathrm{o}$, then $\sum_\mu \left(\sum_\nu x_\nu \alpha_{\nu\mu} \right) \otimes y_\mu = \mathrm{o}$, and so, by (4), $\sum_\nu x_\nu \alpha_{\nu\mu} = \mathrm{o}$; thus $\alpha_{\nu\mu} = 0$ for each $(\nu, \mu) \in \mathsf{N} \times \mathsf{M}$. On the other hand, because of (2), every element of $E \otimes F$ can be written as a finite linear combination of the $x_\nu \otimes y_\mu$.

(6) *If F has dimension d then $E \otimes F$ is isomorphic to the direct sum of d vector spaces isomorphic to E.*

For fixed μ, the terms $x_\nu \otimes y_\mu$ form the basis of a subspace E_μ, by (5), and this is isomorphic to E under the correspondence $(\sum x_\nu \alpha_\nu) \otimes y_\mu \to \sum x_\nu \alpha_\nu$; on the other hand, $E \otimes F = \bigoplus_\mu E_\mu$.

It follows easily from (5) that

(7) *If A and B are linear subspaces of E and F respectively, then $A \otimes B$ is isomorphic to the linear subspace of $E \otimes F$ spanned by the elements $a \otimes b$, $a \in A$, $b \in B$.*

(8) *Every element* $z \neq o$ *of* $E \otimes F$ *has a representation* $z = \sum_1^r x^{(i)} \otimes y^{(i)}$ *for which both the* $x^{(i)}$ *and the* $y^{(i)}$ *are linearly independent.*

Proof. By (5), $z = \sum_{k=1}^{s} \sum_{j=1}^{t} (x_{\nu_k} \otimes y_{\mu_j}) \alpha_{kj}$. If we introduce new bases in $E_s = [x_{\nu_1}, \ldots, x_{\nu_k}]$ and $F_t = [y_{\mu_1}, \ldots, y_{\mu_t}]$, given by $x_{\nu_k} = \sum_l x'_l \beta_{lk}$ and $y_{\mu_j} = \sum_m y'_m \gamma_{mj}$ respectively, then

$$z = \sum_l \sum_m (x'_l \otimes y'_m) \sum_k \sum_j \beta_{lk} \alpha_{kj} \gamma_{mj},$$

so that the coefficient matrix $\mathfrak{A} = ((\alpha_{kj}))$ is replaced by $\tilde{\mathfrak{A}} = \mathfrak{B} \mathfrak{A} \mathfrak{C}'$, with $\mathfrak{B} = ((\beta_{lk}))$, $\mathfrak{C} = ((\gamma_{mj}))$. Since arbitrary invertible matrices can be chosen for $\mathfrak{B}$ and $\mathfrak{C}$, if they are chosen suitably, the matrix $\tilde{\mathfrak{A}}$ has the equivalent canonical form (cf. § 8, 9.(2)). If r is the rank of $\tilde{\mathfrak{A}}$, then $z = \sum_1^r x'_l \otimes y'_l$ is the required representation.

Calling the number r which appears in (8) the rank of z, we have

(9) *The rank of an element* $z \in E \otimes F$ *is equal to the rank of the matrix* $((\alpha_{\nu\mu}))$, *for any representation* $z = \sum_\nu \sum_\mu (x_\nu \otimes y_\mu) \alpha_{\nu\mu}$ *in terms of bases* $\{x_\nu\}$ *and* $\{y_\mu\}$ *of* E *and* F.

When we introduce new bases by $x_\nu = \sum_{\nu'} \tilde{x}_{\nu'} \beta_{\nu'\nu}$ and $y_\mu = \sum_{\mu'} \tilde{y}_{\mu'} \gamma_{\mu'\mu}$, only finitely many non-zero $\tilde{\alpha}_{\nu'\mu'}$ appear in the expression $z = \sum (\tilde{x}_{\nu'} \otimes \tilde{y}_{\mu'}) \tilde{\alpha}_{\nu'\mu'}$; it follows that we can obtain this representation of z from the first by a change of basis within finite-dimensional subspaces of E and F. For this, as in the proof of (8), the rank of the coefficient matrix remains unaltered, and in particular it is equal to the rank of z.

7. Linear mappings of tensor products. A mapping $B(x, y)$ from $E \times F$ into a vector space H which is linear in both variables is called a bilinear mapping from $E \times F$ into H. Thus, for all $x_i \in E$, $y_k \in F$,

$$B\Big(\sum_i x_i \alpha_i, \sum_k y_k \beta_k\Big) = \sum_i \sum_k B(x_i, y_k) \alpha_i \beta_k. \tag{1}$$

If H is the field K of coefficients, we speak of bilinear forms, or bilinear functionals. The set of all bilinear mappings from $E \times F$ into H forms a vector space $\mathfrak{B}(E \times F, H)$. We denote the space of all bilinear functionals on $E \times F$ by $\mathfrak{B}(E \times F)$.

(2) *Every bilinear mapping* B *of* $E \times F$ *into* H *defines a linear mapping* $\hat{B}$ *of* $E \otimes F$ *into* H, *and conversely. This correspondence is an isomorphism between* $\mathfrak{B}(E \times F, H)$ *and* $\mathfrak{S}(E \otimes F, H)$.

Putting $B\Big[\sum_{(x,y)}(x,y)\alpha_{x,y}\Big]=\sum_{(x,y)}B(x,y)\alpha_{x,y}$, B is extended linearly to the whole of $\Lambda(E\times F)$. B vanishes on Λ_0, by (1), and so defines a linear mapping $\hat{B}$ of $E\otimes F$ into H.

If, conversely, $\hat{B}$ is a linear mapping of $E\otimes F$ into H, then we define a mapping B from $E\times F$ into H by the equation $B(x,y)=\hat{B}(x\otimes y)$; this is bilinear by 6.(2).

Since B and $\hat{B}$ are determined by their values at the points (x,y) and $x\otimes y$ respectively, the correspondence is an isomorphism between $\mathfrak{B}(E\times F,H)$ and $\mathfrak{S}(E\otimes F,H)$.

(3) *If E, F and H are vector spaces,* $\mathfrak{S}(E\otimes F,H)\cong\mathfrak{S}(E,\mathfrak{S}(F,H))$.

It is sufficient, by (2), to show that $\mathfrak{B}(E\times F,H)\cong\mathfrak{S}(E,\mathfrak{S}(F,H))$. If $z=B(x,y)$ is a bilinear mapping of $E\times F$ into H, for fixed x let B_x be the linear mapping which sends each $y\in F$ to the element

(4) $$B_x y=B(x,y)=z\in H.$$

In this way B corresponds to the linear mapping $x\to B_x$ from E into $\mathfrak{S}(F,H)$.

Conversely, given a linear mapping $x\to B_x$ from E into $\mathfrak{S}(F,H)$, a bilinear mapping is defined from $E\times F$ into H, using (4) again. This correspondence is clearly one-one and linear.

If $H=\mathsf{K}$, we obtain as a special case

(5) *The vector space* $(E\otimes F)^*$ *is isomorphic to* $\mathfrak{S}(E,F^*)$, *and to* $\mathfrak{B}(E\times F)$.

How is $E^*\otimes F^*$ related to $(E\otimes F)^*$?

(6) $E^*\otimes F^*$ *can always be interpreted as a subspace of* $(E\otimes F)^*$. *The two spaces coincide if and only if E or F is finite-dimensional.*

Proof. If, for $u\in E^*$, $v\in F^*$, $x\in E$, $y\in F$, we put

$$(u,v)(x,y)=(u\,x)(v\,y),$$

we obtain bilinear forms on $E\times F$ and on $E^*\times F^*$, which, by (2), determine linear forms

(7) $$(u\otimes v)(x\otimes y)=(u\,x)(v\,y),$$

on $E\otimes F$ and $E^*\otimes F^*$ respectively. In this way every element of $E^*\otimes F^*$ is interpreted as a linear functional on $E\otimes F$. Different elements of $E^*\otimes F^*$ determine different linear functionals: for this, it is enough to show that an element $\sum u_i\otimes v_i$ with linearly independent v_i and non-zero u_i does not vanish identically.

The linear functional on $E\otimes F$ determined by $\sum_{i=1}^{n}u_i\otimes v_i$ corresponds, by (4) and (5), to the linear mapping

(8) $$x\to\sum_{i=1}^{n}(u_i x)v_i,$$

from E into F^*. This does not vanish identically, because the v_i are linearly independent.

The mappings (8) in $\mathfrak{S}(E,F^*)$ all have finite-dimensional image spaces in F^*. But if E and F are both infinite-dimensional, there are mappings in $\mathfrak{S}(E,F)$ with infinite-dimensional image spaces, so that in this case $E^*\otimes F^*$ is a proper subspace of $(E\otimes F)^*$.

Finally we show that $E^*\otimes F^*=(E\otimes F)^*$ if E is finite-dimensional. Let $\{x_i\}$ and $\{y_\mu\}$ be bases of E and F respectively, so that each $w\in(E\otimes F)^*$ is determined by its values $w(x_i\otimes y_\mu)=\psi_{i\mu}$. For each i, there is a $v^{(i)}\in F^*$, with $v^{(i)}y_\mu=\psi_{i\mu}$ for all μ. If $\{u_i\}$ is the basis dual to $\{x_i\}$, then $\left(\sum_{j=1}^{n} u_j\otimes v^{(j)}\right)(x_i\otimes y_\mu)=\psi_{i\mu}$ for all i,μ, so that $w=\sum_{j=1}^{n} u_j\otimes v^{(j)}\in E^*\otimes F^*$.

Similarly, we show that, provided $E,F,G,H\neq o$,

(9) *$\mathfrak{S}(E,F)\otimes\mathfrak{S}(G,H)$ can always be interpreted as a subspace of $\mathfrak{S}(E\otimes G,F\otimes H)$. Equality holds if and only if either E and F, or G and H, or E and G are finite-dimensional.*

((6) is a special case of (9), obtained by replacing F and H by K, and observing that $\mathsf{K}\otimes\mathsf{K}\cong\mathsf{K}$, by 6.(6).)

Proof. As above, a linear mapping

$$(A\otimes B)(x\otimes z)=(Ax)\otimes(Bz), \tag{10}$$

from $E\otimes G$ into $F\otimes H$ is defined by putting $(A,B)(x,z)=(Ax)\otimes(Bz)$ for $A\in\mathfrak{S}(E,F)$, $B\in\mathfrak{S}(G,H)$, $x\in E$ and $z\in G$, and using (2). To the mapping $\sum_{i=1}^{n} A_i\otimes B_i$ there corresponds, by (4), the mapping

$$x\to\sum_{i=1}^{n}(A_i x)\otimes B_i \tag{11}$$

from E into $\mathfrak{S}(G,F\otimes H)$. If the B_i are linearly independent, then (11) is different from zero, and so we obtain the representation of $\mathfrak{S}(E,F)\otimes\mathfrak{S}(G,H)$ as a subspace of $\mathfrak{S}(E\otimes G,F\otimes H)$.

As x runs through all of E, we obtain on the right-hand side of (11) only those mappings from G into $F\otimes H$ which comprise the same B_i. A mapping $\sum A_i\otimes B_i$ thus can only correspond to a mapping of E into $\mathfrak{S}(G,F\otimes H)$ which sends E into a linear subspace of mappings contained in the linear span of mappings of the form $y\otimes B_i$, where $y\in F$, $i=1,\dots,n$. If G or H is infinite-dimensional, there are infinitely many linearly independent $B^\sigma\in\mathfrak{S}(G,H)$. If E is also of infinite dimension, then there is a linear mapping of E into $\mathfrak{S}(G,F\otimes H)$ whose image space contains infinitely many $y_0 \otimes B^\sigma$, where y_0 is a fixed non-zero element of F. This mapping cannot be produced by a $\sum A_i \otimes B_i$.

If E and at least one of G and H are infinite-dimensional $\mathfrak{S}(E,F)\otimes\mathfrak{S}(G,H)$ is therefore a proper subspace of $\mathfrak{S}(E\otimes G, F\otimes H)$. Since E and G are on an equal footing, it also follows that if G is infinite-dimensional, equality is only possible if E and F are finite-dimensional. The conditions of (9) are therefore necessary.

It remains to show that the relation

$$\mathfrak{S}(E,F)\otimes\mathfrak{S}(G,H)=\mathfrak{S}(E\otimes G, F\otimes H) \tag{12}$$

holds in the specified cases.

a) Let E and G be finite-dimensional, with bases $\{x_n\}$ and $\{z_k\}$. If $\{y_\mu\}$ and $\{t_\lambda\}$ are bases of F and H, a mapping C from $E\otimes G$ into $F\otimes H$ is determined by $C(x_n\otimes z_k)=\sum_\mu\sum_\lambda \gamma_{nk\mu\lambda}(y_\mu\otimes t_\lambda)$; only finitely many γ are different from zero. If $A_{j\mu}$ is the linear mapping in $\mathfrak{S}(E,F)$ which sends x_j to y_μ and $x_{j'}$ to $\mathfrak{o}$, when $j\neq j'$, and if, correspondingly, $B_{l\lambda}\in\mathfrak{S}(G,H)$ sends z_l to t_λ and $z_{l'}$ to $\mathfrak{o}$, when $l\neq l'$, then

$$\Big(\sum_j\sum_l\sum_\mu\sum_\lambda \gamma_{jl\mu\lambda}A_{j\mu}\otimes B_{l\lambda}\Big)(x_n\otimes z_k)=\sum_\mu\sum_\lambda \gamma_{nk\mu\lambda}(y_\mu\otimes t_\lambda),$$

so that (12) holds.

b) Let E and F be finite-dimensional, with bases $\{x_n\}$ and $\{y_m\}$, and let $\{z_\kappa\}$ and $\{t_\mu\}$ be bases for G and H respectively. A mapping $C\in\mathfrak{S}(E\otimes G, F\otimes H)$ is determined by $C(x_n\otimes z_\kappa)=\sum_m\sum_\lambda \gamma_{n\kappa m\lambda}(y_m\otimes t_\lambda)$. Let $A_{nm}\in\mathfrak{S}(E,F)$ be defined by $A_{nm}x_n=y_m$, and $A_{nm}x_{n'}=\mathfrak{o}$, for $n\neq n'$; let $B_{nm}\in\mathfrak{S}(G,H)$ be defined by $B_{nm}z_\kappa=\sum\gamma_{n\kappa m\lambda}t_\lambda$, for all κ. Then

$$\Big(\sum_{n'm} A_{n'm}\otimes B_{n'm}\Big)(x_n\otimes z_\kappa)=\sum_m\sum_\lambda \gamma_{n\kappa m\lambda}(y_m\otimes t_\lambda),$$

so that again (12) holds.

c) The case where G and H are finite-dimensional follows from b) by interchanging E and F with G and H.

If A and B in (10) have the form

$$Ax=\sum_\mu y_\mu\sum_\nu \alpha_{\mu\nu}\xi_\nu,\qquad Bz=\sum_\lambda t_\lambda\sum_\kappa \beta_{\lambda\kappa}\zeta_\kappa,$$

with respect to bases $\{x_\nu\}$, $\{y_\mu\}$, $\{z_\kappa\}$ and $\{t_\lambda\}$, then there are corresponding matrices $\mathfrak{A}=((\alpha_{\mu\nu}))$ and $\mathfrak{B}=((\beta_{\lambda\kappa}))$. With respect to the bases $x_\nu\otimes z_\kappa$ and $y_\mu\otimes t_\lambda$, the mapping $A\otimes B$ is represented by the matrix $((\alpha_{\mu\nu}\beta_{\lambda\kappa}))$, which is called the Kronecker product $\mathfrak{A}\otimes\mathfrak{B}$. This clearly results from replacing the element $\alpha_{\mu\nu}$ of the matrix $\mathfrak{A}$ by the matrix $\alpha_{\mu\nu}\mathfrak{B}$.

§ 10. Linearly topologized spaces

1. Preliminary remarks. There is an evident symmetry between a vector space E over a field K and its dual space E^*, as our earlier considerations have shown. If E is finite-dimensional the symmetry is complete, as E and E^* are isomorphic, and each can be interpreted as the dual of the other. If E is infinite-dimensional, however, E does not comprise all the linear functionals on E^*. In the same way, the correspondence which sends each linear subspace of E to its orthogonal space only produces the orthogonally closed subspaces of E^*. Finally $\mathfrak{S}(E)$ is anti-isomorphic to the algebra $\mathfrak{S}'(E)$ of adjoint mappings, and $\mathfrak{S}'(E)$ is a proper subalgebra of $\mathfrak{S}(E^*)$ if E is infinite-dimensional.

It is natural to ask whether this complete duality between E and E^* in the finite-dimensional case can be obtained in the general case as well, by introducing some concept of continuity, and so some suitable topology, in E and E^*.

This must happen in such a way that the continuous linear functionals on E^* must be just those which are determined by the elements of E. We can use this fact to derive a condition on the topologies on E^* under considerations.

A linear functional $l(u)$ on E^* maps E^* into the field K. In order to be able to speak about a continuous linear functional, a topology must be defined on K. Every Hausdorff topology on a field with finitely many elements is the discrete one, by § 1, 5.(2). In order to obtain a theory valid for general fields, throughout this chapter we shall suppose *that* K *is discrete.*

Now if $l(u)$ is to be continuous at $\mathfrak{o}$, there must be a neighbourhood $U(\mathfrak{o})$ for which $l(u)=0$ for all $u\in U(\mathfrak{o})$. Since $l(u)$ is linear, the linear span of U satisfies this condition, if U does. The continuity of $l(u)$ at a general point $u_0\in E^*$ is then satisfied if we take $U(u_0)$ to be equal to the linear manifold $u_0+U(\mathfrak{o})$.

It will therefore be appropriate for our purpose if we restrict ourselves to topologies for which there is a base of neighbourhoods of $\mathfrak{o}$ consisting of linear subspaces, and for which the neighbourhoods $U(u_0)$ are obtained by taking the translates $u_0+U(\mathfrak{o})$ of the neighbourhoods of zero. Following LEFSCHETZ [1], we call such a topology linear.

2. Linearly topologized spaces. Let L be a vector space over K. Suppose that a topology $\mathfrak{T}_0$ is defined on K. A topology $\mathfrak{T}$ defined on L is said to be compatible with the vector space operations in L if λx and $x+y$ are jointly continuous in both variables. For $x+y$ this means that the mapping $(x,y)\to x+y$ from $L\times L$ into L is continuous, when $L\times L$ is given the product topology. Similarly $(\lambda,x)\to\lambda x$ must be a continuous function on the topological product $\mathsf{K}\times L$, with values in L. If K is discrete, it is sufficient to establish the partial continuity of λx on L.

If $\mathfrak{T}$ is Hausdorff and compatible with the vector space operations, then L is called a topological vector space over K. For brevity, we also write this as $L[\mathfrak{T}]$. A topological vector space over a discrete field, with a topology which is linear in the sense of 1., is called a linearly topologized space. We now establish

(1) *Let* $\{U_\alpha\}$, $\alpha \in \mathsf{A}$, *be a filter base consisting of linear subspaces of the vector space* L, *with* $\bigcap_\alpha U_\alpha = \mathfrak{o}$. *If we introduce the linear topology* $\mathfrak{T}$ *on* L *defined by taking* $\{U_\alpha\}$ *as a base of neighbourhoods of* $\mathfrak{o}$, *then* L *is a linearly topologized space with respect to* $\mathfrak{T}$.

In addition, we show

(2) *The topology* $\mathfrak{T}$ *is produced by a Hausdorff uniformity on* L, *so that, in particular,* L *is regular.*

We take for a base of the uniformity on L the sets N_α of all $(x,y) \in L \times L$ with $x - y \in U_\alpha$. The N_α form a filter base on $L \times L$, since the U_α form a filter base on L. Because $\bigcap_\alpha U_\alpha = \mathfrak{o}$, $\bigcap_\alpha N_\alpha$ is the diagonal, i.e. the set of all (x,x) in $L \times L$. Thus (V 1) and (V 4) of § 5, 1. and 2. are established. (V 2′) is trivial, and finally (V 3) holds, since $N_\alpha^2 = N_\alpha$, for each α.

The uniformity defined in this way is therefore Hausdorff. The topology corresponding to it is $\mathfrak{T}$, and regularity follows from § 5, 2.(3).

Finally $\mathfrak{T}$ is compatible with the vector space operations: λx is continuous in x, since $\lambda x \in U_\alpha$ if $x \in U_\alpha$. If $(x,y) \in (x_0 + U_\alpha, y_0 + U_\alpha)$, then $x + y \in x_0 + y_0 + U_\alpha$, and so $x + y$ is also continuous in both variables.

We shall also denote the neighbourhood $x_0 + U_\alpha$ by $U_\alpha(x_0)$.

(3) *Every linear neighbourhood* U *of* $\mathfrak{o}$ *is open and closed; more generally each set* $M + U$, $M \subset L$, *is open and closed.*

Every linearly topologized space is totally disconnected.

$M + U$ is open, since $x + U$ lies in $M + U$ if x does. But $L \sim (M + U)$ is also open, since no point of $y + U$ lies in $M + U$ if y does not belong to $M + U$.

If the subset S of L contains two points x and y, and if $U(x)$ is a linear neighbourhood of x which does not contain y, then $S \cap U(x)$ is a proper open and closed subset of S. No subset of L with more than one point is connected, therefore, and the result follows from § 1, 6.

The next result is immediately obvious:

(4) *A vector space* L *is a linearly topologized space, when it is given the discrete topology.*

(5) *Every finite-dimensional linearly topologized space is discrete.*

For it follows from $\bigcap_\alpha U_\alpha = \mathfrak{o}$ that finitely many linear U have intersection $\mathfrak{o}$, and so $\mathfrak{o}$ is a neighbourhood of $\mathfrak{o}$.

(6) *Let L_1 be a linear subspace of $L[\mathfrak{T}]$.*

(a) *The closure $\bar{L}_1$ is again a linear subspace.*

(b) *L_1 is a linearly topologized space under the subspace topology induced by $\mathfrak{T}$.*

Proof. (a) Let x_0 and y_0 be two closure points of L_1. For each linear neighbourhood U of $\mathfrak{o}$ there are elements $x, y \in L_1$ with $x \in x_0 + U$, $y \in y_0 + U$; $x\alpha + y\beta \in x_0\alpha + y_0\beta + U$ for arbitrary α, $\beta \in \mathsf{K}$, so that $x_0\alpha + y_0\beta \in \bar{L}_1$.

(b) As U_α runs through a linear base of $\mathfrak{T}$-neighbourhoods of $\mathfrak{o}$, the sets $L_1 \cap U_\alpha$ are linear subspaces of L_1 with intersection $\mathfrak{o}$, and they form a base of neighbourhoods of $\mathfrak{o}$ for the induced topology.

(7) *The topological product $\prod_\beta L_\beta$ of linearly topologized spaces is again a linearly topologized space.*

This follows without difficulty from § 7, 8. and § 1, 8.

If we define a topology $\mathfrak{T}$ on the direct sum $\bigoplus_\beta L_\beta$ of linearly topologized spaces L_β, with topologies $\mathfrak{T}_\beta$, by taking as neighbourhoods of $\mathfrak{o}$ the direct sums $\bigoplus_\beta U_\beta$, where each U_β is a linear neighbourhood of $\mathfrak{o}$ in L_β, we again obtain a linearly topologized space, which we call the topological direct sum $\bigoplus_\beta L_\beta[T_\beta]$ of the L_β. Thus

(8) *The topological direct sum $L[\mathfrak{T}] = \bigoplus_\beta L_\beta[\mathfrak{T}_\beta]$ of linearly topologized spaces is again a linearly topologized space.*

We remark that the topology induced on L_β by $\mathfrak{T}$ is the original topology $\mathfrak{T}_\beta$.

Two linearly topologized spaces $L_1[\mathfrak{T}_1]$ and $L_2[\mathfrak{T}_2]$ are said to be topologically isomorphic, and we write $L_1[\mathfrak{T}_1] \cong L_2[\mathfrak{T}_2]$, if there is a linear one-one correspondence between L_1 and L_2, which is continuous in both directions.

For finitely many L_i, $\prod_{i=1}^n L_i[\mathfrak{T}_i]$ and $\bigoplus_{i=1}^n L_i[\mathfrak{T}_i]$ are topologically isomorphic (§ 7, 8.).

(9) *If a linear mapping A from $L_1[\mathfrak{T}_1]$ into $L_2[\mathfrak{T}_2]$ is continuous at $\mathfrak{o}$, then it is continuous everywhere, and is also uniformly continuous.*

It is enough to establish uniform continuity. By (2), a basic vicinity N of L_2 consists of all (y_1, y_2) with $y_1 - y_2 \in V$, where V is a neighbourhood of zero in L_2. By hypothesis there is a neighbourhood $U \subset L_1$ with $A(U) \subset V$, and so, since $Ax_1 - Ax_2 = A(x_1 - x_2)$, the image of the vicinity consisting of all those (x_1, x_2) with $x_1 - x_2 \in U$ is contained in V.

3. Dual pairs, weak topologies. DIEUDONNÉ and MACKEY have introduced an idea which has proved to be particularly fruitful in the study of topological vector spaces.

Two vector spaces L_1 and L_2 over K form a dual pair or linear system $\langle L_2, L_1\rangle$ when an element of K, denoted by ux or $\langle u,x\rangle$, is associated with every pair $(u,x)\in L_2\times L_1$, in such a way that the following hold:

(D1) ux is a bilinear form, i.e.

$$u(x_1\alpha_1+x_2\alpha_2)=(ux_1)\alpha_1+(ux_2)\alpha_2,\quad (\beta_1 u_1+\beta_2 u_2)x=\beta_1(u_1 x)+\beta_2(u_2 x).$$

(D2′) If, for some $x\in L_1$, $ux=0$ for all $u\in L_2$, then $x=\mathfrak{o}$.

(D2″) If, for some $u\in L_2$, $ux=\mathfrak{o}$ for all $x\in L_1$, then $u=\mathfrak{o}$.

By (D1), each $u\in L_2$ determines a linear functional in L_1^*, and distinct u determine distinct linear functionals, by (D2″); thus L_2 can be interpreted as a subspace of L_1^*, and L_1 as a subspace of L_2^*. The conditions (D2) say that L_1 and L_2 contain "sufficiently many" linear functionals in L_2^* and L_1^*, respectively.

We say that two dual pairs $\langle L_2,L_1\rangle$ and $\langle\tilde{L}_2,\tilde{L}_1\rangle$ are isomorphic, and write $\langle L_2,L_1\rangle\cong\langle\tilde{L}_2,\tilde{L}_1\rangle$, if there are linear one-one correspondences $x\leftrightarrow\tilde{x}$ and $u\leftrightarrow\tilde{u}$ between L_1 and $\tilde{L}_1$ and between L_2 and $\tilde{L}_2$, for which $ux=\tilde{u}\tilde{x}$ always holds.

In these terms, the results of § 9, 1 can be formulated as follows:

(1) *A vector space E and its algebraic dual E^* form a dual pair $\langle E^*,E\rangle$. If E has dimension d over* K, $\langle E^*,E\rangle\cong\langle\omega_d(\mathsf{K}),\varphi_d(\mathsf{K})\rangle$.

As in § 9, 2., we call the set $M^\perp$ of all $u\in L_2$ with $ux=0$ for all $x\in M\subset L_1$ the space orthogonal to M in L_2. $M\subset L_1$ is said to be orthogonally closed with respect to L_2 if the subspace $M^{\perp\perp}$ of L_1 orthogonal to $M^\perp$ is equal to M.

Theorems § 9, 2.(2), (3) and (4), are still valid, and as in § 9. 3. imply

(2) *Let $\langle L_2,L_1\rangle$ be a dual pair. If we make each orthogonally closed subspace of L_1 (respectively L_2) correspond to its orthogonal space, then the correspondence is a dual isomorphism between the complete lattices $\bar{V}(L_1)$ and $\bar{V}(L_2)$.*

Given a dual pair $\langle L_2,L_1\rangle$, we have the following natural construction for a linear topology on L_1 (with a corresponding one for L_2): we take as base of neighbourhoods of $\mathfrak{o}$ all sets $U_{u_1,\dots,u_n}$, $u_i\in L_2$, consisting of all those $x\in L_1$ for which

$$u_i x=0,\quad i=1,\dots,n. \tag{3}$$

A neighbourhood determined by this base of neighbourhoods is called a weak neighbourhood of $\mathfrak{o}$. The weak neighbourhoods of $\mathfrak{o}$ in L_1 are just the spaces $F^\perp$ orthogonal to the finite-dimensional subspaces F

of L_2, and the subsets of L_1 containing them. The topology defined on L_1 by L_2 in this way is called the linear weak topology on L_1 with respect to L_2, and is denoted by $\mathfrak{T}_{ls}(L_2)$. In the same way L_1 determines the weak topology $\mathfrak{T}_{ls}(L_1)$ on L_2.

(4) *If* $\langle L_2, L_1\rangle$ *is a dual pair,* $L_1[\mathfrak{T}_{ls}(L_2)]$ *is a linearly topologized space.*

The intersection condition of 2.(1) is satisfied, because of (D2).

4. The dual space. Given a linearly topologized space $L[\mathfrak{T}]$, we can consider the continuous linear functionals on it. If u_1 and u_2 are continuous linear functionals on $L[\mathfrak{T}]$, so are αu_1, $\alpha \in \mathsf{K}$, and u_1+u_2, and so the collection of continuous linear functionals forms a vector space, which is called the dual space or conjugate space $L[\mathfrak{T}]'$ (or more simply L') of $L[\mathfrak{T}]$ with respect to $\mathfrak{T}$. The algebraic dual L^* is nothing but the dual of L with respect to the discrete topology. Thus we always have $L' \subset L^*$.

As in § 9, 2., the following extension theorem holds:

(1) *If a continuous linear functional* $l(y)$, $y \in F$, *is defined on a linear subspace* F *of the linearly topologized space* L, *then it can be extended continuously to the whole of* L. *Thus there exists* $u \in L'$ *with* $uy = l(y)$, *for* $y \in F$.

(1′) *If* x_0 *is an element of* L *not lying in the closed linear subspace* F *of* L, *there exists a* $u_0 \in L'$ *with* $u_0 x_0 = 1$ *and* $u_0 y = 0$ *for all* $y \in F$.

First we establish (1′). By hypothesis, there exists a linear neighbourhood U of $\mathfrak{o}$ for which $(x_0+U) \cap F$ is empty. $F+U$ is therefore a linear subspace which does not contain x_0. By § 9, 2.(1′), there exists $u_0 \in L^*$ with $u_0 x_0 = 1$ and $u_0 z = 0$, for all $z \in F+U$. Since u_0 vanishes on the whole of U, u_0 is continuous, and so belongs to L'.

It is sufficient to establish (1) for a linear functional $l(y)$ which does not vanish identically on F. If $l(y_0)=1$, F can be written as $F=[y_0]\oplus F_1$, where $l(y_0)=1$ and $l(z)=0$ for all $z \in F_1$ (put $y = \alpha y_0 + (y - \alpha y_0)$, for arbitrary $y \in F$, where $l(y)=\alpha$). Since l is continuous on F, there is a neighbourhood U of $\mathfrak{o}$ in L, for which the induced neighbourhood $(y_0+U)\cap F$ in F has an empty intersection with F_1. But it follows from this that y_0 does not lie in F_1+U, and this is closed, by 2.(3); applying (1′), we obtain a $u_0 \in L'$ which vanishes on F_1+U, and which takes the value $u_0 y_0 = 1$. Thus u_0 coincides with $l(y)$ on F.

(2) *Every linearly topologized space* L, *together with its dual space* L', *forms a dual pair* $\langle L', L\rangle$.

Since $L' \subset L^*$, conditions (D1) and (D2″) are satisfied. (D2′) follows from (1), for if $x \neq 0$, there is a non-zero continuous linear functional on the one-dimensional subspace $[x] \subset L$, by 2.(4), (5) and (6b); using (1), this can be extended to $u \in L'$, with $ux \neq 0$.

(3) *If $\langle L_2, L_1\rangle$ is a dual pair, then $L_1' = L_2$ and $L_2' = L_1$, the duals being taken with respect to the linear weak topologies.*

Proof. It is enough to show that $L_1' = L_2$.

(a) Every $u \in L_2$ defines a weakly continuous linear functional on L_1, since $ux=0$ for each $x \in U_u$, and so u is continuous at $\mathfrak{o}$.

(b) If, conversely, u is a weakly continuous linear functional on L_1, there is a weak neighbourhood $F^\perp = [u_1, \ldots, u_n]^\perp$ of $\mathfrak{o}$, with the u_i independent, on which u vanishes. By § 9, 2.(7a), u is a linear combination of the u_i, and so lies in L_2.

$\langle L', L\rangle$ forms a dual pair, by (2), and so, using 3., the weak topology $\mathfrak{T}_{ls}(L')$ can be introduced on L. Its relation to the original topology $\mathfrak{T}$ is given by

(4) *If $L[\mathfrak{T}]$ is a linearly topologized space, $\mathfrak{T}$ is always finer than the linear weak topology $\mathfrak{T}_{ls}(L')$.*

If $u_0 \in L'$, there is a $\mathfrak{T}$-neighbourhood U of $\mathfrak{o}$ with $u_0 x = 0$, for all $x \in U$. Every hyperplane $u_0 x = 0$ is thus a $\mathfrak{T}$-neighbourhood of $\mathfrak{o}$. Every basic $\mathfrak{T}_{ls}$-neighbourhood of $\mathfrak{o}$ is therefore a $\mathfrak{T}$-neighbourhood of $\mathfrak{o}$, as it is the intersection of finitely many such hyperplanes.

(4) can also be expressed as

(5) *$\mathfrak{T}_{ls}(L')$ is the coarsest linear topology on L for which L' is the dual space.*

(6) *If F is a $\mathfrak{T}$-closed linear subspace of $L[\mathfrak{T}]$, F is orthogonally closed with respect to L'; conversely a linear subspace F of L which is orthogonally closed with respect to L' is $\mathfrak{T}_{ls}(L')$-closed. In particular the $\mathfrak{T}$-closed and $\mathfrak{T}_{ls}(L')$-closed linear subspaces of $L[\mathfrak{T}]$ coincide.*

The closure $\bar{F}$ of an arbitrary linear subspace $F \subset L$ is equal to $F^{\perp\perp}$.

Proof. (1') implies that a $\mathfrak{T}$-closed subspace is orthogonally closed. Suppose conversely that $F^{\perp\perp} = F$, and that x_0 is a weak closure point of F. If u_0 is any point of $F^\perp$, there is at least one $y \in F$ in the weak neighbourhood $U_{u_0}(x_0)$, and so $u_0 x_0 = u_0(x_0 - y) = 0$. Since this holds for each $u_0 \in F^\perp$, x_0 lies in $F^{\perp\perp} = F$. The last assertion now follows from 2.(6a).

(7) *If F is a closed linear subspace of the linearly topologized space L, every linear subspace G in which F has finite codimension is also closed.*

It is enough, by (6), to show that $G = F \oplus [x_0]$ is orthogonally closed. By (1') there exists $u_0 \in F^\perp$ with $u_0 x_0 = 1$. If u is an arbitrary element of $F^\perp$, $v = u - (u x_0) u_0$ belongs to $G^\perp$. If $z \in G^{\perp\perp}$, we must have $vz = uz - (u x_0) u_0 z = 0$, and so $u(z - x_0(u_0 z)) = 0$, for all $u \in F^\perp$. This means that $z - (u_0 z) x_0 \in F$, that $z \in G$, and that $G^{\perp\perp} = G$.

(8) *The linear weak topology of a linearly topologized space* $L[\mathfrak{T}]$ *has as base of neighbourhoods of* $\circ$ *the collection of all* $\mathfrak{T}$*-closed linear subspaces of finite co-dimension.*

If F is an n-dimensional linear subspace of L', $F^{\perp}$ is of codimension n in L, by § 9, 2.(7a). Conversely, a $\mathfrak{T}$-closed linear subspace of codimension n is intersection of n hyperplanes H_i, by § 7, 7.(7), and these are $\mathfrak{T}$-closed, by 7. H_i is the space orthogonal to a one-dimensional subspace $[u_i] \subset L'$, by 4.(1'), and so $H = [u_1, \ldots, u_n]^{\perp}$.

Another proof of (8) will be given in 9.(5).

5. The dual pairs $\langle E^*, E\rangle$. We now inquire whether we can answer the problem in 1., using the results which we have established.

We give E and E^* the linear weak topologies $\mathfrak{T}_{ls}(E^*)$ and $\mathfrak{T}_{ls}(E)$ respectively. Then, by 4.(3), $E^* = E'$ and $E = (E^*)'$, so that the linear functionals on E^* which lie in E are indeed characterized as the weakly continuous ones, whereas conversely all the linear functionals on E are weakly continuous.

By 4.(6), the orthogonally closed subspaces of E^* are precisely the weakly closed ones; on the other hand, by 4.(6) and § 9, 2.(5), all the linear subspaces of E are weakly closed.

Further, we have

(1) *If* $A \in \mathfrak{S}(E,F)$, *the adjoint mapping* A' *is a weakly continuous mapping from* F^* *into* E^*; *conversely every weakly continuous linear mapping from* F^* *into* E^* *is the adjoint of some* $A \in \mathfrak{S}(E,F)$.

(a) It is enough to consider continuity at $\circ$. Let $A \in \mathfrak{S}(E,F)$. Given $U_{x_1,\ldots,x_n}(\circ)$ in E^*, the image $A'v$ of any v in the neighbourhood $U_{Ax_1,\ldots,Ax_n}(\circ)$ in F^* lies in $U_{x_1,\ldots,x_n}(\circ)$, since $(A'v)x_i = v(Ax_i)$; hence A' is weakly continuous.

(b) Conversely, let B be a weakly continuous linear mapping of F^* into E^*. For each $x_0 \in E$, $(Bv)x_0$ is a weakly continuous linear functional on F^*. Since $(F^*)' = F$, there is a uniquely determined $y_0 \in F$ for which $(Bv)x_0 = v y_0$. We put $y_0 = Ax_0$. The mapping defined in this way is clearly linear, and so $A \in \mathfrak{S}(E,F)$. It follows immediately from $(Bv)x = v(Ax)$ that $B = A'$.

(2) *Every* $A \in \mathfrak{S}(E,F)$ *is weakly continuous.*

For given a neighbourhood $U_{v_1,\ldots,v_n}(\circ)$, $v_i \in F^*$, in F, Ax lies in this neighbourhood if $x \in U_{A'v_1,\ldots,A'v_n}(\circ)$.

Using § 9, 4.(6) in the case where $E = F$, (1) and (2) imply that

(3) *The algebras* $L(E)$ *and* $L(E^*)$ *of weakly continuous endomorphisms are anti-isomorphic.*

The weak topology therefore satisfies all the requirements made in 1. The next section shows, however, that it is not satisfactory in every respect.

6. Weak convergence and weak completeness. If E is infinite-dimensional, every basic weak neighbourhood of $\circ$ is a linear subspace of E of finite co-dimension. The topology $\mathfrak{T}_{ls}(E^*)$ on E is therefore different from the discrete topology. Nevertheless, the two topologies give the same criterion for the convergence of sequences, as we shall now show.

A sequence $x_n \subset E$ is said to be almost constant if all the terms of the sequence are the same, from a certain n_0 onwards. A sequence is clearly a Cauchy sequence for the discrete topology if and only if it is almost constant. On the other hand, we also have

(1) *Every weak Cauchy sequence in E is almost constant.*

$\mathfrak{T}_{ls}$ induces the discrete topology on every finite dimensional subspace of E. It is therefore enough to show that a weak Cauchy sequence lies in a finite-dimensional subspace of E. If this were not so, it would be possible to find a subsequence x_{n_j} of linearly independent terms, and it would then be possible to define a linear functional u_0 on E with $u_0 x_{n_j} \neq u_0 x_{n_{j-1}}$, since K has at least two elements. But there would then be no n_0 beyond which the terms $x_{n_j} - x_{n_{j'}}$, would lie in $U_{u_0}(\circ)$.

An immediate consequence is

(2) *E is weakly sequentially complete.*

In order to describe weak convergence in E^*, we establish

(3) *If E has dimension d over* K, E^*, *with the topology* $\mathfrak{T}_{ls}(E)$, *is topologically isomorphic to the topological product* K^d, *with* K *discrete.*

Let $\{x_\nu\}$, $\nu \in \mathsf{N}$, be a basis for E, and let $\{u_\nu\}$ be the dual system. Let $[u_\nu]$ be the one-dimensional subspace of E^*, isomorphic to K, with basic element u_ν. By § 9, 1., E^* is isomorphic, as a vector space, to $\prod_\nu [u_\nu]$. $\mathfrak{T}_{ls}$ induces the discrete topology on $[u_\nu]$, and the sets $U_{x_{\nu_1},\ldots,x_{\nu_n}}(\circ)$ form a base of neighbourhoods of $\circ$ in E^*. The result therefore follows from § 1, 8.

From this there follows immediately

(4) *If* $\{x_\nu\}$, $\nu \in \mathsf{N}$, *is a basis of E, a sequence* $u_n \in E^*$ *is a weak Cauchy sequence if and only if every sequence* $u_n x_\nu$, $n = 1, 2, \ldots$, *is almost constant.*

In particular, a sequence $\mathfrak{u}^{(n)} = \{v_\nu^{(n)}\}$ *in* $\omega_d(\mathsf{K})$ *is a weak Cauchy sequence if and only if every sequence* $v_\nu^{(n)}$, $n = 1, 2, \ldots$, *is almost constant in* K.

It follows that E^* is weakly sequentially complete. We can say even more, however.

(5) *E^* is weakly complete.*

For E^* is topologically isomorphic to $\prod_\nu [u_\nu]$, by (3). Each $[u_\nu]$ is a discrete topological space, and is complete, by § 5, 4.(3). But then $\prod_\nu [u_\nu]$ is complete, by § 5, 7.(2), and so, therefore, is E^*.

In contrast, we have

(6) *If E is infinite dimensional, E is not weakly complete. Considering E as a subspace of E^{**}, E^{**} is the weak closure of E.*

Here E^{**} is to be considered as a linearly topologized space under $\mathfrak{T}_{ls}(E^*)$. E is a proper subspace of E^{**}, by § 9, 5.(2).

The space $E^\perp$ orthogonal to E in E^* is $\mathfrak{o}$, so that the space $E^{\perp\perp}$ orthogonal to $E^\perp$ in E^{**} is E^{**}. Hence E^{**} is the orthogonal closure of E, and the result follows from 4.(6).

The linear weak topology is therefore not completely satisfactory, since E and E^* are not both weakly complete. It is reasonable to try to replace $\mathfrak{T}_{ls}$ by a finer linear topology, which coincides on E with the discrete topology, but which does not alter the results for E^*. The solution of this problem will be given as we investigate linearly topologized spaces further; we turn to the study of these once more.

7. Quotient spaces and topological complements. Let $L[\mathfrak{T}]$ be a linearly topologized space, and let $\mathfrak{T}$ be defined by linear neighbourhoods U_α of $\mathfrak{o}$. Let L_1 be a linear subspace of L. By 2.(6), it is again a linearly topologized space.

A topology can be introduced in a natural way on the quotient space L/L_1, by taking as open sets in L/L_1 the images $\hat{O}$ of the open sets O in L, under the canonical mapping K of L onto L/L_1 (cf. § 8, 2.). If O is open in L, so is $O+L_1$, and so $K^{(-1)}(\hat{O})$ is again open. From this it follows easily that (O1) and (O2) of § 1, 1. are satisfied; for example, $\bigcap_{i=1}^{n} \hat{O}_i$ is again open, since $\bigcap \hat{O}_i = \bigcap K(K^{(-1)}\hat{O}_i) = K(\bigcap K^{(-1)}\hat{O}_i)$.

The topology defined in this way on L/L_1 is called the topology induced by $\mathfrak{T}$; it is again denoted by $\mathfrak{T}$.

This topology can only be Hausdorff if L_1 is closed in L, for if x_0 is a closure point of L_1 which does not lie in L_1, then $\hat{x} \neq \hat{\mathfrak{o}}$ belongs to every neighbourhood of $\hat{\mathfrak{o}}$. On the other hand, we have

(1) *If L_1 is a closed linear subspace of $L[\mathfrak{T}]$, L/L_1 is a linearly topologized space under the induced topology $\mathfrak{T}$.*

Proof. The induced topology is linear, since the images $\hat{U}_\alpha$ of the U_α are linear subspaces of L/L_1, and they form a base of neighbourhoods of $\mathfrak{o}$.

Further, L/L_1 is Hausdorff if there is for each $\hat{x}_0 \neq \hat{\mathfrak{o}}$ a neighbourhood of $\hat{\mathfrak{o}}$ which does not contain $\hat{x}_0$. For if $\hat{U}_\alpha$ is a linear neighbourhood contained in this neighbourhood, $\hat{x}_0+\hat{U}_\alpha$ and the neighbourhood $\hat{U}_\alpha$ of $\hat{o}$ are disjoint. But the set of all $\hat{x}\neq\hat{x}_0$ is open and contains $\hat{o}$, since it is the K-image of the complement of the closed set x_0+L_1.

(2) *If L_1 is a subspace of L which is both open and closed, then L/L_1 is discrete.*

For $\hat{L}_1=\hat{o}$ is a neighbourhood of $\hat{o}$.

A continuous linear mapping A from one linearly topologized space L_1 into another L_2 is called a topological homomorphism if it is open – that is, if every open set in L_1 is mapped into an open set in $A(L_1)$. If A is also one-one A is called a topological monomorphism of L_1 into L_2. A is then a homeomorphism of L_1 and $A(L_1)$ (cf. § 1, 7.). If $A(L_1)=L_2$ as well, A is called a topological isomorphism of L_1 and L_2.

It follows directly from the definition of the induced topology that

(3) *If L_1 is closed, the canonical mapping K of L onto L/L_1 is a topological homomorphism.*

(4) *A continuous linear mapping A from L_1 into L_2 always has a closed null-space $N[A]$. A is the product of the canonical homomorphism K of L_1 onto $L_1/N[A]$, a continuous one-one linear mapping $\breve{A}$ of $L_1/N[A]$ onto $A(L_1)$, and the embedding J of $A(L_1)$ into L_2.*

Proof. The null-space $N[A]$ is closed, since it is the inverse image of the closed set $\{o\}\subset L_2$. By (3), K is therefore a topological homomorphism. The inverse image $A^{(-1)}(O)$ of an open set $O\subset A(L_1)$ is open, and so therefore is $KA^{(-1)}(O)=\breve{A}^{(-1)}(O)$; thus $\breve{A}$ is continuous, and it is clearly one-one.

Further, we have an analogue to § 8, 2.(1)

(5) *Every topological homomorphism A of L_1 onto L_2 is the product of the canonical homomorphism K of L_1 onto $L_1/N[A]$ and a topological monomorphism $\hat{A}$ of $L_1/N[A]$ into L_2. The topological monomorphism $\hat{A}$ of $L_1/N[A]$ into L_2 is the product of a topological isomorphism $\breve{A}$ of $L_1/N[A]$ onto $A(L_1)$, and the embedding J of $A(L_1)$ into L_2, which is a topological monomorphism.*

Conversely, the continuous linear mapping A is a topological homomorphism if $\breve{A}$ is a topological isomorphism.

The simple proof, which uses (4), can be omitted.

Let $F\subset G$ be two closed linear subspaces of $L[\mathfrak{T}]$, and let K_1 be the canonical homomorphism of L onto L/F. The image $K_1(G)$ is G/F. If we apply the canonical homomorphism K_2 of L/F onto $(L/F)/(G/F)$ to L/F, K_2K_1 is a topological homomorphism of L onto $(L/F)/(G/F)$, with $N[K_2K_1]=G$, and so by (5) we have a $\mathfrak{T}$-isomorphism $(L/F)/(G/F)\cong L/G$ (cf. § 7, 6(7)). The other of the two isomorphism theorems, $F/F\cap G\cong(F+G)/G$, is not generally true in the topological sense for closed linear subspaces F and G of $L[\mathfrak{T}]$ (cf. § 13, 6.).

Two complementary closed linear subspaces L_1 and L_2 of L are called $\mathfrak{T}$-complementary if $L[\mathfrak{T}]$ is the topological direct sum (the topological product) of $L_1[\mathfrak{T}]$ and $L_2[\mathfrak{T}]$, in the sense of 2.

(6) *A closed linear subspace L_1 of $L[\mathfrak{T}]$ has a $\mathfrak{T}$-complement L_2 if and only if there is a continuous projection P of L onto L_1. P_1 is then a topological homomorphism, $L_2=N[P_1]$, and L/L_2 is topologically isomorphic to L_1.*

Proof. a) If $L[\mathfrak{T}]=L_1[\mathfrak{T}]\times L_2[\mathfrak{T}]$, the projection P_1 of L onto L_1 is continuous, by § 1, 8.

b) On the other hand, if P_1 is continuous, and if $N[P_1]=L_2$, then L_2 is closed, and by § 8.3 it is an algebraic complement of L_1. Moreover the projection $I-P_1=P_2$ of L onto L_2 is continuous. Each $x=x_1+x_2$, $x_1\in L_1$, $x_2\in L_2$ corresponds in a one-one linear way to an element $\tilde{x}=(x_1,x_2)\in L_1[\mathfrak{T}]\times L_2[\mathfrak{T}]=\tilde{L}$. The mapping $\tilde{x}\to x$ from $\tilde{L}$ onto L is continuous, for given a neighbourhood U of $\mathfrak{o}$ in L, it is sufficient to take $\tilde{x}\in(U\cap L_1)\oplus(U\cap L_2)$. Conversely the mapping $Q_1x=(x_1,\mathfrak{o})$ is continuous, as it is the product of the continuous projection $P_1x=x_1$ and the continuous mapping $x_1\to(x_1,\mathfrak{o})$; the mapping $(Q_1+Q_2)x=\tilde{x}$ is therefore also continuous.

c) The projection of $L_1\times L_2$ onto L_1 is a homomorphism since it is open, by § 1, 8. The last assertion follows from (5).

(7) *Every algebraic complement of a linear neighbourhood U of $\mathfrak{o}$ in a linearly topologized space $L[\mathfrak{T}]$ is a discrete $\mathfrak{T}$-complement of U.*

If H is an algebraic complement of U, $H\cap U=\mathfrak{o}$ is an open set in H, and so H is discrete. The projection of L onto H is continuous, for every subset M of H has an open inverse image $M+U$, by 2.(3). The assertion follows from (6).

(8) *Every finite dimensional linear subspace G of a linearly topologized space $L[\mathfrak{T}]$ has a $\mathfrak{T}$-complement.*

By 2.(5), G is discrete. If $x_1,\dots,x_n$ is a basis of G, then by 4.(1) there are elements $u_1,\dots,u_n\in L'$, with $u_ix_i=1$ and $u_ix_k=0$, if $i\neq k$. Let F be the n-dimensional linear subspace of L' spanned by the u_i. $F^\perp$ is a $\mathfrak{T}$-neighbourhood of $\mathfrak{o}$, and $L=F^\perp\oplus G$, by § 9, 2.(7a). $F^\perp$ and G are $\mathfrak{T}$-complementary, by (7).

(9) *Every linearly topologized space $L[\mathfrak{T}]$ is topologically isomorphic to a linear subspace H of a topological product of discrete spaces.*

Let $\{U_\alpha\}$ be a base of $\mathfrak{T}$-neighbourhoods of $\mathfrak{o}$ consisting of linear subspaces. By 2.(3), each U_α is open and closed, and, by (2), $L/U_\alpha = L_\alpha$ is discrete. We denote the coset of x with respect to U_α by $\hat{x}_\alpha$. Let $\hat{L}$ be the topological product $\prod_\alpha L_\alpha$. If we make each $x \in L$ correspond to the element $\hat{x} = (\hat{x}_\alpha) \in \hat{L}$, L is mapped in a one-one linear way onto a linear subspace H of $\hat{L}$. This mapping is a topological isomorphism, for if $\hat{U}_\beta$ denotes the set of all $y = (y_\alpha)$ in $\hat{L}$ for which $y_\beta = 0$, the neighbourhood U_α of $\mathfrak{o}$ in L is mapped onto the neighbourhood $\hat{U}_\alpha \cap H$ of $\mathfrak{o}$ in H, and the $\hat{U}_\alpha$ form a sub-base of neighbourhoods of $\mathfrak{o}$ for the topology of $\prod_\alpha L_\alpha$.

8. Dual spaces of subspaces and quotient spaces. If H is a linear subspace of $L[\mathfrak{T}]$, every $u \in L'$ defines a continuous linear functional $u^{(0)}$ on H, with the induced topology, and conversely the extension theorem shows that every continuous linear functional $u^{(0)}$ on H can be extended to a continuous linear functional defined on the whole of L. The linear mapping $u \to u^{(0)}$ is called the n a t u r a l homomorphism of L' onto H'. The kernel of this mapping is $H^\perp$, and so

(1) *The natural homomorphism of L' onto H' defines an algebraic isomorphism $L'/H^\perp \cong H'$.*

If $u^{(0)} \in H'$ and $\hat{u} \in L'/H^\perp$ correspond to each other, $u^{(0)} y = \hat{u} y$ for all $y \in$ H, so that

(2) *The induced topology $\mathfrak{T}_{ls}(L')$ and the topology $\mathfrak{T}_{ls}(H')$ coincide on H.*

The natural question, whether the natural homomorphism is a topological one, is answered for the linear weak topology as follows:

(3) *If $H \subset L$ is closed, the natural homomorphism of L' onto H' is a topological homomorphism for the topologies $\mathfrak{T}_{ls}(L)$ and $\mathfrak{T}_{ls}(H)$. The algebraic isomorphism $L'/H^\perp \cong H'$ is therefore a topological isomorphism for these topologies.*

By 7.(5) it is enough to show that the isomorphism $L'/H^\perp \cong H'$, established in (1), is a topological isomorphism, when H' is given the topology $\mathfrak{T}_{ls}(H)$, and $L'/H^\perp$ the topology $\mathfrak{T}_{ls}(L)$ induced by the canonical homomorphism K. The $\mathfrak{T}_{ls}(L)$-neighbourhood $F^\perp$ of L', where F is finite-dimensional in L, is sent by K into $\widehat{F^\perp}$. The isomorphism (1) transforms $\mathfrak{T}_{ls}(H)$ into a topology on $L'/H^\perp$, which has the sets $\widehat{G^\perp}$ as base of neighbourhoods of $\mathfrak{o}$, where G runs through the finite-dimensional linear subspaces of H. $F^\perp$ has the same image under K as $V = F^\perp + H^\perp$. Since $F^\perp$ is a neighbourhood of $\mathfrak{o}$, V is closed, by 2.(3), and is therefore orthogonally closed. Since $V^\perp = F^{\perp\perp} \cap H^{\perp\perp} = F \cap H$, V is of the form $G^\perp$. Thus we have shown that the image of every $\mathfrak{T}_{ls}(L)$-

neighbourhood $F^{\perp}$ is a $\mathfrak{T}_{ls}(H)$-neighbourhood $G^{\perp}$; the topology $\mathfrak{T}_{ls}(L)$ on $L'/H^{\perp}$ therefore coincides with the topology $\mathfrak{T}_{ls}(H)$ induced by (1).

If we apply (3) to $H^{\perp} \subset L'$ instead of to H, and use the fact that $H^{\perp\perp} = H$, we obtain

(3a) *If H is a closed linear subspace of L, $L/H \cong (H^{\perp})'$, for the topologies $\mathfrak{T}_{ls}(L')$ and $\mathfrak{T}_{ls}(H^{\perp})$.*

For the dual space $(L/H)'$ of L/H, with the induced topology, we have

(4) *If H is a closed linear subspace of $L[\mathfrak{T}]$, $(L/H)'$ is algebraically isomorphic to $H^{\perp}$.*

Proof. If $u \in H^{\perp}$, a linear functional u' on L/H is clearly defined by $u'\hat{x} = ux$. $ux = 0$ on a linear neighbourhood U of $\mathfrak{o}$, and so $u'\hat{x} = 0$ on $\hat{U} = KU$, where K is the cannonical homomorphism of L onto L/H. Thus u' is continuous on L/H.

If, conversely, $u'\hat{x}$ is continuous on L/H, a linear functional is defined on L by $ux = u'\hat{x}$. If u' vanishes on $\hat{U}$, u vanishes on $U = K^{(-1)}\hat{U} \supset H$, and so u is continuous and lies in $H^{\perp}$.

(5) *The isomorphism $(L/H)' \cong H^{\perp}$ is a topological isomorphism for the topologies $\mathfrak{T}_{ls}(L/H)$ on $(L/H)'$ and $\mathfrak{T}_{ls}(L)$ on $H^{\perp}$.*

For it follows from the fact that $u'\hat{x} = ux$ for $u \in H^{\perp}$, $u' \in (L/H)'$, $x \in L$ and $\hat{x} \in L/H$ that the $\mathfrak{T}_{ls}(L)$-neighbourhood $U_{x_1,\dots,x_n}$ in $H^{\perp}$ corresponds to the $\mathfrak{T}_{ls}(L/H)$-neighbourhood $U_{\hat{x}_1,\dots,\hat{x}_n}$ in $(L/H)'$, and conversely.

(6) *If $L[\mathfrak{T}]$ is the topological direct sum $L_1 \oplus L_2$ of two closed subspaces, then $L_1^{\perp}$ and $L_2^{\perp}$ are $\mathfrak{T}_{ls}(L)$-complementary to each other in L', and $L_1^{\perp} \cong L_2'$ and $L_2^{\perp} \cong L_1'$, for the linear weak topologies.*

Each $x \in L$ has a decomposition $x = x_1 + x_2$, with $x_1 \in L_1$, $x_2 \in L_2$. If $u \in L'$, a linear functional is defined on L by $u_1 x = u x_1$. It is $\mathfrak{T}$-continuous, for if $ux = 0$ for all $x \in U_1 \oplus U_2$, where U_1 and U_2 are linear neighbourhoods of $\mathfrak{o}$ in L_1 and L_2 respectively, then $u_1 x = u x_1 = 0$, for all these x. Each u_1 lies in $L_2^{\perp}$, and clearly $v = v_1$, if $v \in L_2^{\perp}$. Thus $P^2 = P$ holds for the linear mapping $Pu = u_1$; it is a projection of L' onto $L_2^{\perp}$, and $(I-P)L' = L_1^{\perp}$. P is $\mathfrak{T}_{ls}(L)$-continuous, for given $x^{(1)}, \dots, x^{(n)}$ in L, where $x^{(i)} = x_1^{(i)} + x_2^{(i)}$, $x_1^{(i)} \in L_1$, $x_2^{(i)} \in L_2$, it clearly follows from $u x_1^{(i)} = 0$ that $u_1 x^{(i)} = 0$, for $i = 1, \dots, n$. The decomposition $L' = L_1' \oplus L_2'$ is therefore a $\mathfrak{T}_{ls}(L)$-topological direct sum, by 7.(6). The algebraic isomorphism $L_2' \cong (L/L_1)'$ follows from the isomorphism $L/L_1 \cong L_2$, for the induced topologies $\mathfrak{T}$, which was established in 7.(6). This is also a topological isomorphism for $\mathfrak{T}_{ls}(L_2)$ and $\mathfrak{T}_{ls}(L/L_1)$; applying (5), $L_2' \cong L_1^{\perp}$ for the topologies $\mathfrak{T}_{ls}(L_2)$ and $\mathfrak{T}_{ls}(L)$.

The following example shows the difficulties associated with the idea of complementary spaces.

Let L_1 be the space $\varphi \oplus \varphi$, consisting of all pairs $(\mathfrak{x}, \mathfrak{y})$ of finite vectors over K. Let L_2 be the space of all vectors $(\mathfrak{u}, \mathfrak{v}) = (\mathfrak{x}, \mathfrak{y}) + \alpha(\mathfrak{e}, \mathfrak{e})$, where $(\mathfrak{x}, \mathfrak{y}) \in L_1$, $\mathfrak{e} = \{1, 1, \dots\}$ and $\alpha \in$ K. $\langle L_2, L_1 \rangle$ is a dual pair with respect to the bilinear form $\langle (\mathfrak{u}, \mathfrak{v}), (\mathfrak{x}, \mathfrak{y}) \rangle = \mathfrak{u}\mathfrak{x} + \mathfrak{v}\mathfrak{y}$, where on the right hand side we take the scalar product of the vectors. Let $H_1 \subset L_1$ be the subspace consisting of all $(\mathfrak{x}, \mathfrak{o})$, and let H_2 be the subspace consisting of all $(\mathfrak{o}, \mathfrak{y})$.

$$L_1 = H_1 \oplus H_2 \tag{7}$$

is an algebraic complementary decomposition of L_1 into two $\mathfrak{T}_{ls}(L_2)$-closed subspaces. The spaces orthogonal to H_1 and H_2 in L_2 are H_2 and H_1 respectively, and so $H_1^{\perp} \oplus H_2^{\perp} = L_1$, which is a proper subspace of L_2. By (6), the algebraic decomposition (7) is not $\mathfrak{T}_{ls}(L_2)$-complementary; further the lattice-theoretic complementary decomposition $L_2 = H_1^{\perp} \vee H_2^{\perp}$ in $V(L_2)$ dual to (7) is not an algebraic complementary decomposition.

9. Linearly compact spaces. Following LEFSCHETZ, a linearly topologized space L is said to be l i n e a r l y c o m p a c t if every filter $\mathfrak{F}$ with a base $\{F_\alpha\}$ of linear sub-manifolds F_α of L has an adherent point in L.

Every compact linearly topologized space is clearly linearly compact, and so the concept "linearly compact" is weaker than "compact". However the most important properties of compact sets carry over to linearly compact subspaces.

(1) *A closed linear subspace F of a linearly compact space is again linearly compact.*

The proof of § 3, 2.(5) can be carried over to give

(2) *If A is a continuous linear mapping of a linearly compact space L_1 into a linearly topologized space L_2, then $A(L_1)$ is linearly compact.*

It follows from (1), (2) and 7.(3) that

(3) *If L_1 is a closed linear subspace of the linearly compact space L, then L/L_1 is linearly compact.*

Further, we have

(4) *A discrete linearly compact space L is finite-dimensional.*

P r o o f. Let L be infinite-dimensional, and let $\{x_\nu\}$, $\nu \in$ N, be a basis of L. If L_ν is the linear subspace spanned by the $x_{\nu'}$, $\nu' \neq \nu$, L_ν is closed, since L is discrete. The sets $x_\nu + L_\nu$, $\nu \in$ N, and their finite intersections form a filter base of linear manifolds in L. An adherent point of this filter must be equal to $\sum_\nu x_\nu$, which is not possible as there are infinitely many x_ν.

(5) *The co-dimension of an open and closed linear subspace U of a linearly compact space L is always finite. Conversely, every closed linear subspace of finite co-dimension in a linearly topologized space is open.*

If U is open and closed, $U=\hat{o}$ is open in L/U, and so L/U is discrete. L/U is linearly compact, by (3), and so, by (4), is finite-dimensional. Conversely, if U is closed and L/U is finite-dimensional, L/U is discrete, by 2.(5), and so U is open, as it is the inverse image of $\hat{o}$ under the canonical mapping.

(6) *A linearly compact space L is complete. A linearly compact subspace of a linearly topologized space is therefore always closed.*

Proof. If $\mathfrak{F}=\{F^{\beta}\}$ is a Cauchy filter on L, then for each U_{α} there exists an $F^{\beta(\alpha)}$ with $x-y\in U_{\alpha}$, for all x, $y\in F^{\beta(\alpha)}$, and so $F^{\beta(\alpha)}+U_{\alpha}$ is a linear manifold $x_{\alpha}+U_{\alpha}$. All these manifolds are closed, by 2.(3), and they form a filter base on L. By hypothesis there exists an adherent point x_0 in L; $x_0\in F^{\beta(\alpha)}+U_{\alpha}$, for each α, and so $x_0+U_{\alpha}=F^{\beta(\alpha)}+U_{\alpha}\supset F^{\beta(\alpha)}$. x_0 is therefore an adherent point of $\mathfrak{F}$.

The analogue of TYCHONOFF's theorem is valid:

(7) *The topological product of arbitrarily many linearly compact spaces is linearly compact.*

L is a linearly topologized space, by 2.(7). The proof of (7) is a transcription of the proof of TYCHONOFF's theorem given in § 3, 3.:

Filters with bases of linear manifolds have a property similar to that for arbitrary filters:

(8) *Every filter with a base of linear manifolds can be refined to a maximal such filter.*

This is proved as in § 2, 7.(1).

In place of § 2, 7.(3) we have the following result, proved in exactly the same way.

(9) *The image under a linear mapping of a filter maximal among those with a base of linear manifolds is again a filter maximal among those with a base of linear manifolds.*

(7) now follows from (8) and (9) in the same way as in TYCHONOFF's theorem.

(10) *Every continuous linear mapping A of a linearly compact space $L_1[\mathfrak{T}]$ into a linearly topologized space L_2 is a topological homomorphism with a closed image space.*

By (1), (2) and (6), the image of every closed linear subspace is closed. In particular $A(L_1)$ is closed, and by (3), $L_1/N[A]$ is linearly compact. Under the one-one mapping $\breve{A}$ of $L_1/N[A]$ onto $A(L_1)$ (which is continuous by 7.(4)), every closed linear subspace of finite co-dimension is sent into one with the same property, by (3). It follows from (5) that the $\breve{A}$-image of every linear neighbourhood is a linear neighbourhood, so that $\breve{A}$ is open, and is a topological isomorphism. Thus, by 7.(5), A is a topological homomorphism.

10. E^* as a linearly compact space. If the algebraic dual E^* of a linear space E is given the topology $\mathfrak{T}_{ls}(E)$, E^* is topologically isomorphic to K^d, with K discrete, by 6.(3). Since K, considered as a one-dimensional vector space, is linearly compact, 9.(7) shows that E^* is linearly compact:

(1) *E^* is linearly compact with respect to $\mathfrak{T}_{ls}(E)$.*

We shall establish a converse to this theorem, which will characterise linearly compact spaces.

(2) *If $L[\mathfrak{T}]$ is linearly compact, $\mathfrak{T}$ is the topology $\mathfrak{T}_{ls}(L')$.*

By 4.(4) it is enough to show that every $\mathfrak{T}$-neighbourhood is a $\mathfrak{T}_{ls}$-neighbourhood. The linear $\mathfrak{T}$-neighbourhoods of $\mathfrak{o}$ have finite codimension, by 9.(5), and further are $\mathfrak{T}$-closed subspaces, so that, by 4.(8), they are $\mathfrak{T}_{ls}(L')$ neighbourhoods.

It follows from (2) and 8.(2) that

(2a) *Every linearly compact subspace of a linearly topologized space L is linearly $\mathfrak{T}_{ls}(L')$-compact.*

(3) *If $L[\mathfrak{T}]$ is linearly compact, L is topologically isomorphic to $(L')^*$, so that linearly compact spaces can be characterised as algebraic duals E^* of linear spaces E, equipped with the topology $\mathfrak{T}_{ls}(E)$.*

By 4., L can be considered as a subspace of L'^*. The space orthogonal to L in L' is $[\mathfrak{o}]$, and so by 4.(6) the linear $\mathfrak{T}_{ls}(L')$-closure of L in $(L')^*$ is equal to $(L')^*$. But since L is complete, by 9.(6), $L=(L')^*$. The rest of the assertion follows from (1).

11. The topology $\mathfrak{T}_{lk}$. Besides the weak topology, we can introduce another topology on $L[\mathfrak{T}]$, and this we shall now investigate.

(1) *The sum $F=\sum_{i=1}^{n} F_i$ of finitely many linearly compact subspaces of L is again linearly compact.*

For by 9.(7) the topological product $\prod_{i=1}^{n} F_i$ is linearly compact; if we map this by $(x_1,\dots,x_n)\to x_1+\cdots+x_n$ onto $F\subset L$, we obtain a continuous linear mapping, and so, by 9.(2), F is linearly compact.

(2) *If U is a linear neighbourhood of $\mathfrak{o}$ in $L[\mathfrak{T}]$, $U^\perp$ is linearly weakly compact in L'.*

By 7.(7), $L=U\oplus L_1$, where L_1 is discrete, and so $L_1'=L_1^*$. By 8.(6), $L'=U^\perp\oplus L_1^\perp$ and $U^\perp$ is weakly isomorphic to L_1'; by 10.(1) and 8.(2), $U^\perp$ is therefore linearly weakly compact.

We now define the topology $\mathfrak{T}_{lk}(L')$ in $L[\mathfrak{T}]$, by taking as base of neighbourhoods of $\mathfrak{o}$ in L the spaces $C^\perp$ orthogonal to the linearly $\mathfrak{T}_{ls}(L)$-compact subspaces C of L'. By (1), the $C^\perp$ form a filter base;

the intersection condition of 2.(1) is satisfied, since $\mathfrak{T}_{lk}(L')$ is finer than $\mathfrak{T}_{ls}(L')$, by (2). Thus L is a linearly topologized space under $\mathfrak{T}_{lk}(L')$.

(3) *Let $L[\mathfrak{T}]$ be linearly compact. Then the topology $\mathfrak{T}_{lk}(L')$ on L is equal to $\mathfrak{T}_{ls}(L')$, and $\mathfrak{T}_{lk}(L)$ is the discrete topology on L'. L and L' are $\mathfrak{T}_{lk}$-complete, and are dual to each other under $\mathfrak{T}_{lk}$.*

Proof. $\mathfrak{T}_{lk}(L)$ is the discrete topology on L', as, by 10.(2), L is itself linearly weakly compact, so that $L^\perp=[\mathfrak{o}]$ is a $\mathfrak{T}_{lk}(L)$-neighbourhood in L'. By 10.(3), $(L'[\mathfrak{T}_{lk}(L)])'=L$. On the other hand, the method of proof of 9.(4) shows without difficulty that the linearly weakly compact subspaces of a discrete space are finite-dimensional; thus the topology $\mathfrak{T}_{lk}(L')$ on L is equal to $\mathfrak{T}_{ls}(L')$. Completeness follows from 9.(6) and 10.(2), duality from 10.(3).

If E is a vector space and E^* is its algebraic dual, the topologies $\mathfrak{T}_{lk}$ on E and E^* are therefore topologies which answer the question raised at the end of 6., and which maintain a complete symmetry between E and E^*.

The topology $\mathfrak{T}_{lk}$ is characterised by

(4) *The topology $\mathfrak{T}_{lk}(L')$ is the finest linear topology on $L[\mathfrak{T}]$ which has L' as dual space.*

Proof. a) Let $\mathfrak{T}^*$ be a linear topology on L with the property that $L[\mathfrak{T}^*]'=L[\mathfrak{T}]'=L'$. We must show that every linear $\mathfrak{T}^*$-neighbourhood U is a $\mathfrak{T}_{lk}$-neighbourhood. U is $\mathfrak{T}^*$-closed, by 2.(3), and so is orthogonally closed, by 4.(6). U is therefore the space orthogonal to $U^\perp$, which is linearly compact, by (2); it follows that U is a $\mathfrak{T}_{lk}$-neighbourhood.

b) In the opposite direction, we must show that every $\mathfrak{T}_{lk}$-continuous linear functional u_0 is $\mathfrak{T}_{ls}$-continuous on L. By hypothesis, there exists a linearly weakly compact subspace C of L', with the property that u_0 vanishes on $C^\perp\subset L$. If we form the space $C^{\perp\perp}$ orthogonal to $C^\perp$ in the algebraic dual E^*, then u_0 lies in $C^{\perp\perp}$. By 4.(6), $C^{\perp\perp}$ is the $\mathfrak{T}_{ls}(L)$-closure of C in L^*. On the other hand C is $\mathfrak{T}_{ls}(L)$-complete, by 9.(6), so that $C^{\perp\perp}=C$, and $u_0\in L'$.

The topology $\mathfrak{T}_{lk}$ is the analogue of the Mackey topology considered in § 21, 4., and (4) corresponds to the Mackey-Arens theorem (cf. § 21, 4.).

12. $\mathfrak{T}_{lk}$-continuous linear mappings. We denote the space of continuous linear mappings from $L_1[\mathfrak{T}_1]$ into $L_2[\mathfrak{T}_2]$ by $\mathfrak{L}(L_1[\mathfrak{T}_1], L_2[\mathfrak{T}_2])$.

(1) *Every continuous linear mapping A of $L_1[\mathfrak{T}_1]$ into $L_2[\mathfrak{T}_2]$ is also $\mathfrak{T}_{ls}$-continuous and $\mathfrak{T}_{lk}$-continuous, and we have*

$$(2)\qquad \mathfrak{L}(L_1[\mathfrak{T}_1], L_2[\mathfrak{T}_2])\subset \mathfrak{L}(L_1[\mathfrak{T}_{ls}(L_1')], L_2[\mathfrak{T}_{ls}(L_2')]) = \mathfrak{L}(L_1[\mathfrak{T}_{lk}(L_1')], L_2[\mathfrak{T}_{lk}(L_2')]).$$

The adjoint mappings A' form a linear subspace of

$$\mathfrak{L}(L_2'[\mathfrak{T}_{ls}(L_2)], L_1'[\mathfrak{T}_{ls}(L_1)]) = \mathfrak{L}(L_2'[\mathfrak{T}_{lk}(L_2)], L_1'[\mathfrak{T}_{lk}(L_1)]).$$

Proof. As in § 9, 4., it follows that a linear mapping is defined from L_2' into L_1' by

(3) $$(A'v)x = v(Ax) \quad \text{for all} \quad x \in L_1, v \in L_2'.$$

This is the adjoint mapping A' of A, to which A is adjoint in turn. As in 5.(1), (2) follows from the weak continuity of A and A'. Further A is $\mathfrak{T}_{lk}$-continuous, for given a linearly weakly compact subspace C of L_2', $A'(C)$ is linearly weakly compact, by 9.(2), and the image Ax of each $x \in A'(C)^{\perp}$ lies in $C^{\perp}$, by (3). The $\mathfrak{T}_{lk}$-continuity of each weakly continuous A' follows by interchanging the spaces and their duals. The fact, that we have equality instead of inclusion in (2) is proved by taking the topologies $\mathfrak{T}_{ls}$ (respectively $\mathfrak{T}_{lk}$) as the original topologies.

It follows from (1) and 7.(6) that

(4) *Two closed linear $\mathfrak{T}$-complementary subspaces of $L[\mathfrak{T}]$ are $\mathfrak{T}_{ls}(L')$- and $\mathfrak{T}_{lk}(L')$-complementary.*

If F is a linearly compact subspace of $L[\mathfrak{T}]$, F is linearly weakly compact, by 10.(2a), and $F^{\perp}$ is a $\mathfrak{T}_{lk}(L)$-neighbourhood in L'. By 7.(7), this has a $\mathfrak{T}_{lk}(L)$-complement H. By 11.(4) the $\mathfrak{T}_{lk}(L)$-dual space of L' is L again, and by 8.(6) $H^{\perp}$ is thus a closed space complementary to F in L. The projection P of L onto F is $\mathfrak{T}$-continuous: by 10.(2) a $\mathfrak{T}$-neighbourhood in F is of the form $U = U_{u_1,\ldots,u_n} \cap F$, where the u_i can be chosen in H. An element $x \in U_{u_1,\ldots,u_n}$, with decomposition $x = x_1 + x_2$, $x_1 \in F$, $x_2 \in H^{\perp}$, satisfies the relations $u_i x = u_i x_1 = 0$, so that $Px \in U$. It therefore follows from 7.(6) that

(5) *Every linearly compact subspace F of a linearly topologized space $L[\mathfrak{T}]$ has a $\mathfrak{T}$-complement.*

Information about the $\mathfrak{T}_{lk}$-topology on L/H is given by the following result, corresponding to 8.(3):

(6) *If H is a closed subspace of L, the natural homomorphism of L' onto H' is a topological homomorphism for the topologies $\mathfrak{T}_{lk}(L)$ and $\mathfrak{T}_{lk}(H)$, so that $L'/H^{\perp} \cong H'$ is a topological isomorphism. Similarly, $L/H \cong (H^{\perp})'$ for the topologies $\mathfrak{T}_{lk}(L')$ and $\mathfrak{T}_{lk}(H^{\perp})$.*

The argument is analogous to the proof of 8.(3): a $\mathfrak{T}_{lk}(L)$-neighbourhood of $\mathfrak{o}$ in L' has the form $C^{\perp}$, where C is a linearly $\mathfrak{T}_{ls}(L')$-compact subspace of L. A $\mathfrak{T}_{lk}(H)$-neighbourhood of $\mathfrak{o}$ in $L'/H^{\perp}$ has the form $\widehat{D^{\perp}}$, where D is a linearly $\mathfrak{T}_{ls}(H')$-compact subspace of H. By 8.(2), D is a linearly $\mathfrak{T}_{ls}(L')$-compact subspace of H, so that each $\widehat{D^{\perp}}$ is a $\widehat{C^{\perp}}$. Again, $C^{\perp}$ has the same image as $V = C^{\perp} + H^{\perp}$, and V is $\mathfrak{T}_{lk}$-closed, by 2.(3), since $C^\perp$ is a $\mathfrak{T}_{lk}(L)$-neighbourhood of $\mathfrak{o}$. Thus by 4.(6) V is orthogonally closed with respect to L. It follows from the fact that $V' = C \cap H$ that V has the form $D^\perp$, where D is linearly weakly compact in H.

(7) *If H is a closed linear subspace of $L[\mathfrak{T}]$, the topology $\mathfrak{T}_{lk}(H')$ on H is finer than the topology $\mathfrak{T}_{lk}(L')$.*

This follows by applying (1) to the inclusion mapping of H into E.

The question whether $\mathfrak{T}_{lk}(H')$ and $\mathfrak{T}_{lk}(L')$ coincide on every closed subspace H of $L[\mathfrak{T}]$ depends upon whether every closed linear subspace F of L for which L/F is linearly weakly compact has a linearly weakly compact complement, as does the validity of the result corresponding to 8.(5).

We shall give a counterexample in § 31, number 4.

13. Continuous basis and continuous dimension. Using the idea of Cauchy filter, sums with more than countably many terms can be introduced in a linearly topologized space $L[\mathfrak{T}]$. Let $\{x_\nu\}$, $\nu \in \mathsf{N}$, be a family of elements of L. We form the partial sums $\sum_{i=1}^{n} x_{\nu_i}$, for every finite collection of different ν_i, $i = 1, \ldots, n$. Let $M_{\nu_1, \ldots, \nu_n}$ denote the set of all partial sums $\sum_{i=1}^{n} x_{\nu_i} + \sum_{k=1}^{m} x_{\nu'_k}$, as the ν'_k run through collections of pairwise distinct indices, different from the ν_i. The sets $M_{\nu_1, \ldots, \nu_n}$ clearly form a filter-base on L. If this filter is a Cauchy filter, with limit x, x is called the t o p o l o g i c a l s u m of the x_ν, and we write $x = \sum_\nu x_\nu$. If N is finite, we obtain the usual sum.

(1) *If A is a continuous linear mapping from L_1 into L_2, and if $x = \sum_\nu x_\nu$, then $Ax = \sum_\nu Ax_\nu$.*

Since this holds for finite sums, $A(M_{\nu_1, \ldots, \nu_n})$ is the set $\tilde{M}_{\nu_1, \ldots, \nu_n}$ of all $\sum_{i=1}^{n} Ax_{\nu_i} + \sum_{k=1}^{m} Ax_{\nu'_k}$. Because of the uniform continuity of A (2.(9)), the continuous linear image of a Cauchy filter-base is again a Cauchy filter-base, so that Ax is the limit of the filter defined by the $\tilde{M}_{\nu_1, \ldots, \nu_n}$.

If follows directly from this that

(2) *If $x = \sum x_\nu$, then $x\alpha = \sum (x_\nu \alpha)$, for each $\alpha \in \mathsf{K}$.*

Further, if $x = \sum x_\nu$ and $y = \sum y_\nu$, then the family $\{(x_\nu, y_\nu)\}$, $\nu \in \mathsf{N}$, is summable in $L \times L$ to (x, y); moreover the linear mapping $(z_1, z_2) \to z_1 + z_2$ from $L \times L$ into L is continuous. It follows from this, using (1), that

(3) *If $x = \sum x_\nu$ and $y = \sum y_\nu$ in L, then $x + y = \sum (x_\nu + y_\nu)$.*

A set $\{x_\nu\}$, $\nu\in\mathsf{N}$, is called a **continuous basis** of L if every $x\in L$ can be written in one and only one way as a topological sum $x=\sum x_\nu\xi_\nu$, $\xi_\nu\in\mathsf{K}$, and if the component ξ_ν is a continuous linear function of x, for each ν. The smallest cardinal of a continuous basis of L is called the **continuous dimension** of L. It is invariant under topological isomorphism, by (1).

If $\{x_\mu\}$, $\mu\in\mathsf{M}$, is a basis for the vector space E, and if $\{u_\mu\}$ is the system dual to the x_μ in E^*, $\{u_\mu\}$ clearly forms a continuous basis for E^*, with respect to the weak topology. This is half of

(4) *Every weakly continuous basis of E^* is the dual system of a basis of E, and conversely.*

If the family $\{u_\mu\}$ forms a continuous basis of E^*, let x_μ be the weakly continuous linear functional on E^* defined by $u x_\mu = v_\mu$, where $u=\sum_\mu u_\mu v_\mu$. $u_{\mu'} x_\mu = 0$ if $\mu'\neq\mu$, and $u_\mu x_\mu = 1$. The x_μ are linearly independent, for given a linear combination $\sum_{i=1}^{k} x_{\mu_i}\alpha_i$ (with not all the α_i vanishing), it is easy to find a $\sum_{i=1}^{k} u_{\mu_i}\beta_i \neq \mathrm{o}$, on which the linear combination does not vanish. The linear span of the x_μ has o as orthogonal space in E^*, and so it is equal to E by § 9, 2.(1). The x_μ therefore form a basis of E, whose dual system is $\{u_\mu\}$.

If d is the dimension of E, every weakly continuous basis of E^* therefore consists of d elements; i.e. the continuous dimension of E^* with respect to $\mathfrak{T}_{ls}(E)$ is equal to the dimension of E.

Using 8.(6), the continuous dimension of every closed subspace of E^* is determined, as well. If we give the vector space E the discrete topology, every basis of E is a continuous basis, and by the same token the dimension of a subspace of E can also be considered as the continuous dimension with respect to the discrete topology.

Not every linearly topologized space has a continuous basis, cf. H. Gross, Math. Ann. 194, 313—315 (1971).

§ 11. The theory of equations in E and E^*

1. The duality of E and E^*. Let us gather together the most important of the results of the preceding paragraph.

A vector space E and its algebraic dual E^* form a dual pair. If we give each of them the $\mathfrak{T}_{lk}$-topology, then this coincides with the discrete topology on E, and with the weak topology on E^*. For these topologies, each space is the collection of continuous linear functionals on the other. If we make each closed linear subspace H of one of the two spaces correspond to its orthogonal space $H^\perp$, we obtain a dual isomorphism between the lattices $\bar{V}(E)$ and $\bar{V}(E^*)$ of closed linear subspaces. Under this correspondence we have topological isomorphisms:

If $E=H_1\oplus H_2$,

(1) $$(E/H_1)^*\cong H_1^\perp\cong H_2^*,$$

(2) $$E/H_1\cong(H_1^\perp)'\cong H_2.$$

If $E^*=G_1\oplus G_2$,

(3) $$(E^*/G_1)'\cong G_1^\perp\cong G_2',$$

(4) $$E^*/G_1\cong(G_1^\perp)^*\cong G_2.$$

Proof. The first of the two isomorphisms in (1) and (3) (respectively in (2) and (4)) follows from § 10, 8.(5) (respectively 8.(3a)), and the others follow from § 10, 8.(6). For this it must be borne in mind that if a space F has the discrete topology, then $F'=F^*$.

Remark. It is not necessary to assume that the direct sum $E^*=G_1\oplus G_2$ of closed linear subspaces G_1 and G_2 is a topological sum: it follows from $E^*=G_1\oplus G_2$ that $E=G_1^\perp\oplus G_2^\perp$, by § 9, 3.(5). In the discrete space E algebraic and topological complements are the same, and so it follows from § 10, 8.(6) and § 10, 12.(4) that the direct sum $G_1^{\perp\perp}\oplus G_2^{\perp\perp}=G_1\oplus G_2$ is a $\mathfrak{T}_{ls}$- and $\mathfrak{T}_{lk}$-topological direct sum.

These relations, together with § 10, 13, imply

(5) *If* $H\subset E$ *has dimension d and codimension c, then* $H^\perp$ *has continuous dimension c and continuous codimension d, and conversely.*

There is also complete symmetry between the linear mappings $A\in\mathfrak{S}(E,F)$ and their adjoint mappings $A'\in\mathfrak{S}(E^*,F^*)$. They form the collections $\mathfrak{L}(E,F)$ and $\mathfrak{L}(F^*,E^*)$ of all continuous linear mappings, for either $\mathfrak{T}_{ls}$ or $\mathfrak{T}_{lk}$, from E into F and from F^* into E^* respectively.

From § 10, 9.(10) we have

(6) *Every weakly continuous linear mapping* $A'\in\mathfrak{L}(F^*,E^*)$ *is a topological homomorphism with a closed image space.*

By § 9, 2.(5), the same holds for $A\in\mathfrak{L}(E,F)$. Since $A'(F^*)$ and $N[A']$ have closed complements, by § 9, 3.(5), all the characteristic subspaces of A and A' are orthogonally closed.

(7) *The following relations hold between the characteristic subspaces of A and A':*

(8) $$N[A']=A(E)^\perp\cong(F/A(E))^*\cong C[A]^*,$$

(9) $$A'(F^*)=N[A]^\perp\cong(E/N[A])^*\cong U[A]^*,$$

(10) $$N[A]=A'(F^*)^{\perp}\cong(E^*/A'(F^*))'\cong C[A']',$$

(11) $$A(E)=N[A']^{\perp}\cong(F^*/N[A'])'\cong U[A']',$$

and the isomorphisms are topological.

Proof. It follows from $v(Ax)=(A'v)x$, for $x\in E$, $v\in E^*$, that, for fixed v, $A'v=o$ if and only if $v\in A(E)^{\perp}$, so that $N[A']=A(E)^{\perp}$. Fixing x, we obtain that $N[A]=A'(F^*)^{\perp}$. Going over, in these equations, to the orthogonal spaces, and using (6), we obtain the first equations of (9) and (11) immediately. The isomorphisms of (8) to (11) follow by applying (1) and (3).

If we call the continuous dimensions of $A'(F^*)$, $N[A']$ and $E^*/A'(F^*)$ the rank $r(A')$, the nullity $s(A')$ and the defect $s'(A')$, the next result follows directly from (9), (10) and (8).

(12) *The ranks, nullities and defects of A and A' are related as follows:*

(13) $$r(A)=r(A'),\quad s(A)=s'(A'),\quad s'(A)=s(A').$$

2. The theory of the solutions of column- and row-finite systems of equations. Let A be a linear mapping from the vector space E of dimension d over K into the vector space F of dimension e over K. For given $y^{(0)}\in F$, the relation

(1) $$Ax=y^{(0)}$$

is called a linear equation in $x\in E$. For given $u^{(0)}\in E^*$, the equation

(2) $$A'v=u^{(0)}$$

in $v\in F^*$ is called the transposed equation of (1); conversely, (1) is called the transposed equation of (2).

(1) and (2) are called homogeneous if $y^{(0)}$ and $u^{(0)}$ are equal to o, respectively; otherwise they are inhomogeneous.

The ideas which we have considered up to now allow us to describe the behaviour of the solutions of (1) and (2) in a way modelled on the determinant-free theory of the solution of finitely many linear equations in finitely many unknowns.

By the rank, nullity and defect of (1) and (2) we mean the rank, nullity and defect of A and A' respectively; by considering 1.(8) to (12), we obtain the following theorem on the solution of equations:

(3) *The equations* (1) *and* (2) *have the same rank r, and the defect of either equation is equal to the nullity of its transposed equation.*

The solutions of the homogeneous form of equation (1) *form a linear subspace of dimension $s(A)=s'(A')$, and the solutions of the homogeneous* *form of the transposed equation* (2) *form a closed linear subspace of* F^* *with continuous dimension* $s'(A)=s(A')$.

The inhomogeneous form of equation (1) *is solvable for just those* $y^{(0)}\in F$ *which are orthogonal to all the solutions of the homogeneous transposed equation* (2). *The collection of all these* $y^{(0)}$ *forms a linear subspace of* F *of dimension* $r(A)$ *and codimension* $s'(A)$. *The inhomogeneous form of equation* (2) *is soluble for just those* $u^{(0)}\in E^*$ *which are orthogonal to all the solutions of the homogeneous equation* (1). *The collection of all these* $u^{(0)}$ *forms a closed linear subspace of* E^* *of continuous dimension* $r(A)$ *and continuous codimension* $s'(A')$.

All the solutions of the inhomogeneous equation can be obtained from just one solution, by adding all the solutions of the homogeneous equation.

As a special case, if the homogeneous equation and the homogeneous transposed equation are insoluble, i.e. if they only have the trivial solution o, *then the inhomogeneous equations have a unique solution for each* $y^{(0)}\in F$ *or* $u^{(0)}\in E^*$ *on the right hand side.*

If we introduce a basis $\{x_\nu\}$, $\nu\in\mathsf{N}$, in E and a basis $\{y_\mu\}$, $\mu\in\mathsf{M}$, in F, then, by § 8, 5., (1) is transformed into e linear equations with d unknowns

$$(4)\qquad \sum_\nu \alpha_{\mu\nu}\xi_\nu=\eta_\mu^{(0)},\qquad \mu\in\mathsf{M},\qquad \text{respectively}\qquad \mathfrak{A}\mathfrak{x}=\mathfrak{y}^{(0)},$$

where $\mathfrak{A}=((\alpha_{\mu\nu}))$ is a column-finite matrix and $\mathfrak{x}=\{\xi_\nu\}\in\varphi_d(\mathsf{K})$ and $\mathfrak{y}^{(0)}=\{\eta_\mu^{(0)}\}\in\varphi_e(\mathsf{K})$ are the vectors representing x and $y^{(0)}$ respectively.

Using § 9, 4., the row-finite system of equations

$$(5)\qquad \sum_\mu \alpha_{\mu\nu}\varphi_\mu=v_\nu^{(0)},\qquad \nu\in\mathsf{N},\qquad \text{respectively}\qquad \mathfrak{A}'\mathfrak{v}=\mathfrak{u}^{(0)},$$

with

$$\mathfrak{v}=\{\varphi_\mu\}\in\omega_e(\mathsf{K}),\qquad \mathfrak{u}^{(0)}=\{v_\nu^{(0)}\}\in\omega_d(\mathsf{K}),$$

is obtained from (2) by considering the continuous bases of E^* and F^* dual to $\{x_\nu\}$ and $\{y_\mu\}$ respectively. (4) and (5) are again called systems of equations, each the transpose of the other. Conversely, any column-finite or row-finite system of equations can be interpreted as representing a linear equation (1) or (2) respectively, by means of a suitable basis, so that the theorem on the solution of equations can also be construed as a theorem about the solution of a row- or column-finite system of equations.

3. Formulae for solutions. The theorem on the solution of equations gives complete information about the behaviour of the solutions of column- and row-finite systems of equations over an arbitrary field K. Nevertheless, it does not enable us to represent the solutions of the given system of equations as functions of the matrix of the system of equations and the right-hand side. The theory of equivalence established in § 8 helps us to do this.

If $\mathfrak{A}$ is a column-finite matrix, then by § 8, 8. and 9. there exist invertible square column-finite matrices $\mathfrak{B}$ and $\mathfrak{C}$ for which $\mathfrak{C}\mathfrak{A}\mathfrak{B}=\mathfrak{D}$, where $\mathfrak{D}$ is a matrix for which the elements $\delta_{\mu'(\nu'),\nu'}=1$, while all the other terms vanish. Here ν' runs through a index set N' of $r(\mathfrak{A})$ elements whose complement N'' in the column-index set N contains $s(\mathfrak{A})$ indices, while μ' is a one-one mapping of N' onto a set M' of indices, whose complement M'' in the row-index set M contains $s'(\mathfrak{A})$ terms.

If we now introduce new variables in 2.(4) by putting $\mathfrak{x}=\mathfrak{B}\mathfrak{z}$, and multiply the resulting equation on the left by $\mathfrak{C}$, we obtain

$$\mathfrak{C}\mathfrak{A}\mathfrak{B}\mathfrak{z}=\mathfrak{D}\mathfrak{z}=\mathfrak{C}\mathfrak{y}^{(0)}=\mathfrak{t}^{(0)}. \tag{1}$$

The system (1) of equations is said to be equivalent to 2.(4); the solutions correspond to each other in a one-one way under the transformations $\mathfrak{x}=\mathfrak{B}\mathfrak{z}$, $\mathfrak{z}=\mathfrak{B}^{-1}\mathfrak{x}$. Written out, (1) has the simple form

$$\zeta_{\nu'}=\tau^{(0)}_{\mu'(\nu')}, \quad 0=\tau^{(0)}_{\mu''}, \ \nu'\in\mathsf{N}', \ \mu'(\nu')\in\mathsf{M}', \ \mu''\in\mathsf{M}''. \tag{2}$$

The homogeneous system has for solutions all those $\mathfrak{z}$ which are of the form $\sum\limits_{\nu''}\mathfrak{e}_{\nu''}\zeta_{\nu''}$, and which take arbitrary values on the $\nu''\in\mathsf{N}''$, and these therefore comprise $N[\mathfrak{D}]$. The inhomogeneous system only has a solution if all the $\tau^{(0)}_{\mu''}=0$, for $\mu''\in\mathsf{M}''$. If this condition is satisfied, a solution is given by $\zeta^{(0)}_{\nu'}=\tau^{(0)}_{\mu'(\nu')}$, $\zeta^{(0)}_{\nu''}=0$, $\nu'\in\mathsf{N}'$, $\nu''\in\mathsf{N}''$. All the solutions are obtained from this by adding the solutions of the homogeneous system of equations.

This result can be expressed by formulae. The transposed matrix $\mathfrak{D}'$ provides an inverse of $\mathfrak{D}$ in the sense of § 8, 4., for $\mathfrak{D}\mathfrak{D}'$ is the matrix which has ones in the $(\mu'(\nu),\mu'(\nu'))$ places of the main diagonal, and has zero everywhere else; $\mathfrak{D}\mathfrak{D}'$ is therefore the projection onto the image-space of $\mathfrak{D}$. $\mathfrak{D}'\mathfrak{D}$ is the projection onto the inverse-image space of $\mathfrak{D}$ consisting of all the $\mathfrak{z}$ with $\zeta_{\nu''}=0$, for $\nu''\in\mathsf{N}''$. $\mathfrak{E}_{\mathsf{N}}-\mathfrak{D}'\mathfrak{D}$ is the projection onto $N[\mathfrak{D}]$, where $\mathfrak{E}_{\mathsf{N}}$ is the square unit-matrix defined on N.

All the solutions of the homogeneous system (2) therefore have the form $(\mathfrak{E}_{\mathsf{N}}-\mathfrak{D}'\mathfrak{D})\mathfrak{y}$, $\mathfrak{y}\in\varphi_d(\mathsf{K})$, and a basis for the solutions is formed by the columns of $\mathfrak{E}_{\mathsf{N}}-\mathfrak{D}'\mathfrak{D}$. The inhomogeneous system is only soluble for those $\mathfrak{t}^{(0)}$ for which $(\mathfrak{E}_{\mathsf{M}}-\mathfrak{D}\mathfrak{D}')\mathfrak{t}^{(0)}=\mathfrak{o}$. One solution is then given by $\mathfrak{z}^{(0)}=\mathfrak{D}'\mathfrak{t}^{(0)}$.

Going back to the equivalent system 2.(4), we obtain

(3) *If* $\mathfrak{D}=\mathfrak{C}\mathfrak{A}\mathfrak{B}$ *is the canonical form for* $\mathfrak{A}$, *the non-zero columns of the matrix* $\mathfrak{B}(\mathfrak{E}_{\mathsf{N}}-\mathfrak{D}'\mathfrak{D})$ *form a basis of the space of solutions of the homogeneous system* $\mathfrak{A}\mathfrak{x}=\mathfrak{o}$. *The inhomogeneous system* $\mathfrak{A}\mathfrak{x}=\mathfrak{y}^{(0)}$ *is*

only soluble for those $\mathfrak{y}^{(0)}$ *which satisfy the condition* $(\mathfrak{E}_{\mathsf{M}}-\mathfrak{D}\mathfrak{D}')\mathfrak{C}\mathfrak{y}^{(0)}=\mathfrak{o}$. *If this condition is satisfied,* $\mathfrak{x}^{(0)}=\mathfrak{B}\mathfrak{D}'\mathfrak{C}\mathfrak{y}^{(0)}$ *is a solution.*

By going over to transposed matrices, we obtain the following analogous result:

(4) *If* $\mathfrak{D}=\mathfrak{C}\mathfrak{A}\mathfrak{B}$, $\mathfrak{D}'=\mathfrak{B}'\mathfrak{A}'\mathfrak{C}'$ *is the canonical form for the row-finite matrix* $\mathfrak{A}'$. *The non-zero columns of the matrix* $\mathfrak{C}'(\mathfrak{E}_{\mathsf{M}}-\mathfrak{D}\mathfrak{D}')$ *form a continuous basis of all the solutions of the homogeneous system* $\mathfrak{A}'\mathfrak{v}=\mathfrak{o}$. *The inhomogeneous system* $\mathfrak{A}'\mathfrak{v}=\mathfrak{u}^{(0)}$ *is only soluble for those* $\mathfrak{u}^{(0)}$ *which satisfy the condition* $(\mathfrak{E}_{\mathsf{N}}-\mathfrak{D}'\mathfrak{D})\mathfrak{B}'\mathfrak{u}^{(0)}=\mathfrak{o}$.

If this condition is satisfied, $\mathfrak{v}^{(0)}=\mathfrak{C}'\mathfrak{D}\mathfrak{B}'\mathfrak{u}^{(0)}$ *is a solution.*

These formulae for the solutions are generalizations of Cramer's rule, although to be sure they provide no method for direct calculation, for it is not possible to determine $\mathfrak{B}$ and $\mathfrak{C}$ by an explicit calculation, since their existence is only established with the aid of Zorn's lemma.

4. The countable case. If E has countable dimension, the argument can be simplified considerably, and it is possible to give a constructive method which produces the canonical form in countably many steps.

Let E be equal to $\varphi(\mathsf{K})$. A vector $\mathfrak{x}=\sum_{i=1}^{\infty}\mathfrak{e}_i\xi_i$ is said to be of l e n g t h n if ξ_n is the last non-zero coordinate. If H_1 is a linear subspace of $\varphi(\mathsf{K})$, a subspace complementary to H_1 is formed by the space H_2 of all those $\mathfrak{y}\in\varphi(\mathsf{K})$ for which $\eta_k=0$ whenever k is the length of some vector $x\in H_1$.

This easily established assertion (use a basis of H_1 which contains, for each length appearing in H_1, exactly one vector of this length) gives a construction for a complement in a way which does not depend upon Zorn's lemma.

(1) *In* $\omega(\mathsf{K})$ *each weak accumulation point is the limit of a convergent sequence.*

For if $\mathfrak{u}^{(0)}$ is an accumulation point of the set M, there is for each n a $\mathfrak{u}^{(n)}\in M$ which agrees with $\mathfrak{u}^{(0)}$ in the first n coordinates. Clearly $\mathfrak{u}^{(n)}$ converges weakly to $\mathfrak{u}^{(0)}$ [cf. § 10, 6.(4)].

Thus the use of filters is not necessary in $\omega(\mathsf{K})$.

If v_n is the first non-zero term in $\mathfrak{u}=\{v_i\}\in\omega(\mathsf{K})$, $\mathfrak{u}$ is said to have s h o r t n e s s n. If H_1 is a weakly closed linear subspace of $\omega(\mathsf{K})$, let $\{\mathfrak{u}_j\}$ be a set of vectors in H_1 which contains, for each shortness j appearing in H_1, just one vector $\mathfrak{u}_j$ of this shortness. The $\{\mathfrak{u}_j\}$ form a continuous basis of H_1, and a closed complement of H_1 is obtained by taking the set H_2 of all continuous sums $\sum_k \mathfrak{e}_k\zeta_k$, $\zeta_k\in\mathsf{K}$, where k runs through the set complementary to the set of j's.

We now sketch a method of producing the canonical form of a row-finite matrix $\mathfrak{A}=((\alpha_{ik}))$, $(i,k=1,2,\ldots,)$.

A square matrix with an ordered index set is called a t r i a n g u l a r m a t r i x if the terms beneath the main diagonal vanish. The product of two triangular matrices is again a triangular matrix. If the columns $\mathfrak{c}_1,\mathfrak{c}_2,\ldots,$ of the matrix $\mathfrak{C}$, considered as vectors in φ, have lengths $1,2,\ldots,$ then $\mathfrak{C}$ is an invertible triangular matrix, whose inverse is again a triangular matrix which can be calculated in a simple way.

Let the first non-zero column of $\mathfrak{A}$ have index s_1 and length l_1, so that $\alpha_{l_1 s_1}\neq 0$. Let the next non-zero column be the s_2-nd. We substract a multiple of the s_1-st column from the s_2-nd column, choosing the multiple so that the l_1-st element of the s_2-nd column of the resulting matrix $\mathfrak{A}^{(1)}$ vanishes. If the s_2-nd column of $\mathfrak{A}^{(1)}$ does not vanish, it has a length l_2 different from l_1. For the second step, we subtract from the next non-zero column of $\mathfrak{A}^{(1)}$ (the s_3-rd) suitable multiples of the s_1-st and s_2-nd columns, so that the l_1-st and l_2-nd elements of the s_3-rd column of the corresponding matrix $\mathfrak{A}^{(2)}$ vanish, and so on.

The elementary operation which sends $\mathfrak{A}^{(n-1)}$ to $\mathfrak{A}^{(n)}$ is obtained by multiplying on the right by a matrix $\mathfrak{B}^{(n)}$. This has ones in the main diagonal, and zeros every where else with the exception of the s_{n+1}-th column, which has certain elements of K above the one of the main diagonal. The final result of these infinitely many operations is a matrix $\tilde{\mathfrak{A}}$, in which all the non-zero columns $\tilde{\mathfrak{a}}_j$ have different lengths l_j.

We can adjoin certain $\mathfrak{e}_k$ to the non-zero $\hat{\mathfrak{a}}_j$, to obtain a system of vectors in which each length appears just once. The matrix obtained in this way can be turned into an invertible triangular matrix by permuting its columns, and is therefore itself invertible. We denote its inverse by $\mathfrak{C}$. This sends each $\hat{\mathfrak{a}}_j$ into an $\mathfrak{e}_{l_j}$, so that the matrix $\mathfrak{C}\tilde{\mathfrak{A}}$ has at most one one in each column and row. By permuting the rows or columns if necessary (again by multiplication by invertible column-finite matrices) this can be transformed into a canonical form, which is a unit matrix when the zero rows and columns are deleted.

To complete the proof of our assertion, it remains to show that $\tilde{\mathfrak{A}}=\mathfrak{A}\mathfrak{B}$, where $\mathfrak{B}$ is an invertible triangular matrix. It is natural to take $\mathfrak{B}$ as limit of the sequence $\mathfrak{B}^{(1)}, \mathfrak{B}^{(1)}\mathfrak{B}^{(2)}, \ldots$.

A sequence $\mathfrak{R}^{(n)}$ of matrices in $\mathfrak{L}(\varphi)$ is said to c o n v e r g e to the matrix $\mathfrak{R}\in\mathfrak{L}(\varphi)$, if to each k there corresponds an $n_0(k)$, such that the first k columns of $\mathfrak{R}-\mathfrak{R}^{(n)}$ vanish, for $n>n_0$. To each Cauchy sequence $\mathfrak{R}^{(n)}$, defined in a corresponding way, there clearly exists a column-finite $\mathfrak{R}$ as limit.

For this concept of limit, $\tilde{\mathfrak{A}}=\lim\mathfrak{A}^{(n)}$, and the sequence of products $\mathfrak{B}^{(1)}\ldots\mathfrak{B}^{(n)}$ converges to an invertible triangular matrix $\mathfrak{B}$. The missing relationship

$$\tilde{\mathfrak{A}}=\lim\mathfrak{A}^{(n)}=\lim\mathfrak{A}\mathfrak{B}^{(1)}\ldots\mathfrak{B}^{(n)}=\mathfrak{A}\mathfrak{B}.$$

follows from the easily established rule for column-finite matrices, that if $\lim\mathfrak{R}^{(n)}=\mathfrak{R}$ and $\lim\mathfrak{M}^{(n)}=\mathfrak{M}$, then $\lim\mathfrak{R}^{(n)}\mathfrak{M}^{(n)}=\mathfrak{R}\mathfrak{M}$. The multiplication of $\tilde{\mathfrak{A}}$ by $\mathfrak{C}$ can also be obtained by making infinitely many elementary operations on the rows of $\mathfrak{A}$, and these operations can be made alternately with the column operations. It finally remains to confirm that this infinite product of elementary operations converges in the sense described above.

5. An example. Show that the matrix

$$\mathfrak{A}=\begin{pmatrix} 1 & 2 & 2 & 2 & 2 & 2 & 2 & 2\ldots \\ 1 & 1 & 1 & 1 & 2 & 2 & 2 & 2\ldots \\ 0 & 0 & 0 & 1 & 2 & 2 & 2 & 2\ldots \\ 0 & 0 & 0 & 1 & 1 & 1 & 1 & 2\ldots \\ 0 & 0 & 0 & 0 & 0 & 0 & 1 & 2\ldots \\ 0 & 0 & 0 & 0 & 0 & 0 & 1 & 1\ldots \\ . & . & . & . & . & . & . & \ldots \end{pmatrix}$$

is equivalent to the matrix which is obtained from the unit matrix $\mathfrak{E}$ by inserting zero-columns between the second and third columns, the fourth and fifth, and so on. The matrices $\mathfrak{B}$ and $\mathfrak{C}$ which make the transformation can be determined easily, and so formulae can be obtained for solving the system of equations with matrix $\mathfrak{A}$.

§ 12. Locally linearly compact spaces

1. The structure of locally linearly compact spaces. We now consider a further example related to the theory of linearly topologized spaces developed in § 10.

A linearly topologized space $L[\mathfrak{T}]$ over K is said to be locally linearly compact if it has a linearly compact neighbourhood V of $\mathfrak{o}$. By § 10, 7.(7), there is a discrete $\mathfrak{T}$-complement W of V. The topology $\mathfrak{T}$ on W is the discrete topology, and so also is the topology induced by $\mathfrak{T}_{lk}(V')$, since $\mathfrak{T}_{lk}(V')$ is finer than $\mathfrak{T}$. By § 10, 10.(2) and 11.(3), $\mathfrak{T}$ is the same as the topology $\mathfrak{T}_{lk}(V')$ on V, and this coincides with the topology induced by $\mathfrak{T}_{lk}(L')$, since on the one hand this last topology is finer than that induced by $\mathfrak{T}$, and on the other $\mathfrak{T}_{lk}(V')$ is finer than $\mathfrak{T}_{lk}(L')$ on V, by § 10, 12.(7). By § 10, 12.(4), V and W are also $\mathfrak{T}_{lk}(L')$-complementary, and so both $\mathfrak{T}_{lk}(L')$ and $\mathfrak{T}$ are direct sums of the topologies $\mathfrak{T}_{lk}(L')$ on V and W, so that $\mathfrak{T}=\mathfrak{T}_{lk}(L')$ on L. Using § 10, 10.(3), § 10, 6.(5) and § 5, 7., we obtain

(1) *Every locally linearly compact space $L[\mathfrak{T}]$ is topologically isomorphic to a topological direct sum $\varphi_{d_1}(\mathsf{K})\oplus\omega_{d_2}(\mathsf{K})$. The topology of L is the topology $\mathfrak{T}_{lk}$, and it is equal to the product of the discrete topology on $\varphi_{d_1}(\mathsf{K})$ and the weak topology on $\omega_{d_2}(\mathsf{K})$. Every locally linearly compact space is complete.*

Conversely, we clearly have

(2) *The topological direct sum $L=L_1\oplus L_2$ of a linearly compact space L_1 and a discrete space L_2 is locally linearly compact. L_1 is an open and closed neighbourhood of* 0 *in L.*

A basis of L_2 and a continuous basis of L_1 together comprise a continuous basis of $L_1\oplus L_2$.

The dimensions d_1 and d_2 are not in general uniquely determined, for finite-dimensional subspaces of φ_{d_1} can be taken across to ω_{d_2}, and conversely. If d_1 or d_2 is finite, we can therefore always take a decomsition in which one of the summands vanishes. Nevertheless, we have

(3) *Suppose that L is locally linearly compact, but neither discrete nor linearly compact. If $L=L_1\oplus L_2$ is a decomposition into a discrete space $L_1\cong\varphi_{d_1}$ and a linearly compact space $L_2\cong\omega_{d_2}$, the infinite dimensions d_1 and d_2 are uniquely determined.*

For if $L=\tilde{L}_1\oplus\tilde{L}_2$ is another such decomposition, with $\tilde{L}_1\cong\varphi_{d'_1}$ and $L_2\cong\omega_{d'_2}$, then $(\tilde{L}_2+L_2)/L_2$ is linearly compact in $L/L_2\cong L_1$, by § 10, 9.(2), and is therefore finite-dimensional by § 10, 9.(4), It follows that $d'_2\leqq d_2$, and that $d'_2=d_2$. The image of $\tilde{L}_1$ under the canonical mapping of $L=\tilde{L}_1 \oplus \tilde{L}_2$ onto $L/L_2 \cong L_1$ contains a complementary space of $(\tilde{L}_2 \oplus L_2)/L_2$; thus $d_1' \geqq d_1$, and so $d_1' = d_1$.

$\omega_{d_1} \oplus \varphi_{d_2}$ is the dual of $\varphi_{d_1} \oplus \omega_{d_2}$, and so we obtain the following theorem of LEFSCHETZ:

(4) *The dual space L' of a locally linearly compact space L is again locally linearly compact with respect to the topology $\mathfrak{T}_{lk}$, and $(L')' = L$.*

If $d_1 = d_2$, L is $\mathfrak{T}_{lk}$-isomorphic to L'.

(5) *Every closed linear subspace F of a locally linearly compact space $L[\mathfrak{T}]$ is locally linearly compact, and has a $\mathfrak{T}$-complementary space G.*

Proof. Let $L = L_1 \oplus L_2$, with L_1 discrete and L_2 linearly compact. $F_2 = F \cap L_2$ is a linearly compact neighbourhood of $\mathfrak{o}$ in F in the induced topology, so that F is locally linearly compact. By § 10, 7.(7), $F = F_1 \oplus F_2$, with F_1 discrete. F_2 has a topological complement G_2 in L_2, by § 10, 12.(5), and by § 7, 6.(2) $F + L_2 = F_1 \oplus F_2 \oplus G_2$. Hence by § 10, 7.(7) the sum $F_1 \oplus L_2$ is a topological direct sum. Again by § 10, 7.(7), $F + L_2$ has a discrete topological complement G_1, and $L = G_1 \oplus (F_1 \oplus (F_2 \oplus G_2))$. Thus L is the topological product of these four subspaces. As the order and way in which they are put together does not matter, $L = (F_1 \oplus F_2) \oplus (G_1 \oplus G_2) = F \oplus G$ is the required topological complementary decomposition.

2. The endomorphisms of ψ. In the case where both dimensions d_1 and d_2 are countable, we obtain the semifinite space $\psi = \varphi \oplus \omega$. Its dual space $\psi' = \omega \oplus \varphi$ is topologically isomorphic to ψ.

A full account of the equivalence of the endomorphisms of ψ can also be established. Using this, the theory of the corresponding "semi-finite" systems of equations is at our command, in the same way as for row- and column-finite systems of equations.

First we show that every continuous mapping $A \in \mathfrak{L}(\psi)$ can be taken in canonical form, by choosing two continuous bases suitably.

By 1.(5), the kernel $N[A]$ has a $\mathfrak{T}$-complementary inverse-image space $U[A]$ which is again locally linearly compact. By § 10,7.(6), $\psi/N[A]$ is topologically isomorphic to $U[A]$, and, by § 10, 7.(4), $U[A]$ is mapped continuously by A in a one-one way onto $G = A(\psi)$. Let $\bar{G}$ be the closure of G. By 1.(5), $\bar{G}$ has a $\mathfrak{T}$-complement G_1. We decompose $U[A]$ as $\tilde{F}_1 \oplus F_2$, where $\tilde{F}_1$ is discrete and F_2 is open and linearly compact; F_2 can be taken either as $\mathfrak{o}$ or isomorphic to ω.

The image space $A(F_2) = G_2$ is again linearly compact, by § 10, 9.(2), and by § 10, 9.(10). A establishes a topological isomorphism between F_2 and G_2. By 1.(5), G_2 has a topological complement $\tilde{G}_2$ in $\bar{G}$. The inverse images of those elements of $A(\psi)$ which lie in $\tilde{G}_2$ form a linear subspace $\tilde{F}_2$ of $U[A]$. By § 10, 7.(7), $\tilde{F}_2$ is topologically complementary to F_2 in $U[A]$, and is discrete, so that $U[A]=\tilde{F}_2 \oplus F_2$. Since $G=A(\tilde{F}_2)\oplus A(F_2)$ is dense in $\bar{G}=\tilde{G}_2 \oplus G_2$, and since $A(\tilde{F}_2)\subset \tilde{G}_2$ and $A(F_2)=G_2$, $A(\tilde{F}_2)$ is dense in $\tilde{G}_2$, by § 1, 8.(1). We now consider the continuous one-one mapping from the discrete space $\tilde{F}_2$ into $\tilde{G}_2$. Let G_4 be open and linearly compact in the locally linearly compact space $\tilde{G}_2$. Once again, we can take G_4 either to be $\mathfrak{o}$ or to be isomorphic to ω. By § 7, 6.(5) and § 10, 7.(7), there is a discrete topological complement G_3 to G_4 in $\tilde{G}_2$, and so $A(\tilde{F}_2)=(A(\tilde{F}_2)\cap G_3)\oplus(A(\tilde{F}_2)\cap G_4)=H_3\oplus H_4$. Let F_3 and F_4 be the inverse images of H_3 and H_4, so that $\tilde{F}_2=F_3\oplus F_4$, $H_3=A(F_3)$, $H_4=A(F_4)$. Since $A(\tilde{F}_2)$ is dense in $\tilde{G}_2$, we must again have $\overline{A(F_3)}=G_3$, $\overline{A(F_4)}=G_4$. G_3 is discrete, so that A determines an isomorphism of F_3 onto G_3.

It finally remains to investigate the linear mapping A from the discrete space F_4 into the linearly compact space G_4, in the case where $G_4 \neq \mathfrak{o}$. Equivalently, we may consider a continuous one-one linear mapping A from φ into ω, whose image space is dense in ω.

We now refer to the ideas developed in § 11, 4. Let $x_1, x_2, \ldots$ be a basis of φ. Since $A(\varphi)$ is dense in ω, there is, for each $k=1,2,\ldots$, a $z\in\varphi$ whose image $y=Az$ has shortness k. In particular let z_1 be chosen so that $y_1=Az_1$ has shortness 1. If x_1 is linearly independent of z_1 (otherwise, take x_2), there is a $z_{k_1}=x_1+z_1\lambda_{11}$ in φ whose image y_{k_1} has shortness $k_1>1$. If $k_1>2$, there exist elements $z_2,\ldots,z_{k_1-1}$ in φ whose images $y_2,\ldots,y_{k_1-1}$ have shortness $2,\ldots,k_1-1$ respectively. The elements $z_1,\ldots,z_k$ are linearly independent, since their images clearly are. Let x_{i_2} be the first term in $x_1, x_2, \ldots$ which is linearly independent of $z_1,\ldots,z_{k_1}$. Then there exists a $z_{k_2}=x_{i_2}+z_1\lambda_{21}+\cdots+z_{k_1}\lambda_{2k_1}$, whose image y_{k_2} has shortness $k_2>k_1$. If $k_2>k_1+1$, elements $z_{k_1+1},\ldots,z_{k_2-1}$ of φ can again be inserted so that all the elements $z_1,\ldots,z_{k_2}$ are linearly independent, and so that their images have shortness $1,\ldots,k_2$, respectively. Let x_{i_3} be the first of the terms $x_{i_2+1},\ldots$ which is linearly independent of $z_1,\ldots,z_{k_2}$. By continuing this process we obtain a basis $\{z_i\}$ of φ for which the images $y_i=Az_i$ have all possible different shortnesses, so that the $\{y_i\}$ form a continuous basis of ω.

Combining everything together, corresponding to $A\in\mathfrak{L}(\psi)$ we have constructed two direct topological decompositions of ψ

$$\psi=N[A]\oplus F_2\oplus F_3\oplus F_4, \tag{1}$$

$$\psi=G_1 \oplus G_2\oplus G_3\oplus G_4 \tag{2}$$

with the following properties: the null-space $N[A]$ can be of finite dimension n or can be topologically isomorphic to φ,ω or ψ. The same holds for the complement G_1 of the closure of the image space $A(\psi)$.

F_2 and G_2 are either both $\mathfrak{o}$ or both isomorphic to ω, and A defines a topological isomorphism of F_2 onto G_2 which takes the form $Ax_2^{(i)}=y_2^{(i)}$, $i=1,2,\ldots$, when continuous bases $x_2^{(i)}$ and $y_2^{(i)}$ of F_2 and G_2 are chosen in a suitable way. F_3 and G_3 are both discrete, and have the same dimension, which can be finite or countably infinite. A defines an isomorphism which takes the form $Ax_3^{(i)}=y_3^{(i)}$, when bases $x_3^{(i)}$ and $y_3^{(i)}$ of F_3 and G_3 are chosen in a suitable way. Either F_4 and G_4 both vanish, or they are isomorphic to φ and ω respectively. By choosing a basis $x_4^{(i)}$ of F_4 and a continuous basis $y_4^{(i)}$ of G_4 suitably, A is represented as the mapping $Ax_4^{(i)}=y_4^{(i)}$ from F_4 onto a dense subspace of G_4. If $U[A]$ is infinite-dimensional and F_3 is finite-dimensional, F_3 and G_3 can always be taken to be $\mathfrak{o}$.

If A is a mapping of infinite rank, it follows from this account that the mapping from $U[A]$ onto $A(\psi)$ is of one of the seven following types (we denote by $\tilde{\varphi}$ the space φ considered as a dense subspace of ω, and provided with the topology induced by the weak topology on ω; this in distinction to φ, with the discrete topology):

$$(3)\qquad \begin{aligned} U[A]&:\ \varphi,\ \varphi,\ \varphi\oplus\varphi,\ \omega,\ \omega\oplus\varphi,\ \omega\oplus\varphi,\ \omega\oplus\varphi\oplus\varphi,\\ A(\psi)&:\ \varphi,\ \tilde{\varphi},\ \varphi\oplus\tilde{\varphi},\ \omega,\ \omega\oplus\varphi,\ \omega\oplus\tilde{\varphi},\ \omega\oplus\varphi\oplus\tilde{\varphi}.\end{aligned}$$

3. The theory of equivalence in ψ. A continuous basis of ψ is obtained if ψ is decomposed in some way as a direct sum $\psi_1\oplus\psi_2$, with ψ_1 discrete and ψ_2 open and linearly compact, and if an algebraic basis x_{-1}, $i=1,2,\ldots$, is chosen for ψ_1 and a continuous basis x_i, $i=1,2,\ldots$, is chosen for ψ_2.

If a second continuous basis x'_{-i}, x'_i is chosen, the mapping $Bx_k=x'_k$, $(k=\pm1,\pm2,\ldots)$ is a topological isomorphism of ψ.

Two mappings $A_1, A_2\in\mathfrak{L}(\psi)$ are said to be equivalent if there are topological automorphisms B and C for which $A_2=BA_1C$; the next two results then follow without difficulty from the results of the preceding number and the remark made above:

(1) *Two mappings of finite rank (i.e. $N[A]$ has finite codimension in ψ) are equivalent if they are of equal rank.*

(2) *Two mappings of infinite rank are equivalent if a) their null-spaces and the complements of the closures of their spaces are respectively either of the same finite dimension or of the same type φ,ω or ψ, and b) the dimensions of the spaces F_2, F_3 and F_4 are in each case either both 0 or both ∞.*

This is a complete division into equivalence classes, as the next result shows:

(3) *Two mappings are equivalent if and only if the conditions given in* (1) *and* (2) *are satisfied.*

This is immediately obvious for (1). In case (2), which is clearly not equivalent to (1), we argue as follows: BA_1C has null-space $C^{-1}(N[A_1])$, which is of the same type as $N[A_1]$, and in the same way B sends a complement $G_1^{(1)}$ of $\overline{A_1(\psi)}$ into a complement $BG_1^{(1)}$ of $\overline{BA_1C(\psi)}$, so that these are both of the same type. The inverse-image space contains an infinite-dimensional linearly compact space if F_2 is infinite-dimensional, and is discrete if F_2 is $\mathfrak{o}$; these properties are preserved under equivalence. If F_3 is infinite-dimensional, the mapping sends an infinite-dimensional discrete subspace onto one of the same kind, and if F_3 is zero, this is not the case. If F_4 is infinite-dimensional, $A(\psi)$ is not closed, whereas $A(\psi)$ is closed if F_4 is $\mathfrak{o}$.

By using similar ideas, it is easy to see that of the seven types 2.(3) of image spaces $A(\psi)$ no two are topologically isomorphic; we can therefore express this result about equivalence in the following terms:

(4) *Two continuous linear mappings of ψ into itself are equivalent if and only if their null-spaces, their image spaces, and the complements of the closures of their image spaces are topologically isomorphic.*

If we have an underlying continuous basis $\dots, x_{-2}, x_{-1}, x_1, x_2, \dots$ of ψ, a linear mapping $A \in \mathfrak{L}(\psi)$ is represented by a matrix $\mathfrak{A} = ((\alpha_{ik}))$, which is infinite in all directions, and which has the following form:

$$\mathfrak{A} = \left(\begin{array}{ccccc|ccccc} \cdot & \cdot & & \cdot & & & \cdot & & \cdot & \cdot \\ \cdot & \alpha_{-2-2} & & \alpha_{-2-1} & & & \alpha_{-21} & & \alpha_{-22} & \cdot \\ \cdot & \alpha_{-1-2} & & \alpha_{-1-1} & & & \alpha_{-11} & & \alpha_{-12} & \cdot \\ \hline \cdot & \alpha_{1-2} & & \alpha_{1-1} & & & \alpha_{11} & & \alpha_{12} & \cdot \\ \cdot & \alpha_{2-2} & & \alpha_{2-1} & & & \alpha_{21} & & \alpha_{22} & \cdot \\ \cdot & \cdot & & \cdot & & & \cdot & & \cdot & \cdot \end{array}\right) = \left(\begin{array}{c|c} \mathfrak{A}_1 & \mathfrak{A}_2 \\ \hline \mathfrak{A}_3 & \mathfrak{A}_4 \end{array}\right),$$

where $\mathfrak{A}_1$ is column-finite, $\mathfrak{A}_2$ is finite (and in particular is column- and row-finite), $\mathfrak{A}_3$ is arbitrary and $\mathfrak{A}_4$ is row-finite. Such a matrix is said to be semi-finite. The fact that $\mathfrak{A}_2$ is finite follows easily from the condition that A maps ψ into itself.

In addition to the topological automorphisms of ψ, which are all equivalent to the identity mapping, there is just one equivalence class of mappings A for which both $Ax=0$, $x \in \psi$, and $A'u=0$, $u \in \psi'$, are not soluble. This is the case where $N[A]=\mathfrak{o}$, $G_1=\mathfrak{o}$, and F_1, F_2 and F_3 are all infinite-dimensional, i.e. the last case of 2.(3). By numbering the elements of the bases in a suitable way, we obtain the following normal form:

$$\left(\begin{array}{cccccccccccccc}
\cdot & \cdot & \cdot & \cdot & \cdot & \cdot & \cdot & \cdot & & \cdot & \cdot & \cdot & \cdot \\
\cdot & 1 & 0 & 0 & 0 & 0 & 0 & 0 & & 0 & 0 & 0 & \cdot \\
\cdot & 0 & 0 & 1 & 0 & 0 & 0 & 0 & & 0 & 0 & 0 & \cdot \\
\cdot & 0 & 0 & 0 & 0 & 1 & 0 & 0 & & 0 & 0 & 0 & \cdot \\
\cdot & 0 & 0 & 0 & 0 & 0 & 0 & 1 & & 0 & 0 & 0 & \cdot \\
\hline
\cdot & 0 & 0 & 0 & 0 & 0 & 0 & 0 & & 1 & 0 & 0 & \cdot \\
\cdot & 0 & 0 & 0 & 0 & 0 & 1 & 0 & & 0 & 0 & 0 & \cdot \\
\cdot & 0 & 0 & 0 & 0 & 0 & 0 & 0 & & 0 & 1 & 0 & \cdot \\
\cdot & 0 & 0 & 0 & 1 & 0 & 0 & 0 & & 0 & 0 & 0 & \cdot \\
\cdot & 0 & 0 & 0 & 0 & 0 & 0 & 0 & & 0 & 0 & 1 & \cdot \\
\cdot & 0 & 1 & 0 & 0 & 0 & 0 & 0 & & 0 & 0 & 0 & \cdot \\
\cdot & \cdot & \cdot & \cdot & \cdot & \cdot & \cdot & \cdot & & \cdot & \cdot & \cdot & \cdot
\end{array}\right)$$

We leave it to the reader to transform the theory of equivalence into a theory about infinite equations represented by semi-finite matrices, and with semi-finite vectors as solutions (cf. KÖTHE and TOEPLITZ [1]).

J. DIEUDONNÉ [4] has shown that there is no corresponding simple theory in the uncountable case.

§ 13. The linear strong topology

1. Linearly bounded subspaces. We continue with the general theory of linearly topologized spaces, by introducing an analogue to the concept of bounded set; this concept is particularly important in the later theory.

Let L be a vector space over K with a linear topology $\mathfrak{T}$, which is defined by linear neighbourhoods U_α, $\alpha \in \mathsf{A}$, of $\mathfrak{o}$. A linear subspace F of L is said to be linearly $\mathfrak{T}$-bounded if $(F+U_\alpha)/U_\alpha$ is finite-dimensional for each U_α.

If $\mathfrak{T}$ is the discrete topology, the linearly $\mathfrak{T}$-bounded subspaces are precisely the finite-dimensional subspaces; if $\mathfrak{T}$ is the linear weak topology $\mathfrak{T}_{ls}(L')$, every linear subspace is linearly weakly bounded, since each U_α is of finite codimension in L.

(1) *The closure of a linearly $\mathfrak{T}$-bounded subspace F is again linearly bounded.*

By § 10, 2.(3), $F+U_\alpha$ is closed, so that $F^{\perp\perp}+U_\alpha=F+U_\alpha$ follows from $F+U_\alpha=(F+U_\alpha)^{\perp\perp}\supset F^{\perp\perp}+U_\alpha\supset F+U_\alpha$. Thus if $(F+U_\alpha)/U_\alpha$ is finite-dimensional, so is $(F^{\perp\perp}+U_\alpha)/U_\alpha$.

(2) *The sum of finitely many linearly $\mathfrak{T}$-bounded subspaces of L is again linearly $\mathfrak{T}$-bounded. Every linear subspace of a linearly $\mathfrak{T}$-bounded subspace is again linearly $\mathfrak{T}$-bounded.*

The proof is trivial.

(3) *The continuous linear image in $L_2[\mathfrak{T}_2]$ of every linearly $\mathfrak{T}_1$-bounded subspace of $L_1[\mathfrak{T}_1]$ is linearly $\mathfrak{T}_2$-bounded.*

Let $A \in \mathfrak{L}(L_1, L_2)$. If V is a closed linear neighbourhood of $\circ$ in L_2, there is a neighbourhood U of $\circ$ in L_1 with $A(U) \subset V$. Since $(F+U)/U$ is finite-dimensional, $A(F+U)/A(U) = (A(F)+A(U))/A(U)$ is finite-dimensional, and so therefore is $(A(F)+V)/V$.

(4) *A linear subspace F of $L[\mathfrak{T}]$ is linearly $\mathfrak{T}$-bounded if and only if $\mathfrak{T}$ and the linear weak topology coincide on F.*

a) If $\mathfrak{T} = \mathfrak{T}_{ls}$ on F, then for each $\mathfrak{T}$-neighbourhood U_α there exists a $\mathfrak{T}_{ls}$-neighbourhood V with $F \cap V = F \cap U_\alpha$. Then by § 7, 6.(6), $(F+U_\alpha)/U_\alpha \cong F/(F \cap U_\alpha) = F/(F \cap V) \cong (F+V)/V$. Since L/V is finite-dimensional, so therefore is $(F+U_\alpha)/U_\alpha$.

b) Conversely, if $(F+U_\alpha)/U_\alpha$ is finite-dimensional, we can write $F+U_\alpha = U_\alpha \oplus [x_1, \ldots, x_n]$, where the x_i are linearly independent. Because each $U_\alpha \oplus [x_1, \ldots, x_{i-1}, x_{i+1}, \ldots, x_n]$ is closed, there exist continuous linear functionals $u_1, \ldots, u_n$ in L' which vanish on U_α, and which take the values $u_i x_i = 1$, $u_i(x_k) = \circ$ for $i \neq k$, on the x_i, by § 10, 4.(1'). We than have $F+V = V \oplus [x_1, \ldots, x_n]$ for the weak neighbourhood $V = U_{u_1, \ldots, u_n}(\circ)$, so that $(F+V)/V \cong (F+U_\alpha)/U_\alpha$, and $F/(F \cap V) \cong F/(F \cap U_\alpha)$; since $F/F \cap V$ is finite-dimensional and $F \cap U_\alpha \subset F \cap V$, we finally deduce that $F \cap U_\alpha = F \cap V$.

Starting from a dual pair $\langle L_2, L_1 \rangle$, linear $\mathfrak{T}_{lk}$-boundedness is defined in both L_1 and L_2. This is the most important of the various concepts of boundedness.

(5) *Every linearly weakly compact subspace F of $L[\mathfrak{T}]$ is linearly $\mathfrak{T}_{lk}$-bounded.*

If U is a linear $\mathfrak{T}_{lk}$-neighbourhood of $\circ$ in L, L/U is discrete, by § 10, 7.(2). Since $(F+U)/U$ is the continuous image of F, and is therefore linearly weakly compact, it must be finite-dimensional.

The converse of (5) is not valid. Consider the dual pair $\langle \varphi, \varphi \rangle$. By § 10, 10.(3) and § 9, 5.(3), a linearly weakly compact space never has countable dimension. Thus only the finite dimensional linear subspaces of φ are linearly weakly compact. Hence $\mathfrak{T}_{lk} = \mathfrak{T}_{ls}$, and φ itself is linearly $\mathfrak{T}_{lk}$-bounded.

2. The linear strong topology. Once again, let $\langle L_2, L_1 \rangle$ be a dual pair. As B runs through all the linearly $\mathfrak{T}_{lk}$-bounded subspaces of L_2, the spaces $B^\perp$ satisfy the conditions of § 10, 2.(1), by 1.(2); taken as neighbourhoods of $\mathfrak{o}$ they therefore define a linear topology on L_1, which we call the linear strong topology $\mathfrak{T}_{lb}(L_2)$ on L_1.

It follows immediately from 1.(5) that

(1) *The strong topology* $\mathfrak{T}_{lb}(L_2)$ *is finer than the topology* $\mathfrak{T}_{lk}(L_2)$.

Further, it follows from § 10, 11.(4) and 1.(1) that

(2) *The dual space of* $L_1[\mathfrak{T}_{lb}(L_2)]$ *is equal to* L_2 *if and only if* $\mathfrak{T}_{lb}(L_2) = \mathfrak{T}_{lk}(L_2)$, *and so if and only if every weakly closed linearly* $\mathfrak{T}_{lk}$*-bounded subspace of* L_2 *is linearly weakly compact.*

Let L be a linearly topologized space with topology $\mathfrak{T}$. The dual space L', equipped with the strong topology $\mathfrak{T}_{lb}(L)$, is called the strong dual of L. L is said to be strongly semi-reflexive if the dual $(L'[\mathfrak{T}_{lb}])'$ of the strong dual is equal to L; L is said to be strongly reflexive if, in addition, the strong topology $\mathfrak{T}_{lb}(L')$ on L is the same as the original topology.

(3) *L is strongly semi-reflexive if and only if every closed linearly* $\mathfrak{T}_{lk}$*-bounded subspace of L is linearly weakly compact.*

(4) *L is strongly reflexive if and only if every closed linearly* $\mathfrak{T}_{lk}$*-bounded subspace is linearly weakly compact, in both L and L', and the original topology* $\mathfrak{T}$ *is the same as* $\mathfrak{T}_{lk}(L')$.

Proof. (3) is a consequence of (2). If L is strongly reflexive, then $\mathfrak{T} = \mathfrak{T}_{lb}(L')$, and *a fortiori* $\mathfrak{T}_{lk}(L') = \mathfrak{T}_{lb}(L')$, so that every closed linearly $\mathfrak{T}_{lk}$-bounded subspace is linearly weakly compact in L' as well. The converse is also clear.

As an application, we give a refinement of the theorem of LEFSCHETZ (cf. § 12, 1.(4)):

(5) *Every locally linearly compact space L is strongly reflexive.*

By § 12, 1.(1) and § 12, 1.(4), it is enough to show that every linearly $\mathfrak{T}_{lk}$-bounded subspace B of L is contained in a linearly weakly compact subspace. We can write $L = L_1 \oplus L_2$, with L_1 discrete and L_2 open and linearly weakly compact. By 1.(3), $(B+L_2)/L_2$ is linearly $\mathfrak{T}_{lk}$-bounded in the discrete space L/L_2. Thus $(B+L_2)/L_2$ is finite-dimensional, so that $B+L_2$ is linearly weakly compact.

3. The completion. In § 5, 5, it was shown how a uniquely determined completion can be constructed for every Hausdorff uniform space. Later (§ 15, 3.), we shall see that the completion of a topological vector space constructed in this way is again a topological vector space. In the present case, however, there is a simple way of constructing this completion, which corresponds to the method of GROTHENDIECK (cf. § 21, 9.). This construction is independent of the ideas of § 5.

Let L be a vector space with topology $\mathfrak{T}$ defined by weakly closed linear neighbourhoods U of $\mathfrak{o}$. We denote by $\tilde{L}$ the collection of all those linear functionals y on L' whose restrictions to each $U^\perp$ are weakly continuous. $\tilde{L}$ is clearly a vector space which contains L as a linear subspace.

We define a linear topology $\tilde{\mathfrak{T}}$ on $\tilde{L}$ by taking as neighbourhoods in $\tilde{L}$ the linear subspaces $U^{\perp\perp} \supset U$ orthogonal to the spaces $U^\perp$. It is easy to see that $\tilde{\mathfrak{T}}$ is a linear topology: the intersection of finitely many $U_i^{\perp\perp}$ contains the neighbourhood $\left(\bigcap_i U_i\right)^{\perp\perp}$, so that the $U^{\perp\perp}$ form a filter base. That $\bigcap U^{\perp\perp} = \mathfrak{o}$ follows from the fact that a $y \in \tilde{L}$ which vanishes on all of the $U^\perp$ is identically zero.

Since $U^{\perp\perp} \cap L = U$, $\mathfrak{T}$ is the topology induced on L by $\tilde{\mathfrak{T}}$. That $\tilde{L}$ is the completion of L now follows from

(1) *$\tilde{L}$ is complete with respect to $\tilde{\mathfrak{T}}$, and L is dense in $\tilde{L}$.*

If $\{F_\alpha\}$ is a Cauchy filter on L, then for each $U^\perp$ there is an F_α such that all the $y \in F_\alpha$ have the same restriction $y^{(0)}_{U^\perp}$ to $U^\perp$, and this restriction is a weakly continuous linear functional on $U^\perp$. The $y^{(0)}_{U^\perp}$ defined on all the sets $U^\perp$ in this way determine a linear functional $y^{(0)} \in \tilde{L}$, defined on the whole of L'. For if $U_1^\perp$ and $U_2^\perp$ are distinct, and if $y_1 = y^{(0)}_{U_1^\perp}$ for all $y_1 \in F^{\alpha_1}$ and $y_2 = y^{(0)}_{U_2^\perp}$ for all $y_2 \in F^{\alpha_2}$ respectively, then we clearly have $y^{(0)}_{U_1^\perp} = y = y^{(0)}_{U_2^\perp}$ on $U_1^\perp \cap U_2^\perp$ for all $y \in F^\gamma \subset F^{\alpha_1} \cap F^{\alpha_2}$. The linear functionals $y^{(0)}_{U_1^\perp}$ and $y^{(0)}_{U_2^\perp}$ therefore agree on the intersection of their domains of definition, and $y^{(0)}$ is the limit of the filter $\{F^\alpha\}$. Since $y^{(0)}_{U^\perp}$, being a weakly continuous linear functional on $U^\perp$, can also be defined by an $x_{U^\perp} \in L$, $y^{(0)}$ is a $\tilde{\mathfrak{T}}$-adherent point of the $x_{U^\perp}$, and so L is dense in $\tilde{L}$.

As a first consequence, we obtain a further characterization of the linearly $\mathfrak{T}$-bounded sets.

A linear subspace F of L is said to be linearly $\mathfrak{T}$-precompact if the closure of F in $\tilde{L}$ is linearly $\tilde{\mathfrak{T}}$-compact.

(2) *A linear subspace F is linearly $\mathfrak{T}$-precompact if and only if it is linearly $\mathfrak{T}$-bounded.*

It is sufficient to show this for L itself, for a linear subspace F of $L[\mathfrak{T}]$ is linearly $\mathfrak{T}$-bounded or linearly $\mathfrak{T}$-precompact if and only if it has the same property with respect to the topology induced on F by $\mathfrak{T}$. If L is linearly $\mathfrak{T}$-bounded, L/U is finite-dimensional, and so by § 10, 8.(4) $U^\perp$ is finite-dimensional; the linear functionals in $\tilde{L}$ therefore consist of all the linear functionals on L', i.e. $\tilde{L} = (L')^*$. $\tilde{L}$ is therefore linearly weakly compact, by § 10, 10.(1), and $\tilde{\mathfrak{T}} = \mathfrak{T}_{ls}(L')$.

If, conversely, L is linearly $\mathfrak{T}$-precompact, $\tilde{L}$ is linearly $\mathfrak{T}$-compact, and $\tilde{\mathfrak{T}}$ is the weak topology, by § 10, 10.(2). If therefore follows from the remark before 1.(1) that L is linearly $\mathfrak{T}$-bounded.

If L is $\mathfrak{T}_{lk}$-complete, every closed linearly $\mathfrak{T}_{lk}$-bounded subspace of L is linearly $\mathfrak{T}_{lk}$-compact, and is therefore linearly weakly compact, so that by 2.(3) we have

(3) *If L is $\mathfrak{T}_{lk}$-complete, then L is strongly semi-reflexive.*

Likewise, we have

(4) *If L and L' are $\mathfrak{T}_{lk}$-complete, then $L[\mathfrak{T}_{lk}(L')]$ is strongly reflexive.*

Both these theorems are no longer valid if we only require $\mathfrak{T}_{lb}$-completeness, as the example of the dual pair $\langle\varphi,\varphi\rangle$ shows (cf. the remark at the end of 1.). In both spaces the strong topology is the discrete topology, under which each is complete, while the strong dual spaces are isomorphic to ω.

Let us mention one closely related problem: is a linearly topologized space strongly complete if every closed linearly $\mathfrak{T}_{lk}$-bounded subspace is linearly weakly compact?

4. Topological sums and products. We give some examples to illustrate the theory we have just developed.

Let L_α, $\alpha\in\mathsf{A}$, be linearly topologized spaces with topologies $\mathfrak{T}_\alpha$.

(1) *If $L=\bigoplus_\alpha L_\alpha$ is the topological direct sum of the spaces L_α, then L' is $\mathfrak{T}_{lk}$-isomorphic to $\prod_\alpha L'_\alpha[\mathfrak{T}_{lk}(L_\alpha)]$.*

If $L=\prod_\alpha L_\alpha$ is the topological product of the spaces L_α, L' is $\mathfrak{T}_{lk}$-isomorphic to $\bigoplus_\alpha L'_\alpha[\mathfrak{T}_{lk}(L_\alpha)]$.

Proof. Let us denote by $\mathfrak{T}$ the topology defined on $L=\bigoplus L_\alpha$ by the topologies $\mathfrak{T}_\alpha$, as in § 10, 2. We can consider L_α as a linear subspace of L. Each $u\in L'$ then defines a $\mathfrak{T}_\alpha$-continuous linear functional $u_\alpha\in L'_\alpha$ on L_α, and this we can consider as an element of L', by making it take the value zero on the spaces L_β with $\beta\neq\alpha$. It is now quite clear that the correspondence $u\to\{u_\alpha\}\in\prod_\alpha L'_\alpha$ is an algebraic isomorphism between the spaces L' and $\prod_\alpha L'_\alpha$.

In a similar way, for the second half of the proposition we establish the algebraic isomorphism $L'\cong\bigoplus_\alpha L'_\alpha$ (for a neighbourhood of $\mathfrak{o}$ in $\prod_\alpha L_\alpha$ has the form $\prod_\alpha U_\alpha$, with $U_\alpha=L_\alpha$ for all but finitely many α, so that if $u\in L'$ there are only finitely many $u_\alpha\neq\mathfrak{o}$). We show that the isomorphisms are $\mathfrak{T}_{lk}$-isomorphisms in the following way: the projection C_α of a linearly $\mathfrak{T}_{ls}(L')$-compact set C of $\bigoplus_\alpha L_\alpha$ or of $\prod_\alpha L_\alpha$ onto L_α is linearly $\mathfrak{T}_{ls}(L'_\alpha)$-compact, by § 10, 8.(2). In the case $L=\prod_\alpha L_\alpha$, on the one hand C is contained in the topological product of the linearly

$\mathfrak{T}_{ls}(L'_\alpha)$-compact sets C_α, and on the other hand $\prod_\alpha C_\alpha$ is again linearly $\mathfrak{T}_{ls}(L')$-compact, by § 10, 9.(7), so that the topology $\mathfrak{T}_{lk}(L)$ on L' is equal to the direct sum topology of the spaces $L'_\alpha[\mathfrak{T}_{lk}(L_\alpha)]$.

In the case where $L = \bigoplus_\alpha L_\alpha$, the homeomorphism is established in a similar way, using the fact that only finitely many projections C_α are different from o, so that C is contained in $\bigoplus_i C_{\alpha_i}$. For this, it is sufficient to consider the following case: Let $L = \bigoplus_{k=1}^{\infty} L_k$, and suppose there is a sequence $x^{(i)} = \sum_{k=1}^{\infty} x_k^{(i)}$, $x_k^{(i)} \in L_k$, in C, with $x_i^{(i)} \neq o$, and $x_k^{(i)} = o$ for $k > i$. The linear span H of all the $x^{(i)}$ is infinite-dimensional. We show that H, considered as a subspace of L, is discrete. There is a neighbourhood U_1 of o in L_1 for which $L_1 = U_1 \oplus [x^{(1)}]$. The coset in $(L_1 \oplus L_2)/L_1 = L_2$ determined by $x^{(2)}$ is non-zero, and so there is a neighbourhood U_2 of o in L_2 for which $(L_1 \oplus U_2) \cap [x^{(2)}] = o$. Thus $L_1 \oplus L_2 = U_1 \oplus U_2 \oplus [x^{(1)}, x^{(2)}]$, and so on. Then the neighbourhood $\bigoplus_{k=1}^{\infty} U_k$ of o in L meets H only in the point o.

H, being a closed linear subspace of C, is linearly compact, and is therefore finite-dimensional, by § 10, 9.(4); this gives a contradiction.

We know (§ 5, 7.) that $\prod_\alpha L_\alpha$ is $\mathfrak{T}$-complete if the spaces L_α are complete with respect to $\mathfrak{T}_\alpha$. Similarly we have

(2) *The topological direct sum $L = \bigoplus_\alpha L_\alpha$ of complete spaces L_α is complete.*

Proof. If $\mathfrak{F} = \{F^{(\beta)}\}$ is a Cauchy filter on L, the projection $P_\alpha \mathfrak{F} = \{P_\alpha(F^{(\beta)})\}$ is a Cauchy filter on L_α. Since L_α is complete, $P_\alpha \mathfrak{F}$ has a limit $x_\alpha \in L_\alpha$. We assert that $x_\alpha = o$, except for at most finitely many α_i. Otherwise there would be a sequence $x_{\alpha_i} \neq o$, $i = 1, 2, \ldots$. Let U_{α_i} be a closed linear neighbourhood of o in L_{α_i}, with the property that o does not lie in $x_{\alpha_i} + U_{\alpha_i}$. Let $U = \bigoplus_\alpha U_\alpha$ be a neighbourhood of o in L for which the U_{α_i} are the neighbourhoods which have just been determined, and for which the other U_α are chosen arbitrarily. Further, let $F^{(\beta)}$ be small of order U. If now $z \in F^{(\beta)}$, $P_{\alpha_i} z = z_{\alpha_i}$ lies in $x_{\alpha_i} + U_{\alpha_i}$, and so must be different from zero for all i, which is impossible. Since $F^{(\beta)}$ cannot be empty, we have reached a contradiction.

If $x_{\alpha_1}, \ldots, x_{\alpha_n}$ are the finitely many non-zero limits of the filters $P_\alpha \mathfrak{F}$, then clearly $x_{\alpha_1} + \cdots + x_{\alpha_n}$ is the limit of $\mathfrak{F}$.

(3) *If the spaces L_α are all strongly reflexive, then $\bigoplus_\alpha L_\alpha$ and $\prod_\alpha L_\alpha$ are strongly reflexive.*

Proof. L_α is strongly reflexive, and so the topology of L_α is the $\mathfrak{T}_{lk}$-topology. Applying (1) to $L' = \bigoplus_\alpha L'_\alpha[\mathfrak{T}_{lk}(L_\alpha)]$ instead of to $L = \bigoplus_\alpha L_\alpha$, we obtain, for $L=(L')'$, that $\prod_\alpha L_\alpha$ has the topology $\mathfrak{T}_{lk}(L')$. In the same way we show that the direct sum topology on $\bigoplus_\alpha L_\alpha$ is equal to $\mathfrak{T}_{lk}$. Our assertion therefore follows from 2.(4), provided that we can show that every closed linearly $\mathfrak{T}_{lk}$-bounded subspace of $\prod_\alpha L_\alpha$ (respectively $\bigoplus_\alpha L_\alpha$) is linearly weakly compact, when each L_α has this property.

Let C be closed and linearly $\mathfrak{T}_{lk}$-bounded in $\prod L_\alpha$. $(C+U)/U$ is finite-dimensional, for each $\mathfrak{T}_{lk}$-neighbourhood $U=\prod U_\alpha$ in $\prod L_\alpha$. If C_α is the projection on L_α, it follows that $(C_\alpha+U_\alpha)/U_\alpha$ is finite-dimensional, and by 1.(1) the $\mathfrak{T}_{lk}$-closure $\bar{C}_\alpha$ is also linearly bounded. By hypothesis, $\bar{C}_\alpha$ is linearly weakly compact, by § 10, 9.(7) $\prod_\alpha \bar{C}_\alpha$ is, as well, and finally so is $C\subset\prod_\alpha \bar{C}_\alpha$. If C is closed and linearly $\mathfrak{T}_{lk}$-bounded in $\bigoplus_\alpha L_\alpha$, only finitely many C_α can be non-zero, for otherwise a neighbourhood $U=\bigoplus_\alpha U_\alpha$ could be constructed, as in the proof of (1), for which $(C+U)/U$ would be infinite-dimensional. The linear weak compactness of C then follows as before.

5. Spaces of countable degree. Starting from the spaces φ, ω and ψ we construct further spaces with countable continuous bases, which are no longer locally linearly compact, but which are strongly reflexive.

By 2.(5), φ and ω are strongly reflexive. We denote them by S_1 and S_1' respectively. Let S_2 be the topological direct sum of countably many copies of S_1'.

Clearly all the vectors of S_2 are obtained when in each vector $\{\xi_1,\ldots,\xi_n,0,0,\ldots\}$ of φ the non-zero ξ_i are replaced by arbitrary non-zero vectors $\mathfrak{u}_i$ of ω, and the zeros are replaced by the zero-vector $\mathfrak{o}$ of ω. Because of this, we also denote S_2 by $\varphi\,\omega$.

By 4., S_2' is the topological product of countably many copies of $S_1''=\varphi$. We also write $\omega\varphi$ for S_2', since the vectors in S_2' can be obtained as above by replacing the coordinates of vectors in ω by vectors in φ.

This procedure can be extended. Let α run through the ordinals of the first and second classes. We suppose that S_β has already been defined, for $\beta<\alpha$. If α is not a limit ordinal, we put S_α equal to the topological direct sum of countably many copies of $S_{\alpha-1}'$, so that $S_\alpha=\varphi S_{\alpha-1}'$. If α is a limit ordinal, we put $S_\alpha=\bigoplus_{\beta<\alpha} S_\beta'$.

By transfinite induction, we easily obtain

(1) *S_α is $\mathfrak{T}_{lk}$-complete and strongly reflexive. S_α' is equal to $\omega S_{\alpha-1}$, if α is not a limit ordinal, and S_α' is equal to $\prod_{\beta<\alpha} S_\beta'$, if α is a limit ordinal.*

We say that a linearly topologized space over K is of countable degree if it is obtained from $\varphi(\mathsf{K})$ and $\omega(\mathsf{K})$ by a well-ordered collection (up to ordinals of the second class) of topological sums and products.

It can be shown (cf. KÖTHE [1]) that every space of countable degree can be transformed into one of the spaces S_α, S'_α or $S_\alpha \oplus S'_\alpha$ by a permutation of countably many coordinates of its vectors, and that these canonical forms are all different, in the sense that no two are topologically isomorphic.

Thus the degree α of such a space can be defined; this is an invariant of the space. Only the spaces φ, ω and ψ of the first degree are locally linearly compact. In all these spaces, the unit vectors form a continuous basis.

In $\varphi\omega$ there is no simple theory of equivalence similar to that for φ, ω and ψ, and no full account has yet been given.

6. A counterexample. Circumstances can be considerably more complicated for general linearly topologized spaces than they are for φ and ω. This is made particularly clear by the example of the strongly reflexive $\mathfrak{T}_{lk}$-complete space $\varphi\omega \oplus \omega\varphi$ of second degree (cf. KÖTHE [5]). We assume in the following that K does not have characteristic 2.

By 5., the vectors in $\varphi\omega$ and $\omega\varphi$ have the form

$$\mathfrak{x} = \{\xi_{11}, \xi_{12}, \ldots; \ \xi_{21}, \xi_{22}, \ldots; \ \ldots\}.$$

We denote the vector with $\xi_{ik} = 1$, and $\xi_{jl} = 0$ otherwise, by $\mathfrak{e}_{ik}$. A vector lying in both $\varphi\omega$ and $\omega\varphi$ has only finitely many non-zero ξ_{ik}, and conversely any vector with this property lies in both $\varphi\omega$ and $\omega\varphi$; for this reason, we write $\varphi\omega \cap \omega\varphi = \varphi$, where we consider the coordinates of the vectors in φ to be indexed by pairs (i,k) of positive integers.

Now let H_1 be the linear subspace of $\varphi\omega \oplus \omega\varphi$ consisting of all the $(\mathfrak{x}, \mathfrak{x})$ with $\mathfrak{x} \in \varphi\omega \cap \omega\varphi = \varphi$. It is easy to see that $H_1^\perp$ consists of all $(\mathfrak{x}, -\mathfrak{x})$, $\mathfrak{x} \in \varphi$, in $\omega\varphi \oplus \varphi\omega$. We denote this space by H_2. $H_2^\perp = H_1$, so that H_1 and H_2 are orthogonally closed both in $\varphi\omega \oplus \omega\varphi$ and in $\omega\varphi \oplus \varphi\omega$.

(1) *H_1 and H_2 are orthogonally closed in $\varphi\omega \oplus \omega\varphi$ and $H_1 \cap H_2 = \mathfrak{o}$, but $H_1 \oplus H_2$ is not orthogonally closed.*

$H_1 \oplus H_2$ consists of all the finite vectors in $\varphi\omega \oplus \omega\varphi$ and so the closure of $H_1 \oplus H_2$ is equal to $\varphi\omega \oplus \omega\varphi$. This is in contrast to the behaviour of the closed subspaces of ω (cf. § 9, 3.(2)).

(2) *The quotient space $(\varphi\omega \oplus \omega\varphi)/H_1$ of the $\mathfrak{T}_{lk}$-complete space $L = \varphi\omega \oplus \omega\varphi$ is not complete in the induced topology $\mathfrak{T}_{lk}(L')$.*

Because $H_1^\perp = H_2 \subset \omega\varphi \oplus \varphi\omega$, it is enough, by § 10, 12.(6), to take the topology $\mathfrak{T}_{lk}(H_2)$ on $(\varphi\omega + \omega\varphi)/H_1$. A linearly weakly compact subspace of H_2, however, is finite-dimensional. Given a sequence $\mathfrak{x}^{(n)}=(\mathfrak{y}^{(n)},\mathfrak{z}^{(n)})$ of vectors in $\varphi\omega\oplus\omega\varphi$, the cosets $\hat{\mathfrak{x}}^{(n)}$ with respect to H_1 form a $\mathfrak{T}_{lk}(H_2)$-Cauchy sequence if the sequence is almost constant in each coordinate. In particular if we take the sequence $\mathfrak{x}^{(n)}=\left(\sum_{i,k=1}^{n}\mathfrak{e}_{ik}, -\sum_{i,k=1}^{n}\mathfrak{e}_{ik}\right)$, then $\hat{\mathfrak{x}}^{(n)}$ is a Cauchy sequence. But this has no limit in $(\varphi\omega\oplus\omega\varphi)/H_1$, for there is no $\mathfrak{x}=(\mathfrak{y},\mathfrak{z})\in\varphi\omega\oplus\omega\varphi$ for which $\hat{\mathfrak{x}}-\hat{\mathfrak{x}}^{(n)}$ converges to $\hat{\mathfrak{o}}$; for suitably chosen (j,l), $\eta_{jl}=\zeta_{jl}=0$, and so if $n\geqq j,l$, we have $(\mathfrak{e}_{jl},-\mathfrak{e}_{jl})(\mathfrak{x}-\mathfrak{x}^{(n)})=-2$.

(3) *H_1 has no closed complement in $\varphi\omega\oplus\omega\varphi$.*

For a complement would be $\mathfrak{T}_{lk}$-isomorphic to $(\varphi\omega\oplus\omega\varphi)/H_1$ by § 10, 7.(6), and would therefore not be closed.

The theorem on complementary spaces holds in $\varphi\omega$ and $\omega\varphi$, however (cf. HAGEMANN [1]).

The algebraic isomorphism $H_2=H_2/(H_2\cap H_1)\cong(H_2+H_1)/H_1$ is not an isomorphism for the topologies $\mathfrak{T}_{lk}(L')$. For H_2 is closed, and so complete, whereas $(H_2+H_1)/H_1$ is not complete, by (2). This is the counterexample mentioned in § 10, 7.

7. Further investigations. The theory of linearly topologized spaces has recently been developed systematically in works of FISCHER and GROSS [1], [2], [3], in order to provide a foundation for a theory of quadratic forms in infinite-dimensional vector spaces.

A large part of the theory of linearly topologized spaces developed here applies when a skew field K is used as field of coefficients. In the definition of vector space over K given in § 7.1, it is necessary to restrict the definition in (L2) to $x\xi$; in this way we obtain a right vector-space or right module L over K. For the sake of simplicity, we have not developed the theory in full generality; cf. DIEUDONNÉ [6].

The circumstances are more complicated when K is supposed to be a general topological field. We refer to the investigations of FLEISCHER [1], KÖTHE [7], NACHBIN [1] and VILENKIN [1]. The case of a non-archimedean valued field has had detailed examination; it shows many similarities to the case of the real number field: cf. for example BOURBAKI [6], Vol. 1, FLEISCHER [2], INGLETON [1] and MONNA [1], [2].

Abelian groups, modules and rings with linear topologies have also been investigated; we refer to BALLIER [1], LEPTIN [1] and ZELINSKY [1].

If L is a linearly topologized space, if $A\in\mathfrak{L}(L)$ and if C is a topological automorphism of L, CAC^{-1} is said to be similar to A (cf. § 8, 7.). It is well known that, in the case where L is of finite dimension over the field Γ of complex numbers, the determination of canonical forms for A under "similarity-transformations" is achieved using the theory of elementary divisors. H. ULM [1] has solved the similarity problem for a large class of endomorphisms of $\omega(\Gamma)$, using theorems about infinite abelian groups.

The problem of the equivalence of quadratic forms can also be formulated in linearly topologized spaces. Let $\langle L_2, L_1\rangle$ be a dual pair. If A is a weakly continuous linear mapping from L_1 into L_2, so that $A\in\mathfrak{L}(L_1,L_2)$, then A' maps $L_2'=L_1$ into $L_1'=L_2$, so that A' is also in $\mathfrak{L}(L_1,L_2)$. We call A symmetric when $A=A'$. If A is symmetric, a symmetric bilinear form $(y,x)\to y(Ax)$ is defined on $L_1\times L_1$; the correspondence $(x,x)\to x(Ax)$ is a quadratic form on L_1. The function $y(Ax)$ is continuous on $L_1\times L_1$ for the topologies $\mathfrak{T}_{ls}$ and $\mathfrak{T}_{lk}$, and every continuous symmetric bilinear form on $L_1\times L_1$ is determined by some $A\in\mathfrak{L}(L_1,L_2)$.

If $x=Cz$ is a topological automorphism of L_1, an equivalent form $z((C'AC)z)$ is obtained from the quadratic form $x(Ax)$; the problem then arises of classifying quadratic forms with respect to this equivalence.

No general account of this problem has yet been given. For $\varphi(\mathsf{P})$, $\omega(\mathsf{P})$ and $\psi(\mathsf{P})$, however, it has been completely solved by K. RITZDORFF [1].

CHAPTER THREE

Topological Vector Spaces

§ 14 contains the elementary theory of normed spaces and Banach spaces. A number of classical examples are discussed, to which we shall refer time and again in the later parts of the book.

The concept of topological vector space over the real or complex field is developed in full generality in § 15. In the investigation of these spaces we restrict our attention essentially to those questions which are of significance for the later theory of locally convex spaces. One important result is that every locally compact topological vector space is finite-dimensional, and indeed is topologically isomorphic to some P^n with the usual topology. The spaces L^p, with $0<p<1$, are investigated to give an example to show that the dual space can consist of one element alone. Metrizable spaces are considered in detail. The classical results of the theory developed by BANACH and his colleagues, together with a theorem of BOURBAKI on bilinear mappings from metrizable spaces, bring § 15 to an end.

In the next two paragraphs a detailed account is given of convex sets and the HAHN-BANACH theorem. Provided that necessary care is taken, methods which go back to MINKOWSKI can also be applied to convex sets in vector spaces of infinite dimension. We give three proofs of the HAHN-BANACH theorem. The first two are of a geometrical nature, and are consequences of the separation theorem for convex sets, while the third is the classical analytic proof. The chapter ends with applications to normed spaces, and the theorem of F. RIESZ on the representation of the continuous linear functionals on the space of continuous functions in terms of Stieltjes integrals.

§ 14. Normed spaces

1. Definition of a normed space. In this and the following chapters we shall only deal with real or complex vector spaces. From now on the coefficient field K means either the field P of real numbers or the field Γ of complex numbers. When it is not explicitly mentioned that we are dealing with a real or a complex vector space, then statements are always valid in both cases.

A vector space E is called a normed space if a real number $\|x\|$, the norm of x, is associated with each $x\in E$, with the following properties:

(N 1) $\|x\|\geqq 0$ for all $x\in E$,

(N 2) if $\|x\|=0$, then $x=\mathfrak{o}$,

(N 3) $\|\lambda x\| = |\lambda| \|x\|$ for each $\lambda \in \mathsf{K}$,

(N 4) $\|x+y\| \leqq \|x\| + \|y\|$.

If $\|x\|$ only satisfies (N 1), (N 3) and (N 4), $\|x\|$ is called a semi-norm on E.

As well as $\|x\|$, we shall use the notation $p(x)$ or $q(x)$ for a norm or semi-norm respectively.

The simplest example of a normed space is K itself, with the modulus as norm. As in this case, we have in general that

$$(1) \qquad \big|\|x\| - \|y\|\big| \leqq \|x-y\| \leqq \|x\| + \|y\|.$$

If we introduce a distance in E by putting

$$(2) \qquad |x, y| = \|x-y\|$$

then it follows without difficulty that axioms (D 1) to (D 4) of § 4, 1. are satisfied, so that *every normed space is a metric space*. The properties of metric spaces developed in § 4 are therefore also valid for normed spaces.

The distance defined by (2) is translation-invariant, i.e. $|x+z, y+z| = |x, y|$, for each $z \in E$. The open ball $K_r(x)$ of radius r about x, which consists of all $y \in E$ with $\|y-x\| < r$, is obtained from the ball $K_r(\circ)$ by a translation; $K_r(x) = x + K_r(\circ)$.

(3) *The topology defined by the norm is compatible with the vector space operations, i.e. λx and $x+y$ are continuous in both variables together.*

If $\|x-x_0\| < \frac{\varepsilon}{2}$, $\|y-y_0\| < \frac{\varepsilon}{2}$, then, by (N 4), $\|(x+y)-(x_0+y_0)\| < \varepsilon$, so that $x+y$ is continuous in both variables.

It follows from

$$\begin{aligned} \|\lambda x - \lambda_0 x_0\| &= \|(\lambda-\lambda_0)(x-x_0) + (\lambda-\lambda_0)x_0 + \lambda_0(x-x_0)\| \\ &\leqq |\lambda-\lambda_0| \|x-x_0\| + |\lambda-\lambda_0| \|x_0\| + |\lambda_0| \|x-x_0\| \end{aligned}$$

that we can make $\|\lambda x - \lambda_0 x_0\| < \varepsilon$ if we take $|\lambda-\lambda_0|$ and $\|x-x_0\|$ to be sufficiently small.

In particular it follows from (3) that $\sum_1^m \alpha_k^{(n)} x_k^{(n)}$ converges to $\sum_1^m \alpha_k x_k$ if $\alpha_k^{(n)} \to \alpha_k$, $x_k^{(n)} \to x_k$.

The norm $\|x\|$ is a uniformly continuous function on E, by (1).

A linear subspace H of a normed space E is again a normed space, when the norm of E is restricted to H.

If a semi-norm $\|x\|$ is given on E, a normed space can be obtained in the following way: if $\|x\| = 0$ and $\|y\| = 0$, then it follows from (N 3) and (N 4) respectively that $\|\lambda x\|=0$ and $\|x+y\|=0$. The elements of E with vanishing semi-norm therefore form a linear subspace N of E. On the quotient space E/N let us set $\|\hat{x}\|=\|x\|$, if x lies in the coset $\hat{x}$. $\|\hat{x}\|$ is uniquely defined, since it follows from (1) that $\|x\|=\|x+z\|$, for $z\in N$. All the properties of a norm are immediately satisfied by $\|\hat{x}\|$. Thus we have shown:

(4) *If $\|x\|$ is a semi-norm on E and if N is the null-space of the semi-norm, then E/N is normed by $\|\hat{x}\|=\|x\|$.*

2. Norm isomorphism, equivalent norms. Two normed spces E and F are said to be norm isomorphic if there is a one-one linear mapping of E onto F, under which corresponding elements have equal norms. E and F are then isometric, as metric spaces.

Two norms on the same vector space E are said to be equivalent if they define the same topology on E.

(1) *Two norms $p_1(x)$ and $p_2(x)$ on E are equivalent if and only if there exist two positive numbers m and M such that*

$$m \leqq \frac{p_1(x)}{p_2(x)} \leqq M \tag{2}$$

holds for all non-zero x.

Proof. Let $K_r^{(1)}(\mathrm{o})$ and $K_r^{(2)}(\mathrm{o})$ denote the open balls of radius r about o, with respect to the norms p_1 and p_2 respectively.

If (2) holds, then it follows from the left hand inequality that $K_r^{(1)}(\mathrm{o})$ is contained in $K_{r/m}^{(2)}(\mathrm{o})$, and it follows from the right hand inequality that $K_r^{(2)}(\mathrm{o})$ is contained in $K_{rM}^{(1)}(\mathrm{o})$. The two systems of balls therefore form equivalent bases of neighbourhoods of o. But since the metrics are translation-invariant it is enough to consider the neighbourhoods of o, and so (2) is sufficient.

Conversely, if p_1 and p_2 are equivalent the closed ball $\overline{K_1^{(1)}}(\mathrm{o})$ must contain a closed ball $\overline{K_\delta^{(2)}}(\mathrm{o})$. If $p_2(x)=\delta$, it follows that $p_1(x)\leqq 1$. We therefore have $\delta p_1(x)\leqq p_2(x)$ for all x with $p_2(x)=\delta$. But this inequality must hold for all x, because of (N 3), so that the right hand side of (2) is established, with $M=1/\delta$. The left hand side is established in a similar way.

Two normed spaces E and F are said to be topologically isomorphic if there is an algebraic isomorphism of E onto F, which at the same time is a homeomorphism. If we have a topological isomorphism, and if we transfer the norm of E to F by giving corresponding elements the same norm, then we obtain a norm on F which is equivalent to the original one. (1) can therefore also be expressed in the following way:

(3) *Two normed spaces E and F are topologically isomorphic if and only if there exist an algebraic isomorphism* $x_1 \leftrightarrow x_2$ *of E onto F and two positive numbers m and M, for which*

$$m \leqq \frac{\|x_1\|}{\|x_2\|} \leqq M \tag{4}$$

holds for all non-zero x_1 *in E and corresponding* x_2 *in F.*

3. Banach spaces. A complete normed space is called a Banach space or (B)-space.

(1) *Every normed space can be embedded in a minimal (B)-space* $\tilde{E}$, *the completion of E, and this is unique up to norm isomorphism.*

The closed unit ball in $\tilde{E}$ *is the closure in* $\tilde{E}$ *of both the closed and the open unit ball of E.*

The proof can be established using the corresponding theorem for metric spaces (§ 4, 4.(1)). As a consequence of this, E is contained in a complete metric space $\tilde{E}$. If $\tilde{x}$ and $\tilde{y}$ are two elements of $\tilde{E}$, and if $(x^{(n)})$ and $(y^{(n)})$ are two sequences in E converging to $\tilde{x}$ and $\tilde{y}$ respectively, then $|\tilde{x}, \tilde{y}| = \lim |x^{(n)}, y^{(n)}|$. If we put $\|\tilde{x}\| = |\mathfrak{o}, \tilde{x}|$, we obtain an extension of the norm from E to $\tilde{E}$, which clearly satisfies (N 1) and (N 2). If we put $\lambda \tilde{x}$ equal to the equivalence class of Cauchy sequences to which $(\lambda x^{(n)})$ belongs, and likewise put $\tilde{x} + \tilde{y}$ equal to the class to which $(x^{(n)} + y^{(n)})$ belongs, it is easy to see that unique elements of $\tilde{E}$ are defined in this way (e.g. if $z^{(n)} \to \tilde{x}$ as well, then $(\lambda z^{(n)})$ and $(\lambda x^{(n)})$ are equivalent Cauchy sequences, since $\|\lambda z^{(n)} - \lambda x^{(n)}\| = |\lambda| \, \|z^{(n)} - x^{(n)}\|$). Axioms (L 1) and (L 2) for a vector space, together with axioms (N 3) and (N 4) for the norm, are now established for $\tilde{E}$ by taking limits in the corresponding axioms for E. Thus $\tilde{E}$ is a minimal (B)-space containing E.

Two minimal (B)-spaces containing E are always isometric: because of continuity, sums and scalar products correspond under this isometry, and so the isometry is a norm isomorphism.

Finally, if $x_0 \neq \mathfrak{o}$ belongs to the closed unit ball of $\tilde{E}$ and if $x_n \in E$ and $x_n \to x_0$, then $y_n = \dfrac{n-1}{n} \dfrac{\|x_0\|}{\|x_n\|} x_n \to x_0$ as well, since $\|x_n\| \to \|x_0\| \leqq 1$, and y_n belongs to the open unit ball of E.

Trivially,

(2) *A closed linear subspace of a* (B)-*space is again a* (B)-*space.*

It is often of interest to know if a (B)-space is separable.

(3) *If a normed space E is separable, then every subset, and in particular every linear subspace, is separable, and so is the completion* $\tilde{E}$.

Proof. By § 4, 5.(1), separability is equivalent to the existence of a countable basis of open sets of E. This property is then satisfied by every subset. If, further, x_i is a countable collection of elements which is dense in E, then x_i is also dense in $\tilde{E}$.

4. Quotient spaces and topological products. Suppose that E is a normed space and that H is a closed linear subspace. The induced topology on the quotient space E/H is defined, as in § 10, 7., by the images $K(O)$ of the open subsets O of E, where K is the canonical mapping of E onto E/H.

(1) *If H is a closed linear subspace of a normed space E, and if we introduce a norm on the quotient space E/H by putting*

(2) $\|\hat{x}\| = \inf_{x \in \hat{x}} \|x\|$ (as x runs through all the elements of the coset $\hat{x}$),

then E/H becomes a normed space, whose topology is the induced quotient-space topology.

Proof. If $\hat{x} \neq \mathfrak{o}$ and $x_0 \in \hat{x}$, then $\inf_{z \in H} \|x_0 + z\| > 0$, for otherwise x_0 would be a closure point of the closed subspace H, and would therefore lie in H. Thus (N1) and (N2) are satisfied. (N3) follows immediately from the corresponding (N3) for E, and (N4) follows from

$$\inf_{x+y \in \hat{x}+\hat{y}} \|x+y\| \leqq \inf_{x \in \hat{x},\, y \in \hat{y}} \|x+y\| \leqq \inf_{x \in \hat{x},\, y \in \hat{y}} (\|x\| + \|y\|) = \inf_{x \in \hat{x}} \|x\| + \inf_{y \in \hat{y}} \|y\|.$$

Thus (2) defines a norm on E/H.

If $\|x\| < \rho$, then $\|\hat{x}\| < \rho$; conversely if $\|\hat{x}\| < \rho$, there is an $x \in \hat{x}$ with $\|x\| < \rho$, and so the ball $\|\hat{x}\| < \rho$ is the image of the ball $\|x\| < \rho$ under the canonical mapping. The topology defined by the norm (2) is therefore the quotient-space topology.

(3) *If E is a* (B)*-space, so also is E/H.*

Proof. We must show that E/H is complete.

Preliminary remark: if $\hat{x}$ and $\hat{y}$ are two elements of E/H, and if $x \in \hat{x}$ is given, then, by (2), there exists $y \in \hat{y}$ with $\|x - y\| \leqq 2\|\hat{x} - \hat{y}\|$.

First let $\hat{x}_n$ be a Cauchy sequence in E/H with $\sum_{n=1}^{\infty} \|\hat{x}_n - \hat{x}_{n+1}\| < \infty$. Choose x_1 arbitrarily in the coset $\hat{x}_1$; using the preliminary remark, there is an $x_2 \in \hat{x}_2$ with $\|x_1 - x_2\| \leqq 2\|\hat{x}_1 - \hat{x}_2\|$, and, generally, there is an $x_{n+1} \in \hat{x}_{n+1}$ with $\|x_n - x_{n+1}\| \leqq 2\|\hat{x}_n - \hat{x}_{n+1}\|$. But it follows from $\sum_{n=1}^{\infty} \|x_n - x_{n+1}\| < \infty$ that x_n is a Cauchy sequence in E. If x_0 is its limit, $\hat{x}_n \to \hat{x}_0$ in E/H, since $\|\hat{x}_n - \hat{x}_0\| \leqq \|x_n - x_0\|$.

If $\hat{x}_n$ is an arbitrary Cauchy sequence in E/H, there is a subsequence $\hat{x}_{n_k}$ with $\sum_{k=1}^{\infty} \|\hat{x}_{n_k} - \hat{x}_{n_{k+1}}\| < \infty$. This subsequence has a limit $\hat{x}_0$, and this is also limit of the whole sequence.

(4) *If E is separable, so is E/H.*

We have seen in § 4, 7. that the topological product of countably many metric spaces can be given a metric which induces the same topology as the product topology. The topological product of arbitrarily many normed spaces is again a vector space, on which a topology is defined. We shall see in § 15, 4. that even when there are countably many factors this topology cannot be defined by a norm. We can only assert:

(5) *Let $E_1, \ldots, E_n$ be finitely many normed spaces, and let $E = \prod_{i=1}^{n} E_i$ be the topological product, with elements $x = (x_1, \ldots, x_n)$. If we define norms*

(6a) $\|x\|_1 = \sup_i \|x_i\|$, (6b) $\|x\|_2 = \sqrt{\sum_i \|x_i\|^2}$, (6c) $\|x\|_3 = \sum_i \|x_i\|$,

on E, then under each of these norms E is a normed space whose topology coincides with the product topology.

The simple proof is left to the reader. There are other norms which have the same property. E is a (B)-space if and only if all the E_i are (B)-spaces (cf. § 4, 7.(2)). E is separable if and only if all the E_i are.

5. The dual space. We consider the continuous linear functions with values in K on a normed space E over K. If ux is such a function, there must be a ball $\|x\| \leqq \delta$ such that $|ux| \leqq 1$ for all x in the ball. For general non-zero z $x = \frac{\delta}{\|z\|} z$ lies in the ball, and so $|uz| \leqq \frac{1}{\delta} \|z\|$. This gives one half of

(1) *A linear functional ux on the normed space E is continuous if and only if there exists $M > 0$ for which*

$$|ux| \leqq M \|x\| \tag{2}$$

holds for all $x \in E$.

Since it is enough to establish continuity at o, (2) is clearly also sufficient.

The collection of all the continuous linear functionals on a normed space E forms a vector space. We again call this the dual or conjugate space E' of E.

It is natural to introduce a norm on E' as well, by putting

$$\|u\| = \sup_{\|x\| \leqq 1} |ux|. \tag{3}$$

It is trivial to show that this is a norm. The following inequality follows directly from (3):

(4) $$|u x| \leqq \|u\| \, \|x\|.$$

(5) *The dual E' of a normed space E is a* (B)*-space, when it is given the norm* (3).

Proof. If $u^{(n)}$ is a Cauchy sequence, $\|u^{(n)}-u^{(m)}\| \leqq \varepsilon$ for $n,m \geqq n_0(\varepsilon)$. By (4), $|(u^{(n)}-u^{(m)})x| \leqq \varepsilon\|x\|$, so that the sequence $u^{(n)}x$ converges for each x. We put $v x = \lim u^{(n)} x$.

v is a linear functional on E. Since $|(v-u^{(m)})x| \leqq \varepsilon\|x\|$ for $m \geqq n_0$, $v-u^{(m)}$ is continuous. Since $u^{(m)}$ is also continuous, the sum v is, as well, i.e. v belongs to E'. Finally it follows from $\sup\limits_{\|x\| \leqq 1} |(v-u^{(m)})x| \leqq \varepsilon$ that $u^{(n)}$ converges to v.

We are not yet in a position to show that there always exist non-zero continuous linear functionals on a (B)-space. The HAHN-BANACH theorem (§ 17, 6.) is required before we can do this. In this paragraph, however, we shall determine the duals of some classical examples of (B)-spaces.

We can also form the dual of the space E'; in this way we obtain the bidual space E'' of E. The elements x_0 of E can be interpreted as linear functionals $x_0(u) = u x_0$, $u \in E'$, on E'. If the (B)-space E, considered in this way, is identical with the (B)-space E'', E is said to be reflexive. Later in this paragraph we shall meet examples of both reflexive and non-reflexive (B)-spaces.

If E is separable, E' need not be separable in the norm topology (cf. 7.).

6. Continuous linear mappings. Let E and F be two normed spaces, and let A be a continuous linear mapping from E into F. Just as for linear functionals, we can establish

(1) *A linear mapping A from a normed space E into a normed space F is continuous if and only if there is an $M > 0$ for which*

(2) $$\|A x\| \leqq M \|x\|$$

holds, for all $x \in E$.

Let us recall that the continuity of a mapping between metric spaces can be defined in terms of sequential continuity, i.e. by requiring that $x^{(n)} \to x^{(0)}$ always implies that $A x^{(n)} \to A x^{(0)}$.

The vector space $\mathfrak{L}(E,F)$ of continuous linear mappings from E into F is turned into a normed space, when we put

(3) $$\|A\| = \sup_{\|x\| \leqq 1} \|A x\|.$$

We observe that

$$\|Ax\| \leqq \|A\|\,\|x\|. \tag{4}$$

(5) *If F is a* (B)*-space, then* $\mathfrak{L}(E,F)$ *is a* (B)*-space.*

The proof of 5.(5) carries over directly. The topology defined on $\mathfrak{L}(E,F)$ by (3) is called the uniform norm topology.

(6) *If A is a continuous linear mapping of E into F, and if B is a continuous linear mapping of F into G, then the continuous mapping BA of E into G satisfies*

$$\|BA\| \leqq \|B\|\,\|A\|. \tag{7}$$

For by (4), $\|BAx\| \leqq \|B\|\,\|Ax\| \leqq \|B\|\,\|A\|\,\|x\|$, so that $\sup\limits_{\|x\| \leqq 1} \|BAx\| \leqq \|B\|\,\|A\|$.

If a normed space E is an algebra over K in the sense defined in § 8, 1., and if the product xy of two elements of E always satisfies

$$\|xy\| \leqq \|x\|\,\|y\|, \tag{8}$$

then E is called a normed algebra. If, further, E is a (B)-space, then we call it a Banach algebra.

It follows from (7) that

(9) *The set $\mathfrak{L}(E)$ of continuous endomorphisms of a normed space forms a normed algebra with unit element. If E is a Banach space, then $\mathfrak{L}(E)$ is a Banach algebra.*

It follows from (8) that $\|xy - x_n y_n\| \leqq \|x\| \cdot \|y - y_n\| + \|y_n\| \cdot \|x - x_n\|$, so that the product xy is continuous in both variables together; condition (8) therefore means that the product operation is compatible with the normed space topology.

This idea also applies to sequences of mappings $A_n \in \mathfrak{L}(E,F)$, $B_n \in \mathfrak{L}(F,G)$: if $A_n \to A$ and $B_n \to B$, then $B_n A_n \to BA$ in $\mathfrak{L}(E,G)$.

The theory of Banach algebras will not be considered in this book; we refer the reader to Hille and Phillips [1], Loomis [1] and Naimark [1].

7. The spaces c_0, c, l^1 and l^∞. If we interpret a bounded sequence ξ_k, $k=1,2,\ldots$, of real or complex numbers as a coordinate vector $\mathfrak{x}=(\xi_k)$, and if we define $\mathfrak{x}+\mathfrak{y}$ to be the vector with coordinates $\xi_k+\eta_k$, where $\mathfrak{y}=(\eta_k)$, and define $\lambda\mathfrak{x}$ to be $(\lambda\xi_k)$, then we obtain a real or complex vector space l^∞.

We introduce a norm on l^∞ by setting

$$\|\mathfrak{x}\|_\infty = \|\mathfrak{x}\| = \sup_k |\xi_k|. \tag{1}$$

The norm properties are satisfied. Moreover,

(2) *l^∞ is a* (B)*-space.*

For if $\mathfrak{x}^{(n)}=(\xi_k^{(n)})$ is a Cauchy sequence in l^∞, then $\sup_k |\xi_k^{(n)}-\xi_k^{(m)}| \leqq \varepsilon$ for all $n,m \geqq n_0$. Thus for fixed k each coordinate sequence $\xi_k^{(n)}$ converges to some $\xi_k^{(0)}$, so that $\mathfrak{x}^{(n)}$ is coordinatewise convergent to $\mathfrak{x}^{(0)}=(\xi_k^{(0)})$. Further, $\sup_k |\xi_k^{(n)}-\xi_k^{(0)}| \leqq \varepsilon$, for $n \geqq n_0$, so that $\mathfrak{x}^{(n)} \to \mathfrak{x}^{(0)}$, and $\mathfrak{x}^{(0)}$ is a bounded sequence with $\|\mathfrak{x}^{(n)}-\mathfrak{x}^{(0)}\| \leqq \varepsilon$; i.e. $\mathfrak{x}^{(n)}$ has $\mathfrak{x}^{(0)}$ as limit in l^∞.

We denote by c and c_0 the linear subspace of l^∞ consisting of all convergent sequences and all sequences convergent to 0 respectively. We give both spaces the norm (1). Then we have

(3) *c and c_0 are* (B)*-spaces, and so are closed linear subspaces of l^∞.*

Proof. Let $\mathfrak{x}^{(n)}=(\xi_k^{(n)})$ be a Cauchy sequence in c. We must show that its limit $\mathfrak{x}^{(0)}=(\xi_k^{(0)})=\left(\lim_n \xi_k^{(n)}\right)$, which exists by (2), is again a convergent sequence.

By hypothesis $\lim_{k\to\infty} \xi_k^{(n)}=\xi^{(n)}$ exists, for each n. Given $\varepsilon>0$, there exists an n_0 such that $\sup_k |\xi_k^{(n)}-\xi_k^{(m)}| < \frac{\varepsilon}{3}$ for $n,m \geqq n_0$, so that

$$|\xi^{(n)}-\xi^{(m)}| \leqq |\xi^{(n)}-\xi_k^{(n)}| + |\xi_k^{(n)}-\xi_k^{(m)}| + |\xi_k^{(m)}-\xi^{(m)}| < 3\,\frac{\varepsilon}{3} = \varepsilon,$$

for sufficiently large k, and for $n,m \geqq n_0$. Thus $\lim_{n\to\infty} \xi^{(n)}=\xi^{(0)}$ exists.

We now show that $\lim_{k\to\infty} \xi_k^{(0)}=\xi^{(0)}$. Given $\varepsilon>0$, let n_0 be large enough for $|\xi^{(n)}-\xi^{(0)}| \leqq \frac{\varepsilon}{3}$ and $\sup_k |\xi_k^{(n)}-\xi_k^{(0)}| \leqq \frac{\varepsilon}{3}$ to hold for $n \geqq n_0$. If for a fixed $n_1 \geqq n_0$ we choose k_0 large enough for $|\xi_k^{(n_1)}-\xi^{(n_1)}| \leqq \frac{\varepsilon}{3}$ to hold for all $k \geqq k_0$, then

$$|\xi_k^{(0)}-\xi^{(0)}| \leqq |\xi_k^{(0)}-\xi_k^{(n_1)}| + |\xi_k^{(n_1)}-\xi^{(n_1)}| + |\xi^{(n_1)}-\xi^{(0)}| \leqq \varepsilon$$

for $k \geqq k_0$, as required.

The proof for c_0 is included in this.

We denote by l^1 the vector space of all $\mathfrak{x}=(\xi_k)$ with $\sum_{k=1}^{\infty} |\xi_k| < \infty$. This is normed by

(4) $$\|\mathfrak{x}\|_1 = \|\mathfrak{x}\| = \sum_{k=1}^{\infty} |\xi_k|\,.$$

(5) *l^1 is a* (B)*-space.*

The simple proof is left to the reader. It is also contained in the proof of 8.(7).

If we only consider the vector space structure, we clearly have

$$(6) \qquad l^1 \subset c_0 \subset c \subset l^\infty .$$

The question of separability is easily answered:

(7) *l^1, c_0 and c are separable, and l^∞ is not.*

Proof. The vectors with finitely many non-zero rational coordinates are dense both in l^1 and c_0. The vectors with finitely many arbitrary rational coordinates and with the rest equal to one fixed rational are dense in c.

In l^∞ two vectors whose coordinates are equal to $+1$ or -1 are always distance 2 apart. Since there is a continuum of these, the set of these vectors is not separable, and so by 3.(3) neither is l^∞.

We now attempt to determine the dual spaces precisely.

(8) *l^∞ is the dual of l^1.*

Proof. More precisely, our assertion says that $(l^1)'$ is norm isomorphic to l^∞, so that each continuous linear functional $u(\mathfrak{x})$ can be represented, in a way yet to be defined, by a vector $\mathfrak{u}$ in l^∞. We shall come across this problem of the concrete representation of abstract linear functionals more and more often.

We shall formulate the line of thought in a rather more general setting, as the same idea can be applied in other situations.

A set M of elements of a normed space E is called a fundamental set if its linear span is dense in E. We also say that M is total in E. A continuous linear functional is clearly uniquely determined by its values on a fundamental set.

The unit vectors $\mathfrak{e}_p$, $p=1,2,\ldots$, form a fundamental set in l^1, so that if $u\in(l^1)'$, u is determined by the vector $\mathfrak{u}=(v_p)$, where $v_p=u(\mathfrak{e}_p)$. If $\mathfrak{x}=(\xi_p)\in l^1$, it follows from the fact that the sections $\mathfrak{x}_n=(\xi_1,\ldots,\xi_n,0,0,\ldots)$ converge to $\mathfrak{x}$ that

$$(9) \qquad u(\mathfrak{x}) = \lim_n u(\mathfrak{x}_n) = \lim_n \sum_1^n v_p \xi_p = \sum_1^\infty v_p \xi_p = \mathfrak{u}\,\mathfrak{x} .$$

If we put $\mathfrak{x}^{(p)}=\varepsilon_p^{-1}\mathfrak{e}_p$ when $v_p=\varepsilon_p|v_p|\neq 0$, and put $\mathfrak{x}^{(p)}=\mathfrak{e}_p$ if $v_p=0$, then $\|\mathfrak{x}^{(p)}\|=1$, so that $\mathfrak{u}\,\mathfrak{x}^{(p)}=|v_p|\leqq\|u\|$; i.e. $\mathfrak{u}$ lies in l^∞, and in l^∞ it has norm $\leqq\|u\|$. On the other hand,

$$(10) \qquad |u(\mathfrak{x})| \leqq \left(\sup_p |v_p|\right) \sum_{p=1}^\infty |\xi_p| ,$$

so that $\|u\| \leqq \|\mathfrak{u}\|$. It therefore follows that $\|u\| = \|\mathfrak{u}\|$, which gives the required norm isomorphism.

It follows immediately from (10) that conversely every $\mathfrak{u} \in l^\infty$ defines a continuous linear functional on l^1. We observe that the dual of the separable space l^1 is not separable.

(11) *l^1 is the dual space of c_0.*

The sections $\mathfrak{x}_n$ converge to $\mathfrak{x}$ in c_0 as well, and we again obtain

$$u(\mathfrak{x}) = \lim_n \sum_1^n v_p \xi_p = \mathfrak{u}\,\mathfrak{x},$$

for a continuous linear functional u, where $v_p = u(\mathfrak{e}_p)$. Again, let $v_p = \varepsilon_p |v_p|$ if $v_p \neq 0$, let $\varepsilon_p = 1$ if $v_p = 0$, and let $\mathfrak{x}' = \sum\limits_1^n \varepsilon_p^{-1} \mathfrak{e}_p$. Then $\|\mathfrak{x}'\| = 1$ in c_0, so that $\mathfrak{u}\,\mathfrak{x}' = \sum\limits_1^n |v_p| \leqq \|u\|$. Hence $\mathfrak{u}$ lies in l^1, and $\|\mathfrak{u}\| \leqq \|u\|$. Conversely, $|u(\mathfrak{x})| \leqq \sum\limits_1^\infty |v_p|\,\|\mathfrak{x}\|$, so that $\|u\| \leqq \|\mathfrak{u}\|$.

(8) and (11) show that the bidual of c_0 is equal to l^∞, and is therefore larger than c_0. Hence c_0 is an example of a non-reflexive (B)-space.

We now consider the dual of c. Here we also find that $c' = l^1$, although the representation is a different one.

The space c contains c_0 as a subspace, and c_0 has codimension 1.

$$c = c_0 \oplus [\mathfrak{e}], \tag{12}$$

where $[\mathfrak{e}]$ is the one-dimensional space consisting of the scalar multiples of the vector $\mathfrak{e}$, which has all its coordinates equal to 1. Every convergent sequence $\mathfrak{x} = (\xi_n)$ can be represented in the form

$$\mathfrak{x} = \xi_0 \mathfrak{e} + \mathfrak{x}_0 \quad \text{with} \quad \xi_0 = \lim_{n \to \infty} \xi_n \quad \text{and} \quad \mathfrak{x}_0 \in c_0. \tag{13}$$

A fundamental set in c is formed by $\mathfrak{e}$ and the vectors $\mathfrak{e}_p$, $p = 1, 2, \ldots$.

If $u(\mathfrak{x})$ is a continuous linear functional on c, let us put $v_0' = u(\mathfrak{e})$ and $v_p = u(\mathfrak{e}_p)$, for $p \geqq 1$. It follows from (13) and (11) that

$$u(\mathfrak{x}) = u(\xi_0 \mathfrak{e}) + u(\mathfrak{x}_0) = \xi_0 v_0' + \sum_1^\infty v_p(\xi_p - \xi_0) \quad \text{with} \quad (v_1, v_2, \ldots) \in l^1.$$

If we put $\mathfrak{u} = (v_0, v_1, \ldots)$, where $v_0 = v_0' - \sum\limits_1^\infty v_p$, and put $\mathfrak{x}' = (\xi_0, \xi_1, \ldots)$ then

$$u(\mathfrak{x}) = \left(v_0' - \sum_1^\infty v_p \right) \xi_0 + \sum_1^\infty v_p \xi_p = \mathfrak{u}\,\mathfrak{x}'. \tag{14}$$

Conversely, if $u \in l^1$, (14) always determines a continuous linear functional with $\|u\| \leqq \|\mathfrak{u}\|$, since

$$|\mathfrak{u}\,\mathfrak{x}'| \leqq \left(\sum_0^\infty |v_p|\right) \sup_{p \geqq 0} |\xi_p| = \|\mathfrak{u}\| \sup_{p \geqq 0} |\xi_p| = \|\mathfrak{u}\|_1 \, \|\mathfrak{x}\|_\infty .$$

The inequality $\|\mathfrak{u}\| \leqq \|u\|$, and hence the norm isomorphism, is established in the following way. Again let $v_p = \varepsilon_p |v_p|$ for $v_p \neq 0$, and $\varepsilon_p = 1$ for $v_p = 0$, and let $\mathfrak{x}^{(n)}$ be the convergent sequence, the first n of whose terms are equal to $\varepsilon_1^{-1}, \ldots, \varepsilon_n^{-1}$, and whose subsequent coordinates are all equal to ε_0^{-1}. Then $\|\mathfrak{x}^{(n)}\|_\infty = 1$, and $|u(\mathfrak{x}^{(n)})| = |\mathfrak{u}\,\mathfrak{x}^{(n)\prime}| \geqq |v_0| + \sum_1^n |v_p| - \sum_{n+1}^\infty |v_p|$. It therefore follows from $|u(\mathfrak{x}^{(n)})| \leqq \|u\|$ that $|v_0| + \sum_1^n |v_p| - \sum_{n+1}^\infty |v_p| \leqq \|u\|$. Letting $n \to \infty$, this gives $\|\mathfrak{u}\| \leqq \|u\|$.

We therefore obtain

(15) *The dual space of c is norm isomorphic to l^1, when we make the continuous linear functional u correspond to the vector $\mathfrak{u} = (v_0, v_1, \ldots)$, where $v_p = u(\mathfrak{e}_p)$ for $p = 1, 2, \ldots$ and $v_0 = u(\mathfrak{e}) - \sum_1^\infty v_p$, and when we apply $\mathfrak{u}$ to $\mathfrak{x}$ in the way defined by* (13) *and* (14).

c is therefore not reflexive either. There is no such simple representation of the dual of l^∞, but we shall see later on that l^1 and l^∞ are also not reflexive. We observe that at any rate l^1 is a linear subspace of $(l^\infty)'$, when we interpret each $\mathfrak{u} \in l^1$ as the linear functional $\mathfrak{u}\,\mathfrak{x}$ on l^∞. A detailed investigation of $(l^\infty)'$ is made in § 31, No. 1.

8. The spaces l^p, $1 < p < \infty$. For the sake of completeness, we prove the fundamental inequalities upon which the theory of these spaces depends.

(1) *If* $0 < \alpha < 1$ *and* $a \geqq 0, b \geqq 0$, *then*

(2) $$a^\alpha b^{1-\alpha} \leqq \alpha a + (1-\alpha) b .$$

Proof. If $a = b$ we have equality. We can clearly suppose that $b > a > 0$. By the mean-value theorem,

$$b^{1-\alpha} - a^{1-\alpha} = (1-\alpha)(b-a)\,\xi^{-\alpha} \quad \text{with} \quad a < \xi < b .$$

Since $\xi^{-\alpha} < a^{-\alpha}$, we have

$$b^{1-\alpha} - a^{1-\alpha} < (1-\alpha)(b-a)\,a^{-\alpha} .$$

Multiplying by a^α, we obtain (2).

We remark that the proof shows that equality only holds in (2) when $a=b$.

Using this result, we prove Hölder's inequality

$$(3)\quad \sum_1^\infty |v_k \xi_k| \leqq \left(\sum_1^\infty |v_k|^q\right)^{1/q} \left(\sum_1^\infty |\xi_k|^p\right)^{1/p} \qquad \frac{1}{p}+\frac{1}{q}=1, \qquad 1<p<\infty.$$

Here v_k and ξ_k are arbitrary real or complex numbers for which the sums on the right-hand side converge.

It is clearly sufficient to establish (3) for $\sum_1^n$ instead of for $\sum_1^\infty$.

If we put $\alpha=\frac{1}{p}$, $1-\alpha=\frac{1}{q}$, $a=c_k^p$, $b=d_k^q$, in (2), we obtain

$$(4)\qquad c_k d_k \leqq \frac{1}{p} c_k^p + \frac{1}{q} d_k^q.$$

Let $c_k = \dfrac{|\xi_k|}{\left(\sum_1^n |\xi_l|^p\right)^{1/p}}$, $d_k = \dfrac{|v_k|}{\left(\sum_1^n |v_l|^q\right)^{1/q}}$ for $k=1,\dots,n$. Summing over the resulting inequalities, we obtain

$$\frac{\sum_1^n |v_k \xi_k|}{\left(\sum_1^n |v_k|^q\right)^{1/q} \left(\sum_1^n |\xi_k|^p\right)^{1/p}} \leqq \frac{1}{p}\frac{\sum |\xi_k|^p}{\sum |\xi_k|^p} + \frac{1}{q}\frac{\sum |v_k|^q}{\sum |v_k|^q} = 1$$

which establishes (3), with n in place of ∞.

We remark that equality holds in (3) if and only if it holds in all the inequalities (4), and so if and only if one of the two vectors $(|v_k|^q)$ and $(|\xi_k|^p)$ is a scalar multiple of the other.

From Hölder's inequality we obtain Minkowski's inequality

$$(5)\quad \left(\sum_1^\infty |\xi_k+\eta_k|^p\right)^{1/p} \leq \left(\sum_1^\infty |\xi_k|^p\right)^{1/p} + \left(\sum_1^\infty |\eta_k|^p\right)^{1/p}, \qquad 1<p<\infty.$$

Proof. We have

$$\sum_1^n |\xi_k+\eta_k|^p \leqq \sum_1^n |\xi_k|\,|\xi_k+\eta_k|^{p-1} + \sum_1^n |\eta_k|\,|\xi_k+\eta_k|^{p-1}.$$

Applying (3) to each of the two summands on the right-hand side, and bearing in mind that $(p-1)q=p$, we obtain

$$\begin{aligned}\sum |\xi_k+\eta_k|^p &\leqq \left(\sum |\xi_k|^p\right)^{1/p} \left(\sum |\xi_k+\eta_k|^p\right)^{1/q} + \left(\sum |\eta_k|^p\right)^{1/p} \left(\sum |\xi_k+\eta_k|^p\right)^{1/q} \\ &= \left[\left(\sum |\xi_k|^p\right)^{1/p} + \left(\sum |\eta_k|^p\right)^{1/p}\right] \left(\sum |\xi_k+\eta_k|^p\right)^{1/q}.\end{aligned}$$

If we bring the last factor across to the left-hand side, then because $1-\frac{1}{q}=\frac{1}{p}$ the left hand side becomes $(\sum|\xi_k+\eta_k|^p)^{1/p}$, so that we obtain (5), with n in place of ∞.

Again, equality holds in (5) if and only if one of the vectors (ξ_k) and (η_k) is a non-negative scalar multiple of the other.

We denote by l^p, $1<p<\infty$, the collection of real or complex vectors $\mathfrak{x}=(\xi_1,\xi_2,\ldots)$ for which $\sum_{k=1}^{\infty}|\xi_k|^p$ converges. The expression

$$(6)\qquad \|\mathfrak{x}\|=\|\mathfrak{x}\|_p=\left(\sum_1^{\infty}|\xi_k|^p\right)^{1/p}$$

is used to define a norm on l^p. The norm properties are trivial except for $\|\mathfrak{x}+\mathfrak{y}\|\leqq\|\mathfrak{x}\|+\|\mathfrak{y}\|$, and this is Minkowski's inequality.

(7) l^p, $1<p<\infty$, *is a* (B)*-space.*

Proof. For a Cauchy sequence $\mathfrak{x}^{(n)}$, $\sum_1^{\infty}|\xi_k^{(n)}-\xi_k^{(m)}|^p\leqq\varepsilon^p$ holds for $n,m\geqq n_0$, and this implies the coordinatewise convergence of $\mathfrak{x}^{(n)}$ to some $\mathfrak{x}^{(0)}=(\xi_k^{(0)})$. It also implies that $\sum_1^{r}|\xi_k^{(n)}-\xi_k^{(0)}|^p\leqq\varepsilon^p$ for each r, so that $\sum_1^{\infty}|\xi_k^{(n)}-\xi_k^{(0)}|^p\leqq\varepsilon^p$. But this means that $\mathfrak{x}^{(n)}-\mathfrak{x}^{(0)}$ lies in l^p, and therefore that $\mathfrak{x}^{(0)}$ does too, and that $\mathfrak{x}^{(0)}$ is the limit of the sequence $\mathfrak{x}^{(n)}$.

(8) *Each* l^p, $1<p<\infty$, *is separable, and the dual of* l^p *is* l^q, *where* $\frac{1}{p}+\frac{1}{q}=1$.

Proof. The unit vectors $\mathfrak{e}_k$, $k=1,2,\ldots$, again form a fundamental set, and their rational linear combinations are dense in l^p, so that l^p is separable.

From the fact that the sections $\mathfrak{x}_n$ of $\mathfrak{x}$ converge to $\mathfrak{x}$ it follows once again that a continuous linear functional u can be represented by

$$u(\mathfrak{x})=\lim_{n\to\infty}\sum_1^{n}v_k\xi_k=\sum_1^{\infty}v_k\xi_k=\mathfrak{u}\,\mathfrak{x}\qquad\text{with}\qquad v_k=u(\mathfrak{e}_k).$$

Let $v_k=\varepsilon_k|v_k|$ for $|v_k|\neq 0$, let $\varepsilon_k=1$ for $v_k=0$, and put $\mathfrak{x}^{(n)}=\sum_1^{n}\varepsilon_k^{-1}|v_k|^{q-1}\mathfrak{e}_k$.

The norm of $\mathfrak{x}^{(n)}$ in l^p is equal to

$$\|\mathfrak{x}^{(n)}\|_p=\left(\sum_1^{n}|v_k|^{(q-1)p}\right)^{1/p}=\left(\sum_1^{n}|v_k|^q\right)^{1/p}.$$

Hence $|u(\mathfrak{x}^{(n)})| = \sum_1^n |v_k|^q \leqq \|u\| \left(\sum_1^n |v_k|^q\right)^{1/p}$, so that

$$\left(\sum_1^n |v_k|^q\right)^{1-\frac{1}{p}} = \left(\sum_1^n |v_k|^q\right)^{1/q} \leqq \|u\|,$$

i.e. $\mathfrak{u}$ lies in l^q and $\|\mathfrak{u}\|_q \leqq \|u\|$.

If, conversely, $\mathfrak{u} \in l^q$, then because of Hölder's inequality we have

$$|\mathfrak{u}\,\mathfrak{x}| \leqq \sum_1^\infty |v_k \xi_k| \leqq \|\mathfrak{u}\|_q \|\mathfrak{x}\|_p,$$

so that $\mathfrak{u}$ defines a continuous linear functional u on l^p, and $\|u\| \leqq \|\mathfrak{u}\|_q$.

l^2 is Hilbert space, which is therefore dual to itself. By (8), all the spaces l^p, $1<p<\infty$, are reflexive, in contrast to l^1 and l^∞.

(9) *If* $1 \leqq p_1 < p_2 \leqq \infty$, l^{p_1} *is a proper subspace of* l^{p_2}. *The topology on* l^{p_1} *defined by* l^{p_2} *is coarser than the norm topology of* l^{p_1}, *and* $\|\mathfrak{x}\|_{p_2} \leqq \|\mathfrak{x}\|_{p_1}$, *for each* $\mathfrak{x} \in l^{p_1}$.

Proof. If $\mathfrak{x}$ is an element of the unit sphere of l^{p_1}, so that $\sum_1^\infty |\xi_k|^{p_1} = 1$, then $\sum_1^\infty |\xi_k|^{p_2} \leqq 1$ for $p_1 < p_2 < \infty$, and $\sup_k |\xi_k| \leqq 1$ for $p_2 = \infty$. It follows that $\|\mathfrak{x}\|_{p_2} \leqq \|\mathfrak{x}\|_{p_1}$ for these $\mathfrak{x} \in l^{p_1}$. But this inequality also holds for all $\lambda\mathfrak{x}$, and it therefore holds for all $\mathfrak{x} \in l^{p_1}$.

If $p = \dfrac{p_2}{1+\varepsilon}$, and if $p_1 < p < p_2$ then clearly the sequence $\xi_k = k^{-\frac{1}{p}}$, $k = 1, 2, \ldots$, is an element of l^{p_2} which is not in l^{p_1}.

The spaces l^p $(1 \leqq p \leqq \infty)$ can also be defined for general index sets. If α runs through an index set A with cardinal d, then, for $p < \infty$, let $\mathfrak{x} = (\xi_\alpha)$ denote a vector with d coordinates, of which at most countably many are non-zero. Then (6) has a meaning when the sum is taken over all α; l_d^p is the normed space of all those $\mathfrak{x}$ for which $\|\mathfrak{x}\|_p$ is finite, and $\|\mathfrak{x}\|_p$ is again the norm. Once again l_d^p is a (B)-space with dual l_d^q, for $1 < p < \infty$. The proof we have given holds in this case as well. The dual of l_d^1 is equal to l_d^∞, which consists of all the vectors with $\|\mathfrak{x}\|_\infty = \sup_\alpha |\xi_\alpha| < \infty$. If d is uncountable, the spaces l_d^p are no longer separable. If d is finite, so that $d = n$, we obtain finite-dimensional normed spaces. The dual of l_n^∞ is l_n^1, as can be seen immediately; l_n^∞ is therefore reflexive.

9. (B)-spaces of continuous and holomorphic functions. Let K be a compact space. We have shown in § 6, 4. that there are "sufficiently many" continuous functions defined on the whole of K (both real- and complex-valued), and by § 6, 2.(7) $\sup_{x \in K} |f(x)|$ is always finite. We have

(1) *The vector space* $C(K)$ *of all real-(respectively complex-)valued continuous functions defined on the compact space* K *is a* (B)-*space when we give it the norm*

$$\|f\| = \sup_{x\in K} |f(x)|. \tag{2}$$

Proof. The linear operations are defined in the usual way: $(f_1+f_2)(x)=f_1(x)+f_2(x)$, $(\lambda f)(x)=\lambda(f(x))$. If f_n is a Cauchy sequence in $C(K)$, there is for each $\varepsilon>0$ an n_0 with $|f_n(x)-f_m(x)| \leqq \frac{\varepsilon}{3}$ for $n,m\geqq n_0$, and for all $x\in K$. The existence of a limit function $f_0(x)$ follows from this, and $|f_n(x)-f_0(x)| \leqq \frac{\varepsilon}{3}$ for $n\geqq n_0$ and for all $x\in K$; i.e. the f_n converge to $f_0(x)$ uniformly on K. Finally $f_0(x)$ is continuous, for there is a neighbourhood $U(x_0)$ in which $|f_n(x_0)-f_n(x)| \leqq \frac{\varepsilon}{3}$, so that

$$|f_0(x)-f_0(x_0)| \leqq |f_0(x)-f_n(x)| + |f_n(x)-f_n(x_0)| + |f_n(x_0)-f_0(x_0)| \leqq \varepsilon$$

for all $x\in U(x_0)$.

The dual of $C(K)$ is called the space $\mathfrak{M}(K)$ of measures on K. We shall study this in detail in the second volume. We shall consider the classical result of F. Riesz in the case where $K=I=[0,1]$ as an example, in § 17, 7.

$C(I)$ is separable. In the case where the functions are real-valued this follows directly from the Weierstrass approximation theorem, since any continuous function can clearly also be approximated uniformly by polynomials with rational coefficients. The complex-valued case is then trivial.

Besides these spaces, we shall have to consider spaces of holomorphic functions.

First let $\mathfrak{G}$ be a bounded open domain of arbitrary connectedness in the complex plane. Suppose that the closure $\bar{\mathfrak{G}}$ of $\mathfrak{G}$ consists of $\mathfrak{G}$ together with a system C of finitely or countably many boundary curves C_i. We consider the functions $f(z)$ defined on $\bar{\mathfrak{G}}$, which are holomorphic in $\mathfrak{G}$ and continuous on $\bar{\mathfrak{G}}$. They clearly form a vector space $HB(\mathfrak{G})$, and we have

(3) *If we introduce a norm on* $HB(\mathfrak{G})$ *by setting*

$$\|f\| = \sup_{z\in\bar{\mathfrak{G}}} |f(z)|, \tag{4}$$

$HB(\mathfrak{G})$ *becomes a* (B)-*space.*

For if f_n is a Cauchy sequence with respect to the norm (4), f_n is uniformly convergent to a continuous function f_0 on $\bar{\mathfrak{G}}$, which is also holomorphic throughout $\mathfrak{G}$, by a theorem of Weierstrass.

We shall also consider the case where $\mathfrak{G}$ consists of finitely many open domains $\mathfrak{G}_i$ in the complex sphere Ω, whose closures $\bar{\mathfrak{G}}_i$ are disjoint. $\mathfrak{G}$ must be different from Ω itself. A function $f(z)$ defined on $\mathfrak{G}$ is said to be locally holomorphic in $\mathfrak{G}$ if it is differentiable at each point $z \in \mathfrak{G}$, and if it vanishes at the point ∞, if ∞ belongs to $\mathfrak{G}$. Now let $HB(\mathfrak{G})$ be the space of functions defined on $\bar{\mathfrak{G}}$ which are locally holomorphic in $\mathfrak{G}$ and are continuous on $\bar{\mathfrak{G}}$, with the norm (4). (3) also holds for this space, and it can be seen without difficulty that it is the topological product of the $HB(\mathfrak{G}_i)$, with the norm (6a) of 4.

10. The L^p-spaces ($p \geqq 1$). If, for a continuous function $f(x)$ on $I=[a,b](-\infty<a<b<+\infty)$, we take the expression

$$\|f\|_p = \left[\int_a^b |f(t)|^p \, dt\right]^{1/p}, \quad 1 \leqq p < \infty, \tag{1}$$

as a norm, then the norm properties are satisfied, as we shall see presently; it is easily established however that a Cauchy sequence in this normed space of continuous functions need not have a limit. By 3., this normed space which we have just defined has a completion. We denote this completion by $L^p(I)$, or L^p for short. Rather than being satisfied with this abstract definition, however, we give a concrete representation of L^p, using the theory of Lebesgue measure and integration.

Here we assume that this theory is known, although general measure theory will be developed in the second volume, and this contains the classical theory as a special case. The elementary L^p-spaces considered here will also appear as special cases of general L^p-spaces.

A real- or complex-valued measurable function $f(t)$ defined on $I=[a,b]$ except perhaps on a set of measure zero, is called p-th power summable (or integrable) ($p \geqq 1$) if

$$\int_a^b |f(t)|^p \, dt < \infty. \tag{2}$$

For fixed $p \geqq 1$ let $L^{(p)}$ be the set of all measurable p-th power summable functions on I. For each $f \in L^{(p)}$, (1) takes a finite value. $\|f\|_p = 0$ if and only if f vanishes except on a set of measure zero.

(3) *$L^{(p)}$ is a vector space.*

Clearly λf is in $L^{(p)}$ if f is, and $\|\lambda f\|_p = |\lambda| \, \|f\|_p$. If a and b are arbitrary complex numbers and $p \geqq 1$,

$$|a+b|^p \leqq (2\max(|a|,|b|))^p \leqq 2^p(|a|^p + |b|^p). \tag{4}$$

It follows from this that $\int |f+g|^p dt \leqq 2^p(\int |f|^p dt + \int |g|^p dt)$; i.e. $f+g$ is in $L^{(p)}$ if f and g are.

We again prove Hölder's inequality:

(5) *If* $f(t) \in L^{(p)}$, $g(t) \in L^{(q)}$, $\frac{1}{p} + \frac{1}{q} = 1, p>1$, *then* $f(t)\,g(t)$ *belongs to to* $L^{(1)}$, *and*

(6) $\int\limits_a^b |f(t)g(t)|\,dt \leqq \left(\int\limits_a^b |f(t)|^p dt\right)^{1/p} \cdot \left(\int\limits_a^b |g(t)|^q dt\right)^{1/q} = \|f\|_p \|g\|_q.$

Proof. If $\|f\|_p = 0$ or $\|g\|_q = 0$, then the left hand side is also zero, so that the inequality holds. Otherwise we can put $c = \frac{|f|}{\|f\|_p}$ and $d = \frac{|g|}{\|g\|_q}$ in 8.(4), to obtain

$$\frac{|fg|}{\|f\|_p \|g\|_q} \leqq \frac{1}{p}\,\frac{|f|^p}{\|f\|_p^p} + \frac{1}{q}\,\frac{|g|^q}{\|g\|_q^q}. \tag{7}$$

fg is measurable, and the absolute value of fg is majorized by an integrable function, by (7), so that fg is integrable; i.e. fg lies in $L^{(1)}$. Integration of (7) gives (6).

(8) *Minkowski's inequality. If* $f, g \in L^{(p)}$, $p \geqq 1$, *then*

$$\|f+g\|_p \leqq \|f\|_p + \|g\|_p. \tag{9}$$

For p=1 the assertion is trivial. For $p>1$ it again follows from Hölder's inequality (which can be applied, since $f+g$ lies in $L^{(p)}$ by (3), so that $|f+g|^{p-1}$ lies in $L^{(q)}$, as $(p-1)q = p$):

$$\begin{aligned} \int |f+g|^p dt &\leqq \int |f|\,|f+g|^{p-1} dt + \int |g|\,|f+g|^{p-1} dt \\ &\leqq (\|f\|_p + \|g\|_p)(\int |f+g|^p dt)^{1/q}. \end{aligned}$$

Again we observe that for $p>1$ equality holds in (6) if and only if one of the functions $|f|^p$ and $|g|^q$ is a scalar multiple of the other, almost everywhere, while equality holds in (9) if and only if one the functions f and g is a non-negative scalar multiple of the other, almost everywhere.

Thus we have shown that in $L^{(p)}$ the expression (1) satisfies all the properties of a norm with the exception of (N2). Thus (1) only defines a semi-norm on $L^{(p)}$. The null-space of the semi-norm is the space of all those functions which vanish almost everywhere.

By 1.(4), $L^{(p)}/N$ *is a normed space, under the norm* (1).

It is customary to make no distinction between $f \in L^{(p)}$ and the equivalence class in $L^{(p)}/N$ determined by f. It is readily confirmed that all the results for functions in $L^{(p)}$ contained in this number also hold for the equivalence classes in $L^{(p)}/N$, since these consist of functions which are equal almost everywhere.

(10) *$L^{(p)}/N$ is a* (B)*-space.*

Proof. Let f_n be a Cauchy sequence in $L^{(p)}/N$. We say that the functions f_n form a Cauchy sequence in the p-th mean. Then for each ν there is an $n_\nu > n_{\nu-1}$ for which $\|f_m - f_n\|_p^p < \frac{1}{3^\nu}$ holds, for $m, n \geqq n_\nu$. In particular,

$$\|f_{n_{\nu+1}} - f_{n_\nu}\|_p^p < \frac{1}{3^\nu}. \tag{11}$$

If now M_ν is the set of points for which $|f_{n_{\nu+1}}(t) - f_{n_\nu}(t)| > 2^{-\nu/p}$, then the measure $m(M_\nu) < (\frac{2}{3})^\nu$, for it follows from (11) that

$$2^{-\nu} m(M_\nu) \leqq \int_{M_\nu} |f_{n_{\nu+1}} - f_{n_\nu}|^p \, dt < \frac{1}{3^\nu}.$$

If t does not lie in $\bigcup\limits_{N+1}^{\infty} M_\nu$, then $|f_{n_{\nu+1}}(t) - f_{n_\nu}(t)| \leqq 2^{-\nu/p}$ for $\nu \geqq N+1$, so that $\sum\limits_{\nu=1}^{\infty} |f_{n_{\nu+1}}(t) - f_{n_\nu}(t)| < \infty$ for these t; $\sum\limits_{\nu=1}^{\infty} (f_{n_{\nu+1}}(t) - f_{n_\nu}(t))$ therefore converges for all t which do not lie in $M = \bigcap\limits_{N=1}^{\infty} \bigcup\limits_{N+1}^{\infty} M_\nu$. The set M has measure zero, and $\lim\limits_{\nu \to \infty} f_{n_\nu}(t) = f(t)$ throughout $I \sim M$. On M we put $f(t) = 0$.

We now show that $f(t) \in L^{(p)}$, and that f_n converges to f in p-th mean. $\int_a^b |f_{n_\nu}(t) - f_n(t)|^p \, dt \leqq \varepsilon$, for sufficiently large n_ν and n. The functions $|f_{n_\nu}(t) - f_n(t)|^p$ are non-negative, and as $n_\nu \to \infty$ they converge almost everywhere to $|f(t) - f_n(t)|^p$. Applying FATOU's lemma, $|f(t) - f_n(t)|^p$ is integrable, and

$$\int_a^b |f(t) - f_n(t)|^p \, dt \leqq \varliminf_\nu \int_a^b |f_{n_\nu}(t) - f_n(t)|^p \, dt \leqq \varepsilon.$$

Thus $f - f_n$, and consequently f, lie in $L^{(p)}$, and $\|f - f_n\|_p \to 0$. We now establish

(12) *$L^{(p)}/N$ is identical with L^p.*

The space of continuous functions on $I = [a,b]$ can be considered as a subspace of $L^{(p)}/N$, for two distinct continuous functions never differ by a function in N, so that the embedding into $L^{(p)}/N$ is one-one.

If we can show that the continuous functions are dense in $L^{(p)}/N$, it will follow from (10) and the uniqueness of the completion that $L^{(p)}/N$ and L^p are identical.

Since every real valued function in $L^{(p)}$ can be expressed as the difference of two non-negative functions in $L^{(p)}$, and since every complex-valued function can be expressed as $f_1-f_2+i(f_3-f_4)$, $f_i \geqq 0$, it is enough to show that a non-negative f can be approximated in p-th mean by a continuous function.

As is shown in the theory of real functions, given $\varepsilon>0$, $\delta>0$ and a non-negative measurable function $f(t)$ on I, a continuous function $\varphi(t)$ can be found for which $|f(t)-\varphi(t)|<\delta$ for all t outside some set of measure $<\varepsilon$. If $f(t)\leqq K$, $\varphi(t)$ can also be chosen with $\varphi(t)\leqq K$.

If now $0\leqq f(t)\leqq K$ then $\int_a^b |f(t)-\varphi(t)|^p\,dt \leqq \delta^p(b-a)+K^p\varepsilon$, so that f can be approximated by continuous functions.

If $f(t)$ is unbounded, we introduce the functions $f_n(t)$, which are defined by $f_n(t)=f(t)$ if $f(t)\leqq n$, and $f_n(t)=0$ for $f(t)>n$. If M_n is the set on which f_n vanishes, then $\|f-f_n\| = \left(\int_{M_n} |f(t)|^p\,dt\right)^{1/p} < \frac{\varepsilon}{2}$ for sufficiently large n. Since a continuous function φ can be chosen for which $\|f_n-\varphi\| < \frac{\varepsilon}{2}$, the approximation property also holds for unbounded $f\in L^{(p)}/N$.

(13) L^p, $p\geqq 1$, *is separable.*

For it follows directly from the Weierstrass approximation theorem that $C[I]$ is separable under the norm (1); so therefore is its completion L^p.

We remark without proof that l^2 and L^2 are norm isomorphic. This is essentially the RIESZ-FISCHER theorem.

Analogous to the results about the l^p spaces, we have

(14) *The dual space of* L^p *is* L^q, *where* $\frac{1}{p}+\frac{1}{q}=1$ *and* $1<p,q<\infty$.

We shall give the proof later (§ 26, 7.), using more general ideas.

11. The space L^∞. A measurable function $f(t)$ defined on $I=[a,b]$, except perhaps on a set of measure zero, is said to be essentially bounded if there is an $M\geqq 0$ for which $|f(t)|\leqq M$ for all t outside a set of measure zero. By the essential supremum of such a function, in symbols $\operatorname{ess\,sup}|f(t)|$, we mean the greatest lower bound M_0 of all the essential upper bounds M. M_0 is itself an essential upper bound, for if $|f(t)| \leqq M_i$, $i=1,2,\ldots$, $M_1 > M_2 > \cdots \to M_0$, and if N_i are the sets on which $|f(t)| > M_i$, then $N_0 = \bigcup_{i=1}^{\infty} N_i$ is again of measure zero.

We denote by $L^\infty(I)$ the vector space of all essentially bounded measurable functions on I, where again we identify functions which are equal almost everywhere.

We introduce a norm on $L^\infty(I)$ by putting

(1) $$\|f\|_\infty = \operatorname*{ess\,sup}_{t \in I} |f(t)|.$$

For this we have

(2) *L^∞ is a* (B)*-space under the norm* (1).

The norm properties are trivially satisfied. Let f_n be a Cauchy sequence. The inequality $\|f_n - f_m\|_\infty \leqq \frac{1}{k}$ means that $|f_n(t) - f_m(t)| \leqq \frac{1}{k}$ for all t outside some set $N_{n,m}^{(k)}$ of measure zero. The sequence $f_n(t)$ is therefore uniformly convergent for all t outside the set $N_0 = \bigcup_{m,n,k=1}^{\infty} N_{n,m}^{(k)}$ of measure zero. The limit function $f_0(t)$ defined on $I \sim N_0$ is measurable and bounded, and so lies in L^∞. It is clearly the limit of the sequence f_n.

The (B)-space $C(I)$ with the norm defined in 9.(2) is a proper closed linear subspace of L^∞, and so $C(I)$ is not dense in L^∞.

(3) *L^∞ is not separable.*

If we divide $I = [0,1]$ into a sequence of intervals $I_n = \left[\frac{1}{n+1}, \frac{1}{n}\right)$, and consider functions $f(t)$ which take a fixed value $+1$ or -1 on each I_n, then we obtain uncountably many functions on L^∞ any two of which are distance 2 apart.

The inclusion relations between the various L^p spaces are the exact opposite of the relations between the l^p spaces.

(4) *If $1 \leqq p_1 < p_2 \leqq \infty$, L^{p_2} is a proper subspace of L^{p_1}. The topology induced on L^{p_2} by L^{p_1} is coarser than the norm topology of L^{p_2}, and if $f \in L^{p_2}$ we always have*

(5) $$\|f\|_{p_1} \leqq \|f\|_{p_2} (b-a)^{1/p_1 - 1/p_2}, \qquad 1 \leqq p_1 < p_2 \leqq \infty.$$

If $f \in L^\infty$, $\|f\|_\infty = \lim_{p \to \infty} \|f\|_p$.

Proof. Let $f \in L^{p_2}$, $p_2 < \infty$. For those t for which $|f(t)| \geqq 1$ we have $|f(t)|^{p_1} \leqq |f(t)|^{p_2}$, so that $|f(t)|^{p_1}$ is majorized by the integrable function $g(x) = \max\{1, |f(t)|^{p_2}\}$; thus $|f(t)|^{p_1}$ is itself integrable and $f \in L^{p_1}$. If $p_2 = \infty$, it is trivial that $L^\infty \subset L^{p_1}$.

If $p_2=\infty$, (5) is also trivial. Let $p_2<\infty$. We apply Hölder's inequality to $|f|^{p_1}\in L^{p_2/p_1}$, $g\equiv 1$ and the exponent $p=p_2/p_1$ to obtain

$$\int_a^b |f|^{p_1}\,dt \leqq \left(\int_a^b |f|^{p_2}\,dt\right)^{p_1/p_2}(b-a)^{\frac{p_2-p_1}{p_2}}$$

from which (5) follows.

Finally let $\|f\|_\infty=M>0$. Then given $\varepsilon>0$, there is a set N of positive measure $\mu>0$ on which $|f(t)|\geqq M-\varepsilon$. Hence $\|f\|_p\geqq[\mu(M-\varepsilon)^p]^{1/p}$ $=\mu^{1/p}(M-\varepsilon)$, so that $\varliminf_p \|f\|_p\geqq M$; it follows from (5) that $\|f\|_\infty$ $=\lim_{p\to\infty}\|f\|_p$.

If we take the underlying interval to be $I=(-\infty,\infty)$, then the definition of $L^p(-\infty,\infty)$ must be made differently. In this case we start from the space $C_\infty(-\infty,\infty)$ of all continuous functions on I which vanish outside a finite interval. For these functions

$$(6)\qquad \|f\|_p=\left(\int_{-\infty}^{+\infty}|f(t)^p|\,dt\right)^{1/p}$$

is finite, and we define $L^p(-\infty,\infty)$, $1\leqq p<\infty$, to be the completion of the space C_∞ with the norm (6). On the other hand the definition of $L^{(p)}$ and L^∞ is again the same as for a finite interval $[a,b]$, and the preceding results and proofs remain unaltered, up to the following point:

The assertion of 10.(12) still holds, so that $L^{(p)}/N$ is equal to L^p, but the proof must be changed. We show that C_∞ is dense in $L^{(p)}/N$. Because of the convergence of (6), every $f\in L^{(p)}$ is approximable in p-th mean by an $f\in L^{(p)}$ which vanishes outside some finite interval. Using the proof of 10.(12), this f is approximable by a continuous function vanishing outside the same interval, so that C_∞ is dense in $L^{(p)}/N$; i.e. $L^{(p)}/N=L^p$.

The separability of L^p again follows from the separability of C_∞.

The space of all bounded continuous functions on $(-\infty,\infty)$ is a proper closed linear subspace of L^∞. Theorem (4) of this number is false for $L^p(-\infty,\infty)$; indeed, L^∞ is not contained in any of the spaces L^p.

(7) *The dual space of L^1 is L^∞.*

This theorem will also not be proved until later (cf. § 26, 7.).

§ 15. Topological vector spaces

1. Definition of a topological vector space. In this paragraph we consider vector spaces over the field K of real or complex numbers; this is given its usual topology by means of the modulus.

The ideas are closely related to those of § 10.2., where of course we were concerned with an arbitrary field, with the discrete topology.

A vector space E over K is called a **topological vector space** or a **topological linear space** $E[\mathfrak{T}]$ if a Hausdorff topology $\mathfrak{T}$ is defined on E which is compatible with the vector space structure. Compatibility once more means that the mappings $(x,y)\to x+y$ from $E\times E$ into E and $(\alpha,x)\to\alpha x$ from $\mathsf{K}\times E$ into E are continuous. The concept goes back to A. KOLMOGOROFF [1] and J. VON NEUMANN [1].

The topological isomorphism of two topological vector spaces $E_1[\mathfrak{T}_1]$ and $E_2[\mathfrak{T}_2]$, in symbols $E[\mathfrak{T}_1]\cong E_2[\mathfrak{T}_2]$, is defined as in § 10, 2. as a vector space isomorphism which at the same time is a homeomorphism.

By § 14, 1.(3), a normed space is an example of a topological vector space. Giving a norm, however, provides a richer structure than that provided by the topology which it induces; using § 14, 2. (3) we see: two normed spaces can very well be topologically isomorphic as topological vector spaces without being norm isomorphic.

If follows directly from the definition of topological vector space that $\sum\limits_1^m \alpha_k^{(n)} x_k^{(n)}$ converges to $\sum\limits_1^m \alpha_k x_k$ with respect to $\mathfrak{T}$ if each of the sequences $x_k^{(n)}$, $n=1,2,\ldots$, converges to x_k with respect to $\mathfrak{T}$ and each sequence $\alpha_k^{(n)}\to\alpha_k$ in K.

(1) *The mapping* $x\to x+x_0$ *is a homeomorphism of* $E[\mathfrak{T}]$ *onto itself. The mapping* $x\to\alpha x$, $\alpha\neq 0$, *is a topological automorphism of* $E[\mathfrak{T}]$.

For these mappings are continuous, and linear and continuous, respectively, and have the whole of $E[\mathfrak{T}]$ as image space; the inverse mappings $x\to x-x_0$ and $x\to\dfrac{1}{\alpha}x$ exist and have the same properties.

If $\mathfrak{U}=\{U\}$ is a base of neighbourhoods of $\mathfrak{o}$, the sets x_0+U therefore form a base for the neighbourhood filter of x_0. Further, if U is a neighbourhood of $\mathfrak{o}$, so is αU, for $\alpha\neq 0$.

(2) *If* $E[\mathfrak{T}]$ *is a topological vector space over* K *and* $\mathfrak{U}=\{U\}$ *is a base of neighbourhoods of* $\mathfrak{o}$, *then we have*

(LT 1) *For each* $U\in\mathfrak{U}$ *there is a* $V\in\mathfrak{U}$ *with* $V+V\subset U$.

(LT 2) *For each* $U\in\mathfrak{U}$ *there is a* $V\in\mathfrak{U}$ *for which* $\alpha V\subset U$ *for all* α *with* $|\alpha|\leqq 1$.

(LT 3) *For each* $U\in\mathfrak{U}$ *and each* $x\in E$ *there is a positive integer* $n(x,U)$ *for which* $x\in nU$.

A subset M of E is said to be **absorbent** if a suitable multiple ρx, $\rho>0$, of each element x of E lies in M; (LT 3) says that every neighbourhood of $\mathfrak{o}$ is absorbent.

Proof of (2). (LT1) is nothing but the continuity of $(x,y)\to(x+y)$ at (o,o). It follows from the continuity of αx at $(0,o)$ that there is an $\varepsilon>0$ and a neighbourhood W of o such that $\xi x\in U$ for all $x\in W$ and all $|\xi|\leqq\varepsilon$, so that (LT2) is satisfied with $V=\varepsilon W$.

If U were not absorbent, there would be an $x_0\in E$ which would not lie in any nU. Thus $\frac{1}{n}x_0\notin U$ would hold, which contradicts the convergence of $\frac{1}{n}x_0$ to o.

A set $M\subset E$ is said to be circled if αx_0 belongs to M whenever $|\alpha|\leqq 1$ and $x_0\in M$. The circled cover of a set M consists of all αx, $x\in M$, $|\alpha|\leqq 1$.

(3) *A topological vector space E always has a base of neighbourhoods of o consisting of circled neighbourhoods.*

For the sets $\bigcup_{|\alpha|\leqq 1}\alpha U$, $U\in\mathfrak{U}$, form a base of neighbourhoods of o, by (LT2).

2. A second definition. We now establish the converse of Theorem (2) of the preceding number.

(1) *Let* $\mathfrak{U}=\{U\}$ *be a filter base on a real or complex vector space E, with* $\bigcap U=o$, *which satisfies conditions* (LT1) *to* (LT3). *If a topology* $\mathfrak{T}$ *is defined on E by taking as neighbourhoods the sets* $U(x)=x+U$, $U\in\mathfrak{U}$, *then* $E[\mathfrak{T}]$ *is a topological vector space, with* $\mathfrak{U}$ *as base of neighbourhoods of* o.

By 1.(2), every topological vector space can be defined in this way, for by the remark after 1.(1) the topology on E is always determined by a base of neighbourhoods of o.

The following definition is even simpler:

(2) *Suppose that* $\mathfrak{U}=\{U\}$ *is a filter base of absorbent circled sets U on a real or complex vector space E; suppose that* $\bigcap U=o$, *and that for each U there exists a* $V\in\mathfrak{U}$ *with* $V+V\subset U$. *If we define a topology* $\mathfrak{T}$ *on E by taking as neighbourhoods the set* $U(x)=x+U$, $U\in\mathfrak{U}$, *then* $E[\mathfrak{T}]$ *is a topological vector space, with* $\mathfrak{U}$ *as base of neighbourhoods of* o.

Once again, every topological vector space can be defined in this way, by 1.(2) and 1.(3). It is sufficient to establish (2), for, given a filter base $\mathfrak{U}$ which satisfies the hypotheses of (1), we obtain an equivalent filter base satisfying the hypotheses of (2), just as in the proof of 1.(3).

Proof of (2). Let $\mathfrak{N}$ be the filter generated by $\mathfrak{U}$, then $\mathfrak{N}(x)=x+\mathfrak{N}$ satisfies the properties (N1) to (N4) of § 1, 2. and it follows from § 1, 2. (1) that we have defined a topology $\mathfrak{T}$ on E with $\mathfrak{U}$ as base of neighbourhoods of o. $\mathfrak{T}$ is Hausdorff: suppose that $x\neq y$. Since $\bigcap U=o$ there is a

U with $x-y \notin U$. If $V+V \subset U$, with V circled, then $x+V$ and $y+V$ are disjoint. For if $x+z_1=y+z_2$, with $z_1, z_2 \in V$, then $x-y=z_2-z_1 \in V + V \subset U$, contrary to our assumption.

The continuity of the mapping $(x,y) \to x+y$ at (x_0, y_0) follows from $(x_0+V)+(y_0+V) \subset x_0+y_0+U$.

The continuity of the mapping $(\alpha, x) \to \alpha x$ at (α_0, x_0) is shown as follows: suppose that we are given a neighbourhood $\alpha_0 x_0 + U$, with U circled, and suppose that $|\alpha_0| \leqq n$. We can then find a circled V such that $V + \cdots + V$ ($n+2$ summands) lies in U (this must be distinguished from $(n+2)V$, which only contains the elements $(n+2)x$, $x \in V$). *A fortiori*, $nV+V+V \subset U$. Further, let us choose a positive integer m such that $x_0 \in mV$.

If now $|\alpha-\alpha_0| \leqq \frac{1}{m}$, $x \in x_0+V$, the relation $\alpha x \in \alpha_0 x_0 + V + V + nV \subset \alpha_0 x_0 + U$ follows from

$$\alpha x = \alpha_0 x_0 + (\alpha-\alpha_0) x_0 + (\alpha-\alpha_0)(x-x_0) + \alpha_0 (x-x_0)$$

since $\frac{1}{m}(mV)=V$, $(\alpha-\alpha_0)(x-x_0) \in \frac{1}{m} V \subset V$ and $\alpha_0(x-x_0) \in nV$. Hence αx is continuous.

(3) *Every topological vector space $E[\mathfrak{T}]$ is uniformizable, and is therefore regular (and indeed, by* § 6, 8.(1), *is completely regular).*

The uniformity is uniquely determined, if we require it to have a base of translation-invariant vicinities.

Proof. We call a vicinity $N \subset E \times E$ translation-invariant if $(x+z, y+z)$ lies in N whenever (x,y) does, for arbitrary z in E.

The uniformity on $E[\mathfrak{T}]$ is defined by the vicinities N_U consisting of all pairs $(x,y) \in E \times E$ with $y-x \in U$, $U \in \mathfrak{U}$. We show that the N_U form the base of a Hausdorff uniformity.

The N_U form a filter base on $E \times E$ which clearly satisfies (V 1) of § 5, 1., and also satisfies (V 4) of § 5, No. 2, since $\bigcap U = o$. For circled U, $U=-U$, so that N_U is symmetric, i.e. $N_U = N_U^{-1}$, and (V 2′) is satisfied. Finally it follows from the existence of a circled V with $V+V \subset U$ that $N_V^2 \subset N_U$, so that (V 3) is satisfied. The topology $\mathfrak{T}$ is evidently the topology defined by this uniform structure.

Finally, suppose that some translation-invariant uniformity is given on E for which $\mathfrak{T}$ is the corresponding topology. If U is a neighbourhood of o which is obtained from a translation-invariant vicinity N, so that U is the set of all y with $(o, y) \in N$, then $x+U$ is the set of all z with $(x,z) \in N$, i.e. N is equal to the set N_U defined above, and the uniformity is uniquely determined.

Theorem (3) allows us to apply the results of § 5 about uniform spaces to topological vector spaces.

(4) *If a linear mapping A of $E[\mathfrak{T}_1]$ into $F[\mathfrak{T}_2]$ is continuous at* $\mathfrak{o}$, *it is continuous everywhere, and indeed is uniformly continuous.*

For the proof, cf. § 10, 2.(9).

(5) *The mapping $(x,y)\to x+y$ from $E[\mathfrak{T}]\times E[\mathfrak{T}]$ into $E[\mathfrak{T}]$ is uniformly continuous.*

Proof. Let $V+V\subset U$, and let N_V be the vicinity of all (x,x') with $x'-x\in V$. Then the image of the vicinity $N_V\times N_V$ lies in N_U, since the differences $(x'+y')-(x+y)$ of the $(x,x',y,y')\in N_V\times N_V$ lies in U.

Trivially,

(6) *Every linear subspace H of a topological vector space is again a topological vector space, under the induced topology.*

(7) *The closure $\bar{H}$ of a linear subspace H is again a linear subspace.*

Let x_0 and y_0 be closure points of H and let U be a neighbourhood of $\mathfrak{o}$. There exists $V+V\subset U$, and so if $x,y\in H$ and $x\in x_0+V$, $y\in y_0+V$, then $x+y\in x_0+y_0+U$. Hence x_0+y_0 is a closure point of the $x+y\in H$. Further if x_0 is a closure point of the points $x\in H$, λx_0 is a closure point of the points λx.

3. The completion. Every normed space can be completed to give a (B)-space (§ 14, 3.(1)); in the same way the following result holds for arbitrary topological vector spaces:

(1) *Every topological vector space $E[\mathfrak{T}]$ over K can be embedded in a smallest complete topological vector space $\tilde{E}[\tilde{\mathfrak{T}}]$, the* completion *of $E[\mathfrak{T}]$. This is unique up to topological isomorphism.*

The closures in $\tilde{E}[\tilde{\mathfrak{T}}]$ of the neighbourhoods of a base of neighbourhoods of $\mathfrak{o}$ in $E[\mathfrak{T}]$ form a base of neighbourhoods of $\mathfrak{o}$ in $\tilde{E}[\tilde{\mathfrak{T}}]$. The closure of a circled neighbourhood is again circled.

Proof. $E[\mathfrak{T}]$ is a uniform space, by 2.(3). By § 5, 5.(2), there is a smallest completion $\tilde{E}[\tilde{\mathfrak{T}}]$, uniquely determined up to uniform space isomorphism, whose points consist of equivalence classes of Cauchy nets. $\tilde{E}[\tilde{\mathfrak{T}}]$ is again Hausdorff.

By 2.(5), addition $x+y$ is a uniformly continuous mapping from $E[\mathfrak{T}]\times E[\mathfrak{T}]$ into $E[\mathfrak{T}]$; it is also uniformly continuous as a mapping from $E[\mathfrak{T}]\times E[\mathfrak{T}]$ into $\tilde{E}[\tilde{\mathfrak{T}}]$. By § 5, 4.(4) and § 5, 7. it can be extended uniquely to a uniformly continuous mapping from $\tilde{E}[\tilde{\mathfrak{T}}]\times\tilde{E}[\tilde{\mathfrak{T}}]$. By 2.(4) αx is also uniformly continuous on $E[\mathfrak{T}]$, for fixed α; scalar multiplication can therefore also be extended uniquely, in a uniformly continuous way, to the whole of $\tilde{E}[\tilde{\mathfrak{T}}]$. Axioms (L 1) and (L 2) for a vector space now follow for $\tilde{E}$ by taking limits in the axioms for E.

If $\{N_U\}$ is the vicinity basis derived from a base $\{U\}$ of neighbourhoods of $\circ$ of $E[\mathfrak{T}]$, then by § 5, 5.(4) the closures $\bar{N}_U$ in $\tilde{E}\times\tilde{E}$ form a base of vicinities in $\tilde{E}[\tilde{\mathfrak{T}}]$. We shall show that $\bar{N}_U = N_{\bar{U}}$, where $\bar{U}$ is the closure of U in $\tilde{E}[\tilde{\mathfrak{T}}]$. A point $(\tilde{x},\tilde{y})\in\bar{N}_U$ is a closure point of pairs $(x,y)\in N_U$. But it follows from $y-x\in U$ that $\tilde{y}-\tilde{x}\in\bar{U}$, so that $\bar{N}_U\subset N_{\bar{U}}$. Conversely if $(\tilde{x},\tilde{y})\in N_{\bar{U}}$, so that $\tilde{z}=\tilde{y}-\tilde{x}\in\bar{U}$, then $\tilde{z}$ is a closure point of points $z\in U$ and $\tilde{x}$ is a closure point of points $x\in E$; hence $(\tilde{x},\tilde{y})=(\tilde{x},\tilde{x}+\tilde{z})$ is a closure point of elements $(x,x+z)\in N_U$, implying that $N_{\bar{U}}\subset\bar{N}_{\bar{U}}$. The closures $\bar{U}$ in $\tilde{E}[\tilde{\mathfrak{T}}]$ of the neighbourhoods U therefore form a base of neighbourhoods of $\circ$ in $\tilde{E}[\tilde{\mathfrak{T}}]$.

If the sets U are circled, so also are the sets $\bar{U}$, for from $\alpha U\subset U(|\alpha|\leqq 1)$ it follows by continuity that $\alpha\bar{U}\subset\bar{U}$. From $V+V\subset U$ and the continuity of $(x,y)\to x+y$ it follows that $\bar{V}+\bar{V}\subset\bar{U}$. Finally every $\bar{U}$ is absorbent, as well: given $x_0\in E$ there exists $x\in E$ with $x_0-x\in\bar{V}$; if $x\in nV$, then $x_0=(x_0-x)+x\in\bar{V}+n\bar{V}\subset n\bar{U}$. Applying 2.(2), $\tilde{E}[\tilde{\mathfrak{T}}]$ is therefore a topological vector space.

Two smallest completions of $E[\mathfrak{T}]$ are isomorphic as uniform spaces; because of the continuity of the vector space operations, sums and scalar multiples correspond, so that the uniform space isomorphism is a topological isomorphism of topological vector spaces. This completes the proof of (1).

Remark. The completion of a linearly topologized space (cf. § 13, 3.) can be constructed in exactly the same way. In this case U is a linear subspace; we conclude from this that $\bar{U}$ is also a linear subspace and, using § 10, 2.(1), obtain the result corresponding to (1).

4. Quotient spaces and topological products. As for normed spaces and linearly topologized spaces we define the induced topology $\mathfrak{T}$ on a quotient space E/H of a topological vector space $E[\mathfrak{T}]$ by taking the images $K(O)$ of the open sets O of E as open sets (where K is the canonical mapping of E onto E/H).

(1) *If H is a closed linear subspace of the topology vector space $E[\mathfrak{T}]$, E/H is a topological vector space under the induced topology.*

A similar argument to that of § 10, 7. shows that the sets $K(O)$ determine a Hausdorff topology on E/H. If $U\subset E$ is absorbent and circled, $K(U)$ is also absorbent and circled, and it follows from $V+V\subset U$ that $K(V)+K(V)\subset K(U)$. If, therefore, we start from a base of neighbourhoods of $\circ$ in E which satisfies the requirements of 2.(2), we obtain a base for the induced topology on E/H with the same properties, using the canonical mapping. The assertion now follows from 2.(2).

We remark that E/H need not be complete if E is. For counterexamples see § 23, 5. and § 31, 6.

The concepts of topological homomorphism, topological monomorphism and topological isomorphism are defined as in § 10, 7., and the proofs given there establish the three following results:

(2) *The canonical mapping* K *of* $E[\mathfrak{T}]$ *onto* E/H, *with* H *closed, is a topological homomorphism.*

(3) *A continuous linear mapping* A *of* $E[\mathfrak{T}_1]$ *into* $F[\mathfrak{T}_2]$ *always has a closed null space* $N[A]$. A *is the product of the canonical homomorphism* K *of* E *onto* $E/N[A]$, *a continuous one-one linear mapping* $\check{A}$ *of* $E/N[A]$ *onto* $A(E)$ *and the embedding* J *of* $A(E)$ *into* F.

(4) *Every topological homomorphism of* $E[\mathfrak{T}_1]$ *into* $F[\mathfrak{T}_2]$ *is the product of the canonical homomorphism* K *of* E *onto* $E/N[A]$, *and a topological monomorphism* $\hat{A}$ *of* $E/N[A]$ *into* F. *The topological monomorphism* $\hat{A}$ *is the product of a topological isomorphism* $\check{A}$ *of* $E/N[A]$ *onto* $A(E)$, *and the embedding* J *of* $A(E)$ *into* F, *which is a topological monomorphism.*

Just as in § 10, 7. if $F \subset G$ are two closed linear subspaces of $E[\mathfrak{T}]$ we have a topological isomorphism $(E/F)/(G/F) \cong E/G$. In the converse direction we have

(5) *A linear mapping* A *of* $E[\mathfrak{T}_1]$ *into* $F[\mathfrak{T}_2]$ *with a closed nullspace is continuous (respectively a topological homomorphism) if* $\check{A}$ *is continuous (a topological isomorphism).*

In § 14, 4. we could only show that the topological product of finitely many normed spaces is again normable. Now we have, more generally,

(6) *The topological product* $E[\mathfrak{T}] = \prod_{\alpha} E_\alpha[\mathfrak{T}_\alpha]$ *of arbitrarily many topological vector spaces* $E_\alpha[\mathfrak{T}_\alpha]$ *is again a topological vector space.*

To prove this, we use § 1, 8. and § 7, 8. It is not difficult to verify that the conditions of 2.(2) are satisfied by the base of neighbourhoods $U = \prod_{\alpha} W^\alpha$, where finitely many W^α are circled neighbourhoods U^α of $\mathfrak{o}$ in E_α, and all the other $W^\alpha = E_\alpha$.

(7) *The topological product of infinitely many normed spaces* $\neq (\mathfrak{o})$ *is not normable.*

For a neighbourhood $\|x\| < 1$ must certainly contain a neighbourhood $\prod_{\alpha} W^\alpha$. Now let x be an element all of whose components $x_\alpha = \mathfrak{o}$, except for an x_β which lies in a W^β which is equal to E_β. But then all scalar multiples of x must also lie in $\prod_{\alpha} W^\alpha$; since $\|\lambda x\| = |\lambda| \, \|x\| < 1$ for all $\lambda \in \mathsf{K}$, we obtain $\|x\| = 0$, which is a contradiction.

We have already shown, in § 5, 7., that the completion of a topological product is equal to the topological product of the completions, so that $E[\mathfrak{T}] = \prod_{\alpha} E_\alpha[\mathfrak{T}_\alpha]$ *is complete if the* $E_\alpha[\mathfrak{T}_\alpha]$ *are.*

As in § 7, 8., we denote the topological product of d copies of the field K by $\omega_d(\mathsf{K})$. If $d=\aleph_0$, we simply write $\omega(\mathsf{K})$. Since we no longer take the discrete topology on the field of real or complex numbers, but take the natural topology instead, we obtain a topology on ω_d which is different from the linear topology.

By (7), ω_d is only normable if d is finite.

5. Finite dimensional topological vector spaces. The n-dimensional real or complex space K^n is a topological vector space under its natural topology. The fact that this is the only topology on K^n which is compatible with the vector space operations was established by TYCHONOFF [1]. Thus we have

(1) *Every n-dimensional topological vector space* $E[\mathfrak{T}]$ *over* K *is topologically isomorphic to* K^n *with its natural topology.*

Proof. a) We can identity the vector space E with K^n. Let U be a circled $\mathfrak{T}$-neighbourhood of $\mathfrak{o}$, and let V be a circled $\mathfrak{T}$-neighbourhood of $\mathfrak{o}$ for which the n-fold sum $V+\cdots+V$ is contained in U. Since V is absorbent, there is a $k>0$ such that all the $k\mathfrak{e}_i$, $i=1,\dots,n$, lie in V, where the $\mathfrak{e}_i$ are the unit vectors. But then all the terms $k\sum_{i=1}^{n}\alpha_i\mathfrak{e}_i$ with $\sum|\alpha_i|^2\leqq 1$ lie in U, i.e. U contains the Euclidean ball of radius k determined by taking the $\mathfrak{e}_i$ as unit vectors. The topology $\mathfrak{T}$ is therefore coarser than the natural topology on K^n.

b) In order to show that $\mathfrak{T}$ is finer than the natural topology on K^n it is enough to show that there is a $\mathfrak{T}$-neighbourhood of $\mathfrak{o}$ which is bounded in K^n, for then it lies in some Euclidean ball.

Let $U_0\neq E$ be a circled neighbourhood of $\mathfrak{o}$. U_0 can contain a linear subspace of dimension at most $n-1$. If V is a circled neighbourhood of $\mathfrak{o}$ with $V+V\subset U_0$, V can only contain one linear subspace H_{n-1} of dimension $n-1$, for otherwise U_0 would be equal to E. We now take a circled neighbourhood W of $\mathfrak{o}$ which does not contain some non-zero x in H_{n-1}. Then $U_1=V\cap W$ can contain linear subspaces of dimension at most $n-2$. Continuing this process, we obtain a circled neighbourhood U of $\mathfrak{o}$ which is bounded along every straight line through $\mathfrak{o}$. Because $E[\mathfrak{T}]$ is regular, we can suppose that U is closed. If U were not bounded in K^n, we could find a sequence $\mathfrak{x}^{(p)}$ of vectors with Euclidean lengths $\|\mathfrak{x}^{(p)}\|=1$ and with $\mathfrak{x}^{(p)}\in\frac{1}{p}U$. Some subsequence of this would then converge in K^n to a vector $\mathfrak{x}^{(0)}\neq\mathfrak{o}$, and by a) it would also converge in the topology $\mathfrak{T}$. Because U is closed, $\mathfrak{x}^{(0)}$ would lie in each $\frac{1}{p}U$, and would therefore lie in $\bigcap_p\frac{1}{p}U$. But this intersection is $\mathfrak{o}$, since U is bounded on every straight line through $\mathfrak{o}$. This contradicts the fact that $\mathfrak{x}^{(0)} \neq \mathfrak{o}$.

As an immediate consequence of the fact that K^n is complete we have

(2) *Every finite-dimensional linear subspace of a topological vector space is closed.*

As in § 10, 4.(7), we have

(3) *If F is a closed linear subspace and G a finite dimensional linear subspace of a topological vector space $E[\mathfrak{T}]$, then $F+G$ is closed.*

We cannot apply the theory of duality to prove this result, as we did in § 10. Let K be the canonical mapping of E onto E/F. The image $K(G)$ is finite-dimensional in E/F, and so is closed, by (2). Because K is continuous, the inverse image $K^{(-1)}(K(G))=F+G$ is closed.

We recall that every linearly topologized space is totally disconnected. In contrast, we have

(4) *Every topological vector space $E[\mathfrak{T}]$ over K is connected.*

By (1), every straight line through $\mathfrak{o}$ in $E[\mathfrak{T}]$ is topologically isomorphic to K.

The straight line joining x_0 and y_0 results from a parallel displacement of the line through $\mathfrak{o}$ and y_0-x_0; it is also topologically isomorphic to K, by 1.(1), and it is therefore connected. The connectedness of $E[\mathfrak{T}]$ follows from this, using § 1, 6.

6. Bounded and compact subsets. A subset B of a topological vector space $E[\mathfrak{T}]$ is said to be bounded (or $\mathfrak{T}$-bounded, if we wish to specify the topology) if for each neighbourhood U of $\mathfrak{o}$ there is a $\rho>0$ with $B \subset \rho U$.

In a normed space, this clearly means $\sup_{x\in B} \|x\| < \infty$, in agreement with the definition of § 4, No. 2.

If x_0 is a non-zero element of $E[\mathfrak{T}]$ and if U is a circled neighbourhood of $\mathfrak{o}$ which does not contain x_0, then the only scalar multiples αx_0 which can lie in U are those with $|\alpha|<1$. If B is bounded and $B \subset \rho U$, then the only scalar multiples αx_0 which can lie in B are those with $|\alpha|<\rho$; a bounded set therefore cuts a straight line through $\mathfrak{o}$ in a subset of a finite interval.

Every subset of a bounded set is bounded. Every finite set is bounded. The circled cover of a bounded set is bounded.

(1) *The sum and the union of finitely many bounded sets are bounded.*

We show this for two bounded sets B_1 and B_2. Let U be circled; if $B_1 \subset \rho_1 U$, $B_2 \subset \rho_2 U$, and if $\rho=\max(\rho_1,\rho_2)$ then $B_1 \cup B_2 \subset \rho U$.

Further let $V+V \subset U$, and as before, let $B_1 \subset \rho V$, $B_2 \subset \rho V$. Then $B_1+B_2 \subset \rho V+\rho V \subset \rho U$.

(2) *The closure $\bar{B}$ of a bounded set B is again bounded.*

For $\bar{B} \subset \rho \bar{U}$ follows from $B \subset \rho U$; but the sets $\bar{U}$ form a base of neighbourhoods of $\mathfrak{o}$, since $E[\mathfrak{T}]$ is regular; hence $\bar{B}$ is bounded.

(3) *A subset $B \subset E[\mathfrak{T}]$ is bounded if and only if whenever x_n is a sequence in B and $\alpha_n \to 0$ in K then $\alpha_n x_n \to \mathfrak{o}$ in $E[\mathfrak{T}]$.*

a) Let B be bounded and let U be a circled neighbourhood of $\mathfrak{o}$. There is a $\rho > 0$ such that $x_n \in \rho U$ for all n. Then $\alpha_n x_n \in \alpha_n \rho U = |\alpha_n| \rho U$. If n_0 is chosen sufficiently large, so that $|\alpha_n| \rho \leqq 1$ for $n \geqq n_0$, then $\alpha_n x_n$ lies in U, for $n \geqq n_0$. Since this holds for each U, $\alpha_n x_n$ converges to $\mathfrak{o}$.

b) If B is unbounded, there is a sequence $x_n \in B$ and a circled neighbourhood U of $\mathfrak{o}$ with $x_n \notin n U$. The sequence $\frac{1}{n} x_n$ lies outside U and does not converge to $\mathfrak{o}$.

(4) *In a topological vector space, every Cauchy sequence x_n is bounded.*

Given a circled U, let $V + V \subset U$, with V circled as well. All the terms $x_n - x_{n_0}$ lie in V, for $n \geqq n_0$, so that $x_n \in x_{n_0} + V$. There exists a ρV, $\rho \geqq 1$, which contains all the elements $x_1, \ldots, x_{n_0}$. But then all the x_k, $k = 1, 2, \ldots$, lie in $\rho V + V \subset \rho U$.

(5) *The continuous linear image of a bounded set is bounded.*

Suppose that A is a continuous linear mapping of $E[\mathfrak{T}_1]$ into $F[\mathfrak{T}_2]$. Given a neighbourhood U of $\mathfrak{o}$ in F, there is a neighbourhood V of $\mathfrak{o}$ in E such that $A(V) \subset U$. It then follows from $B \subset \rho V$ that $A(B) \subset \rho A(V) \subset \rho U$.

We now establish some results concerning precompact and compact subsets of a topological vector space $E[\mathfrak{T}]$.

(6) *Every precompact subset K of $E[\mathfrak{T}]$ is bounded; so therefore is every compact subset.*

By § 5, 6.(2), K is totally bounded, so that given a neighbourhood V of $\mathfrak{o}$ there are finitely many sets $x_i + V$, $x_i \in K$, which cover K. If $\rho > 1$ is chosen so that all the x_i lie in ρV, then we have $K \subset \bigcup_{i=1}^{n} (x_i + V) \subset \rho V + V \subset \rho U$, if V and U are circled and $V + V \subset U$.

For compact sets, the following result is contained in § 3, 2.(5):

(7) *The continuous linear image of a precompact set K is precompact.*

Let K be precompact in $E[\mathfrak{T}_1]$, and let A be a continuous linear mapping from $E[\mathfrak{T}_1]$ into $F[\mathfrak{T}_2]$. Given a neighbourhood U of $\mathfrak{o}$ in $F[\mathfrak{T}_2]$, there is a neighbourhood V of $\mathfrak{o}$ in E with $A(V) \subset U$. If $K \subset \bigcup_{i=1}^{n} (x_i + V)$, then $A(K) \subset \bigcup_{i=1}^{n} (A x_i + A(V))$. It follows from this that $A(K)$ is covered by n sets which are small of order N_U, so that $A(K)$ is precompact.

(8) *If* $K_1, \dots, K_n$ *are precompact (respectively compact) subsets of* $E[\mathfrak{T}]$ *and if* $\alpha_1, \dots, \alpha_n$ *are arbitrary constants in* K, *then the set* $\alpha_1 K_1 + \cdots + \alpha_n K_n$ *is also precompact (compact).*

First we show this for compact K_i. Since $\alpha_i K_i$, being the continuous image of K_i, is again compact, it is sufficient to prove the result with $\alpha_i = 1$.

By TYCHONOFF's theorem the topological product $K_1 \times \cdots \times K_n$ is compact. The mapping $(x_1, \dots, x_n) \to \sum_{i=1}^{n} x_i$ from $K_1 \times \cdots \times K_n$ into $E[\mathfrak{T}]$ is continuous, and so the image $K_1 + \cdots + K_n$ is compact.

If the K_i are precompact, we consider the completion $\tilde{E}[\tilde{\mathfrak{T}}]$. The closures $\bar{K}_i$ in $\tilde{E}[\tilde{\mathfrak{T}}]$ are compact, and, as we have just shown, so is $\alpha_1 \bar{K}_1 + \cdots + \alpha_n \bar{K}_n = \overline{(\alpha_1 K_1 + \cdots + \alpha_n K_n)}$; i.e. $\alpha_1 K_1 + \cdots + \alpha_n K_n$ is precompact.

The union of finitely many compact (respectively precompact) sets is compact (precompact).

(9) *If* K *is compact and disjoint from the closed subset* M *of* $E[\mathfrak{T}]$, *there is a neighbourhood* U *of* $\mathfrak{o}$ *for which* $(K+U) \cap (M+U)$ *is also empty.*

Proof. For each $y \in K$ there is an open circled neighbourhood V_y of $\mathfrak{o}$ for which $(y + V_y + V_y + V_y) \cap M$ is empty. $(y + V_y + V_y) \cap (M + V_y)$ is also empty. K is covered by finitely many $y_i + V_{y_i}$. Let $U = \bigcap_i V_{y_i}$. Then for each $y \in K$ we have

$$y + U \subset y_i + V_{y_i} + U \subset y_i + V_{y_i} + V_{y_i},$$

so that $(y+U) \cap (M+U)$ is empty, for all $y \in K$.

(10) *If* K *is compact and* M *is closed in* $E[\mathfrak{T}]$, *then* $K+M$ *is closed.*

If $x \notin K+M$, $(x-M) \cap K$ is empty. But $x-M$ is also closed, so that by (9) there is a U with $((x-M)+U) \cap K$ empty. Hence $(x+U) \cap (K+M)$ is empty, and so the complement of $K+M$ is open.

(11) *If* K *is closed, precompact and disjoint from the complete subset* M *of* $E[\mathfrak{T}]$, *then there is a neighbourhood* U *of* $\mathfrak{o}$ *for which* $(K+U) \cap (M+U)$ *is also empty.*

The closure $\tilde{K}$ of K in the completion $\tilde{E}[\tilde{\mathfrak{T}}]$ is compact, and is still disjoint from M. (9) proves the result for the subsets $\tilde{K}$ and M of $\tilde{E}$; the result for the subsets K and M of E follows from this.

In the same way, it follows from (10) that

(12) *If* K *is precompact and closed, and if* M *is complete in* $E[\mathfrak{T}]$, *then* $K+M$ *is closed.*

As an example of the concept of boundedness, we consider topological products.

(13) *The bounded sets of a topological product* $E[\mathfrak{T}]=\prod_{\alpha} E_\alpha[\mathfrak{T}_\alpha]$ *are just the subsets of sets of the form* $\prod_{\alpha} B_\alpha$, *with* B_α *bounded in* $E_\alpha[\mathfrak{T}_\alpha]$.

If B is bounded in E, then by (5) the projections B_α of B are $\mathfrak{T}_\alpha$-bounded in $E_\alpha[\mathfrak{T}_\alpha]$, and so $B\subset\prod_{\alpha} B_\alpha$. Conversely a set $\prod_{\alpha} B_\alpha$ is bounded in $E[\mathfrak{T}]$; this follows directly from the definition of the product topology.

Since every bounded set in the field K is relatively compact, it follows from TYCHONOFF's theorem that every bounded set in $\omega_d(\mathsf{K})$ is relatively compact. A set B of vectors $\mathfrak{x}=(\xi_\alpha)$ in ω_d is bounded if it is "coordinate-wise" bounded, i.e. if for each α there is an $M_\alpha>0$ for which $|\xi_\alpha|\leqq M_\alpha$ for all $x\in B$.

7. Locally compact topological vector spaces. In 5. we showed that every finite-dimensional topological vector space is topologically isomorphic to K^n, and so is locally compact. The following converse holds:

(1) *Every locally precompact topological vector space is finite dimensional.*

The hypothesis means that $E[\mathfrak{T}]$ has an open precompact circled neighbourhood U of $\mathfrak{o}$. Starting with $H_0=\mathfrak{o}$, we construct a strictly increasing sequence $H_0\subset H_1\subset\cdots$ of finite-dimensional linear subspaces of E. Suppose that H_k has already been constructed, and that it is a proper subspace of E. H_k is closed by 5.(2). By 6.(6), U is bounded, and it is mapped into a bounded subset of E/H_k by the canonical mapping K of E onto E/H_k, by 6.(5). $K(U)$, being bounded, is a proper subset of E/H_k, so that its inverse image $K^{(-1)}(K(U))=H_k+U$ is a proper subset of E. We now show that the closure $\bar{U}$ of U is not contained in H_k+U. If $\bar{U}\subset H_k+U$, then we would have $H_k+\bar{U}=H_k+U$. But by (5.(1) and 6.(12), $H_k+\bar{U}$ is closed. On the other hand, H_k+U is open, so that E would possess a proper open and closed subset, contradicting 5.(4). There therefore exists a point $y_k\in\bar{U}$ which does not lie in H_k+U. We now put $H_{k+1}=H_k\oplus[y_k]$. If the sequence of spaces H_k constructed in this way were not to produce E after finitely many steps, we would obtain a sequence $y_k\in\bar{U}$ with $y_k-y_i\notin U$ for $i\neq k$. There exists a circled V with $V+V\subset U$. Since $\bar{U}$ is totally bounded, it is covered by finitely many x_j+V. But then at least two y_i, y_{i_1} and y_{i_2} say, would lie in the same x_j+V, and so $y_{i_1}-y_{i_2}\in V+V\subset U$, which is impossible.

8. Topologically complementary spaces. We now link up with the ideas of § 10, 7. Two closed algebraically complementary linear subspaces H_1 and H_2 of the topological vector space $E[\mathfrak{T}]$ are said to be **topologically complementary** if the mapping $(x_1, x_2) \to x_1 + x_2$ of the topological product $H_1[\mathfrak{T}] \times H_2[\mathfrak{T}]$ onto $E[\mathfrak{T}]$ is a homeomorphism.

We repeat a result which was also valid before (§ 10, 7.(6)):

(1) *A closed linear subspace H_1 of $E[\mathfrak{T}]$ has a topological complement H_2 if and only if there is a continuous projection P_1 of E onto H_1. P_1 is then a topological homomorphism, $H_2 = N[P_1]$, and E/H_2 is topologically isomorphic to H_1.*

The proof is almost word for word the same as in § 10, 7., and is left to the reader.

(2) *A closed linear subspace H of finite codimension in a topological vector space $E[\mathfrak{T}]$ always has a topological complement. Every algebraic complement of H is also a topological complement.*

Proof. Let M be an algebraic complement of H, so that $E = H \oplus M$. By 5.(2), M is closed and is topologically isomorphic to K^n. The projection P from E onto M with null-space H is the product of the canonical homomorphism K of E onto E/H, a one-one linear mapping $\check{P}$ of E/H onto M and the embedding J of M into E.

But since E/H is also topologically isomorphic to K^n, $\check{P}$ is a topological isomorphism, and so, by 4.(5), $P = J\check{P}K$ is continuous, and the assertion follows from (1).

The obvious conjecture (cf. § 10, 7.(8)) that every finite-dimensional linear subspace of $E[\mathfrak{T}]$ always has a topological complement is false, as we shall see in the next number.

9. The dual space, hyperplanes, the spaces L^p with $0 < p < 1$. We again denote the vector space of all continuous linear functionals on E by E', and E' is again called the **dual** of E.

In this number we shall show, by giving an example, that there are topological vector spaces on which there are no continuous linear functionals other than the trivial one $u(x) \equiv 0$, so that the dual space can consists of the zero element alone.

This pathological possibility prevents us from establishing a really meaningful and useful theory of general topological vector spaces. Such a theory is only obtained when we restrict attention to locally convex spaces, as we shall do in the next chapter.

The following important relation holds between continuous linear functionals and closed hyperplanes.

(1) *If $u(x)$ is a linear functional on $E[\mathfrak{T}]$ which is not identically zero, its null-space is a hyperplane in E; conversely, for each hyperplane through $\mathfrak{o}$ in E there is a linear functional whose null-space is the hyperplane.*

$u(x)$ is continuous if and only if the corresponding hyperplane is closed.

A (closed) hyperplane in $E[\mathfrak{T}]$ is given by an equation

$$u(x)=\gamma, \quad \gamma \in \mathsf{K}, \tag{2}$$

where $u(x)$ is a (continuous) linear functional.

A hyperplane which is not closed is dense in $E[\mathfrak{T}]$.

Proof. a) If $u \neq o$ is a linear functional on E, $H=u^{-1}(o)$ is a vector space, which is closed if u is continuous.

H has codimension 1 in E: there is an $x_0 \in E$ with $u(x_0)=1$; for arbitrary x in E we have a unique decomposition

$$x=u(x)x_0+(x-u(x)x_0), \tag{3}$$

where the second summand lies in H; the scalar multiples of x_0 therefore form a one-dimensional complement to H.

b) Let H be a hyperplane through o. Let $E=[x_0]\oplus H$. Let us put $u(\alpha x_0+y)=\alpha$ for $y \in H$. Then u is a linear functional on E with null-space H. If H is closed u is continuous, for the mapping $\check{u}$ from E/H onto K is the topological isomorphism $\alpha \check{x}_0 \to \alpha$ (cf. 4.(5)).

c) A general hyperplane is of the form x_0+H, where H is a hyperplane through o; it is closed, if and only if H is. If u is a characteristic linear functional for H, then $u(x_0+y)=u(x_0)=\gamma$ holds for all $y \in H$ and for no other elements of E.

d) If H is not closed, H is a proper subspace of $\bar{H}$. Since H has codimension 1 in E, $\bar{H}$ must equal E.

The following example was originally given by M. I [1]; the simple proof of (9) was given by W. ROBERTSON [1].

Once again let L^p be the space of all measurable functions $f(t)$ on $I=[a,b]$ with $\int_a^b |f(t)|^p dt<\infty$ (again, we identify functions which are equal almost everywhere); the case $I=(-\infty,\infty)$ is also valid in what follows. This time, however, let $0<p<1$.

A topology is determined on L^p by taking as base of neighbourhoods of o the sets $U(\varepsilon)$ of all $f \in L^p$ with

$$\|f\|_p=\left(\int_a^b |f(t)|^p dt\right)^{1/p} \leqq \varepsilon. \tag{4}$$

Clearly $\|\alpha f\|_p=|\alpha|\,\|f\|_p$.

The sets $U(\varepsilon)$ are circled and absorbent, and form a filter-base on L^p with $\bigcap_\varepsilon U(\varepsilon)=o$. L^p is a topological vector space, if we can show that for each U there is a V with $V+V \subset U$.

To this end we show that instead of Minkowski's inequality we have the following inequality:

$$\|f+g\|_p \leqq 2^{\frac{1-p}{p}}(\|f\|_p+\|g\|_p). \tag{5}$$

Proof. For $q>1$ the function $\frac{1+x^q}{(1+x)^q}$ has exactly one minimum for $x\geqq 0$, namely where $x=1$, so that $1+x^q\geqq 2^{1-q}(1+x)^q$. Putting $x=c/d$, it follows from this that if $c, d>0$, then

(6) $$c^q+d^q\geqq 2^{1-q}(c+d)^q \quad (q>1).$$

But it follows from $(c+d)^p\leqq c^p+d^p$, for $0<p<1$, and from (6), with $q=1/p$, that

$$\|f+g\|_p\leqq\left(\int|f(t)|^p\,dt+\int|g(t)|^p\,dt\right)^{\frac{1}{p}}\leqq 2^{\frac{1}{p}-1}(\|f\|_p+\|g\|_p).$$

Thus if U is one of the neighbourhoods (4), it follows from (5) that

(7) $$2^{-1/p}U+2^{-1/p}U\subset U;$$

hence

(8) L^p, $0<p<1$, *is a topological vector space.*

Now let $u(f)$ be a continuous linear functional on L^p which does not vanish identically. There exists $g_0\in L^p$ with $u(g_0)=1$. For $a<s<b$ we put $g_s^{(1)}(t)$ equal to the function which is equal to $g_0(t)$ in the interval $a\leqq t\leqq s$, and which vanishes identically for $t>s$; let $g_s^{(2)}(t)=g_0(t)-g_s^{(2)}(t)$. Now $\|g_s^{(1)}\|_p^p=\int_a^s|g_0(t)|^p\,dt$ increases continuously from 0 to $\|g_0\|_p^p$, so that there is an s_0 with $\|g_{s_0}^{(1)}\|_p^p=\|g_{s_0}^{(2)}\|_p^p=\frac{1}{2}\|g_0\|_p^p$. Since $|u(g_0)|=1$, we must have $|u(g_{s_0}^{(i)})|\geqq\frac{1}{2}$ for $i=1$ or 2. Let $g_1(t)=2g_{s_0}^{(i)}(t)$, for this i. Then $|u(g_1)|\geqq 1$, but $\|g_1\|_p=2^{1-\frac{1}{p}}\|g_0\|_p$. Repeating this procedure we obtain a sequence g_n with $|u(g_n)|\geqq 1$, but with $\|g_n\|_p=2^{n\left(1-\frac{1}{p}\right)}\|g_0\|_p\to 0$, contradicting the continuity of u. Thus we have shown:

(9) *Every continuous linear functional on* L^p, $0<p<1$, *vanishes identically.*

Using this, we can answer the problem raised in the preceding number.

(10) *In* L^p, $0<p<1$, *every hyperplane is dense, and no finite-dimensional linear subspace H has a topological complement in* L^p.

Let H_1 be a one-dimensional linear subspace of H. By 5.(1), H_1 has a topological complement H_2 in H under the topology induced by L^p, so that $H=H_1\oplus H_2$. If G were a topological complement of H, $H_2\oplus G$ would be a topological complement of H_1; $H_2\oplus G$ would therefore be a closed hyperplane of L^p, which is not possible, by (1) and (9).

We establish one further general result about dual spaces:

(11) *A topological vector space* $E[\mathfrak{T}]$ *and its completion* $\tilde{E}[\tilde{T}]$ *have the same dual space, so that* $E'=(\tilde{E})'$.

Since E is dense in $\tilde{E}$, and since every continuous linear functional u on E is uniformly continuous, by 2.(4), u can be extended in a unique way to a uniformly continuous functional on $\tilde{E}$; conversely the restriction to E of a continuous linear functional defined on $\tilde{E}$ is also continuous on E.

10. Locally bounded spaces, quasi-norms, p-norms. The expression (4) of the preceding number does not possess all the properties of a norm. It is an example of a quasi-norm $\|x\|$ on a topological vector space $E[\mathfrak{T}]$, which is characterized by the following properties:

(Q1) $\|x\| \geqq 0$,

(Q2) $x = \mathfrak{o}$ if $\|x\| = \mathfrak{o}$,

(Q3) $\|\alpha x\| = |\alpha| \, \|x\|$ for all $\alpha \in \mathsf{K}$,

(Q4) There is a $k \geqq 1$ for which $\|x+y\| \leqq k(\|x\| + \|y\|)$ for all $x, y \in E[\mathfrak{T}]$.

If it possible to take $k=1$, we obtain a norm.

A base of neighbourhoods of $\mathfrak{o}$ in L^p, $0<p<1$, is given by the sets $\|x\|_p \leqq \varepsilon$. Those spaces whose topology can be given by a quasi-norm in this way can be characterized quite simply.

A topological vector space is said to be locally bounded if there is a bounded neighbourhood of $\mathfrak{o}$. This can be taken to be circled. We have (HYERS [1], BOURGIN [1]):

(1) *The topology of a topological vector space $E[\mathfrak{T}]$ can be given by a quasi-norm if $E[\mathfrak{T}]$ is locally bounded. Conversely, a quasi-normed space is always locally bounded.*

Proof. a) Let U be a circled bounded neighbourhood of $\mathfrak{o}$ in $E[\mathfrak{T}]$. The sets αU, $\alpha>0$, form a base of neighbourhoods of $\mathfrak{o}$, for U is bounded, so that for each neighbourhood V of $\mathfrak{o}$ there exists $\beta>0$ for which $\beta U \subset V$.

We introduce the so-called Minkowski functional $q(x)$ of U:

$$q(x) = \inf_{x \in \alpha U} \alpha \qquad (\alpha \geqq 0). \tag{2}$$

$q(x)$ is well-defined for each x, for since U is absorbent there always exists an $\alpha>0$ with $x \in \alpha U$.

We now show that $q(x)$ has the properties of a quasi-norm. (Q1) is trivially satisfied, and (Q2) follows from the fact that a non-zero x cannot lie in every αU, since $\bigcap_{\alpha>0} \alpha U = \mathfrak{o}$. (Q3) follows from the fact that U is circled.

Since the sets αU, $\alpha>0$, form a base of neighbourhoods, there exists $k \geqq 2$ with $U+U \subset kU$, by (LT1). Now let $q(x)=\rho_0$, $q(y)=\sigma_0$. Then if $\rho>\rho_0$ and $\sigma>\sigma_0$ the elements x/ρ and y/σ lie in U, so that

$$\frac{\rho}{\rho+\sigma}\,\frac{x}{\rho} + \frac{\sigma}{\rho+\sigma}\,\frac{y}{\sigma} = \frac{x+y}{\rho+\sigma} \in U+U. \tag{3}$$

It follows from this that $x+y \in k(\rho+\sigma)U$, so that $q(x+y) \leqq k(q(x)+q(y))$; we have therefore established (Q4).

If V is the set of all $x \in E$ with $q(x) \leqq 1$, then we have $U \subset V \subset (1+\varepsilon) U$, and so the quasi-norm topology coincides with $\mathfrak{T}$.

b) If a quasi-norm $\|x\|$ is defined on E, and if we introduce a topology $\mathfrak{T}'$ on E by means of the absorbent circled sets αV, $\alpha > 0$, consisting of all x with $\|x\| \leqq \alpha$, then we obtain a topological vector space, since $\bigcap_{\alpha>0} \alpha V = \mathfrak{o}$ and since $\frac{1}{2k} V + \frac{1}{2k} V \subset V$, by (Q4). Finally, V is clearly $\mathfrak{T}'$-bounded.

In the next paragraph we shall investigate the concept of convex set in detail. For locally bounded spaces, a generalization of this concept, introduced by M. LANDSBERG, has proved to be important.

Let $0 < p \leqq 1$. A subset M of a vector space E over K is said to be *p-convex* if whenever it contains x and y it contains $\tau x + \sigma y$, where $\tau \geqq 0$, $\sigma \geqq 0$ and $\tau^p + \sigma^p = 1$. M is said to be *absolutely p-convex* if whenever it contains x and y it contains all $\tau x + \sigma y$ with $|\tau|^p + |\sigma|^p \leqq 1$, where τ and σ are real or complex, depending on whether E is a real or complex vector space.

If $p=1$, we speak of *convex* and *absolutely convex* sets, respectively.

The *absolutely p-convex cover* $\Gamma_p(M)$ of a set M is the intersection of all the absolutely p-convex sets which contain M. $\Gamma_p(M)$ consists of all the terms of the form $\sum_{i=1}^{n} \alpha_i x_i$, with $\sum_{i=1}^{n} |\alpha_i|^p \leqq 1$. The proof for p with $0 < p < 1$ is completely similar to the one for $p=1$ given in § 16, 1.(1), using the inequality $|\alpha+\beta|^p \leqq |\alpha|^p + |\beta|^p$.

A *p-norm* $|||x|||$, $0 < p \leqq 1$, on a vector space E is characterized by

(P1) $|||x||| \geqq 0$,
(P2) if $|||x||| = 0$, then $x = \mathfrak{o}$,
(P3) $|||\alpha x||| = |\alpha|^p |||x|||$, for all $\alpha \in \mathsf{K}$,
(P4) $|||x+y||| \leqq |||x||| + |||y|||$.

If $p=1$, we obtain the concept of a norm.

The p-th power of the quasi-norm $\|f\|_p$ in L^p (cf. 9.(4)) is a p-norm $|||f|||$ on L^p, and the neighbourhood 9.(4) of $\mathfrak{o}$ is p-convex. The topology on L^p can also be given by the neighbourhoods $|||f||| \leqq \varepsilon$; we say that L^p is a *p-normable* topological vector space.

We now establish

(4) *A topological vector space $E[\mathfrak{T}]$ is p-normable, $0 < p \leqq 1$, if and only if it has a p-convex bounded neighbourhood of $\mathfrak{o}$.*

In the case $p=1$, this is the characterization of normed spaces given by A. KOLMOGOROFF [1].

Proof. It follows directly from (P1) to (P4) that in a p-normed space the set U of all x with $|||x|||\leqq 1$ is absolutely p-convex; further U is bounded, since $|||\rho^{1/p}U|||\leqq\rho$ holds for $\rho>0$, by (P3), so that $\rho^{1/p}U$ lies in the neighbourhood $|||x|||\leqq\rho$ of $\mathfrak{o}$.

Conversely, let V be a bounded p-convex neighbourhood of $\mathfrak{o}$ in $E[\mathfrak{T}]$. V contains a circled neighbourhood W of $\mathfrak{o}$. Let $\Gamma_p(W)$ be the absolutely p-convex cover of W, and let $\sum_{i=1}^{n}\alpha_i x_i$, with $x_i\in W$, $\sum|\alpha_i|^p\leqq 1$, be an element of $\Gamma_p(W)$. We put $\sum|\alpha_i|^p=\rho^p$. The element $\sum\alpha_i x_i$ $=\sum\frac{|\alpha_i|}{\rho}\varepsilon_i x_i$, with $\varepsilon_i=\frac{\alpha_i}{|\alpha_i|}\rho$, is of the form $\sum\frac{|\alpha_i|}{\rho}y_i$, $y_i\in W\subset V$, since W is circled. Now if a p-convex set contains $z_1,\dots,z_n$, it always contains $\sum_{i=1}^{n}\beta_i z_i$, where $\beta_i\geqq 0$ and $\sum_{i=1}^{n}\beta_i^p=1$ (this is proved analogously to § 16, 1.(1)). Since the y_i lie in V, $\sum_{i=1}^{n}\frac{|\alpha_i|}{\rho}y_i$ also lies in V. This means that $\Gamma_p(W)\subset V$. We can therefore assume that V is absolutely p-convex.

We put $|||x|||=\inf_{x\in\rho V}\rho^p$, $\rho\geqq 0$; this is equal to $q(x)^p$, where $q(x)$ is the quasi-norm of V. (P1) to (P3) therefore follow from (Q1) to (Q3).

If x and y are arbitrary points in E, and if $\rho>\rho_0=q(x)$, $\sigma>\sigma_0=q(y)$, then because V is absolutely p-convex the element

$$\frac{\rho}{(\rho^p+\sigma^p)^{1/p}}\cdot\frac{x}{\rho}+\frac{\sigma}{(\rho^p+\sigma^p)^{1/p}}\,\frac{y}{\sigma}=\frac{x+y}{(\rho^p+\sigma^p)^{1/p}}$$

lies in V, and (P4) follows easily from this; $|||x|||$ is therefore a p-norm, and it follows from $V\subset U\subset(1+\varepsilon)V$, where U is the set $|||x|||\leqq 1$, that $\mathfrak{T}$ is defined by the p-norm.

Every p-normed space is locally bounded. We shall show that conversely every locally bounded space $E[\mathfrak{T}]$ is p-normable for a suitable p with $0<p\leqq 1$ (cf. S. Rolewicz [1]).

For every bounded circled neighbourhood U of $\mathfrak{o}$ in $E[\mathfrak{T}]$, there exists at least one $k\geqq 2$ with $U+U\subset kU$. The greatest lower bound of these k is called the module of concavity of U. The greatest lower bound of all these modules of concavity is called the module of concavity $\kappa(E)$ of $E[\mathfrak{T}]$. We now have

(5) *If $\kappa(E)=2^{1/p_0}$ is the module of concavity of the locally bounded space $E[\mathfrak{T}]$, then for each $p<p_0$ there is a p-norm which determines the topology $\mathfrak{T}$.*

By (4), it is sufficient to show that $E[\mathfrak{T}]$ has a bounded absolutely p-convex neighbourhood of $\circ$.

By hypothesis there is a bounded circled neighbourhood U of $\circ$ with $U+U\subset 2^{\frac{1}{p}}U$, i.e. $2^{-\frac{1}{p}}U+2^{-\frac{1}{p}}U\subset U$. More generally we have

(6) $2^{-\frac{k_1}{p}}U+\cdots+2^{-\frac{k_n}{p}}U\subset U$ for $\sum_1^n 2^{-k_i}\leqq 1$, where the k_i are positive integers.

It is sufficient to prove this for $\sum_1^n 2^{-k_i}=1$. By the order of such a decomposition of 1 we mean the maximum of the k_i which appear. The assertion is correct for decompositions of order 1. Every decomposition of order $k+1$ results from one of order k by replacing certain summands 2^{-k} by $2^{-(k+1)}+2^{-(k+1)}$, since it follows from $\sum 2^{-k_i}=1$ that an even number of summands $2^{-(k+1)}$ appear. Making the inductive hypothesis that the relation (6) is true for k, we obtain the relation for $k+1$ when we replace the corresponding summands $2^{-\frac{k}{p}}U$ by $2^{-\frac{k+1}{p}}U+2^{-\frac{k+1}{p}}U$; this is permissible since $2^{-\frac{1}{p}}U+2^{-\frac{1}{p}}U\subset U$. We now show that $\Gamma_p(U)\subset 2^{\frac{1}{p}}U$, which establishes (5). Let $\sum_{i=1}^n |\alpha_i|^p\leqq 1$, and let k_i be chosen so that $2^{-k_i}\leqq|\alpha_i|^p<2^{-k_i+1}$ holds. But then $\sum_1^n|\alpha_i|^p<2\sum_1^n 2^{-k_i}\leqq 2$, so that $\sum_1^n\alpha_i x_i\in 2^{\frac{1}{p}}\sum 2^{-\frac{k_i}{p}}U\subset 2^{\frac{1}{p}}U$ for arbitrary $x_i\in U$, by (6).

It can be shown, by giving examples, that $E[\mathfrak{T}]$ need not be p_0-normable; L^p, $0<p\leqq 1$, is p-normable, however, and $\kappa(L^p)=2^{1/p}$ (cf. Rolewicz [1]).

(7) *If $E[\mathfrak{T}]$ is locally bounded, then so is every closed subspace and every quotient space, and so also is the completion $\tilde{E}[\tilde{\mathfrak{T}}]$.*

We shall only prove the last assertion. If U is a bounded neighbourhood of $\circ$, the sets ρU, $\rho>0$, form a base of neighbourhoods of $\circ$ in E. By 3.(1) the closures $\rho\bar{U}$ in $\tilde{E}[\tilde{\mathfrak{T}}]$ form a base of neighbourhoods of $\circ$ in $\tilde{E}[\tilde{\mathfrak{T}}]$; i.e. $\bar{U}$ is a bounded neighbourhood of $\circ$ in $\tilde{E}[\tilde{\mathfrak{T}}]$.

Examples and further results about locally bounded spaces are contained in the papers of Bourgin [1], Hyers [1], Landsberg [1], [2] and Rolewicz [1].

11. Metrizable spaces. In the classical theory of Banach, not only were normed spaces considered, but also those spaces on which a metric is defined which is compatible with the vector space operations. First we consider the question of when the topology of a topological vector space $E[\mathfrak{T}]$ can be given by a metric. In this case we say that $E[\mathfrak{T}]$ is metrizable.

(1) *A topological vector space is metrizable if and only if it has a countable base of neighbourhoods of* $\mathfrak{o}$, *so that it satisfies the first axiom of countability. If this is the case, there is always a translation-invariant metric, and this defines the uniform structure of* $E[\mathfrak{T}]$ *as well.*

$E[\mathfrak{T}]$ is only metrizable if there exists a countable base of neighbourhoods of $\mathfrak{o}$. If this is the case, then, by § 6, 7.(1), the translation-invariant uniform structure of $E[\mathfrak{T}]$ (and not just the topology) is metrizable. We now show that the metric constructed in § 6, 7.(2) is translation-invariant. A base of vicinities of the uniformity of $E[\mathfrak{T}]$ is given by the sets N_U, as U runs through the circled neighbourhoods of $\mathfrak{o}$. Following through the proof of § 6, 7.(2) once again, we see that, since (x,y) and $(x-y,\mathfrak{o})$ always belong to the same vicinities, the function $f(x,y)$ defined there satisfies $f(x,y)=f(x-y,\mathfrak{o})$; further, if we make the sequences $x=x_1, x_2,\dots,x_n=y$ and $x-y, x_2-y,\dots,y-y$ correspond to each other, it follows that $|x,y|=|x-y,\mathfrak{o}|$. The metric is therefore translation-invariant.

In addition let us remark that, as each U is circled, λx belongs to U for each $|\lambda|\leqq 1$ if x does; hence $f(\lambda x,\mathfrak{o})\leqq f(x,\mathfrak{o})$, so that $|\lambda x,\mathfrak{o}|\leqq|x,\mathfrak{o}|$. In particular it follows from this that $|\tau x,\mathfrak{o}|=|x,\mathfrak{o}|$ if $|\tau|=1$. Once again we write $\|x\|=|x,\mathfrak{o}|$, and obtain

(2) *The uniform structure of a metrizable topological vector space can be given by a function* $\|x\|$ *with the following properties:*

(F1) $\|x\|\geqq 0$,
(F2) $x=\mathfrak{o}$ *if* $\|x\|=0$,
(F3) $\|\lambda x\|\leqq\|x\|$ *if* $|\lambda|\leqq 1$,
(F4) $\|x+y\|\leqq\|x\|+\|y\|$,
(F5) $\|\lambda x_n\|\to 0$ *if* $\|x_n\|\to 0$,
(F6) $\|\lambda_n x\|\to 0$ *if* $\lambda_n\to 0$.

A function $\|x\|$ with properties (F1) to (F6) is called an (F)-norm, and E is said to be (F)-normed. The sets V_ε of all x with $\|x\|<\varepsilon$ form a base of neighbourhoods of $\mathfrak{o}$ for the topology determined by the (F)-norm.

(F5) and (F6) result from the fact that $\lambda x_n\to\mathfrak{o}$ and $\lambda_n x\to\mathfrak{o}$ in the topology $\mathfrak{T}$, so that $|\lambda x_n,\mathfrak{o}|\to 0$ and $|\lambda_n x,\mathfrak{o}|\to 0$.

We note that (F 5) follows easily from (F 1) to (F 4) (write $\lambda=(k+\eta)e^{i\theta}$ with $k\geqq 0$ integer and $0\leqq\eta<1$).

The converse of (2) is also valid:

(3) *Under the topology defined by the* (F)-*norm, an* (F)-*normed space* E *is a topological vector space with a countable base of* $\mathfrak{o}$-*neighbourhoods.*

For by (F3) the sets V_ε are circled, and by (F6) they are absorbent. These sets clearly form a filter-base, whose intersection is $\mathfrak{o}$, because of (F2). Further, by (F4), $V_{\varepsilon/2}+V_{\varepsilon/2}\subset V_\varepsilon$. The sets $V_{1/n}$ form a countable base of neighbourhoods of $\mathfrak{o}$.

We remark that the uniform continuity of $\|x\|$ on E can be established as in § 14.1, and that the (F)-norm of E can be extended continuously to the completion of E. (F1) to (F6) still hold, as is shown by taking limits.

If we define an (F)-norm on a quotient space E/H by setting $\|\hat{x}\|=\inf_{x\in\hat{x}}\|x\|$, as in § 14, 4., then it is easy to confirm that (F1) to (F6) still hold for $\|\hat{x}\|$ provided that H is closed, and that the topology defined by this (F)-norm is the quotient-space topology.

(4) *The quotient space* E/H *of a complete metrizable space by a closed subspace* H *is again complete.*

The proof of § 14, 4.(3) also holds for this more general case.

A subset M of an (F)-normed space is bounded (in the sense of 6.) only if $\sup_{x\in M}\|x\|<\infty$. As we shall see later, the converse is not true. Sets which are bounded in the metric sense (cf. § 4, 2.) need not therefore be bounded in the sense of 6.

BANACH called a complete (F)-normed space an (F)-space, after FRÉCHET. Following the terminology of BOURBAKI we shall reserve this name for complete metrizable locally convex spaces (cf. § 18, 2.).

The p-norms on locally bounded spaces which were introduced in 10. are also (F)-norms, as can be seen immediately.

We have introduced an (F)-norm $\|x\|$ on every metrizable space $E[\mathfrak{T}]$, and the corresponding metric defines the uniquely determined translation-invariant uniformity of $E[\mathfrak{T}]$ (cf. 2.(3)). Two different (F)-norms corresponding to the same topology on $E[\mathfrak{T}]$ are therefore always equivalent, i.e. the sets N_ε of all (x,y) with $\|x-y\|<\varepsilon$ form a base of vicinities for the uniform structure of $E[\mathfrak{T}]$, for both (F)-norms.

Conversely, instead of starting from a topology $\mathfrak{T}$ on E, we can also start from a metric $\mathfrak{M}$. If this metric is compatible with the vector space operations on E, i.e. if the mappings $(x,y)\to x+y$ and $(\alpha,x)\to\alpha x$ are continuous in the sense of the metric $\mathfrak{M}$, then $E[\mathfrak{T}]$ is called a l i n e a r m e t r i c s p a c e. $\mathfrak{M}$ defines a topology $\mathfrak{T}$ on E, and $E[\mathfrak{T}]$ is clearly a metrizable topological vector space.

If $\mathfrak{M}$ is translation-invariant, then the uniform structures on E determined by $\mathfrak{M}$ and by $\mathfrak{T}$ are the same, by 2.(3). If $\mathfrak{M}$ is not translation-invariant, this need not be the case.

We consider the real line P as an example, first with the metric $\mathfrak{M}$ defined by the modulus (and so by an (F)-norm), and secondly with the metric $\mathfrak{M}_1$ defined by $|\alpha,\beta|_1=|\alpha^3-\beta^3|$. The vector space operations are compatible with $\mathfrak{M}_1$ as well, and $\mathfrak{M}_1$ defines the same topology on P as does $\mathfrak{M}$. The uniform structure corresponding to $\mathfrak{M}_1$ with vicinities $|\alpha,\beta|_1<\varepsilon$, however, is strictly finer than the one corresponding to $\mathfrak{M}$, as can easily be seen. P is also complete with respect to $\mathfrak{M}_1$, since the $\mathfrak{M}$-Cauchy and $\mathfrak{M}_1$-Cauchy sequences in P are the same.

A further metric on P is defined by $|\alpha,\beta|_2=|\tan^{-1}\alpha-\tan^{-1}\beta|$; this metric also gives the same topology, although P is no longer complete; e.g. the sequence $\alpha_n=n$ is a Cauchy sequence which does not converge.

The original metric of a linear metric space therefore need not give the uniform structure of the topological vector space; indeed the example given above is a case where $E[\mathfrak{M}]$ is not complete with respect to the metric $\mathfrak{M}$, although it is complete as a topological vector space.

Nevertheless, we have

(5) *If a linear metric space* $E[\mathfrak{M}_1]$ *is* $\mathfrak{M}_1$*-complete, then it is also complete as a topological vector space, and so is complete with respect to any of its* (F)*-norms.*

This conjecture of Banach was proved by V.L. Klee [3]. First we establish a lemma due to W. Sierpiński [1]:

(6) *Let* H *be a subset of a metric space* $E[\mathfrak{M}]$. *Let a second metric* $\mathfrak{M}_1$ *be defined on* H, *which induces the same topology on* H *as does* $\mathfrak{M}$. *If* H *is* $\mathfrak{M}_1$*-complete, then* H *is the intersection of countably many open subsets of* $E[\mathfrak{M}]$.

Proof. Let $x\in H$. Because $\mathfrak{M}$ and $\mathfrak{M}_1$ induce the same topology on H, there exists, for each n, a $\rho_n(x)<\frac{1}{n}$ for which $|x,y|_1<\frac{1}{n}$ holds for all $y\in H$ with $|x,y|<\rho_n(x)$.

If $U_n(x)$ is the set of all $z\in E$ with $|x,z|<\rho_n(x)$, $O_n=\bigcup_{x\in H}U_n(x)$ is an open subset of $E[\mathfrak{M}]$ which contains H. If we put $D=\bigcap_{n=1}^{\infty}O_n$, then clearly $H\subset D$. (6) is established, if we can show that $D\subset H$.

Let $z_0\in D$. For each n there is an $x_n\in H$ with $z_0\in U_n(x_n)$, so that $|z_0,x_n|<\rho_n(x_n)<\frac{1}{n}$. Therefore $z_0=\lim x_n$, in the sense of the metric $\mathfrak{M}$. We shall show that x_n is a Cauchy sequence with respect to $\mathfrak{M}_1$. It follows from $|x_k,x_n|\leqq|x_k,z_0|+|z_0,x_n|<\frac{1}{k}+|z_0,x_n|$ and $|z_0,x_n|<\rho_n(x_n)$ that $|x_k,x_n|<\rho_n(x_n)$ for $k\geqq k_0(n)$, and so $|x_k,x_n|_1<\frac{1}{n}$ for $k\geqq k_0(n)$.

Hence $|x_l,x_k|_1\leqq\frac{2}{n}$ for $k,\ l\geqq k_0(n)$, so that x_n is an $\mathfrak{M}_1$-Cauchy sequence. Since H is $\mathfrak{M}_1$-complete, the sequence x_n has a limit in H, and this must coincide with z_0, so that $z_0\in H$ and $D\subset H$.

Proof of (5). $E[\mathfrak{M}_1]$ and the completion $\tilde{E}$ of E with respect to the (F)-norm satisfy the hypotheses of (6). Thus E is a dense linear subspace of $\tilde{E}$ which is the intersection $E=\bigcap_{n=1}^{\infty} O_n$ of open subsets of $\tilde{E}$. The complement $\tilde{E}\sim E=\bigcup_{n=1}^{\infty}(\tilde{E}\sim O_n)$, being the union of countably many nowhere dense subsets, is meagre in $\tilde{E}$. If $\tilde{E}\sim E$ were non-empty, it would contain a non-zero x_0, and would contain x_0+E, so that E would also be meagre. This is impossible, since $\tilde{E}=E\cup(\tilde{E}\sim E)$ is not meagre in itself (§ 4, 6.(5)). The assertion now follows from $\tilde{E}=E$.

A detailed account of (F)-norms is to be found in Bessaga, Pelczynski and Rolewicz [1].

12. The Banach-Schauder theorem and the closed-graph theorem. We now investigate when a continuous linear mapping from one complete metrizable topological vector space into another is a topological homomorphism. The fundamental Banach-Schauder theorem states that

(1) *A continuous linear mapping A from one complete metrizable topological vector space E into another, F, either is a topological homomorphism or has an image* $A(E)$ *which is meagre (of the first category) in* $\overline{A(E)}$.

(2) *A is a topological homomorphism if and only if* $A(E)$ *is closed.*

Proof of (1). It is sufficient to consider the case where $\overline{A(E)}=F$, since $\overline{A(E)}$ is itself complete and metrizable.

We assume that $A(E)$ is not meagre. Let U_ρ be the open ball of radius ρ about o in E, i. e. the set of all $x\in E$ with $\|x\|<\rho$, where $\|x\|$ is an (F)-norm on E. First we show that the closure $\overline{A(U_\rho)}$ contains a ball V_σ about o in F.

Since $U_{\rho/2}$ is circled and absorbent, $\bigcup_{n=1}^{\infty} n\,U_{\rho/2}=E$, and so $A(E)$ $=\bigcup_{n=1}^{\infty} A(n\,U_{\rho/2})$. Since $A(E)$ is not meagre, one of the sets $A(n\,U_{\rho/2})$ $=n\,A(U_{\rho/2})$, and consequently $A(U_{\rho/2})$ itself, is not nowhere dense in F. There therefore exists a $y\in \overline{A(U_{\rho/2})}$, together with a neighbourhood V of o in F, for which $y+V\subset\overline{A(U_{\rho/2})}$. It follows from this that $V\subset -y+\overline{A(U_{\rho/2})}\subset\overline{A(U_{\rho/2})}+\overline{A(U_{\rho/2})}\subset\overline{A(U_\rho)}$. However, V contains a ball V_σ.

Next we show that if $\rho'>\rho$ then $A(U_{\rho'})$ contains the ball V_σ. Suppose that $y_0\in V_\sigma$. We put $\rho=\rho_1$, $\sigma=\sigma_1$, and choose $\rho_2>\rho_3>\cdots$ such that $\sum_1^\infty \rho_i<\rho'$. For each U_{ρ_n} there is a ball V_{σ_n} in which $A(U_{\rho_n})$ is dense.

We can take $\sigma_n \to 0$.

Since $A(U_{\rho_1})$ is dense in V_{σ_1}, there is a x_1 with $\|x_1\| < \rho_1$ for which the image $A(x_1) = y_1$ satisfies $\|y_0 - y_1\| < \sigma_2$. $A(U_{\rho_2})$ is dense in V_{σ_2}, and so there is an x_2 with $\|x_2\| < \rho_2$ for which the image $A(x_2) = y_2$ satisfies $\|y_0 - y_1 - y_2\| < \sigma_3$, and so on. It follows from $\left\|\sum_N^M x_i\right\| \leqq \rho_N + \cdots + \rho_M$ and the completeness of E that the series $\sum_{i=1}^{\infty} x_i$ converges to an element x_0; since $\sum_{i=1}^{\infty} \rho_i < \rho'$, x_0 lies in $U_{\rho'}$, and further $A(x_0) = \lim_{N \to \infty} \sum_1^N A(x_i)$ $= \lim \sum_1^{\infty} y_i = y_0$, so that $y_0 \in A(U_{\rho'})$.

We have therefore shown that the image of every open neighbourhood of $\mathfrak{o}$ contains an open neighbourhood of $\mathfrak{o}$. This implies that the image of every open set is open; consequently A is a topological homomorphism.

Proof of (2). If $A(E)$ is closed, $A(E)$ is a complete metric space, and so by BAIRE's theorem (§ 4, 6.(5)) it is not meagre in itself. A is therefore a topological homomorphism. If conversely A is a topological homomorphism, then by 4.(4) A determines a topological isomorphism $\breve{A}$ of the quotient space $E/N[A]$ onto $A(E)$. But since $E/N[A]$ is complete (11.(4)), $A(E)$ is also complete, and so it is closed.

By the graph $G(A)$ of a linear mapping A from a topological vector space $E[\mathfrak{T}_1]$ into a topological vector space $E[\mathfrak{T}_2]$ we mean the linear subspace of the topological product $E[\mathfrak{T}_1] \times F[\mathfrak{T}_2]$ which consists of all pairs (x, Ax), $x \in E$. The closed-graph theorem asserts that

(3) *A linear mapping A from a complete metrizable topological vector space E into a complete metrizable topological vector space F is continuous if and only if its graph $G(A)$ is closed in $E \times F$.*

Proof. a) Suppose that A is continuous. A sequence $(x_n, Ax_n) \in G(A)$ is a Cauchy sequence only if x_n is a Cauchy sequence in E. Let x_0 be the limit of x_n. Because A is continuous, Ax_0 is also the limit of the sequence Ax_n. Thus the Cauchy sequence (x_n, Ax_n) has limit (x_0, Ax_0), and $G(A)$ is closed.

b) If conversely $G(A)$ is closed, the projection $(x, Ax) \to x$ is a continuous one-one linear mapping from the complete metric space $G(A)$ onto E, so that by (2) it is a topological isomorphism. It therefore follows from $x_n \to \mathfrak{o}$ in E that $(x_n, Ax_n) \to \mathfrak{o}$ in $E \times F$, and so $Ax_n \to \mathfrak{o}$ in F. A is therefore continuous at $\mathfrak{o}$, and consequently it is continuous everywhere.

The equivalent formulation given by BANACH reads:

(3′) *A linear mapping A from the complete metrizable space E into the complete metrizable space F is continuous if and only if whenever $x_n \to x_0$ and $Ax_n \to y_0$ then $Ax_0 = y_0$.*

The following assertion is more elementary:

(4) *A linear mapping A from a metrisable space E into a metrisable space F is continuous if $x_n \to \mathfrak{o}$ always implies that Ax_n is bounded.*

Proof. It follows from (F4) that

(5) $\|kx\| \leqq k\|x\|$ for each positive integer k.

If $x_n \to \mathfrak{o}$, then $\|x_n\| \to 0$, and there is a sequence of positive integers $k_n \to \infty$ for which $k_n\|x_n\| \to 0$ as well. Applying (5), $\|k_n x_n\| \to 0$, and so the image sequence $A(k_n x_n)$ is bounded. Then $\frac{1}{k_n} A(k_n x_n) = A(x_n) \to \mathfrak{o}$, by 6.(3), which means that A is continuous at $\mathfrak{o}$.

We establish two corollaries of the BANACH-SCHAUDER theorem.

(6) *Two closed algebraically complementary linear subspaces H_1 and H_2 of a complete metrizable topological vector space E are topologically complementary.*

For the topological product $H_1 \times H_2$ is again a complete metrizable space, and the mapping $(x_1, x_2) \to x_1 + x_2$ from $H_1 \times H_2$ onto E is one-one and continuous; it is therefore a topological isomorphism, by the BANACH-SCHAUDER theorem.

(7) *If a coarser metrizable topology $\mathfrak{T}'$ is given on a complete metrizable vector space $E[\mathfrak{T}]$, and if $E[\mathfrak{T}']$ is again complete, then $\mathfrak{T}'$ is equal to $\mathfrak{T}$.*

For the identity mapping of $E[\mathfrak{T}]$ onto $E[\mathfrak{T}']$ is a topological isomorphism, by (2).

13. Equicontinuous mappings, and the theorems of BANACH and BANACH-STEINHAUS. Let E be a topological space, F a uniform space, and $\mathfrak{A}$ a collection of mappings A from E into F. The collection $\mathfrak{A}$ is said to be equicontinuous at x_0 if for each vicinity N in F there is a neighbourhood $U(x_0)$ in E such that (Ax, Ax_0) lies in N for all $x \in U$, $A \in \mathfrak{A}$. $\mathfrak{A}$ is said to be equicontinuous on E if $\mathfrak{A}$ is equicontinuous at each point of E. If $\mathfrak{A}$ is equicontinuous at x_0, then clearly each $A \in \mathfrak{A}$ is continuous at x_0.

If E is also a uniform space, $\mathfrak{A}$ is said to be uniformly equicontinuous on E if for each vicinity N in F there is a vicinity M in E for which $(Ax, Ay) \in N$ whenever $(x, y) \in M$ and $A \in \mathfrak{A}$.

For linear mappings we have

(1) *A set $\mathfrak{A}$ of linear mappings A from the topological vector space E into the topological vector space F is uniformly equicontinuous if and only if it is equicontinuous at* o, *i. e. if and only if for each neighbourhood V of* o *in F there is a neighbourhood U of* o *in E for which* $A(U)\subset V$ *for all* $A\in\mathfrak{A}$.

For given the vicinity N_V of all (y_1,y_2) with $y_1-y_2\in V$ in F, it follows from $(x_1,x_2)\in N_U$, i.e. $x_1-x_2\in U$, that $A(x_1-x_2)=Ax_1-Ax_2\in V$. In other words $(Ax_1,Ax_2)\in N_V$ for all $A\in\mathfrak{A}$.

The uniform equicontinuity of a collection of linear mappings therefore follows from equicontinuity at o in just the same way as the uniform continuity of a single linear mapping follows from continuity at o.

We now prove the theorem of BANACH:

(2) *Let $\mathfrak{A}$ be a collection of continuous linear mappings A from the complete metrizable space E into the topological vector space F. $\mathfrak{A}$ is equicontinuous if and only if the set $\mathfrak{A}(x)$ of all Ax, $A\in\mathfrak{A}$, is bounded in F, for each $x\in E$.*

a) The condition is necessary: if V is a neighbourhood of o in F, there is a neighbourhood U of o in E with $A(U)\subset V$ for all $A\in\mathfrak{A}$. If $x\in\rho U$ then $Ax\in\rho V$ for all $A\in\mathfrak{A}$, so that $\mathfrak{A}(x)$ is bounded.

b) The condition is sufficient. Let V be a closed circled neighbourhood of o in F, and let W be another one with $W+W\subset V$. We form the set $M=\bigcap\limits_{A\in\mathfrak{A}} A^{(-1)}(W)$. Because of the continuity of A, each $A^{(-1)}(W)$ is closed, and so therefore is M. Further M is absorbent, for $\mathfrak{A}(x_0)$ is bounded for each $x_0\in E$, and so there is a $\rho>0$ with $\rho Ax_0\in W$ for all $A\in\mathfrak{A}$; i.e. $\rho x_0\in M$. Since M is absorbent and circled, $E=\bigcup\limits_{n=1}^{\infty} nM$. By BAIRE's theorem (§ 4, 6.(1)) one of the sets nM, and therefore M itself, contains an open set. Consequently the set $M-M$ contains a neighbourhood U of o, and we have $A(U)\subset A(M-M)\subset W-W\subset V$ for all $A\subset\mathfrak{A}$.

For normed spaces the theorem takes the following form:

(2′) *If $\mathfrak{A}$ is a collection of continuous linear mappings A from the* (B)-*space E into the normed space F, and if* $\sup\limits_{A\in\mathfrak{A}}\|Ax\|=M(x)<\infty$ *for each $x\in E$, then* $\sup\|A\|<\infty$.

The BANACH-STEINHAUS theorem now follows easily from this theorem of Banach's.

(3) *Let A_n be a sequence of continuous linear mappings from the complete metrizable space E into the topological vector space F. If the sequence of images $A_n x$ is bounded for each $x\in E$, and if it is a Cauchy*

sequence for the points x of a dense subset M of E, then $A_n x$ is a Cauchy sequence for each x.

If all the Cauchy sequences $A_n x$ have a limit Ax in F, then the mapping A defined in this way is linear and continuous.

Proof. Suppose that $x_0 \in E$. By (2), for each circled neighbourhood V of $\mathfrak{o}$ in F there is a neighbourhood U of $\mathfrak{o}$ in E with $A_n(U) \subset V$ for all n. We choose an $x \in M$ with $x - x_0 \in U$, and write

(4) $$A_n x_0 - A_m x_0 = (A_n x_0 - A_n x) + (A_n x - A_m x) + (A_m x - A_m x_0).$$

If n_0 is chosen large enough for $A_n x - A_m x$ to belong to V for all $n, m \geqq n_0$, then each of the three summands on the right-hand side of (4) lies in V, so that $A_n x_0 - A_m x_0 \in V + V + V$; $A_n x_0$ is therefore a Cauchy sequence.

If $Ax = \lim A_n x$ exists for each $x \in E$, then the mapping A defined in this way is clearly linear. If V is closed, then since $A_n(U) \subset V$ for all n, $A(U) \subset V$, so that A is continuous.

Banach [3] developed the theory of metrizable vector spaces which we have described here under apparently weaker hypotheses. He considered vector spaces E on which a translation-invariant metric $|x, y| = |x - y, \mathfrak{o}| = \|x - y\|$ is given, which is only required to satisfy

a) $\alpha_n x \to \mathfrak{o}$ if $\alpha_n \to 0$,

b) $\alpha x_n \to \mathfrak{o}$ if $x_n \to \mathfrak{o}$,

c) E is complete with respect to the metric.

We now show:

(5) *A vector space E on which a translation invariant metric is defined, with properties* a), b) *and* c), *is a complete metrizable topological vector space.*

We need only show that E is a topological vector space when the sets U_ε, $\varepsilon > 0$, are taken as a base of neighbourhoods of $\mathfrak{o}$, where U_ε is the closed ball given by $\|x\| \leqq \varepsilon$. We shall prove that the conditions of 1.(2) are satisfied.

The sets U_ε clearly form a filter-base with $\bigcap_{\varepsilon > 0} U_\varepsilon = \mathfrak{o}$. (LT 1) is valid, since $U_{\varepsilon/2} + U_{\varepsilon/2} \subset U_\varepsilon$, and (LT 3) is satisfied since $\frac{1}{n} x \in U_\varepsilon$ always holds for a suitable n, by a), so that $x \in n U_\varepsilon$.

The main difficulty is to prove (LT 2), and so to prove that αx is continuous in both variables; it is only assumed in a) and b) that αx is continuous in each variable separately. We prove this using the method of proof of the theorem of Banach; this itself is not directly applicable to the present situation.

For each $x \in E$ there is an integer k for which $\lambda x \in U_\varepsilon$ for all $|\lambda| \leqq \frac{1}{k}$. For otherwise there would be a sequence $\lambda_n \to 0$ with $|\lambda_n x| > \varepsilon$

for all n, contradicting a). We denote by M the set of all $x \in E$ for which $\lambda x \in U_\varepsilon$ for all $|\lambda| \leqq 1$. Since U_ε is closed, and because of b), M is closed. If x is an arbitrary element of E, and if $\lambda x \in U_\varepsilon$ for all $|\lambda| \leqq \frac{1}{k}$, then $\frac{x}{k} \in M$, so that $x \in kM$. Thus $E = \bigcup_{n=1}^{\infty} nM$, and by BAIRE's theorem once again one of the sets kM contains a closed ball $U_\delta(x_0)$. But then M itself contains the ball $U_{\delta/k}\left(\frac{x_0}{k}\right)$, for it follows by 12.(5) from $\left\|\frac{x_0}{k} - z\right\| \leqq \frac{\delta}{k}$ that

$$\|x_0 - kz\| = \left\|k\left(\frac{x_0}{k} - z\right)\right\| \leqq k\left\|\frac{x_0}{k} - z\right\| \leqq \delta.$$

In other words all the kz belong to $U_\delta(x_0) \subset kM$, so that all the z belong to M.

$M - M$ contains $U_{\delta/k}(\mathrm{o})$, so that if $|\lambda| \leqq 1$ and $x \in U_{\delta/k}(\mathrm{o})$ then $\lambda x \in U_{2\varepsilon}(\mathrm{o})$. This establishes (LT2).

We observe that the neighbourhoods used in the proof of (5) need not be circled, and therefore $\|x\|$ does not necessarily satisfy (F3); using the results of (5) and 11.(2), however, we can pass over to an equivalent (F)-norm.

14. Bilinear mappings. We defined bilinear mappings in § 9, 7. Here we consider bilinear mappings $B(x,y)$ from the topological product $E_1 \times E_2$ of two topological vector spaces into a third topological vector space F.

For fixed x, $B(x,y)$ defines a linear mapping B_x from E_2 into F, and for fixed y it defines a linear mapping B_y from E_1 into F.

A bilinear mapping B from $E_1 \times E_2$ into F is said to be continuous if it is continuous as a mapping from $E_1 \times E_2$ into F, and so if it is continuous in both variables together. If this is so, then in particular all the mappings B_x and all the mappings B_y are continuous linear mappings from E_2 and E_1, respectively, into F.

If it is only assumed that all the mappings B_x and all the mappings B_y are continuous, then $B(x,y)$ is said to be separately continuous.

Equicontinuity and separate equicontinuity are defined in a corresponding way for a collection $\mathfrak{B}$ of bilinear mappings.

(1) *A bilinear mapping B (a collection $\mathfrak{B}$ of bilinear mappings) is continuous (equicontinuous) if it is continuous (equicontinuous) at* (o,o).

We prove this for equicontinuity. We write

$$(2) \qquad B(x,y) - B(x_0,y_0) = B(x_0, y - y_0) + B(x - x_0, y_0) + B(x - x_0, y - y_0).$$

Because of the equicontinuity at (o,o), given a neighbourhood W of o and $W_1+W_1+W_1 \subset W \subset F$, there is a neighbourhood $U \times V$ of (o,o) in $E_1 \times E_2$ for which $B(x-x_0, y-y_0) \in W_1$ for all $B \in \mathfrak{B}$ and all $x-x_0 \in U, y-y_0 \in V$. If $\frac{1}{n} x_0 \in U$, then $B(x_0, y-y_0) = B\left(\frac{x_0}{n}, n(y-y_0)\right) \in W_1$ for all B and for all $y-y_0 \in \frac{1}{n} V$. Likewise it follows from $\frac{1}{m} y_0 \in V$ that $B(x-x_0, y_0) \in W_1$ for all B and all $x-x_0 \in \frac{1}{m} U$. Thus we have $B(x,y)-B(x_0,y_0) \in W$ for all B, provided that $(x-x_0, y-y_0) \in \left(\frac{1}{m} U\right) \times \left(\frac{1}{n} V\right)$.

We have the following important theorem (BOURBAKI):

(3) *Every separately continuous bilinear mapping $B(x,y)$ from the product of two complete metrizable spaces E_1, E_2 into a topological vector space F is continuous.*

A collection $\mathfrak{B}$ of bilinear mappings $B(x,y)$ is equicontinuous if and only if the mappings B are separately continuous and the set $\mathfrak{B}(x,y)$ of all values $B(x,y)$ is bounded in F for each fixed $(x,y) \in E_1 \times E_2$.

It is enough to show that the conditions of the second assertion are sufficient.

We establish a preliminary lemma:

(4) *A collection $\mathfrak{A}$ of mappings A from a metric space E into a uniform space F is equicontinuous at x_0 if whenever $x_n \to x_0$ in E then Ax_n converges uniformly to Ax_0.*

Let N be a vicinity in F. It is enough to show that the set V of all $x \in E$ for which $(Ax, Ax_0) \in N$ for all $A \in \mathfrak{A}$ is a neighbourhood of x_0. If this were not the case, there would be a sequence $x_n \notin V$ with $x_n \to x_0$, and the sequence Ax_n would not converge uniformly to Ax_0.

By (1) and (4) it is therefore sufficient to show that whenever $(x_n, y_n) \to (o,o)$ in $E_1 \times E_2$ then the sequence $B(x_n, y_n)$, $B \in \mathfrak{B}$, converges uniformly to o in F.

By hypothesis, for fixed x_0 the set $\mathfrak{B}_{x_0}$ of continuous linear mappings $B_{x_0} = B(x_0, y)$, $B \in \mathfrak{B}$, from E_2 into F maps each y_0 into a bounded set. By the theorem of BANACH $\mathfrak{B}_{x_0}$ is therefore equicontinuous; since $B(x_0, o) = o$, for each neighbourhood W of o in F there is a neighbourhood V of o in E_2 with $B(x_0, y) \in W$ for all $B \in \mathfrak{B}$ and all $y \in V$.

We now show that the set $C = \bigcup_{n=1,2,\ldots} \mathfrak{B}(x_0, y_n)$ is bounded in F. The y_n lie in V from n_0 onwards, and so $\bigcup_{n \geqq n_0} \mathfrak{B}(x_0, y_n) \subset W$; for fixed

$n=1,\ldots,n_0-1$, however, $\mathfrak{B}(x_0,y_n)$ is bounded, so that $C\subset mW$ for some suitable $m>0$; consequently C is bounded.

If we now consider the $B(x,y)$ as mappings B_y from E_1 into F, we have shown that the set of mappings B_{y_n}, $B\in\mathfrak{B}$, $n=1,2,\ldots$, satisfies the hypotheses of the theorem of BANACH. This set is therefore once again equicontinuous, and so there is a neighbourhood U of $\circ$ in E_1 for which $\bigcup_{\substack{x\in U\\ n=1,2,\ldots}}\mathfrak{B}(x,y_n)\subset W$. For $n\geqq n_1$, therefore $B(x_n,y_n)\in W$ for all $B\in\mathfrak{B}$; this completes the proof.

§ 16. Convex sets

1. The convex and absolutely convex cover of a set. Convex sets in arbitrary real or complex vector spaces have properties which are essentially more complicated than in the n-dimensional case. We shall give a brief introduction to the theory of these sets, without striving for completeness. Not everything will be needed later. Reference may be made to the accounts of BOURBAKI [6] and KLEE [2].

Convex and absolutely convex sets have already been defined in § 15, 10. The concept of convex set is the same for both real and complex spaces, while the concept of absolutely convex set is different in the two cases. Because of this, we shall from time to time speak of real or complex absolutely convex sets.

We observe that every complex vector space E can also be interpreted as a real vector space; if $\{x_\alpha\}$ is a complex algebraic basis of E then the vectors x_α and ix_α together form a real algebraic basis of E.

If follows directly from the definitions that the intersection of arbitrarily many convex (respectively absolutely convex) subsets is again convex (absolutely convex).

The convex cover $C(M)$ of an arbitrary set M is the intersection of all the convex subsets of E containing it. $C(C(M))=C(M)$.

If $M=\bigcup_\alpha M_\alpha$, we shall also write $\underset{\alpha}{C}\, M_\alpha$ for $C(M)$.

The (real or complex) absolutely convex cover $\Gamma(M)$ or $\underset{\alpha}{\Gamma}\, M_\alpha$ is defined similarly.

(1) *$C(M)$ consists of all elements of the form* $\sum_1^n \alpha_i x_i$, $\alpha_i\geqq 0$, $\sum_1^n\alpha_i=1$, $x_i\in M$. *Likewise $\Gamma(M)$ consists of all elements of the form* $\sum_1^n \rho_i x_i$, $\rho_i\in K$, $\sum_1^n|\rho_i|\leqq 1$, $x_i\in M$.

Proof. It can be confirmed immediately that the elements of the given form constitute a convex (respectively absolutely convex) set containing M.

Conversely, we shall show that these elements belong to every convex (respectively absolutely convex) set containing M. We shall prove this for $\complement(M)$; the proof for $\Gamma(M)$ is analogous.

Suppose that it has been shown for $n-1$ that every element $\sum_1^{n-1} \alpha_i x_i$ of the above form lies in $\complement(M)$. Given $\sum_1^n \alpha_i' x_i$ we put $\sum_1^{n-1} \alpha_i' = \alpha$. By the inductive hypothesis $y = \sum_1^{n-1} \frac{\alpha_i'}{\alpha} x_i$ lies in $\complement(M)$; so therefore does $\alpha y + (1-\alpha) x_n = \sum_{i=1}^n \alpha_i' x_i$, since $0 < \alpha \leqq 1$.

We recall the concept of the circled cover of a set (§ 15, 1.). This concept is also different in the real and complex case.

(2) *The absolutely convex cover of a set M is the convex cover of the circled cover of M. A set is absolutely convex if and only if it is convex and circled.*

The circled cover of the convex cover of a set M need not be convex.

Corollary. If the sets M_α are circled, $\Gamma_\alpha M_\alpha = \complement_\alpha M_\alpha$.

Proof. a) Since every absolutely convex set is circled, it is sufficient to show that the convex cover of a circled set M is absolutely convex. By (1), $\complement(M)$ consists of all $\sum_1^n \rho_i x_i$ with $\rho_i \geqq 0$, $\sum_1^n \rho_i = 1$, and so it follows from (1) that $\complement(M) \subset \Gamma(M)$. Conversely given $\beta_i \neq 0$ in K with $\sum_1^n |\beta_i| \leqq 1$, if we put $\alpha_i = \beta_i \frac{\sum |\beta_k|}{|\beta_i|}$ and $\rho_i = \frac{|\beta_k|}{\sum |\beta_i|}$, we see that every element $x = \sum \beta_i x_i \in \Gamma(M)$ can be written in the form $\sum \rho_i \alpha_i x_i$, with $\sum \rho_i = 1$, $\rho_i \geqq 0$, $|\alpha_i| \leqq 1$ $\alpha_i x_i \in M$, so that x belongs to $\complement(M)$. Hence we also have $\Gamma(M) \subset \complement(M)$.

b) Neither the real circled cover in P^2 nor the complex circled cover in Γ^2 of the convex cover of the three points (0,0), (1,0) and (1,1) is convex (proof!).

c) The corollary follows from the fact that $\bigcup M_\alpha$ is circled if the M_α are, and so by (2) we have $\complement_\alpha M_\alpha = \complement\left(\bigcup_\alpha M_\alpha\right) = \Gamma\left(\bigcup_\alpha M_\alpha\right) = \Gamma_\alpha M_\alpha$.

(3) *If $C_1, \ldots, C_n$ are convex (respectively absolutely convex) and $\alpha_1, \ldots, \alpha_n$ are arbitrary elements of K, then $\alpha_1 C_1 + \cdots + \alpha_n C_n$ is convex (absolutely convex).*

Since αC is convex or absolutely convex if C is, it is sufficient to establish the assertion for C_1+C_2. For convex C_1 and C_2 this follows from the relation

$$\tau(x+y)+(1-\tau)(x'+y')=[\tau x+(1-\tau)x']+[\tau y+(1-\tau)y']$$

for $x, x' \in C_1$, $y, y' \in C_2$, $0 \leqq \tau \leqq 1$, and the absolutely convex case is proved analogously.

x_0+C is clearly convex if C is. This result, however, does not hold for absolute convexity, for which the point o plays a special role. If C is absolutely convex we say that x_0+C is absolutely convex about x_0.

(4) *The linear image and the linear inverse image of a convex or absolutely convex set are again convex or absolutely convex, respectively.*

Let A be a linear mapping from the vector space E into the vector space F. $A(C)$ is convex if C is, since $A(\tau x+(1-\tau)y)=\tau Ax+(1-\tau)Ay$. Further if M is convex in $A(E)$ and if Ax and Ay belong to M, then $A(\tau x+(1-\tau)y) \in M$, so that $A^{(-1)}(M)$ is convex. The proof for absolutely convex sets is similar.

Now suppose that E is a topological vector space $E[\mathfrak{T}]$.

(5) *The closure $\bar{C}$ of a convex or absolutely convex set C in $E[\mathfrak{T}]$ is again convex or absolutely convex, respectively.*

We shall prove this for absolutely convex sets. The proof for convex sets is similar.

We must show that if x_0 and y_0 belong to $\bar{C}$, then $\alpha x_0+\beta y_0$ lies in $\bar{C}$ too, if $|\alpha|+|\beta| \leqq 1$. Given a neighbourhood U of o, let V be a circled neighbourhood of o with $V+V \subset U$. To x_0 and y_0 there correspond elements x and y in C with $x_0-x \in V$, $y_0-y \in V$. But then

$$(\alpha x_0+\beta y_0)-(\alpha x+\beta y)=\alpha(x_0-x)+\beta(y_0-y) \in V+V \subset U,$$

so that $\alpha x_0+\beta y_0 \in \bar{C}$.

If M is a subset of $E[\mathfrak{T}]$, the intersection of all the closed convex sets containing M is called the closed convex cover of M. The closed absolutely convex cover is defined in a similar way.

(6) *The closed convex cover of M is equal to the closure $\overline{C(M)}$ of the convex cover of M. $\overline{C(M)}$ is also equal to $\overline{C(\bar{M})}$.*

The closed absolutely convex cover is equal to $\overline{\Gamma(M)}=\overline{\Gamma(\bar{M})}$.

Once again we shall prove this for absolutely convex sets. By (5), $\overline{\Gamma(M)}$ is closed and absolutely convex, and it is clearly contained in all the closed absolutely convex sets which contain M, so that it is the closed absolutely convex cover of M. The last assertion follows from the fact that every closed absolutely convex set containing M must also contain $\Gamma(\bar{M})$.

(7) *If M is open, then so are $\complement(M)$ and $\Gamma(M)$.*

An element of $\Gamma(M)$ is of the form $\sum_1^n \alpha_i x_i$, $\sum_1^n |\alpha_i| = \alpha \leqq 1$, $\alpha_i \neq 0$. There is a circled neighbourhood V of o for which all the sets $x_i + V$ are contained in M. But then every element $\sum_1^n \alpha_i x_i + \alpha z$, $z \in V$, lies in $\Gamma(M)$, since it is of the form $\sum_1^n \alpha_i \left(x_i + \frac{|\alpha_i|}{\alpha_i} z \right)$, $z \in V$.

If M is closed, $\complement(M)$ and $\Gamma(M)$ need not be closed. An example of this is given by the real absolutely convex cover of the closed set M in the plane which consists of the points $(-1,0)$, $(1,0)$ and the y-axis.

2. The algebraic boundary of a convex set. Let C be a convex subset of the vector space E.

(1) *The set $C-C$ is real absolutely convex.*

By 1.(3), $C-C$ is certainly convex. If $0<\tau<1$ and $z_1, z_2 \in C$, then $\tau(z_1 - z_2) = z_1 - (\tau z_2 + (1-\tau) z_1)$ lies in $C-C$, since C is convex. Since $C-C = -(C-C)$, it follows from this that $C-C$ is real circled. The assertion now follows from 1.(2).

If C is an arbitrary subset of E, we call the real linear subspace of E spanned by all the elements of $C-C$, i. e. by all the differences of points of C, the real linear space $L(C)$ associated with C.

(2) *If C is convex, $L(C) = \bigcup_{n=1}^{\infty} n(C-C)$, and if C is absolutely convex, $L(C) = \bigcup_{n=1}^{\infty} nC$.*

It follows from (1) that $\bigcup n(C-C)$ is real absolutely convex. Since further every αx, $\alpha \in \mathrm{P}$, lies in $\bigcup n(C-C)$ if x does, $\bigcup n(C-C)$ is a real vector space.

The assertion about absolutely convex sets is proved in a similar way.

(3) *Let C be an arbitrary subset of E. The smallest real linear manifold $M(C)$ containing C is equal to $C + L(C) = x_0 + L(C)$, where x_0 is any point of C.*

If a linear manifold $z + H$ contains the set C, then the linear space H parallel to it contains all the differences $x_1 - x_2$ of elements of C, and so contains $C-C$. Hence $H \supset L(C)$. On the other hand, if $x_0 \in C$ then clearly $x_0 + L(C) \supset C$. Finally it follows from $x = x_0 + (x - x_0)$ that $x + L(C) = x_0 + L(C)$, for arbitrary $x \in C$.

A point x_0 of a set $C \subset E$ is called an internal point of C if every straight line through x_0 which lies in $M(C)$ contains x_0 as an interior point.

When $M(C)=E$, an internal point of C is called an algebraic interior point of C. The collection of all algebraic interior points is called the algebraic kernel C^i of C.

The algebraic hull C^a of C consists of all those points $y \in E$ for which there exists an $x \in C$ for which $[x,y) \subset C$ ($[x,y)$ denotes the real line segment joining x and y, including x and excluding y).

A point of C^a which is not an algebraic interior point of C is called an algebraic boundary point of C. The collection of all the algebraic boundary points of C is called the algebraic boundary of C. A set C is said to be algebraically closed if $C=C^a$, and to be algebraically open if $C=C^i$.

We remark that an algebraic boundary point of a subset of a topological vector space is always a topological boundary point; Example 2 below shows that the converse is not true, even if the set is convex.

In P^n, every convex set C contains an internal point; indeed, if $M(C)=\mathsf{P}^m$, C contains an m-dimensional simplex, and the centroid of this is an internal point of C. For infinite-dimensional spaces this result does not always hold.

Example 1. Let $\{x_\beta\}$, $\beta \in \mathsf{B}$, be an algebraic basis of the infinite-dimensional vector space E, and let C be the convex cover of $\mathfrak{o}$ and all the x_β. $M(C)=E$. We shall show that C has no algebraic interior points, so that it consists of algebraic boundary points alone. By 1.(1) every element z of C is of the form $z=\sum_1^n \alpha_i x_{\beta_i}$, $\alpha_i \geqq 0$, $\sum \alpha_i \leqq 1$. If $\beta \neq \beta_i$, $i=1,\dots,n$, then the straight line $z+\sigma(x_\beta - z)$ through x_β and z meets C in the segment $[x_\beta, z]$, so that z is not an interior point of C.

(4) *If C is convex, the algebraic hull C^a and algebraic kernel C^i are again convex (the empty set is taken to be convex).*

Proof. a) If y_1 and y_2 are two points of C^a, and if x_1 and x_2 are two points of C with $[x_1,y_1) \subset C$ and $[x_2,y_2) \subset C$, then x_1, x_2, y_1 and y_2 are the four vertices of a tetrahedron, all of whose interior points belong to C. But then the boundary points belong to C^a and so therefore does $[y_1,y_2]$.

b) Suppose that x_1 and x_2 lie in C^i, that x is any point between x_1 and x_2, and that g is any straight line through x. Then on the straight line through x_1 (respectively x_2) parallel to g there is an interval $[y_1,z_1]$ (respectively $[y_2,z_2]$) lying in C. The intersection of g with the quadrilateral y_1, z_1, z_2, y_2 is contained in C; it is an interval in g which contains x as an interior point. Thus x belongs to C^i.

In 4. we shall show that $(C^i)^i = C^i$. We now give a counter-example to show that $(C^a)^a = C^a$ does not always hold.

Example 2. Let E be real and infinite-dimensional, and let $\{x_\alpha\}$, $\alpha \in \mathsf{A}$, be an algebraic basis of E. We denote by C the set of all non-zero $x=\sum_\alpha \xi_\alpha x_\alpha$ whose coefficients are non-negative and satisfy $\sum \xi_\alpha \geqq \frac{1}{n(x)}$, where $n(x)$ is the number of α with $\xi_\alpha \neq 0$. C is convex, for if $z=\sum \zeta_\alpha x_\alpha = \tau x+(1-\tau)y$, $y=\sum \eta_\alpha x_\alpha$, with x and y in C, then $n(z) \geqq \max(n(x), n(y)) > 0$; consequently

$$\sum \zeta_\alpha = \tau \sum \xi_\alpha + (1-\tau)\sum \eta_\alpha \geqq \frac{\tau}{n(x)} + \frac{1-\tau}{n(y)} \geqq \frac{1}{n(z)}$$

which shows that $z \in C$.

Let us determine C^a. We assert that C^a is equal to the set of all $x=\sum \xi_\alpha x_\alpha$ with $\xi_\alpha \geqq 0$ and $n(x)>0$. If $x=\sum \xi_\alpha x_\alpha$ is any non-zero element with $\xi_\alpha \geqq 0$, then $\sum \xi_\alpha > 0$, so that there is a positive integer m for which $\sum \xi_\alpha \geqq \frac{1}{m}$. Then element $z = \frac{x_{\alpha_1}+\cdots+x_{\alpha_m}}{m}$ lies in C. If we choose the α_i to be different from all those indices α for which the coefficients ξ_α of x are non-zero, then each $\tau x+(1-\tau)z$ with $0 \leqq \tau < 1$ belongs to C. x therefore belongs to C^a, as it is the end point of the segment $[z,x)$.

Further if y is any element of C, the whole of the segment $[y,\circ)$ is not in C, but only $\left[y, \frac{1}{n(y)\sum \eta_\alpha} y\right]$, so that $\circ$ does not belong to C^a.

Finally $\circ \in C^{aa}$, so that $C^a \neq C^{aa}$.

As we have just seen, C^a need not be algebraically closed. We can however repeat the process of forming the algebraic hull transfinitely. We put $C^a = C^1$; for an ordinal $\gamma+1$ we put $C^{\gamma+1}=(C^\gamma)^a$, and for a limit ordinal β we put $C^\beta = \bigcup_{\gamma<\beta} C^\gamma$. By (4) all the sets $C^{\gamma+1}$ are convex if C is, and so are the sets C^β, and we must have $C^{\mu+1}=C^\mu$ for a sufficiently large ordinal μ. Thus C^μ is the smallest algebraically closed convex set which contains C.

(5) *Every convex set C has an algebraically closed hull C^α, which is again convex.*

O. Nikodym [1] has shown that in fact arbitrarily large ordinals μ of cardinality $\leqq \aleph_0$ can be needed; cf. Klee [6] as well.

The next result gives a further example of the complicated nature of convex sets in infinite dimensional spaces.

(6) *In every infinite-dimensional vector space E there is a proper convex subset C with $C^a = E$.*

Again, let $\{x_\alpha\}$, $\alpha \in \mathsf{A}$, be a real algebraic basis of E, and let the index set be ordered in a way such that there is no final element. Let C be the set of all $\sum_{i=1}^{n} \xi_i x_{\alpha_i} \neq \circ$, $n=1,2,\ldots$, whose last non-zero coefficient is positive. C is clearly a proper convex subset of E. If $z=\sum \zeta_i x_{\alpha_i}$ is an arbitrary point of E and if $\alpha > \alpha_i$ for $i=1,\ldots,n$, then z is the end-point of the segment $[x_\alpha, z)$, which lies in C.

3. Half-spaces. For the present, let E be a real vector space. A hyperplane in E is either a linear subspace H of codimension 1 or a set x_0+H. We know (§ 15, 9.(1)) that corresponding to H there is always a real functional u in the algebraic dual space E^* whose null-space is exactly H. The hyperplane x_0+H is then the set of all the $x\in E$ for which $u(x)=u(x_0)=\gamma$.

The hyperplane x_0+H determines two algebraically open half-spaces, defined by $u(x)<\gamma$ and $u(x)>\gamma$ respectively; similarly it determines two algebraically closed half-spaces, defined by $u(x)\leqq\gamma$ and $u(x)\geqq\gamma$.

It is a simple consequence of the linearity of u that all these half-spaces are convex, that they are respectively algebraically open or closed and that $u(x)\leqq\gamma$ is the algebraic hull of $u(x)<\gamma$. The hyperplane x_0+H is the algebraic boundary of each of the four half-spaces.

A further consequence of § 15, 9.(1) is that, in a topological vector space $E[\mathfrak{T}]$, the half-space $u(x)\leqq\gamma$ is closed if u is continuous. If this is the case, the half-space $u(x)<\gamma$ is open. If u is not continuous, all four half-spaces are dense in $E[\mathfrak{T}]$.

If E is a complex vector space, E^* consists of complex linear functionals on E, and the question arises of how to characterize the half-spaces of a real hyperplane by means of a complex linear functional.

(1) *If a complex vector space E is considered as a real vector space, and if u is a real linear functional on it, then there is a uniquely determined complex linear functional v on the complex vector space E for which* $ux=\Re(vx)$.

If u is continuous on a complex topological vector space $E[\mathfrak{T}]$, *then v is also continuous; conversely, it follows from the continuity of the complex linear functional v that the real linear functional* $ux=\Re(vx)$ *is continuous.*

Proof. If ux is to be equal to the real part of vx, then the imaginary part $\Im(vx)$ must satisfy $\Im(vx)=\Re(-ivx)=-\Re(v(ix))=-u(ix)$. The only possible extension to v is therefore

$$vx=ux-iu(ix). \tag{2}$$

This is certainly a complex-valued linear function on E, when E is considered as a real space. v is however also complex linear, for we have

$$v(ix)=u(ix)-iu(-x)=i[ux-iu(ix)]=ivx.$$

If $|ux|\leqq\varepsilon$ for all x belonging to a complex circled neighbourhood U of $\mathfrak{o}$ in $E[\mathfrak{T}]$, then by (2) $|vx|\leqq 2\varepsilon$ on U, so that v is continuous if u is. Conversely it follows from $|vx|\leqq\varepsilon$ that $|\Re(vx)|\leqq\varepsilon$.

The analytic characterization of real hyperplanes x_0+H and half-spaces in a complex space E follows from (1): corresponding to x_0+H there is a complex linear functional $v\in E^*$ for which the points of x_0+H are characterised by the equation $\mathfrak{R}(v x)=\mathfrak{R}(v x_0)=\gamma$. The algebraically open half-spaces are given by $\mathfrak{R}(v x)\lessgtr\gamma$. These half-spaces are topologically open if and only if v is continuous.

Every real hyperplane H through $\mathfrak{o}$ contains just one complex hyperplane, namely $H\cap iH$. If H is given by $ux=0$, $H\cap iH$ is given by $vx=0$, where v is defined by (2).

4. Convex bodies and the Minkowski functionals associated with them. A particularly important class of convex sets is formed by those which have at least one algebraic interior point. These are called convex algebraic bodies, or convex α-bodies. The paradoxical possibilities of 2. cannot arise for such convex sets, as we shall see.

Every absorbent absolutely convex set is a convex α-body, since $\mathfrak{o}$ is an algebraic interior point.

If the underlying space is a topological vector space $E[\mathfrak{T}]$, we call a set C a convex $\mathfrak{T}$-body if C has an interior point x_0, in the sense of the topology $\mathfrak{T}$, and so if C is a convex $\mathfrak{T}$-neighbourhood of x_0. Every convex $\mathfrak{T}$-body is a convex α-body, but not conversely.

If $C\subset E$ is a convex α-body with $\mathfrak{o}$ as an algebraic interior point, we define (cf. § 15, 10.) the Minkowski functional $q(x)$ by the equation

$$(1) \qquad q(x)=\inf_{x\in\rho C}\rho \quad (\rho\geqq 0)$$

(2) *The Minkowski functional of a convex α-body C with $\mathfrak{o}$ as an algebraic interior point is a non-negative, positive homogeneous, sub-additive function on E; i. e. for all $x, y\in E$ it satisfies the conditions*

$$(\alpha) \qquad q(x)\geqq 0,$$

$$(\beta) \qquad q(\sigma x)=\sigma q(x) \quad \textit{for} \quad \sigma\geqq 0,$$

$$(\gamma) \qquad q(x+y)\leqq q(x)+q(y).$$

Proof. Because C is absorbent, $q(x)$ is defined for all $x\in E$, and (α) and (β) are clearly satisfied. (γ) is proved as in § 15, 10: Let $q(x)=\rho_0$, $q(y)=\sigma_0$. Then if $\rho>\rho_0$ and $\sigma>\sigma_0$, the elements $\frac{x}{\rho}$ and $\frac{y}{\sigma}$ lie in C, and since C is convex, so does $\frac{\rho}{\rho+\sigma}\frac{x}{\rho}+\frac{\sigma}{\rho+\sigma}\frac{x}{\sigma}=\frac{x+y}{\rho+\sigma}$. Hence $x+y\in(\rho+\sigma)C$, so that $q(x+y)\leqq\rho_0+\sigma_0$.

It follows from (β) and (γ) that $q(x)$ is a convex function on E, i. e.

(3) $$q(\tau x+(1-\tau)y)\leqq \tau q(x)+(1-\tau)q(y), \quad 0\leqq\tau\leqq 1,$$

for all $x, y\in E$.

(4) *If $q(x)$ is the Minkowski functional of the convex α-body C with $\circ$ as an algebraic interior point, then C^i consists of all x with $q(x)<1$, and C^a consists of all x with $q(x)\leqq 1$, so that the algebraic boundary of C is given by $q(x)=1$. Further, C^i is an algebraically open convex α-body, and C^a is algebraically closed.*

Proof. By (1), $q(x)\leqq 1$ if $x\in C$. If $q(x)<1$, then, conversely, $x\in C$. We now show that such an x also lies in C^i. Let $q(x)=\tau<1$, and let $0<\sigma<1-\tau$. The convex α-body $x+\sigma C$ has x as an algebraic interior point. It follows from

$$q(x+\sigma z)\leqq q(x)+\sigma q(z)\leqq \tau+\sigma<1,$$

for $z\in C$, that $x+\sigma C\subset C$, so that every point x with $q(x)<1$ lies in C^i.

Indeed, every such point lies in $(C^i)^i$, for it follows from $q(x+\sigma z)<1$ that $x+\sigma C\subset C^i$, so that x is an algebraic interior point of C^i. Hence C^i is also a convex α-body.

If $q(x)=1$, then because of (β), the interval $[\circ, x)$ lies in C, so that all those points with $q(x)\leqq 1$ lie in C^a. Conversely, suppose that $x\in C^a$. There exists an interval $[z,x)\subset C$. Then since $z+\tau(x-z)\in C$ for all $0\leqq\tau<1$, it follows from

$$x=z+\tau(x-z)+(1-\tau)(x-z)$$

that

$$q(x)\leqq q(z+\tau(x-z))+(1-\tau)q(x-z)\leqq 1+(1-\tau)q(x-z).$$

Letting $\tau\to 1$, it follows that $q(x)\leqq 1$. C^a is therefore characterized by $q(x)\leqq 1$.

If $q(x)=1$, then x does not belong to C^i, for otherwise a segment $[\circ,\sigma x]$, $\sigma>1$, would lie in C, and this is not possible, by (2) (β). Consequently, $(C^i)^i$ and C^i only contain points for which $q(x)<1$, and so consist of the collection of all such points; C^i is therefore algebraically open. Finally, if the argument which was given in the preceding paragraph for C is applied to C^a, we obtain $(C^a)^a=C^a$, so that C^a is algebraically closed.

We observe that it also follows from (4) that $(C^i)^a=C^a$. Since $[\circ, y)\subset C$ whenever $y\in C^a$, every algebraic boundary point of C can be reached from every algebraic interior point by a straight line in C^i.

We have the following converse:

(5) *If $q(x)$ is a function on E which satisfies conditions (α), (β) and (γ), then the set $q(x)<1$ (respectively $q(x)\leqq 1$) that it defines is an algebraically open (respectively algebraically closed) α-body with $\mathfrak{o}$ as an interior point, whose Minkowski functional is $q(x)$.*

Proof. The set C of all x with $q(x)\leqq 1$ is clearly absorbent. $\mathfrak{o}$ is an algebraic interior point of C. It follows directly from (3) that C is convex. The Minkowski functional of C is $q(x)$ again, and the remaining assertions follows from (4).

We remark that if $q(x)=0$ for some non-zero x, this means that the whole of the half-line from $\mathfrak{o}$ through x is contained in C. In particular E itself is described by the functional $q(x)\equiv 0$. Starting from a convex body C, and forming its Minkowski functional, we can certainly recover C^i and C^a, but cannot recover C itself, as C can be any arbitrary convex set satisfying $C^i \subset C \subset C^a$.

If we make the additional hypothesis on C that it is an absolutely convex subset of the real (respectively complex) vector space E, then in place of (β) we obtain the relation $q(\alpha x)=|\alpha|\, q(x)$, for arbitrary real (respectively complex) α. In this case, therefore, $q(x)$ is a real (respectively complex) semi-norm (cf. § 14, 1.). Consequently, we have

(6) *The Minkowski functional of an absorbent absolutely convex set is a semi-norm, and conversely the α-bodies defined by a semi-norm are absolutely convex.*

The question of when a convex α-body in a topological vector space $E[\mathfrak{T}]$ is a $\mathfrak{T}$-body is answered by

(7) *A convex α-body C in $E[\mathfrak{T}]$ with $\mathfrak{o}$ as an algebraic interior point is a $\mathfrak{T}$-body if and only if its Minkowski functional is continuous. If this is so, C^i is the interior of C and C^a is the closure of C; the algebraic and topological boundaries of C are therefore the same.*

Proof. a) Suppose that $\mathfrak{o}$ is an interior point of C. Then $q(x)\leqq\varepsilon$ for all x in the neighbourhood εC of $\mathfrak{o}$, i. e. q is continuous at $\mathfrak{o}$.

It follows from the continuity of $q(x)$ at $\mathfrak{o}$, however, that $q(x)$ is continuous at an arbitrary point $y\in E$, and indeed that $q(x)$ is uniformly continuous: Let V be a circled neighbourhood of $\mathfrak{o}$ contained in C. If $z\in\varepsilon V$, then $q(y+z)\leqq q(y)+\varepsilon$; similarly, the inequality $q(y)-\varepsilon\leqq q(y+z)$ follows from $q(y)=q((y+z)-z)\leqq q(y+z)+q(-z)$, since $-z\in\varepsilon V$. Hence $|q(y+z)-q(y)|\leqq\varepsilon$ for $z\in\varepsilon V$.

b) Conversely, if $q(x)$ is continuous, the set of all x with $q(x)<1$, i. e. C^i, is open and contains $\mathfrak{o}$, and the set $q(x)\leqq 1$, i. e. C^a, is closed, so that C^i is the interior of C and C^a is the closure of C.

The geometric properties of convex bodies with $\mathfrak{o}$ as an interior point naturally also hold for convex bodies with a general point x_0 as an interior point. If x_0 is an algebraic interior point of C and if $q(x)$ in the Minkowski functional of $C-x_0$, then C^a is given by $q(x-x_0)\leqq 1$.

We now make some comments indicating how these ideas can be applied to more general convex sets. If C is a convex set for which C^i is empty, then $(C^i)^i$ is also empty, so that we obtain the following general result from (4):

(8) *If C is an arbitrary convex set, then* $(C^i)^i = C^i$.

If C is an arbitrary convex set, with corresponding linear manifold $M(C)$, and if C has at least one internal point, then C is a convex α-body in $M(C)$, and we obtain from (4) that the set of internal points is algebraically open in $M(C)$, and that $(C^a)^a = C^a$. In particular, this result applies to all the convex subsets of an n-dimensional space, since these have internal points, by (2).

If a convex set has at least one internal point, and if $C^a = M(C)$, then it follows from (4) that we must have $C = M(C)$, and the pathological behaviour of example 2.(6) is therefore no longer possible.

5. Convex cones. A subset $K(x_0)$ of a vector space E is called a cone with vertex x_0 if $K(x_0)$ contains every point $x_0 + \rho(x - x_0)$, $\rho > 0$, whenever it contains x. A cone with vertex $\circ$ therefore contains ρx, $\rho > 0$, whenever it contains x. A cone $K(x_0)$ can always be obtained as a translate $x_0 + K(\circ)$ of a cone $K(\circ)$ with vertex $\circ$.

A cone $K(\circ)$ is convex if $K(\circ)$ always contains $x + y$ when it contains x and y. Conversely if a convex cone $K(\circ)$ contains x and y it always contains $\lambda x + \mu y$, for arbitrary positive λ and μ.

If $K(\circ)$ is a cone, so is $-K(\circ)$. More generally, $x_0 - K(\circ)$ is called the cone diametrically opposite to $K(x_0) = x_0 + K(\circ)$. We denote it by $K^*(x_0)$.

The real vector space $L(K(x_0))$ corresponding to a convex cone $K(x_0) = x_0 + K(\circ)$ is equal to $K(\circ) - K(\circ)$, by 2.(2).

A cone is said to be proper if it contains no real line through its vertex. It is said to be truncated if it does not contain its vertex, and to be pointed, if it does. A truncated cone is always proper. By removing the vertex of a proper pointed cone, a truncated cone is obtained, which is convex if the original cone was. Conversely, if the vertex is added to a truncated convex cone, then a proper pointed convex cone results. A cone $K(\circ)$ is proper if and only if either $K(\circ) \cap K^*(\circ) = \circ$ or $K(\circ) \cap K^*(\circ)$ is empty.

Every linear manifold is a convex cone. An algebraically open half-space is a truncated convex cone, and any point of the boundary hyperplane can be chosen as vertex.

The convex set C of 2.(6) is a truncated convex cone with vertex $\circ$ which has no algebraic interior point.

If $K(\circ)$ is a cone, so is $(K(\circ))^a$. The example which we have just mentioned shows that if $K(\circ)$ is a proper cone, $(K(\circ))^a$ need not be proper.

The linear image and the linear inverse image of a cone with vertex $\mathfrak{o}$ are again cones with vertex $\mathfrak{o}$.

The intersection of cones all with the same vertex is again a cone with the same vertex, and the same holds for the union.

(1) *Given a collection* $K_\alpha(\mathfrak{o})$ *of convex cones, the smallest convex cone containing all the* $K_\alpha(\mathfrak{o})$ *is equal to* $\sum_\alpha K_\alpha(\mathfrak{o})$.

For $\sum_\alpha K_\alpha(\mathfrak{o})$ is a convex cone, since the sum of two elements in $\sum_\alpha K_\alpha(\mathfrak{o})$ belongs to $\sum_\alpha K_\alpha(\mathfrak{o})$, and it is clearly the smallest convex cone containing all the cones $K_\alpha(\mathfrak{o})$.

The cone with vertex x_0 generated by a set M is the smallest cone with vertex x_0 which contains all the elements of M.

(2) *If M is convex, the cone with vertex $\mathfrak{o}$ generated by M is convex, and is equal to* $\bigcup_{\rho>0} \rho M$. *If* $\mathfrak{o} \notin M$ *the cone generated by M is truncated, and is therefore proper.*

The set $\bigcup_{\rho>0} \rho M$ is a cone, and it contains $\lambda x + \mu y$ if λ and μ are positive and x and y belong to it. The vertex $\mathfrak{o}$ is not contained in $\bigcup_{\rho>0} \rho M$ if $\mathfrak{o}$ does not belong to M.

If we construct the set C of example 2 of No. 2 in some hyperplane of a vector space and use it to generate a cone with vertex outside the hyperplane, then we obtain a convex cone for which $(K^a)^a \neq K^a$. Convex cones therefore exhibit the same pathological properties as arbitrary convex sets.

The example of the convex set $\xi\eta \geqq 1$, $\xi, \eta > 0$, and the cone with vertex $\mathfrak{o}$ which it generates shows that the cone generated by a closed set need not be closed.

In topological vector spaces, we have

(3) *If K is a convex cone in $E[\mathfrak{T}]$, then so also is its closure $\bar{K}$. K^i is a convex cone as well, and K^i is the interior of K, if K^i contains an interior point.*

Proof. We can take the vertex to be $\mathfrak{o}$. By 1.(5), $\bar{K}$ is convex. If z is a closure point of the points $x \in K$, then ρz, $\rho > 0$, is closure point of the points ρx, which also lie in K, so that $\bar{K}$ is a cone. If K^i is empty, then the second assertion is true. If K^i is not empty, then K is respectively a convex algebraic body or a convex $\mathfrak{T}$-body, and the second assertion follows from 4.(4) and 4.(7), since K^i is again a cone.

6. Hypercones. A maximal convex truncated cone in E with vertex x_0 is called a hypercone at x_0.

We have the following important existence theorem:

(1) *If M is convex, and if x_0 does not belong to M, then there is a hypercone at x_0 which contains M.*

Proof. By 5.(2) there is a truncated convex cone with vertex x_0 which contains M. Since the union of a totally ordered collection of such cones (ordered by inclusion) is again truncated and convex, it follows from ZORN's lemma that there is a maximal such cone.

The diametrically opposite cone $K^*(x_0)$ is a hypercone if $K(x_0)$ is, and we have

(2) *If* $K(x_0)$ *is a hypercone, then* $K(x_0) \cup K^*(x_0) = E \sim \{x_0\}$, *and* $K(x_0) \cap K^*(x_0)$ *is empty. The complement of a hypercone* $K(x_0)$ *is therefore the convex cone* $K^*(x_0) \cup \{x_0\}$.

The assertion that $K(x_0) \cap K^*(x_0)$ is empty is obvious, since a hypercone is convex and truncated, and is therefore proper. The first assertion follows from maximality: we may clearly suppose that $x_0 = \mathrm{o}$. If $K \cup (-K)$ were not equal to $E \sim \{\mathrm{o}\}$, there would be a real line λx not lying in $K \cup (-K)$. But then the collection of points ρx with $\rho > 0$, together with the points $\rho x + y$ with $y \in K$ and $\rho \geqq 0$, would form a truncated convex cone (since $\rho x + y \neq \mathrm{o}$), and this would contain K as a proper subset, which is not possible.

(3) *Conversely, if* $K(x_0)$ *is a convex cone with* $K(x_0) \cup K^*(x_0) = E \sim \{x_0\}$ *and with* $K(x_0) \cap K^*(x_0)$ *empty, then* $K(x_0)$ *is a hypercone.*

The significance of hypercones in the study of convex sets results from

(4) *Every proper convex subset* C *of* E *is the intersection of the hypercones containing it.*

For if $x_0 \notin C$, there is a hypercone at x_0 containing C and not x_0, by (1).

It follows immediately from (3) that

(5) *The intersection of a hypercone* $K(\mathrm{o})$ *with a linear subspace* H *of* E *is a hypercone in* H.

Example 2.(6) is an example of a hypercone whose algebraic hull is equal to E.

If we apply Theorem 2 of § 17, 1., then in particular we obtain from (2) that *the algebraically closed hull of a hypercone is either* E *or an algebraically closed half-space whose bounding hyperplane is the algebraic boundary of the hypercone.*

In the latter case the hypercone is a convex α-body. In the former case the hypercone has no algebraic interior point (cf. the final remark of 4.).

A hypercone $K(\mathrm{o})$ in P^n is a convex α-body, so that it has a hyperplane as algebraic boundary. The points of $K(\mathrm{o})$ lying in this space again form a hypercone, and so on. It is an easy consequence of this that a hypercone in P^n always has the following form: there is a basis $x_1, \ldots, x_n$ of P^n such that $K(\mathrm{o})$ consists of all the points $\sum_{i=1}^{n} \lambda_i x_i \neq \mathrm{o}$ for which the last non-zero co-ordinate is positive.

§ 17. The separation of convex sets. The HAHN-BANACH theorem

1. The separation theorem. Once again, let E be a vector space over K, where K is the real or complex field.

(1) *If A_1 and A_2 are two disjoint convex proper subsets of E, there exist two complementary convex subsets C_1 and C_2 of E with $C_1 \supset A_1$, $C_2 \supset A_2$.*

We shall give two proofs of this important theorem (cf. HAMMER [1] and BOURBAKI [6], Vol. 1, p. 53).

a) $A_1 - A_2$ is a convex subset of E, by § 16, 1.(3), and it does not contain $\mathfrak{o}$. By § 16, 6.(1), there is a hypercone K with vertex at $\mathfrak{o}$ for which $K \supset A_1 - A_2$. We set $C_1 = \bigcap_{x_2 \in A_2} (x_2 + K)$. Since $A_1 - x_2 \subset K$, so that $A_1 \subset x_2 + K$, it follows that $A_1 \subset C_1$. By § 16, 6.(2), the complement of a set $x_2 + K$ is equal to $x_2 + (K^* \cup \{\mathfrak{o}\})$, so that the complement of C_1 is equal to the union $C_2 = A_2 + (K^* \cup \{\mathfrak{o}\})$ of these complements. C_2 is convex, by § 16, 1.(3), and $C_2 \supset A_2$.

b) The following proof is independent of the theory of hypercones. We consider the collection of pairs (B_1, B_2) of disjoint convex sets with $B_1 \supset A_1$ and $B_2 \supset A_2$. The collection of these pairs is partially ordered if we put $(B_1, B_2) \leqq (B_1', B_2')$ when $B_1 \subset B_1'$ and $B_2 \subset B_2'$. Suppose that $B_1 \cup B_2 \neq E$. We shall show that another pair (B_1', B_2') exists with $(B_1, B_2) < (B_1', B_2')$.

Suppose that $x_0 \notin B_1 \cup B_2$. We assert that either the convex cover of x_0 and B_1 has an empty intersection with B_2, or the convex cover of x_0 and B_2 has an empty intersection with B_1. If this were not so, then for some $y_1 \in B_1$ there would be a point $z_2 \in B_2$ in the segment $[x_0, y_1]$, and for some $y_2 \in B_2$ there would be a point $z_1 \in B_1$ in the segment $[x_0, y_2]$. But then the point of intersection of the segments $[y_1, z_1]$ and $[y_2, z_2]$ would lie in both B_1 and B_2, giving a contradiction. The result now follows by applying ZORN's lemma to the collection of pairs (B_1, B_2).

(2) *If C_1 and C_2 are proper complementary convex subsets of E, then $C_1^a \cap C_2^a$ is either equal to E or is equal to a real hyperplane. In the latter case, i.e. if C_1^i and C_2^i are not both empty, C_1^i and C_2^i each coincide with one of the two algebraically open half-spaces defined by this hyperplane.*

Proof. We set $C_1^a \cap C_2^a = H$. $E \sim H = C_1^i \cup C_2^i$, for since $C_1 \cup C_2 = E$, each boundary point of C_1 (respectively C_2) is also a boundary point of C_2 (respectively C_1), and so lies in H. For the same reason, H is not empty. $C_1^a \cap C_2^a$ is convex, by § 16, 2.(4). Further, if H contains two points z_1 and z_2, it contains the whole of the straight line through them. For suppose that this were not the case, and that z was a point of the line, lying outside the segment $[z_1, z_2]$, which did not belong to H. Let us suppose that z_2 lies between z and z_1. z would lie either in C_1^i or in C_2^i. Suppose that $z \in C_1^i$. Then it follows from the remark preceding § 16, 4.(5) that every point of the segment $(z_1, z]$ would lie in C_1^i, so that z_2 would, and this is impossible. Thus we have shown that H is a real linear manifold.

Now suppose that $H \neq E$. We can suppose that $o \in H$, so that H is a linear subspace. Suppose that $x_0 \notin H$, and that $x_0 \in C_1^i$, say. Then $-x_0$ does not belong to H, either, and so $-x_0 \in C_1^i \cup C_2^i$. But $-x_0$ cannot belong to C_1^i, for otherwise, o, being a point of the segment $[x_0, -x_0]$, would lie in C_1^i, since C_1^i is convex (§ 16, 4.(4)). Thus $-x_0$ belongs to C_2^i. We assert that $H \oplus [x_0] = E$, where $[x_0]$ is the space of all real scalar multiples of x_0. If $x \in C_1$, $[x, -x_0]$ contains a point of H, so that $C_1 \subset H \oplus [x_0]$; if $y \in C_2$, $[y, x_0]$ contains a point of H, so that $C_2 \subset H \oplus [x_0]$, as well. Consequently $H \oplus [x_0] = E$.

Since $E \sim H = C_1^i \cup C_2^i$, one of the algebraically open half-spaces must concide with C_1^i, and the other with C_2^i.

We say that two sets M and N are **separated** by a real hyperplane H if they are contained in different algebraically closed half-spaces defined by H. We also say that they lie on opposite sides of H. M and N may have points of H in common. M and N are said to be **strictly separated** by H if they are contained in different algebraically open half-spaces defined by H.

The algebraic form of the **separation theorem** follows from (1) and (2):

(3) *If A_1 is a convex α-body and A_2 is a convex set which contains no algebraic interior point of A_1, then there is a real hyperplane H which separates A_1 and A_2, and which contains no algebraic interior point of A_1.*

If A_1 and A_2 are disjoint algebraically open convex α-bodies, then there exists a strictly separating real hyperplane.

Proof. By hypothesis A_1^i is non-empty and $A_1^i \cap A_2$ is empty. By (1), there exist complementary $C_1 \supset A_1^i$, $C_2 \supset A_2$. Since C_1^i is non-empty, there exists a hyperplane H, by (2), for which C_1^i, and therefore A_1^i, lies in one of the algebraically open half-spaces and A_2 lies in the complementary algebraically closed half-space. This establishes the first part of the theorem.

If $A_1 = A_1^i$ and $A_2 = A_2^i$, then, since $A_2^i \subset C_2^i$, A_2^i is contained in the other open half-space; in this case, therefore, the separation is strict.

In a topological vector space $E[\mathfrak{T}]$ the separation theorem takes the following geometric form (cf. M. Eidelheit [1], S. Kakutani [1]):

(4) *If A_1 is a convex $\mathfrak{T}$-body in $E[\mathfrak{T}]$, and if A_2 is a convex set containing no interior point of A_1, then there is a closed real hyperplane H separating A_1 and A_2, which contains no interior point of A_1.*

If A_1 and A_2 are disjoint convex open $\mathfrak{T}$-bodies, then there exists a strictly separating closed real hyperplane.

Proof. H is either dense in E or closed, by § 15, 9.(1). Since the set A_1^i is open (§ 16, 4.(7)) and disjoint from H, H must be closed.

It is not always possible to separate two disjoint algebraically closed convex α-bodies A_1 and A_2 strictly: in the plane, take A_1 to be the half-plane $\xi \leqq 0$, and take A_2 to be the set of all (ξ, η) with $\xi\eta \geqq 1$, $\xi, \eta > 0$.

2. The HAHN-BANACH theorem. By considering a special case of the algebraic form of the separation theorem we obtain

(1) *If C is a convex α-body in a vector space E and if M is a linear manifold which contains no algebraic interior point of C, then there is a hyperplane H containing M which again contains no algebraic interior point of C.*

For real E, (1) follows as a special case of 1.(3), when we replace the hyperplane separating C and M by a parallel hyperplane through a point of M; this must contain the whole of M. If E is complex, there is a real hyperplane H with the required properties, again by 1.(3). It is sufficient to consider the case where M goes through $\mathfrak{o}$. But then $H \cap iH$ is a complex hyperplane (cf. § 16, 3.) which again contains M, since $M \cap iM = M$, and *a fortiori* it contains no interior point of C.

From the geometric form of the separation theorem we obtain the following theorem, originally established by MAZUR [2], but called the geometric form of the HAHN-BANACH theorem by BOURBAKI:

(2) *If C is a convex $\mathfrak{T}$-body in a topological vector space $E[\mathfrak{T}]$, and if M is a linear manifold which contains no interior point of C, then there is a closed hyperplane H which contains M, and which again contains no interior point of C.*

If we apply the analytic descriptions of convex bodies and half-spaces which were established in the preceding section, (1) and (2) can be expressed in analytic form.

We must now distinguish between the real and the complex case.

(3) *Suppose that a non-negative positive-homogeneous sub-additive function $q(x)$ is given on a real vector space E. If a linear functional $l(z)$, defined on a linear subspace F, satisfies*

(4) $$l(z) \leqq q(z) \quad \text{for} \quad z \in F,$$

then $l(z)$ can be extended to a linear functional u, defined on the whole of E, which satisfies

(5) $$u x \leqq q(x) \quad \text{for} \quad x \in E.$$

If E is a topological vector space and $q(x)$ is continuous, then $u x$ is also continuous.

Proof. By § 16, 4.(5), the inequality $q(x)<1$ defines an algebraically open convex α-body $C \ni o$. We assume that $l(z)$ does not vanish on the whole of F (otherwise the linear functional which vanishes identically is a solution for (5)). There exists a point $z_0 \in F$ with $l(z)=1$, and $F=[z_0] \oplus F_1$, where l vanishes identically on F_1. l is therefore identically one on the linear manifold z_0+F_1. By (1), there exists a hyperplane z_0+H, with $F_1 \subset H$, which contains no point of C. z_0 does not lie in H, so that each point x of E has the form $x=\alpha z_0+y$, $y \in H$. We define u by $ux=u(\alpha z_0+y)=\alpha$. u is an extension of l, and again u takes the value 1 on the whole of z_0+H. The body C lies in the half-space $ux<1$ to which o belongs.

It is now easy to see that inequality (5) holds: since q is positive-homogeneous, it is sufficient to prove (5) for one point of each ray through o. If a ray cuts z_0+H at a point x, then the most x can be is a boundary point of C, i.e. $q(x) \geqq 1=ux$. A ray which does not cut z_0+H only contains points for which $ux \leqq 0$, so that (5) is satisfied, since $q(x) \geqq 0$.

Finally, if $q(x)$ is continuous, C is a convex $\mathfrak{T}$-body, and so the linear functional u defining H is continuous, by § 15, 9.(1).

For complex vector spaces we obtain

(6) *Suppose that a non-negative positive-homogeneous subadditive function $q(x)$ is given on a complex vector space E. If a complex linear functional $l(z)$, defined on a complex linear subspace F, satisfies*

$$\mathfrak{R}\, l(z) \leqq q(z), \quad \textit{for } z \in F, \tag{7}$$

then $l(z)$ can be extended to a complex linear functional v defined on the whole of E, which satisfies

$$\mathfrak{R}\, v x \leqq q(x), \quad \textit{for } x \in E. \tag{8}$$

If E is a topological vector space and $q(x)$ is continuous then v is also continuous.

This follows from (3), since $\mathfrak{R}\, l(z)$ is a real linear functional on the space F, considered as a real vector space; it can be extended to a linear functional which, by § 16, 3.(1), can be written uniquely as the real part of a complex linear functional vx. The continuity of v follows likewise from the continuity of $\mathfrak{R}\, v$.

3. The analytic proof of the HAHN-BANACH theorem. We have given two essentially geometric proofs of the HAHN-BANACH theorem, which follow from properties of convex sets. The classical proof (cf. HELLY [1], [2], HAHN [2], BANACH [2]) is of an analytic nature, and does not use convexity. It provides a somewhat more general result than 2.(3), as the hypothesis that $q(x) \geqq 0$ can be omitted.

(1) (HAHN-BANACH *theorem*). *Suppose that* $q(x)$ *is a positive-homogenenous subadditive function on a real vector space E. If a linear functional* $l(z)$*, defined on a linear subspace F, satisfies*

$$l(z) \leqq q(z), \quad \text{for} \quad z \in F, \tag{2}$$

then $l(z)$ *can be extended to a linear functional u, defined on the whole of E, which satisfies*

$$u x \leqq q(x), \quad \text{for} \quad x \in E. \tag{3}$$

If E is a topological vector space and $q(x)$ *is continuous at* $\circ$, *then u is also continuous.*

Proof. Suppose that $l(x)$ is defined for $F_1 \supset F$, and that (2) is satisfied on F_1. We shall show that if $x_0 \notin F_1$ the linear functional $l(x)$ can be extended to $[x_0] \oplus F_1$ in such a way that (2) is still satisfied.

If z and z' are arbitrary elements of F_1, then because of (2) and the properties of $q(x)$ we have

$$l(z') - l(z) = l(z' - z) \leqq q[(z' + x_0) + (-z - x_0)] \leqq q(z' + x_0) + q(-z - x_0),$$

so that
$$-q(-z - x_0) - l(z) \leqq q(z' + x_0) - l(z').$$

Since this holds for each $z, z' \in F_1$, we have

$$\sup_{z \in F_1} [-q(-z - x_0) - l(z)] \leqq \inf_{z' \in F_1} [q(z' + x_0) - l(z')].$$

Let γ be a number lying between these two values, i.e. let

$$-q(-z - x_0) - l(z) \leqq \gamma \leqq q(z + x_0) - l(z) \quad \text{for all} \quad z \in F_1. \tag{4}$$

We now set $l(\alpha x_0 + z) = \alpha \gamma + l(z)$ for all $z \in F_1$. In this way l is extended to $[x_0] \oplus F_1$. We shall show that (2) is satisfied.

First suppose that $\alpha > 0$. It follows from the right-hand side of (4) that $\gamma \leqq q\left(\frac{z}{\alpha} + x_0\right) - l\left(\frac{z}{\alpha}\right)$, and so that $\alpha \gamma \leqq \alpha q\left(\frac{z}{\alpha} + x_0\right) - \alpha l\left(\frac{z}{\alpha}\right)$ $= q(z + \alpha x_0) - l(z)$; i.e. $l(z + \alpha x_0) = \alpha \gamma + l(z) \leqq q(z + \alpha x_0)$.

On the other hand suppose that $\alpha = -\rho$, $\rho > 0$. It follows from the left-hand side of (4) that $-q\left(\frac{z}{\rho} - x_0\right) + l\left(\frac{z}{\rho}\right) \leqq \gamma$. Multiplying by ρ, we get $-q(z + \alpha x_0) + l(z) \leqq \rho \gamma$, and so $l(z + \alpha x_0) = \alpha \gamma + l(z) \leqq q(z + \alpha x_0)$.

The existence of u follows either by repeating the process, using transfinite induction, or by using ZORN's lemma. If $q(x)$ is continuous at $\circ$, then $q(x) \leqq \varepsilon$ for all x in some circled neighbourhood U of $\circ$. It follows from $ux \leqq q(x) \leqq \varepsilon$ and $-ux = u(-x) \leqq q(-x) \leqq \varepsilon$ that $|ux| \leqq \varepsilon$ for $x \in U$, and this implies that u is continuous.

The complex case results from this, as in the preceding number. It is identical with 2.(6) except that the hypothesis that $q(x) \geqq 0$ is omitted.

Conversely, the geometric form 2.(2) of the HAHN-BANACH theorem can also be derived from the analytic form (1) or 2.(3): let $q(x)$ be the Minkowski functional of a convex α-body C with $\mathfrak{o}$ as an algebraic interior point. C^i is given by $q(x) < 1$. M has the form $x_0 + F$, where F is a real vector space. $q(x_0 + y) \geqq 1$ for all $x_0 + y \in x_0 + F$. We set $l(\alpha x_0 + y) = \alpha$ on $[x_0] \oplus F$. If $\alpha > 0$, $l(\alpha x_0 + y) = \alpha l\left(x_0 + \frac{y}{\alpha}\right) = \alpha \cdot 1 \leqq \alpha q\left(x_0 + \frac{y}{\alpha}\right) = q(\alpha x_0 + y)$. If $\alpha < 0$, $l(\alpha x_0 + y) < 0$, so that $l(\alpha x_0 + y) < q(\alpha x_0 + y)$, since $q(x) \geqq 0$ on E. Hence the hypotheses of (1) are satisfied, and so there is an extension u of l with $ux \leqq q(x)$, $u(x_0) = 1$ and $uy = 0$ for $y \in F$. Consequently C^i is contained in the half-space $ux < 1$, and $x_0 + F$ is contained in the hyperplane $ux = 1$. This again proves 2.(1) in the real case; the complex case and 2.(2) follow as in 2.

For the most part, the HAHN-BANACH theorem is not applied in its general form (1), but in a more special form, which is the same for real and complex vector spaces:

(5) *Let $p(x)$ be a semi-norm on a vector space E. If $l(z)$ is a linear functional on a linear subspace F which satisfies*

$$(6) \qquad |l(z)| \leqq p(z), \quad \textit{for} \quad z \in F.$$

then $l(z)$ can be extended to a linear functional u, defined on the whole of E, which satisfies

$$(7) \qquad |ux| \leqq p(x) \quad \textit{for} \quad x \in E.$$

If E is a topological vector space and $p(x)$ is continuous, then u is also continuous.

Proof. In the real case it follows from $l(z) \leqq p(z)$ on F that there is an extension u with $ux \leqq p(x)$ on the whole of E. But then we also have $-ux = u(-x) \leqq p(-x) = p(x)$, i.e. (7) holds on the whole of E.

In the complex case, let us put $l_1(z) = \Re l(z)$. Then $|l_1(z)| \leqq p(z)$ on F. From what we have just shown, there is a real linear extension u_1 of l_1 with $|u_1 x| \leqq p(x)$ on the whole of E. By § 16, 3., $ux = u_1 x - i u_1(ix)$ is a complex linear extension of $l(z)$ to the whole of E. For arbitrary $x \in E$, let $ux = re^{i\vartheta}$. Then

$$|ux| = e^{-i\vartheta} ux = u(e^{-i\vartheta} x) = u_1(e^{-i\vartheta} x) \leqq p(e^{-i\vartheta} x) = p(x),$$

so that (7) is satisfied.

The next result is a special case of (5) which is frequently used:

(8) *If $p(x)$ is a continuous semi-norm on $E[\mathfrak{T}]$ and if x_0 is an arbitrary fixed point of E, then there is a continuous linear functional u on E with $|ux| \leqq p(x)$ and $ux_0 = p(x_0)$.*

For $l(\alpha x_0)=\alpha p(x_0)$ defines a linear functional on the one-dimensional subspace $[x_0]$ of E, and (5) can be applied to this.

Corresponding to (1), we have

(9) *If $q(x)$ is a positive-homogeneous subadditive function on a real (respectively complex) vector space E and if $x_0 \in E$, then there is a real (respectively complex) linear functional u on E with $ux \leqq q(x)$ on E and $ux_0 = q(x_0)$ (respectively with $\Re(ux) \leqq q(x)$ and $\Re(ux_0)=q(x_0)$).*

If E is a topological vector space, u is continuous if $q(x)$ is continuous at $\circ$.

Proof. In the real case, a linear functional on $[x_0]$ is defined by $l(\alpha x_0)=\alpha q(x_0)$. If $\alpha \geqq 0$, $l(\alpha x_0)=q(x_0)$ it follows from $q(\circ)=\rho q(\circ)$ for $\rho>0$ that $q(\circ)=0$. Since $0=q(\circ)\leqq q(x_0)+q(-x_0)$, $-q(x_0)\leqq q(-x_0)$, and so if $\alpha<0$ we have

$$l(\alpha x_0)=\alpha q(x_0) \leqq -\alpha q(-x_0)=q(\alpha x_0).$$

The linear functional l defined on $[x_0]$ therefore satisfies the hypotheses of (1).

It is easy to derive the complex case from the real case.

Let us remark that (5) need not hold if a complex-valued linear functional is only defined on a real linear subspace; cf. BOHNENBLUST and SOBCZYK [1].

4. Two consequences of the HAHN-BANACH theorem. In § 15, 9. we saw that there are topological vector spaces on which the only continuous linear functional is the one which vanishes identically. We have the following criterion:

(1) *There exist non-trivial continuous linear functionals on a topological vector space $E[\mathfrak{T}]$ if and only if there is a convex neighbourhood of $\circ$ in E which is different from the whole space E.*

Proof. If u is a non-trivial continuous linear functional on $E[\mathfrak{T}]$, the set of all x with $|ux|\leqq 1$ is an absolutely convex neighbourhood of $\circ$ in E which is different from E. Conversely, if U is a convex neighbourhood of $\circ$ the corresponding Minkowski functional $q(x)$ is continuous, by § 16, 4.(7). If x_0 is a point for which $q(x_0)\neq 0$, then it follows from 3.(9) that there exists a continuous linear functional on $E[\mathfrak{T}]$ which does not vanish identically.

It follows from (1) and § 15, 9.(9) that the spaces L^p, $0<p<1$, have no convex neighbourhood of $\circ$ other than L^p itself.

It sometimes happens that we do not have a linear functional defined on a linear subspace of E, but only a function $l(z)$ defined on a subset of E. We shall restrict our attention to the case corresponding to 3.(5).

(2) *Let $p(x)$ be a continuous semi-norm on $E[\mathfrak{T}]$. A function $l(z)$ defined on a subset M can be extended to a continuous linear functional u on $E[\mathfrak{T}]$ which satisfies $|u(x)| \leqq p(x)$, provided that the inequality*

$$(3) \qquad \left| \sum_1^n \alpha_k l(z_k) \right| \leqq p\left(\sum_1^n \alpha_k z_k \right)$$

holds for all n, all $z_k \in M$ and all real (respectively complex) α_k.

Proof. Let F be the linear span of M. If we define a linear functional on F by setting $l\left(\sum_1^n \alpha_k z_k\right) = \sum_1^n \alpha_k l(z_k)$, this definition is unambiguous, since it follows from $\sum \alpha_k z_k = 0$ that $\sum \alpha_k l(z_k) = 0$, by (3). But the linear functional defined on F in this way satisfies the inequality $|l(z)| \leqq p(z)$; the assertion follows from 3.(5).

5. Supporting hyperplanes. We continue with the study of convex sets. If N is a subset of a vector space E, a real hyperplane H is called a supporting hyperplane of N if H contains at least one point of N and N lies in one of the two algebraically closed half-spaces defined by H. A point of N through which a supporting hyperplane passes is called a point of support of N.

(1) *If C is an algebraically closed convex α-body in E, every boundary point is a point of support of C.*

If $E[\mathfrak{T}]$ is a topological vector space, every point of the boundary of a closed convex $\mathfrak{T}$-body C is a point of support of a closed hyperplane, and every supporting hyperplane of C is closed.

Proof. The first assertion is an immediate consequence of 2.(1), since every boundary point of C can be taken as M. If C is a $\mathfrak{T}$-body, every supporting hyperplane must be closed, by § 15, 9.(1), since it cannot be dense in E.

It follows from (1) that

(2) *An algebraically closed convex α-body C is the intersection of the algebraically closed half-spaces which contain it and which are defined by its supporting hyperplanes.*

If C is a closed convex $\mathfrak{T}$-body in $E[\mathfrak{T}]$, C is the intersection of the closed half-spaces which contain it and which are defined by its closed supporting hyperplanes.

For if $y \notin C$ and x_0 is an algebraic interior point of C, there is a boundary point z_0 on the segment between x_0 and y. A supporting hyperplane through z_0 cuts the straight line through x_0 and y in z_0, and y belongs to the algebraically open half-space which does not contain C.

Since a convex set in P^n is either a convex body in P^n or a convex body in some manifold of P^n, every closed convex set is the intersection of the closed half-spaces containing it.

We now investigate the extent to which (2) can be carried over to arbitrary algebraically closed convex sets.

(3) *If E has a countable real algebraic basis, every algebraically closed convex subset C of E is the intersection of the closed half-spaces containing it.*

First we show: *If $o \notin C$, there exists a convex α-body C', with o as an algebraically interior point, which is disjoint from C.*

If $x_1, x_2, \dots$ is a basis of E, let E_n denote the linear subspace spanned by $x_1, \dots, x_n$, and let C_n be the convex algebraically closed intersection of C and E_n. There exists a compact convex subset C_1' of E_1 which contains o as interior point and which is disjoint from C_1. Applying § 15, 6.(9) to the two-dimensional space E_2 with its usual topology, there exists a convex compact neighbourhood U of o in E_2 for which $(C_1' + U) \cap C_2$ is empty. But $C_2' = C_1' + U$ is a convex compact body in E_2, by § 15, 6.(8) and § 16, 1.(3). Repeating this procedure, we obtain a sequence $C_1' \subset C_2' \subset \cdots$ of convex compact subsets of E_n, whose union $C' = \bigcup_{n=1}^{\infty} C_n'$ is a convex α-body in E with o as an interior point. $C' \cap C$ is empty.

Thus if $x \notin C$ there is a convex α-body C', with x as an interior point, which is disjoint from C. The assertion now follows by applying the algebraic form of the separation theorem.

(3) is false for every vector space with an uncountable basis, as was shown by KLEE [2], III:

(4) *If E is any vector space with an uncountable real basis, there exists an algebraically closed convex set C which does not contain o but which has a non-empty intersection with every convex α-body which contains o as an interior point. In particular, every algebraically closed half-space which contains C also contains o, so that C is not the intersection of the algebraically closed half-spaces which contain it.*

Proof. Let $x_\alpha, \alpha \in A$, be a real algebraic basis of E. We denote by M the set of all elements $\frac{1}{n^2} \sum_{i=1}^{n} x_{\alpha_i}$, $n = 1, 2, \dots$, where the x_{α_i} are any n basic elements. Let C be the convex cover of M. The elements of C have non-negative coordinates, and are all different from o. In every linear subspace F spanned by finitely many x_α, the intersection $C \cap F$ is the convex cover of finitely many points and is therefore a closed polyhedron. It follows from this that C is algebraically closed.

Now let C' be a convex α-body with $\circ$ as an interior point. For each x_α there is a positive integer k for which $\frac{1}{k}x_\alpha \in C'$. Since A is uncountable, there exists a k for which $\frac{1}{k}x_{\alpha_i} \in C'$ for a countably infinite set α_i. If we choose k of these x_{α_i}, then $\sum_{i=1}^{k} \frac{1}{k} \cdot \frac{1}{k} x_{\alpha_i} = \frac{1}{k^2} \sum_{1}^{k} x_{\alpha_i}$ is in C', since C' is convex; it is also in C, so that $C \cap C'$ is non-empty.

If C is a closed convex subset of a topological vector space, C need not be the intersection of the closed half-spaces containing it: in L^p, $0<p<1$, there is no closed hyperplane, by § 15, 9.(10). The statement of (2) has no useful application to L^p, since there is no convex $\mathfrak{T}$-body in L^p other than L^p itself, by 4. There are certainly closed convex sets in L^p though, e. g. sets of the form $[\circ, x]$, $x \in L^p$, and none of these can be expressed as the intersection of closed half-spaces. On the other hand, such a representation is always possible in locally convex spaces (cf. § 20, 7.(5)).

We now obtain some more results about cones.

(5) *Every supporting hyperplane of an algebraically closed cone passes through the vertex of the cone.*

Let $\circ$ be the vertex of the cone K. If $ux=\gamma$ is a supporting hyperplane through x_0, and if K is contained in the half-space $ux \geqq \gamma$, then we must have $u\circ = 0 \geqq \gamma$. If γ were negative $u(\rho x_0) = \rho\gamma$ would then be less than γ for $\rho > 1$, which is impossible. But it follows from $\gamma=0$ that the hyperplane is given by $ux=0$, and so it contains $\circ$.

(6) *If an algebraically closed cone lies in one of the half-spaces defined by a hyperplane H, the hyperplane parallel to H through the vertex of the cone is a supporting hyperplane.*

Again let $\circ$ be the vertex of K, and let $ux \geqq \gamma$ for all $x \in K$.

It follows that $\gamma \leqq 0$. We show that $ux \geqq 0$ for all $x \in K$. If ux_0 were negative for some $x_0 \in K$, $u(\rho x_0)$ would be less than γ for a suitably chosen $\rho > 0$, and this is not possible.

We now give an example of a closed convex cone in the real space l^p, $p \geqq 1$, for which not every topological boundary point is a point of support of a closed hyperplane. Let C be the set of all $x=(\xi_n) \in l^p$ with $\xi_n \geqq 0$, $n=1,2,\ldots$. C is a closed convex cone with vertex $\circ$. Since all the closed hyperplanes of support pass through $\circ$, by (5), they must be of the form $ux = \sum_{n=1}^{\infty} v_n \xi_n = 0$, where $u \in l^p, \frac{1}{p}+\frac{1}{q}=1$, or $u \in l^\infty$ if $p=1$. If C lies in $ux \geqq 0$, all the v_i must be non-negative. Consequently only those points x which have at least one zero coordinate can belong to closed supporting hyperplanes. But it is easy to see that C has no topologically interior points. It follows that no closed supporting hyperplane passes through any of the points of C for which all the coordinates are non-zero, and these are boundary points.

6. The HAHN-BANACH theorem for normed spaces. Adjoint mappings. We shall now derive certain consequences from the HAHN-BANACH theorem: in the next chapter we shall establish these results in a somewhat more general form, for arbitrary locally convex spaces; meanwhile they will serve to complement the theory of normed spaces developed in § 14.

In § 14 we evaluated some specific dual spaces; the Hahn-Banach theorem now enables us to show that every general normed space possesses continuous linear functionals which do not vanish identically.

(1) *Given a linear functional $l(z)$ on a linear subspace of a normed space E which satisfies an inequality $|l(z)| \leqq M \|z\|$, l can be extended to a continuous linear funtional u, defined on the whole of E, which satisfies $|u x| \leqq M \|x\|$.*

If x_0 is an element of E, there exists a functional $u_0 \in E'$ with $u_0 x_0 = \|x_0\|$ and $\|u_0\| = 1$.

The first part comes from 3.(5), and the second part follows from 3.(8), and from the definition of the norm in E' given in § 14, 5.(3).

The next statement, which is somewhat sharper, is frequently used:

(2) *If H is a linear subspace of the normed space E and if x_0 is an element of E at distance d from H, then there exists $u_0 \in E'$ with $u_0 x_0 = 1$, $\|u_0\| = 1/d$ and $u_0 z = 0$ for all $z \in H$.*

Proof. By hypothesis $\|x_0 + y\| \geqq d$ for all y in the closure $\bar{H}$ of H. If we define a linear functional on $[x_0] + \bar{H}$ by the formula $l(\alpha x_0 + y) = \alpha$, then we have $\|\alpha x_0 + y\| = |\alpha| \left\| x_0 + \frac{y}{\alpha} \right\| \geqq |\alpha| \cdot d$, so that $|l(\alpha x_0 + y)| = |\alpha| \leqq \frac{1}{d} \|\alpha x_0 + y\|$. Using (1), l can be extended to a continuous linear functional u_0 on the whole of E which satisfies $|u_0 x| \leqq \frac{1}{d} \|x\|$. This shows that $\|u_0\| \leqq \frac{1}{d}$, and clearly $u_0 x_0 = l(x_0) = 1$ and $u_0 z = 0$ for all $z \in H$.

On the other hand, if y_n is a sequence in H with $\|x_0 + y_n\| \to d$, then $1 = u_0(x_0 + y_n) \leqq \|u_0\| \, \|x_0 + y_n\|$, by § 14, 5.(4), and so $\|u_0\| \geqq \frac{1}{d}$, i. e. $\|u_0\| = \frac{1}{d}$.

(3) *The bidual E'' of a normed space E contains E, and on E the norm of the bidual space coincides with the norm of E, i. e. we have*

$$\|x\| = \sup_{\|u\| \leqq 1} |u x|, \quad \textit{for } x \in E. \tag{4}$$

Proof. Each $x_0 \in E$ defines a continuous linear functional $x_0(u) = u x_0$ on E', and we have $|x_0(u)| = |u(x_0)| \leqq \|u\| \, \|x_0\|$, so that $\sup\limits_{\|u\| \leqq 1} |u x_0|$

$\leqq \|x_0\|$. But on the other hand there exists a u_0 with $\|u_0\|=1$ and $u_0 x_0 = \|x_0\|$, by the second part of (1), and so (4) is established.

Thus equation (4), which had previously been established for some of the examples of § 14, is now proved in complete generality.

In § 14, 6. we also introduced a norm on the space of continuous linear mappings, and in this way made the space $\mathfrak{L}(E,F)$ of continuous linear mappings from a normed space E into a normed space F into a normed space. To each mapping $A \in \mathfrak{L}(E,F)$ there corresponds the adjoint mapping A' which maps the algebraic dual space F^* into E^* (cf. § 9, 4.).

(5) *If A is a continuous linear mapping from the normed space E into the normed space F, the adjoint mapping A' maps F' continuously into E', and $\|A'\| = \|A\|$.*

Proof. A' is defined by the relation $(A'v)x = v(Ax)$, where $x \in E$ and $v \in F^*$. We restrict attention to those v which lie in F'. It follows from § 14, 5.(4), and §14, 6.(4) that

$$|(A'v)x| = |v(Ax)| \leqq \|v\| \, \|Ax\| \leqq \|v\| \, \|A\| \, \|x\|.$$

But this means that $A'v$ is bounded by $\|A\| \, \|v\|$, so that $A'v$ is a continuous linear functional on E, and A' maps F' linearly and continuously into E'. Further, the relation $\|A'\| \leqq \|A\|$ follows from

$$\|A'v\| = \sup_{\|x\| \leqq 1} |(A'v)x| \leqq \|A\| \, \|v\|.$$

Given $\varepsilon > 0$, there exists an $x_0 \in E$ with $\|x_0\| \leqq 1$ and $\|Ax_0\| > \|A\| - \varepsilon$. By (1), corresponding to $y_0 = Ax_0$ there exists a $v_0 \in F'$ with $\|v_0\| = 1$ and $v_0 y_0 = \|y_0\| > \|A\| - \varepsilon$. But then $\|A'v_0\| \geqq (A'v_0)x_0 = v_0 y_0 > \|A\| - \varepsilon$, so that $\|A'\| > \|A\| - \varepsilon$; i. e. $\|A'\| = \|A\|$.

(6) *If A is a continuous linear mapping from the normed space E into the normed space F, A'' is an extension of the continuous linear mapping A to a continuous linear mapping from E'' into F''. $\|A''\| = \|A\|$.*

$A'' \in \mathfrak{L}(E'', F'')$ and $\|A''\| = \|A\|$, by (5). By (3), $E'' \supset E$ and $F'' \supset F$. We establish the value of $A''x$, for $x \in E$. The relation $A''x = Ax$ holds, provided that the two terms represent the same linear functionals on F', and so provided that $v(A''x) = v(Ax)$ for each $v \in F'$. But this follows from a double application of the definition of the adjoint mapping:

$$v(A''x) = (A'v)x = v(Ax).$$

7. The dual space of $C(I)$. As a further example of an application of the Hahn-Banach theorem, we shall prove the theorem of F. Riesz characterizing the dual of the (B)-space $C(I)$ of all continuous real- or complex-valued functions on $I=[0,1]$, i. e. characterizing the space of measures on I.

Let $u(f)$ be a continuous linear functional on $C(I)$ which does not vanish identically. If $\|u\|=m$, then $|u(f)|\leqq m\|f\|$, where $\|f\|=\sup_{0\leqq t\leqq 1}|f(t)|$ is the norm of f in $C(I)$.

By § 14, 11., $C(I)$ is a closed linear subspace of $L^\infty(I)$, and the norm of $L^\infty(I)$ coincides on $C(I)$ with the norm of $C(I)$. By the HAHN-BANACH theorem, u can be extended to a continuous linear functional u defined on the whole of $L^\infty(I)$, and with the same bound m.

In particular the value of u is defined on the function $\varphi_c\in L^\infty(I)$ which is equal to 0 for $t>c$ and is equal to 1 for $t\leqq c$, where $0\leqq c\leqq 1$. We set $u(\varphi_c)=g(c)$.

We show that the function $g(c)$ defined in $[0,1]$ in this way is a function of bounded variation.

Suppose that $0=c_0<c_1<\cdots<c_n=1$ is a decomposition of $[0,1]$ into finitely many subintervals, and let $\varepsilon_i=\dfrac{|g(c_i)-g(c_{i-1})|}{g(c_i)-g(c_{i-1})}$ when $g(c_i)-g(c_{i-1})\neq 0$, and otherwise let $\varepsilon_i=0$. Then

$$(1)\quad \sum_{i=1}^{n}|g(c_i)-g(c_{i-1})|=\sum_{i=1}^{n}\varepsilon_i(g(c_i)-g(c_{i-1}))=u\left(\sum_{i=1}^{n}\varepsilon_i(\varphi_{c_i}-\varphi_{c_{i-1}})\right)$$
$$\leqq m\left\|\sum\varepsilon_i(\varphi_{c_i}-\varphi_{c_{i-1}})\right\|\leqq m,$$

for the function $\sum\varepsilon_i(\varphi_{c_i}-\varphi_{c_{i-1}})$ has absolute value $\leqq 1$ in $[0,1]$. It follows from this that the total variation $\overset{1}{\underset{0}{\mathrm{V}}}\, g$ of g, which is the least upper bound of the values of $\sum_{i=1}^{n}|g(c_i)-g(c_{i-1})|$ for all decompositions of $[0,1]$ into finitely many subintervals, is less than or equal to m.

The value of u on an arbitrary $f\in C(I)$ can be expressed in terms of $g(c)$ in the following way: if we divide $[0,1]$ into n equal parts and form the step-function $f_n(t)=\sum_{k=1}^{n}f\left(\frac{k}{n}\right)\left(\varphi_{\frac{k}{n}}(t)-\varphi_{\frac{k-1}{n}}(t)\right)$, then the functions f_n converge uniformly to f, and so converge to f in the sense of the norm. Consequently $u(f)=\lim u(f_n)=\lim_{n\to\infty}\sum_{k=1}^{n}f\left(\frac{k}{n}\right)\left[g\left(\frac{k}{n}\right)-g\left(\frac{k-1}{n}\right)\right]$. But this limit is identical with the Stieltjes integral $\int_0^1 f(t)\,dg$, which exists since f is continuous and g is of bounded variation (cf. NATANSON [1], for example, for the results used here about functions of bounded variation and the Stieltjes integral).

From the inequality

(2) $$|u(f)| = \left| \int_0^1 f(t)\,dg \right| \leqq \sup |f(t)| \overset{1}{\underset{0}{\mathrm{V}}}(g) = \|f\| \overset{1}{\underset{0}{\mathrm{V}}}(g)$$

it follows that $\|u\| \leqq \overset{1}{\underset{0}{\mathrm{V}}}(g)$; on the other hand, we showed above that $\overset{1}{\underset{0}{\mathrm{V}}}(g) \leqq m = \|u\|$. Consequently $\|u\| = \mathrm{V}(g)$, where g is the function of bounded variation corresponding to u.

Conversely, it follows from (2) that every real or complex function g of bounded variation defines a continuous linear functional on $C(I)$, whose norm is less than or equal to $\mathrm{V}(g)$.

Which functions g of bounded variation determine a continuous linear functional on $C(I)$ which vanishes identically?

(3) $\int_0^1 f\,dg = 0$ *for all* $f \in C(I)$ *if and only if* g *is equal to some fixed constant* K *at* 0, 1 *and all its points of continuity.*

It is well-known (cf. NATANSON [1] p. 219) that the set of points of discontinuity of g is at most countable, and that for each point of discontinuity c_0 the limits $g(c_0+0)$ from the right and $g(c_0-0)$ from the left both exist.

If g satisfies the given conditions, g is therefore equal to K at 0, 1 and on a dense set. Choosing the points of dissection for the approximating sums of the Stieltjes integral from this set, we clearly obtain the value 0 for the integral of any function f.

Conversely, suppose that g does not satisfy the conditions. If $g(0) \neq g(1)$, then, putting $f = 1$, the integral $\int_0^1 dg = g(1) - g(0) \neq 0$. We can therefore suppose that $g(0) = g(1) = 0$. Let c_0 be a point of continuity of g at which $g(c_0) \neq 0$. In a sufficiently small interval $[c_0, c_0 + \varepsilon]$ the variation of $g(c)$ is less than $|g(c_0)|$. This is a consequence of the theorem (cf. NATANSON [1], p. 223) which states that the variation $\overset{x}{\underset{0}{\mathrm{V}}}(g)$ is continuous at a point of continuity of g, and the relation $\overset{b}{\underset{a}{\mathrm{V}}}(g) = \overset{c}{\underset{a}{\mathrm{V}}}(g) + \overset{b}{\underset{c}{\mathrm{V}}}(g)$ for $0 \leqq a < c < b \leqq 1$ (NATANSON [1], p. 217). If we consider $\int_0^1 f\,dg$ for a function f which is equal to 1 from 0 to c_0, which decreases linearly from 1 to 0 in the interval $[c_0, c_0 + \varepsilon]$, and which is equal to 0 from $c_0 + \varepsilon$ to 1, then the integral from 0 to c_0 takes the value $g(c_0)$, the integral from

c_0 to $c_0+\varepsilon$ takes a value with modulus less than $|g(c_0)|$, and the integral from $c_0+\varepsilon$ to 1 is equal to 0. Combining these, $\int_0^1 f\,dg \neq 0$. This completes the proof of (3).

A function $h(c)$ of bounded variation on $[0,1]$ is said to be **normalized** if $h(0)=0$ and $h(c)=h(c+0)$ for $0<c<1$.

If $g(c)$ is an arbitrary function of bounded variation, the corresponding normalized function $g^*(c)$ is obtained by setting $g^*(0)=0, g^*(1)=g(1)-g(0)$ and $g^*(c)=g(c+0)-g(c)$, for $0<c<1$. $\int_0^1 f\,dg=\int_0^1 f\,dg^*$, since $g(c)-g^*(c)$ is equal to $g(0)$ at $0,1$, and all points of continuity of g. Thus every continuous linear functional on $C(I)$ is determined by a normalized g. Conversely every normalized g which does not vanish identically determines a linear functional which does not vanish identically, since a normalized function only satisfies the condition of (3) if it vanishes identically: it is equal to 0 at $0,1$ and each point of continuity, and at points of discontinuity it is equal to $g(c+0)$, which being the limit of values of g at points of continuity must be equal to 0.

Finally we shall show that if $h(c)$ is a normalized function the norm of the continuous linear functional which it defines is equal to $\overset{1}{\underset{0}{\mathrm{V}}}(h)$. An arbitrary function $g(c)$ of bounded variation with $g(0)=0$ differs from the corresponding normalized function $g^*(c)$ only at points of discontinuity. If we form the sum $\sum_1^n |g(c_i)-g(c_{i-1})|$ only at points of continuity, then the least upper bound V' of these sums is the same for g and g^*. But because of the definition of g^* at the points of discontinuity, V' is equal to $\overset{1}{\underset{0}{\mathrm{V}}}(g^*)$; consequently the normalized function has the smallest variation of all those functions which define the same normalized function. As we have already shown, every continuous linear functional is determined by a function g of bounded variation whose variation is equal to $\|u\|$, and so the corresponding normalized function g^* can only have a variation $\leq \|u\|$; it then follows from (2) that $\mathrm{V}(g^*)=\|u\|$.

We gather these results together to give the theorem of RIESZ [1]:

(4) *Every continuous linear functional u on the real (respectively complex) (B)-space $C(I)$ can be represented by a real (respectively complex) normalized function h_u of bounded variation on $[0,1]$:*

$$u(f)=\int_0^1 f\,dh_u, \qquad f\in C(I). \tag{5}$$

The correspondence $u \to h_u$ is a norm isomorphism of the space $C(I)'$ of measures on I onto the space $V(I)$ of normalized functions of bounded variation, equipped with the norm

$$\|h\| = \overset{1}{\underset{0}{\mathrm{V}}}(h). \tag{6}$$

We can now answer the question of whether or not $C(I)$ is reflexive (cf. RIESZ-SZ. NAGY [1]):

(7) *$C(I)$ is not reflexive.*

It is sufficient to produce a continuous linear functional $v_0(h)$ on $V(I)$ which cannot be represented in the form $v_0(h) = \int_0^1 f_0\, dh$, $f_0 \in C(I)$. Now every h has only countably many points of discontinuity, and the sum of the jumps $v_0(h) = \sum_{i=1}^{\infty} (h(c_i) - h(c_i - 0))$, $0 < c_i \leqq 1$, clearly satisfies $|v_0(h)| \leqq \overset{1}{\underset{0}{\mathrm{V}}}(h)$. Further $v_0(h)$ is linear, so that it is a continuous linear functional on $V(I)$. If $v_0(h)$ were representable by a function f_0, then in particular for the function $h = \psi_c$, where $\psi_c = 1$ for $x < c$ and $\psi_c(x) = 0$ for $x \geqq c$, we would have $v_0(\psi_c) = -1 = -f_0(c)$, so that $f_0(c)$ would be identically equal to 1. But for continuous functions h, $\int_0^1 dh = h(1) - h(0)$ is in general different from zero whereas $v_0(h)$ is always equal to 0.

CHAPTER FOUR

Locally Convex Spaces. Fundamentals

The first two paragraphs are concerned with methods of producing new locally convex spaces from given ones. Thus subspaces and quotient spaces of locally convex spaces are again locally convex. The same holds for topological products and locally convex direct sums. The completion of a locally convex space is obtained in a simple way by embedding the space in a topological product of Banach spaces.

In § 19, locally convex hulls and locally convex kernels of locally convex spaces are introduced, together with the related ideas of topological inductive and projective limits. The precise distinction between hulls and inductive limits, and likewise between kernels and projective limits, seems to be advisable when giving a systematic account. Some properties of (LF)-spaces are also dealt with in § 19.

§ 20 begins by examining the dual E' of a locally convex space E; the dual pair $\langle E',E\rangle$ and the weak topology are introduced. The theory of duality for closed linear subspaces, complementary decompositions and linear mappings follows in a simple way. The duality of absolutely convex closed subsets leads to the important properties of polarity. The paragraph ends with the ALAOGLU-BOURBAKI theorem on the weak compactness of the polar of a neighbourhood of $\mathfrak{o}$, and the theorems of BANACH-MACKEY and of MACKEY on the identity of weakly and strongly bounded sets.

§ 21 deals with the different topologies of a locally convex space, and above all with the strong topology, the Mackey topology and the topology of precompact convergence. A detailed knowledge of these is indispensible, since the coincidence of two of these topologies leads to important structural properties of the space. In particular the topologies of metrizable locally convex spaces are investigated.

§ 22 applies the theory of duality to the spaces considered in § 18 and § 19. For example we investigate how far it is possible to express the different topologies of a topological product in terms of the corresponding topologies on the factors. We show that these questions do not by any means always have a simple answer in the completely general case.

§ 18. The definition and simplest properties of locally convex spaces

1. Definition by neighbourhoods, and by semi-norms. A topological vector space $E[\mathfrak{T}]$ over K (which once again is the field of real or complex numbers) is said to be l o c a l l y c o n v e x if it has base of neighbourhoods $\mathfrak{U}=\{U_\alpha\}$ of $\mathfrak{o}$ consisting of convex sets U_α.

It follows from § 15, 1.(3) and § 16, 1.(2) that it then also has a base of neighbourhoods of $\mathfrak{o}$ consisting of absolutely convex sets U_α, since the convex cover of a circled neighbourhood of $\mathfrak{o}$ contained in U_α is absolutely convex.

Corresponding to § 15, 2., we have

(1) *Let* $\mathfrak{U}=\{U_\alpha\}$, $\alpha\in A$, *be a filter-base of absolutely convex absorbent subsets* U_α *of a real or complex vector space* E, *with* $\bigcap_{\alpha\in A} U_\alpha=\mathfrak{o}$. *If each set* ρU_α, $\rho>0$, *belongs to* $\mathfrak{U}$ *whenever* U_α *does, then a locally convex space* $E[\mathfrak{T}]$ *is defined by taking* $\mathfrak{U}$ *as base of neighbourhoods of* $\mathfrak{o}$; *every locally convex space can arise in this way.*

This is a simple consequence of § 15, 2.(2), for $\frac{1}{2}U_\alpha+\frac{1}{2}U_\alpha=U_\alpha$, since U_α is absolutely convex, and so (LT 1) is satisfied.

In applications, the topology on E is often defined in a more appropriate way by means of a system of semi-norms.

We saw in § 16, 4. that every absolutely convex absorbent set has a semi-norm as Minkowski functional, and that the algebraic hull U^a is given by $p(x)\leqq 1$, and the algebraic kernel U^i by $p(x)<1$. Conversely, starting from a semi-norm $p(x)$ on E, the sets $p(x)\leqq 1$ and $p(x)<1$ are absolutely convex, absorbent, and algebraically closed and open, respectively.

By § 16, 4., to the absolutely convex neighbourhoods of a locally convex space there correspond continuous semi-norms; conversely, if $p(x)$ is a continuous semi-norm the sets $p(x)<\varepsilon$ and $p(x)\leqq\varepsilon$ are open (respectively closed) $\mathfrak{T}$-neighbourhoods of $\mathfrak{o}$ (§ 16, 4.(7)).

(2) *Every semi-norm defined on* $E[\mathfrak{T}]$ *which is continuous at* $\mathfrak{o}$ *is uniformly continuous on the whole of* $E[\mathfrak{T}]$.

This follows directly from the inequality § 14, 1.(1), which is also valid for semi-norms.

(3) *Let* $\{p_\alpha(x)\}$ *be a system of semi-norms on a vector space* E, *with the property that for each* $x_0\neq\mathfrak{o}$ *there is at least one* p_α *with* $p_\alpha(x_0)\neq 0$. *If we denote by* U_α *the set of all* $x\in E$ *for which* $p_\alpha(x)<1$, *then the system of scalar multiples* ρU, $\rho>0$, *of finite intersections* $U=\bigcap_{i=1}^{n} U_{\alpha_i}$ *of the* U_α *forms a base of neighbourhoods* $\mathfrak{U}$ *of* $\mathfrak{o}$ *for a locally convex topology* $\mathfrak{T}$ *on* E; *this base consists of absolutely convex open sets.*

Every locally convex space can arise in this way.

The set $\rho U=\rho\bigcap_{i=1}^{n} U_{\alpha_i}$ *is given by the inequality* $p(x)<\rho$, *where* $p(x)$ *is the semi-norm* $\sup_{i=1,\ldots,n} p_{\alpha_i}(x)$.

If instead we take the sets defined by $p_\alpha(x) \leqq 1$, then in a corresponding way we obtain a base of closed absolutely convex neighbourhoods of $\mathfrak{o}$.

Proof. The sets ρU are absorbent and absolutely convex in E, by the preceding remarks, and they form a filter-base on E. Since for each non-zero x_0 there exists a p_α with $p_\alpha(x_0) \neq 0$, the intersection of all the sets ρU is equal to $\mathfrak{o}$, and so by (1), $E[\mathfrak{T}]$ is locally convex. By the preceding remarks the sets ρU are open.

Conversely, a locally convex space given by (1) can also be given by a system of semi-norms. If the functions $p_\alpha(x)$ are the semi-norms corresponding to the sets U_α, the open kernels and closed hulls of the sets ρU_α are given by $p_\alpha(x) < \rho$ and $p_\alpha(x) \leqq \rho$ respectively, and in either case these sets form a base of neighbourhoods of $\mathfrak{o}$ for $\mathfrak{T}$.

Finally, the open kernel of the intersection $\bigcap_{i=1}^{n} U_{\alpha_i}$, which is again absorbent and absolutely convex, is given by the inequality $p(x) = \sup_{i=1,\ldots,n} p_{\alpha_i}(x) < 1$; $p(x)$ is therefore a semi-norm as well.

Let us remark that if $\{p_\alpha\}$ is a system of semi-norms defining the topology of $E[\mathfrak{T}]$, then the neighbourhoods $p_\alpha(x) < \rho$ will in general only form a sub-base for the filter of neighbourhoods of $\mathfrak{o}$.

(4) *Let $\mathfrak{T}$ and $\mathfrak{T}'$ be the locally convex topologies defined by two systems $\{p_\alpha(x)\}$ and $\{q_\beta(x)\}$ of semi-norms on E. $\mathfrak{T}$ is finer than $\mathfrak{T}'$ if and only if, given q_β, there exist finitely many p_{α_i}, $i = 1, \ldots, n$, and a $\rho > 0$ for which the inequality*

$$\rho\, q_\beta(x) \leqq \sup_{i=1,\ldots,n} p_{\alpha_i}(x)$$

holds on E.

Proof. If U_1 and U_2 are two algebraically open absorbent absolutely convex sets, the relation $U_1 \subset U_2$ is equivalent to the inequality $p_2(x) \leqq p_1(x)$ between the corresponding Minkowski functionals p_1 and p_2.

$\mathfrak{T}$ is finer than $\mathfrak{T}'$ if and only if every $\mathfrak{T}'$-neighbourhood of $\mathfrak{o}$ of the form $q_\beta(x) < 1$ contains a $\mathfrak{T}$-neighbourhood of $\mathfrak{o}$. By (3), this can be taken in the form $\sup_{i=1,\ldots,n} p_{\alpha_i}(x) < \rho$, and it therefore has $\frac{1}{\rho} \sup_{i=1,\ldots,n} p_{\alpha_i}(x)$ as Minkowski functional. (4) follows from this.

A criterion for the equality of two topologies, each of which is given by a system of semi-norms, follows directly from (4); this corresponds to the Hausdorff criterion for equivalence (§ 2, 4.(1)).

2. Metrizable locally convex spaces and (F)-spaces. Every normed space is locally convex, since its topology is defined as in 1.(3) by a single semi-norm, which is in fact a norm.

Earlier we saw that by no means every metrizable topological vector space is locally convex; indeed the spaces L^p, $0<p<1$, which are metrizable, by § 15, 11., actually have no convex neighbourhood of o other than L^p itself (§ 17, 11.).

A complete metrizable locally convex space is called an (F)-space, after FRÉCHET (cf. the remark of § 15, 11.). Every (B)-space is an (F)-space.

Since a metrizable locally convex space has a countable base U_n, $n=1,2,\ldots$, of neighbourhoods of o, and since each U_n contains an absolutely convex neighbourhood of o, the sets U_n can all be assumed to be absolutely convex. The sets U_1, $U_1 \cap U_2$, $U_1 \cap U_2 \cap U_3, \ldots$ also form a base of neighbourhoods of o, if $U_1, U_2, U_3, \ldots$, do. We therefore obtain

(1) *The topology of a metrizable locally convex space* $E[\mathfrak{T}]$ *can always be given by a decreasing sequence* $U_1 \supset U_2 \supset \cdots$ *of absolutely convex neighbourhoods of* o, *with* $\bigcap_{n=1}^{\infty} U_n = \mathrm{o}$.

The semi-norms $p_n(x)$ corresponding to the neighbourhoods U_n then form an increasing sequence $p_1(x) \leqq p_2(x) \leqq \cdots$.

Conversely, starting from a countable system $\{q_i(x)\}$ of semi-norms on a vector space E, the increasing sequence of semi-norms $p_i(x) = \sup_{k=1,\ldots,i} q_k(x)$ defines the same topology, by 1.(3). The sets U_n consisting of all those $x \in E$ with $p_n(x) < \dfrac{1}{n}$ then form a decreasing fundamental sequence of neighbourhoods of o.

We therefore have

(2) *A locally convex space* $E[\mathfrak{T}]$ *is metrizable if and only if* $\mathfrak{T}$ *can be defined by a countable system of semi-norms. If this is the case* $\mathfrak{T}$ *can also be defined by an increasing sequence* $p_1(x) \leqq p_2(x) \leqq \cdots$ *of semi-norms. The neighbourhoods* $p_n(x) < \dfrac{1}{n}$, $n=1,2,\ldots$, *form a base of neighbourhoods of* o *for* $\mathfrak{T}$.

The construction of an (F)-norm on a metrizable topological vector space given in § 15, 11. depends upon the complicated methods of § 6, 7.(1). In the locally convex case the following simpler construction is possible:

(3) *Suppose that* $E[\mathfrak{T}]$ *is locally convex and metrizable, and that* $\|x\|_1 \leqq \|x\|_2 \leqq \cdots$ *is an increasing sequence of semi-norms which defines* $\mathfrak{T}$. $\mathfrak{T}$ *is also defined by the* (F)-*norm*

$$\|x\| = \sum_{n=1}^{\infty} \frac{1}{2^n} \frac{\|x\|_n}{1+\|x\|_n}; \tag{4}$$

$\mathfrak{T}$ is therefore defined by a translation-invariant metric given by $|x,y| = \|x-y\|$.

Proof. First we establish the (F)-norm properties. (F 1) is trivial, and (F 2) is satisfied, since for each non-zero x_0 there is a semi-norm $\|x\|_n$ with $\|x_0\|_n \neq 0$.

If $0<a\leqq b$, $\dfrac{a}{1+a} \leqq \dfrac{b}{1+b}$.

It follows from $\|\lambda x\|_n = |\lambda|\,\|x\|_n \leqq \|x\|_n$, for $|\lambda|\leqq 1$, that $\|\lambda x\| \leqq \|x\|$ for $|\lambda|\leqq 1$, so that (F 3) is satisfied.

$\|x+y\| \leqq \|x\| + \|y\|$ follows directly from

$$\frac{\|x+y\|_n}{1+\|x+y\|_n} \leqq \frac{\|x\|_n+\|y\|_n}{1+\|x\|_n+\|y\|_n} \leqq \frac{\|x\|_n}{1+\|x\|_n} + \frac{\|y\|_n}{1+\|y\|_n}$$

so that (F 4) is satisfied.

Next we show that the system of $\mathfrak{T}$-neighbourhoods of $\mathfrak{o}$ is equivalent to the metric defined on E by (4).

a) The neighbourhood $\|x\| < \dfrac{1}{2^k}$, $k\geqq 1$, contains the $\mathfrak{T}$-neighbourhood $\|x\|_{k+1} < \dfrac{1}{2^{k+1}}$:

$\dfrac{\|x\|_n}{1+\|x\|_n} \leqq \|x\|_n$. If x satisfies $\|x\|_{k+1} < \dfrac{1}{2^{k+1}}$, then $\|x\|_1 \leqq \cdots \leqq \|x\|_{k+1} < \dfrac{1}{2^{k+1}}$ as well, so that $\|x\| < \dfrac{1}{2^{k+1}} \sum\limits_{1}^{k+1} \dfrac{1}{2^n} + \sum\limits_{k+2}^{\infty} \dfrac{1}{2^n} < \dfrac{1}{2^k}$.

b) The $\mathfrak{T}$-neighbourhood $\|x\|_m < \dfrac{1}{2^k}$ contains the neighbourhood $\|x\| < \dfrac{1}{2^{m+k+1}}$:

If x satisfies $\|x\| < \dfrac{1}{2^{m+k+1}}$, then $\dfrac{1}{2^m}\dfrac{\|x\|_m}{1+\|x\|_m} < \dfrac{1}{2^{m+k+1}}$, so that $\dfrac{\|x\|_m}{1+\|x\|_m} < \dfrac{1}{2^{k+1}}$. It follows from this that $\|x\|_m\left(1-\dfrac{1}{2^{k+1}}\right) < \dfrac{1}{2^{k+1}}$, and so that $\|x\|_m < \dfrac{1}{2^k}$.

Since the two neighbourhood systems are equivalent, it also follows that (F 5) and (F 6) are satisfied, since the corresponding assertions are certainly true for $\mathfrak{T}$-convergence.

We observe that for this metric the distance between two points is always less than or equal to one.

3. Subspaces, quotient spaces and topological products of locally convex spaces. The results obtained in § 15,4. can be extended, in the locally convex case.

(1) *Every linear subspace H of a locally convex space* $E[\mathfrak{T}]$ *is again locally convex, in the induced topology.*

For $U \cap H$ is absolutely convex if H is.

(2) *Every quotient space* E/H *of a locally convex space* $E[\mathfrak{T}]$ *by a closed linear subspace H is locally convex in the induced topology* $\mathfrak{T}$.

If the topology $\mathfrak{T}$ *of* $E[\mathfrak{T}]$ *is given by a system* $\{p_\alpha(x)\}$ *of semi-norms which always contains a* $p_\gamma \geqq \sup(p_\alpha, p_\beta)$ *when it contains* p_α *and* p_β, *then the induced topology of* E/H *is given by the system* $\{\hat{p}_\alpha(\hat{x})\}$, *where* $\hat{p}_\alpha(\hat{x})$ *is the semi-norm*

$$\hat{p}_\alpha(\hat{x}) = \inf_{x \in \hat{x}} p_\alpha(x), \tag{3}$$

and $\hat{x}$ *is a coset in* E/H.

Proof. $K(U)$ is absolutely convex if U is, where K is the canonical mapping of E onto E/H. The quotient space topology therefore has a base of absolutely convex neighbourhoods of $\mathfrak{o}$.

The assertion about semi-norms is proved in the same way as in § 14, 4.(1).

It follows from (2) and § 15, 11.(4) that

(4) *Every quotient space of an* (F)-*space by a closed linear subspace is again an* (F)-*space.*

If $E[\mathfrak{T}]$ is an arbitrary complete locally convex space, E/H need not be complete (cf. § 23, 5. and § 31, 6.).

Similarly for topological products we have

(5) *The topological product* $E[\mathfrak{T}] = \prod_\alpha E_\alpha[\mathfrak{T}_\alpha]$ *of locally convex spaces is again locally convex.*

For the neighbourhoods $\prod_\alpha W_\alpha$, where W_α is an absolutely convex neighbourhood U_α of $\mathfrak{o}$ for finitely many α, and $W_\alpha = E_\alpha$ otherwise, are clearly absolutely convex.

If, for fixed α, $\{p_\beta^\alpha(x_\alpha)\}$ is a system of semi-norms on E_α which defines the topology $\mathfrak{T}_\alpha$, and if we denote by $\hat{p}_\beta^\alpha(x)$ the seminorm on E defined by $\hat{p}_\beta^\alpha(x) = p_\beta^\alpha(x_\alpha)$, where $x = (x_\alpha) \in E$, then the system of all the semi-norms $\hat{p}_\beta^\alpha(x)$ defines the product topology on E.

Let us recall that it was shown in § 5, 7. that the completion of a topological product is equal to the topological product of the completions. It follows from this and from the ideas which have just been used that

(6) *The topological product of metrizable locally convex spaces is metrizable if and only if the product has finitely or countably many factors.*

The topological product of countably many (F)-*spaces is again an* (F)-*space.*

The next result gives a certain way of looking at all possible locally convex spaces (cf. § 10, 7.(9) as well);

(7) *Every locally convex space* $E[\mathfrak{T}]$ *is topologically isomorphic to a linear subspace* $\hat{E}$ *of a topological product of* (B)-*spaces.* E *is complete if and only if* $\hat{E}$ *is closed.*

Proof. Let $\{p_\alpha(x)\}$ be a system of semi-norms on E which defines the topology $\mathfrak{T}$. If N_α is the null-space of $p_\alpha(x)$, then, by § 14, 1.(4), $E/N_\alpha = E_\alpha$ is a normed space under $\hat{p}_\alpha(\hat{x}_\alpha) = p_\alpha(x)$, where $\hat{x}_\alpha$ is the coset of x in E_α. We form the completion $\tilde{E}_\alpha$ of E_α (§ 15, 3.). Let F be the topological product $\prod_\alpha \tilde{E}_\alpha$ of the (B)-spaces $\tilde{E}_\alpha$. The extension of the norm $\hat{p}_\alpha$ to $\tilde{E}_\alpha$ is again denoted by $\hat{p}_\alpha$.

We now map each $x \in E$ to the corresponding element $\hat{x} = (\hat{x}_\alpha)$ of F. This mapping is one-one and linear, and it maps E onto a linear subspace $\hat{E}$ of F. We saw above that the topology of F is defined by the semi-norms $\hat{p}_\alpha(\hat{x}) = \hat{p}_\alpha(\hat{x}_\alpha)$. Since $p_\alpha(x) = \hat{p}_\alpha(\hat{x}_\alpha) = \hat{p}_\alpha(x)$, corresponding semi-norms p_α and $\hat{p}_\alpha$ take the same value on corresponding elements $x \in E$ and $\hat{x} \in \hat{E}$, and so the correspondence is a topological isomorphism.

It follows from the completeness of E that $\hat{E}$ is closed; conversely, if $\hat{E}$ is closed in F, $\hat{E}$ is complete, since F is, and so therefore is $E[\mathfrak{T}]$.

(8) *Every real locally convex space* $E[\mathfrak{T}]$ *can be embedded in a complex locally convex space* $F[\mathfrak{T}']$ *in such a way that* $F = E \oplus iE$ *and the topologies induced on* E *and* iE *by* $\mathfrak{T}'$ *coincide with* $\mathfrak{T}$.

Proof. We set $F[\mathfrak{T}']$ equal to the topological product $E \times E$, and define multiplication by i by the equation $i(x, y) = (-y, x)$. In this way F becomes a complex vector space. We identify E with the real subspace of all (x, o). Then $(x, y) = x + iy$, so that iE is the real subspace consisting of all points (o, y), and $F = E \oplus iE$. By definition, the product topology $\mathfrak{T}'$ induces the topology $\mathfrak{T}$ on E and on iE.

If E is a real normed space with norm $p(x)$, then by § 14, 4.(5) the expressions $\sup(p(x), p(y))$, $p(x) + p(y)$, $\sqrt{p^2(x) + p^2(y)}$, etc., are norms on F whose restrictions to E and to iE coincide with the given norm.

4. The completion of a locally convex space. By § 15,3.(1), every topological vector space has a completion.

(1) *The completion* $\tilde{E}[\tilde{\mathfrak{T}}]$ *of a locally convex space* $E[\mathfrak{T}]$ *is locally convex.*

For by § 15, 3.(1) we obtain a base of neighbourhoods of o in E by forming the closures $\bar{U}_\alpha$ in $\tilde{E}[\tilde{\mathfrak{T}}]$ of a base of absolutely convex neighbourhoods U_α of $\mathfrak{o}$ in E. But these are again absolutely convex, by § 16, 1.(5).

(2) *If the topology of* $E[\mathfrak{T}]$ *is defined by a system* $\{p_\alpha\}$ *of semi-norms, the topology of* $\tilde{E}[\tilde{\mathfrak{T}}]$ *is defined by the system* $\{\tilde{p}_\alpha\}$, *where* $\tilde{p}_\alpha$ *is the uniquely defined continuous extension of* $p_\alpha(x)$ *to* $\tilde{E}$.

Proof. Every semi-norm $p_\alpha(x)$ is uniformly continuous on E, by 1.(2), and so it has a uniformly continuous extension $\tilde{p}_\alpha(y)$ to $\tilde{E}$, by § 5, 4.(4). Properties (N 1), (N 3) and (N 4) of § 14, 1. are still satisfied, so that $\tilde{p}_\alpha(y)$ is a continuous semi-norm on $\tilde{E}$.

We denote the neighbourhood $p_\alpha(x) < \varepsilon$ of $\mathfrak{o}$ in E by $U_{\alpha\varepsilon}$, and its closure in $\tilde{E}$ by $\bar{U}_{\alpha\varepsilon}$. We denote the neighbourhood $\tilde{p}_\alpha(y) < \varepsilon$ of $\mathfrak{o}$ in $\tilde{E}$ by $V_{\alpha\varepsilon}$; it is open, since $\tilde{p}_\alpha$ is continuous. Its closure in $\tilde{E}$ we denote by $\bar{V}_{\alpha\varepsilon}$; it is given by $\tilde{p}_\alpha(y) \leqq \varepsilon$, by § 16, 4.

To prove that $\tilde{\mathfrak{T}}$ is defined by the system $\{\tilde{p}_\alpha\}$ it is sufficient to show that $\bar{U}_{\alpha\varepsilon} = \bar{V}_{\alpha\varepsilon}$. Because $\tilde{p}_\alpha$ is continuous, we clearly have $\tilde{p}_\alpha(z) \leqq \varepsilon$ for each $z \in \bar{U}_{\alpha\varepsilon}$, so that $\bar{U}_{\alpha\varepsilon} \subset \bar{V}_{\alpha\varepsilon}$. Since a point z in the open set $V_{\alpha\varepsilon}$ is a closure point of $U_{\alpha\varepsilon} = V_{\alpha\varepsilon} \cap E$ we have $V_{\alpha\varepsilon} \subset \bar{U}_{\alpha\varepsilon}$, so that $\bar{V}_{\alpha\varepsilon} \subset \bar{U}_{\alpha\varepsilon}$.

The construction of the completion of a topological vector space given in § 15, 3. depends upon the construction of the completion of a uniform space given in § 5, 5., and this is not particularly simple. Using 3.(7), we can give a rather simpler construction which only depends upon the construction of the completion of a normed space given in § 14, 3.

A further construction will be given in § 21, 9.

We recapitulate the assertion:

(3) *Every locally convex space* $E[\mathfrak{T}]$ *can be embedded in a smallest complete locally convex space* $\tilde{E}[\tilde{\mathfrak{T}}]$, *which is unique up to topological isomorphism. The closures in* $\tilde{E}$ *of the members of a base of neighbourhoods of* $\mathfrak{o}$ *in* E *form a base of neighbourhoods of* $\mathfrak{o}$ *in* $\tilde{E}$.

Proof. By 3.(7), we can embed $E[\mathfrak{T}]$ in a topological product F of (B)-spaces. Because F is complete, the closure $\tilde{E}$ of E in F is a complete locally convex space; this establishes the existence of a completion $\tilde{E}[\tilde{\mathfrak{T}}]$.

Next we show that the closures $\bar{U}_\alpha$ in $\tilde{E}$ of a base of open neighbourhoods U_α of E form a base of neighbourhoods of $\mathfrak{o}$ in $\tilde{E}$. Since $\mathfrak{T}$ is the topology induced on E by $\tilde{\mathfrak{T}}$, for each U_α there is an open neighbourhood $\tilde{U}_\alpha$ of $\mathfrak{o}$ in $\tilde{E}$, with $U_\alpha = \tilde{U}_\alpha \cap E$. Since $\tilde{U}_\alpha$ is open and E is dense in $\tilde{U}_\alpha$, U_α is dense in $\tilde{U}_\alpha$. Hence $\tilde{U}_\alpha \subset \bar{U}_\alpha$, $\bar{\tilde{U}}_\alpha \subset \bar{U}_\alpha$, and so $\bar{U}_\alpha = \bar{\tilde{U}}_\alpha$ is a neighbourhood of $\mathfrak{o}$ in $\tilde{E}$. If now $\tilde{V}$ is an arbitrary closed neighbourhood of $\mathfrak{o}$ in $\tilde{E}$, $V = \tilde{V} \cap E$ is a neighbourhood of $\mathfrak{o}$ in E. Thus, for a suitable U_α, $U_\alpha \subset V$, and so $\bar{U}_\alpha \subset \bar{V} \subset \tilde{V}$. The sets $\bar{U}_\alpha$ therefore form a base of neighbourhoods of $\mathfrak{o}$ in $\tilde{E}[\tilde{\mathfrak{T}}]$.

Two different completions of E can be mapped in a one-one way onto each other in a way which makes the limits of the same Cauchy nets correspond to each other. Under this the closures of the sets U_α correspond, and so the two completions of E are topologically isomorphic, since the correspondence also preserves the linear operations.

In many situations the locally convex spaces under consideration are not complete. The two following weaker properties are frequently useful: a locally convex space $E[\mathfrak{T}]$ is said to be sequentially complete if every $\mathfrak{T}$-Cauchy sequence has a limit in $E[\mathfrak{T}]$, and it is said to be quasi-complete if every bounded closed set of $E[\mathfrak{T}]$ is complete.

Clearly every complete space is quasi-complete, and every quasi-complete space is sequentially complete.

Suppose that a second finer locally convex topology $\mathfrak{T}'$ is given on a locally convex space $E[\mathfrak{T}]$. What relation is there between the completions $\tilde{E}[\tilde{\mathfrak{T}}']$ and $\tilde{E}[\tilde{\mathfrak{T}}]$?

The identity mapping I from $E[\mathfrak{T}']$ into $E[\mathfrak{T}]$ is continuous, and so by § 5, 4.(4) there is a uniquely determined continuous extension $\tilde{I}$ which maps $\tilde{E}[\tilde{\mathfrak{T}}']$ into $\tilde{E}[\tilde{\mathfrak{T}}]$. In general this linear mapping need not still be one-one, so that it is only the quotient of $\tilde{E}[\tilde{\mathfrak{T}}']$ by $\tilde{I}^{-1}(\mathfrak{o})$ which is embedded in $\tilde{E}[\tilde{\mathfrak{T}}]$.

For example, let $E[\mathfrak{T}]$ be a (B)-space and let $\mathfrak{T}'$ be the topology defined on E by a finer norm which is not equivalent to the $\mathfrak{T}$-norm. If $\tilde{I}$ were one-one, $\tilde{E}[\tilde{\mathfrak{T}}']$ and $E[\mathfrak{T}]$ would be topologically isomorphic, by the BANACH-SCHAUDER theorem [cf. § 15, 12.(2)], and this is impossible.

However the following important completeness criterion holds, even for arbitrary topological vector spaces:

(4) *Suppose that a second finer topology $\mathfrak{T}'$ is given on the topological vector space $E[\mathfrak{T}]$. Suppose that $\mathfrak{T}'$ has a base of neighbourhoods of $\mathfrak{o}$ consisting of $\mathfrak{T}$-closed sets. Under these hypotheses*

a) *If a $\mathfrak{T}'$-Cauchy filter converges to x_0 as a $\mathfrak{T}$-Cauchy filter on E, then it also converges to x_0 as a $\mathfrak{T}'$-Cauchy filter.*

b) *Every subset of E which is complete (respectively sequentially complete) with respect to $\mathfrak{T}$ is also complete (respectively sequentially complete) with respect to $\mathfrak{T}'$.*

c) *The identity mapping I from $E[\mathfrak{T}']$ onto $E[\mathfrak{T}]$ can be uniquely extended to a continuous embedding $\tilde{I}$ of $\tilde{E}[\tilde{\mathfrak{T}}']$ into $\tilde{E}[\tilde{\mathfrak{T}}]$.*

Proof of a). A $\mathfrak{T}'$-Cauchy filter $\mathfrak{F}$ is certainly a $\mathfrak{T}$-Cauchy filter, since $\mathfrak{T}$ is coarser than $\mathfrak{T}'$. Suppose that x_0 is its $\mathfrak{T}$-limit. To each $\mathfrak{T}$-closed circled $\mathfrak{T}'$neighbourhood U of $\mathfrak{o}$ in E there is an $F^\alpha \in \mathfrak{F}$ which is small of order U. If $y \in F^\alpha$, $F^\alpha \subset y+U$. Since $y+U$ is $\mathfrak{T}$-closed, x_0

lies in $y+U$, as it is a $\mathfrak{T}$-closure point of F^α. It then follows from $F^\alpha \subset x_0+U+U$ that x_0 is also $\mathfrak{T}'$-limit of $\mathfrak{F}$.

b) follows directly from a) and from the corresponding assertion for sequences, which is proved in an analogous way.

Proof of c). We must show that $\tilde{I}$ is one-one. Let z be an element of $\tilde{E}[\tilde{\mathfrak{T}}']$ with $\tilde{I}z=\mathfrak{o}$. There exists a $\mathfrak{T}'$-Cauchy filter $\mathfrak{F}=\{F_\alpha\}$ in E with $\lim \mathfrak{F}=z$. The sets F_α form the base of a $\mathfrak{T}'$-Cauchy filter $\tilde{\mathfrak{F}}'$ in $\tilde{E}[\tilde{\mathfrak{T}}']$ whose limit is again z. Under the continuous mapping $\tilde{I}$, $\tilde{\mathfrak{F}}'$ is mapped into a $\mathfrak{T}$-Cauchy filter $\tilde{\mathfrak{F}}=\tilde{I}(\tilde{\mathfrak{F}}')$ whose limit $\tilde{I}z$ is $\mathfrak{o}$, by hypothesis. The restriction $\tilde{\mathfrak{F}}\cap E$ of $\tilde{\mathfrak{F}}$ to E is coarser than $\mathfrak{F}$: for if $\tilde{I}(M)$ is an arbitrary set of the filter $\tilde{\mathfrak{F}}$, where $M\in\tilde{\mathfrak{F}}'$, and if $F_\alpha\subset M$, then $F_\alpha=\tilde{I}(F_\alpha)\subset\tilde{I}(M)$, so that $F_\alpha\subset\tilde{I}(M)\cap E$. Thus $\tilde{\mathfrak{F}}\cap E$ is a filter on E, and as it is the restriction of $\tilde{\mathfrak{F}}$ it is a $\mathfrak{T}$-Cauchy filter with limit $\mathfrak{o}$. But then the finer $\mathfrak{T}$-Cauchy filter $\mathfrak{F}$ has $\mathfrak{T}$-limit $\mathfrak{o}$. It now follows from a) that $\mathfrak{F}$, considered as a $\mathfrak{T}'$-Cauchy filter, also has $\mathfrak{o}$ as limit, so that $z=\mathfrak{o}$.

(4) is due to N. Bourbaki [6] and W. Robertson [2]. The latter work contains a detailed investigation of this and related problems.

5. The locally convex direct sum of locally convex spaces. In § 7,8. we introduced the direct sum $E=\bigoplus_\alpha E_\alpha$ of vector spaces E_α. This is the subspace of $\prod_\alpha E_\alpha$ consisting of only those elements $x=(x_\alpha)$, $x_\alpha\in E_\alpha$, which have finitely many non-zero x_α.

We denote the embedding of E_α into E by I_α. This is the mapping which sends the element $x_\alpha\in E_\alpha$ to the element $x\in E$ whose α-th coordinate is equal to x_α, and all of whose other coordinates vanish.

The locally convex direct sum $E[\mathfrak{T}]=\bigoplus_\alpha E_\alpha[\mathfrak{T}_\alpha]$ of the locally convex spaces $E_\alpha[\mathfrak{T}_\alpha]$ is defined to be the direct sum E of the spaces E_α equipped with the finest locally convex topology for which each of the embeddings I_α of E_α into E is continuous.

It is easy to give a base of neighbourhoods of $\mathfrak{o}$ for $E[\mathfrak{T}]$.

(1) *If for each fixed α, $\{U_\beta^\alpha\}$ forms a base of neighbourhoods of* 0 *in E_α, then the absolutely convex covers $\Gamma_\alpha I_\alpha(U_\beta^\alpha)$ form a base of neighbourhoods of $\mathfrak{o}$ in $\bigoplus_\alpha E_\alpha[\mathfrak{T}_\alpha]$.*

For if U is an absolutely convex neighbourhood of $\mathfrak{o}$ for some topology on E for which I_α is continuous, then U must contain a set $I_\alpha(U_\beta^\alpha)$. This holds for each α; but then $\Gamma_\alpha I_\alpha(U_\beta^\alpha)$ also lies in U, since U is absolutely convex. The finest of all these locally convex topologies on E is however clearly the one for which all such sets $\Gamma I_\alpha(U_\beta^\alpha)$ form a base of neighbourhoods of $\mathfrak{o}$.

In what follows we shall write U_β^α instead of $I_\alpha(U_\beta^\alpha)$, provided that there is no danger of misunderstanding.

(2) *The topology* $\mathfrak{T}$ *of* $\bigoplus_\alpha E_\alpha[\mathfrak{T}_\alpha]$ *is Hausdorff, so that* $\bigoplus_\alpha E_\alpha[\mathfrak{T}_\alpha]$ *is again a locally convex space.*

If β *runs through a subset of the indices* α, $\bigoplus_\beta E_\beta[\mathfrak{T}_\beta]$ *is a closed subspace of* $\bigoplus_\alpha E_\alpha[\mathfrak{T}_\alpha]$, *and the topology induced on* $\bigoplus_\beta E_\beta[\mathfrak{T}_\beta]$ *by* $\mathfrak{T}$ *is the locally convex direct sum topology.*

In particular $\mathfrak{T}$ *induces the topology* $\mathfrak{T}_\alpha$ *on each* E_α.

For finitely many summands, $\bigoplus_{i=1}^n E_{\alpha_i}[\mathfrak{T}_{\alpha_i}]$ *and* $\prod_{i=1}^n E_{\alpha_i}[\mathfrak{T}_{\alpha_i}]$ *are topologically isomorphic.*

Proof. The topology induced on $H = \bigoplus_\beta E_\beta[\mathfrak{T}_\beta]$ by $\mathfrak{T}$ is determined by the neighbourhoods $\left(\Gamma_\alpha U^\alpha\right) \cap H = \Gamma_\beta U^\beta$ of $\mathfrak{o}$; these also determine the locally convex direct sum topology on H.

The neighbourhood $\Gamma_{i=1}^n U^{\alpha_i}$ of $\mathfrak{o}$ in $\bigoplus_{i=1}^n E_{\alpha_i}[\mathfrak{T}_{\alpha_i}]$ lies in the neighbourhood $\prod_{i=1}^n U^{\alpha_i}$ of $\mathfrak{o}$ in $\prod_{i=1}^n E_{\alpha_i}[\mathfrak{T}_{\alpha_i}]$; on the other hand $\prod_{i=1}^n \left(\frac{1}{n} U^{\alpha_i}\right)$ lies in $\Gamma_{i=1}^n U^{\alpha_i}$, so that the product topology and the sum topology coincide on $\bigoplus_{i=1}^n E_{\alpha_i} = \prod_{i=1}^n E_{\alpha_i}$.

Since $\prod_{i=1}^n E_{\alpha_i}[\mathfrak{T}_{\alpha_i}]$ is Hausdorff, and since each element $x \in \bigoplus_\alpha E_\alpha[\mathfrak{T}_\alpha]$ lies in some $\bigoplus_{i=1}^n E_{\alpha_i}[\mathfrak{T}_{\alpha_i}]$, there is a $\mathfrak{T}$-neighbourhood of $\mathfrak{o}$ separating x and $\mathfrak{o}$, and so $\mathfrak{T}$ is Hausdorff.

$E = H \oplus H'$, with $H' = \bigoplus_\gamma E_\gamma[\mathfrak{T}_\gamma]$, where γ runs through the indices α which are different from the β. By § 15, 8.(1) this decomposition is topological, since the projections onto H and H' are continuous. H and H', being the null spaces of these projections, are closed.

(3) *If the spaces* $E_\alpha[\mathfrak{T}_\alpha]$ *are complete locally convex spaces, their locally convex direct sum is also complete; consequently, we have* $\widetilde{\bigoplus_\alpha E_\alpha[\mathfrak{T}_\alpha]} \cong \bigoplus_\alpha \tilde{E}_\alpha[\tilde{\mathfrak{T}}_\alpha]$.

We shall show that $E[\mathfrak{T}] = \bigoplus_\alpha E_\alpha[\mathfrak{T}_\alpha]$ is complete, when the spaces $E_\alpha[\mathfrak{T}_\alpha]$ are complete.

Let $\mathfrak{F} = \{F^\beta\}$ be a $\mathfrak{T}$-Cauchy filter on E. Since the projection P_α of E onto E_α is continuous, $P_\alpha(\mathfrak{F})$ is also a Cauchy filter on E_α. Since E_α is complete, $P_\alpha(\mathfrak{F})$ has a limit x_α.

We assert that only finitely many x_α can be different from zero. For otherwise there would be a sequence $x_{\alpha_i} \neq \mathfrak{o}$, $i = 1, 2, \ldots$. For each x_{α_i} there exists a closed absolutely convex $\mathfrak{T}_{\alpha_i}$-neighbourhood U^{α_i} of $\mathfrak{o}$ with

$o \notin x_{\alpha_i} + U^{\alpha_i}$. We form a $\mathfrak{T}$-neighbourhood $V = \ulcorner_{\alpha} V^{\alpha}$, with $V^{\alpha_i} = U^{\alpha_i}$, $i = 1, 2, \ldots$. Let F^{β} be small of order V. For an element $z = (z_{\alpha}) \in F^{\beta}$ we have $P_{\alpha_i} z = z_{\alpha_i} \in P_{\alpha_i}(F^{\beta})$. $P_{\alpha_i}(F^{\beta})$ is small of order $P_{\alpha_i}(V) = U^{\alpha_i}$. Because U^{α_i} is closed and x_{α_i} is a closure point of $P_{\alpha_i}(F^{\beta})$ we have $x_{\alpha_i} \in z_{\alpha_i} + U^{\alpha_i}$, or $z_{\alpha_i} \in x_{\alpha_i} + U^{\alpha_i}$, since U^{α_i} is absolutely convex. Consequently $z_{\alpha_i} \neq o$ for $i = 1, 2, \ldots$, so that z is an element of E with infinitely many non-zero components in the spaces E_{α}; this gives a contradiction.

Now let $x = x_{\alpha_1} + \cdots + x_{\alpha_n}$ be the sum of the non-zero x_{α}. We still have to show that $x = \lim \mathfrak{F}$. Let $U = \ulcorner_{\alpha} U^{\alpha}$ be a $\mathfrak{T}$-neighbourhood of o. Further let F^{β} be small of order U and let z be an arbitrary element of F^{β}. There is a finite sum $H = \bigoplus_{i=1}^{m} E_{\alpha_i}[\mathfrak{T}_{\alpha_i}]$ in which both x and z lie. If P is the projection of E onto H, then by (2) $P(U) \subset U$ and $P(U)$ is a neighbourhood of o in H. It follows from $F^{\beta} \subset z + U$ that $P(F^{\beta}) \subset Pz + P(U) = z + P(U)$. Since $P(\mathfrak{F})$ has x as limit in H, $x \in z + 2P(U)$, so that $z \in x + 2P(U) \subset x + 2U$ and $F^{\beta} \subset x + 2U$. Since this holds for each U and for each F^{β} which is small of order U, $x = \lim \mathfrak{F}$.

(4) *Every bounded subset B of $E[\mathfrak{T}] = \bigoplus_{\alpha} E_{\alpha}[\mathfrak{T}_{\alpha}]$ is a bounded subset of a finite sum $\bigoplus_{i=1}^{n} E_{\alpha_i}[\mathfrak{T}_{\alpha_i}]$, and so is contained in a set of the form $\bigoplus_{i=1}^{n} B_{\alpha_i} \subset E$ where B_{α_i} is bounded in $E_{\alpha_i}[\mathfrak{T}_{\alpha_i}]$.*

Proof. The projection $P_{\alpha}(B) = B_{\alpha}$ of B in E_{α} is bounded in $E_{\alpha}[\mathfrak{T}_{\alpha}]$, by (2). We must show that there can be only finitely many α_i with $B_{\alpha_i} \neq o$.

Let us suppose the contrary. It is sufficient to consider the case $\bigoplus_{i=1}^{\infty} E_i[\mathfrak{T}_i]$, with $B_i \neq o$ for each i. We can also suppose that for each i there is an $x^{(i)} = (x_1^{(i)}, x_2^{(i)}, \ldots) \in B$ with $x_i^{(i)} \neq o$ and $x_k^{(i)} = o$ for $k > i$.

Our assumption will be disproved if we can find a neighbourhood U of o which contains none of the points $\frac{1}{i} x^{(i)}$, for then there would be no scalar multiple of U containing all the $x^{(i)}$, so that B would not be bounded.

For each i we determine an absolutely convex neighbourhood U_i of o in E_i which does not contain $\frac{1}{i} x_i^{(i)}$, and we put $U = \ulcorner_{i=1}^{\infty} U_i$. Then we have $P_i(U) = U_i$ and $P_i\left(\frac{1}{i} x^{(i)}\right) = \frac{1}{i} x_i^{(i)} \notin U_i$. But it follows from this that $\frac{1}{i} x^{(i)} \notin U$.

A vector space E can always be represented as the direct sum of one-dimensional spaces E_{α}, $E = \bigoplus_{\alpha} x_{\alpha} \mathsf{K}$, where $\{x_{\alpha}\}$ is a basis for E.

E, considered as the locally convex direct sum of the spaces $x_\alpha \mathsf{K}$, has a topology $\mathfrak{T}$, which, by (1), has as base of neighbourhoods of $\mathfrak{o}$ all absolutely convex absorbent sets, i.e. all absolutely convex α-bodies. This topology is clearly the finest of all the locally convex topologies on E. With the notation introduced in § 7, 5. we therefore have

(5) *The topology on the locally convex direct sum $\varphi_d(\mathsf{K})$ of d one-dimensional spaces topologically isomorphic to K is the finest possible locally convex topology.*

An immediate consequence of (4) is

(6) *Every bounded subset of $\varphi_d(\mathsf{K})$ is finite-dimensional and relatively compact.*

The introduction of the topology $\mathfrak{T}$ of the direct sum given here provides a certain contrast with the procedure in the case of linearly topologized spaces (§ 10, 2.). There we took the sets $\bigoplus_\alpha U^\alpha$ as base of neighbourhoods of $\mathfrak{o}$. In the present case we can also proceed in this way.

If $E_\alpha[\mathfrak{T}_\alpha]$ are arbitrary topological vector spaces, we define the topology $\mathfrak{T}'$ on $E = \bigoplus_\alpha E_\alpha$ by taking as base of neighbourhoods of $\mathfrak{o}$ sets $\bigoplus_\alpha U^\alpha$, where U^α is an arbitrary neighbourhood of $\mathfrak{o}$ in $E_\alpha[\mathfrak{T}_\alpha]$. In the locally convex case we clearly obtain a weaker topology than $\mathfrak{T}$, since $\Gamma_\alpha U^\alpha \subset \bigoplus_\alpha U^\alpha$. We call $E[\mathfrak{T}']$ the topological direct sum of the spaces $E_\alpha[\mathfrak{T}_\alpha]$.

The same results hold for the topological direct sum:

(7) *$E[\mathfrak{T}'] = (\bigoplus_\alpha E_\alpha)[\mathfrak{T}']$ is locally convex for locally convex $E_\alpha[\mathfrak{T}_\alpha]$, $E[\mathfrak{T}']$ is complete for complete $E_\alpha[\mathfrak{T}_\alpha]$, and the bounded sets are contained in the sets $\bigoplus_{i=1}^{n} B_{\alpha_i}$, so that in the locally convex case they coincide with the $\mathfrak{T}$-bounded sets.*

The proof, which is almost word for word the same as for the locally convex direct sum, is left to the reader.

(8) *In the locally convex case, if there are countably many summands $E_i[\mathfrak{T}_i]$, then $\mathfrak{T}'$ and $\mathfrak{T}$ coincide on $\bigoplus_{i=1}^{\infty} E_i$; for more than countably many summands, $\mathfrak{T}'$ can be different from $\mathfrak{T}$.*

In the countable case $\mathfrak{T}'$ is equal to $\mathfrak{T}$ since every $\mathfrak{T}$-neighbourhood $\Gamma_{i=1}^{\infty} U^i$ contains the $\mathfrak{T}'$-neighbourhood $\bigoplus_{i=1}^{\infty} \frac{1}{2^i} U^i$.

$\mathfrak{T}$ and $\mathfrak{T}'$ are different on φ_d if $d > \aleph_0$: in each one-dimensional space $E_\alpha = x_\alpha \mathsf{K}$ let the neighbourhood U^α of $\mathfrak{o}$ be given by $|\xi_\alpha| \leqq 1$. Then $\Gamma_\alpha U^\alpha$ consists of all $\mathfrak{x} = (\xi_\alpha) \in \varphi_d$ with $\sum_\alpha |\xi_\alpha| \leqq 1$. But for every $\mathfrak{T}'$-neighbourhood of $\mathfrak{o}$ $\bigoplus_\alpha \rho_\alpha U^\alpha$ there exists an ε_0 for which $\rho_\alpha \geqq \varepsilon_0$ for uncountable many α, so that there are always elements $\mathfrak{x}$ in $\bigoplus_\alpha \rho_\alpha U^\alpha$ with arbitrarily large $\sum_\alpha |\xi_\alpha|$.

The reason for giving preference to the topology $\mathfrak{T}$ rather than to $\mathfrak{T}'$ will become apparent when we consider the theory of duality (§ 22, 5.).

§ 19. Locally convex hulls and kernels, inductive and projective limits of locally concex spaces

1. The locally convex hull of locally convex spaces. Locally convex direct sums and topological products of locally convex spaces are special cases of more general ideas which we now consider. We begin with the locally convex hull and the topological inductive limit, which are particularly important for applications.

If a vector space E is the linear span of certain linear subspaces E_α, we write $E=\sum_\alpha E_\alpha$. Of particular interest for us is the case where each E_α is given as the linear image $A_\alpha(F_\alpha)$ of a vector space F_α. We then write $E=\sum_\alpha A_\alpha(F_\alpha)$.

A special case of such a linear span is the direct sum $E=\bigoplus_\alpha E_\alpha$; conversely we have

(1) *Every linear span $E=\sum_\alpha A_\alpha(F_\alpha)$ is isomorphic to a quotient $\hat{E}=\left(\bigoplus_\alpha F_\alpha\right)/H$.*

Proof. A linear mapping A from $\bigoplus_\alpha F_\alpha$ onto $E=\sum_\alpha A_\alpha(F_\alpha)$ is defined by $A\left(\sum_\alpha x_\alpha\right)=\sum_\alpha A_\alpha x_\alpha$, $x_\alpha\in F_\alpha$. If H is the null-space $N[A]$ of A in $\bigoplus_\alpha F_\alpha$, then E is isomorphic to $\hat{E}=\left(\bigoplus_\alpha F_\alpha\right)/H$.

Clearly we have the converse:

(2) *Every quotient $\hat{E}=\left(\bigoplus_\alpha F_\alpha\right)/H$ is equal to the linear span $E=\sum_\alpha K_\alpha(F_\alpha)$, where K_α is the restriction to F_α of the canonical homomorphism K from $\bigoplus_\alpha F_\alpha$ onto $\hat{E}$.*

If the F_α are locally convex spaces $F_\alpha[\mathfrak{T}_\alpha]$, we can try to introduce as natural a locally convex topology as possible on the linear span $E=\sum_\alpha A_\alpha(F_\alpha)$. By analogy with the special case of the locally convex direct sum, the finest locally convex topology $\mathfrak{T}$ for which all the A_α are continuous mappings from F_α into E suggests itself. An absolutely convex subset U of E is then a $\mathfrak{T}$-neighbourhood of $\circ$ if $A_\alpha^{(-1)}(U)$ is a $\mathfrak{T}_\alpha$-neighbourhood U^α of $\circ$ in F_α, for each α. All the sets $\Gamma_\alpha A_\alpha(U^\alpha)$, where each set U^α is taken to be a $\mathfrak{T}_\alpha$-neighbourhood of $\circ$ in $F_\alpha[\mathfrak{T}_\alpha]$, therefore form a base of $\mathfrak{T}$-neighbourhoods of $\circ$ in E.

This topology $\mathfrak{T}$ need not however always be Hausdorff, as we shall presently see. But if this is the case, $E[\mathfrak{T}]=\sum_\alpha A_\alpha(F_\alpha[\mathfrak{T}_\alpha])$ is called the locally convex hull of the $A_\alpha(F_\alpha[\mathfrak{T}_\alpha])$, and $\mathfrak{T}$ is called the hull topology on E.

We can say more about the algebraic isomorphism (1):

(3) *Every locally convex hull* $E[\mathfrak{T}] = \sum_\alpha A_\alpha(F_\alpha[\mathfrak{T}_\alpha])$ *is topologically isomorphic to a quotient* $\hat{E} = \left(\bigoplus_\alpha F_\alpha\right)/H$ *of the locally convex sum of the* $F_\alpha[\mathfrak{T}_\alpha]$ *by a closed linear subspace* H.

Proof. Under the mapping $A\left(\sum_\alpha x_\alpha\right) = \sum_\alpha A_\alpha x_\alpha$ of $\bigoplus_\alpha F_\alpha[\mathfrak{T}_\alpha]$ onto $E[\mathfrak{T}]$, the neighbourhood $\Gamma_\alpha U^\alpha$ of o in $\bigoplus_\alpha F_\alpha[\mathfrak{T}_\alpha]$ is mapped onto the $\mathfrak{T}$-neighbourhood $\Gamma_\alpha A_\alpha(U^\alpha)$ of o in E. A is therefore a topological homomorphism. By § 15, 4.(4), $\check{A}$ is a topological isomorphism of $\left(\bigoplus_\alpha F_\alpha\right)/H$ onto $E[\mathfrak{T}]$, where H is the closed null-space of A. The topology on $\left(\bigoplus_\alpha F_\alpha\right)/H$ is therefore the quotient-space topology.

The stronger assertion corresponding to (2) is clearly also true:

(4) *Every quotient* $\hat{E} = \left(\bigoplus_\alpha F_\alpha\right)/H$ *of a locally convex direct sum of locally convex spaces* $F_\alpha[\mathfrak{T}_\alpha]$ *by a closed linear subspace* H *is topologically isomorphic to the locally convex hull* $\sum_\alpha K_\alpha(F_\alpha[\mathfrak{T}_\alpha])$, *where* K_α *is the restriction to* F_α *of the canonical homomorphism* K *of* $\bigoplus_\alpha F_\alpha$ *onto* $\hat{E}$.

Consequently it is clear that the hull topology on a linear span $\sum_\alpha A_\alpha(F_\alpha[\mathfrak{T}_\alpha])$ is not Hausdorff if and only if the null-space H of the mapping A which establishes the algebraic isomorphism (1) is not closed (cf. § 10, 7.).

Remark. By (4), a quotient space $(E/H)[\mathfrak{T}]$ of $E[\mathfrak{T}]$ can be considered as a locally convex hull; we have $(E/H)[\mathfrak{T}] = K(E[\mathfrak{T}])$, where K is the canonical mapping from E onto E/H.

(5) *Suppose on the one hand that* E *is equal to* $\sum_\alpha A_\alpha(F_\alpha[\mathfrak{T}_\alpha])$, *and on the other that it is equal to* $\sum_\beta B_\beta(G_\beta[\mathfrak{T}'_\beta])$. *Let* $\mathfrak{T}$ *be the hull topology corresponding to the first representation, and let* $\mathfrak{T}'$ *be that corresponding to the second. If for each* α *there is a* β *for which* $A_\alpha(F_\alpha) \subset B_\beta(G_\beta)$, *and if the topology* $\mathfrak{T}'_\beta$ *defined on* $B_\beta(G_\beta)$ *by* B_β *induces a coarser topology on* $A_\alpha(F_\alpha)$ *than the topology* $\mathfrak{T}_\alpha$ *defined on* $A_\alpha(F_\alpha)$ *by* A_α, *then* $\mathfrak{T}'$ *is coarser then* $\mathfrak{T}$ *on* E.

For if V is an absolutely convex $\mathfrak{T}'$-neighbourhood of o, $V \cap B_\beta(G_\beta)$ is the B_β-image of a $\mathfrak{T}'_\beta$-neighbourhood of o in G_β, and by hypothesis this contains the A_α-image of a $\mathfrak{T}_\alpha$-neighbourhood of o in F_α. V is therefore also a $\mathfrak{T}$-neighbourhood of o.

If the two representations define the same topology on E, we speak of equivalent defining systems.

(6) *The formation of locally convex hulls is transitive.*

If $E[\mathfrak{T}] = \sum_{\alpha} A_\alpha\Big(\sum_{\beta_\alpha} B_{\beta_\alpha}(F_{\beta_\alpha}[\mathfrak{T}_{\beta_\alpha}])\Big)$, $E[\mathfrak{T}]$ is also equal to $\sum_{\alpha,\beta_\alpha} A_\alpha B_{\beta_\alpha}(F_{\beta_\alpha}[\mathfrak{T}_{\beta_\alpha}])$. This is trivial for the vector space properties, and the fact that the hull topologies coincide follows from $\Gamma_{\alpha} A_\alpha\Big(\Gamma_{\beta_\alpha} B_{\beta_\alpha}(U^{\beta_\alpha}\Big)$ $= \Gamma_{\alpha,\beta_\alpha} A_\alpha B_{\beta_\alpha}(U^{\beta_\alpha})$.

(7) *A linear mapping B from a locally convex hull $E[\mathfrak{T}] = \sum_{\alpha} A_\alpha(E_\alpha[\mathfrak{T}_\alpha])$ (respectively $\sum_{\alpha} E_\alpha[\mathfrak{T}_\alpha]$) into a locally convex space $F[\mathfrak{T}']$ is continuous if and only if all the mappings BA_α (respectively all the restrictions of B to the spaces E_α) are continuous mappings from $E_\alpha[\mathfrak{T}_\alpha]$ into $F[\mathfrak{T}']$.*

Proof. The condition is clearly necessary. On the other hand, if each BA_α (respectively BI_α, where I_α is the embedding of E_α into E) is continuous, then given an absolutely convex neighbourhood V of $\mathfrak{o}$ in F there is always a $\mathfrak{T}_\alpha$-neighbourhood U^α of $\mathfrak{o}$ with $BA_\alpha(U^\alpha) \subset V$. But then $B\Big(\Gamma_{\alpha} A_\alpha(U^\alpha)\Big) \subset V$, so that B is continuous.

An analogous proof shows that a collection M of linear mappings B is equicontinuous if and only if the mappings BA_α are equicontinuous, for each α.

2. The inductive limit of vector spaces. The ideas which we shall consider in this number are of a purely algebraic nature, and they are also valid for vector spaces over arbitrary fields.

Let $E = \sum_{\alpha} A_\alpha(F_\alpha)$ be a linear span. Suppose that the index set A is a directed set, and that for each pair $\alpha < \beta$ there is a linear mapping $A_{\beta\alpha}$ from F_α into F_β for which

$$(1) \qquad A_\alpha = A_\beta A_{\beta\alpha} \quad \text{for} \quad \alpha < \beta.$$

Further, we put $A_{\alpha\alpha}$ equal to the identity mapping from F_α onto itself.

If $\alpha \leqq \beta$, we always have $A_\alpha(F_\alpha) \subset A_\beta(F_\beta)$.

(2) *If the mappings A_α, $\alpha \in \mathsf{A}$, in (1) are one-one, then so are the mappings $A_{\beta\alpha}$, and we have*

$$(3) \qquad A_{\gamma\beta} A_{\beta\alpha} = A_{\gamma\alpha} \quad \text{for} \quad \alpha \leqq \beta \leqq \gamma.$$

To prove this, multiply both sides of (3) by A_γ and apply (1).

This situation arises in many cases. For example if $E = \sum_{\alpha} E_\alpha$, if the indices α form a directed set A and if $E_\alpha \subset E_\beta$ for $\alpha < \beta$, then we need only set A_α equal to the embedding I_α of E_α in E and $A_{\beta\alpha}$ equal to the embedding $I_{\beta\alpha}$ of E_α in E_β, and (1) is satisfied. In this case (3) is also satisfied.

We can even express every linear span $E=\sum_{\alpha} A_\alpha(F_\alpha)$, $\alpha\in\mathsf{A}$, in this way: for every finite subset Δ of A we form the finite direct sum $F_\Delta=\bigoplus_{\delta\in\Delta} F_\delta$ and define the mapping A_Δ from F_Δ into E by setting $A_\Delta\sum_\delta x_\delta=\sum_\delta A_\delta x_\delta$. The set of Δ's becomes a directed set when we put $\Delta_1\leqq\Delta_2$ whenever $\Delta_1\subset\Delta_2$. Finally for $\Delta_1\leqq\Delta_2$ we define $A_{\Delta_2\Delta_1}$ to be the embedding of F_{Δ_1} into F_{Δ_2}. In this way, $E=\sum_\Delta A_\Delta(F_\Delta)$, and (1) is satisfied. (3) also holds for the mappings $A_{\Delta_2\Delta_1}$.

If the situation described at the beginning holds, it is transformed by the isomorphism 1.(1) in the following way:

(4) *If* $E=\sum_\alpha A_\alpha(F_\alpha)$, *and if mappings* $A_{\beta\alpha}$ *from* F_α *into* F_β *are given which satisfy* (1), *then the quotient space* $\hat{E}=\sum_\alpha K_\alpha(F_\alpha)$ *isomorphic to* E *satisfies the following relations:*

$$K_\alpha=K_\beta A_{\beta\alpha} \quad \textit{for} \quad \alpha<\beta. \tag{5}$$

Proof. The mapping A of the proof of 1.(1) determines the one-one mapping $\hat{A}$ from $\hat{E}=K\left(\bigoplus_\alpha F_\alpha\right)$ onto E, and so by 1.(2) $\sum_\alpha A_\alpha x_\alpha=\hat{A}K\left(\sum_\alpha x_\alpha\right)$ $=\hat{A}\sum_\alpha K_\alpha x_\alpha$. The element $A_\alpha x_\alpha$ therefore corresponds in a one-one way to the element $K_\alpha x_\alpha$. It therefore follows from $A_\alpha x_\alpha=A_\beta(A_{\beta\alpha}x_\alpha)$ that $K_\alpha x_\alpha=K_\beta(A_{\beta\alpha}x_\alpha)$ for all $x_\alpha\in F_\alpha$, which implies (5).

So far we have started from a space E, together with subspaces $A_\alpha(F_\alpha)$ and mappings $A_{\beta\alpha}$ from F_α into F_β. It is natural to ask how far E is determined by the F_α and the $A_{\beta\alpha}$, and whether, given F_α and $A_{\beta\alpha}$, we can find a space E and mappings A_α which satisfy (1).

An answer is given by

(6) *Suppose that* F_α *is a directed system of vector spaces, such that for each pair* $\alpha<\beta$ *there is a linear mapping* $A_{\beta\alpha}$ *from* F_α *into* F_β, *and such that*

$$A_{\gamma\beta}A_{\beta\alpha}=A_{\gamma\alpha} \quad \textit{for} \quad \alpha<\beta<\gamma. \tag{7}$$

We denote by H_0 *the linear span in* $\bigoplus_\alpha F_\alpha$ *of all the elements* $x_\alpha-A_{\beta\alpha}x_\alpha$, *with* $x_\alpha\in F_\alpha$ *and* $\alpha<\beta$. *If* $H\supset H_0$ *is a linear subspace of* $\bigoplus_\alpha F_\alpha$ *and if* K *is the canonical homomorphism of* $\bigoplus_\alpha F_\alpha$ *onto* $\left(\bigoplus_\alpha F_\alpha\right)/H$, *then* $\left(\bigoplus_\alpha F_\alpha\right)/H$ $=\sum_\alpha K_\alpha(F_\alpha)$, *and* (5) *holds.*

This is an easy consequence of 1.(2). For it follows from $H\supset H_0$ that $K(x_\alpha-A_{\beta\alpha}x_\alpha)=o$, so that $K_\alpha x_\alpha=K_\alpha A_{\beta\alpha}x_\alpha$ for all $x_\alpha\in F_\alpha$, and so (5) is satisfied.

Taking $H=H_0$, the uniquely determined space $\left(\bigoplus_\alpha F_\alpha\right)/H_0=\sum_\alpha K_\alpha^{(0)}(F_\alpha)$ is called the inductive limit of the spaces F_α with respect to the mappings $A_{\beta\alpha}$, and it is denoted by $\varinjlim A_{\beta\alpha}(F_\alpha)$.

Even when all the mappings $A_{\beta\alpha}$ are non-zero it can happen that $\varinjlim A_{\beta\alpha}(F_\alpha)$ consists of the element $\mathfrak{o}$ alone.

So far we have made no use of (7); we now use it to determine H_0 exactly.

(8) *The element* $\sum_{i=1}^n x_{\alpha_i}$ *of* $\bigoplus_\alpha F_\alpha$ *lies in* H_0 *if and only if there exists a* $\beta\geqq\alpha_i$, $i=1,\dots,n$, *with*

$$\sum_{i=1}^n A_{\beta\alpha_i}x_{\alpha_i}=\mathfrak{o}. \tag{9}$$

Proof. If (9) holds, then $\sum_{i=1}^n x_{\alpha_i}=\sum_{i=1}^n (x_{\alpha_i}-A_{\beta\alpha_i}x_{\alpha_i})\in H_0$. Conversely if $\gamma\geqq\beta$, the relation $A_{\gamma\alpha}x_\alpha-A_{\gamma\beta}(A_{\beta\alpha}x_\alpha)=\mathfrak{o}$ follows from (7), and so we obtain an equation of the form (9) for $x_\alpha-A_{\beta\alpha}x_\alpha$. By choosing a sufficiently large β, (9) also holds for a linear combination of elements $x_{\alpha_i}-A_{\beta_i\alpha_i}x_{\alpha_i}$, and indeed this holds for all $\beta\geqq\beta_i$, $i=1,\dots,n$.

From (8) we obtain a certain converse to (2):

(9) *If the mappings* $A_{\beta\alpha}$ *in* $\varinjlim A_{\beta\alpha}(F_\alpha)=\sum_\alpha K_\alpha^{(0)}(F_\alpha)$ *are all one-one, then so also are the mappings* $K_\alpha^{(0)}$.

For $x_\alpha\in F_\alpha$, (9) is equivalent to a relation $A_{\beta\alpha}x_\alpha=\mathfrak{o}$. The hypothesis that $A_{\beta\alpha}$ is one-one means that this is only possible if $x_\alpha=\mathfrak{o}$. Thus $F_\alpha\cap H_0=\mathfrak{o}$.

In particular it follows from this that if the mappings $A_{\beta\alpha}$ are one-one, the inductive limit is different from zero, provided that the F_α are different from zero.

(11) *If* $E=\sum_\alpha A_\alpha(F_\alpha)$, *and if* $A_{\beta\alpha}$ *is a system of mappings satisfying* (1) *and* (3), *then E is always a homomorphic image of* $\varinjlim A_{\beta\alpha}(F_\alpha)$. *A necessary and sufficient condition for E to be isomorphic to* $\varinjlim A_{\beta\alpha}(F_\alpha)$ *is that an equation* $A_\alpha x_\alpha=\mathfrak{o}$, $x_\alpha\neq\mathfrak{o}$, *holds in E if and only if there is a* $\beta>\alpha$ *(depending upon* x_α*) for which* $A_{\beta\alpha}x_\alpha=\mathfrak{o}$.

Proof. If, using (4), we go over from E to the isomorphic space $\hat{E}=\sum_\alpha K_\alpha(F_\alpha)=\left(\bigoplus_\alpha F_\alpha\right)/H$, then (5) holds, so that $K_\alpha x_\alpha=K_\beta A_{\beta\alpha}x_\alpha$. This means that $K(x_\alpha-A_{\beta\alpha}x_\alpha)=\mathfrak{o}$, and consequently $x_\alpha-A_{\beta\alpha}x_\alpha\in H$. Thus $H\supset H_0$, and so E is the homomorphic image of $\left(\bigoplus_\alpha F_\alpha\right)/H_0=\varinjlim A_{\beta\alpha}(F_\alpha)$.

Further $A_\alpha x_\alpha = o$ holds in E if and only if $K_\alpha x_\alpha = o$ holds in $\hat{E}$. By (8), $K_\alpha^{(0)} x_\alpha = o$, i.e. $x_\alpha \in H_0$, if and only if $A_{\beta\alpha} x_\alpha = o$ for a suitable $\beta > \alpha$. Thus if it always follows from $K_\alpha x_\alpha = o$ that $A_{\beta\alpha} x_\alpha = o$ for a suitable $\beta > \alpha$, then K_α and $K_\alpha^{(0)}$ have the same null-space in F_α. We still have to show that the equation $H = H_0$ follows from this. Since $H \supset H_0$, it follows from $K\left(\sum_{i=1}^{n} x_{\alpha_i}\right) = o$ for $\beta \geqq \alpha_1, \ldots, \alpha_n$ that $K(\sum A_{\beta\alpha_i} x_{\alpha_i}) = o$; from this it follows that

$$K_\beta(\sum A_{\beta\alpha_i} x_{\alpha_i}) = K_\beta^{(0)}(\sum A_{\beta\alpha_i} x_{\alpha_i}) = K^{(0)}(\sum x_{\alpha_i}) = o,$$

i.e. $H_0 \supset H$.

(11) gives conditions for a linear span to be representable as an inductive limit.

A linear span of the special form $E = \sum_\alpha E_\alpha$, *with* $E_\alpha \subset E_\beta$ *for* $\alpha < \beta$, *is always equal to* $\varinjlim I_{\beta\alpha}(E_\alpha)$, *where* $I_{\beta\alpha}$ *is the embedding of* E_α *into* E_β. We also write $\varinjlim E_\alpha$ for this, and speak of the inductive limit of the directed system of spaces E_α.

As a special case, a direct sum $\bigoplus_\alpha E_\alpha$ is the inductive limit of its finite partial sums.

Likewise, by (11) and (2) a linear span $E = \sum_\alpha A_\alpha(F_\alpha)$ with a system of mappings $A_{\beta\alpha}$ satisfying (1) and (3) can be represented as $\varinjlim A_{\beta\alpha}(F_\alpha)$, if the mappings A_α are all one-one.

3. The topological inductive limit of locally convex spaces. If the vector spaces F_α of the preceding number are locally convex, then we can proceed as in 1.

Let $F_\alpha[\mathfrak{T}_\alpha]$ be a directed system of locally convex spaces, and let $\{A_{\beta\alpha}\}$ be a system of continuous mappings from $F_\alpha[\mathfrak{T}_\alpha]$ into $F_\beta[\mathfrak{T}_\beta]$, for $\alpha < \beta$, for which

(1) $$A_{\gamma\beta} A_{\beta\alpha} = A_{\gamma\alpha} \quad \text{for} \quad \alpha < \beta < \gamma.$$

If H_0 is again the linear span of the $x_\alpha - A_{\beta\alpha} x_\alpha \in \bigoplus_\alpha F_\alpha[\mathfrak{T}_\alpha]$, and if H_0 is closed in $\bigoplus_\alpha F_\alpha[\mathfrak{T}_\alpha]$, then the hull topology (respectively quotient topology) on $\varinjlim A_{\beta\alpha}(F_\alpha) = \sum_\alpha K_\alpha^{(0)}(F_\alpha[\mathfrak{T}_\alpha]) = \left(\bigoplus_\alpha F_\alpha[\mathfrak{T}_\alpha]\right)/H_0$ is Hausdorff.

The space $\varinjlim A_{\beta\alpha}(F_\alpha[\mathfrak{T}_\alpha]$, equipped with this topology $\mathfrak{T}$ is called the topological inductive limit of $A_{\beta\alpha}(F_\alpha)$.

By the final remarks of the preceding number, a locally convex hull $E[\mathfrak{T}] = \sum_\alpha E_\alpha[\mathfrak{T}_\alpha]$ with $E_\alpha \subset E_\beta$ for $\alpha < \beta$ is the topological inductive limit $\varinjlim I_{\beta\alpha}(E_\alpha[\mathfrak{T}_\alpha]) = \varinjlim E_\alpha[\mathfrak{T}_\alpha]$ of the spaces $E_\alpha[\mathfrak{T}_\alpha]$ if and only if $\mathfrak{T}_\beta$ induces a weaker topology than $\mathfrak{T}_\alpha$ on E_α, for $\alpha<\beta$.

In particular, the locally convex hull $\sum_{n=1}^{\infty} E_n[\mathfrak{T}_n]$ of an increasing sequence $E_1[\mathfrak{T}_1] \subset E_2[\mathfrak{T}_2] \subset \cdots$ is the topological inductive limit of the spaces $E_n[\mathfrak{T}_n]$ if and only if the topology $\mathfrak{T}_{n+1}$ induces a weaker topology than $\mathfrak{T}_n$ on E_n, for each n.

Every locally convex hull $E[\mathfrak{T}] = \sum_\alpha E_\alpha[\mathfrak{T}_\alpha]$ with an arbitrary index set can be considered as the topological inductive limit of the locally convex hulls of all the finite collections of $E_\alpha[\mathfrak{T}_\alpha]$.

(2) *Suppose that we have an inductive limit* $E[\mathfrak{T}] = \varinjlim A_{\beta\alpha}(F_\alpha[\mathfrak{T}_\alpha])$ *of locally convex spaces* $F_\alpha[\mathfrak{T}_\alpha]$, *and that* Γ *is a cofinal subset of the directed index set* A. *As* γ *and* δ *run through all pairs with* $\gamma<\delta$ *in* Γ, *then* $E_1[\mathfrak{T}_1] = \varinjlim A_{\delta\gamma}(F_\gamma[\mathfrak{T}_\gamma])$ *also exists, and it is topologically isomorphic to* $E[\mathfrak{T}]$.

Proof. Γ is also a directed set. We write E as the linear span $\sum_\alpha K_\alpha^{(0)}(F_\alpha)$, and E_1 as $\sum_\alpha K_\gamma^{(0)\prime}(F_\gamma)$. Since it follows from $\alpha<\gamma$ that $K_\alpha^{(0)}(F_\alpha) \subset K_\gamma^{(0)}(F_\gamma)$, and since, because Γ is cofinal, there always exists a $\gamma \geqq \alpha$ for each $\alpha \in \mathsf{A}$, E is also equal to the linear span $\sum_{\gamma\in\Gamma} K_\gamma^{(0)}(F_\gamma)$. To each element $\sum_{i=1}^{n} K_{\gamma_i}^{(0)} x_{\gamma_i} \in E$ we now make correspond the element $\sum_{i=1}^{n} K_{\gamma_i}^{(0)\prime} x_i \in E_1$. By 2.(8) the element $\sum_{i=1}^{n} K_{\gamma_i}^{(0)} x_{\gamma_i}$ is equal to $\mathfrak{o}$ if and only if there is a $\beta \geqq \gamma_i$, $i=1,\dots,n$, with $\sum_{i=1}^{n} A_{\beta\gamma_i} x_{\gamma_i} = \mathfrak{o}$ in $\bigoplus_\alpha F_\alpha$. But then if $\delta\in\Gamma$ is such that $\delta \geqq \beta$, we also have $\sum_{i=1}^{n} A_{\delta\gamma_i} x_{\gamma_i} = \mathfrak{o}$. Therefore it follows from 2.(8) that $\sum_{i=1}^{n} K_{\gamma_i}^{(0)} x_{\gamma_i} \in E$ is equal to $\mathfrak{o}$ if and only if the corresponding element $\sum_{i=1}^{n} K_{\gamma_i}^{(0)\prime} x_{\gamma_i} \in E_1$ vanishes. The correspondence between E and E_1 is therefore an algebraic isomorphism.

The hull topology $\mathfrak{T}$ on E is the finest locally convex topology for which all the mappings $K_\alpha^{(0)}$, $\alpha \in \mathsf{A}$, are continuous. But it is enough to require that all the mappings $K_\gamma^{(0)}$ are continuous, since it follows from $K_\alpha^{(0)} = K_\gamma^{(0)} A_{\gamma\alpha}$ for $\alpha<\gamma$ and from the continuity of $A_{\gamma\alpha}$ that $K_\alpha^{(0)}$ is continuous. This implies that $E[\mathfrak{T}]$ and $E_1[\mathfrak{T}_1]$ are topologically isomorphic.

We now return to the general situation considered at the beginning of the preceding number. Suppose now that $E[\mathfrak{T}] = \sum_\alpha A_\alpha(F_\alpha[\mathfrak{T}_\alpha])$ is the locally convex hull of a directed system $\{F_\alpha[\mathfrak{T}_\alpha]\}$, and that to each pair $\alpha<\beta$ there corresponds a continuous linear mapping $A_{\beta\alpha}$ of F_α into F_β, for which again

$$(3) \qquad A_\alpha = A_\beta A_{\beta\alpha} \quad \text{for} \quad \alpha<\beta$$

and

$$(4) \qquad A_{\gamma\beta} A_{\beta\alpha} = A_{\gamma\alpha} \quad \text{for} \quad \alpha<\beta<\gamma.$$

We saw (cf. 2.(11)) that E is the homomorphic image of $\varinjlim A_{\beta\alpha}(F_\alpha)$. It is easy to see that this homomorphism is also a topological homomorphism provided that the topological inductive limit of the $A_{\beta\alpha}(F_\alpha)$ exists, i.e. provided that the hull topology is Hausdorff.

4. Strict inductive limits. A topological inductive limit $E[\mathfrak{T}] = \sum_\alpha E_\alpha[\mathfrak{T}_\alpha]$ is said to be strict if $E_\alpha \subset E_\beta$ for $\alpha<\beta$ and if the topology induced by $\mathfrak{T}_\beta$ on the subspace E_α of E_β is equal to $\mathfrak{T}_\alpha$.

In particular, by § 18, 5.(2) every locally convex direct sum $E[\mathfrak{T}] = \bigoplus_\alpha E_\alpha[\mathfrak{T}_\alpha]$ is the strict inductive limit of its finite partial sums.

By § 18, 5.(2) the sum topology $\mathfrak{T}$ induces the topology $\mathfrak{T}_\alpha$ on each E_α. The question of whether this is also true for the hull topology of every strict inductive limit has recently been settled by a counterexample given by Kōmura [2]. In the countable case, however, we have

(1) *Let E be the union of a strictly increasing sequence $E_1[\mathfrak{T}_1] \subset E_2[\mathfrak{T}_2] \subset \cdots$ of locally convex spaces $E_n[\mathfrak{T}_n]$, and suppose that $\mathfrak{T}_{n+1}$ induces the topology $\mathfrak{T}_n$ on E_n. Then the hull topology $\mathfrak{T}$ is Hausdorff, so that $E[\mathfrak{T}]$ is the strict inductive limit of the spaces $E_n[\mathfrak{T}_n]$; further $\mathfrak{T}$ induces the topology $\mathfrak{T}_n$ on each E_n.*

First we establish two lemmas. We shall only need the first of these in the proof of (1).

(2) *If V is an absolutely convex neighbourhood of $\circ$ in a linear subspace H of the locally convex space $E[\mathfrak{T}]$, there is an absolutely convex neighbourhood U of $\circ$ in E for which $U \cap H = V$.*

Using the definition of the induced topology, there exists a neighbourhood W of $\circ$ with $W \cap H \subset V$. We can suppose that W is absolutely convex. We form $U = \Gamma(W \cup V)$. Every $z \in U$ has the form $z = \alpha x + \beta y$, $x \in W$, $y \in V$, $|\alpha| + |\beta| \leqq 1$. If z lies in H, x also lies in H, so that $x \in H \cap W \supset V$. Consequently $z \in V$ and $U \cap H = V$.

(3) *If, further, H is closed in $E[\mathfrak{T}]$ and $x_0 \notin H$, there is an absolutely convex neighbourhood U of $\circ$ with $U \cap H = V$ and $x_0 \notin U$.*

We can choose an absolutely convex neighbourhood W of $\circ$ for which $W \cap H \subset V$ and $(x_0 + W) \cap H$ is empty. Then if $U = \Gamma(W \cup V)$,

$U \cap H = V$, by (2), and x_0 does not belong to U; for if $x_0 = \alpha x + \beta y$ with $x \in W$, $y \in V$, $|\alpha| + |\beta| \leqq 1$, then $x_0 - \alpha x = \beta y$, which contradicts the fact that $(x_0 + W) \cap H$ is empty.

Proof of (1). Let V_k be an absolutely convex neighbourhood of $\mathfrak{o}$ in E_k. By (2), there is a sequence of absolutely convex neighbourhoods of $\mathfrak{o}$ $V_{k+1} \subset V_{k+2} \subset \cdots$, where V_{k+m} is a $\mathfrak{T}_{k+m}$-neighbourhood of $\mathfrak{o}$ in E_{k+m}, for which $V_{k+m} \cap E_k = V_k$. For the $\mathfrak{T}$-neighbourhood of $\mathfrak{o}$ $U = \Gamma_{m=1}^{\infty} V_{k+m} = \bigcup_{m=1}^{\infty} V_{k+m}$ we then have $U \cap E_k = V_k$. This shows that $\mathfrak{T}$ induces the topology $\mathfrak{T}_k$ on E_k.

Every non-zero x in E lies in some E_k, and so there exists a V_k with $x \notin V_k$; $x \notin U$ for a neighbourhood U of $\mathfrak{o}$ constructed in this way, and so the hull topology is Hausdorff.

(4) *Let* $E[\mathfrak{T}] = \sum_{k=1}^{\infty} E_k[\mathfrak{T}_k]$ *be a strict inductive limit, and suppose that for each k E_k is a proper closed subspace of* $E_{k+1}[\mathfrak{T}_{k+1}]$. *A subset B of* $E[\mathfrak{T}]$ *is bounded if and only if it lies in some* $E_k[\mathfrak{T}_k]$ *and is bounded there.*

The condition is sufficient, by (1). Conversely let us suppose that there is a bounded set B in $E[\mathfrak{T}]$ which is not contained in any E_n. There then exists a sequence $x_i \in B$ and a sequence n_i with $x_i \in E_{n_i} \sim E_{n_i - 1}$. Because $E_{n_i - 1}$ is closed in E_{n_i}, there exists, by (3), a sequence of absolutely convex neighbourhoods V_{n_i} of $\mathfrak{o}$ in E_{n_i} with $V_{n_i} \cap E_{n_i - 1} = V_{n_{i-1}}$ and with $\frac{1}{i} x_i \notin V_{n_i}$. $U = \bigcup_{i=1}^{\infty} V_{n_i}$ is an absolutely convex $\mathfrak{T}$-neighbourhood of $\mathfrak{o}$ in E which contains no $\frac{1}{i} x_i$. This contradicts the boundedness of the sequence x_i.

Thus we have proved an analogue of § 18, 5.(4), although § 18, 5.(4) is by no means a special case of (4).

5. (LB)- and (LF)-spaces. Completeness. A locally convex space E is called a (strict) (LB)- (respectively (LF)-) space if it can be represented as the (strict) topological inductive limit of a properly increasing sequence $E_1[\mathfrak{T}_1] \subset E_2[\mathfrak{T}_2] \subset \cdots$ of (B)- (respectively (F)-) spaces.

We shall make a deeper investigation of the properties of these spaces in the second volume; here we investigate the question of their completeness, which can be tackled in a rather more general setting.

(1) *Let* $E[\mathfrak{T}]$ *be the topological inductive limit of a strictly increasing sequence* $E_1[\mathfrak{T}_1] \subset E_2[\mathfrak{T}_2] \subset \cdots$ *of locally convex spaces. For each $\mathfrak{T}$-Cauchy filter $\mathfrak{F}$ there is a coarser $\mathfrak{T}$-Cauchy filter $\mathfrak{F}'$ and a number k for which* $\mathfrak{F}' \cap E_k$ *is a $\mathfrak{T}$-Cauchy filter on* E_k.

Proof. As F runs through a base of $\mathfrak{F}$ and W runs through all the absolutely convex neighbourhoods of $\mathfrak{o}$ in $E[\mathfrak{T}]$, the sets $F+W$ form the base of a Cauchy filter $\mathfrak{F}'$ on $E[\mathfrak{T}]$, which is coarser than $\mathfrak{F}$. For if F is small of order W, $F+W$ is small of order $3W$, so that the collection of all the sets $F+W$ contains sets of arbitrarily small order. Further the set $(F_1+W_1)\cap(F_2+W_2)$ contains the set $F_3+(W_1\cap W_2)$ if $F_3\subset F_1\cap F_2$.

If there exists a k for which all the sets $(F+W)\cap E_k$ are non-empty, then (1) is established. We suppose that this is not the case. Then there is a sequence $W_1\supset W_2\supset\cdots$ of absolutely convex $\mathfrak{T}$-neighbourhoods of $\mathfrak{o}$, together with sets F_k+W_k, where F_k is small of order W_k, for which the sets $(F_k+W_k)\cap E_k$ are empty, for $k=1,2,\ldots$.

We put $W_k^{(n)}=W_k\cap E_n$; $W_k^{(n)}$ is an absolutely convex $\mathfrak{T}_n$-neighbourhood of $\mathfrak{o}$. Consequently the sets

$$V_k=\Gamma\left(\tfrac{1}{2}W_1^{(1)}\cup\tfrac{1}{2}W_2^{(2)}\cup\cdots\cup\tfrac{1}{2}W_{k-1}^{(k-1)}\cup\tfrac{1}{2}W_k\right)$$

are $\mathfrak{T}$-neighbourhoods of $\mathfrak{o}$, for $k=1,2,\ldots$.

Let $F_k'\in\mathfrak{F}$ be small of order V_k. We shall show that $(F_k'+V_k)\cap E_k$ is also empty.

Since $F_k'\cap F_k$ is non-empty, there is an $x_0\in F_k$ in F_k'. Elements y of F_k' and elements z of $F_k'+V_k$ therefore take the form

$$y=x_0+\sum_{i=1}^{k}\alpha_i x_i,\qquad z=x_0+\sum_{i=1}^{k}\alpha_i x_i+\sum_{i=1}^{k}\alpha_i' x_i',$$

with $x_i,\ x_i'\in\frac{1}{2}W_i^{(i)}$ for $i<k$, $x_k,\ x_k'\in\frac{1}{2}W_k$, $\sum|\alpha_i|\leq 1$, $\sum|\alpha_i'|\leq 1$.

Now $\alpha_k x_k+\alpha_k' x_k'\in\frac{1}{2}W_k+\frac{1}{2}W_k=W_k$. Since $x_0\in F_k$ and $(F_k+W_k)\cap E_k$ is empty, the element $x_0+\alpha_k x_k+\alpha_k' x_k'\in F_k+W_k$ cannot lie in E_k. On the other hand $\sum_{i=1}^{k-1}\alpha_i x_i+\sum_{i=1}^{k-1}\alpha_i' x_i'$ lies in E_{k-1}, so that z also does not lie in E_k; hence $(F_k'+V_k)\cap E_k$ is empty.

We now form the neighbourhood $U=\Gamma_{k=1}^{\infty}\frac{1}{2}W_k^{(k)}$ of $\mathfrak{o}$. $U\subset V_k$ for each k, since $W_{k'}^{(k')}\subset W_k$ for each $k'\geq k$. There is a set $F_0\in\mathfrak{F}$ which is small of order U. Let y_0 be an element of F_0; y_0 lies in some E_k. We assert that $F_k'\cap F_0$ is empty, which contradicts the filter properties.

For if $y\in F_k'$, $y+V_k$ contains no element of E_k (for $(F_k'+V_k)\cap E_k$ is empty) and so $y-y_0$ does not lie in $V_k\supset U$; this means that y does not belong to y_0+U, and so does not belong to F_0, i.e. $F_k'\cap F_0$ is empty.

The following completeness criterion is a simple consequence of (1):

(2) *Let $E[\mathfrak{T}]$ be the topological inductive limit of a strictly increasing sequence $E_1[\mathfrak{T}_1]\subset E_2[\mathfrak{T}_2]\subset\cdots$ of locally convex spaces. $E[\mathfrak{T}]$ is*

complete if and only if, for each n, every $\mathfrak{T}$-Cauchy filter on E_n has a limit in E, and so has a limit in some E_{n+k}.

As a special case we have

(3) *Let $E[\mathfrak{T}]$ be the strict inductive limit of the strictly increasing sequence $E_1[\mathfrak{T}_1] \subset E_2[\mathfrak{T}_2] \subset \cdots$. $E[\mathfrak{T}]$ is complete if each $E_k[\mathfrak{T}_k]$ is complete.*

In particular, every strict (LF)*-space is complete.*

The following theorem is due to GROTHENDIECK [13]:

(4) *Let A be a continuous linear mapping from an* (F)*-space F into the* (LF)*-space $E[\mathfrak{T}] = \bigcup_{n=1}^{\infty} E_n[\mathfrak{T}_n]$. There exists a k with $A(F) \subset E_k$, and A is a continuous mapping from F into $E_k[\mathfrak{T}_k]$.*

Proof. Let H_n be the set of all pairs $(y, Ay) \in F \times E_n$, with $Ay \in E_n$. H_n is a closed linear subspace of $F \times E_n$, and so it is an (F)-space. If P_n is the continuous mapping $P_n(y, Ay) = y$ from H_n into F, $P_n(H_n)$ is the set of all y with $Ay \in E_n$. Using BAIRE's category theorem, it follows from $F = \bigcup_{n=1}^{\infty} P_n(H_n)$ that some $P_k(H_k)$ is not meagre in F. It follows from the BANACH-SCHAUDER theorem that $F = P_k(H_k)$, so that $A(F) \subset E_k$. The graph H_k of the mapping A of F into E_k is closed, so that A is continuous, by the closed graph theorem.

(5) *Every absolutely convex, bounded, complete subset M of an* (LF)*-space $E[\mathfrak{T}] = \bigcup_{n=1}^{\infty} E_n[\mathfrak{T}_n]$ is a bounded subset of some $E_k[\mathfrak{T}_k]$.*

The linear subspace E_M of E generated by M is a (B)-space with unit ball M (cf. § 20, 11.(2)). If we apply (4) with $F = E_M$ and A equal to the continuous embedding of E_M in E, we obtain (5).

Finally we remark that the question of the completeness of a topological inductive limit of complete spaces $E_\alpha[\mathfrak{T}_\alpha]$ is identical with the question of whether the quotient space $(\bigoplus_\alpha E_\alpha[\mathfrak{T}_\alpha])/H$, which by 1.(3) is topologically isomorphic to $\varinjlim E_\alpha[\mathfrak{T}_\alpha]$, is complete. Since $\bigoplus_\alpha E_\alpha[\mathfrak{T}_\alpha]$ is complete, by § 18, 5.(3), any example of an incomplete $\varinjlim E_\alpha[\mathfrak{T}_\alpha]$ (with the $E_\alpha[\mathfrak{T}_\alpha]$ complete) at the same time gives an example of an incomplete quotient of a complete space. We shall give such an example in § 31, 6.

6. The locally convex kernel of locally convex spaces. There is a close parallelism between the ideas of this and the following numbers and the ideas of numbers 1. to 3. of this paragraph.

Suppose that E is a vector space and that we are given a collection of vector spaces E_α and a collection of linear mappings A_α from E into E_α, such that for each non-zero x there is at least one E_α in which the image $A_\alpha x$ is non-zero. We then call E the kernel of the $A_\alpha^{(-1)}(E_\alpha)$, and write $E = \underset{\alpha}{\mathsf{K}} A_\alpha^{(-1)}(E_\alpha)$.

As an example, if the spaces E_α are subspaces of a vector space H, their intersection $E = \bigcap_\alpha E_\alpha$ is equal to $\underset{\alpha}{\mathsf{K}} I_\alpha^{(-1)}(E_\alpha)$, where I_α is the embedding of E into E_α. The product $E = \prod_\alpha E_\alpha$ can also be represented as the kernel $\underset{\alpha}{\mathsf{K}} P_\alpha^{(-1)}(E_\alpha)$, where P_α is the projection of E onto E_α. Conversely we have

(1) *Every kernel* $E = \underset{\alpha}{\mathsf{K}} A_\alpha^{(-1)}(E_\alpha)$ *is isomorphic to a linear subspace* $\hat{E}$ *of* $\prod_\alpha E_\alpha$.

The mapping $Ax = \hat{x} = (A_\alpha x)$ of E into $\prod_\alpha E_\alpha$ provides the required embedding, because of the hypothesis that for non-zero x at least one $A_\alpha x$ is non-zero.

On the other hand, we have

(2) *Every linear subspace* $\hat{E}$ *of* $\prod_\alpha E_\alpha$ *can be represented as the kernel* $\underset{\alpha}{\mathsf{K}} \hat{P}_\alpha^{(-1)}(E_\alpha)$, *where* $\hat{P}_\alpha$ *is the restriction to* $\hat{E}$ *of the projection* P_α *of* $\prod_\alpha E_\alpha$ *onto* E_α.

If the spaces E_α are locally convex spaces $E_\alpha[\mathfrak{T}_\alpha]$, then, following the model of the topological product, we introduce as kernel-topology on $E = \underset{\alpha}{\mathsf{K}} A_\alpha^{(-1)}(E_\alpha[T_\alpha])$ the coarsest topology for which all the A_α are continuous mappings from E into $E_\alpha[\mathfrak{T}_\alpha]$.

We now determine a base of $\mathfrak{T}$-neighbourhoods of $\mathfrak{o}$. If U_α is an absolutely convex $\mathfrak{T}_\alpha$-neighbourhood of $\mathfrak{o}$ in E_α, $V_\alpha = A_\alpha^{(-1)}(U_\alpha)$ must be a $\mathfrak{T}$-neighbourhood of $\mathfrak{o}$. The sets V_α are absolutely convex and absorbent. Their finite intersections clearly form a base of $\mathfrak{T}$-neighbourhoods of $\mathfrak{o}$.

Since every non-zero x in E has a non-zero image $A_\alpha x$ in some E_α, the kernel topology is always Hausdorff. $E[\mathfrak{T}] = \underset{\alpha}{\mathsf{K}} A_\alpha^{(-1)}(E_\alpha[\mathfrak{T}_\alpha])$ is thus always a locally convex space; it is called the locally convex kernel of the $A_\alpha^{(-1)}(E_\alpha[\mathfrak{T}_\alpha])$.

The algebraic isomorphism (1) can be sharpened to

(3) *Every locally convex kernel* $E[\mathfrak{T}] = \underset{\alpha}{\mathsf{K}} A_\alpha^{(-1)}(E_\alpha[\mathfrak{T}_\alpha])$ *is topologically isomorphic to a linear subspace* $\hat{E}$ *of the topological product* $\prod_\alpha E_\alpha[\mathfrak{T}_\alpha]$.

Proof. Under the mapping $Ax = \hat{x} = (A_\alpha x)$ of E onto the subspace $\hat{E}$ of $\prod_\alpha E_\alpha$, $V_\alpha = A_\alpha^{(-1)}(U_\alpha)$ is sent to $\hat{P}_\alpha^{(-1)}(U_\alpha)$; the sets $P_\alpha^{(-1)}(U_\alpha)$ form a sub-base of neighbourhoods of $\mathfrak{o}$ in $\hat{E}$ for the topology induced on $\hat{E}$ by the product topology on $\prod_\alpha E_\alpha[\mathfrak{T}_\alpha]$.

The converse is trivial:

(4) *Every linear subspace* $\hat{E}[\mathfrak{T}]$ *of a topological product* $E[\mathfrak{T}]=\prod_\alpha E_\alpha[\mathfrak{T}_\alpha]$ *is topologically isomorphic to the kernel* $\mathsf{K}_\alpha \hat{P}_\alpha^{(-1)}(E_\alpha[\mathfrak{T}_\alpha])$, *where* $\hat{P}_\alpha$ *is the restriction to* $\hat{E}$ *of the projection* P_α *of* E *onto* E_α.

In particular every linear subspace H *of a locally convex space* $E[\mathfrak{T}]$ *can be considered as* $\mathsf{K}J^{(-1)}(E[\mathfrak{T}])$, *where* J *is the embedding of* H *into* $E[\mathfrak{T}]$.

If a locally convex space $E[\mathfrak{T}]$ is represented as a kernel in two different ways, $E[\mathfrak{T}]=\mathsf{K}_\alpha A_\alpha^{(-1)}(E_\alpha[\mathfrak{T}_\alpha])=\mathsf{K}_\beta B_\beta^{(-1)}(F_\beta[\mathfrak{T}'_\beta])$, we again speak of equivalent defining systems.

(5) *The formation of locally convex kernels is transitive.*

For the locally convex kernel $E[\mathfrak{T}]=\mathsf{K}_\alpha A_\alpha^{(-1)}\Big(\mathsf{K}_{\beta_\alpha} B_{\beta_\alpha}^{(-1)}(F_{\beta_\alpha}[\mathfrak{T}_{\beta_\alpha}])\Big)$ can also be considered as the locally convex kernel $\mathsf{K}_{\alpha,\beta_\alpha} A_\alpha^{(-1)} B_{\beta_\alpha}^{(-1)}(F_{\beta_\alpha}[\mathfrak{T}_{\beta_\alpha}])$; in either case a base of neighbourhoods of $\mathfrak{o}$ is formed by the finite intersections of the sets $(B_{\beta_\alpha}A_\alpha)^{(-1)}(U_{\beta_\alpha})$, where each U_{β_α} is a $\mathfrak{T}_{\beta_\alpha}$-neighbourhood of $\mathfrak{o}$ in F_{β_α}.

(6) *A linear mapping* B *from a locally convex space* $F[\mathfrak{T}']$ *into a locally convex kernel* $E[\mathfrak{T}]=\mathsf{K}_\alpha A_\alpha^{(-1)}(E_\alpha[\mathfrak{T}_\alpha])$ *is continuous if and only if* $A_\alpha B$ *is a continuous mapping from* $F[\mathfrak{T}']$ *into* $E_\alpha[\mathfrak{T}_\alpha]$, *for each* α.

Proof. The continuity of $A_\alpha B$ follows from the continuity of A_α and of B. Conversely if all the $A_\alpha B$ are continuous, and if $A_\alpha^{(-1)}(U_\alpha)$ is a neighbourhood of $\mathfrak{o}$ in the sub-basis defining the topology $\mathfrak{T}$ on E, there is a neighbourhood W of $\mathfrak{o}$ in F with $A_\alpha B(W)\subset U_\alpha$, so that $B(W)\subset A_\alpha^{(-1)}(U_\alpha)$, and B is continuous.

(7) *A subset* M *of a locally convex kernel* $E[\mathfrak{T}]=\mathsf{K}_\alpha A_\alpha^{(-1)}(E_\alpha[\mathfrak{T}_\alpha])$ *is bounded (respectively precompact) if and only if* $A_\alpha(M)$ *is bounded (respectively precompact) in* $E_\alpha[\mathfrak{T}_\alpha]$, *for each* α.

Proof. a) $A_\alpha(M)$ is bounded if M is (§ 15, 6 (5)). On the other hand if all the sets $A_\alpha(M)$ are bounded, and if $V=\bigcap_{i=1}^{n} A_{\alpha_i}^{(-1)}(U_{\alpha_i})$ is a $\mathfrak{T}$-neighbourhood of $\mathfrak{o}$, then it follows from $A_{\alpha_i}(M)\subset\rho_i U_{\alpha_i}$ that $M\subset\rho_i A_{\alpha_i}^{(-1)}(U_{\alpha_i})$, so that $M\subset\rho V$, where $\rho=\max\rho_i$.

b) By § 15, 6.(7), $A_\alpha(M)$ is precompact if M is. On the other hand if all the sets $A_\alpha(M)$ are precompact, then each set $A_\alpha(M)$ is covered by finitely many sets $B_\alpha^{(i)}$ which are small of order U_α, and so M is covered by finitely many sets $A_\alpha^{(-1)}(B_\alpha^{(i)})$ which are small of order $A_\alpha^{(-1)}(U_\alpha)$. If $V=A_\alpha^{(-1)}(U_\alpha)\cap A_\beta^{(-1)}(U_\beta)$ then each $M\cap A_\alpha^{(-1)}(B_\alpha^{(i)})$ is covered by finitely many sets which are small of order $A_\beta^{(-1)}(U_\beta)$, and consequently

M is covered by finitely many sets which are small of order V. The corresponding assertion for a V which is intersection of $n>2$ sets $A_\alpha^{(-1)}(U_\alpha)$ follows in exactly the same way. Consequently M is totally bounded, and so it is precompact.

We mention one more consequence of the definition of the kernel topology:

(8) *The locally convex kernel of at most countably many metrizable locally convex spaces is metrizable.*

7. The projective limit of vector spaces. The ideas of this number are also valid for vector spaces over arbitrary fields.

Let $E = \underset{\alpha}{\mathsf{K}} A_\alpha^{(-1)}(E_\alpha)$ be the kernel of $A_\alpha^{(-1)}(E_\alpha)$. Suppose that the set A of indices α is a directed set, and that for each pair $\alpha<\beta$ there is a linear mapping $A_{\alpha\beta}$ of E_β into E_α for which

$$(1) \qquad A_\alpha = A_{\alpha\beta} A_\beta \quad \text{for} \quad \alpha<\beta.$$

Again we write $A_{\alpha\alpha}$ for the identity mapping of E_α onto itself.

If $\alpha \leqq \beta$, $A_\alpha(E)$ is therefore isomorphic to a quotient space of $A_\beta(E)$.

(2) *If each A_α maps E onto E_α then $A_{\alpha\beta}$ also maps the space E_β onto E_α, and we have*

$$(3) \qquad A_{\alpha\beta} A_{\beta\gamma} = A_{\alpha\gamma} \quad \textit{for} \quad \alpha<\beta<\gamma.$$

For an arbitrary kernel $E = \underset{\alpha}{\mathsf{K}} A_\alpha^{(-1)}(E_\alpha)$ such mappings $A_{\alpha\beta}$ are not always given in the first place. A kernel can however always be expressed in this form. For each finite subset Δ of the index set A form the product $E_\Delta = \prod_{\delta\in\Delta} E_\delta$, and define the mapping A_Δ from E into E_Δ as the mapping $A_\Delta x = \prod_{\delta\in\Delta} A_\delta x$. Write $\Delta_1 \leqq \Delta_2$ if $\Delta_1 \subset \Delta_2$, and put $A_{\Delta_1\Delta_2}$ equal to the projection of E_{Δ_2} onto E_{Δ_1}. Then $A_{\Delta_1} = A_{\Delta_1,\Delta_2} A_{\Delta_2}$ and $E = \underset{\Delta}{\mathsf{K}} A_\Delta^{(-1)}(E_\Delta)$. Further $A_{\Delta_1,\Delta_2} A_{\Delta_2,\Delta_3} = A_{\Delta_1,\Delta_3}$, so that (3) holds.

(4) *Suppose that $E = \underset{\alpha}{\mathsf{K}} A_\alpha^{(-1)}(E_\alpha)$ and that mappings $A_{\alpha\beta}$ are given from E_β into E_α, which satisfy* (1). *Then the relations*

$$(5) \qquad \breve{P}_\alpha = A_{\alpha\beta} \breve{P}_\beta \quad \textit{for} \quad \alpha<\beta$$

hold for the space $\breve{E} = \underset{\alpha}{\mathsf{K}} \breve{P}_\alpha^{(-1)}(E_\alpha)$, which is isomorphic to E, by 6.(1) *and* 6.(2).

This follows directly from

$$\breve{P}_\alpha \breve{x} = A_\alpha x = A_{\alpha\beta} A_\beta x = A_{\alpha\beta} \breve{P}_\beta \breve{x} \quad \text{for all} \quad \breve{x} \in \breve{E}.$$

So far we have started from a vector space E, together with mappings A_α of E onto spaces E_α and mappings $A_{\alpha\beta}$ of E_β into E_α. To what extent is E determined by the spaces E_α and the mappings $A_{\alpha\beta}$ alone?

(6) *Suppose that* E_α, $\alpha \in A$, *is a directed system of vector spaces, that a linear mapping* $A_{\alpha\beta}$ *from* E_β *into* E_α *is given for each* $\alpha < \beta$, *and that*

$$A_{\alpha\beta} A_{\beta\gamma} = A_{\alpha\gamma} \quad \text{for} \quad \alpha < \beta < \gamma. \tag{7}$$

We denote by $\hat{E}$ *the linear subspace of* $E = \prod_\alpha E_\alpha = \mathsf{K}_\alpha P_\alpha^{(-1)}(E_\alpha)$ *consisting of all* $\hat{x} = (x_\alpha)$ *with* $x_\alpha = A_{\alpha\beta} x_\beta$ *for* $\alpha < \beta$.

If $\check{E}$ *is a linear subspace of* $\hat{E}$, *and if* $\check{P}_\alpha$ *is the restriction of* P_α *to* $\check{E}$, *then* $\check{E} = \mathsf{K}_\alpha \check{P}_\alpha^{(-1)}(E_\alpha)$ *and (5) is satisfied.*

This follows from 6.(2) and the hypothesis that $x_\alpha = A_{\alpha\beta} x_\beta$ for the components of the elements of $\check{E}$.

$\hat{E}$ *is the largest possible subspace* $\check{E}$ *of* $\hat{E}$. The uniquely determined space $\hat{E} = \mathsf{K}_\alpha \hat{P}_\alpha^{(-1)}(E_\alpha)$ defined in this way is called the projective limit of the space E_α under the mappings $A_{\alpha\beta}$, and it is also denoted by $\varprojlim A_{\alpha\beta}(E_\beta)$.

It can happen that $\varprojlim A_{\alpha\beta}(E_\beta)$ reduces to the single element o.

In contrast to 2.(10), where it follows from the fact that the mappings $A_{\alpha\beta}$ are one-one that the mappings $K_\alpha^{(0)}$ have the same property, if $\hat{E}$ is a projective limit for which the mappings $A_{\alpha\beta}$ map E_β onto E_α it need not follow that the $\hat{P}_\alpha$ map $\hat{E}$ onto E_α. This is so, however, in the following special case:

(8) *Suppose that the directed set* A *is countable. If the* $A_{\alpha\beta}$, $\alpha < \beta$, $\alpha \in \mathsf{A}$, *each map* E_β *onto* E_α, *then in the projective limit* $\hat{E} = \mathsf{K}_\alpha \hat{P}_\alpha^{(-1)}(E_\alpha)$ *the* $\hat{P}_\alpha$ *map* $\hat{E}$ *onto* E_α.

In this case we can therefore be sure that $\hat{E} \neq o$ if the E_α are different from o.

Proof. We must show that given $x_\beta^{(0)} \in E_\beta$ there is an element in $\hat{E}$ whose component in E_β is $x_\beta^{(0)}$. We arrange the indices α as a sequence α_i, $i = 1, 2, \ldots$, with $\alpha_1 = \beta$. If $\alpha_j < \beta$, we put $x_{\alpha_j}^{(0)} = A_{\alpha_j,\beta} x_\beta^{(0)}$. Let α_{i_1} be the first term in the sequence of indices α_i for which $\alpha_i \not< \alpha_1$, and let α_{k_1} be the first α_i with $\alpha_i \geqq \alpha_{i_1}$ and $\alpha_i > \alpha_1$. By hypothesis there is an $x_{\alpha_{k_1}}^{(0)}$ with $x_\beta^{(0)} = A_{\beta,\alpha_{k_1}} x_{\alpha_{k_1}}^{(0)}$. For all $\alpha_j < \alpha_{k_1}$ we again put $x_{\alpha_j}^{(0)} = A_{\alpha j,\alpha_{k_1}} x_{\alpha_{k_1}}^{(0)}$. Because of (7), this is consistent with the terms which have already been fixed. We now repeat the procedure, letting α_{i_2} be the first α_i with $\alpha_i \not< \alpha_{k_1}$ and α_{k_2} the first α_i with $\alpha_i \geqq \alpha_{i_2}$ and $\alpha_i > \alpha_{k_1}$, and so on. In this way we obtain an element $x^{(0)} \in \hat{E}$ with component $x_\beta^{(0)}$ in E_β.

(9) *If* $E = \underset{\alpha}{\mathsf{K}} A_\alpha^{(-1)}(E_\alpha)$, *and if mappings* $A_{\alpha\beta}$ *are given from* E_β *into* $\hat{E}_\alpha$ *which satisfy* (1) *and* (3), *then E is always isomorphic to a linear subspace of* $\varprojlim A_{\alpha\beta}(E_\beta)$.

We have only to go over from E to the space $\check{E} = \underset{\alpha}{\mathsf{K}} \check{P}_\alpha^{(-1)}(E_\alpha)$, which is isomorphic to it by 6.(1) and 6.(2), to obtain a linear subspace of $\varprojlim A_{\alpha\beta}(E_\beta)$, using (4) and (6).

8. The topological projective limit of locally convex spaces. If the spaces E_α are locally convex spaces $E_\alpha[\mathfrak{T}_\alpha]$, and if the mappings $A_{\alpha\beta}$ from E_β into E_α are all continuous, then on $\hat{E} = \varprojlim A_{\alpha\beta}(E_\beta[\mathfrak{T}_\beta])$ we define as topology $\mathfrak{T}$ the topology induced on $\hat{E}$ by the topology of $\prod_\beta E_\beta[\mathfrak{T}_\beta]$, i.e. the kernel topology of $\underset{\alpha}{\mathsf{K}} \hat{P}_\alpha^{(-1)}(E_\alpha[\mathfrak{T}_\alpha])$. $\hat{E}[\mathfrak{T}]$ is called the topological projective limit of the $A_{\alpha\beta}(E_\beta[\mathfrak{T}_\beta])$.

(1) *If* $E[\mathfrak{T}] = \underset{\alpha}{\mathsf{K}} A_\alpha^{(-1)}(E_\alpha[\mathfrak{T}_\alpha])$ *is a locally convex kernel, and if continuous mappings* $A_{\alpha\beta}$ *are given from* E_β *into* E_α *which satisfy* 7.(1) *and* 7.(3), *then* $E[\mathfrak{T}]$ *is topologically isomorphic to a linear subspace* $\mathring{E}$ *of the topological projective limit* $\varprojlim A_{\alpha\beta}(E_\beta[\mathfrak{T}_\beta]$.

A base of $\mathfrak{T}$*-neighbourhoods of* $\circ$ *is given by all sets of the form* $V_\alpha = A_\alpha^{(-1)}(U_\alpha)$, *where* U_α *is a* $\mathfrak{T}_\alpha$*-neighbourhood of* $\circ$ *in* $E_\alpha[\mathfrak{T}_\alpha]$.

Proof. The first part of the assertion follows from 7.(9) and the fact that the topologies on $\mathring{E}$ and $\varprojlim A_{\alpha\beta}(E_\beta)$ are the kernel topologies induced by the product topology on $\prod_\alpha E_\alpha[\mathfrak{T}_\alpha]$.

A $\mathfrak{T}$-neighbourhood of $\circ$ for the kernel topology can be taken in the form $V = \bigcap_{i=1}^{n} A_{\alpha_i}^{(-1)}(U_{\alpha_i})$, where U_{α_i} is a $\mathfrak{T}_{\alpha_i}$-neighbourhood of $\circ$ in E_{α_i}. Since the indices form a directed set, there exists a β with $\alpha_i < \beta$ for $i = 1, \ldots, n$, and there exists a $U_\beta \subset E_\beta$ with $A_{\alpha_i\beta}(U_\beta) \subset U_{\alpha_i}$.

It follows from this, and from 7.(1) and 7.(3), that

$$\begin{aligned} A_\beta^{(-1)}(U_\beta) \subset A_{\alpha_i}^{(-1)} A_{\alpha_i}(A_\beta^{(-1)}(U_\beta)) &= A_{\alpha_i}^{(-1)} A_{\alpha_i\beta} A_\beta(A_\beta^{(-1)}(U_\beta)) \\ &= A_{\alpha_i}^{(-1)} A_{\alpha_i\beta}(U_\beta) \subset A_{\alpha_i}^{(-1)}(U_{\alpha_i}), \end{aligned}$$

i.e. $A_\beta^{(-1)}(U_\beta) \subset V$, and so the second part of the proposition is proved.

It follows from the remark after 7.(2) and from (1) that every locally convex kernel is topologically isomorphic to a linear subspace of a topological projective limit.

In particular the topological product $\prod_\alpha E_\alpha[\mathfrak{T}_\alpha]$ can be represented as the topological projective limit $\varprojlim P_{\Delta_1,\Delta_2}(E_{\Delta_2}[\mathfrak{T}_{\Delta_2}])$ of the finite partial products $E_\Delta[\mathfrak{T}_\Delta] = \prod_{\delta \in \Delta} E_\delta[\mathfrak{T}_\delta]$ under the projections P_{Δ_1,Δ_2} of E_{Δ_2} onto E_{Δ_1}.

(2) *Let E be the topological projective limit* $E[\mathfrak{T}] = \varprojlim A_{\alpha\beta}(E_\beta[\mathfrak{T}_\beta])$ *of locally convex spaces* $E_\alpha[\mathfrak{T}_\alpha]$, *and let* Γ *be a cofinal subset of the directed index set* A. *As* γ *and* δ *run through all pairs* $\gamma < \delta$ *of* Γ, $E^+ = \varprojlim A_{\gamma\delta}(E_\delta[\mathfrak{T}_\delta])$ *is topologically isomorphic to* $E[\mathfrak{T}]$.

Proof. E consists of all $x = (x_\alpha)$, $\alpha \in \mathsf{A}$, with $x_\alpha = A_{\alpha\beta} x_\beta$ for $\alpha < \beta$. If each $x \in E$ we make correspond the element $x^+ = (x_\gamma)$, $\gamma \in \Gamma$, of E^+, we obtain a homomorphism of E into E^+. Conversely suppose that $x^+ = (x_\gamma)$ belongs to E^+. For each $\alpha \in \mathsf{A}$ there is a $\gamma \in \Gamma$ with $\alpha < \gamma$; we set $x_\alpha = A_{\alpha\gamma} x_\gamma$. Because of 7.(3), x_α is independent of the choice of γ, and for the same reason we have $x_\alpha = A_{\alpha\beta} x_\beta$ for each pair $\alpha < \beta$. Consequently each $x^+ \in E^+$ corresponds in a one-one way to an $x \in E$, so that E and E^+ are isomorphic.

This is a topological isomorphism: by (1), we can restrict our attention to neighbourhoods U'_α of $\mathfrak{o}$, which consist of all $x = (x_\alpha) \in E$ for which $x_\alpha \in U_\alpha$, where U_α is a neighbourhood of $\mathfrak{o}$ in E_α. If $\alpha < \delta \in \Gamma$, U'_α also consists of all x with $x_\delta \in A_{\alpha\delta}^{(-1)}(U_\alpha)$. To this there corresponds the neighbourhood of all $x^+ = (x_\gamma) \in E^+$ with $x_\delta \in A_{\alpha\delta}^{(-1)}(U_\alpha)$, and so the isomorphism between E and E^+ is a topological one.

9. The representation of a locally convex space as a projective limit. We return to the topological isomorphism $x \to \hat{\mathfrak{x}} = (\hat{x}_\alpha)$ between a locally convex space $E[\mathfrak{T}]$ and a subspace $\hat{E}$ of a topological product of (B)-spaces which was established in § 18, 3.(7).

(1) *Every locally convex space* $E[\mathfrak{T}]$ *is topologically isomorphic to a dense linear subspace of a topological projective limit of* (B)-*spaces.*

Every complete locally convex space is isomorphic to a topological projective limit of (B)-*spaces.*

Proof. We take a system $\{p_\alpha(x)\}$ of semi-norms on E corresponding to a base of $\mathfrak{T}$-neighbourhoods of $\mathfrak{o}$. The indices α form a directed set when we define $\alpha \leqq \beta$ if $p_\alpha(x) \leqq p_\beta(x)$ for all $x \in E$. If we again denote the coset of $x \in E$ in $E_\alpha = E/N_\alpha$ (respectively $E_\beta = E/N_\beta$) by $\hat{x}_\alpha$ (respectively $\hat{x}_\beta$), then a continuous linear mapping $A_{\alpha\beta}$ is defined from the normed space $E_\beta[\mathfrak{T}_\beta]$ onto the normed space $E_\alpha[\mathfrak{T}_\alpha]$ by setting $\hat{x}_\alpha = A_{\alpha\beta} \hat{x}_\beta$. This mapping can be extended to a continuous linear mapping $\tilde{A}_{\alpha\beta}$ from the completion $\tilde{E}_\beta[\tilde{\mathfrak{T}}_\beta]$ into $\tilde{E}_\alpha[\tilde{\mathfrak{T}}_\alpha]$. The relation $A_{\alpha\beta} A_{\beta\gamma} = A_{\alpha\gamma}$ is satisfied for $\alpha < \beta < \gamma$, and consequently we have $\tilde{A}_{\alpha\beta} \tilde{A}_{\beta\gamma} = \tilde{A}_{\alpha\gamma}$.

As a result, we can form the topological projective limit $\hat{E}[\mathfrak{T}] = \varprojlim \tilde{A}_{\alpha\beta}(\tilde{E}_\beta[\tilde{\mathfrak{T}}_\beta]) = \mathop{\mathsf{K}}\limits_{\alpha} \hat{P}_\alpha^{(-1)}(\tilde{E}_\alpha[\tilde{\mathfrak{T}}_\alpha])$, where $\hat{P}_\alpha$ is the restriction to $\hat{E}$ of the projection P_α of $\prod\limits_\alpha \tilde{E}_\alpha[\tilde{\mathfrak{T}}_\alpha]$ onto $\tilde{E}_\alpha$. Under the mapping $x \to \check{\mathfrak{x}} = (\hat{x}_\alpha)$, $E[\mathfrak{T}]$ is topologically isomorphic to the subspace $\check{E}$ of all such elements $\check{\mathfrak{x}}$ in $\hat{E}$, since $\hat{x}_\alpha = \tilde{A}_{\alpha\beta} \hat{x}_\beta$ if $\alpha < \beta$.

We now show that $\check{E}$ is dense in $\hat{E}$. By the second part of 8.(1) this is so if $\check{P}_\alpha(\check{E})$ is dense in $\hat{P}_\alpha(\hat{E})$ for each α. But $\check{P}_\alpha(\check{E})=E_\alpha$, and $\hat{P}_\alpha(\hat{E})\subset\tilde{E}_\alpha$.

As a special case, we have

(2) *Every* (F)*-space is topologically isomorphic to a projective limit of a sequence of* (B)*-spaces.*

10. A criterion for completeness. In contrast to the behaviour of inductive limits (cf. 5.), the question of the completeness of a projective limit is easy to answer.

Let $E[\mathfrak{T}]$ be a locally convex kernel $\underset{\alpha}{\mathsf{K}} A_\alpha^{(-1)}(E_\alpha[\mathfrak{T}_\alpha])$.

(1) *A filter $\mathfrak{F}$ on $E[\mathfrak{T}]$ is a Cauchy filter if and only if all the filters $A_\alpha(\mathfrak{F})$ are Cauchy filters in the spaces $E_\alpha[\mathfrak{T}_\alpha]$.*

Suppose that $\mathfrak{F}$ is a Cauchy filter. The continuous images $A_\alpha(\mathfrak{F})$ are Cauchy filters in the spaces $E_\alpha[\mathfrak{T}_\alpha]$. Conversely, suppose that each $A_\alpha(\mathfrak{F})$ is a Cauchy filter in $E_\alpha[\mathfrak{T}_\alpha]$. Suppose that $V=\bigcap_{i=1}^{n} A_{\alpha_i}^{(-1)}(U_{\alpha_i})$ is a $\mathfrak{T}$-neighbourhood of $\circ$ in $E[\mathfrak{T}]$, where U_{α_i} is a $\mathfrak{T}_{\alpha_i}$-neighbourhood of $\circ$ in $E_{\alpha_i}[\mathfrak{T}_{\alpha_i}]$. By hypothesis there is an $F_i\in\mathfrak{F}$ whose image $A_{\alpha_i}(F_i)$ is small of order U_{α_i}. Then F_i is small of order $A_\alpha^{(-1)}(U_{\alpha_i})$ and $\bigcap_{i=1}^{n} F_i$ is small of order V, so that $\mathfrak{F}$ is a Cauchy filter in $E[\mathfrak{T}]$.

For arbitrary locally convex kernels the completeness of $E[\mathfrak{T}]$ certainly does not follow from the completeness of the spaces $E_\alpha[\mathfrak{T}_\alpha]$. This follows from 6.(4), taking the $E_\alpha[\mathfrak{T}_\alpha]$ to be complete and $\hat{E}[\mathfrak{T}]$ to be a non-closed subspace of $E[\mathfrak{T}]$.

For projective limits, however, we have

(2) *A topological projective limit* $E[\mathfrak{T}]=\varprojlim A_{\alpha\beta}(E_\beta[\mathfrak{T}_\beta])$ *is complete (respectively quasi-complete) (respectively sequentially complete) if this is so for each space $E_\alpha[\mathfrak{T}_\alpha]$.*

Proof. In the terminology of 7.(6) we have $E[\mathfrak{T}]=\hat{E}[\mathfrak{T}] =\underset{\alpha}{\mathsf{K}}\hat{P}_\alpha^{(-1)}(E_\alpha[\mathfrak{T}_\alpha])$. Suppose that the spaces $E_\alpha[\mathfrak{T}_\alpha]$ are complete, and let $\mathfrak{F}$ be a Cauchy filter on $E[\mathfrak{T}]$. Then each of the Cauchy filters $\hat{P}_\alpha(\mathfrak{F})$ has a limit x_α in $E_\alpha[\mathfrak{T}_\alpha]$. If $\alpha<\beta$, it follows from 7.(5) and the continuity of $A_{\alpha\beta}$ that

$$x_\alpha=\lim\hat{P}_\alpha(\mathfrak{F})=\lim A_{\alpha\beta}\hat{P}_\beta(\mathfrak{F})=A_{\alpha\beta}\lim\hat{P}_\beta(\mathfrak{F})=A_{\alpha\beta}x_\beta.$$

The element $\mathfrak{x}=(x_\alpha)$ therefore satisfies the compatibility conditions of 7.(6) and so it belongs to $\hat{E}=E$. It follows directly from the fact that $P_\alpha(\mathfrak{F})\to x_\alpha$ for each α that $\mathfrak{F}\to\mathfrak{x}$ in $\hat{E}$, so that $E[\mathfrak{T}]$ is complete.

If $\mathfrak{F}$ is a Cauchy filter on a bounded set $M\subset E[\mathfrak{T}]$, the $P_\alpha(\mathfrak{F})$ are Cauchy filters on the bounded sets $P_\alpha(M)$ in $E_\alpha[\mathfrak{T}_\alpha]$, and we finish the argument as before.

The assertion about sequential completeness is obtained in a similar way.

(3) *A topological projective limit* $E[\mathfrak{T}] = \varprojlim A_{\alpha\beta}(E_\beta[\mathfrak{T}_\beta])$ *is a closed linear subspace of the topological product* $\prod_\alpha E_\alpha[\mathfrak{T}_\alpha]$.

For a closure point of $E[\mathfrak{T}]$ in $\prod_\alpha E_\alpha[\mathfrak{T}_\alpha]$ is the limit of a Cauchy filter in $E[\mathfrak{T}]$, and so, as in the proof of (2), it satisfies the compatibility conditions of 7.(6); consequently it belongs to $E[\mathfrak{T}]$.

References to additional results on inductive and projective limits: N. Bourbaki [3], vol. 3., J. Braconnier [1], S. Lefschetz [1], D. A. Raikov [1], J. Sebastião e Silva [4], O. Takenouchi [1], A. Weil [2].

§ 20. Duality

1. The existence of continuous linear functionals. In § 17, 6. we showed, using the Hahn-Banach theorem, that there are sufficiently many continuous linear functionals on any normed space E, and derived results about the dual space E'. It is also possible to do this in the more general locally convex case.

Once again, the extension theorem holds:

(1) *Every continuous linear functional* $l(z)$ *defined on a linear subspace* F *of a locally convex space* $E[\mathfrak{T}]$ *can be extended to a continuous linear functional* u *defined on the whole of* E.

In particular if $|l(z)| \leqq p(z)$ *on* F, *where* $p(x)$ *is a continuous semi-norm on* $E[\mathfrak{T}]$, *there is an extension* u *with* $|u x| \leqq p(x)$ *on the whole of* E.

Proof. If $l(z)$ is continuous on F, then by § 18, 1. there exists a continuous semi-norm $p(x)$ on E with $|l(z)| \leqq p(z)$ on F. The existence of a continuous extension u of l, with $|u x| \leqq p(x)$, follows from § 17, 3.(5).

(2) *If* $x_1, \dots, x_n$ *are linearly independent elements of* $E[\mathfrak{T}]$ *and if* $\alpha_1, \dots, \alpha_n$ *are real (respectively complex) numbers, there exists a* $u \in E'$ *with* $u x_i = \alpha_i$, $i = 1, \dots, n$.

In particular, for each non-zero x_0 *in* E *there exists a* $u_0 \in E'$ *with* $u_0 x_0 = 1$.

Proof. The n-dimensional subspace F of E spanned by $x_1, \dots, x_n$ is topologically isomorphic to K^n, by § 15, 5.(1). The equations $l(x_i) = \alpha_i$, $i = 1, \dots, n$, therefore define a continuous linear functional on F, and this has a continuous extension u to the whole of E, by (1).

A third theorem on the existence of continuous linear functionals which is frequently useful is given by

(3) *If* F *is a closed linear subspace of* $E[\mathfrak{T}]$ *and if* $x_0 \notin F$, *then there is a* $u_0 \in E'$ *with* $u_0 x_0 = 1$ *and* $u_0 y = 0$ *for all* $y \in F$.

Proof. E/F is again locally convex under the induced topology $\mathfrak{T}$, by § 18, 3.(2). By (2) there is a continuous linear functional $\tilde{u}_0$ on E/F with $\tilde{u}_0 \hat{x}_0 = 1$, where $\hat{x}_0$ is the coset $K x_0$ of x_0 in E/F. A linear functional u_0 is defined on E by setting $u_0 x = \tilde{u}_0 \hat{x}$, where x is an arbitrary element of E; clearly $u_0 x_0 = 1$ and $u_0 y = \tilde{u}_0 \mathrm{o} = 0$ for all $y \in F$. u_0, being the product of the continuous mappings K and $\tilde{u}_0$, is a continuous linear functional on E.

2. Dual pairs and weak topologies. We now use the concept of dual pair, defined in § 10, 3., once again.

(1) *If $E[\mathfrak{T}]$ is a locally convex space, E and its dual space E' form a dual pair $\langle E', E \rangle$ over the real (respectively complex) field.*

Proof. As in § 10, 3. we define $B(u,x) = ux$, i.e. the value of the linear functional u at x, as the bilinear functional on $E' \times E$. Condition (D 2″) is trivially satisfied, and condition (D 2′) follows from 1.(2).

Thus we have established a result for locally convex spaces which we obtained for linearly topologized spaces in § 10, 4.

At the same time, the results which we proved in Chapter II for dual pairs over arbitrary fields are also valid for dual pairs consisting of a locally convex space and its dual space.

As in Chapter II, it will become apparent that many of the most important concepts of the duality theory of locally convex spaces do not depend upon the original topology $\mathfrak{T}$ of E, but only on the dual pair $\langle E', E \rangle$.

As in § 10, 3. we shall now start from a dual pair $\langle E_2, E_1 \rangle$ over K, where now K can only be the field of real or complex numbers. We shall denote the bilinear form by ux, or occasionally by $\langle u, x \rangle$.

In § 10, 3. we introduced a linear topology on E_1 which we called the linear weak topology $\mathfrak{T}_{ls}$. As we saw in § 10, 4.(4), this is the coarsest linear topology on E_1 for which all the $u \in E_2$ are continuous linear functionals on E_1. For this, the topology on K is the discrete topology.

For our study of topological vector spaces in Chapter III, however, the underlying topology on K is the topology defined in terms of the modulus. For this reason, we now define the weak topology $\mathfrak{T}_s(E_2)$ on E_1 to be the coarsest topology on E_1 for which each element $u \in E_2$ defines a continuous linear functional $\langle u, x \rangle = ux$ on E_1, when K is given the topology defined in terms of the modulus.

The sets $U_{u_1, \ldots, u_n; \varepsilon}$ form a base of neighbourhoods of o for the weak topology $\mathfrak{T}_s(E_2)$, where $U_{u_1, \ldots, u_n; \varepsilon}$ is the set of all $x \in E_1$ for which

$$\sup_{i=1,\ldots,n} |u_i x| < \varepsilon; \tag{2}$$

the u_i are n arbitrary elements of E_2, and $n = 1, 2, \ldots$. For if $u \in E_2$, the set $U_{u;\varepsilon}$ of all $x \in E_1$ for which $|ux| < \varepsilon$ must be a neighbourhood of o.

On the other hand every intersection of neighbourhoods of $\mathfrak{o}$ of this form contains a neighbourhood of $\mathfrak{o}$ of the form (2).

The topology $\mathfrak{T}_s(E_1)$ is defined on E_2 in a similar way.

As in § 10, the following result holds for these topologies:

(3) *If $\langle E_2, E_1\rangle$ is a dual pair, $E_1[\mathfrak{T}_s(E_2)]$ and $E_2[\mathfrak{T}_s(E_1)]$ are locally convex, and each is the dual of the other.*

The proof proceeds in a completely analogous way: the neighbourhoods (2) are absolutely convex and absorbent and define a filter, and as a result of 1.(2), the weak topology is Hausdorff; thus $E_1[\mathfrak{T}_s(E_2)]$ and $E_2[\mathfrak{T}_s(E_1)]$ are locally convex, by § 18, 1.

Each $u\in E_2$ defines a weakly continuous linear functional on E_1, since $|ux|<\varepsilon$ for each $x\in U_{u;\varepsilon}$.

Conversely if u is a weakly continuous linear functional on E_1 there is a weak neighbourhood $U=U_{u_1,\ldots,u_n;\varepsilon}$ of $\mathfrak{o}$ with $|ux|\leqq \sup_{i=1,\ldots,n}|u_i x|$. In particular $ux=0$ if $u_i x=0$ for $i=1,\ldots,n$. But then by § 9, 2.(7a) u is a linear combination of the u_i, so that $u\in E_2$.

If $E[\mathfrak{T}]$ is a locally convex space, then using the dual pair $\langle E', E\rangle$ we can introduce the weak topology $\mathfrak{T}_s(E')$ on E, and we can likewise introduce the weak topology $\mathfrak{T}_s(E)$ on the dual space E'; this makes E' a locally convex space. We call $E'[\mathfrak{T}_s(E)]$ the w e a k d u a l of $E[\mathfrak{T}]$.

The relation between the original topology $\mathfrak{T}$ and the weak topology is given by

(4) *The original topology $\mathfrak{T}$ on a locally convex space $E[\mathfrak{T}]$ is always finer than the weak topology $\mathfrak{T}_s(E')$.*

For every element of E' is $\mathfrak{T}$-continuous, while $\mathfrak{T}_s(E')$ is the coarsest topology for which this is so.

If E is an infinite-dimensional normed space, the norm topology is certainly different from the weak topology, since none of the semi-norms $\sup_{i=1,\ldots,n}|u_i x|$ is a norm.

We make the following further remark about the relation between locally convex spaces and their corresponding dual pairs (cf. § 10, 3.):

(5) *If the locally convex spaces $E_1[\mathfrak{T}_1]$ and $E_2[\mathfrak{T}_2]$ are isomorphic, then the dual pairs $\langle E_1', E_1\rangle$ and $\langle E_2', E_2\rangle$ are isomorphic; the converse need not be true.*

The fact that the converse is not true follows from the fact that the original topology cannot be retrieved from the dual pair. If two dual pairs are isomorphic, however, it is at least possible to deduce that the spaces are topologically isomorphic when they are given their weak topologies. For these topologies depend on the dual pair alone.

3. The duality of closed subspaces. Let $\langle E_2, E_1\rangle$ be a dual pair. We recall the concept of the space $M^\perp$ orthogonal to a set $M\subset E_1$, which was introduced in § 9, 2. $M^\perp$ consists of all $u\in E_2$ for which $uy=0$ for all $y\in M$.

Just as for linear topologies, the following theorem also holds for locally convex topologies:

(1) *If F is a $\mathfrak{T}$-closed linear subspace of the locally convex space $E[\mathfrak{T}]$, F is orthogonally closed with respect to E'; conversely a linear subspace F of E which is orthogonally closed with respect to E' is $\mathfrak{T}_s(E')$-closed in E.*

The proof proceeds in a way analogous to the linear topology case (§ 10, 4.(6)): it follows from 1.(3) that F is orthogonally closed if it is $\mathfrak{T}$-closed. On the other hand if $F^{\perp\perp}=F$, if x_0 is a weak closure point of F and if $u_0\in F^\perp$, then there is at least one $y_0\in F$ in each weak neighbourhood $U_{u;\varepsilon}(x_0)$. Consequently $|u_0 x_0|=|u_0(x_0-y_0)|<\varepsilon$, so that, as ε is arbitrarily small, $u_0 x_0=0$, and $x_0\in F^{\perp\perp}=F$.

We have thus shown that although the original topology $\mathfrak{T}$ of a locally convex space $E[\mathfrak{T}]$ is certainly not determined by the dual pair $\langle E', E\rangle$, nevertheless the $\mathfrak{T}$-closed linear subspaces are determined by the dual pair alone.

If we call a locally convex (respectively linear) topology on E_1 compatible with the dual pair $\langle E_2, E_1\rangle$ when the dual space of E_1 is E_2, then we can formulate this result as follows:

(2) *If $\langle E_2, E_1\rangle$ is a dual pair, a linear subspace F of E_1 is closed with respect to a compatible locally convex or linear topology if and only if it is orthogonally closed.*

In this sense we can thus speak simply of "closed" linear subspaces. The question of characterising all the compatible linear topologies was answered in § 10, 11.(4), and a characterization of all the compatible locally convex topologies will be given in § 21, 4.(3). So far at any rate we know one compatible topology, namely the weak topology.

If we define $\bigwedge_\alpha F_\alpha$ to be the intersection $\bigcap_\alpha F_\alpha$ of the closed linear subspaces F_α of E_1 and define $\bigvee_\alpha F_\alpha$ to be the smallest closed linear subspace of E_1 containing all the F_α, then the closed linear subspaces of E_1 form a complete lattice $\bar{V}(E_1)$, and the duality properties of closed linear subspaces follow from § 10, 3.(2):

(3) *Let $\langle E_2, E_1\rangle$ be a dual pair. The complete lattices $\bar{V}(E_1)$ and $\bar{V}(E_2)$ of closed linear subspaces of E_1 and E_2 respectively are dually isomorphic; a dual isomorphism is obtained when we make each such subspace correspond to its orthogonal space.*

In particular this duality holds for the $\mathfrak{T}$-closed linear subspaces of a locally convex space $E[\mathfrak{T}]$, and the $\mathfrak{T}_s(E)$-closed linear subspace of the dual space E'.

This duality of closed linear subspaces will prove in 8. to be a special case of a more general duality.

As in § 14, 7. a subset $M \subset E[\mathfrak{T}]$ is said to be total in $E[\mathfrak{T}]$, or a fundamental subset of E, if the closed linear span of M coincides with E. We clearly have

(4) *A set M is total in $E[\mathfrak{T}]$ if and only if it is total with respect to any locally convex or linear topology compatible with the dual pair $\langle E', E\rangle$, i.e. if and only if $M^{\perp} = \mathfrak{o}$.*

Thus the concept of total set depends only on the dual pair, as well.

4. Duality of mappings. Suppose that two dual pairs $\langle E_2, E_1\rangle$ and $\langle F_2, F_1\rangle$ are given. Then by § 10, 3., E_2 and F_2 are linear subspaces of the algebraic dual spaces E_1^* and F_1^* respectively. To every linear mapping A from E_1 into F_1 there corresponds the adjoint mapping A' from F_1^* into E_1^*, defined by

$$v(Ax) = (A'v)x \quad \text{for all} \quad x \in E_1, \quad v \in F_1^*.$$

In what follows, A' will always mean the restriction of A' to $F_2 \subset F_1^*$.

(1) *Suppose that $\langle E_2, E_1\rangle$ and $\langle F_2, F_1\rangle$ are two dual pairs over* K. *A linear mapping A from E_1 into F_1 is weakly continuous if and only if the adjoint mapping maps F_2 into E_2.*

Corollary. A is weakly continuous from E_1 into F_1 if and only if A' is weakly continuous from F_2 into E_2.

Proof. a) Suppose that $A'(F_2) \subset E_2$. If $U = U_{v_1, \ldots, v_n; \varepsilon}$, $v_i \in F_2$, is a weak neighbourhood of $\mathfrak{o}$ in F_1, then, since $v_i(Ax) = (A'v_i)x$ and since $A'v_i \in F_2$, the image Ax of an $x \in U_{A'v_1, \ldots, A'v_n; \varepsilon}$ is contained in U. Thus A is weakly continuous at $\mathfrak{o}$, and consequently it is weakly continuous on E_1.

b) Suppose that A is weakly continuous. Then the linear functional on E_1 defined by $l(x) = v_0(Ax)$, $v_0 \in F_2$, is weakly continuous, since v_0 is weakly continuous, by 2.(3). $l(x)$ is therefore defined by some $u_0 \in E_2$; on the other hand, $v_0(Ax) = (A'v_0)x$, so that $A'v_0 = u_0 \in E_2$, and A' maps F_2 into E_2.

c) If A is weakly continuous, then A' maps F_2 into E_2, by b). The mapping $(A')'$ adjoint to A' is equal to A, and so maps E_1 into F_1. By a), A' is therefore weakly continuous. Since $A'' = A$, the corollary follows by interchanging A and A'.

We observe that this method of proof has already been used in § 10, 5.(1); accordingly, (1) can also be derived for the linear weak topologies in exactly the same way. Consequently

(2) *The $\mathfrak{T}_s$-continuous and $\mathfrak{T}_{ls}$-continuous linear mappings from a locally convex space $E[\mathfrak{T}]$ into a locally convex space $F[\mathfrak{T}']$ are the same.*

If $E_1[\mathfrak{T}_1]$ and $E_2[\mathfrak{T}_2]$ are two locally convex spaces, we denote the vector space of all continuous linear mappings from E_1 into E_2 by $\mathfrak{L}(E_1[\mathfrak{T}_1], E_2[\mathfrak{T}_2])$. The following duality theorem now follows from (1):

(3) *Let $\langle E_2, E_1\rangle$ and $\langle F_2, F_1\rangle$ be two dual pairs over K. If to each weakly continuous linear mapping A from E_1 into F_1 we make correspond its adjoint mapping A' from F_2 into E_2, we obtain a vector space isomorphism between the spaces $\mathfrak{L}(E_1[\mathfrak{T}_s(E_2)], F_1[\mathfrak{T}_s(F_2)])$ and $\mathfrak{L}(F_2[\mathfrak{T}_s(F_1)], E_2[\mathfrak{T}_s(E_1)])$.*

In the case where the two dual pairs coincide, we obtain

(4) *The correspondence which sends each weakly continuous endomorphism A of E_1 into the weakly continuous adjoint endomorphism A' of E_2 maps the algebra $\mathfrak{L}(E_1[\mathfrak{T}_s(E_2)])$ of weakly continuous endomorphisms of E_1 anti-isomorphically onto the algebra $\mathfrak{L}(E_2[\mathfrak{T}_s(E_1)])$ of weakly continuous endomorphisms of E_2.*

In 2. we found that every continuous linear functional on a locally convex space is weakly continuous, and conversely. In general, for linear mappings we can only say

(5) *Every continuous linear mapping A from $E_1[\mathfrak{T}_1]$ into $E_2[\mathfrak{T}_2]$ is also weakly continuous, i.e. $\mathfrak{L}(E_1[\mathfrak{T}_1], E_2[\mathfrak{T}_2])$ is a linear subspace of $\mathfrak{L}(E_1[\mathfrak{T}_s(E_1')], E_2[\mathfrak{T}_s(E_2')])$.*

Proof. The linear functional defined by $l(x)=v_0(Ax)$, $v_0 \in E_2'$, is $\mathfrak{T}_1$-continuous, and so it is defined by some $u_0 \in E_1'$. A' therefore maps E_2' into E_1', so that A is weakly continuous, by (1).

E be an infinite dimensional normed space. The norm topology $\mathfrak{T}$ is finer than, and different from, the weak topology $\mathfrak{T}_s(E')$. The identity mapping of E onto itself is certainly an element of $\mathfrak{L}(E[\mathfrak{T}_s], E[\mathfrak{T}_s])$, but does not belong to $\mathfrak{L}(E[\mathfrak{T}_s], E[\mathfrak{T}])$.

From (5) and the corollary of (1) we obtain

(6) *If A is a continuous linear mapping of $E_1[\mathfrak{T}_1]$ into $E_2[\mathfrak{T}_2]$, A' is a weakly continuous linear mapping of E_2' into E_1'.*

5. Duality of complementary spaces. In § 15, 9.(10) we saw that in arbitrary topological vector spaces not every finite-dimensional linear subspace need have a topological complement. Locally convex spaces have this property, however, as do linearly topologized spaces (cf. § 10, 7.(8)); this we shall now show.

Every weak topological complementary decomposition has a dual decomposition:

(1) *Let $\langle E_2, E_1\rangle$ be a dual pair. To every $\mathfrak{T}_s$-complementary decomposition*

(2) $$E_1 = H_1 \oplus H_2$$

of E_1 into two closed linear subspaces there corresponds the $\mathfrak{T}_s$-complementary decomposition

(3) $$E_2 = H_1^\perp \oplus H_2^\perp.$$

If P_1 and P_2 are the two projections corresponding to the decomposition (2), *with $P_1(E_1) = H_1, P_2(E_1) = H_2$, then their adjoints P_1' and P_2' are the projections corresponding to the decomposition* (3), *with $P_1'(E_2) = H_2^\perp, P_2'(E_2) = H_1^\perp$.*

Proof. A weakly continuous projection of E_1 is a weakly continuous endomorphism P of E_1, with $P^2 = P$. By 4.(4) the adjoint mapping P' satisfies the equation $(P')^2 = P'$, so that P' is a weakly continuous projection of E_2.

The projections corresponding to (2) satisfy the equations $I = P_1 + P_2$, $P_1 P_2 = P_2 P_1 = 0$. Again by 4.(4), the adjoint mappings satisfy the equations $I = P_1' + P_2'$, $P_2' P_1' = P_1' P_2' = 0$.

It follows from § 15, 8.(1) that P_1' and P_2' determine a $\mathfrak{T}_s$-complementary decomposition of E_2, and $E_2 = P_1'(E_2) \oplus P_2'(E_2)$.

We still have to show that $P_1'(E_2) = H_2^\perp$ and $P_2'(E_2) = H_1^\perp$.

$P_1'(E_2)$ consists of all $u \in E_2$ with $P_1' u = u$. This is equivalent to

$$(P_1' u)x = u(P_1 x) = u x = u(P_1 x + P_2 x) = u(P_1 x) + u(P_2 x)$$

for all $x \in E_1$, i.e. $u \in H_2^\perp$.

We show that $P_2'(E_2) = H_1^\perp$ in the same way.

Since every $\mathfrak{T}$-continuous projection of a locally convex space $E[\mathfrak{T}]$ is also weakly continuous, by 4.(5), we have

(4) *Every $\mathfrak{T}$-continuous complementary decomposition $E = H_1 \oplus H_2$ of a locally convex space $E[\mathfrak{T}]$ is also a $\mathfrak{T}_s$-continuous decomposition, and it determines a dual $\mathfrak{T}_s(E')$-continuous complementary decomposition $E' - H_1^\perp \oplus H_2^\perp$.*

In general, however, it is too much to expect a weakly continuous complementary decomposition of $E[\mathfrak{T}]$ to be $\mathfrak{T}$-continuous as well, for not every weakly continuous projection need be $\mathfrak{T}$-continuous.

(5) *Let $E[\mathfrak{T}]$ be locally convex. Every finite-dimensional linear subspace H of E has a $\mathfrak{T}$-complement.*

Proof. H is a closed subspace of E (cf. § 15, 5.(2)) and, applying § 9, 2.(7a) to E' instead of to E, its orthogonal complement $H^\perp$ has codimension n in E', where n is the dimension of H. By § 15, 8.(2), $H^\perp$ has a $\mathfrak{T}_s$-complement G of dimension n. Since $H^{\perp\perp} = H$, to the complementary decomposition $E'=H^{\perp}\oplus G$ there corresponds the $\mathfrak{T}_s$-complementary decomposition $E=H\oplus G^{\perp}$, by (1). But this decomposition is also $\mathfrak{T}$-complementary, by § 15, 8.(2), since H is a finite-dimensional algebraic complement of $G^{\perp}$.

In § 15, 12.(6) we showed that two algebraically complementary closed linear subspaces of a complete metrizable space, and in particular of an (F)-space, are always topologically complementary. This is by no means true for arbitrary locally convex spaces. In § 10.(8) we gave a simple example of a dual pair in which an algebraically complementary decomposition into $\mathfrak{T}_{ls}$-closed subspaces need not be $\mathfrak{T}_{ls}$-continuous. The proof also holds for the weak topologies, because of 3.(2), and because a $\mathfrak{T}_s$-continuous projection is also a $\mathfrak{T}_{ls}$-continuous projection.

By the same token, the example of a closed linear subspace of $\varphi\omega\oplus\omega\varphi$ with no topological complement given in § 13, 6. also provides an example of a locally convex space with a weakly closed linear subspace with no $\mathfrak{T}_s$-complement. In § 31 we go into the question of the existence of a topological complement in more detail.

6. The convex cover of a compact set. We now link up with the ideas of § 15, 6.

From the fact that every locally convex space has a base of neighbourhoods of $\mathfrak{o}$ consisting of absolutely convex sets, it follows immediately that

(1) *If M is a bounded subset of a locally convex space, the closed convex cover $\overline{C(M)}$ and the closed absolutely convex cover $\overline{\Gamma(M)}$ are again bounded.*

We now investigate the corresponding question for compact and precompact subsets of a locally convex space.

We begin with a counterexample. The space φ of all finite vectors is a subspace of the Hilbert space l^2, and φ is a normed space under the norm induced by the norm of l^2. The set M of vectors $\mathfrak{x}^{(n)}=\frac{1}{n}\mathfrak{e}_n$ is a sequence tending to zero in l^2; together with $\mathfrak{o}$ it forms a compact subset of φ. If $\alpha_n>0$, $\sum_{n=1}^{\infty}\alpha_n=1$, the sequence $\mathfrak{y}^{(k)}=\frac{1}{\sum_1^k\alpha_n}\sum_1^k\alpha_n\mathfrak{x}^{(n)}$ is a Cauchy sequence in $C(M)$, whose limit is not a finite vector and so does not belong to φ. Thus $\overline{C(M)}$ is not compact.

In contrast to this, for precompact sets we have

(2) *If M is a precompact subset of the locally convex space $E[\mathfrak{T}]$, $\overline{C(M)}$ and $\overline{\Gamma(M)}$ are again precompact.*

It is sufficient to show that $\Gamma(M)$ is totally bounded, for then $\overline{\Gamma(M)}$ and $\overline{C(M)}\subset\overline{\Gamma(M)}$ are precompact.

Let V be a closed absolutely convex neighbourhood of o. There are finitely many points $x_i \in M$, $i=1,\dots,m$, such that $M \subset \bigcup_{i=1}^{m} (x_i + \frac{1}{2}V)$. Let δ be chosen so that δM, and consequently $\Gamma(\delta M)$, is contained in $\frac{1}{2}V$, and let A be the subset of the m-dimensional space K^m consisting of all $\alpha = (\alpha_1, \dots, \alpha_m)$ with $\sum_{i=1}^{m} |\alpha_i| \leqq 1$. There is a finite set of points $\beta^{(j)} = (\beta_1^{(j)}, \dots, \beta_m^{(j)})$ in A with the property that for each $\alpha \in A$ there is a $\beta^{(j)}$ with $\sum_{i=1}^{m} |\alpha_i - \beta_i^{(j)}| \leqq \delta$.

It is sufficient to show that $\Gamma(M) \subset \bigcup_j \left(\sum_{i=1}^{m} \beta_i^{(j)} x_i + V \right)$. An element of $\Gamma(M)$ has the form $y = \sum_1^N \gamma_k y_k$ with $\sum |\gamma_k| \leqq 1$, $y_k \in M$. For each y_k there is an x_{i_k} with $y_k = x_{i_k} + z_k$, $z_k \in \frac{1}{2}V$, so that $y = \sum_1^N \gamma_k x_{i_k} + \sum_1^N \gamma_k z_k = \sum_1^m \alpha_i x_i + z$, $z \in \frac{1}{2}V$, $\sum |\alpha_i| \leqq 1$.

But then for a suitable $\beta^{(j)}$, $y = \sum_1^m \beta_i^{(j)} x_i + \sum_1^m (\alpha_i - \beta_i^{(j)}) x_i + z = \sum_1^m \beta_i^{(j)} x_i + z' + z$ with $z' \in \Gamma(\delta M) \subset \frac{1}{2}V$, so that $y \in \sum \beta_i^{(j)} x_i + V$ and so $\Gamma(M) \subset \bigcup_j \left(\sum_i \beta_i^{(j)} x_i + V \right)$.

From (2) there follows

(3) *If M is a compact subset of the locally convex space $E[\mathfrak{T}]$, $\overline{C(M)}$ (respectively $\overline{\Gamma(M)}$) is compact if and only if $\overline{C(M)}$ (respectively $\overline{\Gamma(M)}$) is complete.*

This is always the case if $E[\mathfrak{T}]$ is quasi-complete.

Proof. If $\overline{C(M)}$ (respectively $\overline{\Gamma(M)}$) is complete, then every closure point of the precompact set $C(M)$ (respectively $\Gamma(M)$) lies in this set, and so it is compact.

The second part of the assertion follows from the fact that every precompact set is bounded, by § 15, 6. (6).

Later on we shall prove KREIN's theorem, which extends the second part of (3) in a profound way (cf. § 24, 5. and 6.).

(4) *The circled cover of a compact subset K of a topological vector space is again compact.*

Proof. The circled cover of K consists of all αx, $x \in K$, $|\alpha| \leqq 1$. If A is the set of all $\alpha \in \mathsf{K}$ with $|\alpha| \leqq 1$, then the topological product $\mathsf{A} \times K$ is compact, by TYCHONOFF's theorem. The mapping which sends each

$(\alpha, x) \in \mathsf{A} \times K$ to the element $\alpha x \in E[\mathfrak{T}]$ is continuous, and maps $\mathsf{A} \times K$ onto the circled cover of K; this is therefore compact.

(5) *If $K_1, \ldots, K_n$ are convex (respectively absolutely convex) compact subsets of the locally convex space $E[\mathfrak{T}]$, the convex cover $\overset{n}{\underset{i=1}{C}} K_i$ (respectively absolutely convex cover $\overset{n}{\underset{i=1}{\Gamma}} K_i$) is again compact.*

Proof. Let A be the set of all $\alpha = (\alpha_1, \ldots, \alpha_n) \in \mathsf{K}^n$ with $\sum_{i=1}^{n} \alpha_i = 1$, $\alpha_i \geqq 0$. A is a compact subset of K^n. The topological product $A \times K_1 \times \cdots \times K_n$ is again compact, by TYCHONOFF's theorem. If we map each $(\alpha, x_1, \ldots, x_n) \in A \times K_1 \times \cdots \times K_n$ to the corresponding element $\sum_{i=1}^{n} \alpha_i x_i \in E[\mathfrak{T}]$, we obtain a continuous mapping whose image $\overset{n}{\underset{i=1}{C}} K_i$ is compact.

For absolutely convex sets the proof is similar; instead of A we consider the set B of all $\beta = (\beta_1, \ldots, \beta_n)$ with $\sum_{i=1}^{n} |\beta_i| \leqq 1$, $\beta_i \in \mathsf{K}$.

In the case of a *real* locally convex space we have the further result:

(6) *The absolutely convex cover $\Gamma(K)$ of a convex compact set K is compact.*

Proof. All elements of the form $\rho x - \sigma y$ with x and y in K, $\rho \geqq 0$, $\sigma \geqq 0$ and $\rho + \sigma \leqq 1$ lie in $\Gamma(K)$. On the other hand, all the elements of ΓK are of this form: to show this, it is sufficient to show that if two elements $z_1 = \rho_1 x_1 - \sigma_1 y_1$ and $z_2 = \rho_2 x_2 - \sigma_2 y_2$ have this form, then so does $\alpha_1 z_1 + \alpha_2 z_2$, where $|\alpha_1| + |\alpha_2| \leqq 1$. Since $-z = \sigma y - \rho x$ is again of the required form, we may suppose that $\alpha_1 > 0$ and $\alpha_2 > 0$. But then

$$\alpha_1 z_1 + \alpha_2 z_2 = (\alpha_1 \rho_1 + \alpha_2 \rho_2)\left(\frac{\alpha_1 \rho_1}{c} x_1 + \frac{\alpha_2 \rho_2}{c} x_2\right) - (\alpha_1 \sigma_1 + \alpha_2 \sigma_2)\left(\frac{\alpha_1 \sigma_1}{d} x_1 + \frac{\alpha_2 \sigma_2}{d} x_2\right),$$

where $\alpha_1 \rho_1 + \alpha_2 \rho_2 = c$, $\alpha_1 \sigma_1 + \alpha_2 \sigma_2 = d$. Since $0 \leqq c + d \leqq 1$, and since the elements in the brackets lie in K, the assertion now follows.

If now C is the set of all (ρ, σ) with $\rho \geqq 0$, $\sigma \geqq 0$, $\rho + \sigma \leqq 1$, C is compact. The mapping from $C \times K \times K$ into E which sends the element (ρ, σ, x, y) to the element $\rho x - \sigma y$ is continuous, and so ΓK is compact.

For *complex* locally convex spaces, (6) is not true in general; however in this case we have

(7) *The absolutely convex cover ΓK of a compact convex set M is relatively compact, so that $\overline{\Gamma K}$ is compact.*

Proof. By (6), we can suppose that K is real absolutely convex. The set $K_1 = K + iK$ is convex, by § 16, 1.(3), and is compact, by § 15, 6.(8). K_1 contains every set of the form $e^{i\varphi} K$, and so it contains the complex circled cover of K. Since K_1 is convex, K_1 also contains the convex cover of the circled cover of K, i.e. $\Gamma(K)$, so that $\Gamma(K)$ is relatively compact.

Consequently (5) can be sharpened to give

(8) *If $E[\mathfrak{T}]$ is real locally convex and if $K_1, \ldots, K_n$ are convex compact subsets of $E[\mathfrak{T}]$, then $\Gamma_{i=1}^{n} K_i$ is compact.*

If $E[\mathfrak{T}]$ is complex locally convex and if $K_1, \ldots, K_n$ are convex compact subsets of $E[\mathfrak{T}]$, then $\Gamma_{i=1}^{n} K_i$ is relatively compact, so that $\overline{\Gamma_{i=1}^{n} K_i}$ is compact.

The real case follows from (5), when we replace the K_i by their absolutely convex covers, which are compact, by (6). In the complex case we apply (5) to the sets $\overline{\Gamma(K_i)}$, which are compact, by (7), to obtain that $\Gamma_{i=1}^{n} (\overline{\Gamma K_i})$ is compact, so that $\Gamma_{i=1}^{n} K_i$ is relatively compact.

Example. In ω, the topological product of countably many one-dimensional complex spaces, the set K of all $x = (x_n)$ with x_n real and $|x_n| \leqq 1$ is real absolutely convex, and is compact. $\overline{\Gamma(K)}$ consists of all $y = (y_n)$ with y_n complex and $|y_n| \leqq 1$. We shall show that ΓK is a proper subset of $\overline{\Gamma K}$, and is therefore not compact.

The elements of ΓK have the form

$$\mathfrak{z} = \sum_{k=1}^{n} \alpha_k e^{i\varphi_k} \mathfrak{x}^{(k)}, \quad \mathfrak{x}^{(k)} \in K, \quad \sum \alpha_k \leqq 1, \quad \alpha_k \geqq 0.$$

A coordinate

$$z_p = \sum_{k=1}^{n} \alpha_k e^{i\varphi_k} x_p^{(k)} = \sum_{k=1}^{n} \beta_k e^{i\varphi_k}$$

with $\sum |\beta_k| \leqq 1$ is of modulus 1 only if all the β_k vanish except for one $\beta_k = 1$. An element $\mathfrak{y}_0 = (e^{i\varphi_n})$ of $\overline{\Gamma K}$ with infinitely many different φ_n can therefore never be a $\mathfrak{z}$.

7. The separation theorem for compact convex sets. The existence theorems obtained in 1. were sufficient to derive the duality properties that we have obtained so far. We need a further existence theorem, however, for the duality theory of closed convex sets which we shall develop in the next number; this we shall first formulate geometrically as a separation theorem for compact convex sets:

(1) *Let A be a closed convex subset of a locally convex space $E[\mathfrak{T}]$, and let K be a compact convex set which is disjoint from A. Then there exists a closed real hyperplane which separates A and K strictly.*

Proof. By § 15, 6.(9) there exists an absolutely convex open neighbourhood U of $\mathfrak{o}$ for which $A+U$ and $K+U$ are disjoint. By § 17, 1.(4) these two open convex sets are separated strictly by a closed real hyperplane, and consequently so are A and K.

The analytic form follows by characterising the real hyperplane by an equation $\Re(u x)=\gamma$ (cf. § 16, 3.):

(2) *Under the same hypotheses as for* (1), *there exists a continuous linear functional* $u_0 \in E'$ *and a real number* γ *for which*

$$\sup_{y \in A} \Re(u_0 y) \leqq \gamma < \inf_{z \in K} \Re(u_0 z). \tag{3}$$

The strict inequality in (3) follows from the fact that the infimum is attained in K.

If instead of § 15, 6.(9) we apply theorem § 15, 6.(11), then in place of (1) we obtain

(1′) *Let A be a complete convex subset of the locally convex space $E[\mathfrak{T}]$, and let K be a closed precompact convex subset which is disjoint from A. Then there exists a real closed hyperplane which separates A and K strictly.*

Every closed convex $\mathfrak{T}$-body C in a topological vector space is the intersection of the closed half-spaces, containing C, which correspond to the closed supporting hyperplanes of C (§ 17, 5.). The corresponding result for compact convex subsets of a locally convex space now follows from (1):

(4) *In a locally convex space, every compact convex set K is the intersection of the closed real half-spaces containing K which correspond to the supporting hyperplanes of K.*

Proof. If $x_0 \notin K$, then by (1) there exists a hyperplane $\Re(u_0 x)=\gamma$ which separates x_0 and K strictly. Since K is compact, the infimum ρ in (3) is attained for some element $z_0 \in K$. Then z_0 lies in the hyperplane $\Re(u_0 x) = \inf_{z \in K} \Re(u_0 z)=\rho$, and K is contained in the half-space $\Re(u_0 x) \geqq \rho$; thus the hyperplane is a supporting hyperplane at z_0 and it separates K from x_0. The assertion follows from this.

For arbitrary convex sets we have

(5) *Every closed convex proper subset C of a locally convex space $E[\mathfrak{T}]$ is the intersection of the closed real half-spaces containing it.*

We have only to apply (1) to C and a point x_0 which does not lie in C, since x_0 is certainly a compact convex set.

As a result we have shown that forming the closed convex cover of a set M depends only upon the dual space, and does not depend upon the original topology:

(6) *The closed convex cover $\overline{\complement(M)}$ of a set M in a locally convex space $E[\mathfrak{T}]$ can be obtained using any locally convex topology compatible with the dual pair $\langle E', E\rangle$, e.g. the weak topology.*

Any compatible linear topology can be used as well.

As a consequence of (6) we have

(7) *Suppose that $E[\mathfrak{T}]$ is a metrizable locally convex space. If a sequence $x_n \in E$ has x_0 as a weak closure point in E, then there is a sequence of finite linear combinations of the x_n which converges to x_0 with respect to $\mathfrak{T}$.*

For x_0 belongs to the weakly closed convex cover of the set of x_n, and the convex cover of the x_n is dense in this set with respect to the topology $\mathfrak{T}$, by (6).

In § 17, 5. we saw that (5) need not hold in arbitrary topological vector spaces.

We mention one further result about compact sets:

(8) *Suppose that M is a compact subset of a locally convex space $E[\mathfrak{T}]$, and that H is a real closed hyperplane in E. Then M has a supporting hyperplane parallel to H.*

If H is given by $\Re(ux)=0$, and if the maximum α of $\Re(ux)$ on M is attained at z_0, then $\Re(ux)=\alpha$ is a support hyperplane of M, since it contains z_0 and since M lies in the half-space $\Re(ux)\leqq\alpha$.

8. Polarity. Suppose that we are given a dual pair $\langle E_2, E_1\rangle$. If M is a subset of E_1, we call the set of all $u\in E_2$ for which $\Re(ux)\leqq 1$ for all $x\in M$ the polar M° of M in E_2. The polar in E_1 of a subset of E_2 is defined similarly.

If M is circled, M° is also equal to the set of all u for which $|ux|\leqq 1$ for all $x\in M$: for if $|\alpha|\leqq 1$, αx is also in M, and it follows from $\Re(\alpha ux)\leqq 1$, for all $|\alpha|\leqq 1$, that $|ux|\leqq 1$; and clearly if $|ux|\leqq 1$ then $\Re(ux)\leqq 1$.

If we call the set of all $u\in E_2$ for which $|ux|\leqq 1$ for all $x\in M$ the absolute polar of M, then the absolute polar of M is the polar of the circled cover of M. If M is absolutely convex, M° is the absolute polar of M. The absolute polar is always absolutely convex.

The next result follows easily from the definition:

(1) *The following rules hold for forming polars:*

a) $(\alpha M)^\circ = \frac{1}{\alpha} M^\circ$ *for each non-zero* $\alpha\in \mathsf{K}$.

b) *If* $M\subset N$, *then* $N^\circ \supset M^\circ$ *and* $M^{\circ\circ}\subset N^{\circ\circ}$.

c) $M\subset M^{\circ\circ}$.

d) $M^\circ = M^{\circ\circ\circ}$.

We shall only prove d): if we apply c) to M° we obtain $M^\circ \subset M^{\circ\circ\circ}$, and if we apply b) to c) we obtain $M^\circ \supset M^{\circ\circ\circ}$; the two together give d).

$M^{\circ\circ}$ is also called the bipolar of M. Similarly the absolute polar of the absolute polar of M is called the absolute bipolar of M. It is equal to the bipolar of the circled cover of M.

(2) *The polar M° of a set $M \subset E_1$ is convex, weakly closed and contains* $\mathfrak{o}$. *If M is circled, M° is absolutely convex.*

Indeed M° is the intersection of the weakly closed half-spaces $\mathfrak{R}(u x_0) \leqq 1$ in E_2, as x_0 runs through all the elements of M, and so M° is weakly closed and convex; further the element $\mathfrak{o}$ belongs to all these half-spaces. If M is circled, M° is the absolute polar of M, and so it is absolutely convex.

Conversely, we have

(3) *M and the weakly closed convex cover $\overline{\complement(M,\mathfrak{o})}$ of M and $\mathfrak{o}$ have the same polar M°.*

For $\mathfrak{R}(u x) \leqq 1$, for all $u \in M^\circ$ and $x \in M$, and also $\mathfrak{R}(u\mathfrak{o}) = 0 < 1$; further all the elements x of the real line segment joining x_1 and x_2 satisfy $\mathfrak{R}(u x) \leqq 1$, if x_1 and x_2 do.

If M is a linear subspace, $M^\circ = M^\perp$, so that forming an orthogonal space is just a special case of forming a polar set.

If C is a cone with vertex $\mathfrak{o}$, the polar set C° is again a cone with vertex $\mathfrak{o}$, which is given by the equation

$$(4) \qquad \mathfrak{R}(u x) \leqq 0 \quad \text{for all} \quad x \in C.$$

For if C contains x it contains ρx for all $\rho > 0$, and so if $\mathfrak{R}(u x) \leqq 1$ we also have $\rho R(u x) \leqq 1$; this is only possible if $\mathfrak{R}(u x) \leqq 0$.

We now come to the theorem of bipolars:

(5) *The bipolar $M^{\circ\circ}$ of a subset M of E_1 is the weakly closed convex cover $\overline{\complement(\mathfrak{o},M)}$ of $\mathfrak{o}$ and M. If M is convex and weakly closed, and if M contains $\mathfrak{o}$, then $M = M^{\circ\circ}$.*

The absolute bipolar of a subset M of E_1 is the weakly closed absolutely convex cover $\overline{\Gamma(M)}$ of M. If M is circled, $M^{\circ\circ} = \overline{\complement(M)}$ is equal to the absolute bipolar of M.

Proof. By (3) we may suppose that $\mathfrak{o} \in M$, and that M is convex and weakly closed. By (1) c), $M \subset M^{\circ\circ}$. If on the other hand $x_0 \notin M$, then by 7.(1) x_0 and M are strictly separated by a closed real hyperplane. Since this does not pass through $\mathfrak{o}$, by 7.(2) it can be taken in the form $\mathfrak{R}(u_0 x) = 1$, where $\mathfrak{R}(u_0 y) \leqq 1$ for all $y \in M$ and $\mathfrak{R}(u_0 x_0) > 1$. But then x_0 does not belong to $M^{\circ\circ}$, and so $M \supset M^{\circ\circ}$.

The second part of (5) follows directly from the remarks we made about the relation between polars and absolute polars.

We denote by $\mathfrak{C}(E_1)$ the collection of all convex weakly closed subsets of E_1 which contain o. By 7.(6) it is enough to require the sets to be closed with respect to any topology compatible with the dual pair $\langle E_2, E_1\rangle$, instead of requiring them to be weakly closed.

Under the relation $C_1 \subset C_2$, $\mathfrak{C}(E_1)$ forms a complete lattice; $\bigwedge_\alpha C_\alpha$ is the intersection $\bigcap_\alpha C_\alpha$ and $\bigvee_\alpha C_\alpha = \overline{\mathsf{C}_\alpha C_\alpha}$.

(6) *Let $\langle E_2, E_1\rangle$ be a dual pair. The complete lattices $\mathfrak{C}(E_1)$ and $\mathfrak{C}(E_2)$ are dually isomorphic, when each set is made to correspond to its polar.*

Proof. By (2), $M^\circ \in \mathfrak{C}(E_2)$ (respectively $\mathfrak{C}(E_1)$) if $M \in \mathfrak{C}(E_1)$ (respectively $\mathfrak{C}(E_2)$), by (5) the correspondence is one-one, and by (1) b) it reverses the partial order, so that it is a dual lattice isomorphism (cf. § 7, 9.).

In particular, because of the correspondence between $\bigwedge$ and $\bigvee$ we have

(7) *If the sets C_α are weakly closed convex subsets of E_1 containing o, we have*

(8) $$\Big(\bigcap_\alpha C_\alpha\Big)^\circ = \overline{\mathsf{C}_\alpha C_\alpha^\circ}.$$

The dual formula holds for arbitrary subsets M_α of E_1:

(9) $$\Big(\bigcup_\alpha M_\alpha\Big)^\circ = \bigcap_\alpha M_\alpha^\circ.$$

Since under the formation of polars absolutely convex sets, cones and linear subspaces correspond to absolutely convex sets, cones and linear subspaces, respectively, we obtain dually isomorphic lattices of weakly closed absolutely convex sets, of weakly closed convex cones and of weakly closed linear subspaces. This last of these results has already been obtained in 3.

If the sets C_α are weakly closed absolutely convex sets or weakly closed linear subspaces F_α, respectively, (8) takes the form

(10) $$\Big(\bigcap_\alpha C_\alpha\Big)^\circ = \overline{\Gamma_\alpha C_\alpha^\circ}$$

or

(11) $$\Big(\bigcap_\alpha F_\alpha\Big) = \overline{\sum_\alpha F_\alpha^\perp}.$$

respectively.

9. The polar of a neighbourhood of o. We begin by investigating the dual pair $\langle E^*, E\rangle$, where E is a vector space and E^* is its algebraic dual. We have (cf. § 10, 6. (3))

(1) *If E is a vector space of dimension d over K, then $E^*[\mathfrak{T}_s(E)]$ is topologically isomorphic to $\omega_d(\mathsf{K})$, the topological product of d copies of the field K.*

Proof. If $\{x_\alpha\}$ is an algebraic basis of E, each linear functional u on E is given by the corresponding coordinate vector $\mathfrak{u}=\{v_\alpha\}$, where $v_\alpha=u x_\alpha$; conversely, every such coordinate vector $\mathfrak{u}$ defines a linear functional on E, so that E^* is algebraically isomorphic to $\omega_d(\mathsf{K})$. In defining the weak neighbourhoods U of $\mathfrak{o}$ it is sufficient to take finitely many elements x_α of the basis. $\sup_{i=1,\ldots,n} |u x_{\alpha_i}|<\varepsilon$ means however that finitely many coordinates of the elements $\mathfrak{u}$ of the neighbourhood U are bounded in modulus by ε. The weak topology $\mathfrak{T}_s(E)$ therefore coincides with the product topology on $\omega_d(\mathsf{K})$.

(2) *If* $\langle E_2, E_1\rangle$ *is a dual pair, the completion of* $E_1[\mathfrak{T}_s(E_2)]$ *is* $E_2^*[\mathfrak{T}_s(E_2)]$, *and it is topologically isomorphic to* $\omega_d(\mathsf{K})$, *where* d *is the algebraic dimension of* E_2.

Proof. Since $E_1\subset E_2^*$ and $E_2^*[\mathfrak{T}_s(E_2)]$ is topologically isomorphic to $\omega_d(\mathsf{K})$, and further since $\omega_d(\mathsf{K})$, being the topological product of complete spaces, is again complete, it is sufficient to show that E_1 is dense in E_2^* with respect to the weak topology $\mathfrak{T}_s(E_2)$. If this were not the case, the weak closure $\bar{E}_1$ of E_1 in E_2^* would be different from E_2^*; then by 1.(3) there would be a non-zero weakly continuous linear functional v on E_2^* which would vanish on E_1. This contradicts the fact that the weak dual of E_2^* is equal to E_2, by 2.(3), and that there is no such non-zero v in E_2.

Using these results we can characterise the weakly bounded subsets of a locally convex space:

(3) *The weakly bounded subsets of a locally convex space* $E[\mathfrak{T}]$ *are the same as the weakly precompact subsets of* E.

Proof. Every weakly precompact set is weakly bounded, by § 15, 6.(6). On the other hand, if M is $\mathfrak{T}_s(E')$ bounded in E, M is also $\mathfrak{T}_s(E')$ bounded in the weak completion $(E')^*$ of E, and this is topologically isomorphic to some $\omega_d(\mathsf{K})$, by (2). But every bounded subset of $\omega_d(\mathsf{K})$ is relatively compact, by § 15, 6.

We now prove the Alaoglu-Bourbaki theorem:

(4) *If* U *is a neighbourhood of* $\mathfrak{o}$ *in the locally convex space* $E[\mathfrak{T}]$, U° *is a weakly compact subset of* E'.

Proof. Suppose first that U is absolutely convex. To begin with we form the polar of U in $E^*\supset E'$. If the linear functional $v\in E^*$ belongs to this polar, $|v x|\leqq 1$ for all $x\in U$. But this means that v is continuous, and is therefore an element of E'. U therefore has the same polar in E' as it has in E^*.

U° is weakly bounded in E' and in E^*. For given any $x\in E$, U contains a suitable scalar multiple ρx, $\rho>0$, and it follows from this that given finitely many x_i in E, $\sup_{\substack{i=1,\ldots,n\\ v\in U^\circ}} |v x_i|<M$ for a suitable $M>0$.

By (2), $E^*[\mathfrak{T}_s(E)]$ is topologically isomorphic to some $\omega_d(\mathsf{K})$, in which every bounded set is relatively compact. Thus U°, considered as a subset of E^*, is relatively weakly compact; since it is weakly closed, by 8.(2), it is therefore weakly compact.

Since every neighbourhood U contains an absolutely convex V, and since $U^\circ \subset V^\circ$, U° is also weakly compact.

We observe that there is a certain analogy between (4) and § 10, 11.(2).

In § 14, 5.(3) we introduced a norm on the dual space E' of a normed space E. The closed unit ball in E' is simply the polar of the unit ball in E. We therefore obtain as a special case of (4):

(5) *The closed unit ball of the dual of a normed space is weakly compact.*

As an application of (5) we prove:

(6) *The absolute bipolar of a sequence x_n which converges weakly to $\mathfrak{o}$ in the sequentially complete locally convex space $E[\mathfrak{T}]$ is weakly compact, and consists of all* $\sum\limits_{n=1}^{\infty} \xi_n x_n$ with $\sum\limits_{n=1}^{\infty} |\xi_n| \leqq 1$.

Proof. The x_n form a weakly bounded subset N of E, by § 15, 6.(4). By the theorem of MACKEY, which will be proved in the next number but one (11.(7)), N is also $\mathfrak{T}$-bounded in E. By 6.(1) the absolute bipolar $\overline{\Gamma(N)}$ is also bounded. All the partial sums $\sum\limits_{n=1}^{m} \xi_n x_n$ clearly belong to $\overline{\Gamma(N)}$, and are $\mathfrak{T}$-convergent, since $E[\mathfrak{T}]$ is sequentially complete. The collection M of all these $\sum\limits_{n=1}^{\infty} \xi_n x_n$ is therefore a bounded subset of $E[\mathfrak{T}]$ which lies in $\overline{\Gamma(N)}$.

The mapping which sends each $\mathfrak{x}=(\xi_n)$ in l^1 to the element $A\mathfrak{x} = \sum\limits_{n=1}^{\infty} \xi_n x_n \in E$ is a linear mapping from l^1 into $E[\mathfrak{T}]$, under which M is the image of the closed unit ball K of l^1. l^1 is the dual of c_0, by § 14, 7.(11). If $v \in E'$, $v\left(\sum\limits_{n=1}^{\infty} \xi_n x_n\right) = \sum\limits_{n=1}^{\infty} \xi_n (v x_n)$; the sequence $v_n = (v x_n)$ is convergent to zero, and therefore represents an element $\mathfrak{v}=(v_n)$ of c_0. It follows from the relation $v(A\mathfrak{x}) = \mathfrak{v}\mathfrak{x}$ which we have just established that A is a weakly continuous mapping for the topologies $\mathfrak{T}_s(c_0)$ on l^1 and $\mathfrak{T}_s(E')$ on E. The set K is weakly compact, by (5), and so therefore is its continuous image M. Consequently M is absolutely convex and weakly closed, so that it is equal to the absolute bipolar $\overline{\Gamma(N)}$ of the sequence x_n.

10. A representation of locally convex spaces. In § 14, 9. we introduced the (B)-space $C(K)$ of all continuous real- (respectively complex-)

valued functions on the compact space K. The following representation of (B)-spaces follows from 9.(5):

(1) *Every* (B)-*space* E *is norm-isomorphic to a closed linear subspace of some suitable* $C(K)$, *where* K *is compact.*

Proof. We take for K the closed unit ball in the dual space $E'[\mathfrak{T}_s(E)]$. Each x is then a weakly continuous linear functional on K, and so is an element of $C(K)$. The norm $\|x\|$ is equal to $\sup_{u\in K}|ux|$, by § 17, 6.(4), and so it is equal to the norm of x as an element of $C(K)$. In this way E becomes a closed linear subspace of $C(K)$.

In order to represent arbitrary locally convex spaces as function spaces we have to use locally compact spaces.

Let R be a locally compact topological space. We denote by $C(R)$ the vector space of all continuous real-(respectively complex-)valued functions on R. As topology $\mathfrak{T}$ on $C(R)$ we introduce the topology of uniform convergence on the compact subsets of R, which has as base of neighbourhoods of $\mathfrak{o}$ the sets $U_{K;\varepsilon}$ consisting of all $f(x)$ with $\sup_{x\in K}|f(x)|<\varepsilon$, where K is a compact subset of R.

The topology $\mathfrak{T}$ is also defined by the semi-norms $p_K(f)=\sup_{x\in K}|f(x)|$.

By § 6, 4.(5) there are "sufficiently many" continuous functions on R, i.e. for every two points $x\neq y$ there is always a continuous function f with $f(x)\neq f(y)$. $C(R)$ is a locally convex space with respect to $\mathfrak{T}$. If R is compact, we obtain the special case considered at the beginning of this number.

(2) $C(R)$ *is complete. If* R *is countable at infinity,* $C(R)$ *is an* (F)-*space.*

Proof. If $\mathfrak{F}$ is a Cauchy filter of functions in $C(R)$, then for each $x_0\in R$ we can form the corresponding Cauchy filter of values of the functions at x_0. This has a limit $f(x_0)$. The function $f(x)$ defined in this way is the uniform limit on each compact set K of functions in $\mathfrak{F}$, and so it is continuous on each compact set K; it is therefore continuous on R, so that $C(R)$ is complete.

If R is countable at infinity, there is a sequence $K_1\subset K_2\subset\cdots$ of compact sets whose union is the whole of R, and with the property that every compact subset of R is contained in one of the sets K_n. But then the topology $\mathfrak{T}$ is defined by countably many semi-norms $p_{K_n}(x)$, so that $C(R)$ is metrizable; since it is complete, by the first part of the proof, $C(R)$ is an (F)-space.

As a generalisation of (1) we now obtain:

(3) *Every locally convex space* $E[\mathfrak{T}]$ *is topologically isomorphic to a linear subspace of a suitable* $C(R)$, *where* R *is locally compact.*

If, further, $E[\mathfrak{T}]$ is metrizable, R can be taken to be countable at infinity.

Proof. By § 18, 3.(7) it is sufficient to establish the result for a topological product $F[\mathfrak{T}]=\prod_\alpha E_\alpha$ of (B)-spaces E_α. Let K_α be the closed unit ball of E'_α equipped with the topology $\mathfrak{T}_s(E_\alpha)$, under which K_α is compact. The disjoint union $R=\bigcup_\alpha K_\alpha$ becomes a locally compact space when for each $u_\alpha\in K_\alpha$ we take as base of neighbourhoods in R its neighbourhoods in the compact space K_α. Each compact subset of R is contained in the union of finitely many K_α. The topology on $C(R)$ of uniform convergence on the compact sets is therefore given by the semi-norms $p_K(f)=\sup_{u\in K}|f(u)|$, $K=K_{\alpha_1}\cup\cdots\cup K_{\alpha_n}$. If we consider each element $x=(x_\alpha)$ of F as an element of $C(R)$, by considering each $x_\alpha\in E_\alpha$ as a continuous function on K_α, as in (1), then $p_K(x)$ coincides with the semi-norm $p(x)=\sup_{i=1,\ldots,n}\|x_{\alpha_i}\|$ on $F[\mathfrak{T}]$; the assertion follows directly from this.

If $E[\mathfrak{T}]$ is metrizable, $F[\mathfrak{T}]$ is the product of countably many spaces E_α, and R is countable at infinity.

We shall obtain a stronger result than (1) for separable (B)-spaces in § 21, 3.

11. Bounded and strongly bounded sets in dual pairs. We saw in 7. that the formation of the closed absolutely convex cover of a set $M\subset E[\mathfrak{T}]$ depends only on the dual pair $\langle E',E\rangle$, and that every compatible locally convex topology gives the same result. We shall now show that the bounded sets of a locally convex space $E[\mathfrak{T}]$ are also the same for every topology compatible with the dual pair $\langle E',E\rangle$, so that the concept of bounded set depends only on the dual pair.

We shall obtain this result as a special case of a more general result.

Starting from a dual pair $\langle E_2,E_1\rangle$, the following definition of boundedness for a subset of E_1 is particulary natural:

A subset M of E_1 is said to be E_2-bounded if $\sup_{x\in M}|ux|=\mu(u)<\infty$ for each $u\in E_2$.

It can be seen at once that this is not a new concept:

(1) *The E_2-bounded subsets of E_1 are the subsets of E_1 which are bounded in the weak topology $\mathfrak{T}_s(E_2)$.*

The following definition does however produce a new concept: a subset M of E_1 is said to be strongly E_2-bounded if

$$\sup_{u\in B,\, x\in M}|ux|=\mu(B)<\infty \quad \text{for each } E_1\text{-bounded set } B\subset E_2.$$

Clearly every strongly E_2-bounded set is also E_2-bounded. We shall see in § 21, 2. that strong boundedness also coincides with boundedness with respect to a suitable locally convex topology on E_1. We now investigate the question of when a bounded set is strongly bounded.

Let M be an absolutely convex weakly bounded subset of E_1. The linear span of M in E_1 is equal to $E_{1M} = \bigcup_{n=1}^{\infty} nM$. M is an absolutely convex α-body in the space E_{1M}, and its Minkowski functional defines a norm $\|x\|_M$ on E_{1M}. In this way E_{1M} becomes a normed space. We observe that the norm topology on E_{1M} is finer than the induced topology $\mathfrak{T}_s(E_2)$, since each weak neighbourhood of $\mathfrak{o}$ contains a scalar multiple of M.

We say that M is complete in itself if every sequence in M which is Cauchy with respect to the norm has a limit in M. If M is complete in itself then E_{1M} is a (B)-space with M as closed unit ball.

(2) *Let M be an absolutely convex, bounded, closed, sequentially complete subset of the locally convex space $E[\mathfrak{T}]$. Then E_M is a* (B)*-space with closed unit ball M, and $\mathfrak{T}$ induces a topology on E_M which is coarser than the norm topology.*

Proof. $\mathfrak{T}$ is coarser than the norm topology, since each $\mathfrak{T}$-neighbourhood of $\mathfrak{o}$ contains a scalar multiple of M. By hypothesis, M is $\mathfrak{T}$-sequentially complete, and applying § 18, 4.(4) b) to E_M, M is also sequentially complete with respect to the norm, so that E_M is a (B)-space.

We now obtain the Banach-Mackey theorem:

(3) *Let $\langle E_2, E_1 \rangle$ be a dual pair. An absolutely convex weakly bounded subset of E_1 which is complete in itself is strongly E_2-bounded.*

Every absolutely convex, closed, bounded, sequentially complete subset M of a locally convex space $E[\mathfrak{T}]$ is strongly E'-bounded.

We give two proofs of this important theorem.

a) The first proof reduces the theorem to the theorem of Banach (§15, 13.(2′)). If B is an arbitrary weakly bounded subset of E_2, the restrictions of the functionals $u \in B$ to E_{1M} form a set of continuous linear mappings from E_{1M} into K, and this set is bounded on each $x \in E_{1M}$, i.e. $\sup_{u \in B} |ux| < \infty$. But then by § 15, 13.(2′) $\sup_{u \in B, x \in M} |ux| < \infty$, so that M is strongly E_2-bounded. This establishes the first part of (3).

The second part follows from the first, and from (2), since M is complete in itself, by (2), and is weakly bounded.

b) The second proof uses the "sliding hump" technique of H. Lebesgue and O. Toeplitz (cf. F. Hausdorff [1], for example).

We first consider the case where M is complete in itself, and suppose that M is not strongly E_2-bounded.

There then exists a weakly bounded set $B \subset E_2$ and a sequence $u_n \in B$ with $\nu(u_n) = \sup_{x \in M} |u_n x| \to \infty$. On the other hand $\mu(x) = \sup_{u \in B} |u x| < \infty$ for each $x \in E_1$.

For each u_n there exists an $x_n \in M$ with

$$(4) \qquad |u_n x_n| \geqq \tfrac{1}{2} \nu(u_n).$$

We now determine a sequence $n_1 < n_2 < \cdots$ of integers for which

(5)

$$\frac{1}{6.4^k} \nu(u_{n_k}) \geqq \frac{1}{4} \mu(x_{n_1}) + \frac{1}{4^2} \mu(x_{n_2}) + \cdots + \frac{1}{4^{k-1}} \mu(x_{n_{k-1}}) + k = R_k + k.$$

Since $\|x_{n_k}\| \leqq 1$, and since M is absolutely convex and complete in itself, the sequence $\sum_{k=1}^{\infty} \frac{1}{4^k} x_{n_k}$ is convergent in E_{1M} to an element $x_0 \in M$.

We decompose x_0 into three parts:

$$(6) \qquad \begin{cases} x_0 = \left(\frac{1}{4} x_{n_1} + \cdots + \frac{1}{4^{k-1}} x_{n_{k-1}}\right) + \frac{1}{4^k} x_{n_k} + \left(\frac{1}{4^{k+1}} x_{n_{k+1}} + \cdots\right), \\ \quad = \qquad y_k \qquad + \frac{1}{4^k} x_{n_k} + \qquad z_k. \end{cases}$$

Then for each $u \in B$ we have

$$|u y_k| \leqq \frac{1}{4} \mu(x_{n_1}) + \cdots + \frac{1}{4^{k-1}} \mu(x_{n_{k-1}}) = R_k$$

and

$$|u z_k| \leqq \frac{1}{4^{k+1}} \nu(u) + \frac{1}{4^{k+2}} \nu(u) + \cdots = \frac{1}{3.4^k} \nu(u).$$

Thus $|u x_0| \geqq \frac{1}{4^k} |u x_{n_k}| - R_k - \frac{1}{3.4^k} \nu(u)$.

In particular, writing $u = u_{n_k}$, we obtain

$$|u_{n_k} x_0| \geqq \frac{1}{2.4^k} \nu(u_{n_k}) - \frac{1}{6.4^k} \nu(u_{n_k}) - \frac{1}{3.4^k} \nu(u_{n_k}) + k = k,$$

because of (4) and (5); this contradicts the fact that B is bounded.

The proof also holds if we make the hypothesis on M that M is absolutely convex and bounded, and that every $\mathfrak{T}$-Cauchy sequence in M has a limit in M, where $\mathfrak{T}$ is a locally convex topology on E_1 which induces a coarser topology on E_{1M} than the norm $\|x\|_M$. For $\sum_{n=1}^{\infty} \frac{1}{4^k} x_{n_k}$ is then also $\mathfrak{T}$-convergent to an $x_0 \in M$.

The second part of (3) follows from this remark, since on any bounded subset M of a locally convex space $E[\mathfrak{T}]$ the topology $\mathfrak{T}$ is coarser than the norm topology of E_{1M}.

As a first corollary of (3) we derive the theorem of MACKEY which was announced at the beginning of this number:

(7) *Suppose that $\langle E_2, E_1\rangle$ is a dual pair. All the compatible locally convex topologies on E_1 define the same collection of bounded sets.*

In particular the bounded and weakly bounded subsets of a locally convex space $E[\mathfrak{T}]$ are the same.

Proof. It is sufficient to prove the second assertion, and so it is sufficient to show that a weakly bounded subset M of $E[\mathfrak{T}]$ is bounded in the topology $\mathfrak{T}$. If U is a closed absolutely convex $\mathfrak{T}$-neighbourhood of $\mathfrak{o}$, $U = U^{\circ\circ}$, by the theorem of bipolars, and U° is weakly compact in E', by 9.(4). The second part of (3), applied to $U^\circ \subset E'[\mathfrak{T}_s(E)]$, shows that U° is strongly E-bounded, i.e. $\sup_{u\in U^\circ, x\in M} |ux| = \mu < \infty$. This means that $M \subset \mu U^{\circ\circ} = \mu U$, i.e. that M is $\mathfrak{T}$-bounded.

The next result is a special form of the BANACH-MACKEY theorem which is convenient for applications:

(8) *If the locally convex space $E[\mathfrak{T}]$ is sequentially complete, every weakly bounded subset of E (respectively of E') is strongly E'-bounded (respectively E-bounded).*

For by (3) all the absolutely convex bounded closed subsets of $E[\mathfrak{T}]$ are strongly bounded, and, by (7), so are all the weakly bounded subsets B of E. But this means that if B' is a weakly bounded subset of E' $\sup_{u\in B', x\in B} |ux| < \infty$ holds for each B, i.e. B' is strongly bounded.

We end with an example. In the dual pair $\langle\varphi,\varphi\rangle$, where φ is the space of finite coordinate-vectors $\mathfrak{x} = \{\xi_1, \ldots, \xi_n, 0, 0, \ldots\}$, with bilinear form $\mathfrak{u}\mathfrak{x} = \sum_{i=1}^{\infty} v_i \xi_i$, a subset $M \subset \varphi$ is φ-bounded if and only if there are numbers $m_i > 0$ with $|\xi_i| \leq m_i$ for $i = 1, 2, \ldots$, for all $\mathfrak{x} \in M$.

In contrast, M is strongly bounded if and only if in addition there exists an n_0 such that $\xi_n = 0$ for all $\mathfrak{x} \in M$, $n > n_0$.

§ 21. The different topologies on a locally convex space

1. The topology $\mathfrak{T}_{\mathfrak{M}}$ of uniform convergence on $\mathfrak{M}$. In § 20, 2., given a dual pair $\langle E_2, E_1\rangle$, we introduced the weak topology $\mathfrak{T}_s(E_2)$ on the space E_1. The definition given there can also be formulated in the following way:

We consider the class $\mathfrak{F}$ of all finite subsets F of E_2; the absolute polars $(\ulcorner F)^\circ$ of the sets F form a base of neighbourhoods of $\circ$ for $\mathfrak{T}_s(E_2)$; alternatively the semi-norms $p_F(x)=\sup_{u\in F}|ux|$ define the topology $\mathfrak{T}_s(E_2)$.

It is natural to try to find other classes $\mathfrak{M}$ of subsets of E_2 which can be used to introduce locally convex topologies on E_1.

If M is a subset of E_2, we set $p_M(x)=\sup_{u\in M}|ux|$. Using this terminology, we have

(1) *If* $M\subset E_2$, $p_M(x)$ *is a semi-norm on* E_1 *if and only if* M *is weakly bounded in* E_2.

For if M is bounded in E_2, $p_M(x)<\infty$ for each $x\in E_1$, so that $p_M(x)$ is a semi-norm. On the other hand if $p_M(x)$ is a semi-norm, $\sup_{u\in M}|ux|<\infty$ for each x, so that M is bounded in E_2.

Right from the beginning, then, the only suitable classes are classes $\mathfrak{M}$ of bounded sets in E_2. We call such a class $\mathfrak{M}$ total in E_2 if its union $\bigcup_{M\in\mathfrak{M}} M$ is total in E_2 for the topology $\mathfrak{T}_s(E_1)$.

(2) *Suppose that* $\mathfrak{M}=\{M\}$ *is a collection of bounded subsets of* E_2. *The topology* $\mathfrak{T}_{\mathfrak{M}}$ *defined by the system of semi-norms* $\{p_M(x)\}$ *on* E_1 *is a locally convex topology on* E_1 *if and only if* $\mathfrak{M}$ *is total in* E_2.

Proof. By § 18, 1.(3), $\mathfrak{T}_{\mathfrak{M}}$ is a locally convex topology on E_1 if and only if $\mathfrak{T}_{\mathfrak{M}}$ is Hausdorff; this happens if and only if the linear subspace N of all those $x\in E_1$ with $p_M(x)=0$ for all $M\in\mathfrak{M}$ consists of $\circ$ alone.

But if $p_M(x)=0$ for all $M\in\mathfrak{M}$ then $x\in\bigcap_{M\in\mathfrak{M}} M^\perp=(\bigcup M)^\perp$, and conversely for such an x $p_M(x)=0$ for all M. But $\mathfrak{M}$ is total in E_2 if and only if $\left(\bigcup_{M\in\mathfrak{M}} M\right)^\perp=\circ$.

The locally convex topology $\mathfrak{T}_{\mathfrak{M}}$ is also called the topology of uniform convergence on the sets M of $\mathfrak{M}$, or on $\mathfrak{M}$. Here, the $x\in E_1$ are considered as functions, namely as linear functionals on E_2.

A base of $\mathfrak{T}_{\mathfrak{M}}$-neighbourhoods of $\circ$ is obtained when we form the absolute polars $(\ulcorner M)^\circ$ of the sets M of $\mathfrak{M}$, form their scalar multiples $\rho(\ulcorner M)^\circ$, and then form the finite intersections of these.

When do two topologies $\mathfrak{T}_{\mathfrak{M}_1}$ and $\mathfrak{T}_{\mathfrak{M}_2}$ coincide?

A class $\mathfrak{M}=\{M\}$ is said to be saturated if the following conditions hold:

1) If M belongs to $\mathfrak{M}$, so does every subset of M.

2) If M belongs to $\mathfrak{M}$, so does every scalar multiple ρM, $\rho\in K$.

3) If M_1 and M_2 belong to $\mathfrak{M}$, so does their weakly closed absolutely convex cover $\overline{\ulcorner(M_1,M_2)}$.

If $\mathfrak{M}$ is an arbitrary class of bounded subsets of E_2, we can form the smallest saturated class $\tilde{\mathfrak{M}}$ which contains $\mathfrak{M}$. This saturated cover of $\mathfrak{M}$ again clearly consists only of bounded subsets of E_2.

A little earlier we gave a base of $\mathfrak{T}_{\mathfrak{M}}$-neighbourhoods of $\mathfrak{o}$. If $\mathfrak{M}$ is saturated, we have:

(3) *If $\mathfrak{M}$ is total and saturated, the polars M° of the absolutely convex weakly closed $M \in \mathfrak{M}$ form a base of $\mathfrak{T}_{\mathfrak{M}}$-neighbourhoods of $\mathfrak{o}$ in E_1.*

For if M° is the polar of an weakly closed absolutely convex set in $\mathfrak{M}$, so is $\left(\frac{1}{\rho} M\right)^\circ = \rho M^\circ$, and if M_1° and M_2° are, so is $\overline{\Gamma(M_1, M_2)}^\circ = M_1^\circ \cap M_2^\circ$.

The next result answers the question we raised above:

(4) *The topologies $\mathfrak{T}_{\mathfrak{M}}$ and $\mathfrak{T}_{\tilde{\mathfrak{M}}}$ on E_1 are the same; two topologies $\mathfrak{T}_{\mathfrak{M}_1}$ and $\mathfrak{T}_{\mathfrak{M}_2}$ coincide if and only if $\tilde{\mathfrak{M}}_1 = \tilde{\mathfrak{M}}_2$. If $\tilde{\mathfrak{M}}_1 \subset \tilde{\mathfrak{M}}_2$, $\mathfrak{T}_{\tilde{\mathfrak{M}}_1}$ is coarser than $\mathfrak{T}_{\tilde{\mathfrak{M}}_2}$, and the converse also holds.*

Proof. a) In order to show that $\mathfrak{T}_{\mathfrak{M}} = \mathfrak{T}_{\tilde{\mathfrak{M}}}$ it is sufficient to prove that the absolute polars of sets which are obtained from the sets of $\mathfrak{M}$ by the processes of 1), 2) and 3) are again $\mathfrak{T}_{\mathfrak{M}}$-neighbourhoods of $\mathfrak{o}$.

If $M \in \mathfrak{M}$ and $N \subset M$, $(\Gamma N)^\circ \supset (\Gamma M)^\circ$, so that $(\Gamma N)^\circ$ is a $\mathfrak{T}_{\mathfrak{M}}$-neighbourhood of $\mathfrak{o}$. It follows from $(\Gamma \rho M)^\circ = \frac{1}{\rho}(\Gamma M)^\circ$ that $(\Gamma \rho M)^\circ$ is also a $\mathfrak{T}_{\mathfrak{M}}$-neighbourhood of $\mathfrak{o}$. Finally $\overline{\Gamma(M_1, M_2)}^\circ = (\Gamma M_1)^\circ \cap (\Gamma M_2)^\circ$ is a $\mathfrak{T}_{\mathfrak{M}}$-neighbourhood of $\mathfrak{o}$.

b) Suppose that $\mathfrak{N}$ and $\mathfrak{M}$ are saturated, and that $\mathfrak{T}_{\mathfrak{N}} = \mathfrak{T}_{\mathfrak{M}}$. It is sufficient to show that $\mathfrak{N} \subset \mathfrak{M}$. If $N \in \mathfrak{N}$, $(\Gamma N)^\circ$ is a $\mathfrak{T}_{\mathfrak{M}}$-neighbourhood of $\mathfrak{o}$, and so by (3) there is an $M \in \mathfrak{M}$ with $M^{\circ\circ} = M$ and $M^\circ \subset (\Gamma N)^\circ$. It follows from this that $M = M^{\circ\circ} \supset (\Gamma N)^{\circ\circ} \supset N$. N, being a subset of M, is therefore also in $\mathfrak{M}$, so that $\mathfrak{N} \subset \mathfrak{M}$.

We remark that the class $\mathfrak{F}$ of finite subsets of E_2 which we introduced at the beginning is not saturated. The saturated cover $\tilde{\mathfrak{F}}$ clearly consists of all the bounded finite-dimensional subsets of E_2.

2. The strong topology. Among the locally convex topologies $\mathfrak{T}_{\mathfrak{M}}$ on E_1 which can be defined in terms of the dual pair $\langle E_2, E_1 \rangle$ there is a finest one, by 1.(1), namely the topology of uniform convergence on all the weakly bounded subsets of E_2. For brevity, this is called the strong topology $\mathfrak{T}_b(E_2)$ on E_1.

By 1.(3), the collection of all sets B°, where B is a weakly closed bounded absolutely convex subset of E_2, forms a base of neighbourhoods of $\mathfrak{o}$ for the strong topology. It is also given by the system of semi-norms $p_B(x) = \sup_{u \in B} |u x|$.

A subset M of E_1 is bounded with respect to the strong topology if it is strongly bounded in the sense of the definition of § 20, No. 11. The BANACH-MACKEY theorem gives a sufficient condition for the bounded subsets of E_1 to coincide with the strongly bounded sets. This does not always happen, as was shown by an example given in § 20, 11. Since, on the other hand, every locally convex topology compatible with the dual pair defines the same bounded sets as the weak topology, by MACKEY's theorem (§ 20, 11.), there are situations in which the strong topology on E_1 is not compatible, and so has a larger dual space than E_2.

We shall consider this question in more detail in § 23.

If $E[\mathfrak{T}]$ is a locally convex space, we already have an abundance of possible ways of introducing a locally convex topology on the dual space E'. If $\mathfrak{M}$ is any total class of bounded subsets of E with respect to either $\mathfrak{T}$ or $\mathfrak{T}_s(E')$ (cf. § 20, 3.(4)), $E'[\mathfrak{T}_{\mathfrak{M}}]$ is a locally convex space.

In particular we call the space E' with the strong topology $\mathfrak{T}_b(E)$ the s t r o n g d u a l of $E[\mathfrak{T}]$.

In the case of a normed space E, the strong dual has already been introduced in § 14, 5., for the norm introduced there on E' clearly defines the strong topology on E'.

We now give two characterisations of the strong topology.

Following the terminology of BOURBAKI, we call a closed absorbent absolutely convex subset of a locally convex space $E[\mathfrak{T}]$ a b a r r e l. By § 21, 7.(6), we can require the sets to be closed in any compatible topology, e. g. the weak topology.

(1) *Suppose that $\langle E_2, E_1 \rangle$ is a dual pair. The barrels in E_1 form a base of neighbourhoods of* $\mathfrak{o}$ *for the strong topology $\mathfrak{T}_b(E_2)$ on E_1.*

P r o o f. If U is a barrel in E_1, U°, being the polar of an absorbent set, is weakly bounded. By the theorem of bipolars, $U = (U^\circ)^\circ$, so that U is a strong neighbourhood of $\mathfrak{o}$. Conversely every strong neighbourhood B° of $\mathfrak{o}$, where B is weakly bounded and absolutely convex in E_2, is a barrel in E_1.

Following BOURBAKI, a locally convex space $E[\mathfrak{T}]$ is said to be b a r r e l l e d if the barrels form a base of $\mathfrak{T}$-neighbourhoods of $\mathfrak{o}$. From (1), we get:

(2) *The barrelled spaces $E[\mathfrak{T}]$ are those locally convex spaces whose topology $\mathfrak{T}$ coincides with the strong topology $\mathfrak{T}_b(E')$.*

The next result gives the second characterisation of the strong topology:

(3) *If $E[\mathfrak{T}]$ is locally convex, the semi-norms $p(x)$ which are lower semi-continuous with respect to $\mathfrak{T}$ coincide with the $\mathfrak{T}_b(E')$-continuous semi-norms.*

$\mathfrak{T}$ is therefore the strong topology if and only if every lower semi-continuous semi-norm on $E[\mathfrak{T}]$ is continuous.

Proof. By § 6, 2.(3) the lower semi-continuity of a semi-norm $p(x)$ implies that the subsets of $E[\mathfrak{T}]$ defined by $p(x) \leqq \gamma$ are closed. These sets are therefore closed absolutely convex α-bodies, i.e. barrels. Similarly, by § 6, 2.(3) the semi-norm corresponding to a barrel is lower semi-continuous. The assertion therefore follows from (1).

3. The original topology of a locally convex space; separability. We have just given a condition for the original topology $\mathfrak{T}$ of a locally convex space $E[\mathfrak{T}]$ to be the strong topology. We shall now show that in every case $\mathfrak{T}$ is one of the topologies $\mathfrak{T}_{\mathfrak{M}}$ defined in 1.

We recall the concept of an equicontinuous set of mappings (§ 15,13.). The elements u of E', being continuous linear functionals, are continuous mappings from $E[\mathfrak{T}]$ into K. Considering them in this way, we have:

(1) *A subset M of E' is $\mathfrak{T}$-equicontinuous if and only if $M \subset U^\circ$, where U is a suitable absolutely convex neighbourhood of $\mathfrak{o}$ in $E[\mathfrak{T}]$.*

For by § 15, 13.(1) the equicontinuity of M is equivalent to the existence of an absolutely convex $\mathfrak{T}$-neighbourhood U of $\mathfrak{o}$ with $\sup_{u \in M, x \in U} |ux| \leqq 1$, i.e. $M \subset U^\circ$.

(2) *Suppose that $E[\mathfrak{T}]$ is locally convex. Let $\mathfrak{G}$ denote the class of all $\mathfrak{T}$-equicontinuous subsets of E'. $\mathfrak{G}$ is total and saturated in E', and $\mathfrak{T} = \mathfrak{T}_{\mathfrak{G}}$; $\mathfrak{T}$ is therefore equal to the topology of uniform convergence on the $\mathfrak{T}$-equicontinuous subsets of E'.*

For every locally convex topology $\mathfrak{T}_{\mathfrak{M}}$ on E, the saturated cover $\tilde{\mathfrak{M}}$ is the class of all $\mathfrak{T}_{\mathfrak{M}}$-equicontinuous subsets of E'.

Proof. The closed absolutely convex $\mathfrak{T}$-neighbourhoods of $\mathfrak{o}$ form a base of neighbourhoods of $\mathfrak{o}$. Their polars U° are equicontinuous, and, since $U = U^{\circ\circ}$, $\mathfrak{T}$ is the topology of uniform convergence on the sets U°, so that $\mathfrak{T} = \mathfrak{T}_{\mathfrak{G}}$.

$\mathfrak{G}$ is saturated since $\rho U^\circ = \left(\frac{1}{\rho} U\right)$ and $\overline{\Gamma(U_1^\circ, U_2^\circ)} = (U_1 \cap U_2)^\circ$ again belong to $\mathfrak{G}$.

The second assertion follows from 1.(4) and the definition of equicontinuity.

In the special case of normed spaces, this result has already been proved in § 17, 6.(3).

If $\langle E_2, E_1 \rangle$ is a dual pair, every compatible locally convex topology is therefore a topology $\mathfrak{T}_{\mathfrak{M}}$. We saw in 2. that the strong topology need not be compatible, and it is finer than every compatible locally convex topology.

Conversely we know (§ 20, 2.(4)) that the weak topology $\mathfrak{T}_s(E_2)$ is the coarsest compatible locally convex topology on E_1. A locally convex topology on E_1 which is strictly coarser than the weak topology will therefore not be compatible, and will produce a smaller dual than E_2.

We now show that this can happen.

Let N be a weakly total subset of E_2. We denote by $\mathfrak{T}_s(N)$ the topology of uniform convergence on all the finite subsets of N. By 1.(2), $\mathfrak{T}_s(N)$ is locally convex on E_1, and it is coarser than $\mathfrak{T}_s(E_2)$.

In particular if N is a weakly dense linear subspace $H \neq E_2$ of E_2, then $\mathfrak{T}_s(H)$ is simply the weak topology on E_1 defined by the dual pair $\langle H, E_1 \rangle$ (which is a dual pair, since H is dense in E_2). By § 20, 2.(3), the dual of $E_1[\mathfrak{T}_s(H)]$ is H, and so it is smaller than E_2. The topology $\mathfrak{T}_s(H)$ is therefore not compatible with the dual pair $\langle E_2, E_1 \rangle$, and it is coarser than every compatible topology.

This topology can however coincide with the weak topology on certain subsets of E_1:

(3) *Suppose that $E[\mathfrak{T}]$ is locally convex and that M is a $\mathfrak{T}$-equicontinuous subset of E'. Then if N is a weakly total subset of $E[\mathfrak{T}]$ the topologies $\mathfrak{T}_s(E)$ and $\mathfrak{T}_s(N)$ coincide on M.*

Proof. We can take $M = U^\circ$, by (1). By § 20, 9.(4), U° *is* weakly compact. The locally convex, and therefore Hausdorff, topology $\mathfrak{T}_s(N)$ on E' is coarser than the weak topology; it must therefore coincide with it, by § 3, 2.(6).

From (3) we obtain

(4) *If the locally convex space $E[\mathfrak{T}]$ contains a countable total subset N, every equicontinuous subset M of E' is metrizable in the weak topology, and is weakly sequentially separable.*

For $\mathfrak{T}_s(N)$ has a countable base of neighbourhoods of o in E', so that E' is metrizable under $\mathfrak{T}_s(N)$, and, by (3), M is metrizable under the weak topology. Since further M is weakly relatively compact, by § 4, 5.(2) M is weakly separable in the sense that every element of M is limit of a subsequence of some fixed sequence of elements of M.

(5) *If $E[\mathfrak{T}]$ is a separable metrizable locally convex space, E' is weakly sequentially separable.*

For if the absolutely convex sets $U_n, n=1,2,\ldots$, form a countable base of neighbourhoods of o in $E[\mathfrak{T}]$, $E' = \bigcup_{n=1}^{\infty} U_n^\circ$, and each U_n° is weakly sequentially separable, by (4).

Consequently the assertion also holds for E'.

As an application of (5) we prove a result of Banach and Mazur relating to § 20, 10.(1) (cf. Banach [3]):

(6) *Every separable normed space E is norm-isomorphic to a linear subspace of $C(I)$, the space of continuous functions on the interval $I=[0,1]$.*

Proof. The closed unit ball K of E' is a compactum under the weak topology, as in the proof of (4). By a well-known theorem (cf. HAUSDORFF [2], p. 132 or BOURBAKI [5], Vol. 4, p. 31, ex. 18) K is the continuous image of Cantor's ternary set $K_0 \subset I$, which is nowhere dense in I. If the point $t \in K_0$ there corresponds the continuous linear functional $u_t \in K$, we set $x(t)=u_t(x)$. $x(s)$ is defined on the disjoint components of the open set $I \sim K_0$ by linear interpolation.

This function $x(s)$, defined on the whole of I, is continous on I: it is sufficient to prove this on K_0. But if x is a weakly continuous linear functional on K, then, because of the continuity of the mapping from K_0 onto K, x also defines a continuous function on K_0. The correspondence $x \to x(s)$ is clearly linear, and it is also an isometry, since $\|x\| = \sup_{u_t} |u_t x| = \max_{s \in I} |x(s)|$.

4. The Mackey topology. We shall now give a precise characterization of the topologies $\mathfrak{T}_{\mathfrak{M}}$ which are compatible with a dual pair $\langle E', E \rangle$.

By § 20, 9.(4) the polar U° of an absolutely convex neighbourhood U of $\mathfrak{o}$ of a locally convex space $E[\mathfrak{T}]$ is absolutely convex and weakly compact in E'. The class $\mathfrak{G}$ of all $\mathfrak{T}$-equicontinuous subsets of E' is therefore contained in the class $\mathfrak{K}$ of all absolutely convex weakly compact subsets of E', together with their subsets. Since all the sets of $\mathfrak{K}$ are bounded in E', the topology $\mathfrak{T}_k(E')$ of uniform convergence on all the sets of $\mathfrak{K}$ is a locally convex topology on E which is finer than the original topology.

We call $\mathfrak{T}_k(E')$ the Mackey topology on E. Like the strong topology, it depends only on the dual pair $\langle E', E \rangle$, and not on the original topology.

(1) *The class $\mathfrak{K}$ of all absolutely convex weakly compact subsets of E', together with their subsets, is total and saturated, and therefore consists of all the $\mathfrak{T}_k(E')$-equicontinuous subsets of E'.*

A base of neighbourhoods of $\mathfrak{o}$ for the Mackey topology is formed by the sets K°, as K runs through the absolutely convex weakly compact subsets of E'.

The first part follows from the fact that if the sets K_i are weakly compact and absolutely convex then $\Gamma_{i=1}^{n} K_i$ is also weakly compact and absolutely convex, by § 20, 6.(5). The second part follows from 1.(3).

Because of the counterexample in § 20, 6., the class $\mathfrak{K}'$ of all weakly relatively compact subsets of E' is in general larger than $\mathfrak{K}$ and the corresponding topology $\mathfrak{T}_{\mathfrak{K}'}$ is therefore different from the Mackey topology.

We now prove the MACKEY-ARENS theorem (MACKEY [5], ARENS [1]):

(2) *Suppose that* $\langle E_2, E_1\rangle$ *is a dual pair. The Mackey topology* $\mathfrak{T}_k(E_2)$ *is the finest locally convex topology on* E_1 *which has* E_2 *as dual space, and so it is the finest locally convex topology compatible with the dual pair* $\langle E_2, E_1\rangle$.

Proof. We have already shown that every compatible locally convex topology on E_1 is coarser than the Mackey topology.

It remains to show that E_2 is the dual space of $E_1[\mathfrak{T}_k(E_2)]$. We consider the dual pair $\langle E_1^*, E_1\rangle$, where E_1^* is the algebraic dual of E_1. Let u_0 be a $\mathfrak{T}_k$-continuous linear functional on E_1. It lies in E_1^*, and by (1) there is an absolutely convex weakly compact subset K of E_2 for which $\sup_{x\in K^\circ}|u_0 x|\leqq 1$; u_0 therefore belongs to the polar $(K^\circ)^\circ$ of K° in E_1^*. Since $E_2\subset E_1^*$, K is also absolutely convex and weakly compact in E_1^*, and so it is weakly closed. By the theorem of bipolars, applied to the dual pair $\langle E_1^*, E_1\rangle$, we therefore have $K=K^{\circ\circ}$. Thus it follows from $u_0\in K^{\circ\circ}$ that $u_0\in K\subset E_2$. The dual space of $E_1[\mathfrak{T}_k(E_2)]$ is therefore a subspace of E_2. Since conversely every element of E_2 defines a $\mathfrak{T}_k$-continuous linear functional on E_1, (2) is established.

From (2) and § 20, 2.(4) there follows directly

(3) *If* $E[\mathfrak{T}]$ *is a locally convex space, the original topology* $\mathfrak{T}$ *is finer than the weak topology* $\mathfrak{T}_s(E')$ *and coarser than the Mackey topology* $\mathfrak{T}_k(E')$. *The Mackey topology in turn is coarser than the strong topology* $\mathfrak{T}_b(E')$.

All the locally convex topologies on E_1 *which are compatible with the dual pair* $\langle E_2, E_1\rangle$ *are obtained by forming the topologies* $\mathfrak{T}_{\mathfrak{M}}$, *where* $\mathfrak{M}$ *is a class of subsets of* E_2 *with* $\mathfrak{F}\subset\mathfrak{M}\subset\mathfrak{K}$, *where* $\mathfrak{F}$ *is the class of finite subsets of* E_2 *and* $\mathfrak{K}$ *is the class defined above.*

We remark that analogous results were obtained for linearly topologized spaces in § 10; in particular § 10, 11.(4) is the analogue of the MACKEY-ARENS theorem.

(4) *If* $E[\mathfrak{T}]$ *is barrelled,* $\mathfrak{T}$ *coincides with both the strong and the Mackey topologies, and every bounded subset of* E' *is weakly relatively compact.*

For by 2.(2) $\mathfrak{T}$ is the strong topology, and so it is finer than the Mackey topology. On the other hand $\mathfrak{T}$ is compatible with the dual pair $\langle E', E\rangle$, and so $\mathfrak{T}$ must be coarser than the Mackey topology. The identity of the topologies implies the identity of the equicontinuous sets, so that every bounded subset of E' is weakly relatively compact; conversely a weakly relatively compact set is always bounded.

(5) *The completion* $\tilde{E}[\tilde{\mathfrak{T}}]$ *of a locally convex space* $E[\mathfrak{T}]$ *has the same dual space* E' *and the same equicontinuous sets in* E'.

The topologies $\mathfrak{T}_s(E)$ and $\mathfrak{T}_s(\tilde{E})$ coincide on the equicontinuous sets of the dual of a locally convex space $E[\mathfrak{T}]$.

If $\mathfrak{T}$ is the Mackey topology of $E[\mathfrak{T}]$, $\tilde{\mathfrak{T}}$ is the Mackey topology of $\tilde{E}[\tilde{\mathfrak{T}}]$.

Proof. The identity of the duals was proved in § 15, 9.(11).

If U is a closed absolutely convex neighbourhood of o in $E[\mathfrak{T}]$, the closure $\tilde{U}$ in $\tilde{E}$ has the same polar $U^\circ = \tilde{U}^\circ$ in E', and so the $\mathfrak{T}$-equicontinuous and $\tilde{\mathfrak{T}}$-equicontinuous subsets coincide.

By § 20, 9.(4), the polar U° of a neighbourhood U of o in $E[\mathfrak{T}]$ is $\mathfrak{T}_s(E)$-compact, and from what we have just proved it is also $\mathfrak{T}_s(\tilde{E})$-compact; since $\mathfrak{T}_s(\tilde{E})$ is finer than $\mathfrak{T}_s(E)$, the two topologies coincide on U°, by § 3, 2.(6).

Applying this to the Mackey topology $\mathfrak{T}$, the classes of absolutely convex $\mathfrak{T}_s(E)$-compact sets and of absolutely convex $\mathfrak{T}_s(\tilde{E})$-compact sets coincide. But using the first result of (5) this means that $\tilde{\mathfrak{T}}$ is the Mackey topology on $\tilde{E}$.

In analogy to § 10, 12.(2), we have:

(6) *Suppose that $E_1[\mathfrak{T}_1]$ and $E_2[\mathfrak{T}_2]$ are locally convex. A linear mapping from E_1 into E_2 is $\mathfrak{T}_k$-continuous if and only if it is weakly continuous. Every continuous mapping is $\mathfrak{T}_k$-continuous.*

Proof. By § 20, 4.(5) every $\mathfrak{T}_k$-continuous mapping is weakly continuous. Suppose conversely that A is $\mathfrak{T}_s$-continuous. The mapping A' is weakly continuous, by § 20, 4.(6), and it therefore maps every absolutely convex weakly compact subset $K \subset E_2'$ into an absolutely convex weakly compact subset $A'(K) \subset E_1'$. But then it follows from $v(Ax) = (A'v)x$ that the image $A(A'(K)^\circ)$ of the $\mathfrak{T}_k$-neighbourhood $A'(K)^\circ \subset E_1$ lies in the given $\mathfrak{T}_k$-neighbourhood $K^\circ \subset E_2$; A is therefore $\mathfrak{T}_k$-continuous.

Example. It is easy to see that for the dual pair $\langle \varphi, \varphi \rangle$ considered at the end of § 20, 11. $\mathfrak{T}_k(\varphi) = \mathfrak{T}_s(\varphi)$.

5. The topology of a metrizable space. Most of the locally convex spaces met with in applications have a natural original topology $\mathfrak{T}$, which coincides either with the strong or the Mackey topology. Here we shall establish this fact for metrizable spaces.

(1) *Suppose that $E[\mathfrak{T}]$ is locally convex. Every absolutely convex weakly compact subset of E' is strongly bounded.*

This is a special case of the Banach-Mackey theorem (§ 20, 11.).

If we call the topology of uniform convergence on the strongly bounded sets of E_2 the topology $\mathfrak{T}_{b*}(E_2)$, we can formulate (1) in the following way:

(2) *The topology $\mathfrak{T}_{b*}(E_2)$ is always finer than the Mackey topology $\mathfrak{T}_k(E_2)$.*

For metrizable spaces we have

(3) *If the locally convex space* $E[\mathfrak{T}]$ *is metrizable,* $\mathfrak{T}$ *is equal to the Mackey topology* $\mathfrak{T}_k(E')$ *and to the topology* $\mathfrak{T}_{b*}(E')$.

If, further, $E[\mathfrak{T}]$ *is complete, so that it is an* (F)*-space,* $\mathfrak{T}$ *is equal to the strong topology* $\mathfrak{T}_b(E')$*; thus all* (F)*-spaces are barrelled.*

Proof. a) Suppose that $E[\mathfrak{T}]$ is metrizable. By the MACKEY-ARENS theorem, $\mathfrak{T}$ is coarser than $\mathfrak{T}_k(E')$, and $\mathfrak{T}_k(E')$ is coarser than $\mathfrak{T}_{b*}(E')$, by (1). The first part of (3) will therefore be established if we can show that every strongly bounded subset B of E' is $\mathfrak{T}$-equicontinuous, and so if we can show that there is a $\mathfrak{T}$-neighbourhood U of $\mathfrak{o}$ with $B \subset U^\circ$.

The topology $\mathfrak{T}$ is defined by an increasing sequence of semi-norms $\|x\|_n$ on E. We suppose that B is not $\mathfrak{T}$-equicontinuous. Then for each n

$$\sup_{u\in B,\, \|x\|_n \leq 1} |u x| = \infty.$$

There is therefore a sequence $u_n \in B$, $x_n \in E$, $\|x_n\|_n \leq 1$ for which $|u_n x_n| > n$ for $n = 1, 2, \ldots$. It follows from $\sup_{n=1,2,\ldots} \|x_n\|_k \leq \max(\|x_1\|_k, \ldots, \|x_k\|_k, 1) < \infty$ that the x_n form a bounded subset X of E; but then the strong boundedness of B implies that $\sup_{u\in B,\, x_n \in X} |u x_n| < \infty$, which contradicts $|u_n x_n| > n$.

b) If $E[\mathfrak{T}]$ is complete, the strongly bounded and bounded sets in E' coincide, by § 20, 11.(8), so that the two topologies $\mathfrak{T}_b(E')$ and $\mathfrak{T}_{b*}(E')$ are the same.

The second part of (3) can also be expressed in the following way:

(4) *In the dual* E' *of an* (F)*-space* $E[\mathfrak{T}]$ *the following classes of sets coincide: a) the bounded sets, b) the strongly bounded sets, c) the weakly relatively compact sets and d) the equicontinuous sets.*

We now give an example of a normed space E on which the strong and the Mackey topologies are different. By (3) this is the case if there is a bounded set in E' which is not strongly bounded.

As in § 20, 6., let $E[\mathfrak{T}]$ be the space φ equipped with the l^2-norm topology. Then $E' = l^2$, and the strongly bounded sets of E' are the l^2-bounded sets, while the weakly bounded sets of E' are the subsets whose elements $\mathfrak{x} = (\xi_n)$ satisfy inequalities $|\xi_n| \leq M_n$, $n = 1, 2, \ldots$, for arbitrary non-zero M_n.

6. The topology $\mathfrak{T}_c$ of precompact convergence. If $E[\mathfrak{T}]$ is locally convex, the $\mathfrak{T}$-precompact subsets of E form a class of bounded sets which cover E, and so the topology $\mathfrak{T}_c$ of uniform convergence on the precompact subsets of $E[\mathfrak{T}]$ is a locally convex topology on the dual space E'.

In contrast to the topologies $\mathfrak{T}_b(E)$, $\mathfrak{T}_k(E)$ and $\mathfrak{T}_{b*}(E)$, this topology depends upon the original topology, and not just upon the dual pair $\langle E', E\rangle$.

By § 20, 6., the class $\mathfrak{C}$ of all precompact sets in $E[\mathfrak{T}]$ is saturated; by contrast this need not be the case for the class of all relatively compact subsets of E, since the closed absolutely convex cover of a compact set need not be compact.

(1) *If $E[\mathfrak{T}]$ is quasi-complete, the topology $\mathfrak{T}_c(E)$ on E' is coarser than the Mackey topology $\mathfrak{T}_k(E)$, so that $(E'[\mathfrak{T}_c(E)])'$ is again equal to E.*

Proof. If $E[\mathfrak{T}]$ is quasi-complete, every absolutely convex precompact subset of E is relatively compact. Every $\mathfrak{T}$-compact set C is also weakly compact, by § 3, 2.(6), so that $\mathfrak{C} \subset \mathfrak{K}$, where $\mathfrak{K}$ is the class defined in 4. It follows from $\mathfrak{C} \subset \mathfrak{K}$ that $\mathfrak{T}_k$ is finer than $\mathfrak{T}_c$, so that, applying the Mackey-Arens theorem, E is again the dual of $E'[\mathfrak{T}_c]$.

The following theorem gives an important property of the topology $\mathfrak{T}_c$:

(2) *Suppose that $E[\mathfrak{T}]$ is locally convex. On every equicontinuous set $M \subset E'$ the topologies $\mathfrak{T}_s(E)$, $\mathfrak{T}_c(E)$, $\mathfrak{T}_s(\tilde{E})$ and $\mathfrak{T}_c(\tilde{E})$ coincide, where $\tilde{E}[\tilde{\mathfrak{T}}]$ is the completion of $E[\mathfrak{T}]$.*

Proof. a) Suppose that M is equicontinuous and that $u_0 \in M$. In order to establish the identity of $\mathfrak{T}_c(E)$ and $\mathfrak{T}_s(E)$ on M it is sufficient to show that, given a precompact set $C \subset E$, we can find finitely many $x_i \in E$ with the property that, if $u \in M$ and $\sup_{i=1,\dots,n} |(u-u_0)x_i| < 1$, then $\sup_{y \in C} |(u-u_0)y| < 2$.

Because of the equicontinuity of M, and therefore of $M-u_0$, there is a $\mathfrak{T}$-neighbourhood U of $\circ$ with $\sup_{u \in M, z \in U} |(u-u_0)z| < 1$. C is totally bounded, so that there are finitely many $x_i \in C$ with $C \subset \bigcup_{i=1}^{n} (x_i + U)$. Each $y \in C$ therefore takes the form $y = x_i + z$, $z \in U$. For all $u \in M$ with $\sup |(u-u_0)x_i| < 1$ we then have

$$\sup_{y \in C} |(u-u_0)y| \leqq \sup_{i=1,\dots,n} |(u-u_0)x_i| + \sup_{z \in U} |(u-u_0)z| < 2.$$

b) By 4.(5), $\tilde{E}[\tilde{\mathfrak{T}}]$ has the same equicontinuous sets as $E[\mathfrak{T}]$, and $\mathfrak{T}_s(E)$ and $\mathfrak{T}_s(\tilde{E})$ coincide on the equicontinuous sets. The assertion about $\mathfrak{T}_c(\tilde{E})$ follows from this.

The following strengthening of the Alaoglu-Bourbaki theorem follows from (2):

(3) *Every equicontinuous set of the dual E' of a locally convex space $E[\mathfrak{T}]$ is relatively compact with respect to the topologies $\mathfrak{T}_c(E)$ and $\mathfrak{T}_c(\tilde{E})$.*

If U is a $\mathfrak{T}$-neighbourhood of $\circ$, U° is therefore compact with respect to these topologies.

Compare 3.(3), as well!

We know that the strong dual of a normed space is complete (§14, 5. (5)). A more general and stronger result is

(4) *If* $E[\mathfrak{T}]$ *is metrizable and locally convex,* E' *is complete under both the strong topology and the topology* $\mathfrak{T}_c(E)$.

If $E[\mathfrak{T}]$ *is an* (F)*-space,* E' *is also complete under the Mackey topology* $\mathfrak{T}_k(E)$.

Before proving this we establish a lemma due to GROTHENDIECK:

(5) *A linear mapping* A *from a locally convex space* $E[\mathfrak{T}]$ *into a locally convex space* $F[\mathfrak{T}']$ *is uniformly continuous on an absolutely convex subset* $M \subset E$ *if and only if it is continuous at* o *on* M.

Proof. Suppose that A is continuous at o on M; then given an absolutely convex $\mathfrak{T}'$-neighbourhood V of o there is an absolutely convex $\mathfrak{T}$-neighbourhood U of o with $A(M \cap U) \subset \frac{1}{2} V$. If x and y are two elements of M with $x-y \in U$, then $x-y \in 2M$, since M is absolutely convex. From this it follows that $x-y \in (2M) \cap U = 2(M \cap \frac{1}{2} U)$. Consequently $\frac{x-y}{2} \in M \cap U$, so that $A\left(\frac{x-y}{2}\right) \in \frac{1}{2} V$, $A(x-y) \in V$. This means that $A(x-y) \in V$ if $x-y \in U$, so that A is uniformly continuous on M.

In particular a linear functional is $\mathfrak{T}$-continuous on M if it is $\mathfrak{T}$-continuous at o on M.

Proof of (4). a) First we show that $E'[\mathfrak{T}_c(E)]$ is complete. A $\mathfrak{T}_c$-Cauchy filter $\mathfrak{F} = \{F^\alpha\}$ on E' is certainly a $\mathfrak{T}_s(E)$-Cauchy filter, and so by § 20, 9. (2) it has a $\mathfrak{T}_s$-adherent point $u_0 \in E^*$. If $\varepsilon > 0$ and C is a precompact subset of E, then $\sup_{x \in C} |(u'-u)x| \leqq \frac{\varepsilon}{2}$ for all u, u' in a suitable F^α, and so

(6) $$\sup_{x \in C} |(u_0 - u)x| \leqq \frac{\varepsilon}{2} \quad \text{for} \quad u \in F^\alpha.$$

u_0 is therefore also a $\mathfrak{T}_c$-adherent point of $\mathfrak{F}$.

We now show that u_0 is $\mathfrak{T}$-continuous on every absolutely convex precompact set, and so on every precompact set.

Each u is $\mathfrak{T}$-continuous at o in C, and so for any one u in F^α there is a $\mathfrak{T}$-neighbourhood U of o with $\sup_{x \in U \cap C} |ux| \leqq \frac{\varepsilon}{2}$; (6) then shows that $\sup_{x \in U \cap C} |u_0 x| \leqq \varepsilon$, so that u_0 is $\mathfrak{T}$-continuous at o in C. It then follows from (5) that u_0 is $\mathfrak{T}$-continuous on C.

If u_0 were not continuous on E, then since $E[\mathfrak{T}]$ is metrizable, there would be a sequence $x_n \in E$ converging to zero with $u_0(x_n) \not\to 0$. But the subset C_0 of E consisting of o and the x_n is compact, and u_0 would not be continuous on this set C_0, contradicting what we have just proved.

Consequently u_0 belongs to E', and $E'[\mathfrak{T}_c(E)]$ is complete.

b) The strong topology $\mathfrak{T}_b(E)$ is finer than $\mathfrak{T}_c(E)$ on E'. The polars B° of the bounded subsets B of E form a base of neighbourhoods of $\mathfrak{o}$ for the strong topology, and the sets B° are weakly closed, and *a fortiori* $\mathfrak{T}_c(E)$-closed, so that by § 18, 4.(4) b) E' is strongly complete as well.

If $E[\mathfrak{T}]$ is an (F)-space, the topology $\mathfrak{T}_k(E)$ on E' is finer than $\mathfrak{T}_c(E)$, by (1), and so E' is also $\mathfrak{T}_k$-complete.

7. Polar topologies. We have the following theorem of GROTHENDIECK:

(1) *Suppose that $\langle E_2, E_1\rangle$ is a dual pair, that $\mathfrak{M}$ is a saturated collection of bounded subsets M of E_1 which cover E_1, and that $\mathfrak{N}$ is a similar collection in E_2.*

The four following statements are equivalent:

α) Each $M\in\mathfrak{M}$ is $\mathfrak{T}_{\mathfrak{N}}$-precompact.

β) Each $N\in\mathfrak{N}$ is $\mathfrak{T}_{\mathfrak{M}}$-precompact.

γ) The topologies $\mathfrak{T}_{\mathfrak{N}}$ and $\mathfrak{T}_s$ coincide on each $M\in\mathfrak{M}$.

δ) The topologies $\mathfrak{T}_{\mathfrak{M}}$ and $\mathfrak{T}_s$ coincide on each $N\in\mathfrak{N}$.

Proof. The sets $N\in\mathfrak{N}$ are $\mathfrak{T}_{\mathfrak{N}}$-equicontinuous subsets in $(E_1[\mathfrak{T}_{\mathfrak{N}}])'\supset E_2$. By 6.(2) the topology $\mathfrak{T}_s$ coincides with the topology of $\mathfrak{T}_{\mathfrak{N}}$-precompact convergence on N. If α) holds, $\mathfrak{T}_{\mathfrak{M}}$ is coarser than this last topology, and, since $\bigcup_{M\in\mathfrak{M}} M=E_1$, it is finer than $\mathfrak{T}_s(E_1)$; $\mathfrak{T}_{\mathfrak{M}}$ and $\mathfrak{T}_s$ therefore coincide on N. Thus δ) follows from α). We can however derive β) from δ), for the $\mathfrak{T}_{\mathfrak{N}}$-equicontinuous subsets N are weakly bounded in $(E_1[\mathfrak{T}_{\mathfrak{N}}])'$ and are therefore weakly precompact in E_2 by § 20, 9.(3); they are therefore also $\mathfrak{T}_{\mathfrak{M}}$-precompact, since $\mathfrak{T}_s$ and $\mathfrak{T}_{\mathfrak{M}}$ are the same on the sets N.

Because of the symmetry of the situation, γ) now follows from β), and α) from γ).

As before let $\mathfrak{N}$ be a saturated collection of bounded sets covering E_2, and let $\mathfrak{T}$ be the topology $\mathfrak{T}_{\mathfrak{N}}$ on E_1. We call the topology $\mathfrak{T}_c(E_1[\mathfrak{T}])$ on E_2 the topology $\mathfrak{T}^\circ$ polar to $\mathfrak{T}$.

From (1) there follows

(2) a) *The topology $\mathfrak{T}^\circ$ polar to $\mathfrak{T}=\mathfrak{T}_{\mathfrak{N}}$ is the finest locally convex topology of the form $\mathfrak{T}_{\mathfrak{M}}(E_1)$ on E_2 which coincides with $\mathfrak{T}_s(E_1)$ on all the $\mathfrak{T}$-equicontinuous subsets N of E_2.*

b) *The topology $\mathfrak{T}^{\circ\circ}$ bipolar to $\mathfrak{T}$ on E_1 is always finer than $\mathfrak{T}$, and it is equal to $\mathfrak{T}$ if and only if the collection of all the $\mathfrak{T}$-equicontinuous subsets of E_2 consists of all the $\mathfrak{T}^\circ$-precompact sets.*

c) *If $\mathfrak{T}_1$ is finer than $\mathfrak{T}_2$ on E_1, then $\mathfrak{T}_1^\circ$ is coarser than $\mathfrak{T}_2^\circ$.*

d) $\mathfrak{T}^{\circ\circ\circ}=\mathfrak{T}^\circ$.

Proof. a) $\mathfrak{T}^\circ$ is a topology which coincides with $\mathfrak{T}_s(E_1)$ on the $\mathfrak{T}$-equicontinuous subsets $N \in \mathfrak{N}$. On the other hand if $\mathfrak{T}' = \mathfrak{T}_{\mathfrak{M}}$ is a second such topology on E_2, then δ) holds. Theorem (1) then implies that α) holds; i. e. each $M \in \mathfrak{M}$ is $\mathfrak{T}$-precompact, and so $\mathfrak{T}_{\mathfrak{M}}$ is coarser than $\mathfrak{T}^\circ$.

In particular, putting $\mathfrak{T}_{\mathfrak{M}} = \mathfrak{T}^\circ$, the class of $\mathfrak{T}^\circ$-equicontinuous subsets of E_1 consists of the collection of all $\mathfrak{T}$-precompact subsets of E_1.

b) The collection $\mathfrak{N}$ of all the $\mathfrak{T}$-equicontinuous subsets of E_2 consists of $\mathfrak{T}^\circ$-precompact sets, while the collection $\mathfrak{N}'$ of all the $\mathfrak{T}^{\circ\circ}$-equicontinuous sets consists of all the $\mathfrak{T}^\circ$-precompact sets. It therefore follows from $\mathfrak{N} \subset \mathfrak{N}'$ that $\mathfrak{T}^{\circ\circ}$ is finer than $\mathfrak{T}$.

c) This assertion is immediately obvious.

d) On the one hand $\mathfrak{T}^{\circ\circ\circ} = (\mathfrak{T}^\circ)^{\circ\circ}$ is finer than $\mathfrak{T}^\circ$, by b), and on the other $\mathfrak{T}^{\circ\circ}$ is finer than $\mathfrak{T}$, again by b), so that $\mathfrak{T}^{\circ\circ\circ} = (\mathfrak{T}^{\circ\circ})^\circ$ is coarser than $\mathfrak{T}^\circ$, by c).

(3) *Suppose that $\langle E_2, E_1 \rangle$ is a dual pair. The topology polar to the weak topology $\mathfrak{T}_s(E_2)$ is the strong topology $\mathfrak{T}_b(E_1)$, and the topology polar to this is the topology of strongly precompact convergence.*

For since the weakly precompact subsets of E_1 coincide with the bounded sets, $\mathfrak{T}_s(E_2)^\circ$ is equal to $\mathfrak{T}_b(E_1)$.

The topology bipolar to the weak topology is therefore different from the weak topology, in general; in contrast the strong topology, being the topology $\mathfrak{T}_s(E_2)^\circ$, is its own bipolar, by (2)d).

8. The topologies $\mathfrak{T}^f$ and $\mathfrak{T}^{lf}$. Suppose that $\langle E_2, E_1 \rangle$ is a dual pair, and that $\mathfrak{T}$ is a topology $\mathfrak{T}_{\mathfrak{N}}(E_2)$ on E_1 lying between $\mathfrak{T}_s(E_2)$ and $\mathfrak{T}_b(E_2)$. We saw in 7. that $\mathfrak{T}^\circ = \mathfrak{T}_c(E_1[\mathfrak{T}])$ is the finest topology $\mathfrak{T}_{\mathfrak{M}}(E_1)$ on E_2, defined by a saturated class of bounded subsets of E_1 covering E_1, which coincides with the weak topology $\mathfrak{T}_s(E_1)$ on the $\mathfrak{T}$-equicontinuous subsets of E_2.

We now denote by $\mathfrak{T}^f$ (respectively $\mathfrak{T}^{lf}$) the finest topology (respectively the finest locally convex topology) on E_2 which coincides with $\mathfrak{T}_s(E_1)$ on the $\mathfrak{T}$-equicontinuous subsets of E_2.

As we shall see later, $\mathfrak{T}^f$ is generally different from $\mathfrak{T}^{lf}$, and so need not be locally convex. However, we have:

(1) *The topology $\mathfrak{T}^f$ on E_2 is Hausdorff and translation-invariant, and it has a base of circled absorbent neighbourhoods of* $\mathfrak{o}$.

Proof. The topology $\mathfrak{T}^f$ is determined by giving the collection of $\mathfrak{T}^f$-closed subsets of E_2. Let $\mathfrak{A}$ be the collection of all subsets A of E_2 whose intersections $A \cap M$ with all the $\mathfrak{T}$-equicontinuous subsets M of E_2 are weakly closed in M. It is easy to see that this collection $\mathfrak{A}$ satisfies the axioms for the closed sets of a topology $\mathfrak{T}_0$ (§ 1, 1.). $\mathfrak{T}_0$

coincides with $\mathfrak{T}_s(E_1)$ on the sets M, since it gives the same closed sets in M as the weak topology does. On the other hand A is the most extensive class of subsets of E_2 for which this is the case, and so $\mathfrak{T}_0$ is equal to $\mathfrak{T}^f$.

The open $\mathfrak{T}^f$-neighbourhoods of o are the sets U whose intersections $U \cap M$ with the $\mathfrak{T}$-equicontinuous sets M containing o are weak open neighbourhoods of o in M. But then the sets x_0+U are the open $\mathfrak{T}^f$-neighbourhoods of x_0, for x_0+M is $\mathfrak{T}$-equicontinuous if M is, and so $(x_0+U)\cap(x_0+M)=x_0+U\cap M$ is a weak open x_0-neighbourhood in x_0+M. The topology $\mathfrak{T}^f$ is therefore translation-invariant.

The open $\mathfrak{T}^f$-neighbourhoods U of o form a filter base, and the intersection of all the sets U reduces to the point o, since this is already so for the weak neighbourhoods of o. Each U is absorbent, for every $u \in E_2$ belongs to some absolutely convex $\mathfrak{T}$-equicontinuous set M, and so $U \cap M$ contains a suitable scalar multiple of u. Finally each U contains a circled open $\mathfrak{T}^f$-neighbourhood of o. To prove this, let V be the set of all those x for which the set X of all αx, $|\alpha| \leqq 1$, lies in U. If M is an absolutely convex $\mathfrak{T}$-equicontinuous set, $V \cap M$ is a weak neighbourhood of o in M contained in $U \cap M$. We show that $V \cap M$ is weakly open in M, which implies that V is a circled $\mathfrak{T}^f$-open neighbourhood of o. If $x \in V \cap M$, X is weakly compact in the open subset $U \cap M$ of M. By § 15, 6.(9) there is an absolutely convex weak neighbourhood W of o in E_2 with $(X+W)\cap M \subset U \cap M$. But then if $y \in (x+W)\cap M$, αy lies in $U \cap M$ for every $|\alpha| \leqq 1$, so that $y \in V \cap M$; $V \cap M$ is therefore weakly open in M.

Kōmura [2] has given an example to show that $E_2[\mathfrak{T}^f]$ need not be a topological vector space.

A base of neighbourhoods of o for the topology $\mathfrak{T}^{lf}$ is clearly obtained by taking all the absolutely convex $\mathfrak{T}^f$-neighbourhoods of o; a base is also obtained by taking all the convex $\mathfrak{T}^f$-neighbourhoods of o, for by (1) every $\mathfrak{T}^f$-neighbourhood of o contains a circled $\mathfrak{T}^f$-neighbourhood of o.

We now give an example due to Klee [2] III to show that $\mathfrak{T}^f$ and $\mathfrak{T}^{lf}$ can be different.

(2) *Suppose that E_1 is the space ω_d, that E_2 is φ_d and that $\mathfrak{T}$ is the weak topology $\mathfrak{T}_s(\varphi_d)$ on ω_d. $\mathfrak{T}^f$ and $\mathfrak{T}^{lf}$ coincide on E_2 if and only if $d \leqq \aleph_0$.*

Proof. The $\mathfrak{T}$-equicontinuous sets in φ_d are the bounded finite-dimensional subsets of φ_d. Thus $\mathfrak{T}^{lf}$ is the finest locally convex topology on φ_d, and $\mathfrak{T}^f$ is the topology whose open neighbourhoods of o consist of those sets whose intersections with the finite-dimensional bounded sets M containing o are open neighbourhoods of o in M.

The open $\mathfrak{T}^f$-neighbourhoods of $\circ$ can also be defined as the sets U whose intersections with the finite-dimensional linear subspaces H of φ_d are open neighbourhoods of $\circ$ in H.

Suppose that $d=\aleph_0$. Once again we denote by φ_n the linear subspace of φ consisting of the vectors $\mathfrak{x}=\{\xi_1,\ldots,\xi_n,0,0,\ldots\}$. Every $\mathfrak{T}^f$-neigbourhood of $\circ$ contains one of the form $U=\bigcup_{n=1}^{\infty} U_n$, where U_n is an open neighbourhood of $\circ$ in φ_n and $U_n \subset U_{n+1}$. U_1 contains a compact absolutely convex neighbourhood K_1 of $\circ$. Since $K_1 \subset U_1$, there is a compact absolutely convex neighbourhood K_2 of $\circ$ with $K_1+K_2 \subset U_2$. Repeating this process we obtain by § 16, 1.(3) an absolutely convex $\mathfrak{T}^{lf}$-neighbourhood of $\circ$ $K_1+K_2+\cdots \subset U$. $\mathfrak{T}^f$ and $\mathfrak{T}^{lf}$ are therefore the same on φ.

Now suppose that $d>\aleph_0$. In order to show that the topologies $\mathfrak{T}^f$ and $\mathfrak{T}^{lf}$ are different on φ_d, it is sufficient to give a convex $\mathfrak{T}^f$-closed set C which does not contain $\circ$, but for which $\circ$ is a $\mathfrak{T}^{lf}$-closure point.

We have constructed such a set in § 17, 5.(4). We need only set the space E considered there equal to φ_d. The set C constructed there intersects every finite-dimensional linear subspace in a closed convex polyhedron, and so it is $\mathfrak{T}^f$-closed. Further since the intersection $C' \cap C$ with every convex α-body C' containing $\circ$ is non-empty, $\circ$ is a $\mathfrak{T}^{lf}$-closure point of C.

Finally, using an example of COLLINS [1], we show that in general $\mathfrak{T}^{lf}$ is different from $\mathfrak{T}^\circ$.

Let E be a (B)-space, and let E' be its dual. We consider the dual pair $\langle E,E'\rangle$ and the topology $\mathfrak{T}=\mathfrak{T}_s(E')$ on E'. Then by 7.(3) $\mathfrak{T}^\circ$ is equal to $\mathfrak{T}_b(E')$, and so is the norm topology on E. The topology $\mathfrak{T}^{lf}$ is however the finest locally convex topology on E, and so in general it is different from $\mathfrak{T}^\circ$ (cf. § 18, 5.(5)).

We shall determine the topology $\mathfrak{T}^{lf}$ more precisely in the next number.

9. GROTHENDIECK's construction of the completion. So far we have become acquainted with two methods of constructing the completion $\tilde{E}[\tilde{\mathfrak{T}}]$ of a locally convex space $E[\mathfrak{T}]$, one depending upon the construction of the completion of a uniform space (§ 15, 3.) and the other using the embedding of $E[\mathfrak{T}]$ in a topological product of (B)-spaces (§ 18, 4.).

We now consider a third method, due to GROTHENDIECK [1]; this method has already been used in the linear topological case (cf. § 13, 3.).

(1) *Suppose that $\langle E_2,E_1\rangle$ is a dual pair and that $\mathfrak{T}$ is a locally convex topology on E_1 lying between $\mathfrak{T}_s(E_2)$ and $\mathfrak{T}_b(E_2)$. Then the dual space of $E_2[\mathfrak{T}^f]$, and of $E_2[\mathfrak{T}^{lf}]$, consists of all linear functionals on E_2 whose restrictions to the $\mathfrak{T}$-equicontinuous subsets of E_2 are $\mathfrak{T}_s(E_1)$-continuous.*

Proof. Suppose that u is a linear functional on E_2 whose restrictions to the $\mathfrak{T}$-equicontinuous sets M containing $\circ$ are weakly continuous. If U_ε is the set of all $x \in E_2$ with $|ux|<\varepsilon$, $U_\varepsilon \cap M$ is an open $\mathfrak{T}_s(E_1)$-neighbourhood of $\circ$ in M. By the proof of 8.(1), U_ε is therefore an open $\mathfrak{T}^f$-neighbourhood, and, since U_ε is absolutely convex, it is also a

$\mathfrak{T}'^f$-neighbourhood of $\mathfrak{o}$. u is therefore $\mathfrak{T}^f$- and $\mathfrak{T}'^f$-continuous at $\mathfrak{o}$, and consequently continuous throughout E_2. Conversely, the restriction of a $\mathfrak{T}^f$- or $\mathfrak{T}'^f$-continuous linear functional to any M is clearly weakly continuous.

We now have the following theorem of GROTHENDIECK:

(2) *Suppose that* $\langle E_2, E_1\rangle$ *is a dual pair, and that* $\mathfrak{M}$ *is a saturated collection of bounded subsets of* E_2 *which cover* E_2. *Then the completion of* $E_1[\mathfrak{T}_{\mathfrak{M}}]$ *consists of all the linear functionals on* E_2 *whose restrictions to the sets* $M\in\mathfrak{M}$ *are weakly continuous.*

Expressed differently, *if we form the topology* $\mathfrak{T}^f$ *or* $\mathfrak{T}'^f$ *on* E_2 *corresponding to the topology* $\mathfrak{T}=\mathfrak{T}_{\mathfrak{M}}$ *on* E_1, *then* $(E_2[\mathfrak{T}^f])'$ *and* $(E_2[\mathfrak{T}'^f])'$ *are equal to* $\tilde{E}_1[\tilde{\mathfrak{T}}]$ as vector spaces.

Proof. [PELAYO HENRIQUES]. Let $\hat{E}_1$ be the space of all linear functionals on E_2 with weakly continuous restrictions to the sets $M\in\mathfrak{M}$, and let $\tilde{E}_1$ be the completion of $E_1[\mathfrak{T}_{\mathfrak{M}}]$. $\hat{E}_1\subset E_2^*$, and $\tilde{E}_1\subset E_2^*$, by § 20, 9.(2) and § 18, 4.(4).

We show that $\tilde{E}_1\subset\hat{E}_1$. All the $M\in\mathfrak{M}$ are equicontinuous subsets of $E_1'=E_1[\mathfrak{T}_{\mathfrak{M}}]'$ and by 4.(5) the topologies $\mathfrak{T}_s(E_1)$ and $\mathfrak{T}_s(\tilde{E}_1)$ coincide on M. But then every $z\in\tilde{E}_1\subset E_2^*$ is a linear functional on $E_2\subset E_1'$ whose restrictions to all the sets M are weakly continuous, so that z lies in $\hat{E}_1$.

Secondly we show that E_1 is $\mathfrak{T}_{\mathfrak{M}}$-dense in $\hat{E}_1$. This implies that $\hat{E}_1\subset\tilde{E}_1$, and so $\tilde{E}_1=\hat{E}_1$.

$\langle E_2,\hat{E}_1\rangle$ is a dual pair, and from the definition of $\hat{E}_1$ the topologies $\mathfrak{T}_s(E_1)$ and $\mathfrak{T}_s(\hat{E}_1)$ coincide on the sets $M\in\mathfrak{M}$. Thus the sets $M\in\mathfrak{M}$ are also $\mathfrak{T}_s(\hat{E}_1)$-bounded, and $\hat{E}_1[\mathfrak{T}_{\mathfrak{M}}]$ is locally convex. It is sufficient to show that every $v\in\hat{E}_1'=\hat{E}_1[\mathfrak{T}_{\mathfrak{M}}]'$ that vanishes on E_1 vanishes on the whole of $\hat{E}_1$.

Suppose that v is an arbitrary element of $\hat{E}_1'$. Then there is an absolutely convex $M\in\mathfrak{M}$ with $\sup_{z\in M^\circ}|vz|\leqq 1$, i.e. $v\in(M^\circ)^\circ$, where the polar of $M^\circ\subset\hat{E}_1$ is formed in $\hat{E}_1'$. $(M^\circ)^\circ$ is the $\mathfrak{T}_s(\hat{E}_1)$-closure of M in $\hat{E}_1'$. There is therefore a $\mathfrak{T}_s(\hat{E}_1)$-Cauchy filter $\mathfrak{F}$ on M with $v=\lim\mathfrak{F}$. Since $\mathfrak{T}_s(\hat{E}_1)$ and $\mathfrak{T}_s(E_1)$ coincide on M, $\mathfrak{F}$ is also a $\mathfrak{T}_s(E_1)$-Cauchy filter on M. As such it has a limit v_0 in E_1^*, and v_0 is the restriction of v to E_1, since if $x\in E_1$, $v_0x=\lim\mathfrak{F}x=vx$. If now $v_0=\mathfrak{o}$, $\mathfrak{F}$ converges on M to $\mathfrak{o}\in M$. But this also holds for $\mathfrak{F}$ considered as a $\mathfrak{T}_s(\hat{E}_1)$-Cauchy filter on M, and so $v=\mathfrak{o}$ as well.

The following completeness criterion is an immediate consequence of (2):

(4) *The locally convex space* $E[\mathfrak{T}]$ *is complete if and only if every linear functional on* E' *whose restrictions to the equicontinuous sets on* E' *are weakly continuous is weakly continuous on the whole of* E'.

If we apply 3.(4), we obtain from (2):

(5) *If the complete locally convex space* $E[\mathfrak{T}]$ *contains a countable total subset, for example if* $E[\mathfrak{T}]$ *is a separable* (F)*-space, then every weakly sequentially continuous linear functional on* E' *is weakly continuous.*

(4) characterises the completeness of a locally convex space in terms of duality. (4) can be expressed in a somewhat different form which goes back to V. PTÁK [1] and H. S. COLLINS [1]:

(6) *The locally convex space* $E[\mathfrak{T}]$ *is complete if and only if every* $\mathfrak{T}^f$*-closed linear subspace* H *of codimension* 1 *in* E' *is also weakly closed, i. e. if and only if the fact that* $H \cap M$ *is weakly closed for every weakly closed* $\mathfrak{T}$*-equicontinuous subset* M *of* E' *implies that* H *is weakly closed.*

A hyperplane H *in* E' *is* $\mathfrak{T}^f$*-closed if and only if it is* $\mathfrak{T}^{lf}$*-closed.*

We shall prove the last assertion first. Suppose that the $\mathfrak{T}^f$-closed hyperplane H is given by $z(u)=0$, $z \in (E')^*$, $u \in E'$, and suppose that $u_0 \notin H$. By 8.(1) there is a circled $\mathfrak{T}^f$-neighbourhood U of o with $(u_0+U)\cap H$ empty. $z(U)$ is circled and, as it is a subset of P (respectively Γ), it is therefore absolutely convex, so that $z(\ulcorner U)=z(U)$. Since $0 \notin z(u_0+U)$ it follows that $0 \notin z(u_0+\ulcorner U)$, and so the $\mathfrak{T}^{lf}$-neighbourhood $u_0+\ulcorner U$ of u_0 does not meet H; H is therefore also $\mathfrak{T}^{lf}$-closed.

It is therefore sufficient to prove (6) with $\mathfrak{T}^{lf}$ in place of $\mathfrak{T}^f$.

By § 15, 9.(1), the $\mathfrak{T}^{lf}$-closed hyperplanes through o are precisely the null-spaces of the $\mathfrak{T}^{lf}$-continuous linear functionals on E'. By (4) the $\mathfrak{T}^{lf}$-continuous linear functionals are weakly continuous, i. e. their null-spaces are weakly closed, if and only if $E[\mathfrak{T}]$ is complete.

We can now determine the topology $\mathfrak{T}^{lf}$ more precisely:

(7) *Suppose that* $E[\mathfrak{T}]$ *is a locally convex space, and that* $\tilde{E}[\tilde{\mathfrak{T}}]$ *is its completion. The topology* $\mathfrak{T}^{lf}$ *on* E' *corresponding to* $\mathfrak{T}$ *is the topology* $\tilde{\mathfrak{T}}^\circ$ *of uniform convergence on the compact subsets of* $\tilde{E}$.

For it follows by 3.(2) from $(E'[\mathfrak{T}^{lf}])'=\tilde{E}$ that $\mathfrak{T}^{lf}$ is the topology $\mathfrak{T}_{\mathfrak{N}}$ of uniform convergence on some saturated collection $\mathfrak{N}$ of bounded subsets of $\tilde{E}$ which cover $\tilde{E}$. Since by 6.(2) the topologies $\mathfrak{T}_s(E)$ and $\mathfrak{T}_s(\tilde{E})$ coincide on the $\mathfrak{T}$-equicontinuous subsets M of E', $\mathfrak{T}^{lf}$ is also the finest topology $\mathfrak{T}_{\mathfrak{N}}$ which coincides with $\mathfrak{T}_s(\tilde{E})$ on the sets M. But by 7.(2)a) this is $\tilde{\mathfrak{T}}^\circ$.

(8) *Suppose that* $E[\mathfrak{T}]$ *is a locally convex space. The topologies* $\mathfrak{T}^\circ$ *and* $\mathfrak{T}^{lf}$ *on* E' *are the same if and only if every compact subset of* $\tilde{E}[\tilde{\mathfrak{T}}]$ *lies in the closure of a precompact subset of* $E[\mathfrak{T}]$.

A base of $\mathfrak{T}^\circ$-neighbourhoods of o in E' is obtained by taking all the sets C°, where C is absolutely convex and precompact in $E[\mathfrak{T}]$; it is also obtained by taking all the sets $(C^{\circ\circ})^\circ$, where $C^{\circ\circ}$ is the compact cover of C in $\tilde{E}[\tilde{\mathfrak{T}}]$. On the other hand the sets K°, with K compact in

$\tilde{E}[\tilde{\mathfrak{T}}]$, form a base of $\mathfrak{T}'^f$-neighbourhoods of o in E'. The two topologies are therefore equal if and only if each K lies in a $C^{\circ\circ}$.

10. The Banach-Dieudonné theorem. Interest in the topologies $\mathfrak{T}^f$ and $\mathfrak{T}'^f$ goes back to a classical result of Banach about (B)-spaces. A new proof was given by Dieudonné, using a method which can be applied to metrizable spaces.

The Banach-Dieudonné theorem states:

(1) *If $E[\mathfrak{T}]$ is locally convex and metrizable, $\mathfrak{T}^\circ$ and $\mathfrak{T}^f$ coincide on E'.*

Expressed in another way, *the topology $\mathfrak{T}_c(E)$ of precompact convergence is the finest topology on E' which coincides with the weak topology on all the equicontinuous subsets of E'.*

Proof. We must show that every open $\mathfrak{T}^f$-neighbourhood V of o is a $\mathfrak{T}^\circ$-neighbourhood, and so that there is a precompact subset C in E with $C^\circ \subset V$.

Let $U_1 \supset U_2 \supset \ldots$ be a sequence of absolutely convex neighbourhoods of o in $E[\mathfrak{T}]$ which form a base of neighbourhoods of o. We set $U_0 = E$.

We need the following lemma:

(2) *For each $n>0$ there is a finite set $F_{n-1} \subset U_{n-1}$ with the property that if $C_n = \bigcup_{p=0}^{n-1} F_p$ then the set $U_n^\circ \cap C_n^\circ$ is contained in V.*

We prove this using complete induction. Suppose that the sets F_p have been determined for $p<n$ in such a way that $U_n^\circ \cap C_n^\circ \subset V$ for $C_n = \bigcup_{p=0}^{n-1} F_p$.

We set $D_n = U_{n+1}^\circ \cap (E' \sim V)$. U_{n+1}° is weakly compact, and $\mathfrak{T}^f$ induces the weak topology on U_{n+1}°. Since V is open, $E' \sim V$ is $\mathfrak{T}^f$-closed, so that D_n is weakly closed, and is therefore weakly compact.

We suppose that there is no finite subset F_n of U_n with the required properties. Then for every finite subset F of U_n $(C_n \cup F)^\circ \cap U_{n+1}^\circ$ does not lie in V, so that $C_n^\circ \cap F^\circ \cap D_n$ is not empty. All the sets of the form $C_n^\circ \cap F^\circ \cap D_n$ are weakly compact, and they form a filter base on D_n, since the intersection of finitely many sets of this form is again of this form. Since D_n is weakly compact all these sets $C_n^\circ \cap F^\circ \cap D_n$ must have a common element u_0; this must lie in $C_n^\circ \cap U_n^\circ \cap D_n$, which contradicts the relation $C_n^\circ \cap U_n^\circ \subset V$. Thus there exists an $F_n \subset U_n$ with the required property, and so the lemma is proved.

We now prove (1). We form the set $C = \bigcup_{p=1}^{\infty} F_p$. It is relatively compact in $E[\mathfrak{T}]$, since every subsequence in C converges to o, because of the conditions on the sets U_n. Further $C^\circ \subset C_n^\circ$, so that $C^\circ \cap U_n^\circ \subset V$ for all n. Since E' is the union of the sets U_n°, $C^\circ \subset V$.

(3) *Every precompact subset of a metrizable locally convex space $E[\mathfrak{T}]$ lies in the closed absolutely convex cover of a sequence convergent to* $\mathfrak{o}$ *in* $E[\mathfrak{T}]$.

The topology $\mathfrak{T}_c(E)$ therefore coincides with the topology of uniform convergence on the sequences which converge to $\mathfrak{o}$ *in* $E[\mathfrak{T}]$.

This is contained in the proof of (2): $\mathfrak{T}_c(E)$ is coarser than $\mathfrak{T}^f$, and so, by this proof, given a $\mathfrak{T}_c$-neighbourhood K° of $\mathfrak{o}$, with K precompact in E, there is a set C, consisting of a sequence convergent to $\mathfrak{o}$, with $C^\circ \subset K^\circ$; by the theorem of bipolars, $C^{\circ\circ} = \ulcorner\!\!\urcorner(C) \supset K$.

Using the definition of $\mathfrak{T}^f$ and the ideas contained in the proof of 8.(1), we obtain the following direct consequence of (1):

(4) *A subset M of the dual E' of a metrizable locally convex space $E[\mathfrak{T}]$ is $\mathfrak{T}_c$-closed if and only if $M \cap B$ is weakly closed for every weakly closed equicontinuous subset B of E'.*

The next result is a special case of (4):

(5) *A convex subset M of the dual E' of an* (F)*-space $E[\mathfrak{T}]$ is weakly closed if and only if the intersection $M \cap B$ is weakly closed for every weakly closed bounded subset B of E'.*

For, by 6.(1), $\mathfrak{T}_c(E)$ is coarser than $\mathfrak{T}_k(E)$, so that it is compatible with the dual pair $\langle E', E\rangle$; by § 20.7.(6), M is $\mathfrak{T}_c$-closed if and only if it is weakly closed, so that (5) follows from (4).

In addition, we record the classical result for (B)-spaces:

(6) *If E is a* (B)*-space and if H is a linear subspace of E', then H is weakly closed if and only if $H \cap K$ is weakly closed, where K is the closed unit ball in E'.*

In the separable case we have

(7) *If $E[\mathfrak{T}]$ is a separable* (F)*-space, a convex subset of E' is weakly closed if and only if it is weakly sequentially closed.*

This follows from (5), for by 3.(4) every bounded subset of E' is metrizable in the weak topology.

11. Real and complex locally convex spaces. We have developed the theory of locally convex spaces in the real and the complex case simultaneously. It is occasionally desirable to be able to carry a theorem proved for real locally convex spaces over to the complex case. We shall prove a theorem which in many cases makes this possible.

Suppose that $E[\mathfrak{T}]$ is a complex locally convex space. We can also consider it as a real locally convex space (conversely, it is not always possible to consider a real locally convex space as a complex one, cf. DIEUDONNÉ [8]). Let E' be the complex dual, E'_r the real dual. By § 16, 3.(1), $E'_r = \Re(E')$, i.e. E'_r consists of the real parts $u_1 = \Re v$ of the complex linear functionals $v = u_1 + iu_2$ on E. Conversely v is determined by its real part, for by § 16, 3.(2) u_2 is defined by

(1) $$u_2 x = -u_1(ix).$$

Consequently on the one hand we have the complex dual pair $\langle E', E \rangle$, and on the other the real dual pair $\langle E'_r, E_r \rangle$. Topologies can now be defined on E in two ways; the question of how far they coincide is answered as follows for the most important topologies:

(2) *If $E[\mathfrak{T}]$ is a complex locally convex space, E' its complex dual and E'_r its real dual, then the topologies $\mathfrak{T}_s(E')$ and $\mathfrak{T}_s(E'_r)$ coincide, as do $\mathfrak{T}_b(E')$ and $\mathfrak{T}_b(E'_r)$, and $\mathfrak{T}_k(E')$ and $\mathfrak{T}_k(E'_r)$.*

Proof. Using the definition of the polar given in § 20, 8., it follows that the polar in E'_r of $M \subset E$ is equal to $\Re(M^\circ)$, where M° is the polar in E'. Therefore by 3.(1) the $\mathfrak{T}$-equicontinuous subsets of E'_r are the real parts $\Re(N)$ of the $\mathfrak{T}$-equicontinuous subsets N of E'. If $\mathfrak{M}$ is a total saturated collection of bounded subsets of E', the topology $\mathfrak{T}_{\mathfrak{M}}(E')$ on E is therefore equal to the topology $\mathfrak{T}_{\Re(\mathfrak{M})}(E'_r)$. We observe that $\Re(\mathfrak{M})$ is also saturated.

In particular if M is a weakly bounded subset of E', $\Re(M)$ is clearly a weakly bounded subset of E'_r; conversely if N is a weakly bounded subset of E'_r the set of functionals $v = u_1 + iu_2$ constructed as in (1), with $u_1 \in N$, forms a weakly bounded subset of E'. It therefore follows from $\mathfrak{T}_{\mathfrak{M}} = \mathfrak{T}_{\Re(\mathfrak{M})}$ that $\mathfrak{T}_b(E') = \mathfrak{T}_b(E'_r)$.

The $\mathfrak{T}_s(E')$-equicontinuous sets are the (complex) finite-dimensional bounded subsets of E', and their real parts are the (real) finite-dimensional bounded subsets of E'_r; from this it follows that $\mathfrak{T}_s(E') = \mathfrak{T}_s(E'_r)$.

Finally, if K is complex absolutely convex and weakly compact in E', $\Re(K)$ is real absolutely convex and weakly compact in E'_r. Conversely suppose that C_1 is real absolutely convex and weakly compact in E'_r. The same holds for the set C_2 of functionals u_2 which correspond by (1) to the $u_1 \in C_1$. By § 20, 6.(5) the closed real absolutely convex cover K_1 of C_1 and C_2 is weakly compact. If we now again use (1) to form the set K of all $v = u_1 + iu_2$ with $u_1 \in K_1$, K is real absolutely convex, and is weakly compact by § 15, 6.(8). Since, further, by the construction of K, $K_1 = \Re(K) = \Im(K)$, K always contains iv if it contains v; by § 16, 1.(1), $\overline{\Gamma(K)}$ is contained in $2K$, and so $\overline{\Gamma(K)}$ is absolutely convex and weakly compact. Thus C_1 lies in the real part of an absolutely convex weakly compact subset of E'. Thus it follows from the definition of the Mackey topology (cf. 4.) that $\mathfrak{T}_k(E')$ and $\mathfrak{T}_k(E'_r)$ are equal.

§ 22. The determination of various dual spaces and their topologies

1. The duals of subspaces and quotient spaces. We considered the corresponding problem for linearly topologized spaces in § 10, 8.

Suppose that H is a linear subspace of the locally convex space $E[\mathfrak{T}]$. For the sake of clarity, in this and the next number we shall denote the topology which $\mathfrak{T}$ induces on H by $\hat{\mathfrak{T}}$ (up till now it has been denoted by $\mathfrak{T}$). We denote the embedding of $H(\hat{\mathfrak{T}})$ into $E[\mathfrak{T}]$ by I. I is a topological monomorphism of $H[\hat{\mathfrak{T}}]$ into $E[\mathfrak{T}]$.

As in § 10, 8., we also introduce the natural homomorphism of E' onto H; this is the mapping N which sends each $u \in E'$ to its restriction $u^{(0)}$ in H'. It follows from the relation

$$u^{(0)} y = (N u) y = u(I y),$$

for all $y \in H$ and all $u \in E'$, that N is the adjoint of I.

(1) *Let $H[\hat{\mathfrak{T}}]$ be a linear subspace of the locally convex space $E[\mathfrak{T}]$.*

a) *The natural homomorphism N of E' onto H' defines an algebraic isomorphism $\hat{N}$ of $E'/H^{\perp}$ onto H'.*

In this sense, H' can be identified with $E'/H^{\perp}$.

b) *The $\hat{\mathfrak{T}}$-equicontinuous subsets of H' are the N-images of the $\mathfrak{T}$-equicontinuous subsets of E', or, considered as subsets of $E'/H^{\perp}$, the K-images of the $\mathfrak{T}$-equicontinuous subsets of E', where K is the canonical homomorphism of E' onto $E'/H^{\perp}$.*

Proof. a) is proved as in § 10, 8.(1), and $\hat{N}$ is the mapping determined by the decomposition $N = \hat{N} K$. If M is a $\mathfrak{T}$-equicontinuous subset of E', so that $M \subset U^{\circ}$ for some absolutely convex neighbourhood U of $\circ$ in E, then clearly $|u y| = |u^{(0)} y| \leqq 1$ for $u \in M$ and $y \in U \cap H$, so that $N(M) \subset (U \cap H)^{\circ}$; the image $N(M)$ is therefore $\hat{\mathfrak{T}}$-equicontinuous in H'.

If conversely M' is $\hat{\mathfrak{T}}$-equicontinuous in H', then $\sup_{y \in U \cap H} |u^{(0)} y| \leqq 1$ for all $u^{(0)} \in M'$, where U is some suitable absolutely convex neighbourhood of $\circ$ in E. By the HAHN-BANACH theorem each such $u^{(0)}$ has an extension u to the whole of E, with $\sup_{x \in U} |u x| \leqq 1$. The set M of all these extensions u of the $u^{(0)} \in M'$ is a $\mathfrak{T}$-equicontinuous subset of E' for which $N(M) = M'$.

Now suppose that H is a closed linear subspace of $E[\mathfrak{T}]$. We give the quotient space E/H the quotient space topology corresponding to $\mathfrak{T}$; in this and the next number we shall denote this topology by $\hat{\mathfrak{T}}$.

If u' is a continuous linear functional on $(E/H)[\hat{\mathfrak{T}}]$, then the equation $u x = u' \hat{x}$, $\hat{x} \in E/H$, $x \in \hat{x}$, defines a continuous linear functional u on $E[\mathfrak{T}]$, for it follows from $|u' \hat{x}| < \varepsilon$ for $\hat{x} \in \hat{U}$ that $|u x| < \varepsilon$ for $x \in K^{(-1)}(\hat{U}) = U$, where K is the canonical homomorphism of E onto E/H.

We call the mapping $u = Iu'$ which we have just defined the **natural embedding** I of $(E/H)'$ into E'. It follows from the relation

$$u'\hat{x} = u'(Kx) = (Iu')x$$

for all $x \in E$ and all $u' \in (E/H)'$ that I is the adjoint of K. We have (cf. § 10, 8.(4)):

(2) *Suppose that H is a closed linear subspace of the locally convex space $E[\mathfrak{T}]$ and that the quotient space E/H is given the quotient topology $\hat{\mathfrak{T}}$.*

a) *The natural embedding I of $(E/H)'$ into E' is an algebraic isomorphism of $(E/H)'$ onto $H^\perp$.*

In this sense, $(E/H)'$ can be identified with $H^\perp$.

b) *Under this isomorphism the $\hat{\mathfrak{T}}$-equicontinuous subsets of $(E/H)'$ correspond to the $\mathfrak{T}$-equicontinuous subsets of E' which lie in $H^\perp$; i.e. the $\hat{\mathfrak{T}}$-equicontinuous and $\mathfrak{T}$-equicontinuous subsets of $H^\perp$ are the same.*

Proof. a) If $u' \in (E/H)'$, then, by the definition of I, the linear functional $u = Iu'$ vanishes on H, so that u lies in $H^\perp$.

Conversely if $u \in H^\perp$ the equation $ux = u'\hat{x}$ defines a linear functional u' on E/H; this u' is continuous on E/H, since it follows from $|ux| < \varepsilon$ for $x \in U$ that $|u'\hat{x}| < \varepsilon$ for $x \in K(U)$.

b) If M' is a $\hat{\mathfrak{T}}$-equicontinuous subset of $(E/H)'$, $|u'(\hat{x})| \leqq 1$ for $u' \in M'$ and $\hat{x} \in K(U)$, where U is a suitable open absolutely convex $\mathfrak{T}$-neighbourhood of $\mathfrak{o}$ in E. Then if $u = Iu' \in I(M')$, $|ux| \leqq 1$ for all $x \in U + H$. Thus $I(M') \subset U^\circ \cap H^\perp$. Conversely if $M \subset (E/H)'$ satisfies $I(M) \subset U^\circ \cap H^\perp = (U+H)^\circ$, then M is contained in $(K(U))^\circ$.

2. The topologies of subspaces, quotient spaces and their duals. Again let H be a subspace of the locally convex space $E[\mathfrak{T}]$, equipped with the subspace topology $\hat{\mathfrak{T}}$. We consider a further topology $\mathfrak{T}_{\mathfrak{M}}$ on E, where $\mathfrak{M}$ is a saturated class of bounded subsets of E', and ask which class of bounded subsets of $H' = E'/H^\perp$ produces the induced topology $\hat{\mathfrak{T}}_{\mathfrak{M}}$ on H.

We denote by $\hat{\mathfrak{M}}$ the collection of all sets $\hat{M} = K(M)$, where K is the canonical mapping of E' onto $E'/H^\perp$.

(1) *Suppose that H is a linear subspace of the locally convex space $E[\mathfrak{T}]$. If $\mathfrak{M}$ is a saturated class of bounded subsets of E', the induced topology $\hat{\mathfrak{T}}_{\mathfrak{M}}(E')$ on H is the same as the topology $\mathfrak{T}_{\hat{\mathfrak{M}}}(E'/H^\perp)$.*

In particular $\hat{\mathfrak{T}}_s(E')$ is equal to $\mathfrak{T}_s(E'/H^\perp)$ on H.

Proof. a) If $M \in \mathfrak{M}$ is absolutely convex, M defines the $\hat{\mathfrak{T}}_{\mathfrak{M}}$-neighbourhood $M^\circ \cap H$ of $\mathfrak{o}$, which consists of all $y \in H$ with $|uy| \leqq 1$ for $u \in M$. But since $uy = \hat{u}y$, where $\hat{u}$ is the coset of u in $E'/H^\perp$, $M^\circ \cap H$ is also equal to $(K(M))^\circ$. Conversely every $\mathfrak{T}_{\hat{\mathfrak{M}}}$-neighbourhood $(K(M))^\circ$ of $\circ$ in H, with absolutely convex M, can be written as $M^\circ \cap H$.

b) As M runs through the bounded finite dimensional subsets of E', $\hat{M}$ runs through the bounded finite dimensional subsets of $E'/H^\perp$, so that $\hat{\mathfrak{T}}_s(E')$ is equal to $\mathfrak{T}_s(E'/H^\perp)$.

This can also be seen directly, as in § 10, 8.(2).

If the topology $\mathfrak{T}_{\mathfrak{M}}$ in (1) is coarser than the Mackey topology $\mathfrak{T}_k(E')$, so that $E[\mathfrak{T}_{\mathfrak{M}}]' = E'$, then we can also prove (1) using 1.(1)b), by putting $\mathfrak{T} = \mathfrak{T}_{\mathfrak{M}}$ there. We then obtain that every $\mathfrak{T}_{\mathfrak{M}}$-equicontinuous subset in $E'/H^\perp$ is the K-image of a $\mathfrak{T}_{\mathfrak{M}}$-equicontinuous set, so that in this case $\hat{\mathfrak{M}}$ is saturated.

If $\mathfrak{T}_{\mathfrak{M}}$ is finer than $\mathfrak{T}_k$, then we can only deduce from the proof of (1) that the saturated cover of $\hat{M}$ is obtained by taking the weak closures in $E'/H^\perp$ of the sets $K(M)$, $M \in \mathfrak{M}$.

$\mathfrak{T}_b(E'/H^\perp)$ is equal to $\hat{\mathfrak{T}}_b(E')$ if and only if every bounded subset of $E'/H^\perp$ is contained in the closure of a set $K(B)$, with B bounded in E'. Thus $\mathfrak{T}_b(E'/H^\perp)$ can be strictly finer than $\hat{\mathfrak{T}}_b(E')$ (for examples cf. § 27.2 and § 31, 5.).

In the same way $\mathfrak{T}_k(E'/H^\perp)$ is equal to $\hat{\mathfrak{T}}_k(E')$ on H if and only if every absolutely convex weakly compact subset of $E'/H^\perp$ is the K-image of an absolutely convex weakly compact subset of E'. We shall give a counterexample in 4.

In general, therefore, we can only say: *The topology $\mathfrak{T}_b(E'/H^\perp)$ (respectively $\mathfrak{T}_k(E'/H^\perp)$) on $H \subset E[\mathfrak{T}]$ is finer than $\hat{\mathfrak{T}}_b(E')$ (respectively $\hat{\mathfrak{T}}_k(E')$).*

If we apply 1.(2) and interchange E and E', we obtain the following result on the topologies on the dual of a quotient space:

(2) *Suppose that H is a closed linear subspace of the locally convex space $E[\mathfrak{T}]$. If $\mathfrak{M}$ is a saturated collection of bounded subsets of E, the induced topology $\hat{\mathfrak{T}}_{\mathfrak{M}}(E)$ on $(E/H)' = H^\perp$ coincides with the topology $\mathfrak{T}_{\hat{\mathfrak{M}}}(E/H)$.*

In particular $\hat{\mathfrak{T}}_s(E)$ is equal to $\mathfrak{T}_s(E/H)$ on $(E/H)' = H^\perp$.

The remarks following (1) apply in this case, as well.

For the topologies on a quotient space we have:

(3) *Suppose that H is a closed linear subspace of the locally convex space $E[\mathfrak{T}]$. Suppose that $\mathfrak{M}$ is a saturated collection of subsets of absolutely convex weakly compact subsets of E', and that $\bar{\mathfrak{M}}$ is the collection of subsets of $\mathfrak{M}$ lying in $H^\perp \subset E'$. Then the quotient topology $\hat{\mathfrak{T}}_{\mathfrak{M}}(E')$ on E/H coincides with the topology $\mathfrak{T}_{\bar{\mathfrak{M}}}(H^\perp)$. $\bar{\mathfrak{M}}$ is also saturated.*

In particular $\hat{\mathfrak{T}}_s(E')$ and $\mathfrak{T}_s(H^\perp)$ coincide on E/H, and so do $\hat{\mathfrak{T}}_k(E')$ and $\mathfrak{T}_k(H^\perp)$.

Proof. a) By the MACKEY-ARENS theorem, the dual of $E[\mathfrak{T}_{\mathfrak{M}}]$ is also E', and we can therefore apply Theorem 1.(2)b) with $\mathfrak{T}=\mathfrak{T}_{\mathfrak{M}}$; the $\hat{\mathfrak{T}}_{\mathfrak{M}}$-equicontinuous sets in $H^\perp$ are therefore the $\mathfrak{T}_{\mathfrak{M}}$-equicontinuous sets; i.e. $\mathfrak{M}$ is saturated and $\mathfrak{T}_{\bar{\mathfrak{M}}}(H^\perp)$ coincides with $\hat{\mathfrak{T}}_{\mathfrak{M}}(E')$ on E/H.

b) The finite-dimensional bounded subsets of $H^\perp$ are the $\hat{\mathfrak{T}}_s(E')$-equicontinuous sets in $H^\perp$, and they clearly define the topology $\mathfrak{T}_s(H^\perp)$ on E/H; thus $\hat{\mathfrak{T}}_s(E')$ and $\mathfrak{T}_s(H^\perp)$ coincide on E/H.

c) By (2), the topologies $\mathfrak{T}_s(E/H)$ and $\hat{\mathfrak{T}}_s(E)$ coincide on $H^\perp$, so that the absolutely convex $\mathfrak{T}_s(E)$-compact sets which lie in $H^\perp$ are the same as the absolutely convex $\mathfrak{T}_s(E/H)$-compact subsets of $H^\perp$; thus $\hat{\mathfrak{T}}_k(E')$ and $\mathfrak{T}_k(H^\perp)$ coincide on E/H, by a).

Combining (3) and 1.(1)a) we obtain the following result for the topologies of the dual of a subspace:

(4) *Suppose that $H[\hat{\mathfrak{T}}]$ is a linear subspace of the locally convex space $E[\mathfrak{T}]$. Suppose that $\mathfrak{M}$ is a saturated collection of subsets of absolutely convex weakly compact subsets of E, and that $\bar{\mathfrak{M}}$ is the collection of all the sets $M\in\mathfrak{M}$ which lie in $\bar{H}=H^{\perp\perp}$. Then the quotient topology $\hat{\mathfrak{T}}_{\mathfrak{M}}(E)$ on $H'=E'/H^\perp$ is equal to the topology $\mathfrak{T}_{\bar{\mathfrak{M}}}(\bar{H})$. $\bar{\mathfrak{M}}$ is also saturated.*

In particular, $\hat{\mathfrak{T}}_s(E)$ and $\mathfrak{T}_s(\bar{H})$ coincide on H', and so do $\hat{\mathfrak{T}}_k(E)$ and $\mathfrak{T}_k(\bar{H})$.

Theorems (3) and (4) need no longer be true for topologies finer than the Mackey topology, and in particular they need not be true for the strong topology. The reason for this is the following: an open absolutely convex $\hat{\mathfrak{T}}_b(E')$-neighbourhood of $\mathfrak{o}$ in E/H is of the form $K(U)$, where U is an open $\mathfrak{T}_b$-neighbourhood of $\mathfrak{o}$ in E. Then $K(U)^\circ=(U+H)^\circ$ $U^\circ\cap H^\perp$ is a bounded subset B of H'. The polar B° of this in E/H is then the $\mathfrak{T}_s(H^\perp)$-closure $\overline{K(U)}$ of $K(U)$. It follows from this that $\mathfrak{T}_b(H^\perp)$ *is always coarser than* $\hat{\mathfrak{T}}_b(E')$ *on* E/H. The $\mathfrak{T}_s(H^\perp)$-closure of $K(U)$, however, need not be the same as the $\hat{\mathfrak{T}}_b(E')$-closure of $K(U)$, since $\hat{\mathfrak{T}}_b(E')$ need not be compatible with the dual pair $\langle H^\perp, E/H\rangle$. Cf. § 31, 7. for a counterexample.

We now give two theorems on the topology of precompact convergence.

(5) *Suppose that the locally convex space $E[\mathfrak{T}]$ is quasicomplete, and that H is a linear subspace of E. The topologies $\hat{\mathfrak{T}}_c(E)$ and $\mathfrak{T}_c(\bar{H})$ coincide on $H'=E'/H^\perp$.*

Proof. $\mathfrak{T}_c(E)$ is coarser than $\mathfrak{T}_k(E)$, by § 21, 6.(1), so that theorem (4) can be applied. But the $\mathfrak{T}$-compact subsets of $\bar{H}$ are the sets which are compact in the induced topology $\hat{\mathfrak{T}}$, so that $\hat{\mathfrak{T}}_c(E)$ is equal to $\mathfrak{T}_c(\bar{H})$.

(6) *If $E[\mathfrak{T}]$ is an* (F)*-space and if H is a closed linear subspace of E, the topologies $\hat{\mathfrak{T}}_c(E)$ and $\mathfrak{T}_c(E/H)$ coincide on $(E/H)'=H^\perp$.*

For E/H is an (F)-space under the quotient topology $\hat{\mathfrak{T}}$ (cf. § 18, 3.(4)), and by the BANACH-DIEUDONNÉ theorem both $\hat{\mathfrak{T}}_c(E)$ and $\mathfrak{T}_c(E/H)$ are the finest topology which defines the weak topology on the weakly bounded subsets of $H^\perp$, and the weak topologies $\hat{\mathfrak{T}}_s(E)$ and $\mathfrak{T}_s(E/H)$ are the same.

(6) can also be expressed as follows:

(7) *If E is an* (F)*-space, every compact subset of E/H is the canonical image of a compact subset of E.*

This follows from (6), (2) and one of the remarks following (1), according to which $\hat{\mathfrak{C}}$ is saturated, where $\mathfrak{C}$ is the class of relatively compact subsets of E; $\hat{\mathfrak{C}}$ therefore contains all the compact subsets of E/H.

(7) can also be proved directly using § 21, 10.(3). If $U_1 \supset U_2 \supset \dots$ is a fundamental sequence of open neighbourhoods of $\mathfrak{o}$ in E and if $\hat{x}_k$ is a sequence convergent to $\hat{\mathfrak{o}}$ in E/H, then there is an increasing sequence of integers k_n such that $\hat{x}_k \in K(U_n)$ for $k \geqq k_n$, $n=1,2,\dots$. There is then a sequence x_k in E, with $x_k \in \hat{x}_k$, such that $x_k \in U_n$ for $k \geqq k_n$, $n=1,2,\dots$. The sequence x_k defined in this way is therefore convergent to $\mathfrak{o}$ in E.

3. Subspaces and quotient spaces of normed spaces. In normed spaces the circumstances are particulary simple; indeed in this case the topological isomorphisms are norm isomorphisms.

(1) *a) If H is a linear subspace of the normed space E, the embedding I of H into E is a norm isomorphism of H into E, and if $y \in H$*

$$(2) \qquad \|y\|_H = \sup_{\|u^{(0)}\| \leqq 1,\, u^{(0)} \in H'} |u^{(0)} y| = \sup_{\|u\| \leqq 1,\, u \in E'} |u y| = \|y\|_E.$$

In particular the topologies $\hat{\mathfrak{T}}_k(E')$ and $\mathfrak{T}_k(H')$ on H are the same.

b) If E/H is a quotient of a normed space E, the natural embedding I of $(E/H)'$ into E' defines a norm isomorphism of $(E/H)'$ onto $H^\perp$, i.e. if $u' \in (E/H)'$ and $I u' = u \in H^\perp$,

$$(3) \qquad \|u'\| = \sup_{\|\hat{x}\| \leqq 1,\, \hat{x} \in E/H} |u' \hat{x}| = \sup_{\|x\| \leqq 1,\, x \in E} |u x| = \|u\|.$$

In particular the topologies $\hat{\mathfrak{T}}_b(E)$ and $\mathfrak{T}_b(E/H)$ on $(E/H)' = H^\perp$ are the same.

Proof. a) I defines a norm isomorphism, by definition. Considered as an element of E an element $y \in H$ has norm $\sup_{\|u\| \leqq 1,\, u \in E'} |u y|$, by § 17, 6.(4), and considered as an element of H it has norm $\sup_{\|u^{(0)}\| \leqq 1,\, u^{(0)} \in H'} |u^{(0)} y|$, and these must be the same.

b) In (3) the suprema can be taken over all $\|\hat{x}\| < 1$ and all $\|x\| < 1$ respectively. If $\|x\| < 1$, then $\|K x\| = \|\hat{x}\| < 1$; conversely for each $\hat{x}$ with

$\|\hat{x}\|<1$ there is an x with $Kx=\hat{x}$ and $\|x\|<1$. It then follows from the relation $u'\hat{x}=u'(Kx)=(Iu')x=ux$ proved in 1. that the two suprema in (3) are equal.

(4) a) *If H is a closed linear subspace of the normed space E, then the norm on E/H satisfies*

$$\|\hat{x}\|=\inf_{x\in\hat{x}}\|x\|=\sup_{\|u'\|\leqq 1,\,u'\in(E/H)'}|u'\hat{x}|. \tag{5}$$

In particular the topologies $\hat{\mathfrak{T}}_k(E')$ and $\mathfrak{T}_k((E/H)')$ on E/H are the same.

b) *If H is a linear subspace of the normed space E, the natural homomorphism $N=\hat{N}K$ of E' onto H' defines a norm isomorphism $\hat{N}$ from $E'/H^{\perp}$ onto H'; i.e. if $\hat{u}\in E'/H^{\perp}$ and $u^{(0)}=\hat{N}\hat{u}\in H'$ we have*

$$\|\hat{u}\|=\inf_{u\in\hat{u}}\|u\|=\sup_{\|y\|\leqq 1,\,y\in H}|u^{(0)}y|=\|u^{(0)}\|. \tag{6}$$

In particular the topologies $\hat{\mathfrak{T}}_b(E)$ and $\mathfrak{T}_b(H)$ on H' are the same.

a) follows from § 17, 6.(4) and § 14, 4.(1). We shall prove b). Since $\hat{u}$ is the coset of all those $u\in E'$ whose restriction to H is equal to $u^{(0)}$, we have $\|u\|\geqq\sup_{\|y\|\leqq 1,\,y\in H}|u^{(0)}y|$ for all $u\in\hat{u}$, so that $\|\hat{u}\|=\inf\|u\|\geqq\|u^{(0)}\|$. On the other hand, by the Hahn-Banach theorem, for each $u^{(0)}$ with bound $\|u^{(0)}\|$ on H there is an extension u of equal bound on E, so that we also have $\|u^{(0)}\|\geqq\|\hat{u}\|$.

4. The quotient spaces of l^1. In Hilbert space theory it is shown that every quotient space of l^2 is norm isomorphic either to l^2 or to K^n. l^1 exhibits a completely different behaviour:

(1) *Every separable* (B)*-space E is topologically isomorphic to a suitable quotient space of l^1.*

Proof. Let $x_1, x_2, \dots$ be a dense sequence of elements in the unit ball of E. We define a mapping A from l^1 into E by making each vector $\mathfrak{x}=(\xi_i)\in l^1$ correspond to the element $A\mathfrak{x}=\sum_{i=1}^{\infty}\xi_i x_i$ in E. Since $\sum|\xi_i|<\infty$, the series defining $A\mathfrak{x}$ in E is convergent and $\|A\mathfrak{x}\|\leqq\sum_{i=1}^{\infty}|\xi_i| =\|\mathfrak{x}\|$, so that A is a norm-continuous linear mapping.

Its image space is the whole of E: given x with $\|x\|\leqq 1$, we determine a sequence $x_{n_1}, x_{n_2}, \dots$ for which $\|x-x_{n_1}\|<\frac{1}{2}$, $\left\|x-x_{n_1}-\frac{1}{2}x_{n_2}\right\|<\frac{1}{2^2}$, $\left\|x-x_{n_1}-\frac{1}{2}x_{n_2}-\frac{1}{4}x_{n_3}\right\|<\frac{1}{2^3}, \dots$; then $\sum_{k=1}^{\infty}\frac{1}{2^k}x_{n_k}=x$.

It follows from the Banach-Schauder theorem (§ 15, 12.(2)) that A is a topological homomorphism. Consequently E is topologically isomorphic to $l^1/N[A]$, where $N[A]$ is the kernel of A.

Conversely by § 14, 7.(7) every quotient space l^1/H, where H is a closed linear subspace, is a separable (B)-space.

(2) *Every weak Cauchy sequence in l^1 is a strong Cauchy sequence, and l^1 is weakly sequentially complete.*

The weak convergence is with respect to $(l^1)'=l^\infty$. In order to prove (2) we use the method of the "sliding hump" (cf. § 20, 11.). First we consider a sequence $\mathfrak{x}^{(n)}\in l^1$ which is weakly convergent to $\mathfrak{o}$. If this were not strongly convergent to $\mathfrak{o}$ there would be an $\varepsilon>0$ and infinitely many n_j, $j=1,2,\ldots$, for which $\|\mathfrak{x}^{(n_j)}\|=\sum_{i=1}^{\infty}|\xi_i^{(n_j)}|>\varepsilon$. Let N_1 be chosen large enough so that $\sum_{N_1+1}^{\infty}|\xi_i^{(n_1)}|\leqq\frac{1}{5}\varepsilon$ and $\sum_1^{N_1}|\xi_i^{(n_1)}|>\frac{4}{5}\varepsilon$. Numbers $v_1,\ldots,v_{N_1}$ of modulus 1 can then be chosen so that $\sum_1^{N_1}v_i\xi_i^{(n_1)}=\sum_1^{N_1}|\xi_i^{(n_1)}|>\frac{4}{5}\varepsilon$; then, however the v_k are chosen with modulus 1, for $k>N_1$, we have $\left|\sum_1^{\infty}v_i\xi_i^{(n_1)}\right|>\frac{3}{5}\varepsilon>\frac{1}{5}\varepsilon$.

For the next step we choose n_{j_2} large enough, so that $\sum_1^{N_1}|\xi_i^{(n_{j_2})}|\leqq\frac{1}{5}\varepsilon$. Then we determine an $N_2>N_1$ so that $\sum_{N_2+1}^{\infty}|\xi_i^{(n_{j_2})}|\leqq\frac{1}{5}\varepsilon$ and $\sum_1^{N_2}|\xi_i^{(n_{j_2})}|>\frac{4}{5}\varepsilon$. We can then choose $v_{N_1+1},\ldots,v_{N_2}$ of modulus 1 so that $\sum_{N_1+1}^{N_2}v_i\xi_i^{(n_{j_2})}=\sum_{N_1+1}^{N_2}|\xi_i^{(n_{j_2})}|>\frac{3}{5}\varepsilon$; then, however the subsequent v_k are chosen with modulus 1, we have

$$\left|\sum_1^{\infty}v_i\xi_i^{(n_{j_2})}\right|\geqq\left|\sum_{N_1+1}^{N_2}\right|-\left|\sum_1^{N_1}\right|-\left|\sum_{N_2+1}^{\infty}\right|>\frac{1}{5}\varepsilon.$$

Repeating the process, we obtain a vector $\mathfrak{u}=(v_n)\in l^\infty$ with $|\mathfrak{u}\,\mathfrak{x}^{(n_{j_k})}|>\frac{\varepsilon}{5}$ for all $k=1,2,\ldots$; this contradicts the weak convergence of $\mathfrak{x}^{(n)}$ to $\mathfrak{o}$.

For a weak Cauchy sequence, we proceed in an analogous way: if it were not a Cauchy sequence, there would be a sequence of pairs of indices (n_j,m_j) and an $\varepsilon>0$ for which $\|\mathfrak{x}^{(n_j)}-\mathfrak{x}^{(m_j)}\|>\varepsilon, n_j,m_j\to\infty$. Then, in the proof above, we use the sequence $\mathfrak{x}^{(n_j)}-\mathfrak{x}^{(m_j)}$, which is weakly convergent to $\mathfrak{o}$, instead of $\mathfrak{x}^{(n_j)}$.

Since l^1 is a (B)-space, it follows that l^1 is also weakly sequentially complete.

(3) *In l^1 the weakly compact, weakly sequentially compact and strongly compact sets coincide.*

A bounded subset C of l^1 is strongly relatively compact if and only if $\lim_{n\to\infty} \sup_{\mathfrak{x}\in C} \sum_{i=n}^{\infty} |\xi_i| = 0.$

In particular the closed unit ball of l^1 is not weakly compact.

Proof. Suppose that M is a weakly compact subset of l^1. Since l^1 is separable, l^∞ is weakly sequentially separable, by § 21, 3.(5). There is therefore a countable weakly dense subset N in l^∞. The topology $\mathfrak{T}_s(N)$ on l^1 is Hausdorff, and it therefore coincides on M with the topology $\mathfrak{T}_s(l^\infty)$. Consequently M is metrizable under $\mathfrak{T}_s(l^\infty)$. Thus by § 4, 5.(4) every weakly compact subset M of l^1 is weakly sequentially compact. By (2), every weakly sequentially compact subset is strongly sequentially compact, and so it is strongly compact, by § 4, 5.(4). But every strongly compact set is certainly weakly compact. This proves the first part of (3).

Now suppose that C is the subset of all $\mathfrak{x}=(\xi_i)$ in l^1 with $\sum_{i=n}^{\infty} |\xi_i| \leqq d_n$ and suppose that $d_n \to 0$. If $\mathfrak{x}^{(k)}$ is a sequence in C, we can obtain a subsequence which converges in each coordinate, by using a diagonal procedure. Suppose then that $\mathfrak{x}^{(k)}$ is coordinatewise convergent to $\mathfrak{x}^{(0)}$. Then $\sum_{i=n}^{\infty} |\xi_i^{(k)}| \leqq d_n$, so that $\sum_{i=n}^{\infty} |\xi_i^{(0)}| \leqq d_n$, i.e. $\mathfrak{x}^{(0)} \in C$. Further it follows from $\sum_{i=n}^{\infty} |\xi_i^{(k)} - \xi_i^{(0)}| \leqq 2d_n$ that $\mathfrak{x}^{(k)}$ converges weakly to $\mathfrak{x}^{(0)}$. Every such set C is therefore weakly sequentially compact.

On the other hand if C is a bounded subset of l^1 with $\sup_{\mathfrak{x}\in C} \sum_{i=n}^{\infty} |\xi_i| = d_n \not\to 0$, C contains a sequence $\mathfrak{x}^{(n_k)}$ with $\sum_{i=n_k}^{\infty} |\xi_i^{(n_k)}| \geqq m > 0$. It is easy to obtain a subsequence $\mathfrak{x}^{(n_j)}$ of this with $\sum_{i=1}^{\infty} |\xi_i^{(n_j)} - \xi_i^{(n_{j'})}| \geqq \frac{m}{2}$ for all j, j'; C is therefore not strongly compact.

We observe that the assertion about the unit ball of l^1 has a simpler proof, which will be given in § 23, 5.

We now give the counterexample which was announced in 2. By (1) there is a topological homomorphism A of l^1 onto l^2. By § 20, 9.(5), the unit ball K_2 in l^2 is weakly compact but not strongly compact. But, by (3), the image of every weakly compact subset M of l^1 is strongly compact, so that there is no weakly compact set which is mapped onto the weakly compact subset K_2 of l^2. The canonical homomorphism of l^1 onto $l^1/N[A]$ therefore has the property that there are weakly compact sets in the image which are not the image of weakly compact sets. We now give l^∞ the topology $\mathfrak{T}_k(l^1)$. Then l^1 is the dual of l^∞. We take $H = N[A]^\perp$ as linear subspace of l^∞. By the remarks in 2. preceding (2), the topology $\mathfrak{T}_k(H') = \mathfrak{T}_k(l^1/N[A])$ is strictly finer than the topology $\hat{\mathfrak{T}}_k(l^1)$ on H. Since $l^1/N[A]$ is topologically isomorphic to the Hilbert space l^2, the topology $\mathfrak{T}_k(l^1/N[A])$ is the norm topology on H, which makes H norm isomorphic to l^2. It follows from 3.(1)b) that this topology on H is also equal to the topology $\hat{\mathfrak{T}}_b(l^1)$, and so it is equal to the topology induced on H by the norm topology on l^∞.

The last statement is true in general:

(4) *The strong dual of a separable* (B)*-space is topologically isomorphic to a weakly closed linear subspace of* l^∞.

This follows directly from (1) and 3.(1)b).

In § 14, 8. the spaces l_d^1 and l_d^∞ were defined for arbitrary cardinal d. If E is a general (B)-space, and if $M=\{x_\alpha\}$ is a set with cardinal d which is dense in the unit ball of E, then E is topologically isomorphic to a quotient space of l_d^1, giving a generalization of (1), and E' is topologically isomorphic to a weakly closed linear subspace of l_d^∞.

(2) is also true for l_d^1, and the proof is the same.

We observe further that every (B)-space can be embedded in a norm-preserving way in a suitable l_d^∞, and in particular every separable (B)-space can be embedded in l^∞. For this it is sufficient to pick out a weakly dense subset $N=\{u_\alpha\}$ of the unit ball in E', and to make each $x\in E$ correspond to the vector $\mathfrak{x}=(\xi_\alpha)$, with $\xi_\alpha=u_\alpha x$, in the corresponding l_d^∞. In the separable case we can manage with countably many u_α, by § 21, 3.(5).

5. The duality of topological products and locally convex direct sums. Suppose that a collection of dual pairs $\langle F_\alpha, E_\alpha\rangle$, $\alpha\in\mathsf{A}$, is given. The direct sum $F=\bigoplus_\alpha F_\alpha$ and the product $E=\prod_\alpha E_\alpha$ form a dual pair $\langle F,E\rangle$ in a natural way, when we define the bilinear form ux for two elements $u=(u_\alpha)\in F$ and $x=(x_\alpha)\in E$ by setting $ux=\sum_\alpha u_\alpha x_\alpha$, where $u_\alpha x_\alpha$ is the bilinear form on $\langle F_\alpha, E_\alpha\rangle$.

The following rules holds for forming polars in $\langle F,E\rangle$:

(1) *If each M_α is a closed absolutely convex subset of E_α and if M_α° is its polar in F_α, then the polar $\left(\prod_\alpha M_\alpha\right)^\circ$ in F of the subset $\prod_\alpha M_\alpha$ of E is equal to $\Gamma_\alpha M_\alpha^\circ$.*

If each N_α is an absolutely convex subset of F_α and if N_α° is its polar in E_α, then the polar $\left(\Gamma_\alpha N_\alpha\right)^\circ$ in E of the subset $\Gamma_\alpha N_\alpha$ of F is equal to $\prod_\alpha N_\alpha^\circ$.

Proof. The first statement follows from § 20, 8.(10), for $\prod_\alpha M_\alpha$ can be considered as the intersection of the sets $M^{(\beta)}=\prod_\alpha M_\alpha^{(\beta)}$, where $M_\beta^{(\beta)}=M_\beta$, and $M_\alpha^{(\beta)}=E_\alpha$ for $\alpha\neq\beta$, and the polar $(M^{(\beta)})^\circ$ in F is equal to the polar M_β° in F_β.

The second assertion follows in a similar way from § 20, 8.(9).

(2) *The dual of a topological product* $\prod_\alpha E_\alpha[\mathfrak{T}_\alpha]$ *is algebraically isomorphic to the direct sum* $\bigoplus_\alpha E'_\alpha$ *of the duals; the dual of a locally convex direct sum* $\bigoplus_\alpha E_\alpha[\mathfrak{T}_\alpha]$ *is algebraically isomorphic to the product* $\prod_\alpha E'_\alpha$ *of the duals.*

In the dual pairs $\left\langle \bigoplus_\alpha E'_\alpha, \prod_\alpha E_\alpha \right\rangle$ *and* $\left\langle \prod_\alpha E'_\alpha, \bigoplus_\alpha E_\alpha \right\rangle$ *which arise in this way, the bilinear form is given by* $ux = \sum u_\alpha x_\alpha$, $u_\alpha \in E'_\alpha$, $x_\alpha \in E_\alpha$.

Proof. a) If $u \in \left(\prod_\alpha E_\alpha[\mathfrak{T}_\alpha]\right)'$, u is bounded on a neighbourhood W of $\mathfrak{o}$. This can be taken in the form $\prod_\alpha U_\alpha$, where $U_{\alpha_i} \neq E_{\alpha_i}$ for finitely many α and $U_\beta = E_\beta$ for the remaining indices. u clearly vanishes on the subspace $\prod_\beta E_\beta$. If P is the projection of $\prod_\alpha E_\alpha$ onto $\prod_i E_{\alpha_i}$ and Q the projection onto $\prod_\beta E_\beta$, where each projection maps the other space to zero, then the relation $ux = u(Px) + u(Qx) = u(Px)$ holds for each $x = (x_\alpha) \in \prod_\alpha E_\alpha$. Thus u can be considered as a linear functional on $\prod_{i=1}^n E_{\alpha_i}[\mathfrak{T}_{\alpha_i}]$. The restriction u_{α_i} of u to E_{α_i} is an element of $(E_{\alpha_i}[\mathfrak{T}_{\alpha_i}])'$, since the topology induced on E_{α_i} by the product topology of $\prod_\alpha E_\alpha[\mathfrak{T}_\alpha]$ is equal to $\mathfrak{T}_{\alpha_i}$.

Since $\prod_{i=1}^n E_{\alpha_i}[\mathfrak{T}_{\alpha_i}] = \bigoplus_{i=1}^n E_{\alpha_i}[\mathfrak{T}_{\alpha_i}]$, we have

$$ux = u(Px) = u \sum_{i=1}^n x_{\alpha_i} = \sum_{i=1}^n u_{\alpha i} x_{\alpha_i}, \qquad u_{\alpha_i} \in E'_{\alpha_i}.$$

Thus each u is mapped in a one-one way to an element of $\bigoplus_\alpha E'_\alpha$; conversely if $(u_\alpha) \in \bigoplus_\alpha E'_\alpha$ the formula $(u_\alpha)x = \sum u_\alpha x_\alpha$ clearly defines a continuous linear functional on $\prod_\alpha E_\alpha[\mathfrak{T}_\alpha]$.

b) By § 18, 5.(2), the topology $\mathfrak{T}$ of the locally convex direct sum $\bigoplus_\alpha E_\alpha[\mathfrak{T}_\alpha]$ induces the topology $\mathfrak{T}_\alpha$ on each space E_α, so that the restriction u_α of a continuous linear functional $u \in \left(\bigoplus_\alpha E_\alpha[\mathfrak{T}_\alpha]\right)'$ to E_α is an element of E'_α; clearly if $x = (x_\alpha) \in \bigoplus_\alpha E_\alpha$ we have $ux = \sum_\alpha u_\alpha x_\alpha$ (the sum containing only finitely many non-zero x_α). In this way each $u \in \left(\bigoplus_\alpha E_\alpha[\mathfrak{T}_\alpha]\right)'$ is mapped in a one-one way to a $(u_\alpha) \in \prod_\alpha E'_\alpha$. Conversely if $v = (v_\alpha)$ is an arbitrary element of $\prod_\alpha E'_\alpha$, and if $|v_\alpha x_\alpha| \leqq 1$ for each x_α in the $\mathfrak{T}_\alpha$-neighbourhood U_α of $\mathfrak{o}$ in E_α, then $|vx| \leqq 1$ for each x in the neighbourhood $\Gamma_\alpha U_\alpha$ of $\mathfrak{o}$ in $\bigoplus_\alpha E_\alpha[\mathfrak{T}_\alpha]$, so that v is continuous. Consequently $\prod_\alpha E'_\alpha$ is algebraically isomorphic to $\left(\bigoplus_\alpha E_\alpha[\mathfrak{T}_\alpha]\right)'$.

We now determine various topologies on topological products and locally convex sums.

(3) *Suppose that* $E[\mathfrak{T}]=\prod_\alpha E_\alpha[\mathfrak{T}_\alpha]$, *with dual space* $E'=\bigoplus_\alpha E'_\alpha$. *Suppose that a saturated collection* $\mathfrak{M}_\alpha$ *of* E_α*-bounded subsets is given in each* E'_α. *We denote by* $\mathfrak{M}$ *the saturated collection of bounded subsets of* E' *consisting of all the finite direct sums* $\bigoplus_{i=1}^m M_{\alpha_i}, M_{\alpha_i}\in\mathfrak{M}_{\alpha_i}$, *together with their subsets. The space* $E[\mathfrak{T}_{\mathfrak{M}}]$ *is then the topological product of the spaces* $E_\alpha[\mathfrak{T}_{\mathfrak{M}_\alpha}]$.

In particular $E[\mathfrak{T}_s(E')]=\prod_\alpha E_\alpha[\mathfrak{T}_s(E'_\alpha)]$, $E[\mathfrak{T}_b(E')]=\prod_\alpha E_\alpha[\mathfrak{T}_b(E'_\alpha)]$, $E[\mathfrak{T}_{b^*}(E')]=\prod_\alpha E_\alpha[\mathfrak{T}_{b^*}(E'_\alpha)]$ *and* $E[\mathfrak{T}_k(E')]=\prod_\alpha E_\alpha[\mathfrak{T}_k(E'_\alpha)]$.

Proof. a) The fact that $\mathfrak{T}_s(E')$ is equal to the product topology on $\prod_\alpha E_\alpha[\mathfrak{T}_s(E'_\alpha)]$ can be deduced directly from (2) and the definition of the product topology.

b) We can take as base of neighbourhoods of o for the topological product $\prod_\alpha E_\alpha[\mathfrak{T}_{\mathfrak{M}_\alpha}]$ the sets $\prod_\alpha U_\alpha$, where $U_{\alpha_i}=M^\circ_{\alpha_i}, M_{\alpha_i}\in\mathfrak{M}_{\alpha_i}$ for finitely many α_i and $U_\alpha=E_\alpha$ otherwise. By a), $\prod_\alpha U_\alpha$ is weakly closed, since the sets U_α are, so that $\prod_\alpha U_\alpha=\left(\prod_\alpha U_\alpha\right)^{\circ\circ}$. By (1) we have $\left(\prod_\alpha U_\alpha\right)^\circ$ $=\overline{\Gamma_{i=1}^{n} M^{\circ\circ}_{\alpha_i}}$. But the saturated collection of subsets of E' defined by the sets $\Gamma_{i=1}^{n} M^{\circ\circ}_{\alpha_i}$ is equal to the saturated collection defined by the sets $\bigoplus_{i=1}^n M_{\alpha_i}$. Since the sets $\bigoplus_{i=1}^n M_{\alpha_i}$ are absolutely convex and weakly closed when the M_{α_i} are, the collection of subsets of E' which are equicontinuous for the product topology is precisely $\mathfrak{M}$.

c) By § 18, 5.(4) an E-bounded subset of E' is always contained in a set $\bigoplus_{i=1}^n B'_{\alpha_i}$, where B'_{α_i} is E_{α_i}-bounded in E'_{α_i}. Conversely every set $\bigoplus_{i=1}^n B'_{\alpha_i}$ is weakly bounded in E'. The product topology on $\prod_\alpha E_\alpha[\mathfrak{T}_b(E'_\alpha)]$ is therefore equal to $\mathfrak{T}_b(E')$, by b).

d) The topology $\mathfrak{T}_{b^*}$ which was introduced in § 21, 5. is the topology of uniform convergence on the strongly bounded sets. Since every strongly bounded subset M' of E' is weakly bounded, there are only finitely many non-zero projections $M'_\alpha=P_\alpha M'$ in the spaces E'_α. Every set of the form $B=\prod_\alpha B_\alpha$, where B_α is bounded in $E_\alpha[\mathfrak{T}_\alpha]$, is bounded in $\prod_\alpha E_\alpha[\mathfrak{T}_\alpha]$. It follows directly from $\sup_{u\in M', x\in B}|ux|<\infty$ that $\sup_{u_\alpha\in M'_\alpha, x_\alpha\in B_\alpha}|u_\alpha x_\alpha|<\infty$, i.e. M'_α is strongly bounded in E'_α. As a result, every strongly bounded subset of E' is contained in a finite sum $\bigoplus_{i=1}^n M'_{\alpha_i}$, where M'_{α_i} is strongly bounded in E'_{α_i}. Since conversely every such set is strongly bounded in E', it follows that $\mathfrak{T}_{b^*}(E')$ is equal to the product topology of $\prod_\alpha E_\alpha[\mathfrak{T}_{b^*}(E'_\alpha)]$.

e) $\mathfrak{T}_s(E_\alpha)$ coincides with the topology induced on E'_α by $\mathfrak{T}_s(E)$, so that every absolutely convex $\mathfrak{T}_s(E_\alpha)$-compact subset $C_\alpha \subset E'_\alpha$ is $\mathfrak{T}_s(E)$-compact, and conversely. If C is an absolutely convex weakly compact subset of E', its projections C_α in the spaces E'_α are also weakly compact, and only finitely many of these sets C_α are different from $\mathfrak{o}$, since C is weakly bounded. It therefore follows from this that $C \subset \bigoplus_{i=1}^{n} C_{\alpha_i}$, where C_{α_i} is weakly compact in E'_{α_i}; conversely if the sets C_{α_i} are absolutely convex and weakly compact, then the set $\bigoplus_{i=1}^{n} C_{\alpha_i}$ is again absolutely convex (§ 16, 1.(3)), and it is weakly compact, by TYCHONOFF's theorem. It follows from this, by a), that $\mathfrak{T}_k(E')$ is the product topology on $\prod_\alpha E_\alpha[\mathfrak{T}_k(E'_\alpha)]$.

The corresponding results on locally convex direct sums run as follows:

(4) *Suppose that* $E[\mathfrak{T}]$ *is the locally convex direct sum* $\bigoplus_\alpha E_\alpha[\mathfrak{T}_\alpha]$, *with dual space* $E' = \prod_\alpha E'_\alpha$. *Suppose that a saturated collection* $\mathfrak{M}_\alpha$ *of* E_α*-bounded subsets is given in each* E'_α. *Let* $\mathfrak{M}$ *be the saturated collection of subsets of* E' *consisting of all sets of the form* $\prod_\alpha M_\alpha$, $M_\alpha \in \mathfrak{M}_\alpha$, *together with their subsets. Then* $E[\mathfrak{T}_{\mathfrak{M}}]$ *is the locally convex direct sum of the spaces* $E_\alpha[\mathfrak{T}_{\mathfrak{M}_\alpha}]$.

In particular we have $E[\mathfrak{T}_b(E')] = \bigoplus_\alpha E_\alpha[\mathfrak{T}_b(E'_\alpha)]$, $E[\mathfrak{T}_{b^*}(E')] = \bigoplus_\alpha E_\alpha[\mathfrak{T}_{b^*}(E'_\alpha)]$ *and* $E[\mathfrak{T}_k(E')] = \bigoplus_\alpha E_\alpha[\mathfrak{T}_k(E'_\alpha)]$.

For the weak topologies, however, the topology of the locally convex direct sum $\bigoplus_\alpha E_\alpha[\mathfrak{T}_s(E'_\alpha)]$ *is equal to the weak topology of* E *if and only if there are only finitely many summands* $E_\alpha[\mathfrak{T}_s(E'_\alpha)]$.

Proof. a) The sum topology $\mathfrak{T}'$ on $\bigoplus_\alpha E_\alpha[\mathfrak{T}_{\mathfrak{M}_\alpha}]$ has a base of neighbourhoods of $\mathfrak{o}$ of the form $U = \Gamma_\alpha M_\alpha^\circ = C_\alpha M_\alpha^\circ$, where each M_α is an absolutely convex weakly closed subset in $\mathfrak{M}_\alpha$. By the second part of (1), $\left(\Gamma_\alpha M_\alpha^\circ\right)^\circ = \prod_\alpha M_\alpha^{\circ\circ} = \prod_\alpha M_\alpha$. It follows that $\mathfrak{M}$ is the class of $\mathfrak{T}'$-equicontinuous sets, and $\mathfrak{T}' = \mathfrak{T}_{\mathfrak{M}}$, provided that we can show that $U^{\circ\circ} = \left(\prod_\alpha M_\alpha\right)^\circ = U$. Clearly $\left(\prod_\alpha M_\alpha\right)^\circ \supset U$. Suppose that $x = \sum_{i=1}^{n} x_{\alpha_i}$, $x_{\alpha_i} \in E_{\alpha_i}$, is an element of $\left(\prod_\alpha M_\alpha\right)^\circ$. Then at all events $\sup_{u_{\alpha_i} \in M_{\alpha_i}} |u_{\alpha_i} x_{\alpha_i}| = \rho_i \leqq 1$. Since further the u_{α_i} can be chosen independently of each other in such a way that $u_{\alpha_i} x_{\alpha_i} \geqq \rho_i - \varepsilon$, it follows that $\sum_{i=1}^{n} \rho_i \leqq 1$. The element $y_i = \frac{1}{\rho_i} x_{\alpha_i}$ lies in $M^\circ_{\alpha_i}$, and consequently $x = \sum_{i=1}^{n} \rho_i y_i$ belongs to $\underset{\alpha}{C}\, M^\circ_\alpha = U$.

b) If $\underset{\alpha}{\oplus}\, E_\alpha[\mathfrak{T}_s(E'_\alpha)]$ has only finitely many summands, the locally convex direct sum is equal to the topological product; in this case, therefore, it follows from (3) that the weak topology on E is equal to the locally convex direct sum topology.

If there are infinitely many summands and if each $\mathfrak{M}_\alpha$ is the collection of bounded finite-dimensional subsets of E'_α, then $\mathfrak{M}$ contains infinite-dimensional subsets, so that the topology $\mathfrak{T}_{\mathfrak{M}}$ is strictly finer than the weak topology $\mathfrak{T}_s(E')$ on E.

We observe however that the formula $u x = \sum_\alpha u_\alpha x_\alpha$ $(u \in E', x \in E, u_\alpha \in E'_\alpha, x_\alpha \in E_\alpha)$ implies that the topology induced on each E_α by $\mathfrak{T}_s(E')$ is the same as $\mathfrak{T}_s(E'_\alpha)$.

c) If B' is a $\mathfrak{T}_s(E)$-bounded subset of E', each of its projections B'_α in E'_α is a $\mathfrak{T}_s(E_\alpha)$-bounded subset of E'_α, so that $B' \subset \prod_\alpha B'_\alpha$; conversely every such set $\prod_\alpha B'_\alpha$ is weakly bounded in E'. It therefore follows from a) that the topology $\mathfrak{T}_b(E)$ coincides with the topology of the locally convex direct sum of the spaces $E_\alpha[\mathfrak{T}_b(E'_\alpha)]$.

d) Analogously to (3)d), we can apply § 18, 5.(4) and show that the strongly bounded subsets of E' are the subsets of sets of the form $\prod_\alpha M'_\alpha$, where each M'_α is strongly bounded in E'_α. It follows from this that $E[\mathfrak{T}_{b^*}(E')] = \underset{\alpha}{\oplus}\, E_\alpha[\mathfrak{T}_{b^*}(E'_\alpha)]$.

e) Using the remark in b), we can establish the proof for the Mackey topology in a way which corresponds to (3)e).

Theorems (3) and (4) can also be interpreted as statements about the topologies on the duals of locally convex direct sums and topological products.

For example it follows from (3) that *the weak (respectively strong) dual of* $\underset{\alpha}{\oplus}\, E_\alpha[\mathfrak{T}_\alpha]$ *is the topological product of the weak (respectively strong) duals* $E'_\alpha[\mathfrak{T}_s(E_\alpha)]$ *(respectively* $E'_\alpha[\mathfrak{T}_b(E_\alpha)]$*)*.

Likewise it follows from (4) that *the strong dual of* $\prod_\alpha E_\alpha[\mathfrak{T}_\alpha]$ *is the locally convex direct sum of the strong duals* $E'_\alpha[\mathfrak{T}_b(E_\alpha)]$, and that *the dual of* $\prod_\alpha E_\alpha[\mathfrak{T}_\alpha]$, *equipped with the Mackey topology, is equal to the locally convex direct sum of the spaces* $E'_\alpha[\mathfrak{T}_k(E_\alpha)]$.

As an example we consider the spaces ω_d and φ_d which were introduced in § 15, 4. and § 18, 5.(5) respectively. These spaces are respectively the topological product and the locally convex direct sum of d one-dimensional spaces K.

By (2), ω_d and φ_d are dual to each other. By § 18, 5.(6) all the bounded subsets of φ_d are finite-dimensional, and by MACKEY's theorem (§ 20, 11.(7)) so also are all the weakly bounded sets. Consequently the strong and the weak topologies on ω_d coincide, and are both equal to the topology of the topological product.

The topology on φ_d is the topology of the locally convex direct sum, and by § 18, 5.(5) it is the finest possible locally convex topology on φ_d. Consequently it coincides with the strong topology. φ_d and ω_d are therefore barrelled (cf. § 21, 2.).

By (3) and (4), all the spaces formed from φ_d and ω_d by repeatedly forming topological products and locally convex direct sums are also barrelled.

The duality between topological products and locally convex direct sums established here no longer holds when we replace the latter by the topological direct sums introduced in § 18, 5.

(5) *The dual of the topological direct sum* $E[\mathfrak{T}']$ *of locally convex spaces* $E_\alpha[\mathfrak{T}_\alpha]$ *is equal to the subspace of* $\prod_\alpha E'_\alpha$ *consisting of all* $u=(u_\alpha)$, $u_\alpha\in E'_\alpha$ *with at most countably many non-zero* u_α.

Proof. The topology $\mathfrak{T}'$ of the topological direct sum is coarser than the topology $\mathfrak{T}$ of the locally convex direct sum, so that by (2) every continuous linear functional on $E[\mathfrak{T}']$ can be represented by a $u=(u_\alpha)\in\prod_\alpha E'_\alpha$. But such a linear functional u is only $\mathfrak{T}'$-continuous if at most countably many u_α are different from o. Let us suppose the contrary. There must be a $\mathfrak{T}'$-neighbourhood $V=\bigoplus_\alpha U_\alpha$ of o, where each U_α is a $\mathfrak{T}_\alpha$-neighbourhood of o in E_α, on which u is bounded. For each non-zero u_α there is an $x_\alpha\in U_\alpha$ with $u_\alpha x_\alpha=\gamma_\alpha>0$. It follows that there is a positive integer n_0 for which $\gamma_\beta\geqq\frac{1}{n_0}$ for uncountably many β. But then $u\sum_{i=1}^{k}x_{\beta i}\geqq\frac{k}{n_0}$ for each element $\sum_{i=1}^{k}x_{\beta i}$, and these lie in V, for each k. This contradicts the fact that u is bounded on V.

On the other hand, if $u=(u_\alpha)$ has at most countably many non-zero components u_α, u is $\mathfrak{T}'$-continuous on E, by § 18, 5.(8).

6. The duality of locally convex hulls and kernels. We now return to the concepts and results of § 19. Suppose that $E[\mathfrak{T}]=\sum_\alpha A_\alpha(F_\alpha[\mathfrak{T}_\alpha])$ is the locally convex hull of arbitrary locally convex spaces $F_\alpha[\mathfrak{T}_\alpha]$. The A_α are then continuous mappings from $F_\alpha[\mathfrak{T}_\alpha]$ into $E[\mathfrak{T}]$. If we form the locally convex direct sum $\bigoplus_\alpha F_\alpha[\mathfrak{T}_\alpha]$ and make each $x=\sum_\alpha x_\alpha\in\bigoplus_\alpha F_\alpha[\mathfrak{T}_\alpha]$ correspond to the element $Ax=\sum_\alpha A_\alpha x_\alpha$, A is the topological homomorphism of $\bigoplus_\alpha F_\alpha[\mathfrak{T}_\alpha]$ onto $E[\mathfrak{T}]$, by § 19, 1.(3), and $E[\mathfrak{T}]$ is topologically isomorphic to the quotient $\left(\bigoplus_\alpha F_\alpha[\mathfrak{T}_\alpha]\right)/H$, where H is the kernel of A.

(1) *The dual E' of $E[\mathfrak{T}]=\sum_\alpha A_\alpha(F_\alpha[\mathfrak{T}_\alpha])$ can be represented as the kernel $\mathsf{K}_\alpha A_\alpha'^{(-1)}(F_\alpha')$; in this expression A_α' is the adjoint of A_α, mapping E' into F_α'.*

The mapping A' adjoint to A maps E' isomorphically onto the subspace $\hat{E}'$ of $\prod_\alpha F_\alpha'$ consisting of all the elements $(A_\alpha' u)$.

Proof. A maps $\bigoplus_\alpha F_\alpha[\mathfrak{T}_\alpha]$ continuously onto $E[\mathfrak{T}]$, so that A' maps E' into $\left(\bigoplus_\alpha F_\alpha\right)'=\prod_\alpha F_\alpha'$. For each $x=\sum_\alpha x_\alpha\in\bigoplus_\alpha F_\alpha$ and each $u\in E'$ we have

(2) $$u(A\,x)=u\left(\sum_\alpha A_\alpha x_\alpha\right)=\sum_\alpha (A_\alpha' u)\,x_\alpha=(A'u)\,x.$$

From this it follows that $A'u=(A_\alpha' u)\in\prod_\alpha F_\alpha'$, and that A' is one-one. But by § 19, 6. this means that $E'=\mathsf{K}_\alpha A_\alpha'^{(-1)}(F_\alpha')$, and that A' is the mapping of § 19, 6.(1), sending E' onto the subspace $\hat{E}'$ of $\prod_\alpha F_\alpha'$.

A completely analogous result holds for locally convex kernels:

(3) *The dual E' of $E[\mathfrak{T}]=\mathsf{K}_\alpha A_\alpha^{(-1)}(F_\alpha[\mathfrak{T}_\alpha])$ is equal to the span $\sum_\alpha A_\alpha'(F_\alpha')$, where A_α' is the adjoint of A_α, mapping F_α' into E'.*

If A is the topological monomorphism $x\to(A_\alpha x)$ of $E[\mathfrak{T}]$ into $\prod_\alpha F_\alpha[\mathfrak{T}_\alpha]$, with image $\hat{E}$, A' is the mapping $\sum_\alpha u_\alpha\to\sum_\alpha A_\alpha' u_\alpha$ from $\bigoplus_\alpha F_\alpha'$ onto $\sum_\alpha A_\alpha'(F_\alpha')$.

Proof. A' clearly maps $\bigoplus_\alpha F_\alpha'$ into E'. Since each $A_\alpha'(F_\alpha')$ is contained in E', $\sum_\alpha A_\alpha'(F_\alpha')$ is at all events a linear subspace of E'. On the other hand suppose that v is an arbitrary element of E'. Since A is a topological monomorphism, a continuous linear functional is defined on $\hat{E}\subset\prod_\alpha F_\alpha[\mathfrak{T}_\alpha]$ by setting $u(Ax)=vx$. By the HAHN-BANACH theorem this can be extended to give an element of $\left(\prod_\alpha F_\alpha\right)'=\bigoplus_\alpha F_\alpha'$, which we shall again denote by u. Then for this $u=\sum_\alpha u_\alpha$ and all $x\in E$ we have

(4) $$v\,x=u(A\,x)=\sum_\alpha u_\alpha(A_\alpha x)=\sum_\alpha (A_\alpha' u_\alpha)\,x=(A'u)\,x.$$

From this it follows that $v=A'u=\sum_\alpha A_\alpha' u_\alpha$, so that $v\in\sum_\alpha A_\alpha'(F_\alpha')$. Consequently $E'=\sum_\alpha A_\alpha'(F_\alpha')$. The assertion about A' now follows immediately.

Theorem (1) can also be carried over to inductive limits.

(4) *The dual E' of a topological inductive limit* $E[\mathfrak{T}]=\varinjlim A_{\beta\alpha}(F_\alpha[\mathfrak{T}_\alpha])$ *can be represented as the projective limit* $\varprojlim A'_{\beta\alpha}(F'_\beta)$, *where $A'_{\beta\alpha}$ is the adjoint of the continuous mapping $A_{\beta\alpha}$ from $F_\alpha[\mathfrak{T}_\alpha]$ into $F_\beta[\mathfrak{T}_\beta]$, mapping F'_β into F'_α.*

Proof. By § 19, 3.(1), if $\alpha<\beta<\gamma$ we have the relation $A_{\gamma\beta}A_{\beta\alpha}=A_{\gamma\alpha}$. This goes over to the relation $A'_{\beta\alpha}A'_{\gamma\beta}=A'_{\gamma\alpha}$, and so the relation § 19, 7.(7) is satisfied by the mappings $A'_{\beta\alpha}$ from F'_β into F'_α. By § 19, 7.(6) we can therefore form $\varprojlim A'_{\beta\alpha}(F'_\beta)$.

By § 19, 2., $E[\mathfrak{T}]=\left(\bigoplus_\alpha F_\alpha[\mathfrak{T}_\alpha]\right)/H_0$, where H_0 is the set of all finite linear combinations of elements of the form $x_\alpha-A_{\beta\alpha}x_\alpha$, $x_\alpha\in F_\alpha$. We observe that $A_{\beta\alpha}x_\alpha$ is an element of F_β. By 1.(2), E' can be identified with the subspace $H_0^\perp$ of $\prod_\alpha F'_\alpha$. $H_0^\perp$ consists of all $u=(u_\alpha)\in\prod_\alpha F'_\alpha$ with

$$(5)\qquad u(x_\alpha-A_{\beta\alpha}x_\alpha)=u_\alpha x_\alpha-u_\beta(A_{\beta\alpha}x_\alpha)=(u_\alpha-A'_{\beta\alpha}u_\beta)x_\alpha=0$$

for all $\alpha<\beta$ and all $x_\alpha\in F_\alpha$. This means that $u_\alpha=A'_{\beta\alpha}u_\beta$ for all $\alpha<\beta$, and so by § 19, 7.(6) $H_0^\perp=\varprojlim A'_{\beta\alpha}(F'_\beta)$.

Theorem (3) cannot be taken over directly to projective limits. We must first take the projective limit in a normal form.

By definition $E[\mathfrak{T}]=\varprojlim A_{\alpha\beta}(F_\beta[\mathfrak{T}_\beta])$ is the subspace $\hat{E}$ of $\prod_\alpha F_\alpha[\mathfrak{T}_\alpha]$ consisting of all (x_α) with $x_\alpha=A_{\alpha\beta}x_\beta$ for all $\alpha<\beta$. We have denoted by $\hat{P}_\alpha(E)$ the linear subspace of F_α formed by the projections of the elements of $\hat{E}$ in F_α. We say that $\varprojlim A_{\alpha\beta}(F_\beta[\mathfrak{T}_\beta])$ is in reduced form if $\hat{P}_\alpha(E)$ is dense in $F_\alpha[\mathfrak{T}_\alpha]$, for each α.

Every topological projective limit can be taken in reduced form. For this it is only necessary to replace the spaces $F_\alpha[\mathfrak{T}_\alpha]$ by the closures of their subspaces $\hat{P}_\alpha(E)$, and to replace the mappings $A_{\alpha\beta}$ by their restrictions to these closed subspaces. The reduced topological projective limit formed in this way is clearly topologically isomorphic to the original one.

We now obtain the result corresponding to Theorem (3):

(6) *The dual E' of a reduced topological projective limit* $E[\mathfrak{T}]=\varprojlim A_{\alpha\beta}(F_\beta[\mathfrak{T}_\beta])$ *is the inductive limit* $\varinjlim A'_{\alpha\beta}(F'_\alpha)$, *where $A'_{\alpha\beta}$ is the adjoint of the continuous mapping $A_{\alpha\beta}$ from $F_\beta[\mathfrak{T}_\beta]$ into $F_\alpha[\mathfrak{T}_\alpha]$, mapping F'_α into F'_β.*

Since the relation $A'_{\beta\gamma}A'_{\alpha\beta}=A'_{\alpha\gamma}$, follows from the relation $A_{\alpha\beta}A_{\beta\gamma}=A_{\alpha\gamma}$, for $\alpha<\beta<\gamma$, the inductive limit $\varinjlim A'_{\alpha\beta}(F'_\alpha)=\left(\bigoplus_\alpha F'_\alpha\right)/H_0$ exists, by § 19, 2.

$E[\mathfrak{T}]$ is a subspace of $\prod_\alpha F_\alpha[\mathfrak{T}_\alpha]$. By 1.(1), E' is equal to $\left(\bigoplus_\alpha F'_\alpha\right)/E^\perp$; our theorem is therefore established if we can show that $E^\perp = H_0$.

Suppose that $u = \sum_{i=1}^n u_{\alpha_i}$ is an arbitrary element of $\bigoplus_\alpha F'_\alpha$, and that $x = (x_\alpha)$ is an element in E. If $\beta \geqq \alpha_i$, $i = 1, \ldots, n$, we have $x_{\alpha_i} = A_{\alpha_i\beta} x_\beta$ for $i = 1, \ldots, n$. Thus we have

$$ux = \sum_{i=1}^n u_{\alpha_i} x_{\alpha_i} = \sum u_{\alpha_i}(A_{\alpha_i\beta} x_\beta) = \left(\sum_{i=1}^n A'_{\alpha_i\beta} u_{\alpha_i}\right) x_\beta. \tag{7}$$

If $u \in E^\perp$, it therefore follows from (7) and the fact that E is in reduced form that the equation $\sum_{i=1}^n A'_{\alpha_i\beta} u_{\alpha_i} = \mathfrak{o}$ holds. By § 19, 2.(8) this means that u belongs to H_0. Conversely, it follows in a similar way from (7) and § 19, 2.(8) that $H_0 \subset E^\perp$.

7. Topologies on locally convex hulls and kernels. We first obtain a simple lemma about the formation of polars.

(1) *Suppose that A is a weakly continuous linear mapping from $E_1[\mathfrak{T}_1]$ into $E_2[\mathfrak{T}_2]$. Then if M is any subset of E_1*

$$A(M)^\circ = A'^{(-1)}(M^\circ). \tag{2}$$

Proof. A' maps E'_2 into E'_1. By definition $A(M)^\circ$ consists of all the $v \in E'_2$ with $\Re v(Ax) \leqq 1$ for all $x \in M$. This is equivalent to $\Re(A'v)x \leqq 1$ for all $x \in M$, and so to $A'v \in M^\circ$, i.e. $v \in A'^{(-1)}(M^\circ)$.

We now consider the locally convex hull $E[\mathfrak{T}] = \sum_\alpha A_\alpha(F_\alpha[\mathfrak{T}_\alpha]) = A\left(\bigoplus_\alpha F_\alpha[\mathfrak{T}_\alpha]\right)$. It follows directly from (2) and 5.(1) that

(3) *If, for each α, N_α is an absolutely convex subset of F_α and N_α° is its polar in F'_α, then the polar of $A\left(\Gamma_\alpha N_\alpha\right)$ in E' is equal to $A'^{(-1)}\left(\prod_\alpha N_\alpha^\circ\right)$.*

We can obtain information about the hull topology from this:

(4) *Suppose that $E[\mathfrak{T}] = \sum_\alpha A_\alpha(F_\alpha[\mathfrak{T}_\alpha]) = A\left(\bigoplus_\alpha F_\alpha[\mathfrak{T}_\alpha]\right)$. Let $\mathfrak{M}_\alpha$ be the collection of $\mathfrak{T}_\alpha$-equicontinuous subsets of F'_α, and let $\mathfrak{M}$ be the saturated collection of subsets of $\prod_\alpha F'_\alpha$ consisting of all the sets $\prod_\alpha M_\alpha$, $M_\alpha \in \mathfrak{M}_\alpha$, together with their subsets. Then $A'^{(-1)}(\mathfrak{M})$ is the collection of $\mathfrak{T}$-equicontinuous subsets of E', so that $\mathfrak{T} = \mathfrak{T}_{A'^{(-1)}(\mathfrak{M})}$.*

The neighbourhoods of $\mathfrak{o}$ of the form $A\left(\Gamma_\alpha U_\alpha\right)$, where each U_α is a $\mathfrak{T}_\alpha$-neighbourhood of $\mathfrak{o}$ in $F_\alpha[\mathfrak{T}_\alpha]$, form a base of $\mathfrak{T}$-neighbourhoods of $\mathfrak{o}$ in $E[\mathfrak{T}]$. Thus the sets $A\left(\Gamma_\alpha U_\alpha\right)^\circ$ and their subsets form the collection of $\mathfrak{T}$-equicontinuous subsets of E'. The assertion now follows from (3).

We now determine the equicontinuous sets for the kernel topology:

(5) *Suppose that* $E[\mathfrak{T}]=\mathsf{K}_\alpha A_\alpha^{(-1)}(F_\alpha[\mathfrak{T}_\alpha])=A^{(-1)}\left(\prod_\alpha F_\alpha[\mathfrak{T}_\alpha]\right)$. *Let* $\mathfrak{M}_\alpha$ *be the collection of* $\mathfrak{T}_\alpha$*-equicontinuous subsets of* F'_α, *and let* $\mathfrak{M}$ *be the collection of subsets* $M=\bigoplus_{i=1}^n M_{\alpha_i}$, $M_{\alpha_i}\in\mathfrak{M}_{\alpha_i}$ *of* $\bigoplus_\alpha F'_\alpha$, *together with their subsets. We denote by* $A'(\mathfrak{M})$ *the collection of the sets* $A'(M)=\sum_{i=1}^n A'_{\alpha_i}(M_{\alpha_i})$ *in* E'.

Then $A'(\mathfrak{M})$ *is the collection of* $\mathfrak{T}$*-equicontinuous subsets of* E', *so that* $\mathfrak{T}=\mathfrak{T}_{A'(\mathfrak{M})}$.

A base of $\mathfrak{T}$-neighbourhoods of $\mathfrak{o}$ is given by the finite intersections of sets $A_\alpha^{(-1)}(M_\alpha^\circ)$, where we can suppose that the $M_\alpha\in\mathfrak{M}_\alpha$ are weakly compact and absolutely convex. The collection of $\mathfrak{T}$-equicontinuous subsets of E' is therefore formed by the weakly closed convex covers of finitely many $A_\alpha^{(-1)}(M_\alpha^\circ)^\circ$, together with their subsets. The mapping A'_α from F'_α into E' is weakly continuous, by § 20, 4.(6); applying (2) to it, we obtain $A_\alpha^{(-1)}(M_\alpha^\circ)=A'_\alpha(M_\alpha)^\circ$; consequently $A_\alpha^{(-1)}(M_\alpha^\circ)=A'_\alpha(M_\alpha)^{\circ\circ}$, and since M_α is absolutely convex and weakly compact, $A'_\alpha(M_\alpha)^{\circ\circ}=A'_\alpha(M_\alpha)$. Since the sets $A'_\alpha(M_\alpha)$ are equicontinuous, so also are the sets $\sum_{i=1}^n A'_{\alpha_i}(M_{\alpha_i})$, which are absolutely convex and weakly compact, so that every equicontinuous set lies in such a set.

If each $\mathfrak{M}_\alpha$ is the collection of finite-dimensional bounded subsets of F'_α, $A'(\mathfrak{M})$ is the collection of finite-dimensional bounded subsets of E', and so it follows from (5) that

(6) *The weak topology* $\mathfrak{T}_s(E')$ *of a kernel* $\mathsf{K}_\alpha A_\alpha^{(-1)}(F_\alpha[\mathfrak{T}_\alpha])$ *is the kernel topology of* $\mathsf{K}_\alpha A_\alpha^{(-1)}(F_\alpha[\mathfrak{T}_s(F'_\alpha)])$.

Interchanging the two spaces, we obtain

(7) *The weak dual of a locally convex hull* $E[\mathfrak{T}]=\sum_\alpha A_\alpha(F_\alpha[\mathfrak{T}_\alpha])$ *is equal to* $\mathsf{K}_\alpha A_\alpha'^{(-1)}(F'_\alpha[\mathfrak{T}_s(F_\alpha)])$.

There are no corresponding general results for the topologies $\mathfrak{T}_k$, $\mathfrak{T}_b$ and $\mathfrak{T}_{b^*}$ on a kernel. Suppose that we are given a kernel in the form $\mathsf{K}_\alpha P_\alpha^{(-1)}(F_\alpha[\mathfrak{T}_\alpha])$, so that we have a subspace of $\prod_\alpha F_\alpha[\mathfrak{T}_\alpha]$. Then $E'=\left(\bigoplus_\alpha F'_\alpha\right)/H$, and A' is the canonical mapping K from $\bigoplus_\alpha F'_\alpha$ onto E'. If $\mathfrak{M}_\alpha$ is the class of all bounded subsets of F'_α, then $\mathfrak{M}$ is the class of all bounded subsets of $\bigoplus_\alpha F'_\alpha$. The sets M of $\mathfrak{M}$ are mapped by K into bounded subsets of E', but the sets $\overline{K(M)}$ need not determine all the bounded sets of E' (cf. 2.). We can draw similar conclusions about $\mathfrak{T}_k$ and $\mathfrak{T}_{b^*}$.

In general, therefore, we can only say that *the strong topology of a kernel is always finer than the kernel topology of* $\mathsf{K}_\alpha A_\alpha^{(-1)}(F_\alpha[\mathfrak{T}_b(F'_\alpha)])$. *Similar results hold for* $\mathfrak{T}_k$ *and* $\mathfrak{T}_{b^*}$.

There is no result analogous to (6) for locally convex hulls. We saw in 5.(4) that even for locally convex direct sums the weak topology can be strictly coarser than the corresponding sum topology. In general the hull topology of $\sum_\alpha A_\alpha(F_\alpha[\mathfrak{T}_s])$ is finer than the weak topology of the hull. However,

(8) *Suppose that* $E[\mathfrak{T}]=\sum_\alpha A_\alpha(F_\alpha[\mathfrak{T}_\alpha])$ *is a locally convex hull. Then we have* $E[\mathfrak{T}_k(E')]=\sum_\alpha A_\alpha(F_\alpha[\mathfrak{T}_k(F'_\alpha)])$.

We can take $E[\mathfrak{T}]$ in the special form $\left(\bigoplus_\alpha F_\alpha[\mathfrak{T}_\alpha]\right)/H$, where $A=K$, the canonical mapping of $\bigoplus_\alpha F_\alpha[\mathfrak{T}_\alpha]$ onto E. Then E' is equal to the space $H^\perp$ orthogonal to H in $\prod_\alpha F'_\alpha$. The mapping K' is the embedding of E' into $\prod_\alpha F'_\alpha$. Now let $\mathfrak{M}_\alpha$ be the class of absolutely convex weakly relatively compact subsets of F'_α. The collection $\mathfrak{M}$ constructed as in (4) contains all such subsets of $\prod_\alpha F'_\alpha$, by 5.(4). The collection $K'^{(-1)}(\mathfrak{M})$ consists of all $M\cap H^\perp$, with $M\in\mathfrak{M}$. $H^\perp$ is weakly closed in $\prod_\alpha F'_\alpha$, and the topology $\mathfrak{T}_s(E)$ coincides with $\mathfrak{T}_s\left(\bigoplus_\alpha F_\alpha\right)$. $K'^{(-1)}(\mathfrak{M})$ therefore consists of all the absolutely convex relatively $\mathfrak{T}_s(E)$-compact subsets of E'; the assertion now follows by applying (4) to the compatible topology $\mathfrak{T}=\mathfrak{T}_k$.

The corresponding statement about strong topologies does not hold in general, and indeed it is false in the special case of a quotient space (cf. 2.).

The result on the Mackey topology of the dual of a locally convex kernel which corresponds to (8) is true, provided that just one further condition is imposed. We can suppose that the locally convex kernel is a subspace $\hat{E}=\mathsf{K}_\alpha P_\alpha^{(-1)}(F_\alpha[\mathfrak{T}_\alpha])$ of $\prod_\alpha F_\alpha[\mathfrak{T}_\alpha]$. The hull topology of $\sum_\alpha K_\alpha(F'_\alpha[\mathfrak{T}_k])$ is then defined by the neighbourhoods $\left(\hat{E}\cap\prod_\alpha C_\alpha\right)^\circ$, where each C_α is a weakly compact absolutely convex subset of $F_\alpha[\mathfrak{T}_\alpha]$. $\prod_\alpha C_\alpha$ is certainly weakly compact and absolutely convex in $\prod_\alpha F_\alpha$, but $\hat{E}\cap\prod_\alpha C_\alpha$ need not be weakly compact in $\hat{E}$, if $\hat{E}$ is not closed in $\prod_\alpha F_\alpha$. If this is the case, however, then $\mathfrak{T}_k(E)$ is equal to the hull topology of $\sum_\alpha A'(F'_\alpha[\mathfrak{T}_k])$; in general this hull topology is finer than $\mathfrak{T}_k(E)$.

For projective limits, therefore, we have, by § 19, 10.(3) and 6.(6):

(9) *If* $E[\mathfrak{T}]=\varprojlim A_{\alpha\beta}(F_\beta[\mathfrak{T}_\beta])$ *is a topological projective limit in reduced form, the topology* $\mathfrak{T}_k(E)$ *of the dual is equal to the hull topology of* $\varinjlim A'_{\alpha\beta}(F'_\alpha[\mathfrak{T}_k])$.

The fact that the inductive limit is a topological inductive limit follows from the fact that the weakly continuous mappings $A'_{\alpha\beta}$ are also $\mathfrak{T}_k$-continuous, by § 21, 4.(6).

CHAPTER FIVE

Topological and Geometrical Properties of Locally Convex Spaces

We continue with the general theory of locally convex spaces. In § 23 the bidual space is introduced, and the question of the semi-reflexivity or reflexivity of a locally convex space is raised. A number of criteria are established, and we examine the connection with other structural properties.

In § 24 we consider the question of finding conditions for a subset of a locally convex space to be weakly compact. The important theorems of EBERLEIN and of KREIN are obtained in full generality, as are several related criteria. We give two proofs of KREIN's theorem. The first is due to GROTHENDIECK and uses integration theory, while the second was given recently by PTÁK, who succeeded in proving the result without the help of integration theory. We also establish three criteria of KLEE's for semireflexivity, which are of a geometric nature.

§ 25 is devoted to a group of problems centred on the KREIN-MILMAN theorem; it deals with extreme points and extreme rays of compact and locally compact convex sets.

Closely related to this is the investigation, in § 26, of the various refinements of the concept of convexity. These have proved to be important above all for the structure of normed spaces. Thus the reflexivity of a Banach space follows from the uniform convexity of its unit ball. Strict convexity is of importance for questions of approximation. The concepts dual to strict and uniform convexity, namely smoothness and uniform smoothness, are equivalent to differentiability properties of the norm.

§ 23. The bidual space. Semi-reflexivity and reflexivity

1. Quasi-completeness. The locally convex spaces met with in applications are frequently not complete. We have already seen that important results in the general theory can be established under the weaker hypotheses of sequential completeness or quasi-completeness (e.g. the BANACH-MACKEY theorem).

Both these concepts were introduced in § 18, 4. We shall meet examples of quasi-complete spaces which are not complete later on.

We now give an example of a sequentially complete space which is not quasi-complete. Suppose that $d>\aleph_0$, and that H is the linear subspace of ω_d consisting of all vectors $\mathfrak{x}=(\xi_\alpha)$ with only countably many non-zero coordinates ξ_α. H is dense in ω_d in the topology introduced on ω_d in § 15, 4., and indeed every element of ω_d is a closure point of a bounded subset of H (cf. § 15, 6.); H is therefore not quasi-complete, although it is sequentially complete.

Clearly we have

(1) *Every closed linear subspace of a sequentially complete (respectively quasi-complete) locally convex space is again sequentially complete (respectively quasi-complete).*

The corresponding assertion for quotient spaces is not true; indeed, the example $\varphi\omega \oplus \omega\varphi$ of § 13, 6. shows that the quotient of a complete locally convex space need not be sequentially complete. We observe that the proof given for linear topologies in § 13, 6. also holds for the locally convex topology on $\varphi\omega \oplus \omega\varphi$ which is obtained from the topologies of ω and φ by forming the topological product and locally convex direct sum.

(2) *The topological product and the locally convex direct sum of sequentially complete (respectively quasi-complete) locally convex spaces are again sequentially complete (respectively quasi-complete).*

Proof. If $E[\mathfrak{T}] = \prod_\alpha E_\alpha[\mathfrak{T}_\alpha]$, a sequence $x^{(n)} = (x_\alpha^{(n)})$ in $E[\mathfrak{T}]$ is a Cauchy sequence if and only if the $x_\alpha^{(n)}$ form a Cauchy sequence in E_α, for each α. The sequential completeness of E therefore follows from the sequential completeness of the spaces E_α.

A set $B \subset E[\mathfrak{T}]$ is bounded if and only if each of its projections $B_\alpha = P_\alpha B$ is bounded in $E_\alpha[\mathfrak{T}_\alpha]$. If the spaces $E_\alpha[\mathfrak{T}_\alpha]$ are all quasi-complete, the closure $\bar{B}_\alpha$ of B_α in E_α is complete, for each α, and so also is $\prod_\alpha \bar{B}_\alpha$ in E. Since this set contains B, B is also complete, provided that it is assumed to be closed.

Applying § 18, 5.(4), we easily obtain the assertion about locally convex direct sums.

A subset M of a locally convex space $E[\mathfrak{T}]$ is said to be quasi-closed if it contains all the closure points in E of its bounded subsets. The intersection of arbitrarily many, and the union of finitely many, quasi-closed subsets are again quasi-closed. If A is a continuous linear mapping from $E[\mathfrak{T}]$ into $F[\mathfrak{T}']$, the inverse image $A^{(-1)}(M)$ of every quasi-closed subset M of F is quasi-closed.

The quasi-closure $\hat{M}$ of a set $M \subset E[\mathfrak{T}]$ is the intersection of all the quasi-closed subsets of E which contain M. Each point of $\hat{M}$ is called a strict closure point of M. Such a point is naturally also a closure point of M, but the converse is not always true. On the other hand a strict closure point of M need not be closure point of a bounded subset of M (cf. the example at the end of this number).

The continuous image Ax of a strict closure point x of M is a strict closure point of $A(M)$, for if $A(M)$ lies in the quasi-closed set N, M

lies in $A^{(-1)}(N)$, and $A^{(-1)}(N)$ is quasi-closed and therefore contains x, so that Ax lies in N.

The quasi-completion $\bar{E}$ of a locally convex-space $E[\mathfrak{T}]$ is the quasi-closure of E in the completion $\tilde{E}[\tilde{\mathfrak{T}}]$. $\bar{E}$ is quasi-complete under under the topology $\bar{\mathfrak{T}}$ induced on $\bar{E}$ by $\tilde{\mathfrak{T}}$.

(3) *The dual E' of a barrelled space $E[\mathfrak{T}]$ is weakly quasi-complete.*

For the bounded weakly closed subsets of E' are weakly compact, by § 21, 4. (4), and so they are weakly complete.

We shall prove the converse in 6. (4).

Hilbert space is weakly quasi-complete, by (3), but is not weakly complete (cf. § 20, 9. (2)).

(4) *Let A be a continuous linear mapping from $E[\mathfrak{T}]$ into $F[\mathfrak{T}']$. If F is quasi-complete (respectively complete), A can be extended in a unique way to a continuous linear mapping from the quasi-completion (respectively completion) of $E[\mathfrak{T}]$ into $F[\mathfrak{T}']$.*

Proof. Suppose that F is complete. By § 15, 2. (4), A is uniformly continuous, and so by § 5, 4. (4) it can be extended continuously to $\tilde{E}$ in a unique way. The extension $\tilde{A}$ is again linear. We have therefore proved the assertion when $F[\mathfrak{T}']$ is complete.

If F is quasi-complete, $\tilde{A}$ maps the quasi-completion $\bar{E}$ into $\tilde{F}$. But every strict closure point of E in $\bar{E}$ is mapped by $\tilde{A}$ into a strict closure point of $A(\bar{E})$ in $\tilde{F}$. Since, by hypothesis, this lies in F, $\tilde{A}(\bar{E}) \subset F$.

Example. Using the terminology of § 13, 5., the locally convex sum of countably many spaces ω is denoted by $\varphi\omega$. The elements of $\varphi\omega$ are of the form $\mathfrak{x} = (\xi_{ik}) = \sum_{i,k=1}^{\infty} \xi_{ik}\,\mathfrak{e}_{ik}$, with $\xi_{ik} = 0$ for $i \geqq i_0$ and all k. Let H be the linear span of the elements $\mathfrak{a}_{nk} = \mathfrak{e}_{1n} + \mathfrak{e}_{nk}$, $n, k = 1, 2, \ldots$. None of the elements $\mathfrak{e}_{1n}$ lies in H, although $\mathfrak{e}_{1n}$ is the limit of $\mathfrak{e}_{1n} + \mathfrak{e}_{nk}$ as $k \to \infty$, and so belongs to the quasi-closure $\hat{H}$. Every sum $\sum_{1}^{m} \mathfrak{e}_{1n}$ also lies in $\hat{H}$; $\mathfrak{x}_0 = \sum_{n=1}^{\infty} \mathfrak{e}_{1n}$ is limit of the terms $\sum_{1}^{m} \mathfrak{e}_{1n}$, and so $\mathfrak{x}_0$ is an element of $\hat{H}$. $\mathfrak{x}_0$, however, is not a closure point of a bounded subset of H; for such a set can only contain linear combinations of the $\mathfrak{e}_{1n} + \mathfrak{e}_{nk}$ with $n \leqq n_0$. $\mathfrak{x}_0$ is therefore a strict closure point of H, but not a closure point of a bounded subset of H.

2. The bidual space. If we give the dual E' of a locally convex space $E[\mathfrak{T}]$ a topology $\mathfrak{T}_{\mathfrak{M}}$, where $\mathfrak{M}$ is a total saturated class of bounded subsets of $E[\mathfrak{T}]$, then we know from § 21, 4. that the dual of $E'[\mathfrak{T}_{\mathfrak{M}}]$ coincides with E if and only if $\mathfrak{T}_{\mathfrak{M}}$ is coarser than the Mackey topology $\mathfrak{T}_k(E)$ and finer than the weak topology $\mathfrak{T}_s(E)$.

By § 21, 1. (2), $E'[\mathfrak{T}_{\mathfrak{M}}]$ is locally convex, and its dual is a subspace of the algebraic dual $(E')^*$, which, by § 20, 9. (2), can also be considered as the weak completion $\tilde{E}[\mathfrak{T}_s(E')]$ of E.

The dual of $E'[\mathfrak{T}_{\mathfrak{M}}]$ can be determined in the following way:

(1) *If $\mathfrak{M}$ is a total saturated class of bounded subsets of the locally convex space $E[\mathfrak{T}]$, the dual of $E'[\mathfrak{T}_{\mathfrak{M}}]$ is equal to $\bigcup_{M \in \mathfrak{M}} \bar{M}$, where $\bar{M}$ is the weak closure of M in $(E')^* = \tilde{E}[\mathfrak{T}_s(E')]$.*

Proof. Every continuous linear functional z on $E'[\mathfrak{T}_{\mathfrak{M}}]$ is bounded in modulus by 1 on some suitable $\mathfrak{T}_{\mathfrak{M}}$-neighbourhood M° of $\mathfrak{o}$, where M is absolutely convex and belongs to $\mathfrak{M}$. In other words, z lies in the polar $M^{\circ\circ}$ of M° in $(E')^*$. Conversely every element of $M^{\circ\circ}$ is a continuous linear functional on E'. By the theorem of bipolars, $M^{\circ\circ} = \bar{M}$, and so we have $(E'[\mathfrak{T}_{\mathfrak{M}}])' = \bigcup_{M \in \mathfrak{M}} \bar{M}$.

(2) *If $\mathfrak{T}_{\mathfrak{M}}$ is finer than $\mathfrak{T}_s(E)$, E is a subspace of $(E'[\mathfrak{T}_{\mathfrak{M}}])'$.*

For each $x_0 \in E$ is a weakly continuous, and *a fortiori* $\mathfrak{T}_{\mathfrak{M}}$-continuous, linear functional on E.

The finest of the topologies $\mathfrak{T}_{\mathfrak{M}}$ on E' is the strong topology $\mathfrak{T}_b(E)$, and this therefore produces the largest dual space. In conformity with the terminology for normed spaces introduced in § 14, 5., we denote the dual of $E'[\mathfrak{T}_b(E)]$ by E'', and call it the b i d u a l space of $E[\mathfrak{T}]$. In this case, (1) becomes:

(3) *If $E[\mathfrak{T}]$ is a locally convex space, the bidual E'' is the union of the weak closures in $(E')^*$ of the bounded subsets of $E[\mathfrak{T}]$.*

E'' therefore always lies in the $\mathfrak{T}_s(E')$-quasi-completion of E.

Every bounded subset B of $E[\mathfrak{T}]$ is weakly precompact, by § 20, 9. (3). Since $(E')^*$ is weakly complete, by § 20, 9. (2), the weak closure $\bar{\bar{B}}$ of B in E'' is weakly compact; it therefore follows from (3) that

(4) *Every bounded subset of $E[\mathfrak{T}]$ is weakly relatively compact in the bidual E''.*

This result can also be expressed as a statement about the topologies on E':

(5) *The Mackey topology $\mathfrak{T}_k(E'')$ on E' is always finer than the strong topology $\mathfrak{T}_b(E)$ on E'.*

3. Semi-reflexivity. In § 14, 5. we said that a (B)-space E was reflexive if the bidual E'' coincided with E as a (B)-space. This means first that E'' is equal to E as a vector space and secondly that the norm of E'' coincides with the norm of E. Because of § 17, 6. (3), the second statement is a consequence of the first.

In the case of an arbitrary locally convex space the corresponding circumstances are more complicated; for this reason we shall begin by considering the general form of the first statement. Suppose that $E[\mathfrak{T}]$ is locally convex. When we consider E as a space of continuous linear functionals on E', we have $E \subset E''$. $E[\mathfrak{T}]$ is said to be s e m i - r e f l e x i v e if E is equal to E''. We have the following criterion:

(1) *A locally convex space $E[\mathfrak{T}]$ is semi-reflexive if and only if every bounded subset of $E[\mathfrak{T}]$ is relatively weakly compact, i. e. if and only if the topologies $\mathfrak{T}_b(E)$ and $\mathfrak{T}_k(E)$ coincide on E'.*

Proof. If $E''=E$, then the relatively weakly compact subsets are the same as the bounded subsets, by 2.(4), so that $\mathfrak{T}_b(E)$ is equal to $\mathfrak{T}_k(E)$ on E'.

On the other hand, if this is the case, it follows from the MACKEY-ARENS theorem that $E''=(E'[\mathfrak{T}_k(E)])'=E$.

A second criterion is

(2) *A locally convex space $E[\mathfrak{T}]$ is semi-reflexive if and only if it is weakly quasi-complete.*

Proof. Suppose that E is weakly quasi-complete. By § 20, 9.(3), every bounded set is weakly precompact, and so, since it is contained in a bounded weakly complete set, it is weakly relatively compact. $E[\mathfrak{T}]$ is therefore semi-reflexive, by (1).

Conversely every semi-reflexive space is weakly quasi-complete, by (1), or 2.(4).

For (B)-spaces, it follows from (1), (2) and the fact that reflexivity and semi-reflexivity are the same that

(3) *A (B)-space E is reflexive if and only if the closed unit ball of E is weakly compact, or weakly complete.*

From (1) and § 21, 2.(2) it follows at once that

(4) *The strong dual of a semi-reflexive locally convex space is barrelled.*

The converse of this is not true. For example the (B)-space c_0 is not reflexive, by § 14, 7., whereas its strong dual l^1 is barrelled, since it is a (B)-space.

(5) *Every closed linear subspace H of a semi-reflexive space $E[\mathfrak{T}]$ is semi-reflexive.*

The topology $\mathfrak{T}_s(H')$ coincides on H with the topology induced by $\mathfrak{T}_s(E')$, by § 22, 2.(1). Since H is weakly closed in E, every weakly closed bounded subset of H is weakly complete, by (2), so that H is weakly quasi-complete, and is therefore semi-reflexive, again by (2).

A quotient space of a semi-reflexive space, and likewise the strong dual of a semi-reflexive space, are in general not semi-reflexive (cf. 5. and 6.).

(6) *The locally convex direct sum and the topological product of semi-reflexive locally convex spaces are again semi-reflexive.*

Proof. Suppose that $E[\mathfrak{T}]=\bigoplus_{\alpha} E_\alpha[\mathfrak{T}_\alpha]$. By § 18, 5.(4), each bounded subset of $E[\mathfrak{T}]$ is contained in a set of the form $B=\bigoplus_{i=1}^{n} B_{\alpha_i}$, where B_{α_i}

is bounded in $E_{\alpha_i}[\mathfrak{T}_{\alpha_i}]$. By hypothesis we can suppose that the sets B_{α_i} are $\mathfrak{T}_s(E'_{\alpha_i})$-compact. By TYCHONOFF's theorem, B is compact in the product topology of $\bigoplus_{i=1}^{n} E_{\alpha_i}[\mathfrak{T}_s(E_{\alpha_i})]$, and this coincides with the topology induced by $\mathfrak{T}_s(E')$, by § 22, 5.(3). The semi-reflexivity of $E[\mathfrak{T}]$ now follows from (1).

b) The semi-reflexivity of a topological product $\prod_\alpha E_\alpha[\mathfrak{T}_\alpha]$ follows in an analogous way from the fact that every bounded set lies in a product $\prod_\alpha B_\alpha$ of bounded and weakly compact sets, and this is weakly compact in $\prod_\alpha E_\alpha$, by § 22, 5.(3).

(7) *The projective limit* $E[\mathfrak{T}] = \varprojlim A_{\alpha\beta}(F_\beta[\mathfrak{T}_\beta])$ *of semi-reflexive spaces* $F_\beta[\mathfrak{T}_\beta]$ *is semi-reflexive.*

By § 19, 10.(3), $E[\mathfrak{T}]$ is a closed linear subspace of $\prod_\alpha F_\alpha[\mathfrak{T}_\alpha]$, and so the assertion follows from (5) and (6).

4. The topologies on the bidual. The bidual E'' of a locally convex space $E[\mathfrak{T}]$ forms a dual pair $\langle E'', E'\rangle$ with E', and so the usual topologies can be defined on E''. Since we must consider the original dual pair $\langle E', E\rangle$ as well as $\langle E'', E'\rangle$ in what follows, we shall occasionally depart from our present usage by mentioning the second space of the dual pair when we describe a topology; for example, $\mathfrak{T}_b(E', E'')$ denotes the strong topology on E'' with respect to the dual pair $\langle E', E''\rangle$. Its equicontinuous sets are the $\mathfrak{T}_s(E'')$-bounded subsets of E'.

By the strong bidual of $E[\mathfrak{T}]$ we mean the bidual E'' equipped with the topology $\mathfrak{T}_b(E', E'')$; the strong bidual is therefore the strong dual of $E'[\mathfrak{T}_b(E)]$.

E is a subspace of E''. We now investigate the topology which $\mathfrak{T}_b(E, E'')$ induces on E. This need not be the strong topology on E; rather, we have

(1) $\mathfrak{T}_b(E', E'')$ *induces the topology* $\mathfrak{T}_{b*}(E', E)$ *on* E.

Proof. By MACKEY's theorem (§ 20, 11.(7)), applied to $E'[\mathfrak{T}_b(E)]$ and its dual E'', the $\mathfrak{T}_s(E'', E')$-bounded subsets of E' are the same as the bounded subsets of $E'[\mathfrak{T}_b(E)]$, and these are the strongly bounded subsets of E' with respect to the dual pair $\langle E', E\rangle$. Thus the $\mathfrak{T}_b(E', E'')$- and $\mathfrak{T}_{b*}(E', E)$-equicontinuous subsets of E' are the same.

Besides the topology $\mathfrak{T}_b(E', E'')$, there is a second important topology, the natural topology $\mathfrak{T}_n(E')$, on the bidual of $E[\mathfrak{T}]$. This is defined as the topology of uniform convergence on the $\mathfrak{T}$-equicontinuous subsets of E'. A base of neighbourhoods of $\mathfrak{o}$ for $\mathfrak{T}_n(E')$ therefore consists of all the polars $U^{\circ\circ}$ in E'' of the sets $U^\circ \subset E'$, as U runs through the absolutely convex neighbourhoods of $\mathfrak{o}$ of $E[\mathfrak{T}]$.

Thus if $\{U\}$ is a base of $\mathfrak{T}$-neighbourhoods of $\mathfrak{o}$ in E, the $\mathfrak{T}_s(E',E'')$-closed convex covers in E'' of the sets U form a base of $\mathfrak{T}_n(E')$-neighbourhoods of $\mathfrak{o}$ in E''.

Clearly,

(2) *The natural topology $\mathfrak{T}_n(E')$ always induces the original topology $\mathfrak{T}$ on E.*

$\mathfrak{T}_n(E')$ is always coarser than the strong topology $\mathfrak{T}_b(E',E'')$, for every $\mathfrak{T}$-equicontinuous subset of E' is $\mathfrak{T}_b(E)$-bounded, by § 21, 5.(1).

When do the natural and the strong topologies coincide on E''? This happens if and only if the $\mathfrak{T}$-equicontinuous subsets of E' are the same as the sets which are strongly bounded with respect to E, and thus if and only if $\mathfrak{T}$ is equal to $\mathfrak{T}_{b*}(E')$.

A locally convex space $E[\mathfrak{T}]$ is said to be quasi-barrelled if $\mathfrak{T}$ coincides with $\mathfrak{T}_{b*}(E')$. Since $\mathfrak{T}_{b*}(E')$ is always finer than $\mathfrak{T}$ and coarser than $\mathfrak{T}_b(E')$, a barrelled space is always quasi-barrelled; the converse is not always true.

Later we shall investigate barrelled and quasi-barrelled spaces in detail; for the present we shall content ourselves with a simple characterization of quasi-barrelled spaces.

We say that a subset M of a vector space absorbs a set N if $\rho N \subset M$ for a suitable $\rho > 0$. Using this terminology, we have

(3) *A locally convex space $E[\mathfrak{T}]$ is quasi-barrelled if and only if every barrel in E which absorbs all the bounded sets of $E[\mathfrak{T}]$ is a $\mathfrak{T}$-neighbourhood of $\mathfrak{o}$.*

For the polar of such a barrel is strongly bounded in E'; conversely the polar of an absolutely convex strongly bounded subset of E' is a barrel with the given property.

It follows from the remarks made above that

(4) *The natural and strong topologies of a bidual space E'' are the same if and only if $E[\mathfrak{T}]$ is quasi-barrelled, and so if and only if the $\mathfrak{T}$-equicontinuous subsets of E' are the same as the $\mathfrak{T}_b(E)$-bounded sets.*

By § 21, 5.(3), all metrizable locally convex spaces, and in particular all (F)-spaces, are quasi-barrelled. Thus the natural and strong topologies coincide on their biduals.

The strong bidual of a normed space is a (B)-space. We shall prove in § 29, 2. that the strong bidual of a metrizable locally convex space is an (F)-space. For the present we establish the weaker result:

(5) *If $E[\mathfrak{T}]$ is metrizable, the strong bidual E'' is also metrizable.*

For if U_n, $n=1,2,\ldots$, is a base of neighbourhoods of $\mathfrak{o}$ in $E[\mathfrak{T}]$, the bipolars $U_n^{\circ\circ}$ in E'' form a base of neighbourhoods of $\mathfrak{o}$ for the natural, and so for the strong, topology on E''.

As a third topology on E'' we consider the Mackey topology $\mathfrak{T}_k(E', E'')$. It has as equicontinuous sets the absolutely convex $\mathfrak{T}_s(E'')$-relatively compact subsets of E', together with their subsets. Since $E'' \supset E$, the topology $\mathfrak{T}_s(E'')$ is finer than $\mathfrak{T}_s(E)$ on E'. Every $\mathfrak{T}_s(E'')$-compact subset of E' is therefore also $\mathfrak{T}_s(E)$-compact. Consequently we have

(6) *The topology $\mathfrak{T}_k(E', E'')$ induces a topology on E which is coarser than $\mathfrak{T}_k(E', E)$.*

We now show by giving an example that the two topologies can be different. In the strong dual l^1 of c_0 the closed unit ball K is $\mathfrak{T}_s(c_0)$-compact, by § 20, 9.(5), but it is not $\mathfrak{T}_s(l^\infty)$-compact, by § 22, 4.(3). As a result, $\mathfrak{T}_k(l^1, l^\infty)$ induces a weaker topology on c_0 than $\mathfrak{T}_k(l^1, c_0)$.

5. Reflexivity. A locally convex space $E[\mathfrak{T}]$ is said to be reflexive if the bidual E'' is equal to E, and if the topology $\mathfrak{T}_b(E', E'')$ coincides with the original topology; put another way, $E[\mathfrak{T}]$ is reflexive if the strong bidual of $E[\mathfrak{T}]$ coincides with $E[\mathfrak{T}]$.

For (B)-spaces, this agrees with the previous definition.

A first criterion for reflexivity is given by

(1) *A locally convex space is reflexive if and only if it is semi-reflexive and quasi-barrelled.*

This follows directly from 4.(2) and 4.(4).

From 3.(2) we obtain

(2) *A locally convex space is reflexive if and only if it is weakly quasi-complete and quasi-barrelled.*

The next result contains two further criteria:

(3) *A locally convex space $E[\mathfrak{T}]$ is reflexive a) if and only if $\mathfrak{T}$ is the Mackey topology and every bounded set in $E[\mathfrak{T}]$ and in $E'[\mathfrak{T}_k(E)]$ is relatively weakly compact, and b) if and only if $\mathfrak{T}$ is the Mackey topology and $E[\mathfrak{T}]$ and $E'[\mathfrak{T}_k(E)]$ are weakly quasi-complete, or semi-reflexive.*

Proof. Suppose that $E[\mathfrak{T}]$ is reflexive. From the fact that $E''=E$, and from the definition of $\mathfrak{T}_b(E', E'')$, it follows that $\mathfrak{T}$ coincides with the topology $\mathfrak{T}_b(E')$. Consequently the Mackey topology, which lies between $\mathfrak{T}$ and $\mathfrak{T}_b(E')$, coincides with them both. It follows from $\mathfrak{T}_b(E') = \mathfrak{T}_k(E')$ that every bounded subset of $E'[\mathfrak{T}_k(E)]$ is weakly relatively compact, and also that E' is weakly quasi-complete. It follows from 3.(1) and 3.(2) that E also has the same properties.

If, conversely, conditions a) or b) are satisfied, then by 3.(1) and 3.(2) $E[\mathfrak{T}]$ is semi-reflexive, and further $\mathfrak{T}_b(E') = \mathfrak{T}_k(E') = \mathfrak{T}$ on E.

Using the fact that every (F)-space is quasi-barrelled, the following generalization of 3.(3) now follows from (1) and (2):

(4) *An* (F)-*space* $E[\mathfrak{T}]$ *is reflexive if and only if it is semi-reflexive, and if and only if it is weakly quasi-complete, and if and only if every bounded subset of* $E[\mathfrak{T}]$ *is weakly relatively compact.*

A reflexive $E[\mathfrak{T}]$ is by definition barrelled, so we have:

(5) *If* $E[\mathfrak{T}]$ *is reflexive, the strong dual* $E'[\mathfrak{T}_b(E)]$ *is also reflexive.*

In many cases, the reflexivity of E can be deduced from the reflexivity of E':

(6) *Suppose that* $E[\mathfrak{T}]$ *is quasi-complete, and that* $\mathfrak{T}$ *is the Mackey topology. If the strong dual* $E'[\mathfrak{T}_b(E)]$ *is semi-reflexive,* $E[\mathfrak{T}]$ *is reflexive.*

Proof. The $\mathfrak{T}_b(E)$-bounded sets of E' are the E''-bounded sets of E', where E'' is the strong bidual. It follows from the semi-reflexivity of E' that these sets are relatively $\mathfrak{T}_s(E'')$-compact. Since $\mathfrak{T}_s(E)$ is coarser than $\mathfrak{T}_s(E'')$ on E', these sets are also relatively $\mathfrak{T}_s(E)$-compact. But the fact that the $\mathfrak{T}_b(E)$-bounded sets of E' are the same as the relatively $\mathfrak{T}_s(E)$-compact sets means that $E[\mathfrak{T}]$ is quasi-barrelled.

By (1), our theorem is proved if we can show that E is semi-reflexive. By 2.(3), E'' is obtained from E by taking the $\mathfrak{T}_s(E')$-closure points in $(E')^*$ of the bounded subsets of E. Since these can be taken to be absolutely convex, it is sufficient by § 20, 7.(6) to consider the $\mathfrak{T}_k(E',E'')$-closure points. But, as the argument of the first part of the proof shows, $\mathfrak{T}_k(E',E'')$ coincides on E with $\mathfrak{T}_k(E',E)=\mathfrak{T}$. It therefore follows from the assumption that $E[\mathfrak{T}]$ is quasi-complete that $E''=E$.

In particular, because of (5) we have

(7) *An* (F)-*space is reflexive if and only if its strong dual is reflexive.*

From the fact that c_0 is not reflexive (cf. § 14, 7.) it therefore follows that neither l^1 nor l^∞ is reflexive. By 3.(3) the closed unit balls of l^1 and of l^∞ are not weakly compact.

The strong dual of a (B)-space is again a (B)-space, and so from (7) we obtain

(8) *If the* (B)-*space E is not reflexive, all the iterated strong duals are not reflexive, and in the sequences*

$$E \subset E'' \subset E'''' \subset \cdots \quad \textit{and} \quad E' \subset E''' \subset \cdots$$

each space is a proper closed subspace of the one following it.

In § 29, 2. we shall prove that this holds for (F)-spaces as well.

The following analogue of 3.(6) holds:

(9) *The locally convex direct sum and the topological product of reflexive spaces are again reflexive.*

Proof. On each of the reflexive spaces $E_\alpha[\mathfrak{T}_\alpha]$, $\mathfrak{T}_\alpha$ is the Mackey topology, and so the locally convex direct sum topology $\mathfrak{T}$ on $E[\mathfrak{T}] = \bigoplus_\alpha E_\alpha[\mathfrak{T}_\alpha]$ is the Mackey topology on E, by § 22, 5.(4). By 3.(6), $E[\mathfrak{T}]$

is semi-reflexive. The spaces $E'_\alpha[\mathfrak{T}_b(E_\alpha)]$ are semi-reflexive, and by 3.(6) so also is $(E[\mathfrak{T}])' = \prod_\alpha E'_\alpha$. The reflexivity of $E[\mathfrak{T}]$ now follows from (3) b).

The argument for the topological product is similar.

It follows from 3.(5) and (4) that

(10) *Every closed linear subspace of a reflexive* (F)-*space is reflexive.*

For (B)-spaces, we have the further result:

(11) *If E is a reflexive* (B)-*space and if H is a closed linear subspace of E, then the* (B)-*space E/H is also reflexive.*

For under the canonical mapping K from E onto E/H the image $K(B)$ of the weakly compact unit ball B of E is $\mathfrak{T}_s(H^\perp)$-compact, by § 22, 2.(3). On the other hand $K(B)$ is dense in the closed unit ball of E/H, and so must coincide with it. The assertion now follows from 3.(3).

Later (§ 31, 5.) we shall give an example of a reflexive (F)-space which has a non-reflexive quotient space. (10) is not true for arbitary locally convex spaces, either. We now give an example of this.

By (9), the spaces of countable degree, which are obtained from the ground field K by repeatedly forming locally convex direct sums and topological products (cf. § 13, 5.), are all reflexive; in particular the space $\varphi\omega \oplus \omega\varphi$ and its dual $\omega\varphi \oplus \varphi\omega$ are reflexive. In § 13, 6., closed linear subspaces $H_1 \subset \varphi\omega \oplus \omega\varphi$ and $H_2 \subset \omega\varphi \oplus \varphi\omega$ were constructed with $H_2^\perp = H_1$. The quotient space $(\varphi\omega \oplus \omega\varphi)/H_1$ contains a sequence which is certainly a Cauchy sequence for the topology $\mathfrak{T}_s(H_1^\perp)$, but which does not converge. It follows from this that $(\varphi\omega \oplus \omega\varphi)/H_1$ is not even semi-reflexive.

Further H_2 is a closed linear subspace of $\omega\varphi \oplus \varphi\omega$ whose dual space coincides with $(\varphi\omega \oplus \omega\varphi)/H_1$, since $H_2^\perp = H_1$. Since this is not semi-reflexive, H_2 is not reflexive.

In general a non-reflexive (B)-space E has infinite co-dimension in E'', as c_0 does in l^∞, for example. James [1], [2] has given an example of a real (B)-space E with one-dimensional E''/E; further E is norm-isomorphic to E''. Civin and Yood [1] have made a thorough investigation of (B)-spaces with finite dimensional E''/E, which they call quasi-reflexive, relating their properties to ideas in a paper by Dixmier [1], which contains interesting results about the iterated duals of a (B)-space.

6. The relationship between semi-reflexivity and reflexivity. The question of whether a locally convex space $E[\mathfrak{T}]$ is semi-reflexive depends only on the dual pair $\langle E', E\rangle$, since the strong topology $\mathfrak{T}_b(E)$ on E' is determined by the dual pair alone. Thus $\mathfrak{T}$ can be replaced by any topology lying between $\mathfrak{T}_s(E')$ and $\mathfrak{T}_k(E')$ without affecting the assertion that E is semi-reflexive.

Thus if a dual pair $\langle E_2, E_1\rangle$ is given, E_1 can be defined to be semi-reflexive with respect to E_2 if $(E_2[\mathfrak{T}_b(E_1)])' = E_1$. E_1 is then semi-reflexive in the earlier sense with respect to any compatible locally convex topology.

In the same way the concept of reflexivity can be defined for dual pairs: a dual pair $\langle E_2, E_1\rangle$ is said to be reflexive if E_1 is semi-reflexive with respect to E_2 and E_2 is semi-reflexive with respect to E_1. Thus $\langle E_2,E_1\rangle$ is reflexive if $(E_1[\mathfrak{T}_b(E_2)])'=E_2$ and $(E_2[\mathfrak{T}_b(E_1)])'=E_1$.

(1) *A locally convex space* $E[\mathfrak{T}]$, *where* $\mathfrak{T}$ *is the Mackey topology, is reflexive if and only if* $\langle E',E\rangle$ *is a reflexive dual pair.*

This is a direct consequence of 5.(3) b).

From 5.(3) a) we obtain

(2) *A dual pair* $\langle E_2,E_1\rangle$ *is reflexive if and only if* E_1 *and* E_2 *are weakly quasi-complete, and if and only if every bounded set in* E_1 *and in* E_2 *is weakly relatively compact.*

There exist semi-reflexive spaces which are not reflexive. If $E[\mathfrak{T}]$ is reflexive, it is only necessary to replace the Mackey topology $\mathfrak{T}$ by a strictly coarser one which still gives E' as the dual. This is of course trivial, since it results from using a topology on E which is clearly unsuitable. What is the position when $\mathfrak{T}$ is the Mackey topology? Put another way, are there dual pairs $\langle E_2,E_1\rangle$ in which only one of the two spaces is semi-reflexive with respect to the other?

(3) *Suppose that* $\langle E_2,E_1\rangle$ *is a dual pair.*

a) $E_1[\mathfrak{T}_k(E_2)]$ *is barrelled if and only if* $E_2[\mathfrak{T}_k(E_1)]$ *is semi-reflexive.*

b) $\langle E_2,E_1\rangle$ *is reflexive if and only if* $E_1[\mathfrak{T}_k(E_2)]$ *and* $E_2[\mathfrak{T}_k(E_1)]$ *are barrelled.*

c) $E_1[\mathfrak{T}_k(E_2)]$ *is semi-reflexive but not reflexive if and only if* $E_2[\mathfrak{T}_k(E_1)]$ *is barrelled but not semi-reflexive.*

Proof. To say that $E_1[\mathfrak{T}_k(E_2)]$ is barrelled is to say that the bounded sets in $E_1'=E_2$ are relatively $\mathfrak{T}_s(E_1)$-compact. Since $E_2'=E_1$, this is the same as saying that $E_2[\mathfrak{T}_k(E_1)]$ is semi-reflexive. This proves a).

b) follows directly from a).

It follows from a) and b) that if $E_1[\mathfrak{T}_k(E_2)]$ is semi-reflexive and not reflexive then $E_2[\mathfrak{T}_k(E_1)]$ is barrelled and not semi-reflexive. Conversely if $E_2[\mathfrak{T}_k(E_1)]$ is barrelled and not semi-reflexive then $E_1[\mathfrak{T}_k(E_2)]$ is indeed semi-reflexive, by a), but it cannot be reflexive, for it would then be barrelled, which is not possible, by b). Thus c) is also proved.

The following refinement of 1.(3) follows from (3)a) and 3.(2):

(4) *Suppose that* $E[\mathfrak{T}]$ *is locally convex and that* $\mathfrak{T}$ *is the Mackey topology.* $E[\mathfrak{T}]$ *is barrelled if and only if* E' *is* $\mathfrak{T}_s(E)$*-quasi-complete.*

In another form:

The locally convex space $E[\mathfrak{T}]$ *is weakly quasi-complete if and only if* $E'[\mathfrak{T}_k(E)]$ *is barrelled.*

It follows from (3) c) that

(5) *We obtain all the locally convex spaces with the Mackey topology which are semi-reflexive but not reflexive by forming the duals E' of all spaces $E[\mathfrak{T}]$ which are barrelled but not semi-reflexive and giving them the Mackey topology $\mathfrak{T}_k(E)$.*

In particular, from § 21, 5.(3) we get

(6) *If $E[\mathfrak{T}]$ is a non-reflexive* (F)*-space, $E'[\mathfrak{T}_k(E)]$ is semi-reflexive, but not reflexive.*

7. Distinguished spaces. We saw in 6.(3) a) that the semi-reflexivity of $E[\mathfrak{T}]$ corresponds dually to $E'[\mathfrak{T}_k(E)]$ being barrelled. On the other hand we know (3.(4)) that the strong dual $E'[\mathfrak{T}_b(E)]$ of a semi-reflexive space $E[\mathfrak{T}]$ is always barrelled. We shall now determine a property of $E[\mathfrak{T}]$ which corresponds dually to the strong dual being barrelled.

A locally convex space is said to be distinguished if every E'-bounded subset B_1 of the strong bidual E'' lies in the $\mathfrak{T}_s(E', E'')$-closure in E'' of a bounded set B of E. This is the same as saying that for every B_1 there is a B with $B_1 \subset B^{\circ\circ}$, where $B^{\circ\circ}$ is the polar in E'' of $B^\circ \subset E'$. We now have

(1) *A locally convex space $E[\mathfrak{T}]$ is distinguished if and only if the strong dual $E'[\mathfrak{T}_b(E)]$ is barrelled.*

Proof. Suppose that $E[\mathfrak{T}]$ is distinguished. As B runs through the bounded sets of E, the sets B° form a base of $\mathfrak{T}_b(E)$-neighbourhoods of $\mathfrak{o}$ in E', and the sets $(B^{\circ\circ})^\circ = B^\circ$ form a base of $\mathfrak{T}_b(E'')$-neighbourhoods of $\mathfrak{o}$, i.e. $\mathfrak{T}_b(E)$ and $\mathfrak{T}_b(E'')$ are the same topology on E'. $E'[\mathfrak{T}_b(E)]$ is therefore barrelled.

Conversely if $E'[\mathfrak{T}_b(E)]$ is barrelled, so that $\mathfrak{T}_b(E)$ and $\mathfrak{T}_b(E'')$ are the same, then every bounded set B_1 of E'', being a $\mathfrak{T}_b(E'')$-equicontinuous set, lies in the polar $(B^\circ)^\circ$ in E'' of a $\mathfrak{T}_b(E)$-neighbourhood B° of $\mathfrak{o}$, where B is bounded in E.

Every semi-reflexive space is distinguished. Every (B)-space is distinguished, since the strong dual, being a (B)-space, is barrelled. A non-reflexive (B)-space therefore gives an example of a distinguished space which is not semi-reflexive.

As we shall see in § 31, 7., there are (F)-spaces which are not distinguished, and whose strong duals are therefore not barrelled.

(2) *If $E[\mathfrak{T}]$ is distinguished, E'' is the $\mathfrak{T}_s(E')$-quasi-completion of E.*

For every bounded set B_1 of E'' lies in a set $B^{\circ\circ}$ which is $\mathfrak{T}_s(E')$-compact, and so is $\mathfrak{T}_s(E')$-complete.

(3) *If $E[\mathfrak{T}]$ is distinguished and $E'[\mathfrak{T}_b(E)]$ is semi-reflexive, then E' is reflexive and E'' is the $\mathfrak{T}_k(E')$-quasi-completion of E.*

It follows directly from (1) and 5.(1) that E' is reflexive. E'' is the $\mathfrak{T}_s(E')$-quasi-completion of E, by (2). Since E' is the dual of E'', the $\mathfrak{T}_s(E')$-closure and the $\mathfrak{T}_k(E')$-closure in E'' of a bounded absolutely convex set $B \subset E$ coincide. E'' is therefore the $\mathfrak{T}_k(E')$-quasi-closure of E (cf. § 18, 4.(4)).

8. The dual of a semi-reflexive space. As an example at the end of this number shows, the strong dual of a semi-reflexive space need not be quasi-complete. However, we have

(1) *Suppose that* $E[\mathfrak{T}]$ *is semi-reflexive, and let* $\bar{E}'$ *(respectively* $\tilde{E}'$*) be the quasi-completion (respectively completion) of the strong dual* $E'[\mathfrak{T}_b(E)]$. *Then* $E[\mathfrak{T}_k(\bar{E}')]$ *and* $E[\mathfrak{T}_k(\tilde{E}')]$ *are also semi-reflexive, and their strong duals are the quasi-completion and completion, respectively, of the strong dual of* $E[\mathfrak{T}]$.

Proof. By hypothesis $E'[\mathfrak{T}_b(E)]$ has E as dual. By § 15, 9.(11), the completion $\tilde{E}'$, and *a fortiori* the quasi-completion $\bar{E}'$, of E' also have E as dual. Now the E'-bounded subsets of $E[\mathfrak{T}]$ are relatively $\mathfrak{T}_s(E')$-compact. By § 21, 4.(5), they are also relatively $\mathfrak{T}_s(\tilde{E}')$-compact, and *a fortiori* are $\mathfrak{T}_s(\bar{E}')$-compact, and so they are also $\bar{E}'$- and $\tilde{E}'$-bounded. But by 3.(1) it follows from this that $E[\mathfrak{T}_k(\tilde{E}')]$ and $E[\mathfrak{T}_k(\bar{E}')]$ are semi-reflexive. Since the subsets of E which are bounded with respect to E', $\bar{E}'$ and $\tilde{E}'$ are the same, the strong topologies on E' and on $\bar{E}'$ are the restrictions of the strong topology on $\tilde{E}'$. The last assertion follows from this.

By making the topology of a semi-reflexive space E finer, it is therefore always possible to arrange for E to remain semi-reflexive, but for the strong dual to be quasi-complete, or indeed complete.

If we make the further assumption that $\bar{E}'$ is $\mathfrak{T}_s(E)$-quasi-complete, $E[\mathfrak{T}_k(\bar{E}')]$ is reflexive, for both $E[\mathfrak{T}_k(\bar{E}')]$ and E' are semi-reflexive. In this case we can obtain certain information about the iterated strong duals.

Thus suppose that $E[\mathfrak{T}]$ is semi-reflexive, that $\mathfrak{T}$ is the Mackey topology and that $E'[\mathfrak{T}_b(E)]$ is not quasi-complete, so that $E[\mathfrak{T}]$ is not reflexive. Further suppose that $\bar{E}'[\mathfrak{T}_b(E)]$ is semi-reflexive (and so reflexive, for then $E[\mathfrak{T}_k(\bar{E}')]$ is indeed reflexive). Then $E''=E$, and the topology $\mathfrak{T}_b(E')$ on $E''=E$ is strictly finer than $\mathfrak{T}$, so that E''' is strictly larger than E', although contained in $\bar{E}'$; thus $\mathfrak{T}_b(E')$ is coarser than $\mathfrak{T}_b(\bar{E}')=\mathfrak{T}_k(\bar{E}')$ on E.

It follows by (1) from $E' \subset E''' \subset \bar{E}'$ that the strong dual of $E'''[\mathfrak{T}_b(E)]$ is again E, equipped with the topology $\mathfrak{T}_b(E''')$. In this way we obtain a sequence $E' \subset E''' \subset \cdots \subset \bar{E}'$ and a sequence $\mathfrak{T}_b(E')$, $\mathfrak{T}_b(E''')$, ... of topologies on E which get finer and finer, and under each of which E is semi-reflexive.

If $\bar{E}'$ is not reached after finitely many steps, we can form $E^{(\omega)} = E' \cup E''' \cup \cdots$. If $E^{(\omega)}$ is also not equal to $\bar{E}'$, so that $E^{(\omega)}$ is also not $\mathfrak{T}_b(E)$-quasi-complete, then $\mathfrak{T}_b(E^{(\omega)})$ is strictly coarser than $\mathfrak{T}_b(\bar{E}')$ on E, and the method can be extended by transfinite induction, until $\bar{E}'$ is reached.

We give an example in which $E''' = \bar{E}' = \tilde{E}'$. It is not known if there are spaces for which a higher iterated dual is first equal to $\bar{E}'$.

We consider the space φ_d, the direct sum of d spaces $E_\alpha = \mathsf{K}$. Suppose that the number d is uncountable. In § 18, 5., besides the topology of the locally convex direct sum we also considered the strictly coarser topology $\mathfrak{T}'$ of the topological direct sum on φ_d. By § 22, 5.(5), $(\varphi_d[\mathfrak{T}'])'$ is equal to the linear subspace $\omega_d^{(0)}$ of ω_d consisting of all $\mathfrak{u} = (v_\alpha) \in \omega_d$ with only countably many non-zero v_α. By § 18, 5.(7), the bounded subsets of $\varphi_d[\mathfrak{T}]$ and $\varphi_d[\mathfrak{T}']$ are the same, being the bounded finite-dimensional subsets. Since these are relatively $\mathfrak{T}_s(\omega_d)$-compact, they are also relatively $\mathfrak{T}_s(\omega_d^{(0)})$-compact, so that $\varphi_d[\mathfrak{T}']$ is semi-reflexive.

The strong topology $\mathfrak{T}_b(\varphi_d)$ on ω_d coincides with the weak topology $\mathfrak{T}_s(\varphi_d)$, and in particular the strongly bounded and weakly bounded subsets of ω_d coincide. The bounded subsets of $\omega_d^{(0)}$ are the intersections of the bounded subsets of ω_d with $\omega_d^{(0)}$. The set B of all $\mathfrak{u} = (v_\alpha) \in \omega_d$ with $|v_\alpha| \leq 1$ for all α is bounded and $\mathfrak{T}_s(\varphi_d)$-compact in ω_d, so that $B^{(0)} = B \cap \omega_d^{(0)}$ is bounded and weakly closed in $\omega_d^{(0)}$. But $B^{(0)}$ is $\mathfrak{T}_s(\varphi_d)$-dense in B, so that $B^{(0)}$ is weakly precompact, but not weakly compact, in $\omega_d^{(0)}$. But this means that the $\mathfrak{T}_b(\varphi_d)$-bounded subsets of $\omega_d^{(0)}$ are in general not relatively weakly compact; $\omega_d^{(0)}$ is thus not semi-reflexive.

By 2.(3), $(\omega_d^{(0)})'' = \omega_d$. If φ_d is given the topology $\mathfrak{T}_k(\omega_d^{(0)})$, which lies between $\mathfrak{T}$ and $\mathfrak{T}'$, φ_d is semi-reflexive; the strong dual $\omega_d^{(0)}$, whose topology coincides with the weak topology, is not quasi-complete, but $(\varphi_d)'''$ is equal to the completion ω_d of $\omega_d^{(0)}$. The space φ_d, which is semi-reflexive, but not reflexive, under the topology $\mathfrak{T}_k(\omega_d^{(0)})$, becomes reflexive when we pass to the finer topology $\mathfrak{T}_k(\omega_d)$.

Kōmura [2] has given an example of a reflexive locally convex space which is not complete.

9. Polar reflexivity. We can raise the questions which lead to the concepts of semi-reflexivity and reflexivity for topologies other than the strong topology. By the Mackey-Arens theorem these questions are only non-trivial for topologies which are finer than the Mackey topology.

We shall investigate these questions for the polar topology $\mathfrak{T}^\circ$, that is, for the topology of precompact convergence. Thus we call a locally convex space $E[\mathfrak{T}]$ polar semi-reflexive if $(E'[\mathfrak{T}^\circ])' = E$, and polar reflexive if further $\mathfrak{T} = \mathfrak{T}^{\circ\circ}$ (cf. § 21, 6. and 7.).

(1) *Suppose that $E[\mathfrak{T}]$ is locally convex. The space $(E'[\mathfrak{T}^\circ])'$ is equal to $\bigcup_{C \in \mathfrak{C}} \tilde{C}$, where $\mathfrak{C}$ is the class of absolutely convex precompact subsets of $E[\mathfrak{T}]$, and $\tilde{C}$ is the completion of C in $\tilde{E}[\tilde{\mathfrak{T}}]$.*

$E[\mathfrak{T}]$ is polar semi-reflexive if and only if every precompact subset of E is relatively compact.

Proof. $\tilde{E}[\tilde{\mathfrak{T}}] \subset (E')^*$. The set $\tilde{C}$ is compact, and therefore weakly compact, and so it is the weak closure of C in E'^*; the first assertion therefore follows from 2.(1). The second part of the theorem follows directly from this.

The class of polar semi-reflexive spaces is larger than the class of semi-reflexive spaces.

(2) *Every quasi-complete space $E[\mathfrak{T}]$ is polar semi-reflexive. In particular, every semi-reflexive space is polar semi-reflexive.*

The first assertion follows from (1). If $E[\mathfrak{T}]$ is semi-reflexive, every bounded subset of $E[\mathfrak{T}]$ is relatively weakly compact, and *a fortiori* so is every precompact set, and the assertion follows from (1).

Analogously, we have

(3) *Every reflexive space is polar reflexive.*

For $E[\mathfrak{T}]$ is polar semi-reflexive and $\mathfrak{T}$ is the strong topology, and it was shown in § 21, 7. that this coincides with $\mathfrak{T}^{\circ\circ}$.

(4) *If $E[\mathfrak{T}]$ is quasi-complete, if $\mathfrak{T}$ is the Mackey topology and if $E'[\mathfrak{T}^\circ]$ is also quasi-complete, then $E[\mathfrak{T}]$ is polar reflexive.*

Proof. By (2), $E[\mathfrak{T}]$ is polar semi-reflexive. We still have to prove that $\mathfrak{T} = \mathfrak{T}^{\circ\circ}$, and so we must show that the absolutely convex weakly relatively compact subsets K of E' are the same as the $\mathfrak{T}_c$-precompact subsets C of E'. Because $E'[\mathfrak{T}^\circ]$ is quasi-complete, every C is relatively $\mathfrak{T}_c$-compact, and *a fortiori* is weakly relatively compact. Since every set C is contained in an absolutely convex set of the same kind, all the sets C are also sets K. The converse follows from § 21, 6.(2).

It follows directly from (4) and § 21, 6.(4) that

(5) *Every* (F)-*space is polar reflexive.*

These ideas have a simple connection with the duality theory of abelian topological groups (FREUNDLICH-SMITH [1]).

Clearly, with respect to addition alone, every locally convex space $E[\mathfrak{T}]$ is an abelian topological group. A character χ on an abelian topological group G is a continuous group homomorphism from G into the multiplicative group of complex numbers of modulus 1. The character $\chi(x) \equiv 1$ will be denoted by 1. The characters of G form a group $\hat{G}$ under multiplication, with 1 as unit element.

We now consider the special case of a real locally convex space $E[\mathfrak{T}]$. We can assign a character χ_u to each element u of the dual space E', by setting $\chi_u(x) = e^{iux}$. Certainly $\chi_u(x_1 + x_2) = \chi_u(x_1) \cdot \chi_u(x_2)$, and χ_u is continuous since u and the exponential function are both continuous.

We shall now show that all the characters on $E[\mathfrak{T}]$ are obtained in this way.

(6) *The correspondence $u \to \chi_u$ is an algebraic isomorphism of the additive group E' onto the multiplicative character group $\hat{E}$ of $E[\mathfrak{T}]$.*

Proof. The linear functional $u = 0$ corresponds to the character 1. If $u \neq 0$, then u does not vanish identically on any neighbourhood U of $\mathfrak{o}$ with $|u(U)| < 2\pi$, so that $e^{iux} \neq 1$ and the correspondence $u \to \chi_u$ is one-one.

It remains to show that every character χ_0 has the form χ_u. Because χ_0 is continuous, there is a neighbourhood U of $\mathfrak{o}$ with $|\chi_0(x)-1|<\pi$ for all $x\in U$. A continuous additive function is defined on U by setting $u_0 x = \frac{1}{i}\log\chi_0(x) = \operatorname{arc}\chi_0(x)$, where the principal value is taken. It is real linear on U, for linearity with rational coefficients follows from additivity, and linearity for real coefficients follows from this by taking limits in the usual way.

Now if z is an arbitrary element of E, $\frac{1}{n}z\in U$ for suitable n. If we put $u_0 z = n\cdot u_0\left(\frac{1}{n}z\right)$, u_0 is extended in a continuous linear way to the whole of E, so that $u_0\in E'$. Since $\chi_0(x)=e^{iu_0x}$ holds on U, $\chi_0(z)=e^{iu_0z}$ holds on the whole of E.

If therefore we start from a real dual pair $\langle E_2, E_1\rangle$, by setting $\{u,x\}=e^{iux}$ we obtain a dual pairing of the two abelian groups E_2 and E_1, under which each element of one group determines a homomorphism of the other group into the group of complex numbers of modulus 1. If we take the Mackey topology as the topology on each of the groups, then by the MACKEY-ARENS theorem each group is the character group of the other.

The well-known PONTRYAGIN duality theorem says that the character group $\hat{G}$ of a locally compact abelian group G is again locally compact under the topology of uniform convergence on the compact subsets of G, and that conversely G is the character group of $\hat{G}$. Since $E[\mathfrak{T}]$ is only locally compact if E is finite-dimensional, this theorem only includes the trivial case for locally convex spaces. On the other hand to every pairing $\{E',E\}$ corresponding to a polar reflexive $E[\mathfrak{T}]$, there corresponds an analogue of the Pontryagin duality theorem; in particular by (5) a real (F)-space and its $\mathfrak{T}^\circ$-dual are each the character group of the other, with respect to the topology of uniform convergence on compact subsets.

§ 24. Some results on compact and on convex sets

1. The theorems of ŠMULIAN and KAPLANSKY. As in § 3, 4., a subset M of a topological space R is said to be countably compact or relatively countably compact, if every sequence in M has an adherent point in M, or in R, respectively; M is said to be sequentially compact, or relatively sequentially compact, if every sequence in M has a subsequence convergent to an element of M, or of R, respectively. Every (relatively) sequentially compact set is (relatively) countably compact.

We remark that the closure of a relatively countably compact set need not be countably compact; a similar remark applies to relatively sequentially compact sets (cf. GROTHENDIECK [6], for example).

(1) *If M is a weakly relatively countably compact subset of a locally convex space $E[\mathfrak{T}]$, then M is bounded.*

For otherwise there would be a sequence $x_n\in M$ and an element $u\in E'$ with $|ux_n|\to\infty$, and x_n could then have no weakly adherent point.

We now give an example of a non-reflexive (B)-space in whose dual there is a weakly relatively countably compact set which is not weakly relatively sequentially compact. The space l^1 is a subspace of $(l^\infty)'$. The set of elements $\mathfrak{e}_1, \mathfrak{e}_2, \ldots \in l^1$ is bounded in $(l^\infty)'$, and is weakly relatively countably compact, being a subset of the unit ball of $(l^\infty)'$, which is weakly compact, by the BANACH-ALAOGLU theorem. But no subsequence $\mathfrak{e}_{n_j}$ is weakly convergent in $(l^\infty)'$; to see this, it is enough to consider the sequence $\mathfrak{u}\,\mathfrak{e}_{n_j}$, where $\mathfrak{u}$ is an element of l^∞ whose n_j-th coordinates are alternately equal to 0 and 1.

We now obtain ŠMULIAN's theorem:

(2) *Suppose that* $E[\mathfrak{T}]$ *is locally convex. If* E' *is weakly separable, every weakly relatively countably compact subset of* E *is weakly relatively sequentially compact.*

Proof. Let x_n be a sequence in E each of whose subsequences has at least one weakly adherent point in E. We must show that x_n has at least one subsequence which is weakly convergent in E. By hypothesis there is a weakly dense sequence u_m in E'. Since the sequence x_n is bounded, using a diagonal procedure we can find a subsequence $x_{n_k} = y_k$, for which $\lim_{k\to\infty} u_m y_k$ exists, for each m. The sequence y_k has a weakly adherent point y in E. If z is any weakly adherent point of the sequence y_k, then it follows from $u_m y = \lim_{k\to\infty} u_m y_k = u_m z$ that $u_m(y-z) = 0$ for each u_m, and the u_m are dense in E'. Thus $y = z$, and the sequence y_k has only one weakly adherent point. But it follows from this that y is the weak limit of the y_k: for if an infinite subsequence y_{k_j} were not to lie in a weak neighbourhood $U(y)$, y_{k_j} would have a weakly adherent point which would have to be different from y.

It follows directly from § 21, 3.(5) that ŠMULIAN's theorem holds for every separable metrizable locally convex space.

The following generalization of (2), due to DIEUDONNÉ and SCHWARTZ [1], is substantially stronger:

(3) *Suppose that* $E[\mathfrak{T}]$ *is locally convex. If there is a metrizable locally convex topology* $\mathfrak{T}'$ *on* E *which is coarser than* $\mathfrak{T}$, *then every weakly relatively countably compact subset of* E *is weakly relatively sequentially compact.*

In particular this holds for every metrizable locally convex space.

Proof. Since $\mathfrak{T}'$ is metrizable, there is a sequence $U_1 \supset U_2 \supset \cdots$ of absolutely convex $\mathfrak{T}'$-neighbourhoods of $\mathfrak{o}$ in E, with $\bigcap_{n=1}^{\infty} U_n = \mathfrak{o}$. By hypothesis the U_n are also $\mathfrak{T}$-neighbourhoods of $\mathfrak{o}$. Suppose that x_n is a sequence in E with the property that every subsequence has a weakly adherent point, and let H be the closed linear span of the x_n in E. Then the set of x_n is total in H, and the sets $V_n = U_n \cap H$ form a sequence of

$\mathfrak{T}$-neighbourhoods of $\mathfrak{o}$ in H, with $\bigcap_{n=1}^{\infty} V_n = \mathfrak{o}$. Thus $\bigcup_{n=1}^{\infty} V_n^\circ$ is weakly dense in H'. Using exactly the same argument as in the proof of § 21, 3.(5), it follows that H' is weakly separable. Since all the weakly adherent points of the subsequence of x_n lie in H, x_n has a $\mathfrak{T}_s(H')$-convergent subsequence x_{n_j}, by ŠMULIAN's theorem. But by § 22, 2.(1), the topologies $\mathfrak{T}_s(H')$ and $\mathfrak{T}_s(E')$ coincide on H, so that x_{n_j} is weakly convergent in E, and consequently the set $\{x_n\}$ is weakly relatively sequentially compact.

We observe that the hypothesis of (3) is more general than the hypothesis of (2). For if N is a countable weakly dense subset of E', then by § 21, 1.(2) $\mathfrak{T}_s(N)$ is a metrizable locally convex topology on E, and it is coarser than $\mathfrak{T}$, by § 20, 2.(4).

(4) *Every weakly relatively countably compact subset of a strict* (LF)*-space is also weakly relatively sequentially compact.*

For the weak closure of a weakly relatively countably compact set M is bounded, every bounded subset of a strict (LF)-space $E[\mathfrak{T}] = \sum_{n=1}^{\infty} E_n[\mathfrak{T}_n]$ always lies in an (F)-space E_n, by § 19, 4.(4), and the weak topology on E_n coincides with the weak topology of E. The assertion therefore follows from (3).

(5) *Suppose that $E[\mathfrak{T}]$ is locally convex and that it satisfies the hypotheses of* (2) *or* (3). *If x_n is a sequence in E each of whose subsequences has at least one weakly adherent point in E, then each of these adherent points is also the weak limit of a suitable subsequence.*

Thus the weakly sequentially compact and the weakly countably compact subsets of E are the same.

Proof. If y is a weakly adherent point of x_n then, in the case where the hypotheses of (2) hold, a diagonal procedure produces a subsequence x_{n_k} with $u_m x_{n_k} \to u_m y$, and as before it can be shown that y is the weak limit of the sequence x_{n_k}. Case (3) can again be reduced to case (2).

We now obtain the following theorem of KAPLANSKY's (cf. BOURBAKI [6], Vol. 2, p. 82)

(6) *Suppose that $E[\mathfrak{T}]$ is locally convex and that E' is the union of countably many weakly compact subsets. If M is an arbitrary subset of E, every weak closure point x_0 of M is always a weak closure point of a countable subset of M.*

Proof. Suppose that E' is the union of the weakly compact sets $C_1 \subset C_2 \subset \cdots$. We consider a weak neighbourhood of x_0 of the form

$|u_i(x_0-x)| < \frac{1}{m}$, $i=1,\ldots,k$, with k elements $u_i \in C_n$. By hypothesis there is a $y \in M$ in this neighbourhood. Since x_0 and y are weakly continuous on E', and so also on C_n, there are weak neighbourhoods V_i of the u_i, for which $|v_i(x_0-y)| < \frac{1}{m}$ holds for all $v_i \in V_i$. We consider such a system of neighbourhoods for each k-tuple $u_1,\ldots,u_k$ and a corresponding $y \in M$. By TYCHONOFF's theorem the k-times topological product C_n^k is compact, so that C_n^k is covered by finitely many $V_1 \times \cdots \times V_k$; there is therefore a finite subset $M_{n,k,m}$ of M with the property that at least one element of $M_{n,k,m}$ lies in each neighbourhood of the form $|u_i(x_0-x)| < \frac{1}{m}$, $i=1,\ldots,k$, $u_i \in C_n$. The set $\bigcup_{n,k,m=1}^{\infty} M_{n,k,m}$ has the required properties.

By § 20, 7.(6), every convex weakly sequentially closed subset of a metrizable locally convex space is always weakly closed. The theorems of ŠMULIAN and KAPLANSKY enable us to make another assertion in this direction:

(7) *Suppose that* $E[\mathfrak{T}]$ *is a metrizable locally convex space or a strict* (LF)*-space. If* x_0 *is a weak closure point of a weakly relatively compact subset* M *of* E, x_0 *is the weak limit of a sequence in* M.

A subset of E *is therefore weakly compact if and only if it is weakly relatively compact and weakly sequentially closed.*

Proof. If $E[\mathfrak{T}]$ is metrizable, and if $U_1 \supset U_2 \supset \cdots$ is a base of neighbourhoods of $\mathfrak{o}$ for $E[\mathfrak{T}]$, then E' is the union of the weakly compact sets U_n°. By (6), x_0 is therefore an adherent point of a sequence in M, and the assertion follows from (5).

The case of an (LF)-space can be reduced to this one as in (4).

2. EBERLEIN's theorem. We have already given an example in § 3, 4. of a sequentially compact (and therefore countably compact) set which is not compact. We can also describe this example in the following way: in l_d^∞, with $d > \aleph_0$, let H be the subspace consisting of all vectors $\mathfrak{x} = (\xi_\alpha)$ with at most countably many non-zero coordinates ξ_α; then the unit ball of H is $\mathfrak{T}_s(l_d^1)$-sequentially compact, but is not $\mathfrak{T}_s(l_d^1)$-compact.

The following theorem of EBERLEIN's, proved first by EBERLEIN [2] for (B)-spaces, and later generalized by GROTHENDIECK [6], gives a far-reaching condition for the identity of relatively countably compact and relatively compact sets:

(1) *If the locally convex space* $E[\mathfrak{T}]$ *is quasi-complete under the Mackey topology* $\mathfrak{T}_k(E')$, *every relatively countably* $\mathfrak{T}$*-compact subset* M *of* E *is relatively compact.*

Indeed, we have

(1′) *If $E[\mathfrak{T}]$ is locally convex, if M is relatively countably compact and if $\complement(M)$ is $\mathfrak{T}_k$-complete, then M is relatively compact.*

Proof. a) First we show that it is sufficient to prove (1′) under the hypothesis that $E[\mathfrak{T}]$ is $\mathfrak{T}_k$-complete.

Suppose therefore that M is a relatively countably compact subset of $E[\mathfrak{T}]$. We form the $\mathfrak{T}_k$-completion $\tilde{E}[\tilde{\mathfrak{T}}_k]$ of E. By § 21, 4.(5), $\tilde{\mathfrak{T}}_k$ is the Mackey topology on $\tilde{E}$. The extension of $\mathfrak{T}$ to $\tilde{E}$ will again be denoted by $\mathfrak{T}$. M is then also relatively countably compact in $\tilde{E}[\mathfrak{T}]$.

We assume that it has been proved that M is relatively compact in $\tilde{E}$; we must then show that M is already relatively compact in E. For this it is enough to show that the closure $\bar{M}$ of M in $\tilde{E}$ is a subset of E.

But the closed convex cover of M in E is $\mathfrak{T}_k$-complete, by hypothesis, and so it is also $\tilde{\mathfrak{T}}_k$-closed in $\tilde{E}$. As it is a convex set, it is also $\mathfrak{T}$-closed in $\tilde{E}$, and consequently it contains $\bar{M}$.

b) By § 5, 6.(3) every relatively countably compact set is precompact. It is therefore sufficient to show that the closure $\bar{M}$ of M in $E[\mathfrak{T}]$ is complete, since a complete precompact set is compact.

We may also assume that $\mathfrak{T}$ is the weak topology. For M is also relatively countably compact with respect to the weak topology, and if the weak closure of M is weakly complete, then by § 18, 4.(4) the $\mathfrak{T}$-closure of M is also $\mathfrak{T}$-complete. Consequently (1′) is reduced to the following assertion:

(2) *Suppose that $E[\mathfrak{T}]$ is $\mathfrak{T}_k$-complete, and that M is a weakly relatively countably compact subset of E. Then every weak closure point z of M in $(E')^*$ always belongs to E.*

Proof. By Grothendieck's theorem (§ 21, 9.(4)) it is sufficient to show that z is weakly continuous on every absolutely convex weakly compact subset K of E'.

We suppose that this is not the case, so that there exists a K on which z is not weakly continuous. By § 21, 6.(5), z is not continuous at $\mathfrak{o}$ on K, so that there is an $\varepsilon>0$ with the property that in each weak neighbourhood of $\mathfrak{o}$ there is a $u\in K$ with $|uz|\geqq\varepsilon$.

We now inductively construct two sequences $u_n\in K$, $x_k\in M$ for which the following inequalities hold:

$$(3)\qquad |u_i x_n - u_i z| \leqq \frac{1}{n}, \qquad 1\leqq i\leqq n-1,$$

$$(4)\qquad |u_n x_i| \leqq \frac{1}{n}, \qquad 1\leqq i\leqq n,$$

$$(5)\qquad |u_n z| \geqq \varepsilon.$$

For if $x_1, \ldots, x_{n-1}$, $u_1, \ldots, u_{n-1}$ have already been constructed, we determine $x_n \in M$ in such a way that (3) is satisfied. This is possible, since z is a weak closure point of M. We then determine u_n in such a way that (4) and (5) are satisfied. This is possible because z is not continuous at $\mathfrak{o}$.

The sequence $x_n \in M$ now has a weakly adherent point x_0 in E, and likewise the sequence $u_n \in K$ has a weakly adherent point $u_0 \in K$. As far as these adherent points are concerned, it follows from (3), by letting n tend to infinity, that $u_i x_0 = u_i z$, and likewise it follows from (4) that $u_0 x_i = 0$. As x_0 is an adherent point of the sequence x_n, it follows that $u_0 x_0 = 0$. But $x_0 \in E$ is weakly continuous on K, so that $u_0 x_0 = 0$ is adherent to the sequence of values $u_i x_0 = u_i z$; this contradicts (5). Thus (2) and (1′) are established.

New criteria for reflexivity follow directly from EBERLEIN's theorem and theorems § 23, 3.(1), (3) and § 23, 5.(4):

(6) *A $\mathfrak{T}_k$-quasi-complete locally convex space is semi-reflexive if and only if every bounded subset is weakly relatively countably compact.*

(7) *An* (F)*-space is reflexive if and only if every bounded subset is weakly relatively countably compact.*

(8) *A* (B)*-space is reflexive if and only if its closed unit ball is weakly relatively countably compact.*

Remark. By ŠMULIAN's theorem, 1.(3), "weakly countably compact" can be replaced in (7) and (8) by "weakly sequentially compact".

It follows directly from 2.(1) and 1.(7) that

(9) *If M is a weakly relatively countably compact subset of an* (F)*-space, the set of all the limits of weakly convergent sequences in M is weakly compact.*

By EBERLEIN's theorem, the weakly closed weakly countably compact subsets of a $\mathfrak{T}_k$-quasi-complete locally convex space are the same as the weakly compact sets. *A fortiori* this also holds for the weakly closed $\aleph_\alpha$-compact sets (cf. § 3, 4.).

3. Further criteria for weak compactness. A subset M of a Hausdorff topological space R is said to be pseudo-compact if every continuous real-valued function $f(x)$ on M is bounded. M is said to be relatively pseudo-compact if for each unbounded continuous function f on M there exists a point x_0 in the closure $\bar{M}$, in all of whose neighbourhoods $f(x)$ is unbounded. The closure of a relatively pseudo-compact set is pseudo-compact.

(1) *Every weakly relatively pseudo-compact subset M of a locally convex space $E[\mathfrak{T}]$ is bounded.*

Every weakly (relatively) countably compact set N is weakly (relatively) pseudo-compact.

Proof. If M is unbounded, there is an element $u \in E'$ with the property that $|ux|$ is unbounded on M. However $|ux|$ is bounded on the weak neighbourhood $x+U_{n;\varepsilon}$ of any $x \in E$.

If $f(x)$ is an unbounded function on N, and if $|f(x_n)| \geqq n$ on the sequence $x_n \in N$, then f is unbounded on every weak neighbourhood of a weak accumulation point x_0 of the sequence x_n.

We now have the following lemma of Pták [3]:

(2) *If M is a weakly (relatively) pseudo-compact subset of a locally convex space $E[\mathfrak{T}]$, if z is a weak closure point of M in E'^* and if u_i is a sequence of elements of E', then there exists a point x_0 in M (in $\bar{M}$) with $u_i z = u_i x_0$ for all $i = 1, 2, \ldots$.*

Proof. The functions $f_i(x) = |u_i(x-z)|$ are weakly continuous on the weak closure $\bar{M}$ of M, and they are each bounded, by (1). Suppose that $|f_i(x)| \leqq k_i$ on $\bar{M}$. The function $f(x) = \sum_{i=1}^{\infty} \frac{1}{2^i k_i} f_i(x)$ is the limit of a uniformly convergent series of continuous functions, and it is therefore continuous on $\bar{M}$. Since we can make $|f_1(x)| \leqq \varepsilon, \ldots, |f_n(x)| \leqq \varepsilon$ by choosing x in a suitable way, $\inf f(x) = 0$ on M. If $f(x)$ were strictly positive on the whole of M, $1/f(x) = g(x)$ would be continuous and unbounded on M. But then by hypothesis there would be an $x_0 \in \bar{M}$ in all of whose weak neighbourhoods $g(x)$ would be unbounded. Then $f(x)$ would have infimum 0 in each of these neighbourhoods, so that, since f is continuous at x_0, $f(x_0)$ would have to be zero. But this means that $u_i x_0 = u_i z$ for all i.

A subset M of a locally convex space $E[\mathfrak{T}]$ is said to be weakly (relatively) convex-compact if the following holds: suppose that $K_1 \supset K_2 \supset \cdots$ is a sequence of closed convex subsets of E for which all the intersections $K_n \cap M$ are non-empty; then the sequence $K_n \cap M$ has a weakly adherent point in M (in E). This concept is due to Šmulian [1], [3].

(3) *Every weakly relatively convex-compact subset M of a locally convex space $E[\mathfrak{T}]$ is bounded.*

Every weakly (relatively) countably compact set N is weakly (relatively) convex-compact.

If M were unbounded, there would be a point $u_0 \in E'$ and a sequence $x_i \in M$ with $|u_0 x_i| \geqq i$. The sequence $K_1 \supset K_2 \supset \cdots$ of closed convex covers K_n of the sets $\{x_n, x_{n+1}, \ldots\}$ could then have no weakly adherent point, since $|u_0 y| \geqq n$ for all $y \in K_n$.

If $K_1 \supset K_2 \supset \cdots$ and if $K_n \cap N$ is non-empty for all n, then a sequence $x_n \in K_n \cap N$ has a weakly adherent point in N (respectively E).

(4) *If M is a weakly (relatively) convex-compact subset of a locally convex space $E[\mathfrak{T}]$, if z is a weak closure point of M in E'^*, and if u_i is a sequence of elements of E, then there is an x_0 in M (in $\bar{M}$) with $\lim_{i\to\infty} u_i(z-x_0)=0$.*

Here $\bar{M}$ means the weak closure of M. To prove this, consider the sequence of closed convex weak neighbourhoods $U_n(z)$ defined by the inequalities $|u_i(z-x)| \leqq \frac{1}{n}$, for $i=1,\ldots,n$. A weakly adherent point x_0 of the sequence $U_n(z)\cap M$ has the required properties.

A bounded subset M of a locally convex space $E[\mathfrak{T}]$ is said to be weakly (relatively) partially compact if the following holds: if a sequence of elements of M has a weakly adherent point z in E'^*, then for each sequence u_i contained in an absolutely convex weakly compact subset of E' there exists an x_0 in M (in $\bar{M}$) with $\lim_{i\to\infty} u_i(z-x_0)=0$. This concept was introduced by Day [8].

(5) *Suppose that M is a subset of the locally convex space $E[\mathfrak{T}]$. If M is weakly (relatively) countably compact or weakly (relatively) pseudo-compact or weakly (relatively) convex-compact, then M is also weakly (relatively) partially compact.*

This follows from (1), (2) and (4).

Eberlein's theorem can be strengthened in the following way:

(6) *Suppose that $E[\mathfrak{T}]$ is a locally convex space and that M is a subset whose closed convex cover $\overline{\complement(M)}$ is $\mathfrak{T}_k$-complete. M is weakly relatively compact if and only if it is weakly relatively partially compact.*

The proof proceeds analogously to that of 2.(1):

If M is weakly relatively compact then it follows from (5) that M is weakly relatively partially compact.

Suppose conversely that M is weakly relatively partially compact, and that z is a weak closure point of M in $(E')^*$. It is sufficient to show that z belongs to the $\mathfrak{T}_k$-completion $\tilde{E}$ of E, for then z, being a weak closure point of M, lies in $\overline{\complement(M)}$, which is weakly closed in $\tilde{E}$, since it is $\mathfrak{T}_k$-complete.

We proceed as in 2.(2) and suppose that z is not weakly continuous on some absolutely convex weakly compact subset K of E'. Then again there are sequences $u_k\in K$ and $x_k\in M$ for which the inequalities (3), (4) and (5) of No. 2 hold.

Since M is bounded, it is weakly precompact, so that the sequence $x_n\in M$ has a weakly adherent point $\tilde{x}_0$ in $(E')^*$, and likewise $u_n\in K$ has a weakly adherent point $u_0\in K$. As in the proof of 2.(2) we obtain the relations $u_i x_0=u_i z$, $u_0 x_i=0$ and $u_0 x_0=0$. Applying the hypothesis that M is weakly relatively partially compact to the sequence u_1, u_0, u_2,

$u_0, \ldots$ in K and to $\tilde{x}_0$, it follows that there is an $x_0 \in \bar{M}$ with $\lim u_i(\tilde{x}_0 - x_0) = 0$ and with $u_0 x_0 = u_0 \tilde{x}_0 = 0$. But $u_0 x_0 = 0$ is adherent to the sequence of values $u_i x_0$, so that 0 is a closure point of the values $u_i \tilde{x}_0 = u_i z$; this contradicts 2.(5).

Consequently z is weakly continuous on each K, so that it belongs to $\tilde{E}$, by GROTHENDIECK's theorem.

(7) *Suppose that M is a subset of a locally convex space $E[\mathfrak{T}]$, and that the closed convex cover $\overline{\complement}(M)$ is $\mathfrak{T}_k$-complete. In particular this hypothesis is satisfied by any bounded subset of a $\mathfrak{T}_k$-quasi-complete space.*

Then the following properties of M are equivalent. a) weak relatively countable compactness, b) weak relative pseudo-compactness, c) weak relative convex-compactness, d) weak relative partial compactness and e) weak relative compactness.

By (5) and (6), e) follows from each of the properties a) to d). Conversely a) and d) follows from e); b) then follows from a), by (1), and c) follows from a), by (3).

The equivalence of weak relative compactness of M with b) was proved by PTÁK [2], [3], the equivalence with c) by DIEUDONNÉ [11] and the equivalence with d) by DAY [8]. Further criteria for weak compactness are given in DAY's book [8] and the work of GROTHENDIECK [6]; we shall consider one of these in No. 6.

Applying 1.(3), we obtain the following special case of (7):

(8) *The following properties of a subset M of an* (F)*-space are equivalent: a) weak relative countable compactness, b) weak relative pseudo-compactness, c) weak relative convex-compactness, d) weak relative partial compactness, e) weak relative compactness and f) weak relative sequential compactness.*

As we saw at the beginning of No. 2, a weakly countably compact set need not be weakly compact; the assertion of (7) is therefore not true in general if the word 'relative' is omitted throughout. For (F)-spaces, however, we have

(9) *In an* (F)*-space E the weakly compact subsets are the same as the weakly countably compact sets, the weakly sequentially compact sets and the weakly convex-compact sets.*

First we show that a weakly convex-compact subset M of E is weakly sequentially closed. Suppose that $x_n \in M$ is weakly convergent to x_0, and let H be the closed linear subspace of E defined by the x_n. Since H is separable, there is a weakly dense countable set N in H', by § 21, 3.(5). The topology $\mathfrak{T}_s(N)$ on H is therefore Hausdorff and metrizable. There is therefore a sequence $K_1 \supset K_2 \supset \cdots$ of closed absolutely convex

$\mathfrak{T}_s(N)$-neighbourhoods of x_0 in H with $\bigcap_{n=1}^{\infty} K_n = \{x_0\}$. Since $K_n \cap M$ is non-empty for each n (indeed it contains infinitely many x_k) x_0 belongs to M, which is assumed to be weakly convex-compact.

By (8) a set M which satisfies one of the conditions of (9) is weakly relatively compact. Since it is also weakly sequentially closed, it follows from 1.(7) that it is weakly compact.

(8) *and* (9) *also hold for strict* (LF)*-spaces.*

The assertion corresponding to (8) follows from (7) and 1.(4), and the one corresponding to (9) from (9) and § 19, 5.(4).

By § 23, 3.(1), new criteria for reflexivity are given by (7) and (8); let us state the one produced by the equivalence of weak relative compactness and property c):

(10) *Suppose that* $E[\mathfrak{T}]$ *is locally convex and* $\mathfrak{T}_k$*-quasi-complete (respectively an* (F)*-space). E is semi-reflexive (respectively reflexive) if and only if every decreasing sequence of closed bounded non-empty convex subsets of E has a non-empty intersection.*

4. Convex sets in spaces which are not semi-reflexive. The theorems of Klee. If a $\mathfrak{T}_k$-quasi-complete space $E[\mathfrak{T}]$ is not semi-reflexive, then by 2.(6) there is a bounded sequence in E with no weakly adherent point. This sequence defines a closed separable linear subspace of $E[\mathfrak{T}]$ which cannot be semi-reflexive. Thus we obtain the following criterion for reflexivity from 2.(6):

(1) *A* $\mathfrak{T}_k$*-quasi-complete locally convex space is semi-reflexive if and only if every closed separable linear subspace is semi-reflexive.*

If E is not semi-reflexive and not separable, E therefore contains a closed linear subspace of infinite codimension in E which is also not semi-reflexive. We now show that the same also holds in the separable case:

(2) *If* $E[\mathfrak{T}]$ *is a* $\mathfrak{T}_k$*-quasi-complete locally convex space which is not semi-reflexive, then there exists a closed linear subspace of infinite co-dimension which is also not semi-reflexive.*

Proof. By 2.(6), E contains a bounded sequence $x^{(n)}$ with no weakly adherent point. We form the closed linear span H of the $x^{(n)}$ in E. If B is the closed absolutely convex cover of the $x^{(n)}$ in E, H also contains the normed space $E_B = \bigcup_{n=1}^{\infty} nB$; the norm topology on E_B is finer than the induced topology, by § 20, 11. Let $z_1, z_2, \dots$ be a sequence of linearly independent elements of $E_B \subset H$. We set $F_k = [z_1, \dots, z_k]$. The sequence

$F_k^\perp$ in H' is decreasing, and each $F_k^\perp$ has codimension 1 in $F_{k-1}^\perp$. Corresponding to the spaces $F_k^\perp$ we can find an increasing sequence of algebraic complements G_k in H', and each G_k has dimension k. By § 15, 8.(2), the decomposition $H' = F_k^\perp \oplus G_k$ is $\mathfrak{T}_s$-complementary. Then by § 20, 5.(1) and § 15, 8.(2) the spaces $H_k = G_k^\perp$ form a sequence of topological complements to the spaces $F_1 \subset F_2 \subset \cdots$, with $H_1 \supset H_2 \supset \cdots$, H_k of codimension 1 in H_{k-1}, and H_1 of codimension 1 in H.

Each $x^{(n)}$ then has the unique representation

$$x^{(n)} = \xi_1^{(n)} z_1 + \cdots + \xi_k^{(n)} z_n + \tilde{x}_k^{(n)}, \qquad \tilde{x}_k^{(n)} \in H_k \cap E_B.$$

Using a diagonal procedure, we now choose a subsequence of the $x^{(n)}$, again denoted by $x^{(n)}$, with the property that $\lim_{n\to\infty} \xi_k^{(n)} = \gamma_k$ exists for all k. Then for each k we can determine an n_k and a $y^{(k)} = \gamma_1 z_1 + \cdots + \gamma_k z_k + \gamma'_{k+1} z_{k+1} + \tilde{x}_{k+1}^{(n_k)} \in E_B$ with $\gamma'_{k+1} \neq \gamma_{k+1}$ in such a way that the E_B-norm distance between $y^{(k)}$ and $x^{(n_k)}$ satisfies $\|y^{(k)} - x^{(n_k)}\| \leq 1/k$. It follows from $\gamma'_{k+1} \neq \gamma_{k+1}$ that $y^{(1)} - y^{(k)}, \ldots, y^{(k-1)} - y^{(k)}$ are linearly independent, and so therefore are $y^{(1)}, \ldots, y^{(k-1)}$, modulo H_k. It follows from this that the closed linear subspace L spanned by $y^{(2)}, y^{(4)}, \ldots$ has infinite codimension in H, and thus in E as well. Since $\|y^{(2k)} - x^{(n_{2k})}\| \leq 1/2k$, the $y^{(2k)}$ form a bounded sequence in E_B, and *a fortiori* in L. A point weakly adherent to the sequence $y^{(2k)}$ must also be weakly adherent to $x^{(n_{2k})}$. From this it follows that L is not semi-reflexive.

We now give a further criterion for reflexivity:

(3) *A $\mathfrak{T}_k$-quasi-complete locally convex space $E[\mathfrak{T}]$ is semi-reflexive if and only if every bounded closed absolutely convex subset K has a supporting hyperplane parallel to each real closed hyperplane H.*

Proof. It is sufficient to prove the theorem for real E (cf. § 21, 11.(2)).

a) Suppose that E is semi-reflexive. Then K is weakly compact, and the existence of the supporting hyperplane follows from § 20, 7.(8).

b) Suppose that E is not semi-reflexive. Given any hyperplane $H \ni o$, we shall produce a bounded closed absolutely convex set which has no supporting hyperplane parallel to H. By § 23, 3.(6), H is not semi-reflexive.

By 3.(10) there is a decreasing sequence of bounded non-empty convex $\mathfrak{T}$-closed subsets C_n of H with empty intersection. Suppose that x_0 is an element of E not lying in H. We form the closed absolutely convex cover K of the sets $\left(1 - \dfrac{1}{n}\right) x_0 + C_n$. Then the only possible support hyperplanes parallel to H are $x_0 + H$ and $-x_0 + H$. Because K is symmetric about o it is sufficient to show that $K \cap (x_0 + H)$ is empty.

Let x_0+y be an element of x_0+H. There exists a C_{k_0} with $y \notin C_{k_0}$ and $k_0 \geqq 2$. Since C_{k_0} is closed, there is an absolutely convex $\mathfrak{T}$-neighbourhood U of $\mathfrak{o}$ in H with $y \notin C_{k_0}+U$. In order to show that $x_0+y \notin K$, it is sufficient to show that there is a $\mathfrak{T}$-neighbourhood V of $\mathfrak{o}$ in E for which x_0+y+V contains no element of the form $K_0 = \operatorname*{\Gamma}_{n=1}^{\infty}\left(\left(1-\frac{1}{n}\right)x_0+C_n\right)$.

The elements of K_0 have the form $z=(1-\rho)x_0+z'$ with $1-\rho = \sum_1^m \alpha_n\left(1-\frac{1}{n}\right)$ and $z'=\sum_1^m \alpha_n y_n$, $y_n \in C_n$, $\sum_1^m |\alpha_n| \leqq 1$. We determine a $\delta>0$ for which $\delta C_1 \subset \frac{U}{3}$ holds. We can then determine a $\rho(\delta)>0$ in such a way that $z=(1-\rho)x_0+z'$, with $\rho \leqq \rho(\delta)$, only lies in K_0 if in the expression $z'=\sum_1^m \alpha_n y_n$, the sum $\sum |\alpha_w|$ over $\alpha_1,\dots,\alpha_{k_0-1}$ and any subsequent negative α_n has magnitude less than δ, and the remaining sum $\sum \alpha_{n''}$ lies between $1-\delta$ and 1. It is easy to confirm that we can choose a number $<\frac{\delta}{k_0-1}$ for $\rho(\delta)$.

Now let V be the neighbourhood of $\mathfrak{o}$ in E consisting of all αx_0+x, $|\alpha| \leqq \rho(\delta)$, $x \in \frac{U}{3} \subset H$. Elements $(1-\rho)x_0+z'$ of K_0 only lie in x_0+y+V if $\rho \leqq \rho(\delta)$. But then z' has the form $z'=\sum \alpha_{n'} y_{n'}+\sum \alpha_{n''} y_{n''}$ with $\sum \alpha_{n'} y_{n'} \in \delta C_1 \subset \frac{U}{3}$ and $\sum \alpha_{n''} y_{n''} \in C_{k_0}+\frac{U}{3}$, so that $z' \in C_{k_0}+\frac{2}{3}U$. However all the elements of x_0+y+V have components in H which lie in $y+\frac{U}{3}$. Therefore $z' \in y+\frac{U}{3}$ must hold, which contradicts $y \notin C_{k_0}+U$. Thus $K_0 \cap (x_0+y+V)$ is empty, and so (3) is established.

We now use this example to give some further constructions. Let k be chosen so that $\mathfrak{o} \notin C_k$. If K is our set, constructed as above, and M is the closed convex set x_0-C_k, then we assert that $\complement(K \cup M)$ is not closed.

To show this, we prove that $x_0 \notin \complement(K \cup M)$, but that x_0 is a closure point of $\complement(K \cup M)$. We have already shown that every element of K has the form $(1-\rho)x_0+z'$, with $\rho>0$ and z' in H. An element of $\complement(K \cup M)$ therefore has the form

$$x=[\alpha_1(1-\rho)+(1-\alpha_1)]x_0+\alpha_1 z'-(1-\alpha_1)z'',$$
$$z' \in H, \quad z'' \in C_k, \quad 0 \leqq \alpha_1 \leqq 1.$$

Because $\rho>0$ and $x_0 \notin H$, x could be equal to x_0 only if $\alpha_1=0$; in this case, though, $x \in M$, and so it is different from x_0. On the other hand, however, if $n \geqq k$ and $y_n \in C_n \subset C_k$, $\left(1-\frac{1}{n}\right)x_0+y_n \in K$ and $x_0-y_n \in M$, so that

$$\frac{1}{2}\left[\left(1-\frac{1}{n}\right)x_0+y_n\right]+\frac{1}{2}(x_0-y_n)=\left(1-\frac{1}{2n}\right)x_0 \in \complement(K \cup M),$$

hence x_0 is in the closure of $\complement(K \cup M)$.

In exactly the same way it can be proved that $K+M$ is not closed, since $2x_0$ is not in $K+M$, but belongs to the closure of $K+M$.

Applying § 20, 6.(5) and § 15, 6.(10), we obtain

(4) *The following two conditions are each necessary and sufficient for a $\mathfrak{T}_k$-quasi-complete locally convex space to be semi-reflexive:*

a) *the convex cover of two bounded closed convex sets is always closed.*

b) *the sum of two bounded closed convex sets is always closed.*

(3) and (4) were proved for (B)-spaces by KLEE [1]; the following theorem can also be found, for (B)-spaces, in KLEE [2] II.

(5) *A $\mathfrak{T}_k$-quasi-complete locally convex space $E[\mathfrak{T}]$ is semi-reflexive if and only if every two disjoint bounded closed convex sets can be separated by a closed real hyperplane.*

Proof. a) It follows from the separation theorem for compact sets (§ 20, 7.(1)) that in semi-reflexive spaces it is even possible to separate the sets strictly. We must therefore construct a counterexample in any non-semi-reflexive $\mathfrak{T}_k$-quasi-complete space $E[\mathfrak{T}]$.

b) Once more we can take $E[\mathfrak{T}]$ to be real and not semi-reflexive, and can take $\mathfrak{T}=\mathfrak{T}_k$. Using the ideas of the proof of (2), we can find a bounded sequence $y^{(k)}$ in E with the following properties: the $y^{(2k)}$ define a closed linear subspace L of infinite codimension in E, and have no weakly adherent point in L, while the $y^{(2k+1)}$ are linearly independent modulo L. Let H be the separable closed linear subspace of E defined by all the $y^{(k)}$. Let B be the closed absolutely convex cover of the $y^{(k)}$, and let x_n be a $\mathfrak{T}$-dense sequence in B. The space $H_B=\bigcup_{n=1}^{\infty} nB \subset H$ with B as unit ball is a (B)-space, since $E[\mathfrak{T}]$ is quasi-complete, and its norm topology is finer than $\mathfrak{T}$. L has infinite codimension in H, and is not semi-reflexive.

We shall now construct the required counterexample in H.

c) The set formed by $\mathfrak{o}$ and the sequence x_n/n is compact in H_B and, by § 20, 6.(3), so also is its closed absolutely convex cover A in H_B. Since A is also $\mathfrak{T}$-compact, the set $K_0=A+L$ is absolutely convex and $\mathfrak{T}$-closed, by § 15, 6.(10), and so also is $K_1=K_0 \cap B \subset H_B$.

d) We show that $\mathfrak{o}$ is a boundary point of $K_0'=K_0 \cap H_B$ in H_B. To do this, we form the closed linear subspace $L^{(n)}$ of H determined by

L and $x_1, \ldots, x_n$; L has finite co-dimension in this space, by § 15, 5.(3). Thus $H_B/L^{(n)} \cap H_B$ is infinite-dimensional. Since the image $\hat{K}_0'$ of K_0' under the canonical mapping of H_B onto $H_B/L^{(n)} \cap H_B$ lies in $\frac{1}{n+1} \hat{B}$, there is an element $y_n \in H_B \sim K_0'$ whose H_B-norm satisfies $\|y_n\| < \frac{1}{n}$. Thus we have established the existence of a sequence in $H_B \sim K_0'$ which converges to $\mathfrak{o}$.

e) There is no $\mathfrak{T}$-closed supporting hyperplane of K_1 through $\mathfrak{o}$. Such a hyperplane must take the form $ux = 0$, with $u \in H'$. But since K_1 contains both x_n/n and $-x_n/n$, we must have $ux_n = 0$ for all n. But the set of all x_n is total in $H[\mathfrak{T}]$, and so it follows that $u \equiv 0$.

f) There is an $x_0 \in H_B$ for which the ray ρx_0, $\rho > 0$, is disjoint from K_0. If there were no such ray, K_0' would be absorbent about $\mathfrak{o}$; it would therefore be a barrel in H_B, and so by § 21, 2.(1) it would be a neighbourhood of $\mathfrak{o}$ in H_B, contradicting d). Since K_0 contains L, x_0 does not lie in L.

g) We now construct the second closed convex set K_2. Since $B_1 = L \cap B$ is not weakly compact, by 3.(7) there is a decreasing sequence C_n of bounded convex $\mathfrak{T}$-closed subsets of $\frac{1}{2} B_1$ which have an empty intersection. We set K_2 equal to the closed convex cover of the sets $\frac{x_0}{n} + C_n$. As in the proof of (3) we see that K_2 only contains elements of the form $\rho x_0 + z$, with $\rho > 0$, $z \in L$. By f), K_1 and K_2 are disjoint.

h) Suppose that $ux = \gamma$ is a closed hyperplane in H which separates K_1 and K_2; suppose that K_1 lies in $ux \leqq \gamma$. Since $\mathfrak{o} \in K_1$, $\gamma \geqq 0$. Let $ux_0 = \alpha$. It follows from $C_n \subset \frac{1}{2} B_1$ and $B_1 \subset K_1$ that $u(C_n) \leqq \frac{\gamma}{2}$. Then letting $n \to \infty$ in $u\left(\frac{1}{n} x_0 + C_n\right) = \frac{\alpha}{n} + u(C_n) \geqq \gamma$, it follows that $\gamma = 0$.

The hyperplane separating K_1 and K_2 must therefore pass through $\mathfrak{o}$, contradicting e).

5. KREIN's theorem. In what follows we shall need certain results about the space $C(K)$ of continuous functions on a compact topological space K. By § 14, 9., $C(K)$ is a (B)-space under the norm $\|f\| = \sup_{x \in K} |f(x)|$.

By the topology $\mathfrak{T}_p$ of pointwise convergence on $C(K)$ we mean the locally convex topology defined by the neighbourhoods $\sup_{i=1,\ldots,n} |f(x_i) - f_0(x_i)| < \varepsilon$, $x_i \in K$. We can relate this topology in a simple way to the weak topology on $C(K)$. Each $x_0 \in K$ determines a continuous linear functional $\delta_{x_0}(f) = f(x_0)$ on $C(K)$. To each $x_0 \in K$ there therefore corresponds an element of $C(K)' = \mathfrak{M}(K)$, which we call the point measure δ_{x_0} corresponding to x_0.

This embedding $\hat{K}$ *of* K *in* $\mathfrak{M}(K)$ *is a homeomorphic mapping, when* $\mathfrak{M}(K)$ *is given the weak topology:* the weak topology on $\hat{K}$ corresponds to the coarsest Hausdorff topology on $\hat{K}$ for which all the $f \in C(K)$ are continuous; the weak topology on $\hat{K}$ is coarser than the one corresponding to the topology of K; since K is compact, it must therefore coincide with it.

If we denote the linear span of $\hat{K}$ in $\mathfrak{M}(K)$ by H, then $\mathfrak{T}_p$ is equal to $\mathfrak{T}_s(H)$. Since $H \subset \mathfrak{M}(K)$, it is coarser than the weak topology $\mathfrak{T}_s(\mathfrak{M}(K))$ on $C(K)$. We observe further that $\hat{K}$ is weakly total in $\mathfrak{M}(K)$.

(1) *Every relatively countably* $\mathfrak{T}_p$*-compact subset of* $C(K)$ *is relatively sequentially* $\mathfrak{T}_p$*-compact.*

Proof. To prove this theorem, we reduce it to an application of Šmulian's theorem. It is sufficient to show that a relatively countably $\mathfrak{T}_p$-compact sequence $f_n \in C(K)$ has a $\mathfrak{T}_p$-convergent subsequence. Let L be the $\mathfrak{T}_p$-closed linear span of the f_n in $C(K)$. Then $\{f_n\}$ is also relatively countably compact in $L[\mathfrak{T}_p]$, and $L[\mathfrak{T}_p]$ is separable. By § 22, 1.(1), the dual of $L[\mathfrak{T}_p]$ is the quotient space $H/L^{\perp}$, where H is the linear span of $\hat{K}$. The image $\hat{\hat{K}}$ of $\hat{K}$ under the canonical mapping from H onto $H/L^{\perp}$ is weakly compact, by § 22, 2.(4). If $N = \{f_n\}$, N is total in $L[\mathfrak{T}_p]$, so that $\mathfrak{T}_s(N)$ is a Hausdorff topology on $\hat{K}$. $\hat{K}$ is therefore also $\mathfrak{T}_s(N)$-compact. Since N is countable, $\hat{K}$ is therefore metrizable under the topology $\mathfrak{T}_s(N) = \mathfrak{T}_s(L)$, and so it is separable (§ 4, 5.(2)); consequently $H/L^{\perp} = L'$ is weakly separable. Since $\mathfrak{T}_p = \mathfrak{T}_s(H)$, the assertion follows by applying Šmulian's theorem (1.(2)) to $f_n \in L[\mathfrak{T}_p]$.

(2) *A sequence* $f_n \in C(K)$ *converges weakly to* $f_0 \in C(K)$ *if and only if the* f_n *are uniformly bounded and converge pointwise to* f_0.

a) Necessity. It follows directly from $\delta_{x_0}(f_n) \to \delta_{x_0}(f_0)$ that f_n converges pointwise to f_0. Since the sequence f_n is a strongly bounded subset of $C(K)$, it must also be uniformly bounded on K, so that the f_n are uniformly bounded.

b) Sufficiency. We use the following theorem of Lebesgue: let μ be a positive measure on K and let L^1_μ be the space of absolutely integrable functions with norm $\|f\| = \int |f|\, d\mu$. If the sequence $h_n \in L^1_\mu$ converges μ-almost everywhere to h_0 and if $|h_n| \leqq g$ μ-almost everywhere, for some g in L^1_μ, then $h_0 \in L^1_\mu$ and $\int |h_n - h_0|\, d\mu \to 0$. Since $C(K)$ is a subspace of every L^1_μ, the weak convergence of f_n follows for every positive μ, and so for every measure on K.

For Lebesgue's theorem cf. Bourbaki [7] Vol. 1, p. 140, for example.

The following theorem, proved by Grothendieck [6], follows from (1) and (2):

(3) *A subset M of $C(K)$ is weakly relatively compact if and only if it is bounded and relatively $\mathfrak{T}_p$-compact.*

Proof. The condition is necessary, for a relatively weakly compact set is also relatively compact under the coarser topology $\mathfrak{T}_p$.

On the other hand suppose that M is bounded and relatively $\mathfrak{T}_p$-compact. In order to show that M is relatively weakly compact it is sufficient, by EBERLEIN's theorem, to show that every sequence f_n in M contains a weakly convergent subsequence. By (1) f_n has a $\mathfrak{T}_p$-convergent subsequence, and this converges weakly, by (2).

In § 20, 6.(3) we showed that the closed convex (respectively absolutely convex) cover $\overline{\complement(M)}$ ($\overline{\Gamma(M)}$) of a compact set M is compact if $\overline{\Gamma(M)}$ is complete. We now obtain the following sharper result, KREIN's theorem:

(4) *The closed convex (respectively absolutely convex) cover $\overline{\complement(K)}$ ($\overline{\Gamma(K)}$) of a compact subset K of the locally convex space $E[\mathfrak{T}]$ is compact if and only if $\overline{\complement(K)}$ ($\overline{\Gamma(K)}$) is $\mathfrak{T}_k$-complete.*

Thus if $E[\mathfrak{T}]$ is $\mathfrak{T}_k$-quasi-complete, the closed absolutely convex cover of a compact set is always compact.

(4) is clearly equivalent to

(4′) *If K is a weakly compact subset of a locally convex space, $\overline{\complement(K)}$ (respectively $\overline{\Gamma(K)}$) is weakly compact if and only if $\overline{\complement(K)}$ ($\overline{\Gamma(K)}$) is $\mathfrak{T}_k$-complete.*

(4′) was proved by M. KREIN [1] for weakly sequentially compact subsets of a separable (B)-space; PHILLIPS [2] proved (4′) for (B)-spaces, and the general case, and the proof given here, are due to GROTHENDIECK.

Proof. a) First we show that it is sufficient to prove the theorem for $\mathfrak{T}_k$-complete spaces.

Suppose that K is weakly compact in $E[\mathfrak{T}]$, and that $\overline{\complement(K)}$ (respectively $\overline{\Gamma(K)}$) is $\mathfrak{T}_k$-complete. We can suppose that $\mathfrak{T}=\mathfrak{T}_k$. By § 21, 4.(5), the topology $\tilde{\mathfrak{T}}_k$ of the completion $\tilde{E}[\tilde{\mathfrak{T}}_k]$ is the Mackey topology on $\tilde{E}$. Since E and $\tilde{E}$ have the same dual, K is also weakly compact in $\tilde{E}$. Because of the assumption that $\overline{\complement(K)}$ (respectively $\overline{\Gamma(K)}$) is $\mathfrak{T}_k$-complete, $\overline{\complement(K)}$ (respectively $\overline{\Gamma(K)}$) is also equal to the closed convex (respectively absolutely convex) cover of K in $\tilde{E}$. If we assume that (4′) has been proved for $\tilde{E}[\tilde{\mathfrak{T}}_k]$, then it follows that $\overline{\complement(K)}$ (respectively $\overline{\Gamma(K)}$) is weakly compact in $\tilde{E}$, and consequently in E.

b) Now suppose that K is a weakly compact subset of the complete space $E[\mathfrak{T}_k]$. It is sufficient to show that $\overline{\Gamma(K)}$ is weakly compact. If $Au=\tilde{u}$ is the restriction of an element $u\in E'$ to K, then A is a linear mapping from E' into $C(K)$. Since all $u\in(\overline{\Gamma(K)})^\circ$ have a restriction $\tilde{u}$

with $\|\tilde{u}\| \leq 1$ in $C(K)$, A is continuous for the strong topology on E' and the norm topology on $C(K)$. Now $\mathfrak{T}_s(E)$ is finer than $\mathfrak{T}_s(K)$ on E', so that A is also a continuous mapping from $E'[\mathfrak{T}_s(E)]$ into $C(K)[\mathfrak{T}_p]$. If B is a closed $\mathfrak{T}_k$-equicontinuous subset of E', and thus a strongly bounded $\mathfrak{T}_s(E)$-compact set, it therefore follows that $A(B)$ is a bounded $\mathfrak{T}_p$-compact subset of $C(K)$. It follows from (3) that $A(B)$ is weakly compact in $C(K)$.

The adjoint mapping A' maps $C(K)' = \mathfrak{M}(K)$ into E''. A neighbourhood of $\mathfrak{o}$ for the natural topology $\mathfrak{T}_n(E')$ on E'' has the form B°, where B is closed and $\mathfrak{T}_k$-equicontinuous in E' (cf. § 23, 4.). By the preceding argument, $(A(B))^\circ$ is a $\mathfrak{T}_k$-neighbourhood of $\mathfrak{o}$ in $\mathfrak{M}(K)$. Since it follows from $u \in (A(B))^\circ$ that $A'u \in B^\circ$, A' is continuous for the $\mathfrak{T}_k$-topology on $\mathfrak{M}(K)$ and the natural topology on E'', and this coincides on E with $\mathfrak{T} = \mathfrak{T}_k$, by § 23, 4.(2).

It follows from $(A'\delta_{x_0})u = \delta_{x_0}(Au) = ux_0$ that A' maps the set $\hat{K}$ of point measures onto $K \subset E''$. $\hat{K}$ is weakly total in $\mathfrak{M}(K)$, and is therefore $\mathfrak{T}_k$-total as well. The image $A'(\mathfrak{M}(K))$ therefore lies in the $\mathfrak{T}_n$-complete linear span of K in E'', and so lies in E, since E is $\mathfrak{T}_k$-complete. Now $\hat{K}$ lies in the weakly compact unit ball C of $\mathfrak{M}(K)$; the mapping A', which is weakly continuous, by § 20, 4.(6), maps C into an absolutely convex weakly compact subset of E, which contains $\overline{\Gamma(K)}$. $\overline{\Gamma(K)}$ is therefore weakly compact.

6. PTÁK's theorem. We now give a second proof of KREIN's theorem. This does not use the theory of integration; instead it uses EBERLEIN's theorem and a combinatorial theorem of PTÁK.

We say that a subset M of a locally convex space $E[\mathfrak{T}]$ has *interchangeable double limits* if for every sequence $x_i \in M$ and every sequence $u_j \in N$, with N absolutely convex and weakly compact in E', the existence of both the double limits $\lim_i \lim_j u_j x_i$ and $\lim_j \lim_i u_j x_i$ ensures that they are equal.

We now have the following variant of EBERLEIN's theorem, due to GROTHENDIECK [6]:

(1) *Suppose that $E[\mathfrak{T}]$ is $\mathfrak{T}_k$-complete. A bounded subset M of E is weakly relatively compact if and only if it has interchangeable double limits.*

Proof. a) Suppose that M is weakly relatively countably compact, that $x_i \in M$ and $u_j \in N$, with N absolutely convex and weakly compact, and that $\alpha = \lim_i \lim_j u_j x_i$ and $\beta = \lim_j \lim_i u_j x_i$ both exist. Let x_0 be weakly adherent to the sequence x_i and let u_0 be weakly adherent to the sequence u_j. Then $\lim_j u_j x_i = u_0 x_i$ and $\lim_i u_j x_i = u_j x_0$. It then follows from $\lim_i u_0 x_i = u_0 x_0$ and $\lim_j u_j x_0 = u_0 x_0$ that $\alpha = \beta$.

b) Suppose that M is bounded and has interchangeable double limits. If M were not weakly relatively compact there would be a weak closure point z of M in E'^* which does not lie in E. As in the proof of 2.(2), this implies the existence of two sequences $x_i \in M$ and $u_k \in N$, where N is absolutely convex and weakly compact in E', with properties 2.(3), 2.(4) and 2.(5). It follows from 2.(3) that $\lim_i u_j x_i = u_j z$ for all j, and it follows from 2.(4) that $\lim_j u_j x_i = 0$, so that $\lim_i \lim_j u_j x_i = 0$.

The sequence $u_j z$ need have no limit, but since $M \subset \rho N^\circ$, $\rho > 0$, the moduli remain bounded. We can therefore choose a subsequence, denoted by u_j again, with the property that $\lim_j u_j z = \beta$ exists. It then follows from 2.(5) that $|\beta| \geqq \varepsilon$, so that $\lim_j \lim_i u_j x_i = \beta \neq 0 = \lim_i \lim_j u_j x_i$. This contradicts the fact that M has interchangeable double limits.

Suppose now that $\mathsf{A} = \{\alpha\}$ is an infinite set, and that $\varphi(\mathsf{A})$ is the space of coordinate vectors $\mathfrak{x} = (\xi_\alpha)$ with only finitely many non-zero ξ_α. We denote by $C(\mathsf{A})$ the set of all $\mathfrak{l} = (\lambda_\alpha) \in \varphi(\mathsf{A})$ with $\lambda_\alpha \geqq 0$, $\sum_{\alpha \in \mathsf{A}} \lambda_\alpha = 1$. If $\mathsf{B} \subset \mathsf{A}$, $C(\mathsf{B})$ denotes the set of all $\mathfrak{l}$ in $C(\mathsf{A})$ with $\lambda_\alpha = 0$ for $\alpha \notin \mathsf{B}$.

Let $\mathfrak{G}$ be a collection of finite subsets Γ of A. We denote by $C(\mathsf{B}, \mathfrak{G}, \varepsilon)$ the set of all $\mathfrak{l} \in C(\mathsf{B})$ with $\sum_{\alpha \in \Gamma} \lambda_\alpha < \varepsilon$ for all $\Gamma \in \mathfrak{G}$. With these definitions, PTÁK's theorem [5] reads

(2) *The two following conditions on $\mathfrak{G}$ are equivalent:*

α) There exists a strictly increasing sequence $\mathsf{A}_1 \subset \mathsf{A}_2 \subset \cdots$ of finite subsets of A, and a sequence $\Gamma_n \in \mathfrak{G}$ with $\mathsf{A}_n \subset \Gamma_n$ for all n;

β) There exists an infinite subset B of A and an $\varepsilon > 0$ for which $C(\mathsf{B}, \mathfrak{G}, \varepsilon)$ is empty.

Proof. a) Suppose that α) holds. Let $\mathsf{B} = \bigcup_{n=1}^{\infty} \mathsf{A}_n$ and $\mathfrak{l} \in C(\mathsf{B}, \mathfrak{G}, \varepsilon)$, $\varepsilon < 1$. Now and in what follows let $\mathsf{N}(\mathfrak{l})$ denote the set of indices $\alpha \in \mathsf{A}$ for which $\lambda_\alpha \neq 0$. There exists an n_0 with $\mathsf{N}(\mathfrak{l}) \subset \mathsf{A}_{n_0}$. But then $\mathsf{N}(\mathfrak{l}) \subset \Gamma_{n_0}$, so that $\sum_{\alpha \in \Gamma_n} \lambda_\alpha = \sum_{\alpha \in \mathsf{A}} \lambda_\alpha = 1$, which contradicts the fact that $\mathfrak{l} \in C(\mathsf{B}, \mathfrak{G}, \varepsilon)$ and that $\varepsilon < 1$.

b) If Δ is a subset of A, let $\mathfrak{G}(\Delta)$ denote the collection of all $\Gamma \in \mathfrak{G}$ with non-empty $\Gamma \cap \Delta$. In particular let $\mathfrak{G}(\alpha)$ denote the collection of all $\Gamma \in \mathfrak{G}$ which contain α. We prove the following lemma:

(3) *Suppose that $C(\mathsf{B}, \mathfrak{G}, \varepsilon)$ is empty and that B is infinite. If Φ is a finite subset of A and if $\varepsilon_1 < \varepsilon$, then there is a non-empty finite subset Δ_1 of A which is disjoint from Φ and for which $C(\mathsf{B} \sim \Phi, \mathfrak{G}(\Delta_1), \varepsilon_1)$ is empty. It then follows from the definition of $C(\mathsf{B}, \mathfrak{G}(\Delta), \varepsilon)$ that $\mathfrak{G}(\Delta_1)$ is non-empty.*

Proof. Suppose that $C(\mathsf{B} \sim \Phi, \mathfrak{G}(\Delta_1), \varepsilon_1)$ is empty for given Φ, given $\varepsilon_1 < \varepsilon$ and every finite non-empty subset $\Delta \subset \mathsf{A}$ which is disjoint from Φ. We choose a finite non-empty subset $\mathsf{M}_1 \subset \mathsf{B} \sim \Phi$. Let $\mathfrak{l}^{(1)}$ be an element of the set $C(\mathsf{B} \sim \Phi, \mathfrak{G}(\mathsf{M}_1), \varepsilon_1)$, which is non-empty, by hypothesis. We then form $\mathsf{M}_2 = \mathsf{M}_1 \cup \mathsf{N}(\mathfrak{l}^{(1)})$. Then $\mathsf{M}_2 \subset \mathsf{B} \sim \Phi$, and we can choose an $\mathfrak{l}^{(2)}$ in $C(\mathsf{B} \sim \Phi, \mathfrak{G}(\mathsf{M}_2), \varepsilon_1)$, and so on. In this way we obtain a sequence $\mathfrak{l}^{(n)} = (\lambda_\alpha^{(n)}) \in C(\mathsf{B} \sim \Phi, \mathfrak{G}(\mathsf{M}_n), \varepsilon_1)$ with $\mathsf{M}_n = \mathsf{M}_1 \cup \mathsf{N}(\mathfrak{l}^{(1)}) \cup \cdots \cup \mathsf{N}(\mathfrak{l}^{(n-1)})$.

Suppose that $\Gamma \in \mathfrak{G}$. We form the sequence $\alpha_n = \sum_{\alpha \in \Gamma} \lambda_\alpha^{(n)} \leqq 1$, and assert that it contains at most one $\alpha_n \geqq \varepsilon_1$. Suppose that $\alpha_p \geqq \alpha_n$ is the first of these α_n. Since $\mathsf{N}(\mathfrak{l}^{(p)}) \subset \mathsf{M}_{p+1}$, $M_{p+1} \cap \Gamma$ is non-empty, so that $\Gamma \in \mathfrak{G}(\mathsf{M}_{p+1})$. Now if $q > p$, $\Gamma \in \mathfrak{G}(\mathsf{M}_{p+1}) \subset \mathfrak{G}(\mathsf{M}_q)$. Since $\mathfrak{l}^{(q)} \in C(\mathsf{B} \sim \Phi, \mathfrak{G}(\mathsf{M}_q), \varepsilon_1)$, $\alpha_q < \varepsilon_1$ for all $q > p$.

Consequently if we form the mean $\frac{1}{n}(\mathfrak{l}^{(1)} + \cdots + \mathfrak{l}^{(n)})$ for sufficiently large n, then this lies in $C(\mathsf{B} \sim \Phi, \mathfrak{G}, \varepsilon)$, which contradicts the hypothesis of (3).

c) Now suppose that β) is satisfied. Then by (3) if $\varepsilon_1 < \varepsilon$, $C(\mathsf{B}, \mathfrak{G}(\Delta_1), \varepsilon_1)$ is also empty. If we apply (3) again, then we obtain for $\varepsilon_2 < \varepsilon_1$ a non-empty subset Δ_2, disjoint from Δ_1, for which $C(\mathsf{B} \sim \Delta_1, \mathfrak{G}(\Delta_1) \cap \mathfrak{G}(\Delta_2), \varepsilon_2)$ is empty. Repeating this procedure, we obtain a sequence $\Delta_1, \Delta_2, \ldots$ of pairwise disjoint non-empty sets for which the intersections $\mathfrak{G}(\Delta_1) \cap \cdots \cap \mathfrak{G}(\Delta_n)$ are non-empty. Now each $\mathfrak{G}(\Delta_k)$ is the union of finitely many $\mathfrak{G}(\alpha_k)$, $\alpha_k \in \Delta_k$; there is therefore an α_1 for which $\mathfrak{G}(\alpha_1) \cap \mathfrak{G}(\Delta_2) \cap \cdots \cap \mathfrak{G}(\Delta_m)$ is non empty, for all m, and corresponding to it an α_2 for which $\mathfrak{G}(\alpha_1) \cap \mathfrak{G}(\alpha_2) \cap \mathfrak{G}(\Delta_3) \cap \cdots \cap \mathfrak{G}(\Delta_m)$ is non-empty, for all m, and so on. Then $\mathfrak{G}(\alpha_1) \cap \cdots \cap \mathfrak{G}(\alpha_n)$ is non-empty, for all n. Now let $\mathsf{A}_n = \{\alpha_1, \ldots, \alpha_n\}$. The sequence A_n is properly increasing. For $\Gamma_n \in \mathfrak{G}(\alpha_1) \cap \cdots \cap \mathfrak{G}(\alpha_n)$, however, $\mathsf{A}_n \subset \Gamma_n$, and so α) is established.

The proof of Krein's theorem can be reduced to the case where $E[\mathfrak{T}]$ is $\mathfrak{T}_k$-complete, as was shown in 5. By § 20, 6.(4), we can restrict ourselves to showing that $\complement(M)$ is relatively weakly compact if $M \subset E$ is. Using (1), Krein's theorem therefore follows from

(4) *If a bounded subset M of a real locally convex space has interchangeable double limits, then so does $\complement(M)$.*

Proof. Suppose that the assertion is false. Then there is an absolutely convex weakly compact subset N of E' and sequences $x_i \in C(M)$, $u_j \in N$, with

$$(5) \qquad \left| \lim_i \lim_j u_j x_i - \lim_j \lim_i u_j x_i \right| = \varepsilon > 0.$$

We denote by μ the finite value of $\sup_{u \in N, x \in M} |u x|$.

There is a countable subset T of M with the property that all the x_i lie in $\complement(T)$. Suppose that u_0 is weakly adherent to the sequence u_j.

Using a diagonal procedure, we can choose a subsequence of the u_j, denoted by u_j once again, for which $\lim_j u_j z = u_0 z$ holds for all $z \in T$. We also have $\lim_j u_j x_i = u_0 x_i$ for all i.

Since (5) holds, we can find $\sigma = \pm 1$ and an infinite set B of indices k for which

$$\sigma\left(\lim_i u_k x_i - \lim_i u_0 x_i\right) \geqq \tfrac{3}{4}\varepsilon \quad \text{for all} \quad k \in \mathsf{B}. \tag{6}$$

For each $z \in T$ let $\Gamma(z)$ be the finite set of all indices j with

$$|(u_j - u_0) z| \geqq \frac{\varepsilon}{4}. \tag{7}$$

Let $\mathfrak{G}$ be the collection of all $\Gamma(z)$, $z \in T$. We now assert that $C\left(\mathsf{B}, \mathfrak{G}, \dfrac{\varepsilon}{8\mu}\right)$ is non-empty. Let us assume the contrary. Then by (2) there are index sets $\mathsf{A}_n = \{j_1, \ldots, j_n\}$ and sets $\Gamma(z_n)$ with $\mathsf{A}_n \subset \Gamma(z_n)$ for all n. By (7) it follows from this that $|(u_{j_q} - u_0) z_n| \geqq \dfrac{\varepsilon}{4}$ for $n \geqq q$. Now let y_n be a subsequence of the sequence z_n with the property that $\lim_n (u_{j_q} - u_0) y_n$ exists for each q. For the sequence $y_n \in M$ and the sequence $v_q = \frac{1}{2}(u_{j_q} - u_0) \in N$ it now follows that $\lim_q v_q y_n = 0$ for each n, and that $\left|\lim_n v_q y_n\right| \geqq \dfrac{\varepsilon}{8}$ for each q. By choosing a subsequence of the sequence v_q, we now obtain $\left|\lim_q \lim_n v_q y_n\right| \geqq \dfrac{\varepsilon}{8}$ and $\lim_n \lim_q v_q y_n = 0$. This contradicts the assumption that M has interchangeable double limits.

Thus $C\left(\mathsf{B}, \mathfrak{G}, \dfrac{\varepsilon}{8\mu}\right)$ is non-empty. Let $\mathrm{l} = (\lambda_n)$ be a vector in this set, so that $\mathsf{N}(\mathrm{l}) \subset \mathsf{B}$. We set $u = \sum_{k \in \mathsf{N}(\mathrm{l})} \lambda_k (u_k - u_0)$. By the definitions of $C\left(\mathsf{B}, \mathfrak{G}, \dfrac{\varepsilon}{8\mu}\right)$ and $\Gamma(z)$, we then have, for each $z \in T$,

$$|uz| = \left|\sum_{k \in \mathsf{N}(\mathrm{l})} \lambda_k (u_k - u_0) z\right| \leqq \left|\sum_{k \in \Gamma(z)}\right| + \left|\sum_{k \in \mathsf{N}(\mathrm{l}) \sim \Gamma(z)}\right|$$

$$\leqq \left(\sum_{\Gamma(z)} \lambda_k\right) 2\mu + \left(\sum_{k \in \mathsf{N}(\mathrm{l})} \lambda_k\right) \frac{\varepsilon}{4} \leqq \frac{\varepsilon}{2}.$$

From this it follows that $|u x_i| \leqq \dfrac{\varepsilon}{2}$ for all i.

Since $\mathsf{N}(\mathsf{I}) \subset \mathsf{B}$, we finally obtain the following contradiction from (6):

$$\frac{\varepsilon}{2} \geqq \lim |u x_i| = \left|\sum_{\mathsf{N}(\mathsf{I})} \lambda_k \left(\lim_i u_k x_i - \lim_i u_0 x_i\right)\right|$$
$$= \left(\sum_{\mathsf{N}(\mathsf{I})} \lambda_k\right) \sigma \left(\lim_i u_k x_i - \lim_i u_0 x_i\right) \geqq \tfrac{3}{4}\varepsilon.$$

For the results of this number I am indebted to a written communication from Professor PTÁK, who has recently (cf. PTÁK [6]) given a more extended account of his combinatorial method.

§ 25. Extreme points and extreme rays of convex sets

1. The KREIN-MILMAN theorem. In the following, E may be either a real or a complex vector space. By (x, y), however, we always mean the real open interval $\tau x + (1-\tau) y$, $0 < \tau < 1$. Suppose that M is a subset of E. A point z of M is called an extreme point of M if it belongs to no open interval $(x, y) \subset M$. If M is convex, this means that z never lies between two points of M.

More generally, we clearly have: if z is an extreme point of the convex set M, and if z lies in the convex cover of the points $x_1, \ldots, x_n$ of M, then z coincides with one of the points x_i.

If M is absolutely convex and if z is an extreme point of M, then so is every point σz with $|\sigma| = 1$.

We denote the set of all extreme points of M by $Ep(M)$. The concept of extreme point is a special case of a more general concept. A real linear manifold $H \subset E$ is called a support manifold of M if $H \cap M$ is non-empty and if every open interval $(x, y) \subset M$ which contains a point of H lies entirely in H, and thus lies in $H \cap M$.

The zero-dimensional support manifolds of M are the extreme points; the support manifolds which are hyperplanes are the supporting hyperplanes of M (cf. § 17, 5.), provided that M is convex. It should be pointed out, however, that not every point of support (cf. § 17, 5.) need be a zero-dimensional support manifold, i.e. an extreme point.

(1) *Suppose that H is a support manifold of M in E. A real linear manifold $H_1 \subset H$ is a support manifold of M if and only if it is a support manifold of $H \cap M$.*

Proof. If H_1 is a support manifold of M, it is also a support manifold of $H \cap M$, for $H_1 \cap (H \cap M) = H_1 \cap M$ is non-empty, and by hypothesis every open interval in $H \cap M$ which meets H_1 lies in $H_1 \cap M = H_1 \cap (H \cap M)$.

Conversely, if H_1 is a support manifold of $H \cap M$, $H_1 \cap M$ is non-empty; further if (x,y) is an interval in M which meets $H_1 \subset H$, then (x,y) is contained in $H \cap M$, since H is a support manifold of M. But then by the hypothesis on H_1, (x,y) lies in H_1.

We now investigate the closed support manifolds of a compact subset of a locally convex space.

(2) *Suppose that M is a compact subset of a locally convex space. Further suppose that $\{H_\alpha\}$ is a system of closed support manifolds of M, which contains an $H_\beta \subset H_{\alpha_1} \cap H_{\alpha_2}$ whenever it contains H_{α_1} and H_{α_2}. Then $H = \bigcap_\alpha H_\alpha$ is a closed support manifold of M.*

Proof. The compact sets $M_\alpha = M \cap H_\alpha$ form a filter-base on M whose intersection $H \cap M = \bigcap_\alpha M_\alpha$ is non-empty. H is a closed support manifold of M, for H is closed, and if $(x,y) \subset M$ has the point z in common with H then z belongs to all the H_α, so that $(x,y) \subset H_\alpha$, and consequently $(x,y) \subset \bigcap_\alpha H_\alpha = H$.

(3) *Every closed support manifold H of a compact subset M of a locally convex space contains at least one extreme point of M.*

Proof. By (2), the closed support manifolds of M contained in H satisfy the hypotheses of ZORN's lemma, so that there is a minimal closed support manifold H_0 of M contained in H.

Suppose that H_0 is not a point. We may suppose that $o \in H_0$ (this can always be obtained by a translation). By § 20, 7.(8) the compact subset $H_0 \cap M$ of H_0 has a closed supporting hyperplane H_1 in H_0. By (1), H_1 is also a closed support manifold of M, which contradicts the minimality of H_0.

The KREIN-MILMAN theorem now reads

(4) *Every compact convex subset M of a locally convex space is the closed convex cover of the set $Ep(M)$ of its extreme points.*

We shall prove the rather more general statement

(5) *A compact subset M of a locally convex space has the same closed convex cover as the set $Ep(M)$ of its extreme points.*

Proof. We assume that $N_1 = \overline{\complement(Ep(M))}$ is a proper subset of $N = \overline{\complement(M)}$. If $z_0 \in N \sim N_1$, then by § 20, 7.(1) there is a closed real hyperplane H which separates z_0 and N_1 strictly. There are points of M which belong to the same open half-space R_1 of H as z_0. By § 20, 7.(8), M has a closed supporting hyperplane $H_1 \subset R_1$ parallel to H. By (3), H_1 contains an extreme point of M, and this cannot lie in N_1, which is impossible.

One half of the next theorem follows from (4):

(6) *Suppose that K is a compact convex subset of a locally convex space. If $M \subset K$, then $\overline{\complement(M)} = K$ if and only if $\bar{M} \supset Ep(K)$.*

The other half results from MILMAN's theorem [2]:

(7) *Suppose that M is a set whose closed convex cover $\overline{\complement(M)}$ is compact. The extreme points of $\overline{\complement(M)}$ lie in $\bar{M}$, and so they are also extreme points of $\bar{M}$.*

For the proof (H. v. WEIZSÄCKER) we use the following

Lemma. Let N, R be subsets of a locally convex space, $N \subset R$, N compact, R convex and locally compact. Then $\overline{\complement(N)}$ is compact and $\overline{\complement(N)} \subset R$.

Proof. Every element of R has a neighbourhoodbase of compact convex subsets of R. Since N is compact there exist compact convex sets $K_1, \dots, K_n$ in R such that $N \subset \bigcup_{i=1}^{n} K_i \subset R$. The set $K = \complement(\bigcup_{i=1}^{n} K_i)$ is convex and compact by § 20, 6. (5). Since R is convex we have $K \subset R$ and finally $\overline{\complement(N)} \subset \overline{\complement(K)} = K \subset R$.

Proof of (7). Let x be extreme point of $\overline{\complement(M)}$ which is not contained in $\bar{M}$. Then $R = \overline{\complement(M)} \sim x$ is locally compact and convex since $\overline{\complement(M)}$ is compact and x is one of its extreme points. If we put $N = \bar{M}$ and apply the lemma we get the contradiction $\overline{\complement(\bar{M})} \subset \overline{\complement(M)} \sim x$.

On going from $\bar{M}$ to $\overline{\complement(M)}$, extreme points can be lost, as the example of the set consisting of three collinear points shows. If M is compact and $\complement(M)$ is not, then $\complement(M)$ can have extreme points which do not lie in M. (For an example, cf. BOURBAKI [6], Vol. 1, p. 85, Ex. 3.)

The next result is a corollary of MILMAN's theorem:

(8) *Suppose that M is a set whose closed absolutely convex cover $\overline{\Gamma(M)}$ is compact. Then every extreme point of $\overline{\Gamma(M)}$ has the form αz, $|\alpha| = 1$, where z is an extreme point of $\bar{M}$.*

Proof. Let $\tilde{M} = \bigcup_{|\alpha|=1} \alpha \bar{M}$. $\tilde{M}$ is compact, since it is the continuous image of the topological product of $\bar{M}$ and the set of all α with $|\alpha| = 1$. Since the circled cover of M lies in the convex cover of $\tilde{M}$, $\overline{\Gamma(M)} = \overline{\complement(\tilde{M})}$, by § 16, 1. (2). By (7), every extreme point of $\overline{\Gamma(M)}$ is therefore an extreme point of $\tilde{M}$, and *a fortiori* of some $\alpha \bar{M}$; it is therefore of the form αz, where z is an extreme point of $\bar{M}$.

The set $Ep(M)$ of a compact convex M is not in general closed. Suppose that in P^3 we are given a circle in the ξ, η-plane through the origin, together with the segment of the ζ-axis between $\zeta = -1$ and $\zeta = +1$, which meets it at o. The closed convex cover of this set has as extreme points the end-points of the segment and all points of the circle with the exception of $\mathfrak{o}$.

Suppose that $E_1[\mathfrak{T}_1]$ and $E_2[\mathfrak{T}_2]$ are locally convex spaces, and that A is a continuous linear mapping from E_1 into E_2. The image Ax_0 of an extreme point x_0 of a set $M \subset E_1$ need not be an extreme point of $A(M)$. On the other hand, it follows easily from the definition of support manifold that the inverse image $A^{(-1)}(H)$ of a closed support manifold H of $A(M)$ in $A(E_1)$ is a closed support manifold of M.

Applying (3), we obtain

(9) *Every extreme point of $A(M)$, where M is compact in $E_1[\mathfrak{T}_1]$, is the image of an extreme point of M.*

2. Examples and applications. The closed unit ball K of c_0 has no extreme points: if $\mathfrak{x}=(\xi_n)\in c_0$ and $\|\mathfrak{x}\| = \sup_n |\xi_n| = 1$, then we may replace some coordinate ξ_k, with $|\xi_k|<1$, alternately by $\xi_k+\varepsilon$ and $\xi_k-\varepsilon$, with ε sufficiently small; $\mathfrak{x}$ then lies between the two points of K obtained in this way.

The closed unit ball K of l^1 has the set of all $\sigma\mathfrak{e}_i$, $i=1,2,\dots$, $|\sigma|=1$, as $Ep(K)$. For if $\mathfrak{x} \neq \sigma\mathfrak{e}_i$, we can divide the set of non-zero coordinates of $\mathfrak{x}$ into two non-empty parts, to give two vectors $\mathfrak{x}_1$ and $\mathfrak{x}_2$. $\mathfrak{x}$ then lies between $\frac{1}{\|\mathfrak{x}_1\|}\mathfrak{x}_1$ and $\frac{1}{\|\mathfrak{x}_2\|}\mathfrak{x}_2$.

The unit ball of l^∞ has as extreme points all $x=(\xi_i)$ with $|\xi_i|=1$ for all i.

If $1<p<\infty$, then every boundary point of the unit ball of l^p (respectively L^p) is an extreme point. This follows easily from the remarks on Minkowski's inequality in § 14, 8. (respectively 10.).

On the other hand, the unit ball of L^1 has no extreme points: suppose that $f\in L^1$, $\int_a^b |f|\,dx=1$; we determine c in such a way that $\int_a^c |f|\,dx=\frac{1}{2}$. If we put f_1 equal to $2f$ on $[a,c)$ and equal to 0 on $[c,b]$, and put f_2 equal to 0 on $[a,c)$ and equal to $2f$ on $[c,b]$, then f is the mid-point of the segment $[f_1,f_2]$, whose end-points both belong to K.

The extreme points of the unit ball of L^∞ are the functions $f(t)$ with $|f(t)|=1$ almost everywhere.

If R is a compact space and $C(R)$ is the space of continuous real- (respectively complex-)valued functions on R (cf. § 14, 9.), then the extreme points of the unit ball K consist of all functions $f(t)$ with $|f(t)|\equiv 1$. If, in addition, R is connected and $C(R)$ is real, then it follows that $f_1\equiv 1$ and $f_2\equiv -1$ are the only two extreme points.

We now investigate $\mathfrak{M}(R)$, the space of real measures on the compact space R.

By § 24, 5., the set $\hat{R}$ of point-measures δ_x, $x \in R$, is a subset of the weakly compact unit ball K of $\mathfrak{M}(R)$ which is homeomorphic to R. A measure $u \in \mathfrak{M}(R)$ is said to be positive if $u(f) \geqq 0$ whenever $f \geqq 0$ on the whole of R. Let $\mathfrak{M}^+(R)$ be the set of positive measures, and let K_1^+ be the set of positive measures v with $\|v\| = 1$.

If $v \in K_1^+$, $v(1) = 1$, so that K_1^+ lies in the closed hyperplane H of $\mathfrak{M}(R)$ given by $u(1) = 1$. It follows from $\|u\| \leqq 1$ that $u(1) \leqq 1$, so that H is a supporting hyperplane of K.

If $u \in K \cap H$, $u \in K_1^+$, so that $K_1^+ = H \cap K$. For if $u(f) = -k < 0$ for $u \in K \cap H$ and for some $f \geqq 0$, we could choose $\rho > 0$ so that $1 - \rho f \geqq 0$; we would then obtain $u(1 - \rho f) = 1 + \rho k > 1$, contradicting $\|u\| = 1$.

We now show that the points δ_x are extreme points of K_1^+. We assume that there exist $v_1, v_2 \in K_1^+$ with

(1) $$\delta_{x0} = \tfrac{1}{2}(v_1 + v_2).$$

Suppose that $g(x) \in C(R)$ and that $g(x_0) = 0$. We assert that $v_1(g) = v_2(g) = 0$. Since every such g is the difference of two non-negative functions vanishing at x_0, it is sufficient to prove this for $g \geqq 0$. In this case, however, the assertion follows from $v_1(g) \geqq 0$, $v_2(g) \geqq 0$ and $\delta_{x_0}(g) = 0 = \frac{1}{2}(v_1(g) + v_2(g))$.

Now every $f \in C(R)$ has the form $f = f(x_0) \cdot 1 + g$, with $g(x_0) = 0$. But then

$$v_1(f) = f(x_0)\, v_1(1) + v_1(g) = f(x_0) = \delta_{x0}(f),$$

so that $v_1 = \delta_{x0}$ and $v_2 = \delta_{x0}$. The points δ_x are therefore extreme points of K_1^+.

The absolute polar of $\hat{R}$ in $C(R)$ is the closed unit ball, and so $K = \overline{\Gamma(\hat{R})}$, by the theorem of bipolars. By 1.(8), the only possible extreme points of K are the elements of $\hat{R}$ and of $-\hat{R}$. The points of $\hat{R}$ are extreme points of K_1^+. Since $K_1^+ = K \cap H$, they are also extreme points of K, by 1.(1). A corresponding argument applies to $-\hat{R}$. Thus we have proved

(2) *Suppose that $\mathfrak{M}(R)$ is the space of real measures on the compact space R. The extreme points of the unit ball K of $\mathfrak{M}(R)$ are the point-measures δ_x and their mirror-images $-\delta_x$ about $\mathfrak{o}$.*

The extreme points are shared between the two closed supporting hyperplanes $u(1) = 1$ and $u(1) = -1$ of K.

Remark. Similarly, the measures $\sigma \delta_x$, with $|\sigma| = 1$, are found to be the extreme points of the unit ball of the space of complex measures on R.

The Banach-Stone theorem is now a simple consequence of (2):

(3) *Two compact spaces R_1 and R_2 are homeomorphic if and only if the real spaces $C(R_1)$ and $C(R_2)$ are norm-isomorphic.*

Proof. The condition is clearly necessary. Conversely, suppose that $C(R_1)$ and $C(R_2)$ are norm-isomorphic. The dual spaces $\mathfrak{M}(R_1)$ and $\mathfrak{M}(R_2)$ are then both weakly isomorphic and norm-isomorphic. In particular the sets of extreme points of the two unit balls are mapped onto each other in a weakly isomorphic way. By (2), every $\delta_y \in \hat{R}_2$ has as image either δ_x or $-\delta_x$ for some $x \in R_1$. It follows from this that the function 1 in $C(R_1)$ corresponds to a function φ in $C(R_2)$ with values $+1$ and -1. Because φ is continuous, R_2 divides into two compact open subsets M_1 and M_2, on which φ is equal to $+1$ and -1 respectively. The mapping which sends every $f(x)$ in $C(R_2)$ to the function $\varphi(x)f(x)$ is therefore a norm-isomorphism of $C(R_2)$, under which φ is sent to the function 1. We may therefore suppose that $1 \in C(R_1)$ is mapped into $1 \in C(R_2)$. $\hat{R}_1$ is then mapped in a weakly isomorphic way onto $\hat{R}_2$. By § 24, 5., R_1 and R_2 are therefore homeomorphic.

The properties of the extreme points of the unit ball of $\mathfrak{M}(R)$ given in (2) characterise such spaces, as the following theorem of Arens and Kelley [1] (cf. Kadison [1], as well), which we give without proof, shows:

(4) *A real* (B)*-space E is norm-isomorphic to some $C(R)$, with compact R, if and only if the following hold:*

a) *the extreme points of the closed unit ball K of the dual space lie in two supporting hyperplanes of the form $u(a)=1$, $u(b)=1$, where a and b are elements of E of norm 1;*

b) *if the weak closure of a set of extreme points contains no two points symmetric about o, then it is contained in some supporting hyperplane of K.*

For further characterisations of $C(R)$, cf. Kadison [1] and Day [8].

The next result follows directly from the Krein-Milman theorem and the weak compactness of every closed equicontinuous set:

(5) *Every weakly closed convex equicontinuous subset of the dual E' of a locally convex space $E[\mathfrak{T}]$ is the weakly closed convex cover of its extreme points.*

In particular the unit ball of the dual of a (B)*-space has this property.*

From this it follows that

(6) *If E is a reflexive* (B)*-space, the closed unit ball of E is the closed convex cover of its extreme points.*

It follows immediately from (6) and the examples considered above that c_0, L^1 and the real space $C(R)$, R compact, infinite and connected, are not reflexive.

Even more can be deduced from (5), however.

(7) *The spaces c_0, L^1 and the real space $C(R)$, R compact, connected and infinite, are not norm-isomorphic to a dual* (B)*-space.*

For since their unit balls contain either no extreme points or only finitely many of them, the closed convex cover of the extreme points is at most finite-dimensional, for any Hausdorff topology, and so it cannot be the unit ball.

The papers of DIXMIER [1] and RUSTON [6] deal with the question of when a (B)-space is norm-isomorphic to the dual of a (B)-space. SCHATTEN [1] considers a further example, using the method of extreme points.

3. Variants of the KREIN-MILMAN theorem. We begin with an example. Let M_0 consist of a square with corners A, B, C, D, together with the coplanar semi-circular disc with diameter AB. A, B, C, D and the points of the open semi-circular arc R are the extreme points of M_0. We can clearly leave out A and B, and still obtain the whole of M_0 as the closed convex cover of C, D and R.

Quite generally, as has indeed been made explicit in 1.(6), $Ep(K)$ can be replaced in the KREIN-MILMAN theorem by any subset M of K, provided only that $\bar{M} \supset Ep(K)$. Variants of the KREIN-MILMAN theorem are therefore obtained when the set $Ep(K)$ is replaced by some such suitable set M.

Suppose that K is a closed convex subset of a locally convex space $E[\mathfrak{T}]$. If x_0 is a point of support of K, i.e. a point through which at least one closed supporting hyperplane passes, let $D(x_0)$ denote the intersection of all the closed supporting hyperplanes through x_0. $D(x_0)$ is a closed support manifold. We call the intersection $S(x_0) = D(x_0) \cap K$ the supporting facet of K through x_0. The supporting facet through a point is always a closed convex subset of K consisting only of points of support. If $y \in S(x_0)$, then $S(y) \subset S(x_0)$.

A supporting facet of K is said to be minimal if it contains no proper subset which is also a supporting facet of K.

If K is compact, every decreasing ordered set $\{S(x_\alpha)\}$, with $S(x_\alpha) \subset S(x_\beta)$ if $\alpha > \beta$, has a non-empty intersection, which contains the supporting facet through each of its points. Thus by applying ZORN's lemma we obtain

(1) *If K is a compact convex subset of a locally convex space $E[\mathfrak{T}]$, there is at least one minimal supporting facet in each closed supporting hyperplane of K.*

Two minimal supporting facets of K either coincide or are disjoint. We now obtain the MILMAN-RUTMAN theorem [1]:

(2) *Every compact convex subset K of a locally convex space $E[\mathfrak{T}]$ is equal to $\complement(M)$, where M is any subset which contains one arbitrary point from each minimal supporting facet of K.*

The proof proceeds in a completely analogous way to the proof of 1.(5), and will be left to the reader.

Since every supporting facet of K is the intersection of a support manifold of K with K, by 1.(1) every extreme point of a minimal supporting facet is also an extreme point of K. Thus by (2) $K=\overline{\mathrm{C}}(M)$, where M is any subset which contains one extreme point from each minimal supporting facet of K.

In the example M_0 above, the extreme points C, D and the points of the open semi-circle R are minimal supporting facets, while the supporting facet through A is the segment AD; it is therefore quite possible for the supporting facet through an extreme point to fail to be minimal.

A minimal supporting facet need by no means consist of a single point. If we add another semicircular disc, with diameter CD, to the set M_0, we obtain another compact convex set M_1 for which the two segments AD and BC are minimal supporting facets.

An essentially stronger result has been obtained by KLEE [9] for normed spaces.

Following STRASZEWICZ [1], a point x_0 of a subset M of a locally convex space $E[\mathfrak{T}]$ is called an exposed point of M if x_0 is a point of support of a closed supporting hyperplane H of M, for which $H \cap M=\{x_0\}$. Every exposed point of M is thus an extreme point, but not conversely, as for example the point A of the set M_0 shows. In particular, exposed points of a set are zero-dimensional supporting facets. The following theorem of KLEE's [9] is thus an extension both of 1.(4) and of (2), for normed spaces.

(3) *Suppose that K is a compact convex subset of a normed space. The set of exposed points of K is dense in $Ep(K)$, so that K is the closed convex cover of the set of exposed points of K.*

Reference must be made to KLEE [9] for the proof, and for a collection of further results about exposed points.

4. The extreme rays of a cone. An analogue of the KREIN-MILMAN theorem can be proved for cones, once a further concept has been introduced. Suppose that K is a convex subset of a vector space E. Let h be a real open half-line contained in K; it is called an extreme ray of K if every open interval in K which meets h lies wholly in h.

It follows from this that the straight line g which h defines meets K in a set which consists only of h and possibly the end point a of h. If a also belongs to K, then a is an extreme point of K.

If C is a convex cone in E with vertex x_0, and if x is a point of the cone other than x_0, then x lies on one of the generators of the cone, namely the open half-line from x_0 through x. It follows directly from this that x_0 is the only possible extreme point of C, and that x_0 is an extreme point of C if and only if C is proper and pointed.

The question of which generators of C are extreme rays can be answered in the following way:

(1) *Suppose that C is a proper convex cone with vertex x_0. If a generator h is an extreme ray, and if a hyperplane H cuts the generator h in one point $y \neq x_0$, then y is an extreme point of $H \cap C$.*

Conversely if there is a hyperplane H which cuts the generator h in an extreme point of $H \cap C$, then h is an extreme ray of C.

Corollary. *If M is a convex set contained in a hyperplane H, and if C is the cone generated by M with vertex x_0 outside H, then the generators through the extreme points of M are precisely the extreme rays of C.*

Proof. The first part of (1) follows directly from the definition of extreme point and extreme ray.

Conversely suppose that h meets H in an extreme point y_0 of $H \cap C$. Suppose that h is not an extreme ray. Then there is an open interval $(x_1, x_2) \subset C$ which meets h but which does not lie in h. It is easy to see that we can suppose that the point of intersection is y_0. But then the open planar surface bounded by the generators $x_0 x_1$ and $x_0 x_2$ is contained in C, contains h, and meets H in an open interval containing y_0; y_0 is therefore not an extreme point of $H \cap C$, contrary to hypothesis.

We now consider cones in a locally convex space $E[\mathfrak{T}]$.

(2) *A proper pointed convex cone C in $E[\mathfrak{T}]$ is locally compact if and only if it is the pointed cone generated by a compact convex set M which does not contain the vertex. M can be chosen to be a subset of a closed hyperplane.*

Such a cone is always closed.

In the proof we can suppose that the vertex is the point $\mathfrak{o}$.

a) Suppose that C is proper, pointed, convex and locally compact. Then there is a closed convex neighbourhood U of $\mathfrak{o}$ in $E[T]$ for which $U \cap C$ is convex and compact. Let M be the intersection of the boundary of U with C, and let $K = \overline{C(M)}$. From $M \subset U \cap C$ it follows that $K \subset U \cap C$, so that K is convex and compact.

If $y \in C$, $y \neq \mathfrak{o}$, then the generator through y contains a boundary point of U, and so a point of M; C is therefore the cone generated by M, and *a fortiori* is the cone generated by K. We still have to show that $\mathfrak{o}$ does not belong to K. If $\mathfrak{o}$ belonged to K, $\mathfrak{o}$ would be an extreme point of K, since C is assumed to be proper. By Milman's theorem (1.(7)), $\mathfrak{o}$ would then have to be an extreme point of M, and this is not possible.

b) Suppose that C is the pointed cone generated by the convex compact set K with vertex $\mathfrak{o} \notin K$. By § 20, 7.(1), there is a real closed hyperplane H which separates $\mathfrak{o}$ and K strictly. H intersects every generator of C in a point which is different from $\mathfrak{o}$. Let $\tilde{K} = H \cap C$. $\tilde{K}$ is a subset of the set C_1 of all ρy, $\rho \in [0,1]$, $y \in K$. This set, being the continuous image of the compact set $[0,1] \times K$, is also compact. Thus $\tilde{K} = H \cap C_1$ is convex and compact, and C is the cone generated by $\tilde{K}$.

c) Let C be the cone generated by $\tilde{K}=H\cap C$, as in b), and let $z\neq o$ be a closure point of C. The ray from o through z meets H, for otherwise z would lie in the hyperplane H_0 through o parallel to H, and z would therefore be a closure point of the part $\tilde{C}_1$ (the set of all ρy, $\rho\in[0,1]$, $y\in\tilde{K}$) of C which lies on the same side of H; but since $\tilde{C}_1$ is compact, $H_0\cap\tilde{C}_1=o$. Since therefore the ray from o through z meets H, there is a hyperplane H_1, parallel to H, such that z is an interior point of the closed half-space R_1 defined by H_1 and containing o. z must be a closure point of the part of C lying in R_1, and this has the form $\tilde{C}_l$ (the set of all σy, $\sigma\in[0,l]$, $y\in\tilde{K}$) and is therefore compact. Hence $z\in C$, and so every proper pointed convex locally compact cone is closed, by a) and b).

d) Finally suppose that C is the cone generated by a compact convex set M which does not contain o. By b), we can replace M by a set $\tilde{K}$ lying in a closed hyperplane H. Every point of C has the form ρy, $\rho\geqq 0$, $y\in\tilde{K}$. If we construct the hyperplane H_2 parallel to H through $(\rho+1)y$, ρy is an interior point of the closed half-space R_2 defined by H_2 and containing o, and so $R_2\cap C$ is a compact neighbourhood of ρy in C. Thus C is locally compact.

We now have the following analogue of the KREIN-MILMAN theorem.

(3) *Every proper pointed convex locally compact cone C in a locally convex space is the closed convex cover $\overline{\complement(Es(C))}$ of the set $Es(C)$ of its extreme rays.*

Proof. By (2), C can be represented as the pointed cone with vertex o generated by a compact convex set $\tilde{K}$ lying in a hyperplane H. By the corollary of (1), the extreme points of $\tilde{K}$ lie on extreme rays of C; by the KREIN-MILMAN theorem, $\tilde{K}$ is contained in $\overline{\complement(Es(C))}$. Any hyperplane parallel to H which cuts C other than in o can be used to produce C, and so every non-zero point of C lies in $\overline{\complement(Es(C))}$; from this it follows that $C\subset\overline{\complement(Es(C))}$. C is closed, by (2), and so we also have the converse inclusion $\overline{\complement(Es(C))}\subset C$.

The set $\mathfrak{M}^+(R)$ of positive measures on the compact space R forms a proper pointed convex weakly closed cone in $\mathfrak{M}(R)$. $\mathfrak{M}^+(R)$ is the cone generated by K_1^+, the weakly compact convex subset of positive measures of norm 1. We have shown that the point-measures δ_x are extreme points of K_1^+. Since $K_1^+=K\cap H$, and H is a closed support hyperplane of the unit ball K, K_1^+ has no further extreme points, by 1.(1) and 3.(2). The extreme rays of $\mathfrak{M}^+(R)$ are thus the generators defined by the points δ_x, and $\mathfrak{M}^+(R)$ is the weakly closed convex cover of these extreme rays.

5. Locally compact convex sets. The KREIN-MILMAN theorem and its analogue for cones, which we have just obtained, are special cases of a more general theorem of KLEE's [8]:

(1) *Every closed convex locally compact subset K of a locally convex space which contains no straight line is the closed convex cover of the set of its extreme points and extreme rays:* $K = \overline{\complement(Ep(K) \cup Es(K))}$.

Before proving this, we establish two lemmas:

(2) *If K is closed, convex and locally compact, K is compact if and only if it contains no half-line.*

We need only show that if K is not compact then it contains a half-line.

We shall prove rather more, namely that through each point of K there is a half-line contained in K. We can take this point to be $\mathfrak{o}$. There exists a closed convex neighbourhood U of $\mathfrak{o}$ for which $U \cap K$ is compact. The intersection B_1 of the boundary of U with K is also compact, and it does not contain $\mathfrak{o}$. Let B_n be the intersection of nU with K, multiplied by $1/n$. Clearly $B_1 \supset B_2 \supset \cdots$. We shall show that no B_k is empty. If B_k were empty, we would have $K = (kU) \cap K$. It would then follow from $(kU) \cap K \subset k(U \cap K)$ and the compactness of $U \cap K$, and thus of $k(U \cap K)$, that K would be relatively compact; since K is closed, K would therefore be compact, contrary to hypothesis.

The sequence $B_1 \supset B_2 \supset \cdots$ of compact sets B_i therefore has a non-empty intersection. If $z \neq \mathfrak{o}$ is an element of this intersection, every nz belongs to K, and so the ray from $\mathfrak{o}$ through z lies in K.

(3) *If K is closed, convex and locally compact, and if K contains no straight line, then K has at least one extreme point.*

We may assume that K is not compact. By (2) there is at least one point x_0 from which there is a half-line contained in K. The collection of all half-lines from x_0 which are contained in K forms a proper, closed, convex cone C. This cone is also locally compact, since each point of C has a compact neighbourhood in K, and therefore also in $K \cap C$. By 4.(2) there is a closed hyperplane H_1 which has only the point x_0 in common with C. Let R_1 be the closed half-space defined by H_1 which does not contain C. R_1 has a locally compact intersection K_1 with K. If K_1 were not compact, then there would be a half-line through x_0 lying in K_1, by the proof (2). But this contradicts the definition of C. If the compact convex subset K_1 of K lies in H_1, then every extreme point of K_1 is also an extreme point of K. If $K_1 \not\subset H_1$ then K_1 has an extreme point which does not lie in H_1, and which is therefore an extreme point of K.

Proof of (1). We assume that $A = \overline{\complement(Ep(K) \cup Es(K))}$ is a proper subset of K. Then by § 20, 7.(1) (applied to a point of $K \sim A$, and A), there is a closed hyperplane H which meets K, but not A. $K \cap H$ satisfies the hypotheses of (3), and so it possesses an extreme point x_0, which is not an extreme point of K, since it does not lie in A. There is therefore a straight line g, cutting H in x_0, whose intersection D with K contains x_0 as an interior point. D is either a segment or a half-line.

Suppose first that D is a segment $[x_1, x_2]$. We assert that x_1 and x_2 are extreme points of K. If x_1 were not an extreme point, there would be two points $y_1, y_2 \in K$, not lying on g, with $x_1 \in (y_1, y_2)$. If z is a point of D on the side of H which does not contain x_1, then the triangle z, y_1, y_2, which lies in K, would meet H in an interval containing x_0 as an interior point, and this is not possible, since x_0 is an extreme point of $K \cap H$. Since x_0 lies between the two extreme points x_1 and x_2 of K, x_0 must belong to A; this gives a contradiction.

If D is a half-line, then the same argument shows that every open interval in K which contains a point of D must lie in D. But then D is an extreme ray of K. Since x_0 lies in D, x_0 must again belong to A, which is impossible. Thus (1) is proved.

Suppose that M is a subset of a locally convex space $E[\mathfrak{T}]$ and that h is a half-line $x_0 + \rho y, \rho \geqq 0$, in E. The half-line H is called an asymptote of M if for each $\rho > 0$ and each $\mathfrak{T}$-neighbourhood U of $x_0 + \rho y$ there exists a $z \in M$ for which $[x_0, z] \cap U$ is non-empty.

Suppose that K is convex, locally compact and closed, and suppose that K contains no straight line. Suppose further that M is a subset of K with $\bar{M} \supset Ep(K)$, and suppose that every extreme ray of K is an asymptote of M. Since the end point of an extreme ray of K is an extreme point of K, and so lies in $\bar{M}$, it follows easily that $\overline{\mathrm{C}(M)}$ contains every extreme ray of K as well. Thus $\overline{\mathrm{C}(M)}$ is equal to K, by (1).

This establishes one half of the following generalization of 1.(6):

(4) *Suppose that K is a closed convex locally compact subset of a locally convex space, which contains no straight line. A subset M of K satisfies the relation $\overline{\mathrm{C}(M)} = K$ if and only if $\bar{M} \supset Ep(K)$ and all the extreme rays of K are asymptotes of M.*

Reference must be made to KLEE [8] for the proof of the other half, which is a not completely straightforward generalisation of MILMAN's theorem.

In 3. we introduced the concept of an exposed point. An exposed ray H of a closed convex subset K of a locally convex space is a closed half-line which is the intersection with K of a closed supporting hyperplane H of K. The corresponding open half-line is an extreme ray of K.

KLEE [9] has proved the following stronger form of (1), corresponding to 3.(3):

(5) *Suppose that K is a closed, convex, locally compact subset of a normed space which contains no straight line. K is the closed convex cover of the set of exposed points and exposed rays of K.*

§ 26. Metric properties of normed spaces

1. Strict convexity. Although every normed space is a topological vector space, it has an even richer structure. There is a collection of properties which are defined in terms of the metric, with the result that they are preserved under norm-isomorphism, but not under topological isomorphism. These properties are important above all for various applications in analysis. In this paragraph we shall investigate some of the concepts of this kind.

First suppose that $E[\mathfrak{T}]$ is locally convex and that K is a closed convex $\mathfrak{T}$-body with $\mathfrak{o}$ as interior point, and with boundary S. K is said to be strictly convex, or rotund, if every boundary point of K is an extreme point. A normed space, and its norm, are said to be strictly convex if the closed unit ball is strictly convex.

(1) *If K is a closed convex $\mathfrak{T}$-body, the following conditions are equivalent:*

a) *K is strictly convex;*

b) *S contains no line segment;*

c) *every supporting hyperplane meets K in at most one point;*

d) *distinct boundary points have distinct supporting hyperplanes;*

e) *every boundary point of K is an exposed point.*

If K is the closed unit ball of a normed space, we can also add

f) *if $\|x\| = \|y\| = 1$ and $x \neq y$, then $\|\frac{1}{2}(x+y)\| < 1$;*

g) *if $\|x+y\| = \|x\| + \|y\|$ and $y \neq \mathfrak{o}$, then $x = \alpha y$ for some non-negative α.*

Proof. b) follows from a). Suppose that b) is satisfied, but not a). Then there is a $z \in S$ and a segment $[x, y] \subset K$ with z as interior point. If x were an interior point of K, z would be an interior point of the convex cover of y and a neighbourhood $U \subset K$ of x, and this is contained in K. Thus $[x, y] \subset S$, contradicting b).

c) is necessary; for if a hyperplane of support H met K in more than one point, $H \cap K \subset S$ would contain a line segment, contradicting b). Conversely if z is a boundary point of K, z is point of support of a closed hyperplane H (cf. § 17, 5.(1)) and it follows from c) that z is a minimal support manifold, and so is an extreme point.

d) follows from c). b) follows from d), since there is a supporting hyperplane through every segment in S (§ 17, 2.(2)).

Since there is a supporting hyperplane through each point of S, c) and e) are equivalent.

If K is the closed unit ball of a normed space, and if f) holds, then b) follows. On the other hand, f) follows directly from a).

Suppose that g) is satisfied. If $x,\ y\in E$ and if $\|x\|=\|y\|=\|\frac{1}{2}(x+y)\|=1$, then $\|x+y\|=\|x\|+\|y\|$; it follows from g) that $x=\alpha y$, and we must then have $x=y$. f) therefore follows from g).

We now assume that f) holds. If $\|x+y\|=\|x\|+\|y\|$ for $x,\ y\in E$, and if $\|y\|\geqq\|x\|$ and $x,y\neq o$, then

$$\left\|\frac{x}{\|x\|}+\frac{y}{\|y\|}\right\|\geqq\left\|\frac{x}{\|x\|}+\frac{y}{\|x\|}\right\|-\left\|\frac{y}{\|x\|}-\frac{y}{\|y\|}\right\|=\frac{\|x\|+\|y\|}{\|x\|}$$
$$-\|y\|\left(\frac{1}{\|x\|}-\frac{1}{\|y\|}\right)=2.$$

It now follows from f) that $x/\|x\|=y/\|y\|$, i.e. g) is satisfied.

For the examples discussed in § 25, 2. we have the following results: the spaces l^p and L^p are strictly convex for $1<p<\infty$, while the spaces c_0, l^1, l^∞, L^1, L^∞, $C(K)$ and $\mathfrak{M}(K)$ are not strictly convex.

2. Shortest distance. Suppose that M is a subset of the normed space E, and that x_0 is a point of E. The quantity $|x_0,M|=\inf_{y\in M}\|x_0-y\|$ is the distance of x_0 from M. If the infimum is attained for some $y_0\in M$, then we call this point a nearest point to x_0 in M.

We shall be concerned with the question of when a nearest point exists, and when it is unique.

(1) *Suppose that E is a normed space, and that M is a closed convex locally compact subset of E. Then there is always at least one nearest point to x_0 in M.*

Proof. The closed unit ball K of E and M are weakly closed. For sufficiently large $\rho\geqq 0$ $(x_0+\rho K)\cap M$ is non-empty. This set is weakly closed and contained in M, so that it is weakly locally compact. By § 25, 5.(2) it is therefore weakly compact. The intersection of all the non-empty $(x_0+\rho K)\cap M$ is thus non-empty, and is equal to $(x_0+\rho_0 K)\cap M$, where ρ_0 is the least of these ρ. All the points of $(x_0+\rho_0 K)\cap M$ (and no others) are nearest points to x_0 in M.

(2) *If E is a reflexive* (B)*-space and if M is a closed convex subset of E, then there exists at least one nearest point to x_0 in M.*

This follows in the same way as (1), for the closed bounded sets $(x_0+\rho K)\cap M$ are all weakly compact.

The next result shows the significance of strict convexity for this problem:

(3) *If E is strictly convex, and if M is a convex subset of E, then each x_0 has at most one nearest point in M.*

If there were two nearest points y_1, y_2 at distance ρ_0 from x_0, then $\frac{1}{2}(y_1+y_2)\in M$, being the midpoint of two boundary points of the strictly convex set $x_0+\rho_0 K$, would, by 1.(1)f, have distance less than ρ_0 from x_0, and this is not possible. Clearly strict convexity is also necessary for (3) to hold.

If E is strictly convex and if M satisfies the hypotheses of (1) we can define a mapping from E onto M, the nearest-point mapping $P_M(x)$, which sends each $x\in E$ to its uniquely determined nearest point in M.

(4) *If E is strictly convex and if M is a closed convex locally compact subset of E, then the nearest-point mapping $P_M(x)$ is continuous.*

We must show that $P_M(x_n)\to P_M(x_0)$ always follows from $x_n\to x_0$. If $\|x_n-x_0\|\leqq\varepsilon$, then $d_n=\inf\limits_{y\in M}\|x_n-y\|\leqq\inf\|x_0-y\|+\|x_0-x_n\|=d_0+\varepsilon$. From this it follows that $\|x_0-P_M(x_n)\|\leqq\|x_n-P_M(x_n)\|+\|x_0-x_n\|\leqq d_n+\varepsilon\leqq d_0+2\varepsilon$. Thus $P_M(x_n)$ lies in $C(\varepsilon)=(x_0+(d_0+2\varepsilon)K)\cap M$. This set is compact. The diameters of these sets $C(\varepsilon)$ must tend to 0 as $\varepsilon\to 0$, for otherwise there would be a nearest point to x_0 in M different from $P_M(x_0)$. The $P_M(x_n)\in C(\varepsilon)$ therefore converge, as $\varepsilon\to 0$, to $P_M(x_0)=\bigcap\limits_{\varepsilon>0} C(\varepsilon)$.

In particular, if M is a finite-dimensional linear subspace of a normed space E there is always at least one best approximation of $x\in E$ by elements of M, by (1). If, further, M is strictly convex, there is only one such best approximation, and it depends continuously on x.

If M is a closed linear subspace of l^2, the nearest-point mapping is the orthogonal projection on M, and so it is linear. In the general case the nearest-point mapping need not be linear, even when M is a closed linear subspace.

We now investigate the behaviour of $|x,M|=f(x)$ with regard to the weak topology. We first make two general remarks about normed spaces.

Suppose that E is a complex normed space, and that E_r is the same space considered as a real normed space. If $u\in E'$, then $\Re u\in(E_r)'$, and we assert that

$$\|u\|=\|\Re u\|. \tag{5}$$

Once again, let K be the closed unit ball of E and of E_r. If $ux=re^{i\varphi}$, then $|ux|=(\Re u)(e^{-i\varphi}x)$. It follows from this that

$$\|u\|=\sup_{x\in K}|ux|\leq\sup_{y\in K}|(\Re u)y|=\|\Re u\|,$$

and since $\|\Re u\|\leqq\|u\|$, (5) follows.

We now prove a generalization of § 17, 6.(2):

(6) *Suppose that M is a closed convex subset of the normed space E, and that x_0 has distance $|x_0,M|=d$ from M. Then there is a $u_0 \in E'$ with $\|u_0\|=1$ and $\mathfrak{R}(u_0(x_0-y)) \geqq d$ for all $y \in M$.*

Because of (5) and § 16, 3.(1), it is sufficient to prove this for a real normed space. Further, we can take $x_0=o$. We denote the open ball $\|x\|<\rho$ in E by K_ρ^i. By hypothesis the two convex sets K_d^i and M have no point in common. By § 17, 1.(4) they are separated by a closed hyperplane given by $u_0 x=\rho, \rho>0$, with $\|u_0\|=1$. In particular $\sup_{x \in K_d^i} u_0 x=\|u_0\| d \leqq \rho$, so that $\rho \geqq d$. But then $u_0 y \geqq d$ for all $y \in M$.

(7) *If M is a closed convex subset of a normed space E, $f(x)=|x,M|$ is a weakly lower semi-continuous function on E.*

Proof. Suppose that u_0 is an element of E' which corresponds to x_0, as in (6). Then

$$f(x_0)=d=\inf_{y \in M}\|x_0-y\| \geqq \inf_{y \in M} \mathfrak{R}(u_0(x_0-y)) \geqq d,$$

so that $f(x_0)=\inf \mathfrak{R}(u_0(x_0-y))$. Now if x is in the weak neighbourhood $|u_0(x_0-y)|<\varepsilon$, we have

$$\begin{aligned} f(x)=|x,M| \geqq \inf \mathfrak{R}(u_0(x-y)) &\geqq \inf \mathfrak{R}(u_0(x_0-y))-|\mathfrak{R}(u_0(x-x_0))| \\ &\geqq f(x_0)-\varepsilon, \end{aligned}$$

so that $f(x)$ is weakly lower semi-continuous at x_0.

The distance between two subsets M_1 and M_2 of E is defined by $\inf_{x \in M_1, y \in M_2}\|x-y\|=|M_1,M_2|$. From (7), we obtain

(8) *Suppose that M_1 is a weakly compact subset, and M_2 a closed, convex, weakly locally compact subset of the normed space E. Then there are two points $x_1 \in M_1$ and $x_2 \in M_2$ with $\|x_1-x_2\|=|M_1,M_2|$.*

Proof. The function $|x,M_2|$, which is weakly lower semi-continuous, by (7), takes its minimum value $|M_1,M_2|$ at a point x_1 of M_1, by § 6, 2.(6). By (1) there is an $x_2 \in M_2$ with $\|x_1-x_2\|=|M_1,M_2|$.

Remark. If E is a reflexive (B)-space, M_1 can be taken to be an arbitrary bounded closed set, and M_2 to be an arbitrary closed convex set.

3. Points of smoothness. We obtain a concept which corresponds dually to the concept of extreme point, by making the following definition. A boundary point x_0 of a convex subset K of a vector space E is called an algebraic point of smoothness of K if at most one supporting hyperplane passes through x_0.

If $E[\mathfrak{T}]$ is locally convex, and if at most one closed supporting hyperplane passes through x_0, then x_0 is called a point of smoothness of K.

We again consider a closed convex $\mathfrak{T}$-body $K\ni o$ in $E[\mathfrak{T}]$ with boundary S; let K° be the convex weakly compact subset of E' polar to K. Every supporting hyperplane of $x_0\in S$ is closed, and has the form $\Re(ux)=1$, with $u\in K^\circ$. By § 17, 5.(1), there is a closed supporting hyperplane through each point x_0 of S. If x_0 is a point of smoothness of K, there is therefore exactly one supporting hyperplane through x_0. In this case we speak of a t a n g e n t h y p e r p l a n e. The points of smoothness of S are therefore the points at which a tangent hyperplane exists.

If to each point of S we make correspond all those points $u\in K^\circ$ which define supporting hyperplanes $\Re(ux)=1$ through the point, we obtain a many-valued function from S into K°. For this we have

(1) *a) To every point of smoothness of S there corresponds just one point of K°, which is extreme;*

b) To every point S which is not a point of smoothness there corresponds at least one non-extreme point;

c) to every point of S which is not an extreme point there corresponds at least one point of K° which is not a point of smoothness.

P r o o f. a) If the point u_0 corresponding to the point of smoothness x_0 were not an extreme point, then there would be an interval $[u_1,u_2]\subset K^\circ$ which contains u_0 as an interior point. But then $\Re(u_0x_0)=1$ would be a point of $[\Re(u_1x_0),\Re(u_2x_0)]$. It would then follow from $\Re(u_1x_0)\leqq 1$ and $\Re(u_2x_0)\leqq 1$ that $\Re(u_1x_0)=\Re(u_2x_0)=1$, so that $\Re(u_1x)=1$ and $\Re(u_2x)=1$ would be two further supporting hyperplanes through x_0, and this is not possible.

b) If x_0 is not a point of smoothness, let $\Re(u_1x)=1$ and $\Re(u_2x)=1$ be two distinct supporting hyperplanes through x_0. Then $[u_1,u_2]$ is contained in K°, and for each interior point u_0 of this line segment we also have $\Re(u_0x)=1$, i.e. u_0 defines a supporting hyperplane through x_0; u_0 is not an extreme point of K°.

c) Suppose that x_0 is an interior point of $[x_1,x_2]\subset S$. There is a supporting hyperplane $\Re(u_0x)=1$ containing $[x_1,x_2]$. But then all supporting hyperplanes of K^0 of the norm $\Re(ux)=1$ with $x\in[x_1,x_2]$ pass through u_0, so that u_0 is not an algebraic point of smoothness of K°; u_0 is also not a $\mathfrak{T}_s(E)$-point of smoothness of K°.

A convex $\mathfrak{T}$-body is said to be s m o o t h if all the points of its boundary are points of smoothness. In particular a normed space, and its norm, are said to be smooth, if the closed unit ball is smooth.

From (1) b) and c) we obtain immediately that

(2) *If the strong dual E' of a normed space is strictly convex (respectively smooth), then E itself is smooth (respectively strictly convex).*

Complete duality clearly holds in the reflexive case:

(3) *A reflexive* (B)-*space is strictly convex (respectively smooth) if and only if its strong dual is smooth (respectively strictly convex).*

DAY [6] has given an example of a non-reflexive strictly convex (B)-space whose dual is not smooth (cf. 9.).

(4) *Linear subspaces of strictly convex (respectively smooth) normed spaces are again strictly convex (respectively smooth).*

This is clear for strict convexity. If H is a subspace of E and if y_0 is a boundary point of the unit ball of H, then it follows from the HAHN-BANACH theorem that if y_0 is a point smoothness of the unit ball of E then y_0 is also a point of smoothness of the unit ball of H. The second part of (4) follows from this.

(5) *If E is a reflexive strictly convex (respectively smooth)* (B)-*space, then every quotient space E/H is also strictly convex (respectively smooth).*

By § 22, 3.(1 b), $(E/H)'$ is norm-isomorphic to a closed linear subspace of E', which is smooth (respectively strictly convex), by (3) and (4). The assertion now follows from (2).

(5) does not hold for quotients of arbitrary normed spaces: cf. KLEE [10].

It follows directly from (2) and the examples in 1. that none of the spaces $l^1=(c_0)'$, $l^\infty=(l^1)'$, $L^\infty=(L^1)'$, $\mathfrak{M}(K)=C(K)'$ is smooth. Further examples follow in number 5.

4. Weak differentiability of the norm. The question of when a boundary point of the unit ball of a normed space has a tangent hyperplane is connected with the differentiability of the norm at this point (cf. MAZUR [2], [3]).

Let $q(x)$ be the Minkowski functional of a convex $\mathfrak{T}$-body $C\ni o$ in the locally convex space $E[\mathfrak{T}]$. $q(x)$ is continuous and has properties (α), (β) and (γ) of § 16, 4.(2).

Let $\Delta(x_0,y,t)=\dfrac{q(x_0+ty)-q(x_0)}{t}$ be the difference quotient of $q(x)$ at x_0 in the direction y.

(1) *$\Delta(x_0,y,t)$ is monotonic increasing for $t>0$.*

Proof. If $0<t_1<t_2$, we have

$$t_2 q(x_0+t_1 y)=q(t_2 x_0+t_2 t_1 y)\leqq t_1 q(x_0+t_2 y)+(t_2-t_1)q(x_0),$$

so that

$$t_2(q(x_0+t_1 y)-q(x_0))\leqq t_1(q(x_0+t_2 y)-q(x_0)).$$

(2) *$\Delta(x_0,y,t)$ is bounded below, for $t>0$, and we have $\Delta(x_0,y,t)\geqq -q(-y)$.*

This follows directly from

$$q(x_0)=q(x_0+ty-ty)\leqq q(x_0+ty)+tq(-y).$$

The existence of the right derivative $q'_+(x_0,y)=\lim_{t\to 0^+}\Delta(x_0,y,t)$ of $q(x_0+ty)$ at $t=0$ now follows from (1) and (2). From (2) we obtain

(3) $$q'_+(x_0,y)\geqq -q(-y).$$

It follows from $\Delta(x_0,y,-t)=-\Delta(x_0,-y,t)$ that $\Delta(x_0,y,t)$ is also monotonic increasing and bounded above for negative t; for the left derivative $q'_-(x_0,y)=\lim_{-t\to 0^-}\Delta(x_0,y,-t)$ we have

(4) $$q'_-(x_0,y)=-q'_+(x_0,-y).$$

It follows from $2q(x_0)\leqq q(x_0+ty)+q(x_0-ty)$ that $\Delta(x_0,y,-t)\leqq\Delta(x_0,y,t)$, and, as $t\to 0^+$, this gives $q'_-(x_0,y)\leqq q'_+(x_0,y)$. From this we obtain

(5) $\Delta(x,y,t)$ *is monotonic increasing for all* t, *and we have the inequality*

(6) $$\frac{q(x_0-ty)-q(x_0)}{-t}\leqq q'_-(x_0,y)\leqq q'_+(x_0,y)\leqq\frac{q(x_0+ty)-q(x_0)}{t},\quad t>0.$$

The next proposition gives further properties of q'_+.

(7) *For fixed* x_0, $q'_+(x_0,y)$ *is a continuous positive homogeneous subadditive function on* $E[\mathfrak{T}]$.

Proof. It follows directly by taking limits in $\Delta(x_0,\sigma y,t)=\sigma\Delta(x_0,y,\sigma t)$, for $\sigma>0$, that $q'_+(x_0,\sigma y)=\sigma q'_+(x_0,y)$.

By subtracting $2q(x_0)$ from the inequality

$$2q\left(x_0+\frac{t}{2}(y_1+y_2)\right)\leqq q(x_0+ty_1)+q(x_0+ty_2)$$

and dividing by t, we obtain

$$\Delta\left(x_0,y_1+y_2,\frac{t}{2}\right)\leqq\Delta(x_0,y_1,t)+\Delta(x_0,y_2,t).$$

As $t\to 0$, this gives $q'_+(x_0,y_1+y_2)\leqq q'_+(x_0,y_1)+q'_+(x_0,y_2)$.

Finally the inequality

(8) $$q'_+(x_0,y)\leqq q(y)$$

follows from $\Delta(x_0,y,t)\leqq q\left(\frac{x_0}{t}\right)+q(y)-q\left(\frac{x_0}{t}\right)=q(y)$, and this, together with (3), establishes the continuity of $q'_+(x_0, y)$ at $\circ$; the ideas of the proof of § 16, 4.(7) then give continuity for all y.

We observe that at x_0 itself we clearly have $q'_+(x_0, x_0) = q'_-(x_0, x_0) = 1$.

If the two derivatives $q'_-(x_0, y)$ and $q'_+(x_0, y)$ coincide, the common value $q'(x_0, y) = \lim_{t \to 0} \Delta(x_0, y, t)$ is called the weak or Gâteaux derivative of q at the point x_0 in the direction y. In particular $q(x)$ is said to be weakly differentiable at x_0 if $q'(x_0, y)$ exists for each $y \in E$. For this it is necessary and sufficient, by (5), for

$$\lim_{t \to 0^+} \frac{q(x_0 + t y) + q(x_0 - t y) - 2 q(x_0)}{t} = 0 \tag{9}$$

to hold for all $y \in E$.

(10) *If $q(x)$ is weakly differentiable at x_0, the function defined by $u y = q'(x_0, y)$ is a real continuous linear functional on $E[\mathfrak{T}]$.*

The linearity and continuity of u follows without difficulty from $q'_-(x_0, y) = q'_+(x_0, y)$, (4) and (7).

The next theorem gives the connection with supporting hyperplanes:

(11) *Suppose that $C \ni \circ$ is a closed convex $\mathfrak{T}$-body in the locally convex space $E[\mathfrak{T}]$, with Minkowski functional $q(x)$. A boundary point x_0 of C is a point of smoothness if and only if $q(x)$ is weakly differentiable at x_0. The tangent hyperplane through x_0 is then given by $q'(x_0, y) = 1$.*

More generally we have: if $u x = 1$ is a real supporting hyperplane through x_0, u satisfies the inequality

$$-q'_+(x_0, -y) \leqq u y \leqq q'_+(x_0, y) \tag{12}$$

for all $y \in E$. Conversely if $y_0 \in E$ and if $-q'_+(x_0, -y_0) \leqq \gamma \leqq q'_+(x_0, y_0)$ then there is a real supporting hyperplane $v x = 1$ through x_0 for which $v y_0 = \gamma$.

It is sufficient to prove the second part of the assertion. Suppose that $u x = 1$ is a supporting hyperplane through x_0. Then $u x \leqq q(x)$, and it follows from

$$1 + t u y = u(x_0 + t y) \leqq q(x_0 + t y), \qquad t > 0,$$

that $u y \leqq \dfrac{q(x_0 + t y) - q(x_0)}{t}$, so that $u y \leqq q'_+(x_0, y)$. It then follows from $u(-y) \leqq q'_+(x_0, -y)$ that $-u(-y) = u(y) \geqq -q'_+(x_0, -y)$. Thus (12) is established.

We now define a linear functional on the real linear subspace H of E spanned by x_0 and y_0 by setting $l(z)=l(\alpha x_0+\beta y_0)=\alpha+\beta\gamma$. For sufficiently small $t>0$ we have

$$\frac{q(x_0+tz)-q(x_0)}{t}=\alpha+\frac{1+\alpha t}{t}\left[q\left(x_0+\frac{t}{1+\alpha t}\beta y_0\right)-1\right],$$

from which it follows that $q'_+(x_0,z)=\alpha+q'_+(x_0,\beta y_0)$. From the definition of γ, we have $\beta\gamma\leqq q'_+(x_0,\beta y_0)$, and we therefore obtain

$$l(z)=\alpha+\beta\gamma\leqq\alpha+q'_+(x_0,\beta y_0)=q'_+(x_0,z).$$

It now follows from (8) and the HAHN-BANACH theorem that there exists a real linear functional v with $vy\leqq q'_+(x_0,y)\leqq q(y)$ for all $y\in E$.

As a special case of (11) we have

(13) *A normed space is smooth if and only if its norm is weakly differentiable at every point except the origin.*

Since weak differentiation always takes place in a two-dimensional subspace, we have

(14) *A normed space is smooth if and only if every two dimensional linear subspace is smooth, and if and only if the norm is weakly differentiable (except at the origin) in each of these subspaces.*

5. Examples. 1. We now determine all the points of smoothness of the unit ball of $C(R)$, R compact (cf. BANACH [3], p. 169).

In § 25, 2. we determined the extreme points of the unit ball of $\mathfrak{M}(R)$. These are the points $\sigma\delta_x$, with x arbitrary in R and $|\sigma|=1$. By 3.(1)a) the supporting hyperplane through a point of smoothness f_0 must have the form $\mathfrak{R}(\sigma\delta_x(f))=1$. But there is just one hyperplane of the form $\mathfrak{R}(\sigma\delta_x(f))=1$ through a point f_0, with $\|f_0\|=1$, if and only if the function f_0 takes a value of modulus 1 at just one point x_0.

The norm $p(f)=\sup\limits_{t\in R}|f(t)|$ is therefore differentiable only at these points f_0, and we have

$$p'(f_0,g)=\mathfrak{R}(\overline{f_0(x_0)}\,\delta_{x_0}(g)). \tag{1}$$

2. A similar idea can be used for L^1. The extreme points of the unit ball of L^∞ are the functions $h(t)$ with $|h(t)|=1$ almost everywhere. If now $\int|f(t)|\,dt=1$, $f(t)$ can only lie on the hyperplane $\mathfrak{R}(\int h(t)f(t)\,dt)=1$ if $f(t)$ is non-zero almost everywhere and if $h(t)=\overline{\dfrac{f(t)}{|f(t)|}}$ holds almost everywhere. From this it follows that the points of smoothness of the unit ball of L^1 are those functions $f_0(t)$ with $\|f_0\|=1$ and $f_0(t)\neq 0$

almost everywhere, and that at these points f_0 the weak derivative of the norm has the form

$$p'(f_0,g)=\int \Re\left(\frac{\overline{f_0(t)}}{|f_0(t)|}g(t)\right)dt. \tag{2}$$

This argument assumes that it has already been shown that $(L^1)'=L^\infty$ (cf. 7.).

A proof can be given without making this assumption by using an argument similar to that of the next example.

3. We now consider L^p, $1<p<\infty$, (cf. MAZUR [3]). We calculate the derivative of the norm directly.

$$\frac{d}{dh}[(\int|f_0(t)+hg(t)|^p dt)^{1/p}]_{h=0}=\frac{1}{p}\cdot(\int|f_0|^p dt)^{-\frac{1}{q}}\frac{d}{dh}(\int|f_0+hg|^p dt)_{h=0},$$

whenever the derivative on the right hand side exists. The difference quotients of the convex function $|f_0+hg|^p$ are monotonic increasing with h; by LEBESGUE's theorem, we can therefore take limits under the integral sign as $h\to-0$ and $h\to+0$; a derivative therefore exists as $h\to 0$, provided that $|f_0+hg|^p$ is differentiable at $h=0$.

If we put $|f_0+hg|^p=[(f_0+hg)(\bar{f}_0+h\bar{g})]^{p/2}$, we obtain

$$\frac{d}{dh}|f_0+hg|^p|_{h=0}=p|f_0|^{p-2}\Re(\bar{f}_0 g)=p|f_0|^{p-1}\Re\left(\frac{\bar{f}_0}{|f_0|}g\right).$$

Combining these results we see that the weak derivative of $\|f\|_p=p(f)$ $=(\int|f|^p dt)^{1/p}$ is equal to

$$p'(f_0,g)=\|f_0\|_p^{1-p}\int|f_0|^{p-1}\Re\left(\frac{\bar{f}_0}{|f_0|}g\right)dt, \tag{3}$$

and it therefore exists at every point except the origin.

Consequently every boundary point of the unit ball is a point of smoothness, and (3) gives its supporting hyperplane.

This result can be proved much more simply if we use the fact, not proved until 7., that $(L^p)'=L^q$, where $\frac{1}{p}+\frac{1}{q}=1$. Since L^q is strictly convex, L^p is smooth, by 3.(3). If $f_0\in L^p$ and $\|f_0\|_p=1$ then $\frac{\bar{f}_0}{|f_0|}|f_0|^{p-1}$ is in L^q and has L^q-norm 1, so that $\int|f_0|^{p-1}\Re\left(\frac{\bar{f}_0}{|f_0|}g\right)dt=1$ is the supporting hyperplane through f_0.

4. The points of smoothness of l^1 can again be found using the extreme points of l^∞. From the results of § 25, 2. we find that the l^1-points

of smoothness are just those points $\mathfrak{x}=(\xi_n)$ for which all the coordinates ξ_n are non-zero. At such a point the derivative of the norm is given by

$$(4)\qquad p'(\mathfrak{x},\mathfrak{y})=\Re\left(\sum_{n=1}^{\infty}\varepsilon_n\eta_n\right),\ \text{where}\ \varepsilon_n=\frac{\bar{\xi}_n}{|\xi_n|}.$$

5. In a similar way it follows that l^p, $1<p<\infty$, is smooth, and that the derivative of the norm is given by

$$(5)\quad p'(\mathfrak{x},\mathfrak{y})=\|\mathfrak{x}\|_p^{1-p}\sum_{n=1}^{\infty}|\xi_n|^{p-1}\Re(\varepsilon_n\eta_n),\ \text{where}\ \varepsilon_n=\frac{\bar{\xi}_n}{|\xi_n|}.$$

6. We shall show that no boundary point of the unit ball in L^∞ is a point of smoothness. We can restrict our attention to $L^\infty(I)$, with $I=[0,1]$, as the case $I=(-\infty,\infty)$ can easily be reduced to it.

If $\|f_0\|=\operatorname{ess\,sup}|f_0(t)|=1$, there is at least one σ, with $|\sigma|=1$, for which $|\sigma-f_0(t)|\leqq\varepsilon$ holds on a set of positive measure, for each $\varepsilon>0$. There is then a sequence I_n of pairwise disjoint measurable subsets of I, each of positive measure and with $|\sigma-f_0(t)|\leqq\frac{1}{n}$ for all $t\in I_n$. The functions $u_n(f)=\dfrac{\bar{\sigma}}{\mu(I_n)}\displaystyle\int_{I_n}f(t)\,dt$ are continuous linear functionals on L^∞ with $|1-u_n(f_0)|\leqq\frac{1}{n}$ and $\|u_n\|=1$. Each of the two sequences u_{2n-1} and u_{2n} has a weakly adherent point $u^{(1)}$ (respectively $u^{(2)}$) in the weakly compact unit ball of $(L^\infty)'$. It follows from $u^{(1)}(f_0)=u^{(2)}(f_0)=1$ that they define two supporting hyperplanes through f_0. These are distinct: if χ_n is the characteristic function of I_n then $u_{2k}\left(\sum_{n=1}^{\infty}\chi_{2n}\right)=\bar{\sigma}$, so that $u^{(2)}(\sum\chi_{2n})=\bar{\sigma}$, while $u_{2k-1}(\sum\chi_{2n})=0$, so that $u^{(1)}(\sum\chi_{2n})=0$.

7. The space l^∞. Suppose that $\mathfrak{x}\in l^\infty$ and that $\|\mathfrak{x}\|=1$. If there is a subsequence (ξ_{n_k}) of coordinates which converges to some σ with $|\sigma|=1$, then it can be divided into two such subsequences, and we can define $u^{(1)}$ and $u^{(2)}$ as weakly adherent points of the sequences $\mathfrak{e}_{n_{2k-1}}$ and $\mathfrak{e}_{n_{2k}}$; as before we obtain two different supporting hyperplanes through $\mathfrak{x}$. If $\mathfrak{x}$ has two coordinates of modulus 1, then there are clearly two supporting hyperplanes through $\mathfrak{x}$. Finally if $\mathfrak{x}$ has the form $\sigma\,\mathfrak{e}_k+\mathfrak{x}'$, with $|\sigma|=1$ and $\|\mathfrak{x}'\|<1$, then for sufficiently small t we have $p(\mathfrak{x}+t\,\mathfrak{y})=|\sigma+t\eta_k|$, so that the norm is weakly differentiable at $\mathfrak{x}$, with $p'(\mathfrak{x},\mathfrak{y})=\Re(\bar{\sigma}\eta_k)$. The points of the form $\sigma\,\mathfrak{e}_k+\mathfrak{x}'$ with $\|\mathfrak{x}'\|<1$ are therefore the points of smoothness of the unit ball of l^∞.

8. Suppose that $\mathfrak{G}$ is a bounded domain in the complex plane, and that $HB(\bar{\mathfrak{G}})$ is the (B)-space, defined in § 14, 9., consisting of functions

which are holomorphic in $\mathfrak{G}$ and continuous on the boundary S. Since $\|f(z)\| = \sup_{z\in\bar{\mathfrak{G}}}|f(z)| = \sup_{z\in S}|f(z)|$, $HB(\bar{\mathfrak{G}})$ is norm-isomorphic to a closed linear subspace H of $C(S)$. By 3.(4) and the first example of this number, all the functions $f(z)$ which attain their maximum modulus 1 at just one point of S are points of smoothness of the unit ball. We shall assume that S is a homeomorphic image of the circumference of the unit circle. Then using results in the theory of conformal mappings, to each point t of S there is an $f(z)\in HB(\bar{\mathfrak{G}})$ which attains its maximum modulus at t. and at t alone.

By § 22, 3.(4) the dual $HB(\bar{\mathfrak{G}})'$ is norm-isomorphic to the quotient space $\mathfrak{M}(S)/H^{\perp}$, where $H^{\perp}$ is the space of all measures which vanish on H. By 3.(1)a) the points $\sigma\hat{\delta}_t$, where $|\sigma|=1$ and $\hat{\delta}_t$ is the coset in $\mathfrak{M}(S)/H^{\perp}$ corresponding to the point measure δ_t, $t\in S$, are extreme points of the unit ball $\hat{K}'$ of $\mathfrak{M}(S)/H^{\perp}$. We shall show that these are the only extreme points of $\hat{K}'$. Since the unit ball K' of $\mathfrak{M}(S)$ is weakly compact, its image under the canonical mapping from $\mathfrak{M}(S)$ onto $\mathfrak{M}(S)/H^{\perp}$ is also weakly compact, and so it is equal to $\hat{K}'$. It then follows from § 25, 1.(9) and § 25, 2. that the points $\sigma\hat{\delta}_t$ are the only possible extreme points of $\hat{K}'$. If we now apply 3.(1)a) again, it follows that the points of smoothness of the unit ball of $HB(\bar{\mathfrak{G}})$ are precisely those $f(z)$ which attain their maximum modulus 1 at just one point of S.

The extreme points of the unit ball of $HB(\bar{\mathfrak{G}})$ are more difficult to determine. For the case where $\mathfrak{G}$ is the unit disc in the complex plane cf. DE LEEUW and RUDIN [1].

6. Uniform convexity. Following CLARKSON [1], a normed space E, and its closed unit ball, are said to be uniformly convex if for each ε with $0<\varepsilon\leqq 2$ there exists a $\delta(\varepsilon)>0$ for which it always follows from $\|x\|\leqq 1$, $\|y\|\leqq 1$ and $\|x-y\|\geqq\varepsilon$ that $\|\frac{1}{2}(x+y)\|\leqq 1-\delta(\varepsilon)$. Such a function $\delta(\varepsilon)$ is called a module of convexity for E.

By 1.(1)f) every uniformly convex space is strictly convex.

(1) *E is uniformly convex if and only if it always follows from* $\|x_n\|\leqq 1$, $\|y_n\|\leqq 1$ *and* $\lim\limits_{n\to\infty}\|\frac{1}{2}(x_n+y_n)\|=1$ *that* $\lim\limits_{n\to\infty}\|x_n-y_n\|=0$.

Proof. The condition is necessary, for a contradiction of uniform convexity follows immediately from $\|x_{n_k}-y_{n_k}\|\geqq\varepsilon$ and $\lim\limits_{n\to\infty}\|\frac{1}{2}(x_{n_k}+y_{n_k})\|=1$.

On the other hand, if the condition were satisfied, and if for some $\varepsilon>0$ there were no $\delta(\varepsilon)$ with the required properties, then there would be sequences x_n, y_n with $\|\frac{1}{2}(x_n+y_n)\|\to 1$ and $\|x_n-y_n\|\geqq\varepsilon$, which would contradict $\lim\|x_n-y_n\|=0$.

(2) *Every linear subspace and every quotient space of a uniformly convex normed space is again uniformly convex.*

If E is uniformly convex, so is its completion.

We need only prove this for quotient spaces E/H. Suppose $\varepsilon>0$ is given, and suppose that $\|\hat{x}\|\leqq 1$, $\|\hat{y}\|\leqq 1$ and $\|\hat{x}-\hat{y}\|\geqq\varepsilon$. Then there exist $x\in\hat{x}$, $y\in\hat{y}$ with $\|x\|\leqq 1+\lambda$, $\|y\|\leqq 1+\lambda$, $\lambda>0$, with $\lambda>0$ arbitrarily small. It then follows from $\|x-y\|\geqq\|\hat{x}-\hat{y}\|\geqq\varepsilon$ that $\left\|\frac{1}{2}(\hat{x}+\hat{y})\right\| \leqq\left\|\frac{1}{2}(x+y)\right\|\leqq\left[1-\delta\left(\frac{\varepsilon}{1+\lambda}\right)\right]\cdot(1+\lambda)$. Consequently $\lim_{\lambda\to 0^+}\delta\left(\frac{\varepsilon}{1+\lambda}\right) \geqq\delta\left(\frac{\varepsilon}{2}\right)$ is a module of convexity for E/H.

(3) *Suppose that E is a uniformly convex normed space, that $u_0\in E'$ and that $\|u_0\|=1$. If x' and x'' are two elements of E, with $\|x'\|\leqq 1$ and $\|x''\|\leqq 1$, and if they satisfy the inequality $|u_0 x-1|<\frac{\delta(\varepsilon)}{2}$, then $\|x'-x''\|<\varepsilon$.*

This follows directly from the inequality

$$\|\tfrac{1}{2}(x'+x'')\|\geqq\tfrac{1}{2}|u_0(x'+x'')|\geqq|u_0 x'|-\tfrac{1}{2}|u_0(x''-x')|>1-\delta(\varepsilon)$$

and the assumption of uniform convexity.

If x_0 is a point of the unit sphere $\|x\|=1$ in E, and if u_0 is chosen in such a way that $u_0 x_0=1$, then (3) means that the weak neighbourhood of x_0 in the unit ball K given by $|u_0 x-1|<\frac{\delta(\varepsilon)}{2}$ is contained in the strong neighbourhood of x_0 in K given by $\|x_0-x\|<\varepsilon$. Thus the weak and the strong topologies coincide on the unit sphere of a uniformly convex space.

It should be observed that it does not follow from this that the corresponding uniform structures are the same; the unit sphere, being bounded, is certainly weakly precompact, but in general it is not strongly precompact, as for example the uniformly convex space l^2 shows (cf. the next number).

We now obtain MILMAN's [1] important criterion for reflexivity:

(4) *Every uniformly convex* (B)-*space is reflexive.*

Proof. (cf. DIEUDONNÉ [2]). If is sufficient to show that every element $z\in E''$ with $\|z\|=1$ is contained in the closed unit ball K of E. z belongs to the polar $K^{\circ\circ}$ of K° in E'', and so, by the theorem of bipolars, it is a weak closure point of K. As V runs through the weak neighbourhoods of z in E'', the sets $V\cap K$ form a weak Cauchy filter $\mathfrak{F}$ on K with limit z. If we can show that sets of arbitrarily small diameter, in the sense of the norm on E, are contained in this filter, then $\mathfrak{F}$ will be a Cauchy filter with respect to the strong topology on K; since K is strongly complete, $\mathfrak{F}$ will therefore have its limit in K, so that z will belong to K.

Since $\|z\|=1$, given $\delta>0$ there exists a u_0 with $\|u_0\|=1$ and $|u_0 z-1| < \frac{\delta}{4}$. The set W_δ of all $x \in K$ with $|u_0(z-x)| < \frac{\delta}{4}$ belongs to the filter $\mathfrak{F}$, and if $x \in W_\delta$ then $|u_0 x-1| < \frac{\delta}{2}$. If $\delta=\delta(\varepsilon)$, then it follows from (3) that $\|x'-x''\|<\varepsilon$ for $x', x'' \in W_\delta$; the diameter of W_δ is therefore at most ε.

(4) can be generalized to arbitrary locally convex spaces:

(5) *A sequentially complete locally convex space $E[\mathfrak{T}]$ is semi-reflexive if every bounded subset of E is contained in a bounded closed absolutely convex uniformly convex set B.*

B uniformly convex means that the normed space E_B is uniformly convex.

Proof. By § 20, 11.(2), E_B is a uniformly convex B-space, and $\mathfrak{T}$ defines a weaker topology than the norm topology of E_B. By (4), E_B is reflexive, so that B is $\mathfrak{T}_s(E'_B)$-compact. Since every $u \in E'$ defines a $\mathfrak{T}_B$-continuous linear functional on E_B, $\mathfrak{T}_s(E')$ is coarser than $\mathfrak{T}_s(E'_B)$ on E_B; B is therefore $\mathfrak{T}_s(E')$-compact. The assertion now follows from § 23, 3.(1).

7. The uniform convexity of the l^p and L^p spaces. The Hilbert space l^2 is easily seen to be uniformly convex. In l^2 we have $\|\mathfrak{x}\|^2=\mathfrak{x}\bar{\mathfrak{x}}$, and from this follows the identity

(1) $$\|\mathfrak{x}+\mathfrak{y}\|^2+\|\mathfrak{x}-\mathfrak{y}\|^2=2(\|\mathfrak{x}\|^2+\|\mathfrak{y}\|^2).$$

Thus if $\|\mathfrak{x}\| \leqq 1$, $\|\mathfrak{y}\| \leqq 1$ and $\|\mathfrak{x}-\mathfrak{y}\| \geqq \varepsilon$, then $\left\|\frac{1}{2}(\mathfrak{x}+\mathfrak{y})\right\|^2 \leqq 1-\frac{\varepsilon^2}{4}$, so that l^2 is uniformly convex.

The proof of the next result is also comparatively simple.

(2) *l^p and L^p are uniformly convex for $p \geqq 2$.*

This depends upon the following inequality, valid for all complex α, β and for $p \geqq 2$:

(3) $$|\alpha+\beta|^p+|\alpha-\beta|^p \leqq 2^{p-1}(|\alpha|^p+|\beta|^p).$$

Proof. Using (1) it follows from § 14, 8.(9) that

$$(|\alpha+\beta|^p+|\alpha-\beta|^p)^{\frac{1}{p}} \leqq (|\alpha+\beta|^2+|\alpha-\beta|^2)^{\frac{1}{2}}=\sqrt{2}(|\alpha|^2+|\beta^2|)^{\frac{1}{2}}.$$

Using Hölder's inequality for $\frac{2}{p}+\frac{p-2}{p}=1$, we get

$$|\alpha|^2+|\beta|^2 \leqq (|\alpha|^p+|\beta|^p)^{\frac{2}{p}}(1+1)^{\frac{p-2}{p}}=(|\alpha|^p+|\beta|^p)^{\frac{2}{p}}2^{\frac{p-2}{p}}.$$

Combining these, we have

$$(|\alpha+\beta|^p+|\alpha-\beta|^p)^{\frac{1}{p}} \leqq 2^{\frac{p-1}{p}} (|\alpha|^p+|\beta|^p)^{\frac{1}{p}},$$

which gives (3).

By summing over the components of vectors in l^p or integrating over functions of L^p, we obtain the relation

(4) $$\|x+y\|_p^p+\|x-y\|_p^p \leqq 2^{p-1}(\|x\|_p^p+\|y\|_p^p), \qquad p \geqq 2,$$

from (3), for any two elements of l^p or L^p, respectively.

Now if $\|x_n\|_p \leqq 1$, $\|y_n\|_p \leqq 1$, $\lim \|x_n+y_n\|_p=2$, then it follows from (4) that $\lim \|x_n-y_n\|_p=0$, and uniform convexity follows from 6.(1).

The case $1<p<2$ is substantially harder to deal with, as we shall shortly see. It follows from (2), 6.(4) and § 14, 10. that the L^p-spaces are reflexive for $p \geqq 2$. Using this, we can now derive a result annonced in § 14, 10.(14):

(5) *If* $1<p<\infty$ *and,* $\frac{1}{p}+\frac{1}{q}=1$, *the spaces* L^p *and* L^q *are each the dual of the other.*

It is sufficient to show that $(L^p)'=L^q$ for $p \geqq 2$. For then $(L^q)'=L^p$ for $1<q\leqq 2$, since L^p is reflexive.

If $g(t)$ is an element of L^q, and if we define a linear functional for all $f\in L^p$ by setting $\langle g, f\rangle = \int_a^b f(t)g(t)\,dt$ then it follows from Hölder's inequality that $\int_a^b |f(t)g(t)|\,dt \leqq \|f\|_p \|g\|_q$, so that $\langle g, f\rangle$ is continuous on L^p. So L^q can be identified with a subspace of $(L^p)'$.

We have

(6) $$\sup_{\|f\|_p\leqq 1} |\langle g, f\rangle| = \|g\|_q.$$

Hölder's inequality implies $\leqq$ in (6). If we put $f(t)=|g(t)|^{q-1}\varepsilon_g(t)$, with $\varepsilon_g(t)=\dfrac{g(t)}{|g(t)|}$ for $g(t)\neq 0$ and $\varepsilon_g(t)=0$ for $g(t)=0$, then $f\in L^p$, since $\int |f|^p dt = \int |g|^{p(q-1)} dt = \int |g|^q dt = \|g\|_q^q$. It is immediately verified that $\left\langle g, \dfrac{f}{\|f\|_p}\right\rangle = \|g\|_q$, which establishes (6).

This means that on L^q the norm of $(L^p)'$ coincides with the norm of L^q. Since L^q is norm-complete, L^q is therefore a closed linear subspace of $(L^p)'$. If L^q were a proper subspace, there would be a continuous linear functional on $(L^p)'$ which vanished on L^q without vanishing on the whole of $(L^p)'$. Because L^p is reflexive, this functional must be given by an $f\in L^p$. But it follows from $\int fg\,dt=0$, for all $g\in L^q$, that in particular, putting $g=|f|^{p-1}\varepsilon_f\in L^q$, we obtain $\int |f|^p dt=0$. This means that f is the zero element in L^p, which gives a contradiction.

The usual proof of (5) uses the theory of differentiation of real functions, for example the LEBESGUE-NIKODYM theorem (cf. BANACH [3] p. 61 ff. or BOURBAKI [7] Vol. 2 p. 55, for example), while the present proof makes do with the results already used in § 14, 10.

It follows from the remarks of § 14, 10. and the proof of (5) that (5) is also valid for $L^p(-\infty, +\infty)$.

The dual of L^1 can now be determined, as well.

(7) *The dual of* L^1 *is* L^∞.

The underlying interval of the real line may again be either $[a,b]$ or $(-\infty, +\infty)$.

We first investigate the case $L^1[a,b]$. Once again, $L^\infty \subset (L^1)'$. Suppose that u is a continuous linear functional on L^1. By § 14, 11.(4), $L^1 \supset L^2$, and u is also a continuous linear functional on L^2. By (5), there is therefore a function $g(t) \in L^2$ with $\langle u, f\rangle = \int g f\, dt$ for all $f \in L^2$. If M is a measurable subset of $[a,b]$, and if $\varphi(t)$ is its characteristic function, then $\varepsilon_g(t)\varphi(t)$ belongs to L^2, and we therefore have $\langle u, \varepsilon_g \varphi\rangle = \int\limits_M |g|\, dt \leqq \|u\|\, \mu(M)$. But it follows directly from this that $\operatorname{ess\,sup} |g| \leqq \|u\|$, i.e. $g \in L^\infty$. The formula $\langle u, f\rangle = \int g f dt$ therefore holds for all $f \in L^1$, since L^2 is dense in L^1.

The case $L^1(-\infty, +\infty)$ can be reduced to this one: $L^1[-n, n]$ can be considered as a closed subspace of $L^1(-\infty, +\infty)$. On $L^1[-n, n]$ the continuous linear functional u is determined by a function $g_n(t)$ which is essentially bounded by $\|u\|$. These functions $g_n(t)$ are the restrictions to $[-n, n]$ of a function $g(t)$ defined on $(-\infty, \infty)$ and essentially bounded by $\|u\|$, and this functions satisfies $\langle u, f\rangle = \int\limits_{-\infty}^{+\infty} g(t) f(t) dt$. The exact details of the proof are left to the reader.

We shall now show that l^p and L^p are uniformly convex for $1<p<2$, as well. We follow a method due to MCSHANE [1], which works for all p with $1<p<\infty$, and which can be applied to more general classes of (B)-spaces. Theorem (12) was first proved by CLARKSON [1].

(8) *Suppose that* E *is a uniformly convex normed space, and that* $1<p<\infty$. *Then given* $\varepsilon>0$, *there exists a* $\delta_p(\varepsilon)>0$ *such that it always follows from* $\|x\| \leqq 1$, $\|y\| \leqq 1$ *and* $\|x-y\| \geqq \varepsilon$ *that*

$$\left\|\frac{1}{2}(x+y)\right\|^p \leqq (1-\delta_p)\left(\frac{\|x\|^p + \|y\|^p}{2}\right). \tag{9}$$

For arbitrary $x, y \in E$, *we therefore have*

$$\left\|\frac{1}{2}(x+y)\right\|^p \leqq \left(1-\delta_p\left(\frac{\|x-y\|}{\sup(\|x\|, \|y\|)}\right)\right)\left(\frac{\|x\|^p + \|y\|^p}{2}\right). \tag{9'}$$

Proof. By 15, 9.(6) we have

(10) $$(\tfrac{1}{2}(1+c))^p \leqq \tfrac{1}{2}(1+c^p) \quad \text{for} \quad c \geqq 0.$$

Since $\dfrac{1+t^p}{(1+t)^p}$ attains its minimum at $t=1$, we have strict inequality in (10) for $0 \leqq c < 1$.

It is sufficient to prove (9) for $\|x\|=1$, $\|y\| \leqq 1$ and $\|x-y\| \geqq \varepsilon$. We assume that (9) is false. Then there exists $\varepsilon > 0$ and elements x_n, y_n with $\|x_n\|=1$, $\|y_n\| \leqq 1$, $\|x_n - y_n\| \geqq \varepsilon$ and

(11) $$\frac{\|\frac{1}{2}(x_n+y_n)\|^p}{\frac{1}{2}(\|x_n\|^p + \|y_n\|^p)} \to 1.$$

We first show that this is only possible if $\|y_n\| \to 1$. If not, there would be y_n with $\|y_n\| \leqq q < 1$, with (11) holding. But it would follow from this and the remark following (10) that we would have

$$\left\|\frac{1}{2}(x_n+y_n)\right\|^p \leqq \left(\frac{1}{2}(1+\|y_n\|)\right)^p \leqq \frac{\rho}{2}(1+\|y_n\|^p) = \frac{\rho}{2}(\|x_n\|^p + \|y_n\|^p)$$

for all n and for some $\rho < 1$, which contradicts (11).

If we now put $z_n = \dfrac{y_n}{\|y_n\|}$, then $\|z_n - y_n\| \to 0$. Consequently if $n \geqq n_0$, $\|x_n - z_n\| \geqq \dfrac{\varepsilon}{2}$. Further, because of (11), $\lim \left\|\dfrac{1}{2}(x_n+z_n)\right\| = 1$. But by 6.(1) this contradicts the uniform convexity of E.

(12) *l^p and L^p are uniformly convex for all $1 < p < \infty$.*

We first prove this for L^p. Suppose that f, g are in L^p, and that $\|f\|_p \leqq 1$, $\|g\|_p \leqq 1$ and $\|f-g\|_p \geqq \varepsilon$. I may be either $[a,b]$ or $(-\infty, +\infty)$. Let M be the set of all $t \in I$ for which

(13) $$|f(t)-g(t)|^p \geqq \frac{\varepsilon^p}{4}(|f(t)|^p + |g(t)|^p) \geqq \frac{\varepsilon^p}{4}\sup(|f(t)|^p, |g(t)|^p)$$

Applying (9′) to $E = \mathsf{K}$, we obtain the following inequality on the measurable set M:

(14) $$\left|\frac{1}{2}(f(t)+g(t))\right|^p \leqq \left(1-\delta_p\left(\frac{\varepsilon}{4^{1/p}}\right)\right)\left(\frac{1}{2}(|f(t)|^p + |g(t)|^p)\right).$$

On $N = I \sim M$ we have $\displaystyle\int_N |f(t)-g(t)|^p\,dt \leqq \frac{\varepsilon^p}{4}\int_I (|f|^p + |g|^p)\,dt \leqq \frac{\varepsilon^p}{2}$.

It follows from this that $\displaystyle\int_M |f-g|^p\,dt \geqq \frac{\varepsilon^p}{2}$. If f_M and g_M denote the restrictions of f and g to M, we therefore have $\|f_M - g_M\|_p \geqq \frac{\varepsilon}{2^{1/p}}$, so that $\sup(\|f_M\|_p, \|g_M\|_p) \geqq \frac{\varepsilon}{2 \cdot 2^{1/p}}$, i.e.

$$(15) \qquad \sup\left(\int_M |f|^p dt, \int_M |g|^p dt\right) \geqq \frac{\varepsilon^p}{2^{p+1}}.$$

It now follows from (14), (15) and the fact that the integrand of the first integral is non-negative, by (9), that

$$\int_I \left\{\frac{1}{2}(|f|^p + |g|^p) - \left(\frac{1}{2}|f+g|\right)^p\right\} dt \geqq \int_M \geqq \int_M \delta_p\left(\frac{\varepsilon}{4^{1/p}}\right)\left(\frac{1}{2}(|f|^p + |g|^p) dt\right)$$

$$\geqq \delta_p\left(\frac{\varepsilon}{4^{1/p}}\right)\frac{\varepsilon^p}{2^{p+2}}.$$

Finally it follows from this that

$$\left\|\frac{1}{2}(f+g)\right\|_p \leqq \left(1 - \delta_p\left(\frac{\varepsilon}{4^{1/p}}\right)\frac{\varepsilon^p}{2^{p+2}}\right)^{1/p}$$

for all f, g with $\|f\|_p \leqq 1$, $\|g\|_p \leqq 1$ and $\|f-g\|_p \geqq \varepsilon$.

The proof for l^p is similar; it is sufficient to replace the integral over I by the sum over $n = 1, 2, \ldots$.

Let us observe that these results are also valid for the spaces l^p_d, where d is any cardinal, and that the proofs are the same.

8. Further examples. If E_n, $n = 1, 2, \ldots$, are (B)-spaces, let $l^p(E_n)$, with $1 \leqq p < \infty$, denote the space of all sequences $x = (x_n)$, $x_n \in E_n$, with $\sum_{n=1}^{\infty} \|x_n\|^p < \infty$. Arguing as for the special case l^p in § 14, 8., it is not difficult to see that $l^p(E_n)$ is again a (B)-space, under the norm $\left(\sum_{n=1}^{\infty} \|x_n\|^p\right)^{1/p}$. The norm induced on E_n is identical with the original norm. If all the spaces E_n are equal to the same (B)-space E, we write $l^p(E)$.

Similarly $l^\infty(E_n)$ is the (B)-space of all sequences $x = (x_n)$, $x_n \in E_n$, with $\|x\| = \sup_n \|x_n\| < \infty$.

(1) *The dual of* $l^p(E_n)$, $p > 1$, *is* $l^q(E'_n)$ $\left(\frac{1}{p} + \frac{1}{q} = 1\right)$, *and the dual of* $l^1(E_n)$ *is* $l^\infty(E'_n)$.

We shall prove this for $p > 1$. Every $u = (u_n) \in l^q(E'_n)$ defines a continuous linear functional on $l^p(E_n)$, as is clear from the inequality

$$|ux| = \left|\sum_{n=1}^{\infty} u_n x_n\right| \leqq \sum \|u_n\| \, \|x_n\| \leqq \left(\sum \|u_n\|^q\right)^{1/q} \left(\sum \|x_n\|^p\right)^{1/p} < \infty.$$

On the other hand every continuous linear functional must be of the form $ux=\sum_{n=1}^{\infty} u_n x_n$, with $u_n \in E_n'$. For each $x=(x_n)$ we can find an $\tilde{x}=(\tilde{x}_n)$ with $\|x_n\|=\|\tilde{x}_n\|$ and $u_n\tilde{x}_n=\|u_n\|\,\|x_n\|-\varepsilon_n$ for arbitrary $\varepsilon_n>0$. It then follows from $\sum u_n\tilde{x}_n<\infty$ that $\sum\|u_n\|\,\|x_n\|<\infty$. Since this holds for all $x\in l^p(E_n)$, we have $(\|u_n\|)\in l^q$, so that $u\in l^q(E_n')$.

Similarly, it is easy to confirm that the norm of $l^q(E_n')$ is the same as the dual norm of $l^p(E_n)$.

It follows directly from (1) that

(2) *$l^p(E_n)$, $p>1$, is reflexive if and only if all the spaces E_n are reflexive.*

We now establish

(3) *$l^p(E_n)$ is strictly convex if and only if all the spaces E_n are strictly convex.*

If E_n is not strictly convex, neither is $l^p(E_n)$, for the boundary of the unit ball of E_n is a subset of the boundary of the unit ball of $l^p(E_n)$.

Conversely suppose that all the spaces E_n are strictly convex, and suppose that $x=(x_n)$ and $y=(y_n)$ are two distinct elements on the boundary of the unit ball of $l^p(E_n)$. Then the vectors $\mathfrak{x}=(\|x_n\|)$ and $\mathfrak{y}=(\|y_n\|)$ in l^p also satisfy $\|\mathfrak{x}\|_p=\|\mathfrak{y}\|_p=1$. If $\mathfrak{x}\neq\mathfrak{y}$, then $\|\frac{1}{2}(\mathfrak{x}+\mathfrak{y})\|_p<1$, since l^p is strictly convex. It follows from this that $\|\frac{1}{2}(x+y)\|<1$. If $\mathfrak{x}=\mathfrak{y}$, there is at least one n_0 with $\|x_{n_0}\|=\|y_{n_0}\|$ and $x_{n_0}\neq y_{n_0}$. Then $\|\frac{1}{2}(x_{n_0}+y_{n_0})\|<\|x_{n_0}\|$, and again it follows that $\|\frac{1}{2}(x+y)\|<1$.

(4) *$l^p(E_n)$, $p>1$, is separable if and only if all the spaces E_n are separable.*

The simple proof is left to the reader.

(5) *$l^p(E)$ is uniformly convex if and only if E is uniformly convex.*

It is easy to see that the condition is necessary. On the other hand if E is uniformly convex the proof of uniform convexity can be followed through in exactly the same way as for l^p, with the difference that 7.(8) must be applied to E, and not to K.

Day [3] has shown that, more generally, $l^p(E_n)$, $p>1$, is uniformly convex if and only if the spaces E_n have a common module of convexity $\delta(\varepsilon)$.

The generalization of the L^p-spaces to spaces of the form $L^p(E)$, which is essentially more important in applications, will be investigated in greater depth in the second volume.

9. Invariance under topological isomorphisms. The properties investigated in this paragraph are of a metric kind. The question arises of whether concepts can be obtained from them which are invariant under topological isomorphisms. Now two normed spaces are topologically isomorphic if and only if they are norm-isomorphic under some suitable equivalent norms (cf. § 14, 2.).

We therefore call a normed space E strictly, smoothly or uniformly normable if an equivalent norm can be defined under which E is respectively strictly convex, smooth or uniformly convex; in this way we define concepts with the required properties.

Theorem 6.(4) can now be expressed in the stronger form: *every uniformly normable* (B)-*space is reflexive.*

DAY [2] has shown that the converse is not true; a counterexample is given by

(1) $l^p(E_n)$, $E_n = l_n^\infty$, $p>1$, *is reflexive, but not uniformly normable.*

Proof. l_n^∞ is the space of all $(\xi_1, \ldots, \xi_n)$ with $\|(\xi_1, \ldots, \xi_n)\| = \sup|\xi_i|$. That $E = l^p(E_n)$ is reflexive follows from 8.(2). If E were uniformly normable there would be a second norm $\|x\|'$ on E with

$$\|x\| \leqq \|x\|' \leqq M\|x\|, \tag{2}$$

for which E is uniformly convex. It then follows from $\|x\|' \leqq 1$, $\|y\|' \leqq 1$ and $\|x-y\|' \geqq \varepsilon$ that $\delta(\varepsilon) = 1 - \sup\|\frac{1}{2}(x+y)\|' > 0$.

Suppose that $z = (\alpha_1, \ldots, \alpha_n)$, $|\alpha_i| = \dfrac{1}{M} = \alpha$. Then z belongs to $l_n^\infty \subset E$, and, since $\|z\| = 1/M$, it follows from (2) that $\|z\|' \leqq 1$. We set $\tilde{z} = (\alpha_1, \ldots, \alpha_{n-1}, -\alpha_n)$. Then $z - \tilde{z} = (0, \ldots, 0, 2\alpha_n)$ and $\frac{1}{2}(z+\tilde{z}) = (\alpha_1, \ldots, \alpha_{n-1}, 0)$. It follows from $\|z-\tilde{z}\|' \geqq \|z-\tilde{z}\| = 2\alpha$ that $\|\frac{1}{2}(z+\tilde{z})\|' \leqq 1 - \delta(2\alpha)$. This means that

$$\left\| \left(\frac{\alpha_1}{1-\delta(2\alpha)}, \ldots, \frac{\alpha_{n-1}}{1-\delta(2\alpha)}, 0 \right) \right\|' \leqq 1.$$

If we now apply the same idea to $\dfrac{1}{1-\delta(2\alpha)}(\alpha_1, \ldots, \alpha_{n-2}, \alpha_{n-1}, 0)$ and $\dfrac{1}{1-\delta(2\alpha)}(\alpha_1, \ldots, \alpha_{n-2}, -\alpha_{n-1}, 0)$, and so on, we finally obtain that

$$\|y_n\|' = \left\| \left(\frac{\alpha_1}{(1-\delta(2\alpha))^{n-1}}, 0, \ldots, 0 \right) \right\|' \leqq 1.$$

Since on the other hand $\|y_n\| = \dfrac{1}{M(1-\delta(2\alpha))^{n-1}}$, and since $(1-\delta(2\alpha))^{n-1}$ tends to 0 as $n \to \infty$, we obtain the contradiction $\|y_n\|' \geqq \|y_n\| > 1$, for sufficiently large n.

The uniformly normable (B)-spaces therefore form a proper subset of the collection of reflexive (B)-spaces.

It is not known if every reflexive (B)-space is strictly normable. There are however non-reflexive (B)-spaces which are strictly normable, as we shall see. The situation is the same for smooth normability.

(3) *A normed space E is strictly normable if and only if there is a continuous one-one linear mapping A from E into a strictly convex normed space F. The norm* $\|x\|'=\|x\|+\|Ax\|$ *is then a strictly convex norm on E, equivalent to* $\|x\|$.

Proof. The equivalence of the two norms follows from $\|x\| \leqq \|x\|' \leqq (M+1)\|x\|$, where $\|A\|=M$. Suppose now that $\|x_1\|'=\|x_2\|'=1$ and $\|x_1+x_2\|'=\|x_1\|'+\|x_2\|'$. This is only possible if $\|x_1+x_2\|=\|x_1\|+\|x_2\|$ and $\|Ax_1+Ax_2\|=\|Ax_1\|+\|Ax_2\|$. It then follows from the strict convexity of F and 1.(1) g) that $Ax_1=\alpha Ax_2$, with $\alpha \geqq 0$, so that $x_1=\alpha x_2$. Since $\|x_1\|'=\|x_2\|'$, this is only possible if $x_1=x_2$, so that E is strictly convex under $\|x\|'$.

It is trivial that the condition is necessary.

(4) *Suppose that E is smoothly normable. If E has a continuous one-one linear image in a space F which is simultaneously strictly and smoothly normable, then E is also simultaneously strictly and smoothly normable.*

We can suppose that E is smooth and that F is both smooth and strictly convex. By (3), the norm $\|x\|'=\|x\|+\|Ax\|$ is strictly convex. But it is also smooth, for by 4.(13) both $\|x\|$ and $\|Ax\|$ are weakly differentiable norms on E, and so also is their sum $\|x\|+\|Ax\|$. Applying 4.(13) again, it follows that $\|x\|'$ is smooth as well.

(5) *Every separable normed space E is simultaneously smoothly and strictly normable.*

The dual of every separable (B)*-space is strictly normable.*

Proof. Let $\{x_i\}$ be a sequence dense in the sphere $\|x\|=1$ of E. A continuous one-one linear mapping from the strong dual E' into l^2 is defined by setting $Au=\left(\frac{ux_1}{2},\ldots,\frac{ux_i}{2^i},\ldots\right)$. If we again set $\|u\|_1 =\|u\|+\|Au\|$ on E', then by (3) $\|u\|_1$ is a strictly convex norm on E' which is equivalent to $\|u\|$. Thus the second part of the theorem is proved.

The mapping A sends every weakly convergent sequence in E' into a norm-convergent sequence in l^2. It follows directly from this that the new unit ball K_1 of E', given by $\|u\|+\|Au\| \leqq 1$, is weakly sequentially closed, and so it is weakly closed, by § 21, 10.(7). By taking K_1° as unit ball, therefore, we obtain a new norm $\|x\|_1$ which is equivalent to the original norm and which gives $\|u\|_1$ as the dual norm, since $K_1=K_1^{\circ\circ}$. $\|x\|_1$ is therefore smooth, by 3.(2).

If we now map E, with norm $\|x\|_1$, into l^2 by setting $Bx=\left(\frac{u_1 x}{2},\dots\frac{u_i x}{2^i},\dots\right)$, where u_i is a weakly dense sequence in $\|u\|_1=1$, then the theorem follows from (4).

In this result of DAY's [6], the separability condition cannot be relaxed. None of the spaces l_d^∞ with $d>\aleph_0$ is smoothly normable (cf. DAY [6]).

By (5), l^1 is both strictly and smoothly normable; its dual l^∞ is strictly normable, by (5), but is not smoothly normable, as DAY [6] has also shown.

10. Uniform smoothness and strong differentiability of the norm. There is a further concept of convexity, which is dual to uniform convexity, and which provides a refinement of the concept of smoothness.

A normed space, and its norm, are said to be u n i f o r m l y s m o o t h if for each $\varepsilon>0$ there is an $\eta(\varepsilon)>0$ for which $\|x\|\geqq 1$, $\|y\|\geqq 1$ and $\|x-y\|\leqq\varepsilon$ always implies $\|x+y\|\geqq\|x\|+\|y\|-\varepsilon\|x-y\|$.

(1) *A normed space E is uniformly smooth if and only if for each $\varepsilon>0$ there is a $\rho(\varepsilon)>0$ for which $\|x\|=1$, $\|y\|\leqq\rho$ always implies that $\|x+y\|+\|x-y\|\leqq 2+\varepsilon\|y\|$.*

a) Suppose that E is uniformly smooth, and that there are sequences $x_n, y_n\in E$ with $\|x_n\|=1$, $\|y_n\|\to 0$ and $\|x_n+y_n\|+\|x_n-y_n\|>2+\varepsilon_0\|y_n\|$.

Setting $x_n+y_n=v_n$, $x_n-y_n=w_n$, we obtain $\|v_n\|+\|w_n\|>2+\frac{\varepsilon_0}{2}\|v_n-w_n\|$. Since $\|v_n+w_n\|=2$, this means that

$$\|v_n+w_n\|<\|v_n\|+\|w_n\|-\frac{\varepsilon_0}{2}\|v_n-w_n\|.$$

The inequality remains valid when we replace v_n and w_n by $v_n'=\frac{v_n}{1-\|y_n\|}$ and $w_n'=\frac{w_n}{1-\|y_n\|}$, respectively. Since $\|v_n'\|\geqq 1$, $\|w_n'\|\geqq 1$ and $\|v_n'-w_n'\|\leqq\frac{2\|y_n\|}{1-\|y_n\|}$, we obtain a contradiction of the uniform smoothness by letting $n\to\infty$.

b) Suppose that E satisfies the conditions of (1), and suppose that there exist $\varepsilon>0$ and sequences x_n, y_n with $\|x_n\|\geqq 1$, $\|y_n\|\geqq 1$, $\|x_n-y_n\|\to 0$ and $\|x_n+y_n\|<\|x_n\|+\|y_n\|-\varepsilon\|x_n-y_n\|$. Setting $x_n+y_n=s_n$, $x_n-y_n=t_n$, we obtain

$$\|s_n+t_n\|+\|s_n-t_n\|>2\|s_n\|+2\varepsilon\|t_n\|.$$

For $s_n'=\frac{s_n}{\|s_n\|}$ and $t_n'=\frac{t_n}{\|s_n\|}$ this gives

$$\|s_n'+t_n'\|+\|s_n'-t_n'\|>2+2\varepsilon\|t_n'\|,$$

and this provides a contradiction to the conditions of (1), since $\|s_n'\|=1$ and $\|t_n'\|\to 0$.

As we saw in 4., smoothness is equivalent to the weak differentiability of the norm. Uniform smoothness can also be expressed in terms of differentiability.

Once again let $q(x)$ be the Minkowski functional of a closed convex $\mathfrak{T}$-body $C\ni o$ in a locally convex space $E[\mathfrak{T}]$. The function $q(x)$ is said to be strongly differentiable at a point x_0 if there is a continuous real linear functional u on $E[\mathfrak{T}]$, and a monotonically increasing function $\delta_{x_0}(\rho)$, $\rho>0$, with $\lim_{\rho\to 0}\delta_{x_0}(\rho)=0$, for which

$$(2)\qquad |q(x_0+y)-q(x_0)-u\,y|\leqq q(y)\,\delta_{x_0}(q(y))$$

holds for all $y\in E$. u is called the strong or Fréchet derivative of $q(x)$ at the point x_0.

If $q(x)$ is a norm, (2) can be written more simply as

$$(3)\qquad \lim_{\|y\|\to 0}\frac{1}{\|y\|}(\|x_0+y\|-\|x_0\|-u\,y)=0.$$

(4) *If the strong derivative of $q(x)$ exists at x_0, then it is equal to the weak derivative $q'(x_0,y)$. The function $q(x)$ is strongly differentiable at x_0 if and only if it is weakly differentiable at x_0 and* $\dfrac{q(x_0+t\,y)-q(x_0)}{t}$ *converges uniformly in $q(y)\leqq 1$ to $q'(x_0,y)$.*

Proof. Uniform weak differentiability at x_0 means the existence of a function $\delta_{x_0}(t)$ with $\lim_{t\to 0^+}\delta_{x_0}(t)=0$ for which

$$(5)\qquad |q(x_0+t\,y)-q(x_0)-q'(x_0,t\,y)|\leqq |t|\,\delta_{x_0}(|t|)$$

holds for all y with $q(y)\leqq 1$. But this is equivalent to (2).

$q(x)$ is said to be uniformly strongly differentiable in $E[\mathfrak{T}]$ if $q(x)$ is strongly differentiable at every point except the origin, and if there is a $\delta(\rho)$ such that (2) (or (5)) is satisfied, with $\delta_{x_0}(\rho)=\delta(\rho)$, for every x_0 with $\|x_0\|=1$.

(6) *A normed space E is uniformly smooth if and only if its norm is uniformly strongly differentiable.*

a) Suppose that E is uniformly strongly differentiable. The inequality 4.(6) gives

$$(7)\qquad \frac{\|x_0-t\,y\|-\|x_0\|}{-t}\leqq q'(x_0,y)\leqq\frac{\|x_0+t\,y\|-\|x_0\|}{t}.$$

From (7) and (5) we obtain

(8) $$\|x_0+ty\|+\|x_0-ty\|\leqq 2+2|t|\,\delta(|t|)$$

for all $\|x_0\|=1$ and all $\|y\|\leqq 1$; but by (1) this means that E is uniformly smooth.

b) Conversely if E is uniformly smooth, then (8) holds for some suitable function $\delta(\rho)$. By 4.(9) the weak derivative $q'(x_0,y)$ exists, and by arguing back from (7) and (8), we obtain uniform strong differentiability.

In particular, we have also shown that *every uniformly smooth normed space is smooth.*

(9) *If the* (B)*-space E is uniformly convex, then E' is uniformly smooth.*

We assume that E' is not uniformly smooth. Then by (1) there exist $\varepsilon_0>0$ and sequences u_n, $v_n\in E'$ with $\|u_n\|=1$, $\|v_n\|\to 0$ and

(10) $$\|u_n+v_n\|+\|u_n-v_n\|>2+\varepsilon_0\|v_n\|.$$

Since E is reflexive, by 6.(4), there are elements x_n, x_n' in E with $\|x_n\|=\|x_n'\|=1$, $\|u_n+v_n\|=(u_n+v_n)x_n$ and $\|u_n-v_n\|=(u_n-v_n)x_n'$. Now $|\,\|u_n+v_n\|-1\,|\leqq\|v_n\|$, so that $|(u_n+v_n)x_n-1|\leqq\|v_n\|$; consequently $|u_nx_n-1|\leqq 2\|v_n\|$. Similarly $|u_nx_n'-1|\leqq 2\|v_n\|$.

Now if $2\|v_n\|<\dfrac{\delta(\varepsilon)}{2}$ it follows from 6.(3) that $\|x_n-x_n'\|<\varepsilon$. It follows from this that

$$\begin{aligned}\|u_n+v_n\|+\|u_n-v_n\|&=|(u_n+v_n)x_n+(u_n-v_n)x_n'|\\&\leqq|u_nx_n+u_nx_n'|+\|v_n\|\,\|x_n-x_n'\|\\&\leqq 2+\varepsilon\|v_n\|,\end{aligned}$$

which contradicts (10), for $\varepsilon<\varepsilon_0$.

(11) *If the* (B)*-space E is uniformly smooth, then E' is uniformly convex.*

Suppose that E' is not uniformly convex. Then by 6.(1) there are sequences u_n, v_n with $\|u_n\|\leqq 1$, $\|v_n\|\leqq 1$, $\lim\|u_n+v_n\|=2$ and $\|u_n-v_n\|\geqq\varepsilon_0$. We may assume that $\|u_n+v_n\|>2-\dfrac{\varepsilon_0}{4n}$.

For each n there exist x_n and x_n' in E with $\|x_n\|=\|x_n'\|=1$ and

$$(u_n+v_n)x_n\geqq\|u_n+v_n\|-\frac{\varepsilon_0}{8n},\quad (u_n-v_n)x_n'\geqq\|u_n-v_n\|-\frac{\varepsilon_0}{8n}.$$

Then we have

$$\left\| x_n + \frac{x'_n}{n} \right\| + \left\| x_n - \frac{x'_n}{n} \right\| \geqq u_n\left(x_n + \frac{x'_n}{n}\right) + v_n\left(x_n - \frac{x'_n}{n}\right)$$

$$= (u_n+v_n)x_n+(u_n-v_n)\frac{x'_n}{n}$$

$$\geqq \|u_n+v_n\| + \frac{1}{n}\|u_n-v_n\| - \frac{\varepsilon_0}{4n}$$

$$> 2 - \frac{\varepsilon_0}{2n} + \frac{\varepsilon_0}{n} = 2 + \frac{\varepsilon_0}{2}\left\|\frac{x'_n}{n}\right\|.$$

Since $\left\|\frac{x'_n}{n}\right\| = \frac{1}{n} \to 0$, this contradicts (1).

It follows from (11) that every uniformly smooth space is reflexive, since its dual is. From this fact, (9) and (11) we obtain

(12) *A normed space is uniformly smooth (respectively uniformly convex) if and only if its strong dual is uniformly convex (respectively uniformly smooth).*

The theorems of this number are due to DAY [5], ŠMULIAN [3], [4], [6] and BOURBAKI [6], Vol. 2 pp. 144-5.

11. Further ideas. Uniform convexity can be weakened in the following way: a normed space is said to be locally uniformly convex if to each $\varepsilon>0$ and each $x\in E$ with $\|x\|=1$ there corresponds a $\delta(\varepsilon,x)$ for which $\|\frac{1}{2}(x+y)\|\leqq 1-\delta(\varepsilon,x)$ for all $\|y\|\leqq 1$ with $\|y-x_0\|\geqq\varepsilon$. This concept has been investigated in detail by LOVAGLIA [1]. If E' is locally uniformly convex, then the norm of E is strongly differentiable. If E is locally uniformly convex, then E' has a strongly differentiable norm, provided that the additional assumption is made that for each $u\in E'$ there is an x in E with $\|x\|=1$ and $ux=\|u\|$.

Every reflexive space has this property. The question of whether conversely every (B)-space with the property that each $u\in E'$ attains its supremum on the unit ball of E is reflexive originated with MAZUR. JAMES [4] first answered this question in the affirmative for separable (B)-spaces, and recently (JAMES [5]) gave the same answer for arbitrary (B)-spaces.

FAN and GLICKSBERG [1], [2] have introduced and investigated further concepts of convexity which are weaker than uniform convexity. In this way they have obtained a stronger form of EBERLEIN's theorem for normed spaces.

CUDIA [2] has recently obtained important results by refining the convexity and differentiability properties further; these results complement and round off the results described here. The survey article of CUDIA [1] gives a full account of the present state of the theory.

CHAPTER SIX

Some Special Classes of Locally Convex Spaces

If the topology of a locally convex space is the same as the strong topology, the space is said to be barrelled. If the topology is the same as the Mackey topology, and if every linear functional which is bounded on the bounded subsets of the space is continuous, then the space is said to be bornological. Every (F)-space has these properties, and each of these properties entails a sequence of important consequences. The investigation of the properties of barrelled and of bornological spaces, which generalises the theory of (F)-spaces, and which goes back to MACKEY and to BOURBAKI, forms an important part of the general theory of locally convex spaces. These two classes of spaces are considered in depth in § 27 and § 28.

The theory of (F)- and (DF)-spaces which is given in §29 provides an application of these results. (DF)-spaces were introduced by GROTHENDIECK to provide a class of spaces which contains the duals of (F)-spaces as a special case.

In § 30, the theory of perfect spaces, which was originated by TOEPLITZ and the author, is developed as a special case of the general theory of locally convex spaces. The special structure of sequence spaces to some extent allows simplifications to be made, and to some extent enables further results to be established.

Some of the questions of the general theory which have been left open until now are answered in § 31 by giving counterexamples, which are perfect spaces. Besides this, § 31 contains a discussion of the problem of the existence of complements; although a collection of counterexamples is known, there are only a few general results on this topic.

§ 27. Barrelled spaces and Montel spaces

1. Quasi-barrelled spaces and barrelled spaces. As in § 21, 2., a locally convex space $E[\mathfrak{T}]$ is said to be barrelled if every barrel in E is a $\mathfrak{T}$-neighbourhood of $\mathfrak{o}$. This is equivalent to $\mathfrak{T}=\mathfrak{T}_b(E')$. As a result, every weakly bounded subset of E' is relatively weakly compact, by § 21, 4.(4).

Every (F)-space is barrelled (§ 21, 5.(3)). By § 23, 6.(4), a locally convex space $E[\mathfrak{T}_k(E')]$ is barrelled if and only if its dual E' is $\mathfrak{T}_s(E)$-quasi-complete. By § 23, 3.(4), the strong dual of every semi-reflexive space is barrelled.

As in § 23, 4., $E[\mathfrak{T}]$ is said to be quasi-barrelled if every barrel in E which absorbs all the bounded sets of E is a $\mathfrak{T}$-neighbourhood of $\mathfrak{o}$.

This is the same as saying that $\mathfrak{T}=\mathfrak{T}_{b^*}(E')$, by § 23, 4.(3). The $\mathfrak{T}$-equicontinuous subsets of E' are then the strongly bounded subsets of E'.

Every metrizable locally convex space is quasi-barrelled (§ 21, 5.(3)).

We now develop some further properties of barrelled and quasi-barrelled spaces.

Every barrelled space is quasi-barrelled. In the opposite direction, we have

(1) *If a quasi-barrelled space is sequentially complete, then it is barrelled.*

For it then follows from the BANACH-MACKEY theorem that $\mathfrak{T}_b(E')$ and $\mathfrak{T}_{b^*}(E')$ coincide on E.

It follows from (1) and § 21, 4.(5) that

(2) *The quasi-completion and the completion of a quasi-barrelled space are both barrelled.*

The following results give further hereditary properties.

(3) *Every locally convex hull of quasi-barrelled (respectively barrelled) spaces is again quasi-barrelled (barrelled).*

Suppose that $E[\mathfrak{T}]=\sum_\alpha A_\alpha(F_\alpha[\mathfrak{T}_\alpha])$, where the $F_\alpha[\mathfrak{T}_\alpha]$ are barrelled. If T is a barrel in $E[\mathfrak{T}]$, then, because of the continuity of the maps A_α, each set $T_\alpha=A_\alpha^{(-1)}(T)$ is again absolutely convex, absorbent and closed, and is therefore a barrel in F_α. By hypothesis, T_α is a $\mathfrak{T}_\alpha$-neighbourhood of $\mathfrak{o}$ in F_α. But then (cf. § 19, 1.) since T is absolutely convex and contains all the sets $A_\alpha(T_\alpha)$, T is a $\mathfrak{T}$-neighbourhood of $\mathfrak{o}$ in E, so that $E[\mathfrak{T}]$ is barrelled.

If B_α is a bounded subset of $F_\alpha[\mathfrak{T}_\alpha]$, $A_\alpha(B_\alpha)$ is bounded in $E[\mathfrak{T}]$. If T absorbs every bounded subset of E, T_α therefore absorbs every bounded subset of F_α. Arguing as above, it follows that if the $F_\alpha[\mathfrak{T}_\alpha]$ are quasi-barrelled, so is $E[\mathfrak{T}]$.

Since a quotient space $(E/H)[\mathfrak{T}]$ can be considered as the locally convex hull $K(E[\mathfrak{T}])$, where K is the canonical map of E onto E/H (cf. § 19, 1.), the next result is a special case of (3):

(4) *Every quotient of a quasi-barrelled (respectively barrelled) space is quasi-barrelled (barrelled).*

In particular (3) holds for inductive limits and locally convex direct sums. Thus all (LB)-spaces and all (LF)-spaces are barrelled.

(5) *The topological product of quasi-barrelled (barrelled) spaces is again quasi-barrelled (barrelled).*

This is a direct consequence of § 22, 5.(3).

There exist spaces which are complete, but are not quasi-barrelled. For, applying § 23, 6.(6) to the case where E is a non-reflexive (F)-space, $E'[\mathfrak{T}_k(E)]$ is certainly semi-reflexive, but is not reflexive, and so it is not quasi-barrelled (§ 23, 5.(1)). But $E'[\mathfrak{T}_k(E)]$ is complete, by § 21, 6.(4).

Every complete space which is not quasi-barrelled is a closed linear subspace of a topological product of (B)-spaces, by § 18. 3.(7), and this space is barrelled, by (5). A closed subspace of a barrelled space therefore need not be quasi-barrelled. Every (not necessarily closed) linear subspace of finite co-dimension is, however, barrelled (for the proof, cf. DIEUDONNÉ [9]).

A topological projective limit of barrelled spaces need not be quasi-barrelled, for by § 19, 9.(1) each of the above spaces $E'[\mathfrak{T}_k(E)]$ is the topological projective limit of (B)-spaces.

A normed space is quasi-barrelled, but need not be barrelled, as the example of § 21, 5. shows. On the other hand there are normed spaces which are barrelled, but not complete: a sequence k_n of positive integers is said to have density zero if $\lim \frac{n}{k_n} = 0$. Let E be the dense linear subspace of l^1 which comprises all $\mathfrak{x} = (\xi_i) \in l^1$ whose coordinates ξ_i vanish except on a set of density zero. It is easy to verify that the $\mathfrak{T}_s(E)$-bounded subsets of $E' = l^\infty$ coincide with the norm-bounded subsets of l^∞. It follows from this that the norm topology of $E \subset l^1$ coincides with the strong topology $\mathfrak{T}_b(E')$.

The strong dual of a barrelled space need not be barrelled, for there exist (F)-spaces which are not distinguished (cf. § 23, 7.(1) and § 31, 7.).

The strong dual and strong bidual of a barrelled space need not be complete, since KŌMURA [2] has shown that there are (M)-spaces which are not complete.

2. (M)-spaces and (FM)-spaces Besides (F)-spaces there is a further class of barrelled spaces which is of particular interest. A barrelled space $E[\mathfrak{T}]$ is called a Montel space, or (M)-space, if every bounded subset of E is relatively compact.

Every (M)-space is clearly quasi-complete.

A normed space which is also an (M)-space is locally compact, and so by § 15, 7.(1) it is finite-dimensional; an infinite dimensional (B)-space is therefore never an (M)-space.

On the other hand there are (F)-spaces which are also (M)-spaces (the space ω is an example of this). We call these spaces (FM)-spaces.

It follows from the definition and § 23, 5.(1) that

(1) *Every* (M)-*space is reflexive.*

The topologies $\mathfrak{T}_b(E)$ and $\mathfrak{T}_c(E)$ on the dual of an (M)-space coincide.

(2) *The strong dual of an* (M)-*space* $E[\mathfrak{T}]$ *is again an* (M)-*space.*

Proof. $E'[\mathfrak{T}_b(E)]$ is barrelled, since it is reflexive. The bounded subsets of E' are the same as the $\mathfrak{T}$-equicontinuous subsets. By § 21, 6.(3)

these are relatively $\mathfrak{T}_c(E)$-compact, and so, since $\mathfrak{T}_b(E)$ and $\mathfrak{T}_c(E)$ are the same, they are relatively $\mathfrak{T}_b(E)$-compact.

Thus the weak and the strong topologies coincide on the bounded subsets of an (M)-space. In particular, we have

(3) *Every weakly convergent sequence in an* (M)-*space is also strongly convergent, to the same limit.*

We have the following hereditary properties:

(4) *The topological product and the locally convex direct sum of* (M)-*spaces are again both* (M)-*spaces.*

The strict inductive limit of a sequence of complete (M)-*spaces is again an* (M)-*space.*

Proof. It follows from 1.(3) and 1.(5) that the spaces are barrelled; and from the structure of bounded sets in topological products (§ 15, 6.(13)), locally convex direct sums (§ 18, 5.(4)) and strict inductive limits (§ 19, 4.(1) and (4)), and from TYCHONOFF's theorem, it follows that every bounded set is relatively compact.

As we shall see in § 31, 5., there are quotient spaces of (FM)-spaces and closed linear subspaces of their dual (M)-spaces which are not (M)-spaces.

For arbitrary (M)-spaces, we can obtain a counterexample as follows: It follows from (4) that every space of countable degree is an (M)-space. Thus $\varphi\omega \oplus \omega\varphi$ and its dual $\omega\varphi \oplus \varphi\omega$ are (M)-spaces. Once again let H_1 and H_2 be the closed subspaces of $\varphi\omega \oplus \omega\varphi$ and $\omega\varphi \oplus \varphi\omega$, respectively, which were introduced in § 13, 6. By § 23, 5., $(\varphi\omega \oplus \omega\varphi)/H_1$ is not semi-reflexive, and so it is not an (M)-space; further H_2 is not reflexive, and so it is not an (M)-space, either.

We further observe that H_2 is semi-reflexive, by § 23, 3.(5), and so it is not barrelled, by § 23, 5.(1). H_2 therefore provides an example of a closed subspace of a locally convex space E on which the strong topology on E does not induce the strong topology of H_2.

There are (M)-spaces which are not separable, for example φ_d, $d > \aleph_0$. However DIEUDONNÉ [13] has proved that

(5) *Every* (FM)-*space E is separable.*

Proof. By § 18, 3.(7) we can consider E as subspace of a topological product $\prod_{n=1}^{\infty} E_n$ of normed spaces E_n, where the projection $P_n(E)$ of E in E_n is equal to E_n, for each n.

Suppose that E is not separable. If all the spaces E_n were separable, $\prod_{n=1}^{\infty} E_n$ would also be separable, and so would E, by § 4, 5.(1). We may therefore assume that E_1 is not separable. There is then a bounded uncountable subset N of E_1, whose elements are all at a distance $\geqq \delta > 0$ from each other. For each $x_1 \in N$ we can determine an $x \in E$ whose projection in E_1 is equal to x_1. Let M be the collection of these x. We put

$M=M_1$. There is then a proper uncountable subset M_2 of M_1 whose projection $P_2(M_2)$ is bounded in E_2 (E_2 is the union of countably many bounded sets). If we continue in this way, we obtain a properly decreasing sequence M_n of uncountable subsets, with each $P_n(M_n)$ bounded in E_n.

Suppose that $x_k \in M_{k+1} \sim M_k$. The sequence $P_n x_k$, $k=1,2,\ldots$, is then bounded in E_n, since $x_k \in P_n(M_n)$ for $k \geqq n-1$, so that x_k is bounded in E. Since E is an (FM)-space, x_n has a subsequence which is a Cauchy sequence in E. The same holds for the projections $P_1(x_n)$ in E_1. But this is a contradiction, since the $P_1(x_n)$ are pairwise at a distance $\geqq \delta$ apart.

Likewise it follows from (2), (3), (5) and § 21, 3.(5) that

(6) *The dual of an* (FM)-*space is strongly sequentially separable.*

We have the following criterion (cf. Dieudonné and Gomes [1]):

(7) *A separable* (F)-*space* $E[\mathfrak{T}]$ *is an* (M)-*space if and only if weak and strong sequential convergence coincide in* E'.

Proof. Necessity follows from (2) and (3). Conversely suppose that E is separable, and that every weakly convergent sequence in E' is strongly convergent. Since E is barrelled, every weakly closed bounded subset M of E' is weakly compact. Since E is separable, the weak topology on M is metrizable (cf. § 21, 3.(4)); by § 4, 5.(4), M is therefore weakly sequentially compact, and so it is strongly sequentially compact. By § 5, 6.(3) every strongly sequentially compact set is strongly precompact, so that M is strongly precompact. Finally if we apply § 21, 6.(3) to the bounded closed subsets B of $E \subset E''$, it follows that each such B is relatively strongly compact in E''. But B is strongly complete, as it is a closed subset of the (F)-space $E[\mathfrak{T}]$, and so B is strongly compact. E is therefore an (M)-space.

The strong duals of (FM)-spaces can be characterised more simply. A locally convex space $E[\mathfrak{T}]$ has a countable fundamental system of absolutely convex compact sets if there is a sequence $K_1 \subset K_2 \subset \cdots$ of absolutely convex compact sets such that each absolutely convex compact subset of E is contained in some K_n.

Following Dieudonné [15], we have

(8) *A barrelled space* $E[\mathfrak{T}]$ *is the strong dual of an* (FM)-*space if and only if* $E[\mathfrak{T}]$ *has a countable fundamental system of absolutely convex compact subsets.*

Proof. If $\{U_n\}$ is a base of absolutely convex neighbourhoods of $\mathfrak{o}$ in the (FM)-space $E[\mathfrak{T}]$, the sets U_n° form a fundamental system of bounded subsets of $E'[\mathfrak{T}_b(E)]$. By (2), the sets U_n° are absolutely convex and compact, and further $E'[\mathfrak{T}_b]$ is barrelled. The conditions are therefore necessary.

Conversely suppose that $E[\mathfrak{T}]$ is barrelled, and that $K_1 \subset K_2 \subset \cdots$ is a countable fundamental system of absolutely convex compact subsets of $E[\mathfrak{T}]$. If $\mathfrak{T}_0$ is the topology of uniform convergence on the absolutely convex compact subsets of E, $E'[\mathfrak{T}_0]$ is metrizable, and so it is quasi-barrelled. Since $\mathfrak{T}=\mathfrak{T}_b(E')$, the bounded subsets of E coincide with the $\mathfrak{T}_0$-equicontinuous subsets, and so coincide with the subsets of the absolutely convex compact subsets of $E[\mathfrak{T}]$. $E[\mathfrak{T}]$ is therefore an (M)-space, and $\mathfrak{T}_b(E)$ and $\mathfrak{T}_0$ coincide on E'. $E'[\mathfrak{T}_0]$ is therefore an (M)-space, by (2). Since $E'[\mathfrak{T}_0]$ is then quasi-complete and metrizable, $E'[\mathfrak{T}_0]$ is an (FM)-space. The assertion now follows from the reflexivity of $E[\mathfrak{T}]$ (cf. (1)).

If $E[\mathfrak{T}]$ is barrelled and if $E[\mathfrak{T}]$ has a countable fundamental system of compact subsets, then $E[\mathfrak{T}]$ is a dense subspace of the strong dual of an (FM)-space (Dieudonné [15]). Recently Garling [1] proved that $E[\mathfrak{T}]$ is in fact the strong dual of an (FM)-space. This result and (8) have analogues for reflexive spaces (Garling [1]).

3. The space $H(\mathfrak{G})$. Some of the most important spaces of analysis are (M)-spaces, e.g. the spaces of infinitely differentiable functions, and of distributions, on a compact interval in P^n (cf. Schwartz [1]). A detailed investigation of these spaces will be made in the second volume. Here we discuss a class of examples which motivated the name "Montel space".

Suppose that $\mathfrak{G}$ is a domain, i.e. an open connected proper subset of the complex sphere Ω. The holomorphic functions $x(z)$ defined on $\mathfrak{G}$ which, in the case where $\mathfrak{G}$ contains ∞, vanish at ∞ form a complex vector space $H(\mathfrak{G})$.

For every compact subset $\mathfrak{K}$ of $\mathfrak{G}$ we define the norm

$$p_{\mathfrak{K}}(x)=\sup_{z\in\mathfrak{K}}|x(z)|. \tag{1}$$

This system of norms defines a locally convex topology $\mathfrak{T}$ on $H(\mathfrak{G})$, the topology of uniform convergence on the compact subsets of $\mathfrak{G}$.

If $\mathfrak{G}_1 \subset \mathfrak{G}_2 \subset \cdots$ is a sequence of open connected subsets of $\mathfrak{G}$, with $\bar{\mathfrak{G}}_n \subset \mathfrak{G}_{n+1}$ and $\bigcup_{n=1}^{\infty} \mathfrak{G}_n=\mathfrak{G}$, then $\mathfrak{T}$ is also defined by the countable collection of norms

$$\|x\|_n=p_{\bar{\mathfrak{G}}_n}(x) \tag{2}$$

so that $H(\mathfrak{G})$ is metrizable.

$\mathfrak{G}$ need not be simply connected, but the $\mathfrak{G}_n$ can be chosen in such a way that each is bounded by finitely many simple closed rectifiable curves. This system of curves is denoted by C_n.

By WEIERSTRASS' theorem, a $\mathfrak{T}$-Cauchy sequence in $H(\mathfrak{G})$ always has a holomorphic function as limit (vanishing at ∞), so that $H(\mathfrak{G})$ is an (F)-space. In addition, we have

(3) *$H(\mathfrak{G})$ is an* (FM)-*space.*

Proof. A subset B of $H(\mathfrak{G})$ is bounded if and only if $p_{\mathfrak{K}}(x) \leqq M(\mathfrak{K})$ for all $x \in B$. If $\mathfrak{G}$ is a domain which does not contain the point ∞, (3) is just the statement of MONTEL's theorem.

The case where $\infty \in \mathfrak{G}$ can be reduced to this by means of a transformation $y(z) = x\left(\dfrac{1}{z-a}\right)$, $a \notin \mathfrak{G}$.

By 2.(2), the strong dual of $H(\mathfrak{G})$ is an (M)-space. This space has a concrete representation, which we shall now derive.

Suppose that $\langle u, x\rangle$ is a continuous linear functional on $H(\mathfrak{G})$. Then for a suitable choice of n we have

$$|\langle u, x\rangle| \leqq M\|x\|_n. \tag{4}$$

Now $\|x\|_n$ is the norm of the (B)-space $HB(\bar{\mathfrak{G}}_n)$ of all functions which are analytic on $\mathfrak{G}_n$ and continuous on $\bar{\mathfrak{G}}_n$ (cf. § 15, 9.). By the HAHN-BANACH theorem, u can be extended from $H(\mathfrak{G})$ to $HB(\bar{\mathfrak{G}}_n)$, with the same bound.

Consequently u is defined for all functions $\dfrac{1}{\lambda - z}$, $\lambda \in \mathfrak{O}_n = \Omega \sim \bar{\mathfrak{G}}_n$, which, considered as functions of z, are elements of $HB(\bar{\mathfrak{G}}_n)$.

By defining

$$\tilde{u}(\lambda) = \left\langle u, \frac{1}{\lambda - z}\right\rangle, \tag{5}$$

we obtain a function of λ defined in $\mathfrak{O}_n$ which, following FANTAPPIÈ, is called the indicatrix of u. If $\lambda = \infty$, $1/(\lambda - z)$ is put equal to 0, so that $\tilde{u}(\infty) = 0$, if $\infty \in \mathfrak{O}_n$.

We observe that $\mathfrak{O}_n$ is open, but that it need not be connected.

(6) *The indicatrix $\tilde{u}(\lambda)$ is locally holomorphic in $\mathfrak{O}_n = \Omega \sim \bar{\mathfrak{G}}_n$.* If $\lambda_0 \neq \infty$, then for $\lambda, \lambda_0 \in \mathfrak{O}_n$ we have

$$\frac{\tilde{u}(\lambda) - \tilde{u}(\lambda_0)}{\lambda - \lambda_0} = \left\langle u, \frac{1}{\lambda - \lambda_0}\left(\frac{1}{\lambda - z} - \frac{1}{\lambda_0 - z}\right)\right\rangle.$$

Now the difference quotient of $\dfrac{1}{\lambda - z}$ converges to $\dfrac{-1}{(\lambda_0 - z)^2}$ with respect to the norm of $HB(\bar{\mathfrak{G}}_n)$. Since u is continuous, we therefore have $\tilde{u}'(\lambda_0) = \left\langle u, \dfrac{-1}{(\lambda_0 - z)^2}\right\rangle$. Thus the indicatrix is analytic at λ_0.

If $\lambda_0=\infty$, $\tilde{u}(\lambda)$ is analytic in a neighbourhood of ∞, and $\tilde{u}(\lambda)$ converges to 0 as $\lambda\to\infty$, since $\frac{1}{\lambda-z}$ converges to 0 with respect to the norm of $HB(\bar{\mathfrak{G}}_n)$ as $\lambda\to\infty$. $\tilde{u}(\lambda)$ is therefore also analytic at ∞, and vanishes there.

Conversely, the linear functional $\langle u,x\rangle$ on $H(\mathfrak{G})$ can be expressed in terms of the indicatrix:

(7) *If* $\tilde{u}(\lambda)$ *is the indicatrix of* u *in* $\mathfrak{O}_n$, *then*

$$\langle u,x\rangle = \frac{1}{2\pi i}\oint_{C_{n+1}} \tilde{u}(t)x(t)\,dt \tag{8}$$

for each $x\in H(\mathfrak{G})$.

Proof. For each (possibly infinite) $z\in\bar{\mathfrak{G}}_n$ and each $x(z)\in H(\mathfrak{G})$ we have

$$x(z) = \frac{1}{2\pi i}\oint_{C_{n+1}} \frac{x(t)}{t-z}\,dt.$$

The system of curves C_{n+1} is here described in such a way that $\bar{\mathfrak{G}}_n$ always lies to the left. We consider a sequence $x_k(z) = \frac{1}{2\pi i}\sum_j \frac{x(t_j^{(k)})\,\Delta t_j^{(k)}}{t_j^{(k)}-z}$ of approximating sums for the integral $x(z)$. Since there is a strictly positive least distance δ between the points z of $\bar{\mathfrak{G}}_n$ and the points of C_{n+1}, the sequence $x_k(z)$ converges to $x(z)$ uniformly for all $z\in\bar{\mathfrak{G}}_n$. Since u is continuous on $HB(\bar{\mathfrak{G}}_n)$, it then follows that

$$\langle u,x\rangle = \lim_{k\to\infty}\frac{1}{2\pi i}\sum x(t_j^{(k)})\left\langle u,\frac{1}{t_j^{(k)}-z}\right\rangle \Delta t_j^{(k)} = \frac{1}{2\pi i}\oint_{C_{n+1}} x(t)\tilde{u}(t)\,dt.$$

We observe that $\mathfrak{O}_n$ is an open set containing $\mathfrak{A}=\Omega\sim\mathfrak{G}$ and that $\mathfrak{O}_n$ comprises finitely many components, each of which contains at least one point of $\mathfrak{A}$; a set with these properties will be called an open neighbourhood $\mathfrak{U}$ of $\mathfrak{A}$, for short. Now suppose that $\tilde{u}(\lambda)$ is a locally holomorphic function in $\mathfrak{U}$, which vanishes at ∞ if ∞ lies in $\mathfrak{U}$. There is then an n for which $\Omega\sim\mathfrak{U}\subset\mathfrak{G}_{n+1}$ holds. The right hand side of (8) is defined for all $x\in H(\mathfrak{G})$, and, as is easy to see, it defines a continuous linear functional $\langle u,x\rangle$ on $H(G)$.

The value of (8) is independent of n, provided that C_{n+1} lies in $\mathfrak{U}$; this follows directly from Cauchy's theorem. For the same reason, two locally holomorphic functions, defined on neighbourhoods $\mathfrak{U}_1$ and $\mathfrak{U}_2$ respectively, define the same linear functional if they agree on a neighbourhood of $\mathfrak{A}$ contained in $\mathfrak{U}_1\cap\mathfrak{U}_2$.

We say that two such functions are equivalent, and call a class of equivalent functions (vanishing at ∞) a **locally holomorphic function on $\mathfrak{A}$**; we also use $\tilde{u}(\lambda)$ to denote the class to which $\tilde{u}(\lambda)$ belongs. The set of all locally holomorphic functions on $\mathfrak{A}$ forms a vector space $H(\mathfrak{A})$.

We have already seen that every $u \in H(\mathfrak{G})'$ is given by a $\tilde{u} \in H(\mathfrak{A})$, and that conversely every $\tilde{u} \in H(\mathfrak{A})$ defines a $u \in H(\mathfrak{G})'$, using (8). This last correspondence is clearly linear. In order to show that $H(\mathfrak{G})'$ and $H(\mathfrak{A})$ are isomorphic, we must show that only one equivalence class $\tilde{u}$ corresponds to each u. Now $\tilde{u}(\lambda)$, considered as a locally holomorphic function on $\mathfrak{U} \supset \mathfrak{A}$, is determined by the values of its derivatives at one point in each of the components of $\mathfrak{U}$. But by hypothesis each component of $\mathfrak{U}$ contains at least one point λ_0 of $\mathfrak{A} = \Omega \sim \mathfrak{G}$.

Thus it is sufficient to show that if the continuous linear functional defined by (8) vanishes identically on $H(\mathfrak{G})$, then $\tilde{u}(\lambda)$, together with all its derivatives, vanishes on $\mathfrak{A}$. If $\lambda \in \mathfrak{A}$ and $\lambda \neq \infty$, then all the functions $x(z) = \dfrac{1}{(z-\lambda)^n}$, $n = 1, 2, \ldots$, lie in $H(\mathfrak{G})$. Substituting these in (8), we get $\tilde{u}^{(n)}(\lambda) = 0$ for $n = 0, 1, \ldots$.

If $\lambda = \infty \in \mathfrak{A}$, then the functions $x(z) = z^n$, $n = 0, 1, \ldots$, lie in $H(\mathfrak{G})$, and (8) shows further that all the coefficients of the Laurent expansion of $\tilde{u}(\lambda)$ at ∞ vanish.

We have therefore proved the following theorem:

(9) *The dual of $H(\mathfrak{G})$ is isomorphic to the space $H(\mathfrak{A})$ of locally holomorphic functions $\tilde{u}(\lambda)$ on $\mathfrak{A} = \Omega \sim \mathfrak{G}$, and the continuous linear functional u corresponding to a $\tilde{u}(\lambda) \in H(\mathfrak{A})$ is given by* (8).

The last part of our proof shows that, in spite of the fact that we used the Hahn-Banach theorem to construct the indicatrix $\tilde{u}(\lambda)$, this is uniquely determined, up to equivalence, by the linear functional $u \in H(\mathfrak{G})'$.

4. (M)-spaces of locally holomorphic functions. It is natural to consider an arbitrary proper open subset $\mathfrak{O}$ of Ω, instead of a domain $\mathfrak{G}$. $\mathfrak{O}$ consists of at most countably many open components $\mathfrak{H}_i$, and we again denote the space of all locally holomorphic functions $x(z)$ on $\mathfrak{O}$ by $H(\mathfrak{O})$. Clearly $H(\mathfrak{O})$ is equal to the Cartesian product of the spaces $H(\mathfrak{H}_i)$, and we define the topology $\mathfrak{T}$ on $H(\mathfrak{O})$ to be that of the topological product of the $H(\mathfrak{H}_i)$. It follows directly from this, using 2.(4), that $H(\mathfrak{O})$ is also an (FM)-space.

It is easily confirmed that $\mathfrak{T}$ coincides with the topology of uniform convergence on the compact subsets $\mathfrak{K}$ of $\mathfrak{O}$. In the present case, of course, the expressions 3.(1) defining $\mathfrak{T}$ need only be semi-norms. The sequence $\mathfrak{G}_1 \subset \mathfrak{G}_2 \subset \cdots$ can be chosen exactly as in 3., the only difference being that the set $\mathfrak{G}_n$ can no longer be assumed to be connected; instead it can be taken to consist of just one connected component in each of $\mathfrak{H}_1, \dots, \mathfrak{H}_n$. This has the consequence that the Banach spaces $HB(\mathfrak{G}_n)$ of locally holomorphic functions are of the type already introduced in § 14, 9.

If 3.(4) holds, then u vanishes on all functions $x(z)$ which vanish on $\mathfrak{H}_1, \dots, \mathfrak{H}_n$. Then $H(\mathfrak{H}_1 \cup \cdots \cup \mathfrak{H}_n) \subset HB(\bar{\mathfrak{G}}_n)$, and we obtain the indicatrix of u as in 3.

The evaluation of the dual can now be carried through as in 3.: a functional $u \in H(\mathfrak{D})'$ satisfies an inequality 3.(4), and so it vanishes identically on $H\left(\bigcup_{n+1}^{\infty} \mathfrak{H}_i\right)$. If we consider u as a linear functional on $H\left(\bigcup_{i=1}^{n} \mathfrak{H}_i\right)$, then $H\left(\bigcup_{i=1}^{n} \mathfrak{H}_i\right) \subset HB(\bar{\mathfrak{G}}_n)$, and we obtain the representation 3.(8) on $H\left(\bigcup_{i=1}^{n} \mathfrak{H}_i\right)$. This is however valid on the whole of $H(\mathfrak{D}) = H\left(\bigcup_{i=1}^{n} \mathfrak{H}_i\right) \oplus H\left(\bigcup_{n+1}^{\infty} \mathfrak{H}_i\right)$, since (8) vanishes identically on $H\left(\bigcup_{n+1}^{\infty} \mathfrak{H}_i\right)$.

The fact that the correspondence between the functionals u and the locally holomorphic functions $\tilde{u}(\lambda)$ on $\mathfrak{A} = \Omega \sim \mathfrak{D}$ is one-one is obtained as in 3.

We therefore have the following general result:

(1) *Using* 3.(5) *and* 3.(8), *the dual of* $H(\mathfrak{D})$ *can be represented as the space* $H(\mathfrak{A})$ *of locally holomorphic functions on* $\mathfrak{A} = \Omega \sim \mathfrak{D}$.

If we give $H(\mathfrak{A})$ its strong topology $\mathfrak{T}_b(H(\mathfrak{D}))$, $H(\mathfrak{A})$ becomes a complete reflexive (M)-space by 2. and § 21, 6.(4). Our aim is now to give a simpler definition of the topology $\mathfrak{T}_b(H(\mathfrak{D}))$, which will enable the structure of the space $H(\mathfrak{A})$ to be seen more clearly.

Let $\mathfrak{A}$ be an arbitrary proper closed subset of Ω. Its complement $\mathfrak{D} = \Omega \sim \mathfrak{A}$ is then an arbitrary proper open subset of Ω. Let $\mathfrak{D}_1 \supset \mathfrak{D}_2 \supset \cdots$ be a sequence of open neighbourhoods of $\mathfrak{A}$ with $\bigcap_{n=1}^{\infty} \mathfrak{D}_n = \mathfrak{A}$. We recall that the sets $\mathfrak{D}_n$ each have only finitely many components, each of which contains at least one point of $\mathfrak{A}$. We further suppose that the boundary of $\mathfrak{D}_n$ is a system C_n of finitely many rectifiable closed curves.

If $\mathfrak{G}_n$ is the sequence determined above for $\mathfrak{D} = \bigcup_{i=1}^{\infty} \mathfrak{H}_i$, we can take $\mathfrak{D}_n = \Omega \sim \bar{\mathfrak{G}}_n$.

For each $\mathfrak{D}_n$ we form the (B)-space $HB(\bar{\mathfrak{D}}_n)$ of functions which are locally holomorphic in $\mathfrak{D}_n$ and continuous on $\bar{\mathfrak{D}}_n$. We now identify each function in $HB(\bar{\mathfrak{D}}_n)$ with its class of equivalent functions with respect to $\mathfrak{A}$, in other words, with a locally holomorphic function on $\mathfrak{A}$. In this way $HB(\bar{\mathfrak{D}}_n)$ becomes a linear subspace of $HB(\bar{\mathfrak{D}}_{n+1})$, and $H(\mathfrak{A})$ becomes the union of the spaces $HB(\bar{\mathfrak{D}}_n)$.

It is now natural to try to interpret $H(\mathfrak{A})$ as topological inductive limit of the spaces $HB(\bar{\mathfrak{D}}_n)$.

Since the norm $\|u\|_{n+1} = \sup_{\lambda \in \bar{\mathfrak{D}}_{n+1}} |u(\lambda)|$ of $HB(\bar{\mathfrak{D}}_{n+1})$ induces a coarser Hausdorff topology on $HB(\bar{\mathfrak{D}}_n)$ than the norm $\|u\|_n$, we can define the hull topology $\mathfrak{T}$ on $H(\mathfrak{A}) = \bigcup_{n=1}^{\infty} HB(\bar{\mathfrak{D}}_n)$. We still have to show that

(2) *The hull topology on $H(\mathfrak{A})$ is Hausdorff.*

First we obtain a simple lemma.

(3) *Every bounded subset M of $HB(\bar{\mathfrak{D}}_n)$ is relatively compact in $HB(\bar{\mathfrak{D}}_{n+1})$, and its closure $\bar{M}$ in $HB(\bar{\mathfrak{D}}_{n+1})$ is compact in each of the spaces $HB(\bar{\mathfrak{D}}_{n+m})$, $m \geqq 1$.*

M consists of functions which are uniformly bounded on $\bar{\mathfrak{D}}_{n+1}$; by MONTEL's theorem, M is therefore relatively compact in $HB(\bar{\mathfrak{D}}_{n+1})$. The rest of (3) follows from § 3, 2.(6).

In order to prove (2) we must show that for each non-zero $u_0 \in H(\mathfrak{A})$ there exists a $\mathfrak{T}$-neighbourhood of $\mathfrak{o}$ which does not contain u_0. Without loss of generality we may suppose that $u_0 \in HB(\bar{\mathfrak{D}}_1)$. There is then a closed ball K_1' about $\mathfrak{o}$ in $HB(\bar{\mathfrak{D}}_2)$ which does not contain u_0. If K_1 is the ball of equal radius in $HB(\bar{\mathfrak{D}}_1)$, u_0 does not belong to the closure $\bar{K}_1$ of K_1 in $HB(\bar{\mathfrak{D}}_2)$, since $\bar{K}_1 \subset K_1'$. By (3), $\bar{K}_1$ is compact in $HB(\bar{\mathfrak{D}}_n)$, $n \geqq 2$. Since u_0 does not belong to $\bar{K}_1$, there is a closed ball K_2' about $\mathfrak{o}$ in $HB(\bar{\mathfrak{D}}_3)$ such that u_0 does not belong to $\Gamma(\bar{K}_1 \cup K_2')$. Again, let $\bar{K}_2$ be the closure in $HB(\bar{\mathfrak{D}}_3)$ of the ball K_2 of equal radius in $HB(\bar{\mathfrak{D}}_2)$. u_0 does not belong to $\Gamma(\bar{K}_1 \cup \bar{K}_2)$, and this is compact in $HB(\bar{\mathfrak{D}}_n)$, $n \geqq 3$, by (3) and § 20, 6.(5). Repeating this procedure, we obtain a $\mathfrak{T}$-neighbourhood $\Gamma_{i=1}^{n} \bar{K}_i$ which does not contain u_0, since u_0 belongs to no $\Gamma_{i=1}^{\infty} \bar{K}_i$.

The topological inductive limit $\varinjlim HB(\bar{\mathfrak{D}}_n)$ therefore exists, and it defines the hull topology $\mathfrak{T}$ on $H(\mathfrak{A})$.

(4) *$H(\mathfrak{D})$ is the dual of $H(\mathfrak{A})[\mathfrak{T}]$.*

Proof. If $\langle x,u \rangle$ is a continuous linear functional on $H(\mathfrak{A})$, we have $|\langle x,u \rangle| \leqq M$ for a suitable $\mathfrak{T}$-neighbourhood $U = \Gamma_{n=1}^{\infty} K_n$, where each K_n is a closed ball in $HB(\bar{\mathfrak{D}}_n)$. The linear functional x is bounded on the $u_n \in K_n$, so that x is continuous on each $HB(\bar{\mathfrak{D}}_n)$.

If, corresponding to 3.(5) (but with the roles of λ and z interchanged), we introduce an indicatrix for x by writing

$$(5) \qquad \tilde{x}(z) = \left\langle x, \frac{1}{\lambda - z} \right\rangle, \qquad z \in \mathfrak{O},$$

then $\tilde{x}(z)$ is locally holomorphic in each $\mathfrak{G}_n = \Omega \sim \bar{\mathfrak{O}}_n$ and so it is locally holomorphic in $\bigcup_{n=1}^{\infty} \mathfrak{G}_n$. If $u \in HB(\bar{\mathfrak{O}}_n)$, we have

$$(6) \qquad \langle x, u \rangle = \frac{1}{2\pi i} \oint_{C_{n+1}} \tilde{x}(t) u(t) \, dt.$$

The system of curves C_{n+1} is described in the same sense as in 3.(8), so that the domain $\mathfrak{G}_{n+1}$ is on the left.

Conversely, using (6), every $\tilde{x}(z) \in H(\mathfrak{O})$ defines a linear functional on $H(\mathfrak{A})$, provided that n is suitably chosen for each u. This linear functional is $\mathfrak{T}$-continuous; for if u belongs to the neighbourhood K_n of all u with $\|u\|_n \leqq \dfrac{2\pi}{|C_{n+1}| \, \|\tilde{x}\|_{n+1}}$, where $\|\tilde{x}\|_{n+1} = \sup\limits_{z \in C_{n+1}} |\tilde{x}(z)|$ and $|C_{n+1}|$ is the total length of the system of curves C_{n+1}, then $|\langle x, u \rangle| \leqq 1$, by (6). Thus $|\langle x, u \rangle| \leqq 1$ through the whole $\mathfrak{T}$-neighbourhood $\Gamma_{n=1}^{\infty} K_n$.

(7) *The hull topology $\mathfrak{T}$ on $H(\mathfrak{A})$ is identical with the strong topology $\mathfrak{T}_b(H(\mathfrak{O}))$.*

By (4), $\mathfrak{T}$ is a locally convex topology on $H(\mathfrak{A})$ for which the dual is $H(\mathfrak{O})$. By the Mackey-Arens theorem (§ 21, 4.(2)), $\mathfrak{T}_b(H(\mathfrak{O}))$ is the finest topology with this property. (7) will therefore be established when we have shown that $\mathfrak{T}$ is finer than $\mathfrak{T}_b(H(\mathfrak{O}))$.

Let B be a bounded subset of $H(\mathfrak{O})$. There exist numbers $M_n \geqq 0$ such that $|x(z)| \leqq M_n$ for all $x \in B$ and $z \in \bar{\mathfrak{G}}_n$. If now K_n is the set of all $u \in HB(\bar{\mathfrak{O}}_n)$ with $\|u\|_n \leqq \dfrac{2\pi}{M_{n+1} |C_{n+1}|}$, then $|\langle x, u \rangle| \leqq 1$ for all $u \in K_n$ and all $x \in B$, by (6); thus the $\mathfrak{T}$-neighbourhood $\Gamma_{n=1}^{\infty} K_n$ is contained in the strong neighbourhood B° of $\mathfrak{o}$.

In particular if $\mathfrak{O}$ is the whole complex plane Γ, we obtain the space $H(\Gamma)$ of entire functions. Its dual is $H(\infty)$, the space of functions holomorphic at ∞.

In the case where $\mathfrak{O}$ is an open circular disc, this duality, and the results relating to it, were first obtained by Toeplitz [2] in the setting of the theory of perfect spaces (cf. § 30). For the general case, and for further generalisations, see Grothendieck [8], [9], Köthe [11], [12], Sebastião e Silva [1], [3], Silva Dias [1] and Tillmann [1] to [4].

§ 28. Bornological spaces

1. Definition. A linear functional on a normed space is continuous if it is bounded on the unit ball. This can also be expressed by saying that every linear functional on a normed space which is bounded on the bounded sets is continuous.

Expressed in this form, this property need no longer hold for arbitrary locally convex spaces, as we shall presently see. If we say that a linear functional $u \in E^*$ is locally bounded when its values remain bounded on any bounded subset of E, then the problem is to characterise those locally convex spaces for which every locally bounded linear functional is continuous. For this we can always suppose that the topology $\mathfrak{T}$ is the Mackey topology, since we are concerned with a property which depends only upon the dual pair $\langle E', E\rangle$, and not upon the original topology $\mathfrak{T}$.

Following BOURBAKI, a locally convex space $E[\mathfrak{T}]$ is said to be bornological if every absolutely convex set M which absorbs all the bounded sets of $E[\mathfrak{T}]$ is a $\mathfrak{T}$-neighbourhood of $\circ$.

Comparing this definition with the characterisation of quasi-barelled spaces given in § 23, 4.(3), we see that a definition of quasi-barrelled space is obtained from the definition of bornological space by adding the further requirement that M be closed. Thus we have

(1) *Every bornological space is quasi-barrelled.*

It therefore follows from § 27, 1.(1) that

(2) *Every sequentially complete bornological space is barrelled.*

The topology $\mathfrak{T}$ of a bornological space always coincides with $\mathfrak{T}_k(E')$, and indeed with $\mathfrak{T}_{b^*}(E')$, since this is so for quasi-barrelled spaces (§ 27, 1.).

Our original question is answered by

(3) *A locally convex space* $E[\mathfrak{T}]$ *has the property that every locally bounded linear functional on* E *is continuous if and only if* $E[\mathfrak{T}_k(E')]$ *is bornological.*

a) Sufficiency. If $E[\mathfrak{T}]$ is bornological, $\mathfrak{T} = \mathfrak{T}_k(E')$, as we have just seen. If $u \in E^*$ is locally bounded, and thus is bounded on every bounded subset B of $E[\mathfrak{T}]$, then the absolutely convex set M consisting of all $x \in E$ with $|ux| \leq 1$ contains a scalar multiple of each set B. It is therefore a $\mathfrak{T}$-neighbourhood of $\circ$, so that u is continuous.

b) Necessity. We consider the locally convex topology $\mathfrak{T}^\times$ on E which is obtained by taking the absolutely convex sets which absorb every bounded set as a base of neighbourhoods of $\circ$. $\mathfrak{T}^\times$ is clearly finer than $\mathfrak{T}_k(E')$. By hypothesis, every $\mathfrak{T}^\times$-continuous linear functional is continuous, and is therefore an element of E'. It now follows from the MACKEY-ARENS theorem that $\mathfrak{T}^\times$ is coarser than $\mathfrak{T}_k(E')$. The fact that $\mathfrak{T}_k(E')=\mathfrak{T}^\times$ implies that $E[\mathfrak{T}_k]$ is bornological.

This characterisation of bornological spaces was used as a definition by MACKEY [5] (who called them "relatively strong with a boundedly closed linear system").

Not every locally convex space $E[\mathfrak{T}]$ with the property that $\mathfrak{T}=\mathfrak{T}_k$ is bornological, for not every such space is quasi-barrelled (cf. the example in § 27, 1.). The class of bornological spaces is however very extensive, as we shall see. As a first result, we have

(4) *Every metrizable locally convex space* $E[\mathfrak{T}]$ *is bornological.*

Before proving this, we establish the following lemma:

(5) *If* $x_n \to o$ *in a metrizable locally convex space* $E[\mathfrak{T}]$, *there exist positive numbers* ρ_n, *with* $\rho_n \to \infty$, *such that* $\rho_n x_n \to o$ *as well.*

Proof. Let $U_1 \supset U_2 \supset \cdots$ be a base of neighbourhoods of o in $E[\mathfrak{T}]$. For each k there exists an $n_k > n_{k-1}$ such that $x_n \in \frac{1}{k} U_k$ for $n \geqq n_k$. We obtain the required sequence by setting $\rho_n = k$ for $n_k \leqq n < n_{k+1}$.

Proof of (4). Suppose that u is a locally bounded linear functional on E, and that $x_n \to o$. Then $\rho_n x_n \to o$ as well, so that $|u(\rho_n x_n)| = \rho_n |u x_n| \leqq M$. But it follows from this that $u x_n \to 0$, so that u is continuous. The assertion now follows from (3).

It follows from (4) that not every bornological space is barrelled, since the example of § 21. 5. gives a normed, and therefore bornological, space which is not barrelled.

Conversely, not every barrelled space is bornological, as NACHBIN [4] and SHIROTA [1] have shown.

2. The structure of bornological spaces. We are also led to bornological spaces by asking another question. Suppose that $E[\mathfrak{T}]$ is locally convex. Let us consider the finest locally convex topology $\mathfrak{T}^\times$ which has the same bounded sets in E as does $\mathfrak{T}$. Its absolutely convex neighbourhoods of o must absorb all the bounded sets of $E[\mathfrak{T}]$. On the other hand, the collection of all the absolutely convex sets which absorb all the bounded sets forms a base of neighbourhoods of o for a locally convex topology on E, and this must be the required topology $\mathfrak{T}^\times$. $E[\mathfrak{T}^\times]$ is clearly bornological, and the dual of $E[\mathfrak{T}^\times]$ consists of all locally bounded linear functionals on $E[\mathfrak{T}]$.

The locally convex space $E[\mathfrak{T}^\times]$ which is uniquely defined in this way is called the bornological space associated with $E[\mathfrak{T}]$. Clearly we have

(1) *$E[\mathfrak{T}]$ is bornological if and only if $E[\mathfrak{T}]$ coincides with its associated bornological space $E[\mathfrak{T}^{\times}]$.*

For each bounded, closed, absolutely convex subset B of a bornological space $E[\mathfrak{T}]$ let us form the normed space $E_B \subset E[\mathfrak{T}]$, as in § 20, 11. $\mathfrak{T}$ induces a coarser topology on E_B than the norm topology. If we form the locally convex hull $E[\mathfrak{T}'] = \sum_B E_B$ of all these spaces E_B, then the hull topology $\mathfrak{T}'$ is finer than $\mathfrak{T}$, by the definition of $\mathfrak{T}'$. But on the other hand every $\mathfrak{T}'$-neighbourhood of o $U = \Gamma_B \rho_B B$, $\rho_B > 0$, is also a $\mathfrak{T}$-neighbourhood of o, since it is absolutely convex, and absorbs all the bounded subsets of $E[\mathfrak{T}]$. Thus we have shown

(2) *Every bornological space is the locally convex hull $E[\mathfrak{T}] = \sum_B E_B$ of normed spaces E_B. If, further, $E[\mathfrak{T}]$ is sequentially complete, $E[\mathfrak{T}]$ is the locally convex hull of* (B)*-spaces.*

The last assertion follows from § 20, 11.(2).

By § 19, 3., $\sum_B E_B$ can also be interpreted as a topological inductive limit.

From (2) there follows an important result, due to MACKEY [5], which generalises 1.(3).

A linear mapping A from a locally convex space E into a locally convex space F is said to be locally bounded if it maps the bounded sets of E into bounded sets of F.

(3) *A locally convex space $E[\mathfrak{T}]$ is bornological if and only if every locally bounded map from $E[\mathfrak{T}]$ into any locally convex space $F[\mathfrak{T}']$ is continuous.*

Proof. a) Necessity. Suppose that $E[\mathfrak{T}]$ is bornological, so that it is equal to $\sum_B E_B$. If A is locally bounded, then the restriction of A to E_B is a mapping which sends the unit ball B of E_B into a bounded set in $F[\mathfrak{T}']$. But then A is continuous, by § 19, 1 (7)

b) Sufficiency. Putting $F[\mathfrak{T}'] = K$, the condition means that every locally bounded linear functional is continuous. Since, further, the identity mapping is a locally bounded mapping from $E[\mathfrak{T}]$ into $E[\mathfrak{T}_k(E')]$, $\mathfrak{T}$ coincides with $\mathfrak{T}_k(E')$. The assertion follows from 1.(3).

If we now consider the associated bornological space $E[\mathfrak{T}^{\times}]$, we obtain the following characterisation of locally bounded mappings from general locally convex spaces from (3):

(4) *A linear mapping from a locally convex space $E[\mathfrak{T}]$ into a locally convex space $F[\mathfrak{T}']$ is locally bounded if and only if it is a continuous mapping from $E[\mathfrak{T}^{\times}]$ into $F[\mathfrak{T}']$.*

3. Local convergence. Sequentially continuous mappings. Criterion 2.(3) for the continuity of linear mappings from bornological spaces can be expressed in another form which is particularly convenient for applications.

We introduce the concept of local convergence (which is also called Mackey convergence). A sequence x_n of elements of a locally convex space $E[\mathfrak{T}]$ is said to be locally convergent to x_0 if there is a bounded, closed absolutely convex subset B of $E[\mathfrak{T}]$ such that x_n and x_0 lie in E_B and such that x_n converges to x_0 with respect to the norm of E_B. A sequence which is locally convergent to o is called a local null-sequence.

The concept of local convergence clearly depends only upon the dual pair $\langle E', E\rangle$.

A locally convergent sequence in $E[\mathfrak{T}]$ is always $\mathfrak{T}$-convergent.

(1) a) *A sequence $x_n \in E[\mathfrak{T}]$ is locally convergent to x_0 if and only if $x_n - x_0$ is a local null-sequence.*

b) *x_n is a local null-sequence if and only if there exist positive numbers ρ_n, with $\rho_n \to \infty$, such that $\rho_n x_n$ is $\mathfrak{T}$-convergent to o.*

c) *If $E[\mathfrak{T}]$ is a metrizable locally convex space, every $\mathfrak{T}$-convergent sequence is locally convergent.*

Proof. a) If x_n converges to x_0 in E_B, $x_n - x_0$ converges to o in E_B. Conversely, if $x_n - x_0$ converges to o in E_B, and if B_1 is a bounded, closed, absolutely convex set containing B, the x_n and x_0, then x_n converges to x_0 in E_{B_1}.

b) If x_n is a local null-sequence, then $x_n \in \varepsilon_n B$, for some suitable B and some suitable sequence $\varepsilon_n \to 0$. But then $\frac{1}{\sqrt{\varepsilon_n}} x_n$ converges to o in E_B, and so it converges in $E[\mathfrak{T}]$. Conversely, if $\rho_n x_n$ is $\mathfrak{T}$-convergent to o, and if B is the closed absolutely convex cover of the $\rho_n x_n$, then x_n converges to o in E_B.

c) follows from a), b) and 1.(5).

The problem of giving a precise characterisation of those locally convex spaces in which every convergent sequence is locally convergent is still unanswered.

(2) *A locally convex space $E[\mathfrak{T}]$ is bornological if and only if every absolutely convex set M which absorbs all the local null-sequences is a $\mathfrak{T}$-neighbourhood of o.*

It is enough to show that such a set M absorbs all the bounded sets. Suppose that M absorbs all local null-sequences, but not the bounded set B. Then for each n^2 there is an $x_n \in B$ with $\frac{x_n}{n^2} \notin M$. But then x_n/n is a local null-sequence, while on the other hand no scalar multiple of the set of all x_n/n is contained in M, and this is impossible.

A linear mapping from E into F is said to be locally continuous if it maps every local null-sequence into a local null-sequence.

(3) *If A is a linear mapping from a locally convex space $E[\mathfrak{T}]$ into a locally convex space $F[\mathfrak{T}']$, the following properties are equivalent: a) A is locally bounded, b) A is locally continuous, c) A maps every local null-sequence into a sequence which is $\mathfrak{T}'$-convergent to $\mathfrak{o}$, and d) A maps every local null-sequence into a bounded sequence.*

Clearly a) implies b), b) implies c), and c) implies d). We must therefore show that d) implies that A is locally bounded. If A were not locally bounded, there would be a bounded set B in E and an absolutely convex neighbourhood V of $\mathfrak{o}$ in F such that $A(B)$ is contained in no $n^2 V$. Thus there exist points $x_n \in B$ with $\frac{A x_n}{n^2} \notin V$. But then $\frac{x_n}{n}$ is a local null-sequence and the $A\left(\frac{x_n}{n}\right)$ are unbounded, contradicting d).

Since continuity of A implies sequential continuity, and since d) follows from this, we have the following result for bornological spaces:

(4) *A linear mapping A from a bornological space into a locally convex space is continuous if and only if it is sequentially continuous, and if and only if it satisfies one of the conditions* a), b), c) *or* d) *of* (3).

4. Hereditary properties. The class of bornological spaces is stable under operations similar to those under which the class of barrelled spaces is stable (cf. § 27, 1.).

(1) *Every locally convex hull of bornological spaces is bornological.*

Suppose that $E[\mathfrak{T}] = \sum_\alpha A_\alpha(F_\alpha[\mathfrak{T}_\alpha])$, where each $F_\alpha[\mathfrak{T}_\alpha]$ is bornological. If the absolutely convex set U absorbs all the bounded subsets of $E[\mathfrak{T}]$, then $A_\alpha^{(-1)}(U)$ is also absolutely convex, and it absorbs all the bounded subsets of $F_\alpha[\mathfrak{T}_\alpha]$. Thus by hypothesesis $A_\alpha^{(-1)}(U)$ is a $\mathfrak{T}_\alpha$-neighbourhood U_α of $\mathfrak{o}$ in $F_\alpha[\mathfrak{T}_\alpha]$. But then $U \supset \Gamma_\alpha A_\alpha(U_\alpha)$, so that, by definition of the hull topology, U is a $\mathfrak{T}$-neighbourhood of $\mathfrak{o}$.

As a special case, as in § 27, 1. we have

(2) *Every quotient of a bornological space is bornological.*

Further,

(3) *The completion $\tilde{E}[\tilde{\mathfrak{T}}]$ of a bornological space $E[\mathfrak{T}]$ is bornological if and only if every locally bounded linear functional on $\tilde{E}[\tilde{\mathfrak{T}}]$ which vanishes on E is identically zero.*

By § 21, 4.(5), $\mathfrak{T}$ is again the Mackey topology, and $E'=(\tilde{E}[\tilde{\mathfrak{T}}])'$. The restriction of any locally bounded linear functional on $\tilde{E}[\tilde{\mathfrak{T}}]$ to E is certainly locally bounded on $E[\mathfrak{T}]$. By hypothesis it lies in E', so that the assertion follows from 1.(3).

The question of whether the topological product of bornological spaces is again bornological leads to some special difficulties. For the time being, we consider the case of countably many factors, and postpone the investigation of the general case until number 8 of this paragraph.

(4) *The topological product of at most countably many bornological spaces is again bornological.*

Suppose that the spaces $F_i[\mathfrak{T}_i]$, $i=1,2,\ldots$, are bornological. The topology $\mathfrak{T}$ of $E[\mathfrak{T}]=\prod_{i=1}^{\infty} F_i[\mathfrak{T}_i]$ is again the Mackey topology on E, by § 22, 5.(3). By 1.(3), we must also show that every locally bounded linear functional on $E[\mathfrak{T}]$ is continuous.

Suppose that u is locally bounded on $E[\mathfrak{T}]$. We assert that there exists an n_0 such that $uy=0$ for all $y=(y_i)$, $y_i\in F_i$, whose first n_0 components y_i all vanish. If this were not the case, there would be a sequence $x^{(k)}=(x_i^{(k)})\in E$ with $x_i^{(k)}=0$ for $i=1,\ldots,k$, and $ux^{(k)}=k$. But the sequence $x^{(k)}$ is bounded in $E[\mathfrak{T}]$; this contradicts the local boundedness of u.

It is therefore sufficient to establish the result for a product $E[\mathfrak{T}]=F_1[\mathfrak{T}]\times F_2[\mathfrak{T}_2]$. If u is locally bounded on $E[\mathfrak{T}]$, so are the linear functionals on E defined by $u_1x=u_1(x_1,x_2)=u(x_1,\mathrm{o})$ and $u_2x=u(\mathrm{o},x_2)$. But by hypothesis the restrictions of u_1 to F_1 and u_2 to F_2 are continuous. so that u_1 and u_2 are continuous on E, and so also is $u=u_1+u_2$.

Since every metrizable locally convex space is bornological (1.(4)), we obtain a very extensive class of bornological spaces by repeated application of (1) to (4). For example all (LF)-spaces belong to this class, and so do all spaces of countable degree.

Not every closed linear subspace H of a bornological space $E[\mathfrak{T}]$ is bornological. For example the subspace H_2 of the bornological space $\omega\varphi\oplus\varphi\omega$ considered in § 27.2 is not bornological. For, by § 27, 2., H_2 is complete and not barrelled, and so by 1.(2) it is not bornological.

Dieudonné [9] has shown that every linear subspace of a bornological space of finite co-dimension is again bornological.

The dual of a bornological space need not be bornological, either; indeed there exist (F)-spaces whose duals are not bornological, as is shown by an example in § 31, 7.

Amemiya [1] has given an example of a bornological space whose strong bidual is not bornological.

5. The dual, and the topology $\mathfrak{T}_{c_0}$. In § 21, 6.(4) we showed that the dual E' of every metrizable locally convex space $E[\mathfrak{T}]$ is $\mathfrak{T}_c$-complete.

By § 21, 10.(3) the topology $\mathfrak{T}_c$ on E' coincides with the topology of uniform convergence on the sequences $\mathfrak{T}$-convergent to o. But by 3.(1)c) these are the same as the local null-sequences in E.

We denote the topology of uniform convergence on all the local null-sequences of a locally convex space $E[\mathfrak{T}]$ by $\mathfrak{T}_{co}(E)$; this topology depends only upon the dual pair $\langle E', E\rangle$, and does not depend on $\mathfrak{T}$. Thus the dual E' of a metrizable locally convex space E is always $\mathfrak{T}_{co}(E)$-complete. More generally, we now have

(1) *The dual E' of a bornological space $E[\mathfrak{T}]$ is $\mathfrak{T}_{co}$-complete.*

In particular E' is always $\mathfrak{T}_c(E)$ and $\mathfrak{T}_b(E)$-complete, and, in the case where E is sequentially complete, E' is also $\mathfrak{T}_k(E)$-complete.

Proof. a) Suppose that $\mathfrak{F}$ is a $\mathfrak{T}_{co}$-Cauchy filter on E'. It is also a $\mathfrak{T}_s(E)$-Cauchy filter, and so it has a limit $u_0 \in E^* = \tilde{E}'[\tilde{\mathfrak{T}}_s]$, and by § 18, 4.(4) this is also the $\mathfrak{T}_{co}$-limit of $\mathfrak{F}$. By 3.(4) it is enough to show that u_0 remains bounded on every local null-sequence x_n.

But there exists a $u \in E'$ with $\sup_n |(u-u_0)x_n| \leq \varepsilon$; since u is bounded on the sequence x_n, it follows that u_0 is also bounded on the sequence x_n. Thus E' is $\mathfrak{T}_{co}$-complete; *a fortiori* it is $\mathfrak{T}_c(E)$- and $\mathfrak{T}_b(E)$-complete, by § 18, 4.(4).

b) Suppose that $E[\mathfrak{T}]$ is sequentially complete. We must show that $\mathfrak{T}_k(E)$ is finer than $\mathfrak{T}_{co}(E)$, i.e. that the closed absolutely convex covers $\overline{\Gamma(C)}$ in $E[\mathfrak{T}]$ of local null-sequences $C = \{x_n\}$ are weakly compact. But this follows from § 20, 9.(6).

Before proving the converse of (1), we consider some preliminary ideas.

If K is a compact subset of $E[\mathfrak{T}']$, every weaker Hausdorff topology $\mathfrak{T}$ coincides with $\mathfrak{T}'$ on K. If K is only precompact with respect to $\mathfrak{T}'$, and if $\mathfrak{T}$ is a weaker Hausdorff on K, then K is still precompact with respect to $\mathfrak{T}$, but $\mathfrak{T}$ need no longer coincide with $\mathfrak{T}'$ on K.

We give an example. Suppose that $E[\mathfrak{T}]'$ and $E[\mathfrak{T}]$ are two normed spaces, and that $\mathfrak{T}$ is weaker than $\mathfrak{T}'$. Suppose that the mapping $\tilde{I}$ from $\tilde{E}[\tilde{\mathfrak{T}}']$ into $\tilde{E}[\tilde{\mathfrak{T}}]$ is not one-one (cf. the example of § 18, 4.(1)). Suppose that $z \in \tilde{I}^{(-1)}(o)$, $z \neq o$, and that x_n is a sequence of elements in E with $\mathfrak{T}'$-limit z. This sequence has $\mathfrak{T}$-limit o; the set $\{o, x_1, x_2, \ldots\}$ is therefore compact in $E[\mathfrak{T}]$, but not in $E[\mathfrak{T}']$.

However, we have

(2) *Suppose that a second, finer, topology $\mathfrak{T}'$ is given on the topological vector space $E[\mathfrak{T}]$. If $\mathfrak{T}'$ has a base of $\mathfrak{T}$-closed neighbourhoods of o, then $\mathfrak{T}$ and $\mathfrak{T}'$ coincide on every $\mathfrak{T}'$-precompact subset M of E.*

Under the given hypotheses, $\tilde{E}[\tilde{\mathfrak{T}}']$ can be considered as a linear subspace of $\tilde{E}[\tilde{\mathfrak{T}}]$, by § 18, 4.(4), and so the $\tilde{\mathfrak{T}}'$-compact closure $\bar{M}$ of M in $\tilde{E}[\tilde{\mathfrak{T}}']$ is a subset of $\tilde{E}[\tilde{\mathfrak{T}}]$. Consequently $\tilde{\mathfrak{T}}$ induces the same topology on $\bar{M}$ as does $\tilde{\mathfrak{T}}'$, and so $\mathfrak{T}$ and $\mathfrak{T}'$ coincide on M.

(3) *Suppose that two locally convex topologies $\mathfrak{T}_1$ and $\mathfrak{T}_2$ are given on the vector space E. If $\mathfrak{T}_1$ and $\mathfrak{T}_2$ coincide on an absolutely convex subset M of E, then the uniformities induced on M by $\mathfrak{T}_1$ and $\mathfrak{T}_2$ are the same.*

Proof. A base for the uniformity induced by $\mathfrak{T}_1$ on M is given by the sets N_U of all $(x,y), x, y \in M, x-y \in U$, where U is an absolutely convex $\mathfrak{T}_1$-neighbourhood of $\mathfrak{o}$. Let W be an absolutely convex $\mathfrak{T}_2$-neighbourhood of $\mathfrak{o}$ for which $M \cap \frac{W}{2} \subset M \cap \frac{U}{2}$, and let N_W be the vicinity corresponding to W. If $x', y' \in M, x'-y' \in W$, then $x'-y' \in 2M$, $\frac{x'-y'}{2} \in M \cap \frac{W}{2} \subset M \cap \frac{U}{2}$, so that $x'-y' \in U$, and consequently $N_W \subset N_U$. The uniformity induced on M by $\mathfrak{T}_2$ is therefore finer than the one induced by $\mathfrak{T}_1$.

We now establish the following dual characterisation of bornological spaces:

(4) *A locally convex space $E[\mathfrak{T}]$ is bornological if and only if $\mathfrak{T}$ is the Mackey topology and E' is $\mathfrak{T}_{c_0}(E)$-complete.*

One half has been proved in (1). Suppose that E' is $\mathfrak{T}_{c_0}$-complete. By 1.(3) it is sufficient to show that every locally bounded linear functional u_0 on $E[\mathfrak{T}]$ belongs to E'. According to GROTHENDIECK's theorem (§ 21, 9.(2)) this is the case if the restrictions of u_0 to the sets $\overline{\Gamma(C)}$, where $C=\{x_n\}$ is a local null-sequence in $E[\mathfrak{T}]$, are weakly continuous.

Every local null-sequence $C=\{x_n\}$ is a precompact subset of a suitable E_B, where B is a bounded, closed, absolutely convex subset of $E[\mathfrak{T}]$. E_B is a normed space; let the norm topology obtained by taking B as unit ball be $\mathfrak{T}_B$. By § 20, 6.(2), $\Gamma(C)$ is also $\mathfrak{T}_B$-precompact in E_B. If we apply (2) to E_B and to the two topologies $\mathfrak{T}_B$ and $\mathfrak{T}_s(E')$, it follows that the topologies $\mathfrak{T}_B$ and $\mathfrak{T}_s(E')$ coincide on $\Gamma(C)$. From this it follows, using (3), that the $\mathfrak{T}_B$-closure of $\Gamma(C)$ in E_B is the same as the $\mathfrak{T}_s(E')$-closure in E_B. Since $\Gamma(C) \subset kB$, and since B is weakly closed in $E[\mathfrak{T}]$, it follows that this is the same as the closure $\overline{\Gamma(C)}$ in E. Thus $\overline{\Gamma(C)}$ is also precompact in E_B, and $\mathfrak{T}_B$ and $\mathfrak{T}_s(E')$ coincide on $\overline{\Gamma(C)}$. The restriction of u_0 to E_B is bounded, and is therefore $\mathfrak{T}_B$-continuous, and so the restriction of u_0 to $\overline{\Gamma(C)} \subset E_B$ is $\mathfrak{T}_s(E')$-continuous.

6. Boundedly closed spaces. If $E[\mathfrak{T}]$ is a locally convex space, the linear subspace of $(E')^*$ consisting of all the linear functionals on E' which are bounded on the weakly bounded subsets of E' is called the bounded closure $\check{E}$ of E. $E[\mathfrak{T}]$ is boundedly closed if $E=\check{E}$.

The bounded closure of $E[\mathfrak{T}]$ clearly depends only upon the dual pair $\langle E', E\rangle$, and $\mathfrak{T}$ can be replaced by any other topology compatible with the dual pair $\langle E', E\rangle$.

MACKEY [4] called a dual pair $\langle E', E\rangle$ for which $E=\check{E}$ a "boundedly closed linear system".

As a direct consequence of the definition of the associated bornological topology $\mathfrak{T}^{\times}$ (cf. 2.), we have

(1) *The bounded closure* $\check{E}$ *of a locally convex space* $E[\mathfrak{T}]$ *is the dual of the bornological space* $E'[\mathfrak{T}_k^{\times}(E)]$.

$\mathfrak{T}_k(E)$ can be replaced by the weak topology, or any other topology which defines the same bounded sets.

There is a simple connection between bornological spaces and boundedly closed spaces; this is made explicit in the next two results.

(2) $E[\mathfrak{T}]$ *is boundedly closed if and only if* $E'[\mathfrak{T}_k(E)]$ *is bornological.*

This follows from the fact that $(E'[\mathfrak{T}_k])'=E$, and from the definition of "boundedly closed" and 1.(3).

Interchanging E and E' in (2), we get

(3) $E[\mathfrak{T}]$ *is bornological if and only if* $\mathfrak{T}$ *is the Mackey topology and* $E'[\mathfrak{T}_k(E)]$ *is boundedly closed.*

Here again, $\mathfrak{T}_k(E)$ can be replaced by any other compatible topology.

We denote the linear span $\bigcup_{n=1}^{\infty} nC$ of an absolutely convex subset C of $(E')^*$ by $L(C)$. We now have (cf. DIEUDONNÉ [9])

(4) *The bounded closure* $\check{E}$ *of a locally convex space* $E[\mathfrak{T}]$ *is equal to* $\bigcap_T L(\bar{T})$ *where* T *runs through all the barrels of* $E[\mathfrak{T}]$, *and* $\bar{T}$ *stands for the weak closure of* T *in* $(E')^*$.

Every locally bounded linear functional z on $E'[\mathfrak{T}_s(E)]$ lies in $(E')^*$. By § 21, 2.(1), the barrels T are the closed absolutely convex $\mathfrak{T}_b(E')$-neighbourhoods of $\mathfrak{o}$ in E. Every weakly bounded subset of E' therefore lies in some set T°. Since z is bounded on T°, $z\in nT^{\circ\circ}$ for some suitable n, where $T^{\circ\circ}$ is the polar of T° in $(E')^*$. By the theorem of bipolars, $T^{\circ\circ}=\bar{T}$. It follows from this that $z\in\bigcap_T L(\bar{T})$. Conversely if $z\in\bigcap_T L(\bar{T})$, z is bounded on all the sets $\bar{T}^{\circ}=T^{\circ}$.

(5) *If* $E[\mathfrak{T}]$ *is sequentially complete,* $\check{E}\supset E''$. *A boundedly closed sequentially complete locally convex space is always semi-reflexive.*

For in these circumstances the weakly bounded sets of E' are strongly bounded (MACKEY's theorem) and every $u\in E''$ is bounded on such a set. It follows that if $\check{E}=E$, then $E=E''$.

$\check{E}$ can also be characterised as follows:

(6) *The bounded closure* $\check{E}$ *of a locally convex space* $E[\mathfrak{T}]$ *is equal to the* $\mathfrak{T}_{co}(E'[\mathfrak{T}_k])$*-completion of* E.

It follows from (1) and 5.(1) that $\check{E}$ is $\mathfrak{T}_{co}(E'[\mathfrak{T}_k^{\times}])$-complete. Since $\mathfrak{T}_k$ and $\mathfrak{T}_k^{\times}$ have the same bounded sets, the local null-sequences in E' are the same for either topology; E is therefore $\mathfrak{T}_{co}(E'[\mathfrak{T}_k])$-complete.

We still have to show that E is $\mathfrak{T}_{co}$-dense in $\check{E}$. By GROTHENDIECK's theorem it is enough to show that every linear functional which is bounded on the weakly bounded subsets of E' is weakly continuous on the sets $\overline{\Gamma(C)}$; here C stands for a local null-sequence $\{u_n\}$ in $E'[\mathfrak{T}_k]$. But this can be proved in exactly the same way as the proof of 5.(4); instead of $E[\mathfrak{T}]$ and E', one considers $E'[\mathfrak{T}_k]$ and its dual E.

(7) *A locally convex space* $E[\mathfrak{T}]$ *is boundedly closed if and only if it is* $\mathfrak{T}_{co}(E'[\mathfrak{T}_k])$*-complete.*

Necessity follows from (6). Conversely if E is $\mathfrak{T}_{co}(E')$-complete, $E'[\mathfrak{T}_k(E)]$ is bornological, by 5.(4), and so $E[\mathfrak{T}]$ is boundedly closed by (2).

7. Reflexivity and completeness. We now obtain some further conclusions from the results proved in 5. and 6.

(1) *If the strong dual* $E'[\mathfrak{T}_b]$ *of a locally convex space* $E[\mathfrak{T}]$ *is bornological, the bidual* E'' *is* $\mathfrak{T}_{co}(E'[\mathfrak{T}_b])$*-complete, and so it is also strongly complete.*

It follows from 5.(4) that E'' is $\mathfrak{T}_{co}(E'[\mathfrak{T}_b])$-complete. The subsets of E' which are strongly bounded with respect to E are the equicontinuous subsets of E' for the strong topology $\mathfrak{T}_b(E',E'')$ on E''. But the sets $\overline{\Gamma(C)}$, where $C=\{u_n\}$ is a local null-sequence in $E'[\mathfrak{T}_b(E)]$, belong to this collection, and these sets form a fundamental collection of $\mathfrak{T}_{co}$-equicontinuous subsets of E'. Hence $\mathfrak{T}_b(E',E'')$ is finer than $\mathfrak{T}_{co}$; from this it follows that E'' is strongly complete.

(2) *Suppose that* $E[\mathfrak{T}]$ *is sequentially complete. If the strong dual* $E'[\mathfrak{T}_b]$ *is bornological, then* $E''=\check{E}=\tilde{E}[\tilde{\mathfrak{T}}_{co}(E')]$.

Suppose that $E'[\mathfrak{T}_b]$ is bornological. By § 20, 11.(8), the sequential completeness of $E[\mathfrak{T}]$ ensures that $\mathfrak{T}_b$, $\mathfrak{T}_b^\times$ and $\mathfrak{T}_k^\times$ coincide on E'. It follows from 6.(1) that $\check{E}=E''$, and from 6.(6) that $\check{E}=\tilde{E}[\tilde{\mathfrak{T}}_{co}(E')]$.

In this case the terminology $\mathfrak{T}_{co}(E')$ is unambiguous, since the weakly and strongly bounded subsets of E' with respect to E are the same, and so the topologies $\mathfrak{T}_{co}(E'[\mathfrak{T}_k])$ and $\mathfrak{T}_{co}(E'[\mathfrak{T}_b])$ are the same.

If $E'[\mathfrak{T}_b]$ is not bornological, and if $\mathfrak{T}_b(E)=\mathfrak{T}_k(E'')$, then by 1.(3) there is at least one locally bounded linear functional on $E'[\mathfrak{T}_b]$ which does not belong to E''; by 6.(1), E'' is a proper subspace of $\check{E}$.

If the topology $\mathfrak{T}_b(E)$ on E' is strictly coarser than $\mathfrak{T}_k(E'')$, $E'[\mathfrak{T}_k(E'')]$ can be bornological. In this case we have $E''=\check{E}=\tilde{E}[\tilde{\mathfrak{T}}_{co}(E')]$. KŌMURA [2] has shown that, even if E is an (F)-space, we need not always have $\mathfrak{T}_b(E)=\mathfrak{T}_k(E'')$.

(3) *If the strong dual of a quasi-barrelled space* $E[\mathfrak{T}]$ *is bornological, then* $\tilde{E}[\tilde{\mathfrak{T}}]\subset E''$.

If $E[\mathfrak{T}]$ is quasi-barrelled, $\mathfrak{T}=\mathfrak{T}_{b*}(E')$. This topology has the same equicontinuous sets in E' as $\mathfrak{T}_b(E',E'')$. The assertion therefore follows from (1).

A special case of (2) and (3) is given by

(4) *If* $E[\mathfrak{T}]$ *is reflexive and* $E'[\mathfrak{T}_b]$ *bornological, then* $E[\mathfrak{T}]$ *is complete, boundedly closed, and* $\mathfrak{T}_{co}(E')$*-complete.*

We now consider (B)- and (F)-spaces briefly.

(5) *The bidual of a* (B)*-space E is the bounded closure of E, and is its* $\mathfrak{T}_c(E')$*-completion.*

$E''=\check{E}=\tilde{E}[\tilde{\mathfrak{T}}_{co}(E')]$ follows directly from (2). But the topology $\mathfrak{T}_{co}(E')$ on the dual of the (B)-space E' is the same as the topology $\mathfrak{T}_c(E')$ of uniform convergence on the compact subsets of E'.

The assertion $E''=\check{E}$ can also be derived simply from the definition of the bounded closure.

From (4) and (5) we obtain the following criterion for reflexivity:

(6) *A* (B)*-space E is reflexive if and only if it is* $\mathfrak{T}_c(E')$*-complete.*

A corresponding result also holds for (F)-spaces, but in order to prove it we need the result that the strong dual of a reflexive (F)-space is bornological; this we obtain in § 29.4.

In the example of the (FM)-space $H(\mathfrak{D})$ considered in § 27, 4., we showed directly that $(H(\mathfrak{D}))'=H(\mathfrak{A})$ is bornological, since we showed (§ 27, 4.(7)) that the strong topology on $H(\mathfrak{A})$ is the topology of the locally convex hull of a sequence of (B)-spaces.

From (2) and 6.(6) there follows directly

(7) *The bidual* E'' *of an* (F)*-space* $E[\mathfrak{T}]$ *is always contained in* $\check{E}=\tilde{E}[\tilde{\mathfrak{T}}_{co}(E')]$.

If, further, the strong dual of $E[\mathfrak{T}]$ *is bornological, then* $E''=\check{E}=\tilde{E}[\tilde{\mathfrak{T}}_{co}(E')]$.

The converse of this last statement is false (see the remark before (3)).

8. The MACKEY-ULAM theorem. We now return to the question of when a topological product of bornological spaces is bornological.

First we prove two lemmas.

(1) *If the topological product* $E[\mathfrak{T}]=\prod_\alpha E_\alpha[\mathfrak{T}_\alpha]$ *of bornological spaces* $E_\alpha[\mathfrak{T}_\alpha]$ *is not bornological, there exists a discontinuous locally bounded linear functional u on E which vanishes on the linear subspace* $\bigoplus_\alpha E_\alpha$.

Since $\mathfrak{T}$ is the Mackey topology, by § 22, 5.(3), there exists a discontinuous locally bounded linear functional u on E, by 1.(3). There can only be finitely many $E_\alpha\subset E$ on which this does not vanish, for otherwise, as in the proof of 4.(4), there would be a sequence $x_i\in E_{\alpha_i}$ which is bounded in E, but on which u is unbounded.

The restriction of u to the product $\prod_{k=1}^{n} E_{\alpha_k}$ of the finitely many E_{α_k} on which u does not vanish is, by hypothesis, a continuous linear functional on $\prod E_{\alpha_k}$. If we extend this linear functional to a linear functional v on the whole of E which vanishes on the product of the other spaces E_α, then v is continuous on E, and $u-v$ has the desired properties.

(2) *Suppose that the spaces $E_\alpha[\mathfrak{T}_\alpha]$ are bornological. If $\prod_\alpha E_\alpha[\mathfrak{T}_\alpha]$ is bornological, so is every subproduct $\prod_\beta E_\beta[\mathfrak{T}_\beta]$, where the β run through a subset B of the set A of indices α.*

For every discontinuous locally bounded linear functional on $\prod E_\beta$ can be extended to a linear functional with the same properties on $\prod E_\alpha$, by setting it equal to zero on the product of the other spaces E_α.

Next we show that the question of whether $\prod E_\alpha$ is bornological or not depends only on the cardinal d of the collection of spaces E_α.

(3) *A product $E[\mathfrak{T}]=\prod_\alpha E_\alpha[\mathfrak{T}_\alpha]$ of d bornological spaces $E_\alpha[\mathfrak{T}_\alpha]$ is bornological if and only if ω_d is bornological.*

a) Suppose that ω_d is bornological. Let us suppose that E is not bornological. By (1) there is a linear functional u with the properties described there. Let x be an arbitrary element of E with components $x_\alpha \in E_\alpha$. The topological product $\prod [x_\beta]$ of the one-dimensional subspaces $[x_\beta]$ of E_β defined by the non-zero x_β is a subspace of E which, under the induced topology $\mathfrak{T}$, is topologically isomorphic to a space $\omega_{d'}$, $d' \leqq d$. The restriction u_0 of u to $\prod [x_\beta]$ is locally bounded and vanishes on $\bigoplus_\beta [x_\beta]$. But $\omega_{d'}$ is bornological, by (2), and so u_0 is continuous. But, by § 22, 5.(2), a continuous linear functional on $\omega_{d'}$ is identically zero if it vanishes on all the elements of $\omega_{d'}$ which have at most finitely many non-zero components. Thus $ux=u_0x=0$ for all $x\in E$, which gives a contradiction.

b) Suppose that $E[\mathfrak{T}]=\prod_\alpha E_\alpha[\mathfrak{T}_\alpha]$ is a product of d bornological spaces $E_\alpha[\mathfrak{T}_\alpha] \neq (\mathrm{o})$, and that $E[\mathfrak{T}]$ is bornological. As we have just seen, ω_d is topologically isomorphic to a space $\prod_\alpha [x_\alpha] \subset E$, where $x_\alpha \neq \mathrm{o}$ for each α. A locally bounded linear functional u_0 on $\prod_\alpha [x_\alpha]$ can be extended to a locally bounded linear functional u on E: for each E_α can be written as a topological product $[x_\alpha]\times F_\alpha$, so that $\prod [x_\alpha]$ has a topological complement in $\prod_\alpha E_\alpha$, and it is sufficient to set u equal to zero on this complement. Since E is bornological, u is continuous on E, and so u_0 is continuous on $\prod [x_\alpha] \cong \omega_d$.

We denote the topological product K^{A} by $\omega(\mathsf{A})$ (cf. § 1, 8.), where K is either the real or the complex field, and A is an index set. $\omega(\mathsf{A})$ is topologically isomorphic to ω_d, where d is the cardinal of A. If $\mathsf{B}\subset\mathsf{A}$,

$\omega(\mathsf{B})$ is identified with the subspace of $\omega(\mathsf{A})$ consisting of all vectors (ξ_α) with $\xi_\alpha=0$ for $\alpha\notin\mathsf{B}$. We say that a linear functional u defined on $\omega(\mathsf{A})$ vanishes on $\mathsf{B}\subset\mathsf{A}$ if its restriction to $\omega(\mathsf{B})$ vanishes.

We now prove a stronger form of (1) for $\omega(\mathsf{A})$.

(4) *If $\omega(\mathsf{A})$ is not bornological, there is a discontinuous locally bounded linear functional on $\omega(\mathsf{A})$ with the following properties:*

a) u vanishes on all the finite subsets of A, and b) if A is decomposed into two disjoint sets A_1 and A_2, then u vanishes on just one of the two sets.

By (1), there is a u which satisfies a). We assert that there is a decomposition of A into finitely many pairwise disjoint subsets A_i such that the restrictions of u to the corresponding spaces $\omega(\mathsf{A}_i)$ also have property b).

Let us suppose that this is not so. There is then a decomposition of A into two sets B_1 and B_2 on which u does not vanish. One of the two sets must be decomposable into two sets on which u does not vanish. Repeating this procedure, we obtain a sequence of pairwise disjoint subsets A_i of A on which u does not vanish. But using a sequence $x_i\in\omega(\mathsf{A}_i)$ for which $ux_i\neq 0$ we can again construct a bounded sequence in $\omega(\mathsf{A})$ on which u is not bounded.

One of the sets A_i has the same cardinal as A; $\omega(\mathsf{A}_i)$ is topologically isomorphic to $\omega(\mathsf{A})$, and the restriction of u to $\omega(\mathsf{A}_i)$ has properties a) and b).

(5) *If ω_d is bornological, so is ω_{2^d}.*

Let us suppose that $\omega(\mathsf{A})$ is not bornological, where A has cardinal 2^d, and let u be a locally bounded linear functional on $\omega(\mathsf{A})$ with properties a) and b) of (4).

The elements of A can be written as vectors $\alpha=(\alpha_\beta)$, where the indices β form a set B with cardinal d, and where α_β takes the values 0 or 1.

Each β determines a decomposition of A into two sets, one consisting of those α for which $\alpha_\beta=0$ and the other those α for which $\alpha_\beta=1$. By b), u vanishes on exactly one of these two sets, which we shall denote by A_β. Let A'_β denote its complement.

We form the set $\mathsf{M}=\bigcup_\beta \mathsf{A}_\beta$. We can also express M as the union of at most d pairwise disjoint subsets $\Gamma_\gamma\in\mathsf{A}_\gamma$, on each of which u vanishes. We assert that u also vanishes on M. Suppose that $x=(x_\gamma)\in\omega(\mathsf{M})$, $x_\gamma\in\omega(\Gamma_\gamma)$. The topological product $\prod[x_\delta]$ of the one-dimensional spaces corresponding to the non-zero x_δ is then bornological, by hypothesis. Since u vanishes on the vectors x_γ, u therefore vanishes on each x, and so it vanishes on M.

By b), u does not vanish on the complement $\bigcap_{\beta} \mathsf{A}'_{\beta}$ of M. But, by the construction of the A'_{β}, the set $\bigcap_{\beta} \mathsf{A}'_{\beta}$ consists of just one element, and so we have obtained a contradiction to a).

A cardinal d is said to be strongly inaccessible if a) $d>\aleph_0$, b) every sum $\sum_{\gamma} d_{\gamma}$ of $d'<d$ cardinals $d_{\gamma}<d$ is a cardinal less than d, and c) if $f<d$, $2^f<d$.

It is in fact not known if any strongly inaccessible cardinals exist. If so, there is a smallest one.

We now come to the MACKEY-ULAM theorem:

(6) *The topological product of d bornological spaces is bornological, if d is smaller than the smallest strongly inaccessible cardinal.*

By (3) it is sufficient to prove this for ω_d. If there is a non-bornological ω_d, there is a smallest cardinal d_0 for which ω_{d_0} is non-bornological. This cardinal has properties a) and c), by 4.(4), and by (5). Since $\omega_{d'}$ is bornological for $d'<d_0$, the topological product of d' spaces $\omega_{d_{\gamma}}$, with $d_{\gamma}<d_0$, is also bornological, by (3), and so d_0 satisfies b) as well.

MACKEY [3] showed that the question of when $\prod_{\alpha} E_{\alpha}$ is bornological is equivalent to a problem in measure theory which ULAM [1] had considered. Theorem (5) is due to ULAM.

It is not known if (6) is true for all cardinals d.

§ 29. (F)- and (DF)-spaces

1. Fundamental sequences of bounded sets. Metrizability. An (F)-space is barrelled and bornological. As a result, theorems about such spaces hold for (F)-spaces. Besides this, we have considered (F)-spaces in detail, within the framework of the general theory. Up to now, however, our knowledge of the properties of the strong dual of an (F)-space has been very slight. In this paragraph we shall develop some of the recent results in the theory of (F)-spaces and their duals, which originated with GROTHENDIECK [2], [10] and with DONOGHUE and K. T. SMITH [1].

We say that a locally convex space $E[\mathfrak{T}]$ possesses a fundamental sequence of bounded sets if there exists a sequence $B_1 \subset B_2 \subset \cdots$ of bounded sets in $E[\mathfrak{T}]$ such that every bounded set B is contained in some B_k.

In a (B)-space such a fundamental sequence always exists. We now investigate whether this is also possible in (F)-spaces which are not (B)-spaces.

(1) *Suppose that the locally convex space* $E[\mathfrak{T}]$ *has a fundamental sequence* $B_1 \subset B_2 \subset \cdots$ *of bounded sets. If none of the sets* B_k *absorbs all the other sets* B_i, *there is a countable subset* M *whose closure is strictly larger than the set obtained by taking all the limits of the Cauchy sequences in* M.

Proof. We can suppose that the sets B_k are absolutely convex, and that no B_k absorbs the next set B_{k+1}. Let x_n be a sequence of non-zero terms in B_1 which converges to $\mathfrak{o}$. For each (n,k) let us choose a $z_{n,k} \in \frac{1}{k} B_n$ with $z_{n,k} \notin (k+1) B_{n-1}$. Let M be the set of all $x_n + z_{n,k}$, $n,k = 1,2,\ldots$. For fixed n, x_n is obtained as a limit as $k \to \infty$. x_n does not belong to M, so that $\mathfrak{o}$ is a limit of limits of Cauchy sequences in M. But $\mathfrak{o}$ is not itself limit of a sequence in M. For on the one hand such a sequence would have to contain elements $x_n + z_{n,k}$ with unbounded n; on the other hand the terms would all have to belong to a fixed set B_m, and this would mean that $z_{n,k} \in B_1 + B_m \subset 2 B_m$, so that $n \leqq m$.

(2) *If* $E[\mathfrak{T}]$ *is metrizable and locally convex, and if* $E[\mathfrak{T}]$ *has a fundamental sequence of bounded sets, then* $E[\mathfrak{T}]$ *is normable.*

For, by (1), $E[\mathfrak{T}]$ is only metrizable if there is a bounded set B which absorbs all the other bounded sets. But then $E'[\mathfrak{T}_b(E)]$ is normable with unit ball B°, so that $E[\mathfrak{T}_{b*}(E')]$ is normable, with unit ball $B^{\circ\circ}$. But $\mathfrak{T} = \mathfrak{T}_{b*}(E')$, by § 21, 5.(3).

As a special case, we have

(3) *An* (F)-*space is a* (B)-*space if and only if it has a fundamental sequence of bounded sets.*

A further consequence of (1) is

(4) *A locally convex space* $E[\mathfrak{T}]$ *with a fundamental sequence of bounded sets is metrizable if and only if its strong dual is normable.*

The next result deals with the bounded subsets of a metrizable locally convex space:

(5) *If* $E[\mathfrak{T}]$ *is locally convex and metrizable, and if* B_n *is a sequence of bounded subsets of* E, *then there always exist positive scalars* ρ_n *such that* $\bigcup_{n=1}^{\infty} \rho_n B_n$ *is also bounded.*

If $U_1 \supset U_2 \supset \cdots$ is a base of absolutely convex neighbourhoods of $\mathfrak{o}$ in $E[\mathfrak{T}]$, and if $\rho_n B_n \subset U_n$, then $\bigcup_{n=m}^{\infty} \rho_n B_n \subset U_m$ for each m, so that $\bigcup_{n=1}^{\infty} \rho_n B_n$ is bounded.

As far as the strongly bounded subsets of the dual of a metrizable locally convex space are concerned, we have

(6) *Suppose that* $E[\mathfrak{T}]$ *is locally convex and metrizable. If* $U_1 \supset U_2 \supset \cdots$ *is a base of neighbourhoods of* $\circ$ *in* $E[\mathfrak{T}]$, *the sets* U_n° *form a fundamental sequence of strongly bounded subsets of* E'.

For by § 21, 5.(3), every strongly bounded subset M of E' is contained in an equicontinuous set U°, where U is a neighbourhood of $\circ$, and, since $U \supset U_k$ for some suitable k, it follows that $M \subset U^\circ \subset U_k^\circ$.

(7) *If* $E[\mathfrak{T}]$ *is locally convex and metrizable,* $E'[\mathfrak{T}_b]$ *is metrizable if and only if* $E[\mathfrak{T}]$ *is normable.*

For if $E'[\mathfrak{T}_b]$ is metrizable, the bidual E'' is normable in the strong topology $\mathfrak{T}_b(E',E'')$, by (4), and so therefore is $E[\mathfrak{T}]$, since $\mathfrak{T}=\mathfrak{T}_{b^*}(E')$ (§ 21, 5.(3)), and since the topology $\mathfrak{T}_b(E',E'')$ induces the topology $\mathfrak{T}_{b^*}(E')$ on E, by § 23, 4.(4).

(8) *Suppose that* $E[\mathfrak{T}]$ *is locally convex and metrizable, but not normable. Then there is a fundamental sequence* $B_1 \subset B_2 \subset \cdots$ *of strongly bounded absolutely convex weakly closed subsets of* E', *none of which absorbs the next, and such that each* E'_{B_n} *is a proper subspace of* $E'_{B_{n+1}}$.

Here E'_{B_n} denotes the linear span of B_n, and $E' = \bigcup_{n=1}^{\infty} E'_{B_n}$.

Proof. It follows from (6) and § 21, 5.(3) that there exists a sequence (B_n) such that no B_n absorbs the next set B_{n+1}. By § 21, 6.(4) and § 20, 11.(2), E'_{B_n} is a (B)-space when B_n is taken as unit ball. If E'_{B_n} were equal to $E'_{B_{n+1}}$, E'_{B_n} would also be a (B)-space under the strictly coarser norm topology of $E'_{B_{n+1}}$, and this contradicts § 15, 12.(7).

(9) *An* (F)-*space* E *is a* (B)-*space if and only if* E *contains a bounded absorbent set.*

The condition is clearly necessary. Conversely if E contains a bounded absorbent set, E also contains a bounded barrel B. Since E is barrelled, B is a bounded neighbourhood of $\circ$ in E, and E is therefore a (B)-space, by § 15, 10.(4).

2. The bidual. Suppose that $U_1 \supset U_2 \supset \cdots$ is a base of absolutely convex neighbourhoods of $\circ$ for the metrizable locally convex space $E[\mathfrak{T}]$. Then the sequence $U_1^{\circ\circ} \supset U_2^{\circ\circ} \supset \cdots$ of polars in E'' of the sets U_n° forms a base of neighbourhoods of $\circ$ for the natural topology $\mathfrak{T}_n(E')$ on the bidual E'' (cf. § 23,4.). Since $E[\mathfrak{T}]$ is metrizable, $\mathfrak{T}_n(E')$ is the same as the strong topology $\mathfrak{T}_b(E',E'')$ *on* E''.

(1) *Suppose that* $E[\mathfrak{T}]$ *is locally convex and metrizable, with strong dual* $E'[\mathfrak{T}_b(E)]$ *and strong bidual* $E''[\mathfrak{T}_n(E')]$. *If the union* $M = \bigcup_{n=1}^{\infty} M_n$ *of countably many* $\mathfrak{T}_b(E')$-*equicontinuous subsets* M_n *of* E'' *is* $\mathfrak{T}_s(E')$-*bounded in* E'', *then* M *is also* $\mathfrak{T}_b(E')$-*equicontinuous.*

Proof. We can suppose that $M_n \subset M_{n+1}$, and that the sets M_n and M are absolutely convex. The equicontinuity of M_n means that $M_n \subset B_n^{\circ\circ}$, where B_n is an absolutely convex bounded subset of $E[\mathfrak{T}]$, and $B_n^{\circ\circ}$ is the polar of $B_n^\circ \subset E'$ in E''. The set M is weakly bounded, by hypothesis; since $E'[\mathfrak{T}_b]$ is complete, it is bounded in $E''[\mathfrak{T}_n]$, by the BANACH-MACKEY theorem. Thus for each $\mathfrak{T}_n$-neighbourhood $U_k^{\circ\circ}$ there is a $c_k > 0$ with $M \subset c_k U_k^{\circ\circ}$. Further, for each k and n, there exists an $a_{nk} > \circ$ with $B_n \subset a_{nk} U_k$. Let $b_k = \max\limits_{n \leq k}(c_k, a_{nk})$. Then $\bigcap\limits_{k=n+1}^{\infty} a_{nk} U_k \subset \bigcap\limits_{n+1}^{\infty} b_k U_k$, so that $\left(\bigcap\limits_{n+1}^{\infty} b_k U_k\right)^\circ \subset \left(\bigcap\limits_{n+1}^{\infty} a_{nk} U_k\right)^\circ \subset B_n^\circ$. It follows from this that $M_n^\circ \supset \left(\bigcap\limits_{n+1}^{\infty} b_k U_k\right)^\circ$.

On the other hand it follows from $M_n \subset M \subset b_k U_k^{\circ\circ}$ that $M_n^\circ \supset \dfrac{1}{b_k} U_k^\circ$, so that $M_n^\circ \supset \Gamma_{k=1}^{n}\left(\dfrac{1}{b_k} U_k^\circ\right) = \left(\bigcap\limits_{k=1}^{n} b_k U_k\right)^\circ$; this last equation follows from § 20, 6.(5) and § 20, 8.(10). Consequently $2M_n^\circ \supset \left(\bigcap\limits_{k=1}^{n} b_k U_k\right)^\circ + \left(\bigcap\limits_{n+1}^{\infty} b_k U_k\right)^\circ$. But now $\left(\bigcap\limits_{k=1}^{n} b_k U_k\right)^\circ$ is the polar of a neighbourhood of $\circ$, and so it is weakly compact in E'. Thus $\left(\bigcap\limits_{k=1}^{n} b_k U_k\right)^\circ + \left(\bigcap\limits_{n+1}^{\infty} b_k U_k\right)^\circ$ is weakly closed, by § 15, 6.(10). But it follows from this that $\left(\bigcap\limits_{k=1}^{\infty} b_k U_k\right)^\circ \subset \left(\bigcap\limits_{k=1}^{n} b_k U_k\right)^\circ + \left(\bigcap\limits_{n+1}^{\infty} b_k U_k\right)^\circ$, and consequently $2M_n^\circ \supset \left(\bigcap\limits_{k=1}^{\infty} b_k U_k\right)^\circ$.

Thus for each n we have $\frac{1}{2} M_n \subset \left(\bigcap\limits_{k=1}^{\infty} b_k U_k\right)^{\circ\circ}$, and so $\frac{1}{2} M \subset \left(\bigcap\limits_{k=1}^{\infty} b_n U_k\right)^{\circ\circ}$. But since $\bigcap\limits_{k=1}^{\infty} b_k U_k$ is a bounded subset of E, our assertion is proved.

(1) can also be expressed as a property of $E'[\mathfrak{T}_b]$:

(2) *Suppose that $E[\mathfrak{T}]$ is locally convex and metrizable, with strong dual $E'[\mathfrak{T}_b]$. If V_n is a sequence of absolutely convex strongly closed neighbourhoods of $\circ$ in $E'[\mathfrak{T}_b]$ whose intersection V is absorbent, then V is also a neighbourhood of $\circ$.*

Proof. If we put $M_n = V_n^\circ$, M_n is a $\mathfrak{T}_b$-equicontinuous subset of E''. $V^\circ = \left(\bigcap\limits_{n=1}^{\infty} V_n\right)^\circ \supset \bigcap\limits_{n=1}^{\infty} V_n^\circ = M$. Since V is absorbent, V° is weakly bounded in E'', so that M is weakly bounded. If we now apply (1), we obtain (2). Similarly it can be shown that (1) follows from (2).

It follows from (1) that every countable bounded subset of $E''[\mathfrak{T}_n]$ is $\mathfrak{T}_b(E)$-equicontinuous, and so is relatively $\mathfrak{T}_s(E')$-compact. In particular every weak Cauchy sequence and every $\mathfrak{T}_n$-Cauchy sequence is a set of this kind; such sequences therefore always have a limit in E''. Since $E''[\mathfrak{T}_n]$ is metrizable, we therefore have

(3) *If* $E[\mathfrak{T}]$ *is locally convex and metrizable, its bidual* $E''[\mathfrak{T}_n(E')]$ *is a weakly sequentially complete* (F)*-space.*

In particular the strong bidual of an (F)*-space is again an* (F)*-space.*

Further, we have

(4) *If* $E[\mathfrak{T}]$ *is a non-reflexive* (F)*-space, the strong bidual* $E''[\mathfrak{T}_n(E')]$ *is again a non-reflexive* (F)*-space.*

Proof. Let us suppose that E'' is reflexive. By § 23, 2.(5), the topology $\mathfrak{T}_k(E'')$ on E' is finer than $\mathfrak{T}_b(E)$. But $E'[\mathfrak{T}_b(E)]$ is complete, so that $E'[\mathfrak{T}_k(E'')]$ is also complete. If we apply § 23, 5.(6) to $E'[\mathfrak{T}_k(E'')]$, it follows that $E'[\mathfrak{T}_k(E'')]$ is reflexive. Consequently $E'[\mathfrak{T}_b(E)]$ is semi-reflexive. Another application of § 23, 5.(6) shows that $E[\mathfrak{T}]$ is reflexive, which gives a contradiction.

From (4) and § 23, 5.(7), we obtain the following result, analogous to the one for (B)-spaces:

(5) *If E is a non-reflexive* (F)*-space, the iterated strong duals are all non-reflexive, and in each of the sequences*

$$E \subset E'' \subset E'''' \subset \cdots \quad \textit{and} \quad E' \subset E''' \subset \cdots$$

each space is a proper subspace of its successor.

3. (DF)-spaces. In 1.(6) and 2.(1) we obtained two special properties of the strong dual of a metrizable locally convex space; these were used by Grothendieck to define a new class of locally convex spaces.

A locally convex space $E[\mathfrak{T}]$ is said to be a (DF)-space if a) it has a fundamental sequence of bounded sets, and b) every strongly bounded subset M of E' which is the union of countably many equicontinuous sets is also equicontinuous.

As in 2.(2), it is easy to see that b) can be replaced by the dual property b′): if U_n is a sequence of closed absolutely convex neighbourhoods of o in $E[\mathfrak{T}]$, and if $U = \bigcap_{n=1}^{\infty} U_n$ absorbs every bounded set, then U is a $\mathfrak{T}$-neighbourhood of o.

A quasi-barrelled locally convex space with a fundamental sequence of bounded sets is always a (DF)-space. Thus every normed space is a (DF)-space.

By 2.(1), the strong dual of a metrizable space is a complete (DF)-space.

As in 2, it follows from a) and b) that

(1) *If* $E[\mathfrak{T}]$ *is a* (DF)-*space,* $E'[\mathfrak{T}_b]$ *is an* (F)-*space.*

The next result, about the neighbourhoods of $\mathfrak{o}$ in a (DF)-space, is due to GROTHENDIECK [10]:

(2) *Suppose that* $E[\mathfrak{T}]$ *is a* (DF)-*space, and that* (B_n) *is a fundamental sequence of closed, absolutely convex bounded subsets of E. An absolutely convex subset W of E is a* $\mathfrak{T}$-*neighbourhood of* $\mathfrak{o}$ *if and only if* $W \cap B_n$ *is a* $\mathfrak{T}$-*neighbourhood of* $\mathfrak{o}$ *in* B_n, *for each n.*

Proof. We have to show that the condition is sufficient. For this, we construct a sequence α_n of positive numbers and a sequence U_n of closed absolutely convex neighbourhoods of $\mathfrak{o}$, such that

$$(3) \quad \alpha_n B_n \subset \tfrac{1}{3} W, \qquad (4) \quad \alpha_n B_n \subset U_k, \qquad (5) \quad U_n \cap B_n \subset W$$

for all n and all k. The proof of (2) follows from this construction; for if we put $U = \bigcap_{n=1}^{\infty} U_n$, then by (4) U absorbs every bounded set, and so by b′) U is a $\mathfrak{T}$-neighbourhood of $\mathfrak{o}$; finally it follows from $U \cap B_n \subset W$ that $U \cap \left(\bigcup_{n=1}^{\infty} B_n\right) = U \subset W$.

Suppose that α_n and U_n have been defined for $n \leqq m$ in such a way that (3), (4) and (5) are satisfied for $n, k \leqq m$. By hypothesis, there is a neighbourhood U of $\mathfrak{o}$ with $U \cap B_{m+1} \subset W$. We choose α_{m+1} in such a way that $\alpha_{m+1} B_{m+1} \subset \tfrac{1}{3} U$ and $\alpha_{m+1} B_{m+1} \subset \tfrac{1}{3} B_{m+1}$. But then $\alpha_{m+1} B_{m+1} \subset \tfrac{1}{3}(B_{m+1} \cap U) \subset \tfrac{1}{3} W$, so that (3) is satisfied for $m+1$. Further α_{m+1} can clearly be chosen small enough for (4) to hold for $n = m+1$ and $k \leqq m$.

We set $B^{(m+1)} = \Gamma_{i=1}^{m+1} \alpha_i B_i$. If we can find an absolutely convex neighbourhood V of $\mathfrak{o}$ such that $U_{m+1} = \overline{B^{(m+1)} + V}$ satisfies condition (5) for $m+1$, then since $\alpha_n B_n \subset B^{(m+1)} \subset U_{m+1}$, (4) is also satisfied for $n \leqq m+1$ and $k = m+1$.

Since $U_{m+1} \subset 2B^{(m+1)} + 2V$, it is sufficient to show that $(2B^{(m+1)} + 2V) \cap B_{m+1} \subset W$. If we set $M = B_{m+1} \cap (E \sim W)$, we must show that $(2B^{(m+1)} + 2V) \cap M$ is empty for a suitable choice of V, i.e. that $2V \cap (M + 2B^{(m+1)})$ is empty. This means that we must show that $N = M + 2B^{(m+1)}$ cannot have $\mathfrak{o}$ as a closure point.

We prove this in the following way. Since $B^{(m+1)} \subset \tfrac{1}{3} W$, we have $\tfrac{1}{3} W + 2B^{(m+1)} \subset W$. Since $W \cap M$ is empty, $(\tfrac{1}{3} W + 2B^{(m+1)}) \cap M$ is also empty, and so $N \cap \tfrac{1}{3} W$ is empty. The set $3N$ is bounded, and so it is contained in some B_k; also $W \cap B_k$ is a neighbourhood of $\mathfrak{o}$ in B_k.

Since $(3N) \cap W$ is empty, $\mathfrak{o}$ is not a closure point of $3N$, and so it is not a closure point of N.

Let us state the result for the special case of the strong dual of a metrizable locally convex space explicitly:

(6) *Suppose that* $E[\mathfrak{T}]$ *is locally convex and metrizable, and that* $U_1 \supset U_2 \supset \cdots$ *is a base of absolutely convex neighbourhoods of* $\mathfrak{o}$. *An absolutely convex subset* W *of* E' *is a strong neighbourhood of* $\mathfrak{o}$ *in* E' *if and only if* $W \cap U_n^\circ$ *is a strong neighbourhood of* $\mathfrak{o}$ *in* U_n°, *for each* n.

As a simple consequence of (2), we have

(7) *A linear mapping* A *from a* (DF)-*space* E *into a locally convex space* F *is continuous if and only if its restrictions to the sets* B_n *of a fundamental sequence of bounded sets are continuous.*

For if A is continuous on the sets B_n, and if V is an absolutely convex neighbourhood of $\mathfrak{o}$ in F, $A^{(-1)}(V) \cap B_n$ is a neighbourhood of $\mathfrak{o}$ in B_n; by (2), the absolutely convex set $A^{(-1)}(V)$ is thus a neighbourhood of $\mathfrak{o}$ in E.

The following result, which is also due to GROTHENDIECK, relates the topology of a (DF)-space to the topology $\mathfrak{T}_{b*}(E')$:

(8) *If* M *is a separable subset of a* (DF)-*space* $E[\mathfrak{T}]$, *the topologies* $\mathfrak{T}$ *and* $\mathfrak{T}_{b*}(E')$ *coincide on* M.

Proof. $\mathfrak{T}_{b*}$ is finer than $\mathfrak{T}$. We must therefore show that given a closed absolutely convex $\mathfrak{T}_{b*}$-neighbourhood V of $\mathfrak{o}$ there is always an open $\mathfrak{T}$-neighbourhood U of $\mathfrak{o}$ with $M \cap U \subset V$. This is equivalent to the assertion that $(U \cap (E \sim V)) \cap M$ is empty. Since $U \cap (E \sim V)$ is open, it is sufficient to show that if x_i is a dense sequence in M, no element of the sequence lies in $U \cap (E \sim V)$. In other words, we must show that there is a U which contains none of the x_{i_k} which lie in $E \sim V$. We denote the sequence x_{i_k} by $x_1, x_2, \ldots$ again. There is clearly nothing to prove, unless this sequence does not terminate.

For the proof, we construct sequences $\alpha_n > 0$ and U_n (closed absolutely convex $\mathfrak{T}$-neighbourhoods of $\mathfrak{o}$) such that

(9) $\quad \alpha_n B_n \subset U_k, \quad$ (10) $\quad \alpha_n B_n \subset V, \quad$ (11) $\quad x_n \notin U_n$

for all n and k; the sets B_n are a fixed fundamental sequence of bounded closed absolutely convex subsets of E.

Suppose that these have been found for $k, n \leqq m$. It is clearly possible to choose α_{m+1} in such a way that (10) holds for $m+1$ and (9) holds for $n = m+1$ and $k \leqq m$.

We set $B^{(m+1)} = \overline{\Gamma_1^{m+1} \alpha_n B_n}$. We now have to choose a closed absolutely convex $U_{m+1} \supset B^{(m+1)}$ in such a way that $x_{m+1} \notin U_{m+1}$. But this can be done, since $B^{(m+1)} \subset V$ and $x_{m+1} \notin V$ (cf. § 15, 6.(9)).

If we put $U = \bigcap_{n=1}^{\infty} U_n$, then because of (9) we obtain an absolutely convex set which absorbs all the bounded sets, and which is therefore a $\mathfrak{T}$-neighbourhood, by b′). Because of (11), U contains none of the $x_n \in E \sim V$.

Using (8), sufficient conditions can now be given for a (DF)-space to be quasi-barrelled, i. e. for $\mathfrak{T}$ and $\mathfrak{T}_{b*}(E')$ to coincide.

(12) a) *Every separable* (DF)-*space is quasi-barrelled.*

b) *If the bounded subsets of a* (DF)-*space* $E[\mathfrak{T}]$ *are metrizable under* $\mathfrak{T}$, $E[\mathfrak{T}]$ *is quasi-barrelled.*

c) *An* (F)-*space is distinguished if the bounded subsets of the strong dual are metrizable.*

Proof. a) follows directly from (8).

By (2), $\mathfrak{T}_{b*}(E')$ is the same as $\mathfrak{T}$ if $\mathfrak{T}_{b*}(E')$ and $\mathfrak{T}$ coincides on any bounded set B. If $\mathfrak{T}$ is metrizable on B, $\mathfrak{T}$ is determined by the collection of sequentially closed subsets of B. But the $\mathfrak{T}$-convergent sequences are the same as the $\mathfrak{T}_{b*}$-convergent sequences, by (8). Since every $\mathfrak{T}$-closed set is $\mathfrak{T}_{b*}$-closed, the $\mathfrak{T}$-closed and $\mathfrak{T}_{b*}$-closed sets of B are the same, so that $\mathfrak{T}$ and $\mathfrak{T}_{b*}$ are the same on B. Thus b) is proved.

By § 23, 7.(1), an (F)-space E is distinguished if and only if $E'[\mathfrak{T}_b]$ is barrelled. Since E' is complete, c) follows from b).

4. Bornological (DF)-spaces. Suppose that $E[\mathfrak{T}]$ is a metrizable locally convex space, that $E'[\mathfrak{T}_b(E)]$ is its strong dual, and that $E''[\mathfrak{T}_n]$ is its strong bidual. We now investigate the bornological space associated with $E'[\mathfrak{T}_b]$.

(1) *Let* B_n *be a fundamental sequence of absolutely convex bounded closed subsets of* $E'[\mathfrak{T}_b]$. *The algebraic hull* V^a *of any set* V *of the form* $V = \Gamma_{n=1}^{\infty} \alpha_n B_n$, $\alpha_n \geqq 0$, *is always* $\mathfrak{T}_b$-*closed.*

Thus $E'[\mathfrak{T}_b^{\times}]$ *has a base of* $\mathfrak{T}_b$-*closed neighbourhoods of* o.

Proof. Let $V_k = \Gamma_{n=1}^{k} \alpha_n B_n$, so that $V = \bigcup_{k=1}^{\infty} V_k$. Since the sets B_k are $\mathfrak{T}_s(E)$-compact (cf. § 21, 5.(3)), the sets V_k are also $\mathfrak{T}_s(E)$-compact, by § 20, 6.(5), and so they are strongly closed. Suppose that u does not belong to V^a. Then by § 16, 4.(4) there exists a $\beta > 1$ with $u \notin \beta V$. Since $u \notin \beta V_k$, there exists a $z_k \in V_k^{\circ} \subset E''$ with $z_k u = \beta$. The sequence z_k is bounded in $E''[\mathfrak{T}_n]$, and so by 2.(1) it is an equicontinuous, and therefore relatively $\mathfrak{T}_s(E')$-compact, subset of E''. If $z_0 \in E''$ is weakly adherent to the sequence z_k, $z_0 u = \beta$, and $z_0 \in \bigcap_{k=1}^{\infty} V_k^{\circ} = V^{\circ}$. Thus u does not belong to $V^{\circ\circ}$, and the $\mathfrak{T}_b$-closure $V^{\circ\circ}$ of V is equal to V^a. Finally the sets V^a form a base of neighbourhoods of $\circ$ for $\mathfrak{T}_b^\times$, by § 28, 2.

(1) now enables us to determine the topology $\mathfrak{T}_b^\times$ on E'.

(2) *Suppose that* $E[\mathfrak{T}]$ *is metrizable and locally convex. The bornological topology* $\mathfrak{T}_b^\times$ *associated with the strong topology* $\mathfrak{T}_b(E)$ *on* E' *is equal to* $\mathfrak{T}_b(E'')$.

$E'[\mathfrak{T}_b(E'')]$ *is thus always a complete bornological* (DF)-*space.*

Proof. By § 21, 2.(1), the closed absolutely convex absorbent subsets of $E'[\mathfrak{T}_b(E)]$ form a base of $\mathfrak{T}_b(E'')$-neighbourhoods of $\circ$ in $E'[\mathfrak{T}_b(E'')]$. But, by (1) and the definition of bornological space (§ 28, 1.), these sets also form a base of $\mathfrak{T}_b^\times$-neighbourhoods of $\circ$ in E'.

The completeness of $E'[\mathfrak{T}_b(E'')]$ follows from the $\mathfrak{T}_b(E)$-completeness of E', using (1) and § 18, 4.(4).

We now use (2) to give a criterion for the strong dual of an (F)-space to be bornological, i. e. for it to be an (LB)-space.

(3) *Suppose that* $E[\mathfrak{T}]$ *is an* (F)-*space. The following are equivalent:*

a) $E[\mathfrak{T}]$ *is distinguished;*

b) $E'[\mathfrak{T}_b(E)]$ *is bornological;*

c) $E'[\mathfrak{T}_b]$ *is barrelled, or quasi-barrelled.*

Proof. Since $E'[\mathfrak{T}_b]$ is complete, a) and c) are equivalent, by § 23, 7.(1). Since every bornological space is quasi-barrelled (§ 28, 1.), c) follows from b). On the other hand, b) follows from c), by (2).

It follows from 3.(12) that *if* $E'[\mathfrak{T}_b]$ *is separable or if the bounded sets of* $E'[\mathfrak{T}_b]$ *are metrizable, then* $E'[\mathfrak{T}_b]$ *is bornological.*

Further we have (cf. § 28, 7.):

(4) *If E is a reflexive* (F)-*space, its strong dual is bornological.*

An (F)-*space E is reflexive if and only if it is boundedly closed, and if and only if it is* $\mathfrak{T}_{co}(E')$-*complete.*

Proof. If $E[\mathfrak{T}]$ is reflexive, $E'[\mathfrak{T}_b]$ is barrelled, and so by (3) it is bornological. It follows from § 28, 7.(4) that a reflexive space E is boundedly closed and $\mathfrak{T}_{co}$-complete. On the other hand if $E=\check{E}$ $=\tilde{E}[\tilde{\mathfrak{T}}_{co}(E')]$, then $E=E''$, by § 28, 7.(7).

In § 31, 7. we shall give an example of an (F)-space which is not distinguished. Thus there are (F)-spaces E with the property that not every bounded set M of E'' is contained in the bipolar $B^{\circ\circ}$ of a bounded subset B of E.

If however M is weakly separable, it follows from the definition of (DF)-space that there always exists such a B, with $M\subset B^{\circ\circ}$.

(5) *If* $E[\mathfrak{T}]$ *is an* (F)-*space, and if* $\check{E}$ *is its bounded closure, then every* $\mathfrak{T}_s(E')$-*bounded subset of* $\check{E}$ *lies in the bipolar* $B^{\circ\circ}$ *in* $\check{E}$ *of a bounded subset* B *of* E''.

By (2), $E'[\mathfrak{T}_b(E'')]$ is barrelled, and by § 28, 6.(1) its dual is $\check{E}$; thus the bounded sets of $\check{E}$ are the same as the $\mathfrak{T}_b(E'')$-equicontinuous sets. These are the subsets of sets $B^{\circ\circ}$, where B° is a $\mathfrak{T}_b(E'')$-neighbourhood of o.

(6) *Suppose that* $E[\mathfrak{T}]$ *is locally convex and metrizable.* $E''=\check{E}=\tilde{E}[\tilde{\mathfrak{T}}_{co}(E')]$ *if and only if* $\mathfrak{T}_k(E'')=\mathfrak{T}_b(E'')$, *and if and only if every* $\mathfrak{T}_s(E')$-*bounded subset of* E'' *is relatively* $\mathfrak{T}_s(E')$-*compact.*

This follows from (2), since $\check{E}=(E'[\mathfrak{T}_b^{\times}])'$ and $E''=(E'[\mathfrak{T}_k(E'')])'$.

In 3. we gave criteria for the topology $\mathfrak{T}$ of a (DF)-space to be equal to $\mathfrak{T}_{b*}(E')$, and in the present number we considered situations in which $\mathfrak{T}$ is bornological. The following simple example shows that in general $\mathfrak{T}$ can be coarser than the Mackey topology.

Suppose that E is a reflexive (B)-space which is not separable. We give E the topology of uniform convergence on the separable bounded subsets of the strong dual E'. Then $E[\mathfrak{T}]$ is a semi-reflexive (DF)-space, and $\mathfrak{T}$ is strictly coarser than the topology $\mathfrak{T}_k(E')=\mathfrak{T}_b(E')$.

5. Hereditary properties of (DF)-spaces.

(1) *If* $E[\mathfrak{T}]$ *is a* (DF)-*space and if* H *is a closed linear subspace, then* $(E/H)[\mathfrak{T}]$ *is also a* (DF)-*space, and the two strong topologies* $\mathfrak{T}_b(E)$ *and* $\mathfrak{T}_b(E/H)$ *on* $H^{\perp}=(E/H)'$ *are the same.*

First we prove the second part of the theorem. The topology $\mathfrak{T}_b(E)$ on $H^{\perp}\subset E'$ is coarser than $\mathfrak{T}_b(E/H)$, so that the identity mapping I from $H^{\perp}[\mathfrak{T}_b(E/H)]$ to $H^{\perp}[\mathfrak{T}_b(E)]$ is continuous. We must show that its inverse is also continuous; then I is a topological isomorphism. $H^{\perp}[\mathfrak{T}_b(E)]$ is metrizable, as it is a subspace of E', and so it is bornological; by § 28, 3.(4) it is sufficient to show that every sequence u_n in $H^{\perp}$ which is $\mathfrak{T}_b(E)$-convergent to o is bounded in $H^{\perp}[\mathfrak{T}_b(E/H)]$. Since $E[\mathfrak{T}]$ is a (DF)-space, though, the set $\{u_n\}$ is $\mathfrak{T}$-equicontinuous in $H^{\perp}$. As such, it is relatively $\mathfrak{T}_s(E)$-compact, and so it is also relatively $\mathfrak{T}_s(E/H)$-compact; consequently it is $\mathfrak{T}_b(E/H)$-bounded. Thus we have shown that on $H^{\perp}$, $\mathfrak{T}_b(E)=\mathfrak{T}_b(E/H)$.

From this it now follows that every bounded set in $(E/H)[\mathfrak{T}]$ is contained in the closure of the canonical image $K(B)$ of a bounded set B of $E[\mathfrak{T}]$. Thus $(E/H)[\mathfrak{T}]$ also has a fundamental sequence of bounded sets. Finally, condition b) of the definition of (DF)-space is satisfied by E/H, since it holds for E.

A closed linear subspace of a (DF)-space need not be a (DF)-space (cf. the counterexample in § 31, 5.).

(2) *If the linear subspace* $H[\mathfrak{T}]$ *of a locally convex space* $E[\mathfrak{T}]$ *is a* (DF)-*space, the strong dual of* $H[\mathfrak{T}]$ *is topologically isomorphic to* $E'[\mathfrak{T}_b(E)]/H^{\perp}$.

We must show that the algebraic isomorphism of $H'[\mathfrak{T}_b(H)]$ onto $E'[\mathfrak{T}_b(E)]/H^{\perp}$ is a topological isomorphism. Since $\mathfrak{T}_b(H)$ is coarser than $\mathfrak{T}_b(E)$ on H', it is sufficient to show that the isomorphism is continuous. Since $H'[\mathfrak{T}_b(H)]$ is metrizable, and so is bornological, we need only show that any sequence $\hat{u}_n$ which converges to $\mathfrak{o}$ in $H'[\mathfrak{T}_b(H)]$ is bounded in the topology $\mathfrak{T}_b(E)$. Once again, the $\hat{u}_n$ form a $\mathfrak{T}$-equicontinuous subset of H', and this set is the image of a $\mathfrak{T}$-equicontinuous subset of E', by § 22, 1.(1). This set is $\mathfrak{T}_b(E)$-bounded, and so also is its image $\{\hat{u}_n\}$.

(3) *a) A* (DF)*-space* $E[\mathfrak{T}]$ *is complete if and only if it is quasi-complete.*

b) The completion of a (DF)*-space is again a* (DF)*-space.*

c) Every semi-reflexive (DF)*-space is complete.*

Proof. a) We apply (2) to E and to its completion $\tilde{E}$. The two strong topologies $\mathfrak{T}_b(E)$ and $\mathfrak{T}_b(\tilde{E})$ coincide on $E'=\tilde{E}'$. Thus every bounded set in $\tilde{E}$ lies in the closure of a bounded set in E. If E is quasi-complete, it follows that $E=\tilde{E}$.

b) If B_n is a fundamental sequence of bounded sets in E, the closures $\bar{B}_n$ in $\tilde{E}$ form a fundamental sequence of bounded sets in $\tilde{E}$. Condition b′) for a (DF)-space is satisfied by $\tilde{E}$, if it is satisfied by E.

c) Every semi-reflexive space is quasi-complete, so that c) follows from a).

(4) *The locally convex hull* $E[\mathfrak{T}]=\sum A_n(E_n[\mathfrak{T}_n])$ *of a sequence of* (DF)*-spaces* $E_n[\mathfrak{T}_n]$ *is again a* (DF)*-space.*

Every bounded subset of E lies in the closed absolutely convex cover of finitely many $A_n(B_n)$, *where* B_n *is bounded in* $E_n[\mathfrak{T}_n]$.

The strong dual $E'[\mathfrak{T}_b(E)]$ *is the locally convex kernel of the spaces* $A_n'^{(-1)}(E_n'[\mathfrak{T}_b(E_n)])$.

First we show that the locally convex direct sum $F=\bigoplus_{n=1}^{\infty} E_n[\mathfrak{T}_n]$ of (DF)-spaces is again a (DF)-space.

As $B_i^{(n)}$ runs through a fundamental sequence of bounded sets in $E_n[\mathfrak{T}_n]$, the sets $\bigoplus_{n=1}^{N} B_{i_n}^{(n)}$, taken in a suitable order, form a fundamental sequence in F, by § 18, 5.(4).

The fact that F satisfies condition b′) for a (DF)-space follows from the fact that the spaces $E_n[\mathfrak{T}_n]$ satisfy b′), and from the definition of the locally convex direct sum topology.

Since $\sum A_n(E_n[\mathfrak{T}_n])$ is topologically isomorphic to a quotient F/H, the first assertion of (4) follows from (1).

Likewise it follows from (1) that every bounded subset of F/H lies in the closure of the canonical image of a bounded set in F, and so it lies in a set of the form $\overline{\Gamma_{i=1}^{n} A_i(B_i)}$. The last part of (4) follows directly from this, when we apply § 22, 7.(5).

In particular (4) says that every topological inductive limit $E[\mathfrak{T}]$ of an increasing sequence $E_1[\mathfrak{T}_1] \subset E_2[\mathfrak{T}_2] \subset \cdots$ of normed spaces is a (DF)-space. By § 28, 4.(1), $E[\mathfrak{T}]$ is then bornological.

We remark that $E[\mathfrak{T}]$ need not be complete, even when the spaces $E_n[\mathfrak{T}_n]$ are (B)-spaces (cf. § 31, 6.).

(5) *A locally convex space* $E[\mathfrak{T}]$ *is a bornological* (DF)-*space if and only if it is the topological inductive limit of an increasing sequence of normed spaces.*

The simple proof that the condition is also necessary is left to the reader.

(6) *The locally convex hull of a sequence of semi-reflexive (respectively reflexive)* (DF)-*spaces* E_n *is again a semi-reflexive (respectively reflexive)* (DF)-*space.*

If the spaces $E_n[\mathfrak{T}_n]$ are semi-reflexive (DF)-spaces, every bounded set $B_n \subset E_n[\mathfrak{T}_n]$ is relatively $\mathfrak{T}_s(E_n')$-compact. But then the absolutely convex cover of finitely many $A_n(B_n)$ is relatively $\mathfrak{T}_s(E')$-compact, so that $\sum A_n(E_n[\mathfrak{T}_n])$ is semi-reflexive, by (4).

If the spaces $E_n[\mathfrak{T}_n]$ are reflexive, they are barrelled. By § 27, 1.(3), $\sum A_n(E_n[\mathfrak{T}_n])$ is also barrelled. The assertion now follows from § 23, 5.(1).

6. Further results, and open questions. The properties of linear mappings between (F)- and (DF)-spaces will be investigated in the second volume, and important special classes of these spaces will also be considered there. For the present, we make a few additional remarks.

If E is a normed space with completion $\tilde{E}$, every bounded subset of $\tilde{E}$ is contained in the completion of a bounded subset of $\tilde{E}$. The question of whether this is also true for metrizable locally convex spaces was settled quite recently.

(1) *If* $E[\mathfrak{T}]$ *is separable, metrizable and locally convex, every bounded subset of the* (F)-*space* $\tilde{E}[\tilde{\mathfrak{T}}]$ *is contained in the completion of a bounded subset of* E.

As GROTHENDIECK [10] showed, this is a simple consequence of the theorems of number 2. $\tilde{E} \subset E''$, by 2.(3). A bounded subset B of $\tilde{E}$ is thus a separable bounded subset of E''. A countable dense subset of B is $\mathfrak{T}_b(E)$-equicontinuous, by 2.(1); thus B, being contained in its closure, is $\mathfrak{T}_b(E)$-equicontinuous. Thus $B \subset M^{\circ\circ} \cap \tilde{E}$, where $M^{\circ\circ}$ is the polar of M° in E'', and M is bounded and absolutely convex in $E[\mathfrak{T}]$. But $M^{\circ\circ} \cap \tilde{E}$ is the completion of M in $\tilde{E}$, by the theorem of bipolars.

AMEMIYA [1] has given the following example, which shows that (1) need no longer be true in the non-separable case.

Let $E_1[\mathfrak{T}]$ be a real (F)-space which is not a (B)-space, and let B_α, $\alpha\in\mathsf{A}$, be a fundamental system of absolutely convex bounded subsets of E_1, which must be uncountable, by 1.(3). By 1.(9), for each α there is a discontinuous linear functional $u_\alpha\in E_1^*$ which vanishes on the elements of B_α.

Let $p_n(x)$ be a sequence of semi-norms on E_1 which define the topology $\mathfrak{T}$. We denote by $E=l^1_{\mathsf{A}}(E_1)$ the space of all $x=(x_\alpha)$, $x_\alpha\in E_1$, for which all the sums $q_n(x)=\sum_\alpha p_n(x_\alpha)$ are finite. E is an (F)-space under the topology defined by the seminorms $q_n(x)$ on E.

We denote by ω_{A} the space of all real vectors $\mathfrak{x}=(\xi_\alpha)$, $\alpha\in\mathsf{A}$. The equation $A(x)=(u_\alpha x_\alpha)$ defines a linear mapping from E into ω_{A}. Let $F=l^1_{\mathsf{A}}$ denote the subspace of all $\mathfrak{x}\in\omega_{\mathsf{A}}$ with $\sum_\alpha|\xi_\alpha|<\infty$. F is a (B)-space under the norm $\sum_\alpha|\xi_\alpha|$. Finally let E_0 be the subspace of E which is mapped into F by A. E_0 is dense in E, since all the terms $x=(x_\alpha)$ with only finitely many non-zero x_α belong to E_0. For the same reason $\hat{A}(E_0)$ is dense in F, where $\hat{A}$ denotes the restriction of A to E_0.

Now let B be a bounded subset of E_0. We assert that $\overline{\hat{A}(B)}$ contains no neighbourhood of $\mathfrak{o}$ in F. If $q_n(x)\leqq M_n$ for all $x\in B$, then $p_n(x_\alpha)\leqq M_n$ for each α; all the components x_α of the elements x of B thus lie in a fixed bounded set B_{α_0}. But then $u_{\alpha_0}x_{\alpha_0}=0$, for all $x\in B$; thus the α_0-th co-ordinate vanishes for all elements of $\hat{A}(B)$. It follows from this that $\overline{\hat{A}(B)}$ can contain no scalar multiple of the unit ball of F.

We now consider the graph $G(\hat{A})$ of the mapping $\hat{A}$ from E_0 into F as a linear subspace of the (F)-space $E\times F$. Since E_0 is dense in E, since $\hat{A}(E_0)$ is dense in F, and since, because the functionals u_α are discontinuous, $\hat{A}^{(-1)}(\mathfrak{o})$ is dense in E_0, $G(\hat{A})$ is dense in $E\times F$. We assert that no bounded subset M of $G(\hat{A})$ is dense in a bounded subset of $E\times F$ of the form $M_1\times K$, where K is the unit ball of F. The set M is contained in a set of the form $B\times\hat{A}(B)$, where B is bounded in E_0. But we have previously shown that $\hat{A}(B)$ is not dense in K, so that M is not dense in $M_1\times K$; this establishes the counterexample.

KŌMURA [2] has given an example, assuming the continuum hypothesis to be true, of an (F)-space which is not separable, but all of whose bounded sets are separable; earlier DIEUDONNÉ [14] had given an example, which again used the continuum hypothesis, of a metrizable locally convex space with the same properties.

The question, raised by GROTHENDIECK [10], of whether the topologies $\mathfrak{T}_b(E)$ and $\mathfrak{T}_k(E'')$ on the strong dual E' of an (F)-space are always the same (cf. § 28, 7.) was also answered in the negative by KŌMURA [2].

The properties "quasi-barrelled" and "bornological" are equivalent for the strong duals of (F)-spaces, by 4.(3). KŌMURA [1] has shown however that there exist barrelled (DF)-spaces which are not bornological.

AMEMIYA [1] has given an example of a reflexive (F)-space, in which no bounded set is total: let R be the set of monotonically increasing sequences $\mathfrak{y}=(\eta_n)$, $\eta_n>0$. Let us consider the vector space E of all real functions $f(\mathfrak{y})$ on R which satisfy $p_i(f)=\left(\sum_{\mathfrak{y}\in R}|f(\mathfrak{y})|^2\eta_i\right)^{\frac{1}{2}}<\infty$, for $i=1,2,\ldots$. Since each one of these conditions defines a (B)-space which is norm-isomorphic to a space l_d^2, the space E, equipped with the norms $p_i(f)$, $i=1,2,\ldots$, is a reflexive (F)-space, as it is the intersection of reflexive (B)-spaces.

It is sufficient to show that the elements f of a bounded set $B\subset E$ all vanish at some fixed $\mathfrak{y}_0$. If $\sup_{f\in B}p_i(f)=M_i$, and if $\mathfrak{x}=(\xi_i)$ is chosen in R in such a way that $\lim_{i\to\infty}M_i^2\xi_i^{-1}=0$, then $f(\mathfrak{x})$ must be zero for each $f\in B$, since we have the inequalities $|f(\mathfrak{x})|^2\xi_i\leqq M_i^2$, for $i=1,2,\ldots$.

(2) *If E is a reflexive* (F)*-space in which no bounded set is total, the strong dual possesses bounded subsets which are not metrizable under the strong topology.*

Let U_n, $n=1,2,\ldots$, be a base of neighbourhoods of $\mathfrak{o}$ in E. It is sufficient to show that not all the sets U_n° are metrizable. If they were all metrizable, there would be bounded sets B_{nk} in E such that the sets $B_{nk}^\circ\cap U_n^\circ$ form a strong base of neighbourhoods of $\mathfrak{o}$ in U_n°. By 1.(5), there is a single bounded set B in E which absorbs all the sets B_{nk}. If u is any non-zero element of E', $u\in U_n^\circ$ for some n, and there is a B_{nk} with $u\notin B_{nk}^\circ$. In particular, u does not vanish identically on B_{nk}, and so it does not vanish identically on B. Thus B would be total in E, which contradicts the assumption made about E.

This result of AMEMIYA's answers a question posed by GROTHENDIECK [10].

§ 30. Perfect spaces

1. The α-dual. Examples. In this paragraph we consider sequence spaces (which are also called coordinate spaces); these are vector spaces λ whose elements are sequences $\mathfrak{x}=(x_i)=(x_1,x_2,\ldots)$ of real or complex numbers. The vector space operations are given by the usual operations on the coordinates.

We can always consider such a sequence space λ as a linear subspace of the space ω of all sequences, $\lambda\subset\omega$.

A sequence space is said to be normal if whenever it contains $\mathfrak{x}=(x_i)$ it also contains all vectors $\mathfrak{y}=(y_i)$ with $|y_i|\leqq|x_i|$ for $i=1,2,\ldots$.

For example the space φ of all sequences with only finitely many non-zero coordinates is normal.

To each sequence space λ we assign another sequence space $\lambda^\alpha=\lambda^\times$, its α-dual. $\lambda^\times$ is defined to be the set of all sequences $\mathfrak{u}=(u_i)$ for which the scalar products $\mathfrak{u}\mathfrak{x}=\sum_{i=1}^{\infty} u_i x_i$ converge absolutely, for all $\mathfrak{x}\in\lambda$. For example $\omega^{\times}=\varphi$ and $\varphi^{\times}=\omega$.

It follows directly from the definition that

(1) *a) If* $\lambda\subset\mu$, *then* $\mu^{\times}\subset\lambda^{\times}$.

b) We always have $\lambda^{\times\times}=(\lambda^{\times})^{\times}\supset\lambda$.

A sequence space λ is said to be perfect if $\lambda^{\times\times}=\lambda$. By the remarks above, φ and ω are perfect spaces.

(2) *The α-dual* $\lambda^{\times}$ *of an arbitrary sequence space is always perfect. For any sequence space* λ, *we have* $\lambda^{\times}=\lambda^{\times\times\times}$. $\lambda^{\times\times}$ *is the smallest perfect space containing* λ.

Proof. By (1) a) it follows from $\lambda^{\times\times}\supset\lambda$ that $(\lambda^{\times\times})^{\times}\subset\lambda^{\times}$. On the other hand, by (1) b), $(\lambda^{\times})^{\times\times}\supset\lambda^{\times}$, so that $\lambda^{\times}=\lambda^{\times\times\times}$; $\lambda^{\times}$ is perfect.

If μ is perfect and $\mu\supset\lambda$, $\mu=\mu^{\times\times}\supset\lambda^{\times\times}$. But $\lambda^{\times\times}$ is perfect, and so it is the smallest perfect space containing λ.

(3) *If* λ *is perfect,* λ *is normal and* $\lambda\supset\varphi$.

This is immediately obvious for $\lambda^{\times}$, and so it also holds for $\lambda^{\times\times}=(\lambda^{\times})^{\times}$. Thus every perfect space λ satisfies $\varphi\subset\lambda\subset\omega$.

We now give some simple examples.

(4) l^1 *and* l^{∞} *are perfect; we have* $(l^1)^{\times}=l^{\infty}, (l^{\infty})^{\times}=l^1$.

Since $\mathfrak{e}=(1,1,\ldots)\in l^{\infty}$, it follows that $\sum_{i=1}^{\infty}|u_i|<\infty$ for each $\mathfrak{u}\in(l^{\infty})^{\times}$, so that $(l^{\infty})^{\times}\subset l^1$. On the other hand, clearly $l^1\subset(l^{\infty})^{\times}$.

Likewise it is trivially true that $l^{\infty}\subset(l^1)^{\times}$. On the other hand given an unbounded sequence $\mathfrak{v}=(v_i)$ there is an $\mathfrak{x}\in l^1$ for which $\mathfrak{v}\,\mathfrak{x}=\sum v_i x_i$ diverges, so that $l^{\infty}=(l^1)^{\times}$.

(5) l^2 *is perfect, and it is the only self α-dual sequence space.*

a) The inequality $|u_n x_n|\leqq|u_n|^2+|x_n|^2$ implies that $\mathfrak{u}\mathfrak{x}$ is absolutely convergent, when both $\mathfrak{x}$ and $\mathfrak{u}$ belong to l^2. Thus $(l^2)^{\times}\supset l^2$. Let us suppose that $(l^2)^{\times}$ is bigger than l^2. Then there is an element $\mathfrak{v}=(v_i)$ in $(l^2)^{\times}$ for which $\sum|v_i|^2=\infty$. We can therefore find $0=n_1<n_2<\cdots$ such that $|v_{n_i+1}|^2+\cdots+|v_{n_{i+1}}|^2=M_i^2\geqq 1$. If we put $x_j=\frac{1}{iM_i}v_j$ for $n_i+1\leqq j\leqq n_{i+1}$, we get $\sum_{j=1}^{\infty}|x_j|^2=\sum\frac{1}{i^2}<\infty$, so that $\mathfrak{x}=(x_i)\in l^2$. But on the other hand $\sum|v_j x_j|\geqq\sum_{i=1}^{\infty}\frac{1}{i}=\infty$, which contradicts the assumption that $\mathfrak{v}\in(l^2)^{\times}$.

b) Suppose that $\lambda=\lambda^{\times}$. λ is therefore perfect and normal. If $\mathfrak{x}=(x_i)\in\lambda$, then $\bar{\mathfrak{x}}=(\bar{x}_i)\in\lambda=\lambda^{\times}$ as well, and so we must have $\mathfrak{x}\bar{\mathfrak{x}}=\sum|x_i|^2<\infty$.

From this it follows that $\lambda \subset l^2$; on the other hand it follows from (1) a) that $\lambda = \lambda^{\times} \supset (l^2)^{\times} = l^2$.

(6) *l^p is perfect, for* $1 < p < \infty$, *and* $(l^p)^{\times} = l^q, \frac{1}{p} + \frac{1}{q} = 1$.

It follows from Hölder's inequality that $(l^p)^{\times} \supset l^q$. For a $\mathfrak{v} \in (l^p)^{\times}$ with $\sum_{i=1}^{\infty} |v_i|^q = \infty$ we proceed in the same way as in (5): we determine sections of $\mathfrak{v}$ satisfying $|v_{n_i+1}|^q + \cdots + |v_{n_{i+1}}|^q = M_i^q \geqq 1$, put $x_j = \frac{1}{i} \frac{|v_j|^{q-1}}{M_i^{q-1}}$ for $n_i + 1 \leqq j \leqq n_{i+1}$, and arrive at a contradiction, as before.

2. The normal topology of a sequence space. It is easy to see how the ideas which have just been introduced can be fitted into the theory of locally convex spaces.

If λ is a sequence space, and if we further suppose that λ contains φ, then λ and its α-dual $\lambda^{\times}$ form a dual pair $\langle \lambda^{\times}, \lambda \rangle$, with bilinear form $\langle \mathfrak{u}, \mathfrak{x} \rangle = \mathfrak{u}\mathfrak{x} = \sum_{i=1}^{\infty} u_i x_i$. The condition $\lambda \supset \varphi$ is necessary in order to make sure that condition (D2″) of § 10, 3. is satisfied.

The construction of $\lambda^{\times}$ therefore enables us to pass from a space $\lambda \supset \varphi$ to a dual pair $\langle \lambda^{\times}, \lambda \rangle$, in a unique way. Consequently if λ is a sequence space, all those concepts which only depend upon the dual pair, such as the weak topology, the Mackey topology and the strong topology, are also uniquely defined.

On the other hand, the way that the α-dual was defined makes it natural to consider the locally convex topology defined by the seminorms

$$p_{\mathfrak{u}}(\mathfrak{x}) = \sum_{i=1}^{\infty} |u_i|\,|x_i|, \qquad \mathfrak{u} \in \lambda^{\times}, \tag{1}$$

on the sequence space λ. We call this the normal topology $\mathfrak{T}$ of the sequence space λ.

A coordinate vector $\mathfrak{x} = (x_i)$ is said to be positive, and we write $\mathfrak{x} > \mathfrak{o}$, if all the x_i are greater than or equal to 0, and if $\mathfrak{x} \neq \mathfrak{o}$. The sets $U_{\mathfrak{u},\varepsilon}$ consisting of all $\mathfrak{x} \in \lambda$ with $p_{\mathfrak{u}}(\mathfrak{x}) \leqq \varepsilon$ form a base of $\mathfrak{T}$-neighbourhoods of $\mathfrak{o}$, as $\mathfrak{u}$ runs through the positive elements of $\lambda^{\times}$. For $\bigcap_{i=1}^{n} U_{\mathfrak{u}_i,\varepsilon}$ contains the neighbourhood $U_{\mathfrak{u},\varepsilon}$ where $\mathfrak{u} = \sum_{i=1}^{n} \mathfrak{u}_i$.

If M is a collection of vectors in ω, the set of all $\mathfrak{y} = (y_i) \in \omega$ with $|y_i| \leqq |x_i|$, $i = 1, 2, \ldots$, for some $\mathfrak{x} = (x_i)$ in M is called the normal cover

M^n of M. If $M=M^n$, M is said to be normal. If a normal space contains M, it contains M^n.

If we denote by $\mathfrak{N}$ the class of normal covers $\{\mathfrak{u}\}^n$ of sets consisting of just one positive element $\mathfrak{u}$ of $\lambda^\times$, together with their subsets, we clearly have

(2) *The normal topology $\mathfrak{T}$ of a sequence space λ is the topology $\mathfrak{T}_{\mathfrak{N}}$ of uniform convergence on the collection $\mathfrak{N}$ of normal covers of the positive elements of $\lambda^\times$, together with their subsets.*

Further,

(3) *If $\lambda \supset \varphi$, $\mathfrak{N}$ is the collection of all $\mathfrak{T}$-equicontinuous subsets of $\lambda^\times$.*

It is sufficient to show that $U^\circ_{\mathfrak{u},1}=\{\mathfrak{u}\}^n$, for the collection of $\mathfrak{T}$-equicontinuous sets consists of the polars of a base of neighbourhoods of $\mathfrak{o}$, together with their subsets.

If the i-th coordinate u_i of $\mathfrak{u}$ is non-zero, every $x_i\mathfrak{e}_i$ with $|x_i| = \dfrac{1}{|u_i|}$ lies in $U_{\mathfrak{u},1}$. From this it follows that if $\mathfrak{v}\in U^\circ_{\mathfrak{u},1}$, then $|v_i|\leqq|u_i|$. If $u_i=0$, then any multiple of $\mathfrak{e}_i$ lies in $U_{\mathfrak{u},1}$, so that v_i must be zero, as well. This proves (3).

If $\lambda=\omega$, then $\lambda^\times=\varphi$, and the normal topology on ω is the same as both the weak and the strong topologies (with respect to φ), and so it is the same as the topology of the product $\prod_{n=1}^{\infty} E_n$, $E_n=\mathsf{K}$.

If $\lambda=\varphi$, then $\lambda^\times=\omega$, and the normal topology on φ is the same as the strong topology, and so it is the same as the locally convex direct sum topology on φ.

If $\lambda=l^1$, then $\lambda^\times=l^\infty$ by 1.(4), and the normal topology on l^1 is the same as the topology given by the norm $\|\mathfrak{x}\|=\sum_{i=1}^{\infty}|x_i|$, so that it is the same as the strong topology.

In the case where $\lambda=l^\infty$, so that $\lambda^\times=l^1$, the collection of sets $\mathfrak{N}$ does not consist of all the norm-bounded subsets of l^1. Indeed, by § 22, 4.(3), $\mathfrak{N}$ is a proper subset of the collection of all $\mathfrak{T}_s(l^\infty)$-compact subsets of l^1. Thus the normal topology on l^∞ is strictly coarser than $\mathfrak{T}_k(l^1)$.

On the l^p spaces, $1<p<\infty$, the normal topology $\mathfrak{T}$ is also strictly coarser than the Mackey topology, as can easily be seen.

If λ is a sequence space and if $\mathfrak{c}=(c_i)$ is a vector with non-zero coordinates c_i, $i=1,2,\ldots$, the set of all $\mathfrak{y}=(c_ix_i)$, where $\mathfrak{x}=(x_i)\in\lambda$ again forms a sequence space μ, which we call a diagonal transform of λ. Its α-dual $\mu^\times$ is obtained from $\lambda^\times$ by making the diagonal transformation by the vector $\mathfrak{d}=\left(\dfrac{1}{c_i}\right)$. The mapping $\mathfrak{x}\to\mathfrak{y}$ is clearly a topological isomorphism of $\lambda[\mathfrak{T}]$ onto $\mu[\mathfrak{T}]$. The $\mathfrak{T}$-equicontinuous sets $\{\mathfrak{u}\}^n$ in $\lambda^\times$ correspond to the $\mathfrak{T}$-equicontinuous sets $\{\mathfrak{v}\}^n$, where $\mathfrak{v}=\left(\frac{u_i}{c_i}\right)$. We now assert:

(4) *The normal topology $\mathfrak{T}$ of a sequence space $\lambda \supset \varphi$ is always coarser than the Mackey topology $\mathfrak{T}_k(\lambda^\times)$.*

The topological dual of a sequence space $\lambda[\mathfrak{T}] \supset \varphi$ is therefore always the same as its α-dual $\lambda^\times$.

It is sufficient to show that every set $\{\mathfrak{u}\}^n \subset \lambda^\times$ is $\mathfrak{T}_s(\lambda)$-compact. Since such a set is certainly absolutely convex, we shall then have $\mathfrak{N} \subset \mathfrak{K}$, where $\mathfrak{K}$ is the collection of $\mathfrak{T}_k(\lambda)$-equicontinuous subsets of $\lambda^\times$.

By the remarks made above, we can restrict our attention to the case where $\mathfrak{u}=\mathfrak{e}=(1,1,\ldots)$, provided that all the coordinates of $\mathfrak{u}$ are non-zero. But then λ is a subspace of l^1. Since $\{\mathfrak{u}\}^n$ is the unit ball of l^∞, and is therefore $\mathfrak{T}_s(l^1)$-compact, $\{\mathfrak{u}\}^n$ is *a fortiori* $\mathfrak{T}_s(\lambda)$-compact.

If we delete coordinates on which $\mathfrak{u}$ vanishes from the vectors of λ and of $\lambda^\times$, we reduce the general case to the one which we have just considered.

3. Sums and products of sequence spaces. If the coordinates of the vectors of a sequence space λ are permuted, so that $\mathfrak{x}=(x_1,x_2,\ldots)$ is transformed to $\mathfrak{x}'=(x_{n_1},x_{n_2},\ldots)$, we obtain a space $\mu[\mathfrak{T}]$ which is topologically isomorphic to $\lambda[\mathfrak{T}]$, and whose dual is obtained by making the same permutation of $\lambda^\times$. For since $\sum u_i x_i$ is absolutely convergent for $\mathfrak{u}\in\lambda^\times$, $\mathfrak{x}\in\lambda$, we have $\mathfrak{u}\mathfrak{x}=\mathfrak{u}'\mathfrak{x}'=\sum_{i=1}^{\infty} u_{n_i} x_{n_i}$.

Sequence spaces which are transformed into themselves by any permutation are said to be s y m m e t r i c. Examples are φ,ω and the l^p-spaces, $1 \leqq p \leqq \infty$.

We can also rearrange the coordinates in another way, where, instead of using the positive integers, we use any countable partially ordered index set. The vectors can, for example, be written as double sequences. If we make the same rearrangement of the α-dual, we again obtain a sequence space which is topologically isomorphic to the original one.

If $\lambda_1,\lambda_2,\ldots$ are sequence spaces, then the cartesian product $\prod_{i=1}^{\infty} \lambda_i$, which consists of all double sequences $(\mathfrak{x}^{(1)},\mathfrak{x}^{(2)},\ldots)$, $\mathfrak{x}^{(i)}\in\lambda_i$, is again a sequence space, in the sense which we have just described. In the same way the direct sum $\bigoplus_{i=1}^{\infty} \lambda_i$ consists of the double sequences $(\mathfrak{x}^{(1)},\ldots,\mathfrak{x}^{(n)},\mathfrak{o},\mathfrak{o},\ldots)$, where $\mathfrak{o}$ stands for the zero-vector in $\lambda^{(n+1)}$, $\lambda^{(n+2)},\ldots$, and n is an arbitrary positive integer. It is not difficult to see that

(1) *If the spaces* $\lambda_i \supset \varphi$ *are normal (respectively perfect), then the spaces* $\bigoplus_{i=1}^{\infty} \lambda_i$ *and* $\prod_{i=1}^{\infty} \lambda_i$ *are also normal (perfect).*

For arbitrary $\lambda_i \supset \varphi$ *we have* $\left(\bigoplus_{i=1}^{\infty} \lambda_i\right)^{\times} = \prod_{i=1}^{\infty} \lambda_i^{\times}$ *and* $\left(\prod_{i=1}^{\infty} \lambda_i\right)^{\times} = \bigoplus_{i=1}^{\infty} \lambda_i^{\times}$.

If, further, the spaces λ_i are given their normal topologies $\mathfrak{T}$, we have

(2) *The normal topologies of* $\bigoplus_{i=1}^{\infty} \lambda_i[\mathfrak{T}]$ *and* $\prod_{i=1}^{\infty} \lambda_i[\mathfrak{T}]$ *are respectively the topologies of the locally convex direct sum, and of the topological product.*

This is easy to prove either directly or by using § 22, 5.(3) and (4).

By starting from φ and ω, and repeatedly forming sums and products, we obtain the spaces of countable degree described in § 13, 5. These are all perfect spaces, and their normal topology is the same as the strong topology, since this is the case for φ and ω, by 2., and since this property is carried over to the spaces formed from them, by § 22, 5.

If $n_1 < n_2 < \cdots$ is a sequence of positive integers, and if for each vector $\mathfrak{x} = (x_1, x_2, \ldots)$ of a sequence space λ we form the vector $\mathfrak{x}' = (x_{n_1}, x_{n_2}, \ldots)$, we obtain another sequence space μ, which we call a s e c t i o n a l s u b s p a c e of λ. Clearly,

(3) *If* $\lambda \subset \varphi$ *is normal or perfect, this is also the case for any sectional subspace. The dual of a sectional subspace is obtained by leaving out the same coordinates of* $\lambda^{\times}$.

μ and the sectional subspace μ' defined on the set complementary to $\{n_1, n_2, \ldots\}$ together form a complementary decomposition of λ: $\lambda = \mu \oplus \mu'$.

4. Unions and intersections of sequence spaces. A linear subspace of a sequence space is again a sequence space, whereas the quotient by a closed linear subspace need not be a sequence space.

Two important ways of constructing new sequence spaces from given ones are special cases of forming hulls and kernels.

If $\{\lambda_\alpha\}$ is a family of sequence spaces, all of which are considered as subspaces of the same space ω, the linear span $\sum_\alpha \lambda_\alpha$ in ω is again a sequence space. Similarly the intersection $\bigcap_\alpha \lambda_\alpha$ in ω is a sequence space.

(1) a) *If all the spaces* λ_α *contain* φ *and are normal, the same holds for* $\sum_\alpha \lambda_\alpha$ *and* $\bigcap_\alpha \lambda_\alpha$.

b) *The intersection* $\bigcap_\alpha \lambda_\alpha$ *of perfect spaces* λ_α *is perfect.*

c) *If all the spaces λ_α contain φ, we have* $\left(\sum_\alpha \lambda_\alpha\right)^\times = \bigcap_\alpha \lambda_\alpha^\times$; *if the spaces λ_α are perfect, we also have* $\left(\bigcap_\alpha \lambda_\alpha\right)^\times = \left(\sum_\alpha \lambda_\alpha^\times\right)^{\times\times}$.

Proof. a) and c) follows easily from the definitions. b) follows from c): if the spaces λ_α are perfect, $(\sum \lambda_\alpha^\times)^\times = \bigcap \lambda_\alpha^{\times\times} = \bigcap \lambda_\alpha$, and so $\bigcap \lambda_\alpha$ is perfect, as it is an α-dual.

The following example shows that the equation $(\bigcap \lambda_\alpha)^\times = \sum \lambda_\alpha^\times$ need not always hold, even for perfect spaces λ_α (cf. § 13, 5.):

Let $\lambda_1 = \varphi\omega$ and $\lambda_2 = \omega\varphi$, where both are considered as subspaces of the same space ω, written as a space of double sequences $\omega\omega$. Then $\lambda_1 \cap \lambda_2$ is equal to $\varphi\varphi$, the space of double sequences with only finitely many non-zero terms. Thus $(\lambda_1 \cap \lambda_2)^\times = \omega\omega$. On the other hand, since $(\varphi\omega)^\times = \omega\varphi$, and $(\omega\varphi)^\times = \varphi\omega$, $\lambda_1^\times + \lambda_2^\times = \omega\varphi + \varphi\omega$, and this is a proper subspace of $\omega\omega$, as can be seen directly.

For the finer theory of sequence spaces there is a special representation in terms of intersections which is particularly important, and which we shall now develop.

Let $\mathfrak{a} = (a_i)$ be an arbitrary coordinate vector. We denote the perfect space consisting of all those $\mathfrak{x} \in \omega$ which satisfy $\sum_{r=1}^{\infty} |a_i|\,|x_i| < \infty$ by $\lambda_\mathfrak{a}$. If all but finitely many a_i are non-zero, $\lambda_\mathfrak{a}$ is a diagonal transform of l^1, as we have already seen in 2. If there are only finitely many non-zero a_i, $\lambda_\mathfrak{a}$ is equal to ω. If both infinitely many a_i are non-zero and infinitely many a_i are zero, $\lambda_\mathfrak{a}$ *divides into two sectional subspaces, of which one is a diagonal transform of l^1, and the other is equal to ω.*

It follows from 1. that the α-dual $\lambda_\mathfrak{a}^\times$ of $\lambda_\mathfrak{a}$ in general divides into two sectional subspaces, of which one is a diagonal transform of l^∞, and the other is equal to φ. $\lambda_\mathfrak{a}^\times$ is the perfect cover of the sequence space consisting of the vector $\mathfrak{a}$ and its scalar multiples.

We observe that for positive $\mathfrak{a}_i$ we clearly have the relation

$$\lambda_{\mathfrak{a}_1+\cdots+\mathfrak{a}_k}^\times = \sum_{j=1}^{k} \lambda_{\mathfrak{a}_j}^\times. \tag{2}$$

Every perfect space λ consists of all those $\mathfrak{x} \in \omega$ for which $\sum |u_i|\,|x_i| < \infty$ for all positive $\mathfrak{u} \in \lambda^\times$. From this, and from (2), we get

(3) *Suppose that λ is perfect. Then $\lambda = \bigcap_\mathfrak{u} \lambda_\mathfrak{u}$ and $\lambda^\times = \bigcup_\mathfrak{u} \lambda_\mathfrak{u}^\times$, where $\mathfrak{u}$ runs through all the positive vectors in $\lambda^\times$.*

If $\lambda \supset \varphi$ is normal, $\lambda = \bigcup_\mathfrak{x} \lambda_\mathfrak{x}^\times$, where $\mathfrak{x}$ runs through all the positive vectors in λ.

Using the terminology of § 19, 6., we can also express the fact that $\lambda = \bigcap_\mathfrak{u} \lambda_\mathfrak{u}$ by writing $\lambda = \mathsf{K}_\mathfrak{u}\, I_\mathfrak{u}^{(-1)}(\lambda_\mathfrak{u})$, where $I_\mathfrak{u}$ is the embedding of λ

in $\lambda_{\mathfrak{u}}$. If we equip each space $\lambda_{\mathfrak{u}}$ with its normal topology, then we can give λ the kernel topology, as in § 19, 6. It follows directly from the definition of this topology and from the definition of the normal topology on λ that these two topologies on λ are the same; thus it follows from (3) that

(4) *Every perfect space* $\lambda[\mathfrak{T}]$, *where* $\mathfrak{T}$ *is the normal topology, is the locally convex kernel* $\underset{\mathfrak{u}}{\mathsf{K}}\, I_{\mathfrak{u}}^{(-1)}(\lambda_{\mathfrak{u}}[\mathfrak{T}])$ *of the spaces* $\lambda_{\mathfrak{u}}[\mathfrak{T}]$ *defined by the positive vectors* $\mathfrak{u}$ *in* $\lambda^{\times}$.

If $\mathfrak{u}$ and $\mathfrak{v}$ are positive vectors in $\lambda^{\times}$, we set $\mathfrak{u}<\mathfrak{v}$ if $\mathfrak{v}-\mathfrak{u}>\mathfrak{o}$. In this way the positive vectors $\mathfrak{u}$ in $\lambda^{\times}$ form a directed set. If for $\mathfrak{u}<\mathfrak{v}$ we define $I_{\mathfrak{u}\mathfrak{v}}$ to be the embedding of $\lambda_{\mathfrak{v}}$ in $\lambda_{\mathfrak{u}}$, then $I_{\mathfrak{u}\mathfrak{v}}$ is a continuous mapping from $\lambda_{\mathfrak{v}}[\mathfrak{T}]$ into $\lambda_{\mathfrak{u}}[\mathfrak{T}]$. Using § 19, 7.(6) and § 19, 8.(1), we obtain the following stronger form of (4):

(5) *Every perfect space* $\lambda[\mathfrak{T}]$ *is topologically isomorphic to the projective limit* $\varprojlim I_{\mathfrak{u}\mathfrak{v}}(\lambda_{\mathfrak{v}}[\mathfrak{T}])$, *where* $\mathfrak{u}$ *and* $\mathfrak{v}$ *run through all pairs of positive vectors in* $\lambda^{\times}$ *satisfying* $\mathfrak{u}<\mathfrak{v}$.

The most important properties of sequence spaces can now be obtained from the topological properties of the spaces $\lambda_{\mathfrak{u}}[\mathfrak{T}]$, and from the general theory of locally convex kernels and projective limits.

5. Topological properties of sequence spaces. As before, let $\mathfrak{e}_i$ denote, the vector all of whose coordinates vanishes except the i-th, which is equal to one. The n-th section of a vector $\mathfrak{x}=(x_i)\in\lambda[\mathfrak{T}]$ is the vector

$$\mathfrak{x}_n=\sum_{i=1}^{n} x_i\,\mathfrak{e}_i=(x_1,\ldots,x_n,0,0,\ldots).$$

If follows directly from the convergence of $\mathfrak{u}\mathfrak{x}=\sum_{i=1}^{\infty} u_i x_i$ for each $\mathfrak{u}\in\lambda^{\times}$ that

(1) *The sequence* $\mathfrak{x}_n$ *of n-th sections of a vector* $\mathfrak{x}$ *in* $\lambda\supset\varphi$ *converges weakly to* $\mathfrak{x}$.

The mapping which is obtained by sending each vector $\mathfrak{x}\in\lambda$ to its i-th coordinate x_i is given by $\mathfrak{e}_i\mathfrak{x}=x_i$; since $\mathfrak{e}_i\in\lambda^{\times}$, this mapping is weakly continuous, and *a fortiori* it is a continuous linear functional on λ for the finer topologies $\mathfrak{T}_k(\lambda^{\times})$ and $\mathfrak{T}$.

(2) *If a sequence* $\mathfrak{x}^{(n)}$ *of vectors in a sequence space* λ *is weakly* $\mathfrak{T}$- *or* $\mathfrak{T}_k$-*convergent to* $\mathfrak{x}^{(0)}\in\lambda$, $\mathfrak{x}^{(n)}$ *is coordinatewise convergent to* $\mathfrak{x}^{(0)}$.

For if $\mathfrak{x}^{(n)}\to\mathfrak{x}^{(0)}$ with respect to one of these topologies, it follows that $\mathfrak{e}_i\mathfrak{x}^{(n)}=x_i^{(n)}\to\mathfrak{e}_i\mathfrak{x}^{(0)}=x_i^{(0)}$, by the remarks made above.

A corresponding result holds for filters $\mathfrak{F}$ which are Cauchy with respect to one of these topologies. If it exists, the limit $\mathfrak{x}^{(0)}$ of $\mathfrak{F}$ has, as i-th coordinate, the limit of the filter $\mathfrak{e}_i\mathfrak{F}$.

In the case where $\lambda=\omega$, all these modes of convergence coincide with coordinatewise convergence; this is the convergence which corresponds to the normal topology on ω.

We now give the first topological characterisation of perfect spaces.

(3) *A sequence space $\lambda\supset\varphi$ is perfect if and only if it is weakly sequentially complete.*

$\lambda^{\times\times}$ is obtained from λ by taking the coordinatewise limits of the weak Cauchy sequences in λ.

Proof. a) Suppose that λ is perfect. ω is weakly complete, and *a fortiori* it is weakly sequentially complete. By § 22, 4.(2), l^1 is also weakly sequentially complete; so therefore is each $\lambda_{\mathfrak{u}}$, $\mathfrak{u}>\mathfrak{o}$, $\mathfrak{u}\in\lambda^{\times}$, by 4.

From 4.4 it follows that

$$(4)\qquad \lambda[\mathfrak{T}_s]=\varprojlim I_{\mathfrak{u}\mathfrak{v}}(\lambda_{\mathfrak{u}}[\mathfrak{T}_s]).$$

This can either be confirmed directly, or we can use § 22, 7.(6). It then follows from § 19, 10.(2) that λ is weakly sequentially complete.

b) Suppose that $\lambda\supset\varphi$. By (1), $\mathfrak{x}\in\lambda^{\times\times}$ is the $\mathfrak{T}_s(\lambda^{\times\times\times})$-limit of its sections $\mathfrak{x}_n$, Since $\lambda^{\times\times\times}=\lambda^{\times}$, and since $\mathfrak{x}_n\in\lambda$, $\mathfrak{x}$ is the $\mathfrak{T}_s(\lambda^{\times})$-limit of the $\mathfrak{x}_n\in\lambda$.

It follows directly from (3) and the BANACH–MACKEY theorem (§ 20, 11.(8)) that

(5) *In every perfect space λ the weakly and strongly bounded sets are the same, and so therefore are the topologies $\mathfrak{T}_b(\lambda^{\times})$ and $\mathfrak{T}_{b^*}(\lambda^{\times})$.*

The next result makes it easier to determine the bounded subsets of a sequence space:

(6) *If $\lambda\supset\varphi$ and if λ is normal, the normal cover of every bounded subset M of λ is bounded.*

Since every bounded subset of λ is also bounded in $\lambda^{\times\times}$, and since λ contains M^n if it contains M, we may suppose that λ is perfect.

Now in l^1 and in ω the normal cover of a bounded set is bounded. This therefore holds in each $\lambda_{\mathfrak{u}}$, and so by 4.(4) and § 19, 6.(7) it also holds in λ.

We now give a second topological characterisation of perfect spaces:

(7) *A sequence space $\lambda\supset\varphi$ is perfect if and only if it is $\mathfrak{T}$-complete, where $\mathfrak{T}$ is the normal topology.*

$\lambda^{\times\times}$ is the $\mathfrak{T}$-completion of $\lambda\supset\varphi$.

Proof. a) If λ is perfect, $\lambda[\mathfrak{T}]$ is the projective limit of the spaces $\lambda_{\mathfrak{u}}[\mathfrak{T}]$, by 4.(5). Since these are complete, $\lambda[\mathfrak{T}]$ is also complete, by § 19, 10.(2).

b) As in (3)b), the fact that $\lambda^{\times\times}$ is the completion of $\lambda[\mathfrak{T}]$ follows from the following result, which strengthens (1):

(8) *The sequence* $\mathfrak{x}_n$ *of n-th sections of a vector* $\mathfrak{x}$ *of a sequence space* $\lambda \supset \varphi$ *is* $\mathfrak{T}$*-convergent to* $\mathfrak{x}$.

It is sufficient to prove this for perfect λ. The assertion is true in ω and in l^1, and so it is true in all the spaces $\lambda_{\mathfrak{u}}$, $\mathfrak{u} \in \lambda^\times$. But it follows from $p_{\mathfrak{u}}(\mathfrak{x} - \mathfrak{x}_n) \to 0$ for all $\mathfrak{u} \in \lambda^\times$ that $\mathfrak{x}_n \to \mathfrak{x}$ with respect to the normal topology on λ.

Since $\mathfrak{T}_k(\lambda^\times)$ is finer that $\mathfrak{T}$ (cf. 2.(4)), it follows from (7) that

(9) *Every perfect space is complete under its Mackey topology.*

If we restrict ourselves to normal spaces λ containing φ, (8) can also be improved:

(10) *If* $\lambda \supset \varphi$ *and if* λ *is normal, the sections* $\mathfrak{x}_n$ *of an* $\mathfrak{x} \in \lambda$ *converge to* $\mathfrak{x}$ *with respect to the Mackey topology.*

This time we use the representation $\lambda = \bigcup_{\mathfrak{x}} \lambda_{\mathfrak{x}}^\times$, $\mathfrak{x} \in \lambda$, of 4.(3).

The assertion is true for φ, and it is also true for l^∞; this can easily be seen from the form that the weakly compact subsets of l^1 take (§ 22, 4.(3)). As a result, the assertion holds in each $\lambda_{\mathfrak{x}}^\times$. But since $\lambda_{\mathfrak{x}} \supset \lambda^\times$, so that $\mathfrak{T}_k(\lambda^\times)$ is coarser than $\mathfrak{T}_k(\lambda_{\mathfrak{x}})$, the fact that $\mathfrak{x}_n \to \mathfrak{x}$ with respect to $\mathfrak{T}_k(\lambda_{\mathfrak{x}})$ implies that $\mathfrak{x}_n \to \mathfrak{x}$ with respect to $\mathfrak{T}_k(\lambda^\times)$.

(11) *Every sequence space* $\lambda[\mathfrak{T}] \supset \varphi$ *is sequentially separable. If further,* $\lambda[\mathfrak{T}]$ *is normal,* λ *is also sequentially separable under the Mackey topology.*

We shall show that the set N of vectors $\sum_{i=1}^{n} \rho_i \mathfrak{e}_i$, with ρ_i rational and n arbitrary, are sequentially dense in λ for the topologies $\mathfrak{T}$ and $\mathfrak{T}_k$. Given $\mathfrak{x} \in \lambda$, we can find, for each n, an element $\mathfrak{x}^{(n)} = \sum_{i=1}^{n} \rho_i^{(n)} \mathfrak{e}_i$ in N for which

$$|x_i - \rho_i^{(n)}| \leqq \frac{|x_i|}{n}, \qquad i = 1, \ldots, n. \tag{12}$$

Now if B is a $\mathfrak{T}$- or $\mathfrak{T}_k$-equicontinuous subset of $\lambda^\times$,

$$\sup_{\mathfrak{u} \in B} |\mathfrak{u}(\mathfrak{x} - \mathfrak{x}^{(n)})| \leqq \sup |\mathfrak{u}(\mathfrak{x} - \mathfrak{x}_n)| + \sup |\mathfrak{u}(\mathfrak{x}_n - \mathfrak{x}^{(n)})|. \tag{13}$$

The first summand of the right-hand side is less than or equal to $\frac{\varepsilon}{2}$ for $n \geqq n_0$, by (8), and by (10), respectively. But the second summand can also be made arbitrarily small: B is a bounded set, and by (6) the normal cover of B is bounded. Consequently $\sup_{\mathfrak{u} \in B} \sum_{i=1}^{\infty} |u_i|\,|x_i| = K < \infty$. It therefore follows from (12) that

$$\sup |\mathfrak{u}(\mathfrak{x}_n - \mathfrak{x}^{(n)})| \leqq \sup \sum_{i}^{n} |u_i|\,|x_i - \rho_i^{(n)}| \leqq \frac{K}{n} \leqq \frac{\varepsilon}{2}$$

for sufficiently large n; the assertion now follows, using (13).

6. Compact subsets of a perfect space. Suppose that λ is perfect and that $\mathfrak{T}'$ is a locally convex topology on λ; the only assumption we make about $\mathfrak{T}'$ is that it is finer than $\mathfrak{T}_s(\lambda^\times)$. We have (cf. KÖTHE [9]):

(1) *Suppose that M is a subset of the perfect space λ. The following properties of M are equivalent:*

a) M is $\mathfrak{T}'$-compact;

b) M is countably $\mathfrak{T}'$-compact;

c) M is sequentially $\mathfrak{T}'$-compact;

d) M is bounded, and every sequence $\mathfrak{x}^{(n)}\in M$ which is coordinatewise convergent to a vector $\mathfrak{x}^{(0)}$ in ω is $\mathfrak{T}'$-convergent to $\mathfrak{x}^{(0)}$, and $\mathfrak{x}^{(0)}$ lies in M.

Proof. b) follows trivially from a).

c) follows from b): Suppose that M is countably $\mathfrak{T}'$-compact, and suppose that $\mathfrak{x}^{(n)}$ is a sequence in M. Since M is bounded, it is coordinatewise bounded, and so using a diagonal procedure we can choose a coordinatewise convergent subsequence of the $\mathfrak{x}^{(n)}$. Let this be denoted by $\mathfrak{x}^{(n)}$ again, and let $\mathfrak{x}_0$ be its coordinatewise limit. By hypothesis $\mathfrak{x}^{(n)}$ has a $\mathfrak{T}'$-adherent point $\mathfrak{y}_0$ in M. By the remark preceding 5.(2), however, this can only be the coordinatewise limit $\mathfrak{x}_0$. For the same reason, $\mathfrak{x}_0$ is the only possible adherent point of any subsequence of $\mathfrak{x}^{(n)}$, so that it must be the $\mathfrak{T}'$-limit of $\mathfrak{x}^{(n)}$. This is essentially the method of proving ŠMULIAN's theorem (cf. § 24, 1.(2)).

d) follows from c): Suppose that M is $\mathfrak{T}'$-sequentially compact. M is then bounded. A sequence $\mathfrak{x}^{(n)}$ which is coordinatewise convergent to $\mathfrak{x}_0$ can only have $\mathfrak{x}_0$ as $\mathfrak{T}'$-limit. If $\mathfrak{x}_0$ were not the $\mathfrak{T}'$-limit of $\mathfrak{x}^{(n)}$, a subsequence would have to be convergent to some $\mathfrak{y}_0 \neq \mathfrak{x}_0$, and this is not possible.

a) follows from d): Suppose that M satisfies hypothesis d), and that $\mathfrak{F}=\{F^\alpha\}$ is a filter on M. If to each $\mathfrak{x}=(x_1,x_2,\ldots)\in M$ we make correspond the element $(x_1,\ldots,x_n)\in \mathsf{K}^n$, we obtain from $\mathfrak{F}=\{F^\alpha\}$ a filter $\mathfrak{F}_n=\{F_n^\alpha\}$ on K^n. Let G_n be the bounded set of adherent points of $\mathfrak{F}_n$ in K^n. Let $\mathfrak{y}^{(n)}\in\lambda$ be chosen in such a way that $(y_1^{(n)},\ldots,y_n^{(n)})$ belongs to G_n. We can pick out a subsequence of the $\mathfrak{y}^{(n)}$ which is coordinatewise convergent; this we again denote by $\mathfrak{y}^{(n)}$. Let its coordinatewise limit be $\mathfrak{x}^{(0)}\in\omega$. In each set F^α there is now an element $\mathfrak{z}^{(n)}$ with $|z_i^{(n)}-y_i^{(n)}|\leq \dfrac{1}{n}$, for $i=1,\ldots,n$. The sequence $\mathfrak{z}^{(n)}$ clearly converges coordinatewise to $\mathfrak{x}^{(0)}$. By d), $\mathfrak{x}^{(0)}$ belongs to M, and it is the $\mathfrak{T}'$-limit of the sequence $\mathfrak{z}^{(n)}$ in F^α. Thus $\mathfrak{x}^{(0)}$ is a $\mathfrak{T}'$-closure point of each F^α, so that $x^{(0)}$ is a $\mathfrak{T}'$-adherent point of the filter $\mathfrak{F}$ on M.

The fact that a) and b) are equivalent contains EBERLEIN's theorem for perfect spaces; in the present case, therefore, this can be proved in a much simpler way.

The next result is important for determining weakly compact sets:

(2) *The weakly closed normal cover* $\overline{M^n}$ *of a weakly compact subset* M *of a perfect space* λ *is always weakly compact.*

This is true for $\lambda=\omega$, since every bounded subset of ω is weakly relatively compact. It is also true for $\lambda=l^1$, as can be seen from the structure of the weakly compact subsets of l^1 (cf. § 22, 4.(3)). Thus (2) holds for all the spaces $\lambda_{\mathfrak{u}}$, $\mathfrak{u}\in\lambda^{\times}$. Now if M is weakly compact in λ, it is *a fortiori* weakly compact in each space $\lambda_{\mathfrak{u}}$. If $\mathfrak{x}^{(n)}\in\overline{M^n}$ is coordinatewise convergent to $\mathfrak{x}_0$, $\mathfrak{x}^{(n)}$ converges weakly to $\mathfrak{x}_0$ in each $\lambda_{\mathfrak{u}}$, by (1) d), so that $\mathfrak{x}_0$ belongs to $\lambda=\bigcap\lambda_{\mathfrak{u}}$, and $\mathfrak{x}^{(n)}$ also converges weakly in λ to $\mathfrak{x}_0$. $\overline{M^n}$ is therefore weakly compact, by (1) d).

(3) *A subset* M *of the perfect space* $\lambda[\mathfrak{T}]$ *(where* $\mathfrak{T}$ *is the normal topology) is weakly compact if and only if it is* $\mathfrak{T}$*-compact. Weakly convergent and* $\mathfrak{T}$*-convergent sequences are therefore always the same.*

Suppose that M is weakly compact in $\lambda[\mathfrak{T}]=\bigcap_{\mathfrak{u}}\lambda_{\mathfrak{u}}[\mathfrak{T}]$. M is weakly compact in each space $\lambda_{\mathfrak{u}}$. By § 22, 4.(3) M is strongly sequentially compact in $\lambda_{\mathfrak{u}}$, and *a fortiori* it is $\mathfrak{T}$-sequentially compact in $\lambda_{\mathfrak{u}}$. If $\mathfrak{x}^{(n)}\in M$ is coordinatewise convergent to $\mathfrak{x}_0$, $\mathfrak{x}^{(n)}$ is $\mathfrak{T}$-convergent to $\mathfrak{x}_0$ in each space $\lambda_{\mathfrak{u}}$, by (1) d), and by 4.(4) $\mathfrak{x}^{(n)}$ is therefore also $\mathfrak{T}$-convergent to $\mathfrak{x}_0$ in λ. Consequently M is $\mathfrak{T}$-compact, by (1).

Applying 5.(7), § 21, 7. and § 21, 9.(7), it follows from (3) that

(4) *The Mackey topology* $\mathfrak{T}_k(\lambda^{\times})$ *on the perfect space* λ *is the polar topology* $\mathfrak{T}^{\circ}$ *of the normal topology* $\mathfrak{T}$ *on* $\lambda^{\times}$.

$\mathfrak{T}_k(\lambda)$ *is the finest locally convex topology which coincides on the* $\mathfrak{T}$*-equicontinuous sets of* $\lambda^{\times}$ *with the weak topology.*

The polar topology of $\mathfrak{T}_k$ can also be characterised in a simple way:

(5) *The topology* $\mathfrak{T}_k^{\circ}$, *the topology of uniform convergence on the* $\mathfrak{T}_k$*-compact subsets of the perfect space* λ, *is the finest locally convex topology* $\mathfrak{T}'$ *on* $\lambda^{\times}$ *which gives the same convergent sequences as the weak topology.*

By § 21, 9.(7), $\mathfrak{T}_k^{\circ}$ is the finest locally convex topology on $\lambda^{\times}$ which coincides with the weak topology on the weakly compact subsets of $\lambda^{\times}$. It follows from this that every weakly convergent sequence is $\mathfrak{T}_k^{\circ}$-convergent. Conversely if every weakly convergent sequence is $\mathfrak{T}'$-convergent, the weakly compact and $\mathfrak{T}'$-compact sets are the same, by (1), and so $\mathfrak{T}_k^{\circ}$ is finer than $\mathfrak{T}'$.

By (3) and (5), $\mathfrak{T}_k^{\circ}$ is always finer than $\mathfrak{T}$.

The next result is analogous to (2):

(6) *The weakly closed normal hull* $\overline{M^n}$ *of every* $\mathfrak{T}_k$*-compact subset* M *of a perfect space is again* $\mathfrak{T}_k$*-compact.*

If $\overline{M^n}$ were not $\mathfrak{T}_k$-compact, there would be a weakly convergent sequence $\mathfrak{u}^{(n)} \to \mathfrak{o}$ in $\lambda^\times$ for which $\sup\limits_{\mathfrak{x} \in \overline{M^n}} |\mathfrak{u}^{(n)}\mathfrak{x}| > m > 0$, by (5). There would thus be a sequence $\mathfrak{x}^{(n)} \in M$ with $\sum\limits_1^\infty |u_i^{(n)}|\,|x_i^{(n)}| \geqq m$. Since $\mathfrak{u}^{(n)}$ converges weakly to $\mathfrak{o}$, every sequence $\mathfrak{v}^{(n)}$ with $|v_i^{(n)}| = |u_i^{(n)}|$ converges weakly to $\mathfrak{o}$ as well, by (3). If $\mathfrak{v}^{(n)}$ is chosen in such a way that $\sum\limits_i^\infty |u_i^{(n)}|\,|x_i^{(n)}| = \mathfrak{v}^{(n)}\mathfrak{x}^{(n)}$, $\sup\limits_{\mathfrak{x} \in M} |\mathfrak{v}^{(n)}\mathfrak{x}|$ would not converge to 0, in contradiction to (5).

The normal cover of a strongly compact subset of a perfect space need not be strongly compact, as the example of the set consisting of the vector $\mathfrak{e} = (1,1,\ldots)$ in l^∞ shows.

7. Barrelled spaces and (M)-spaces. We saw in 5.(10) that in every perfect space λ the sections $\mathfrak{x}_n$ of a vector $\mathfrak{x}$ converge to $\mathfrak{x}$ in the topology $\mathfrak{T}_k(\lambda^\times)$. As the example of l^∞ shows, this need not be so for the strong topology. The class of perfect spaces for which the sections converge strongly can be characterised in the following way:

(1) *The following properties of a perfect space are equivalent:*

a) $\lambda[\mathfrak{T}_k]$ *is barrelled;*

b) $\mathfrak{T}_k$*-convergent and strongly convergent sequences are the same;*

c) every $\mathfrak{T}_k$*-compact subset of* λ *is strongly compact;*

d) the weakly closed normal cover of every strongly compact subset of λ *is strongly compact;*

e) the sections of every $\mathfrak{x} \in \lambda$ *converge strongly to* $\mathfrak{x}$*;*

f) $\lambda[\mathfrak{T}_b]$ *is sequentially separable.*

Proof. b) and c) follow from a), and are equivalent, by 6.(1), d) follows from c) by 6.(6). If d) holds, the weakly closed normal cover of the set consisting of a single vector $\mathfrak{x} \in \lambda$ is strongly compact. From this it follows by 6.(1) that the sections of $\mathfrak{x}$ converge strongly to $\mathfrak{x}$, so that e) holds. f) follows from e) as in 5.(11). In order to be able to derive a) from f) we must show that every weakly closed bounded subset M of $\lambda^\times$ is weakly sequentially compact. Let $\mathfrak{u}^{(n)}$ be a sequence in M, and let $\mathfrak{x}^{(i)}$, $i = 1,2,\ldots$, run through a strongly sequentially dense sequence in λ. By using a diagonal procedure, we obtain a subsequence, which we again denote by $\mathfrak{u}^{(n)}$, for which $\lim\limits_{n\to\infty} \mathfrak{u}^{(n)}\mathfrak{x}^{(i)}$ exists, for each i. We therefore have

$$(2) \qquad |(\mathfrak{u}^{(n)} - \mathfrak{u}^{(m)})\mathfrak{x}^{(i)}| \leqq \frac{\varepsilon}{2} \quad \text{for} \quad n,m \geqq n_0(\varepsilon, i).$$

Let $\mathfrak{x} \in \lambda$, and let $\mathfrak{x}^{(j)}$ be a subsequence of $\mathfrak{x}^{(i)}$ which converges strongly to $\mathfrak{x}$. Since the set of all expressions $\mathfrak{u}^{(n)} - \mathfrak{u}^{(m)}$ is bounded in $\lambda^\times$, there is a j_0 for which

(3) $|(\mathfrak{u}^{(n)}-\mathfrak{u}^{(m)})(\mathfrak{x}-\mathfrak{x}^{(j)})|\leqq\frac{\varepsilon}{2}$ for all n,m and for $j\geqq j_0(\varepsilon)$.

From (2) and (3) it follows that $|(\mathfrak{u}^{(n)}-\mathfrak{u}^{(m)})\mathfrak{x}|\leqq\varepsilon$ for $n,m\geqq n_0(\varepsilon,j_0)$. Thus the $\mathfrak{u}^{(n)}$ form a weak Cauchy sequence in M, whose limit lies in M, by hypothesis, and by 5.(3); thus M is weakly sequentially compact.

Since a perfect space $\lambda[\mathfrak{T}]$ is semi-reflexive if and only if $(\lambda^\times[\mathfrak{T}_b(\lambda)])'=\lambda$, and so if and only if $\lambda^\times[\mathfrak{T}_k(\lambda)]$ is barrelled, the next result follows directly from (1):

(4) *A perfect space $\lambda[\mathfrak{T}]$ (where $\mathfrak{T}$ is the normal topology) is semi-reflexive if and only if $\lambda^\times$ satisfies one of the conditions a) to f) of* (1).

Likewise, we have

(5) *A perfect space $\lambda[\mathfrak{T}_k]$ is reflexive if and only if both λ and $\lambda^\times$ satisfy one of the conditions a) to f) of* (1).

For the Mackey topology, we have

(6) *If λ is a perfect space, the following properties are equivalent:*

a) *weakly convergent and $\mathfrak{T}_k$-convergent sequences in λ are the same;*

b) *every weakly compact subset of λ is $\mathfrak{T}_k$-compact;*

c) *the topologies $\mathfrak{T}_k(\lambda)$ and $\mathfrak{T}_k^\circ(\lambda)$ on $\lambda^\times$ are the same.*

A perfect space λ has these properties if and only if its α-dual $\lambda^\times$ has them.

Proof. a) and b) are equivalent, by 6.(1) a) and c). Further b) is equivalent to c). By 6.(5) the weak topology and the topology $\mathfrak{T}_k^\circ$ coincide on a weakly compact subset of $\lambda^\times$. Thus if c) holds, every weakly compact subset of $\lambda^\times$ is $\mathfrak{T}_k$-compact, i. e. b) is satisfied, for $\lambda^\times$.

Let us remark that in (6) weak convergence and the weak topology can also be replaced by normal convergence and the normal topology, using 6.(3). In particular λ has the properties described in (6) if the normal topology $\mathfrak{T}$ is the same as the Mackey topology.

(7) *Weakly and strongly convergent sequences in a perfect space λ are the same if and only if the bounded sets of $\lambda^\times$ are relatively $\mathfrak{T}_k$-compact. Such a space is always barrelled.*

This follows directly from (1) and (6).

We now give a criterion for a perfect space to be an (M)-space. We observe that, by (1), every perfect (M)-space is sequentially separable.

(8) *A perfect space $\lambda[\mathfrak{T}_k]$ is an* (M)*-space if and only if the sections of each vector $\mathfrak{x}$ in λ converge strongly to $\mathfrak{x}$ and the weakly (or $\mathfrak{T}$-) convergent sequences in $\lambda^\times$ are the same as the strongly convergent ones.*

By (1), the first condition means that $\lambda[\mathfrak{T}_k]$ is barrelled and that every $\mathfrak{T}_k$-compact subset of λ is strongly compact. The second condition means that every bounded closed subset is $\mathfrak{T}_k$-compact, by (7). The two together give the result.

T. and Y. KŌMURA [1] have given an example of a barrelled perfect space which is not bornological.

8. Echelon and co-echelon spaces. In 4. we introduced the special perfect spaces $\lambda_{\mathfrak{a}}$ and their α-duals $\lambda_{\mathfrak{a}}^{\times}$. If we are given countably many vectors $\mathfrak{a}^{(k)}$, $k=1,2,\ldots$, in ω, which we call steps, we call the linear span $\sum_{k=1}^{\infty} \lambda_{\mathfrak{a}^{(k)}}^{\times}$ the co-echelon space, and the intersection $\bigcap_{k=1}^{\infty} \lambda_{\mathfrak{a}^{(k)}}$ the echelon space, corresponding to them.

We know (cf. 4.(1)) that $\bigcap \lambda_{\mathfrak{a}^{(k)}}$ is perfect, and that $\sum \lambda_{\mathfrak{a}^{(k)}}^{\times}$ is normal. The steps can all be supposed to be positive. By possibly going over from the steps $\mathfrak{a}^{(k)}$ to the steps $\mathfrak{a}^{(1)}+\cdots+\mathfrak{a}^{(k)}$, we can obtain a monotonic increasing system of steps, without altering the spaces. Further we can suppose that the system of steps is complete, in the sense that for each index i there is an $\mathfrak{a}^{(k)}$ for which $\mathfrak{a}_i^{(k)}$ is non-zero. In what follows we shall always make these assumptions. Then $\sum \lambda_{\mathfrak{a}^{(k)}}^{\times}$ is the union of the spaces $\lambda_{\mathfrak{a}^{(1)}}^{\times} \subset \lambda_{\mathfrak{a}^{(2)}}^{\times} \subset \cdots$.

The projective limit topology $\mathfrak{T}'$ on $\lambda=\bigcap_k \lambda_{\mathfrak{a}^{(k)}}$ is defined by the semi-norms $q_k(\mathfrak{x})=\sum_{i=1}^{\infty} |a_i^{(k)}|\,|x_i|$, $k=1,2,\ldots$, so that $\lambda[\mathfrak{T}']$ is an (F)-space. The dual λ' of $\lambda[\mathfrak{T}']$ is equal to the co-echelon space $\bigcup_k \lambda_{\mathfrak{a}^{(k)}}^{\times}$. λ', being the dual of an (F)-space, is weakly sequentially complete. Since $(\lambda')^{\times}=\lambda$ (cf. 4.(1)), λ' is perfect, by 5.(3), so that $\lambda'=\bigcup \lambda_{\mathfrak{a}^{(k)}}^{\times}=\lambda^{\times}$, and $\mathfrak{T}'$ is the same as the normal topology $\mathfrak{T}$ on λ.

Thus we have shown

(1) *The spaces $\bigcap \lambda_{\mathfrak{a}^{(k)}}$ and $\bigcup \lambda_{\mathfrak{a}^{(k)}}^{\times}$ are perfect and α-dual to each other. The normal topology $\mathfrak{T}$ on $\lambda=\bigcap \lambda_{\mathfrak{a}^{(k)}}$ is given by the semi-norms $q_k(\mathfrak{x})$, and $\lambda[\mathfrak{T}]$ is an* (F)*-space.*

On any (F)-space, the topology is the strong topology. Thus it follows from $\mathfrak{T}=\mathfrak{T}_b(\lambda^{\times})$ and 6.(3) that weak and strong convergence of sequences coincide in every echelon space. Further, $\lambda[\mathfrak{T}]$ is sequentially separable (7.(1)). In addition it follows from $\mathfrak{T}=\mathfrak{T}_b(\lambda^{\times})$ that every bounded subset of the co-echelon space $\lambda^{\times}=\bigcup \lambda_{\mathfrak{a}^{(k)}}^{\times}$ lies in the normal cover of a vector $\rho\,\mathfrak{a}^{(k)}$, and so it is bounded in some $\lambda_{\mathfrak{a}^{(k)}}^{\times}$. A weakly convergent sequence $\mathfrak{u}^{(n)}\to\mathfrak{u}$ in $\lambda^{\times}$ is therefore bounded in some $\lambda_{\mathfrak{a}^{(k)}}^{\times}$, and is coordinatewise convergent. It is therefore also weakly convergent in some $\lambda_{\mathfrak{a}^{(k)}}^{\times}$.

The definition of co-echelon space was generalised by DIEUDONNÉ and GOMES [1]. Corresponding to a positive vector $\mathfrak{a}$, with $a_i>0$ for all i, we form the set $\lambda_{\mathfrak{a}}^{p}$ of all $\mathfrak{x}$ with $\sum_{i=1}^{\infty} |a_i|\,|x_i|^p<\infty$, for some $p\geqq 1$. $\lambda_{\mathfrak{a}}^{p}$, equipped with the norm $q(\mathfrak{x})=(\sum |a_i|\,|x_i|^p)^{1/p}$, is a diagonal transform of l^p, and so it is perfect. Its dual (and at the same time its α-dual) is the space $(\lambda_{\mathfrak{a}}^p)^\times$ of all $\mathfrak{u}$ with $\sum |a_i|^{-\frac{q}{p}} |u_i|^q < \infty$, $\frac{1}{p} + \frac{1}{q} = 1$. $(\lambda_{\mathfrak{a}}^p)^\times$ is a diagonal transform of l^q, and has norm $q(\mathfrak{u}) = (\sum |a_i|^{-\frac{q}{p}} |u_i|^q)^{\frac{1}{q}}$. If infinitely many a_i are zero, $\lambda_{\mathfrak{a}}^p$ decomposes into a sectional subspace which is a diagonal transform of l^p and one which is isomorphic to ω; corresponding to this there is a sectional subspace φ of $(\lambda_{\mathfrak{a}}^p)^\times$.

Once again let $\mathfrak{a}^{(1)} \leqq \mathfrak{a}^{(2)} \leqq \cdots$ be a complete sequence of positive steps. The intersection $\bigcap_{k=1}^{\infty} \lambda_{\mathfrak{a}^{(k)}}^p$, with the topology $\mathfrak{T}_0$ given by the semi-norms $q_k(\mathfrak{x}) = (\sum |a_i^{(k)}| \, |x_i|^p)^{1/p}$, $k = 1, 2, \ldots$, is called the echelon space of p-th order corresponding to the $\mathfrak{a}^{(k)}$; it is an (F)-space. Similarly, the space $\bigcup_{n=1}^{\infty} (\lambda_{\mathfrak{a}^{(k)}}^p)^\times$ is called the co-echelon space of p-th order. When $p=1$, we obtain the echelon spaces considered above.

Using the same method of proof as before, we obtain

(2) *The spaces* $\bigcap_{k=1}^{\infty} \lambda_{\mathfrak{a}^{(k)}}^p$ *and* $\bigcup_{k=1}^{\infty} (\lambda_{\mathfrak{a}^{(k)}}^p)^\times$ *are perfect and α-dual to each other.*

The topology $\mathfrak{T}_0 = \mathfrak{T}_b(\lambda^\times)$ is of course in general strictly finer than the normal topology for $p>1$. However $\lambda = \bigcap \lambda_{\mathfrak{a}^{(k)}}^p$ is always sequentially separable. It follows from the fact that $\mathfrak{T}_0 = \mathfrak{T}_b$ that every bounded subset of the co-echelon space $\lambda^\times = \bigcup (\lambda_{\mathfrak{a}^{(k)}}^p)^\times$ is contained in a scalar multiple of the unit ball of some $(\lambda_{\mathfrak{a}^{(k)}}^p)^\times$, and that weak sequential convergence in $\lambda^\times$ means the same as weak sequential convergence in a suitable $(\lambda_{\mathfrak{a}^{(k)}}^p)^\times$ (for a bounded coordinatewise convergent sequence in l^p, $p>1$, is weakly convergent in l^p).

It follows from § 23, 3.(7) and the fact that the spaces $\lambda_{\mathfrak{a}^{(k)}}^p$ are reflexive for $p>1$ that

(3) *Every co-echelon space $\lambda^\times[\mathfrak{T}_k(\lambda)]$ and every echelon space $\lambda[\mathfrak{T}_0]$ of order $p>1$ is reflexive.*

Example. If the vectors $\mathfrak{a}^{(k)} = (1, k, k^2, \ldots)$, $k = 1, 2, \ldots$, are taken as steps, the corresponding echelon space λ can be considered as the space of entire transcendental functions: to the vector $\mathfrak{x} = (x_0, x_1, \ldots) \in \lambda$ we assign the function $x(z) = \sum_{i=0}^{\infty} x_i z^i$. If, in a corresponding way, we assign to each $\mathfrak{u} \in \lambda^\times$ the function $u(z) = \frac{1}{z}\left(u_0 + u_1 \frac{1}{z} + u_2 \frac{1}{z^2} + \cdots\right)$, we obtain the space of all functions which are analytic in a neighbourhood of ∞ and which vanish at ∞. The scalar product can now be interpreted in the following way:

$$(4) \qquad \frac{1}{2\pi i} \oint u(z) x(z) \, dz = \sum_{i=0}^{\infty} u_i x_i = \mathfrak{u}\mathfrak{x}.$$

For on multiplying the two series together we obtain the Laurent expansion of $u(z)x(z)$ in an annulus about 0, and $\mathfrak{u}\mathfrak{x}$ is the coefficient of $1/z$. The integral is taken along a circular contour about 0 which lies within the annulus.

In the terminology of § 27,4., $H(\Gamma)$ is the space of entire functions, with the topology defined there. Using (4), we obtain an isomorphism of the dual pairs $\langle H(\infty), H(\Gamma)\rangle$ and $\langle\lambda^\times,\lambda\rangle$; in particular $\lambda[\mathfrak{T}]$ and $\lambda^\times[\mathfrak{T}_k]$ are topologically isomorphic to $H(\Gamma)$ and $H(\infty)$ respectively, when these spaces are given the topologies defined in § 27,4. Consequently $\lambda[\mathfrak{T}]$ and $\lambda^\times[\mathfrak{T}_k]$ are (M)-spaces, each being the other's dual. For this, cf. Toeplitz [2].

9. Co-echelon spaces of type (M). A precise criterion can be given for a co-echelon space to be an (M)-space. We have (cf. Köthe [6], Dieudonné and Gomes [1]):

(1) *Suppose that* $\mathfrak{a}^{(k)}=(a_i^{(k)})$, $k=1,2,\ldots$, *is a complete monotonic increasing system of steps, and that* $\lambda[\mathfrak{T}_0]$ *and* $\lambda^\times[\mathfrak{T}_k]$ *are the corresponding echelon and co-echelon spaces of order* $p\geqq 1$. *These two spaces are both* (M)-*spaces if and only if there is no infinite index set* $\{j_n\}$ *such that for some suitable* k_0, *and for suitable* $M_k>0$, *we have*

(2) $$0<a_{j_n}^{(k)}\leqq M_k a_{j_n}^{(k_0)}$$

for all $k\geqq k_0$.

Expressed in another way: λ *and* $\lambda^\times$ *are both* (M)-*spaces if and only if there is no sectional subspace of* λ *(respectively* $\lambda^\times$*) which is a diagonal transform of* l^p *(respectively* l^q*)* $\left(\frac{1}{p}+\frac{1}{q}=1,\ q=\infty \text{ for } p=1\right)$.

An immediate corollary is: *An echelon or co-echelon space of order* 1 *is reflexive if and only if it is an* (M)-*space.*

Proof of (1). By 7.(8), λ is an (M)-space (and so, therefore, is $\lambda^\times$) if and only if weak and strong sequential convergence coincide in $\lambda^\times$. This does not happen if $\lambda^\times$ has a sectional subspace which is a diagonal transform of l^q, $q>1$, or of l^∞.

We must therefore show that we can deduce (2) from the fact that there is a sequence $\mathfrak{u}^{(n)}$ in $\lambda^\times$ which converges weakly but not strongly to $\mathfrak{o}$. By 8., we can suppose that $\mathfrak{u}^{(n)}$ converges weakly to $\mathfrak{o}$ in some $(\lambda^p_{\mathfrak{a}^{(k_0)}})^\times$. By making a diagonal transformation, and restricting attention to the sectional subspace defined by the non-zero $a_i^{(k_0)}$, we may suppose that $\mathfrak{a}^{(k_0)}=\mathfrak{e}=(1,1,\ldots)$, so that $(\lambda^p_{\mathfrak{a}^{(k_0)}})^\times=l^q$.

Suppose therefore that $\|\mathfrak{u}^{(n)}\|_q=(\sum|u_i^{(n)}|^q)^{1/q}=1$, and that $\mathfrak{u}^{(n)}$ converges coordinatewise to $\mathfrak{o}$. Further suppose that M is a normal bounded subset in λ for which

(3) $$\sup_{\mathfrak{x}\in M}\sum_{i=1}^{\infty}|u_i^{(n)}|\,|x_i|\geqq 2c>0$$

for all n.

For each $\mathfrak{u}^{(n)}$ there therefore exists an $\mathfrak{x}^{(n)}\in M$ for which

$$c<\sum|u_i^{(n)}|\,|x_i^{(n)}|\leqq\|\mathfrak{u}^{(n)}\|_q\,\|\mathfrak{x}^{(n)}\|_p=\|\mathfrak{x}^{(n)}\|_p.$$

From this, from the fact that M is normal and from the fact $\mathfrak{u}^{(n)}$ converges coordinatewise to $\mathfrak{o}$ it follows that there exist infinitely many vectors $\mathfrak{y}^{(n)}=(0,\ldots,0,y_{r_n}^{(n)},\ldots,y_{s_n}^{(n)},0,0,\ldots)$ in M, with $\|\mathfrak{y}^{(n)}\|_p>c$ and $s_n<r_{n+1}$ for $n=1,2,\ldots$.

Since M is bounded, there exists $m_k<\infty$ such that

$$(4)\qquad (q_k(\mathfrak{x}))^p=\sum_{i=1}^{\infty}|a_i^{(k)}|\,|x_i|^p\leqq m_k,\qquad k=1,2,\ldots,$$

for all $\mathfrak{x}\in M$.

Suppose that $d_k>0$ and $\sum_{k=0}^{\infty}\frac{1}{d_k}=\frac{1}{2}$. Let t_{kn} run through all indices $r_n\leqq t_{kn}\leqq s_n$ for which

$$(5)\qquad a_{t_{kn}}^{(k)}>\frac{m_k d_k}{c^p}.$$

For the corresponding coordinates of $\mathfrak{y}^{(n)}$ (with k,n fixed) we have

$$(6)\qquad \sum_{t_{kn}}|y_{t_{kn}}^{(n)}|^p<\frac{c^p}{d_k},$$

since otherwise we would have $\sum_{t_{kn}}|a_{t_{kn}}^{(k)}|\,|y_{t_{kn}}^{(n)}|^p>\frac{m_k d_k}{c^p}\cdot\frac{c^p}{d_k}=m_k$, which would contradict the fact that $\mathfrak{y}^{(n)}\in M$, and (4).

If we form the sum over $k_0,k_0+1,\ldots$, it follows from (6) that

$$\sum_{k=k_0}^{\infty}\sum_{t_{kn}}|y_{t_{kn}}^{(n)}|^p<c^p\sum\frac{1}{d_k}=\frac{c^p}{2},\quad\text{for each } n.$$

But $\|\mathfrak{y}^{(n)}\|_p^p=\sum_{l=r_n}^{s_n}|y_l^{(n)}|^p>c^p$. Consequently for each n there is at least one index j_n with $r_n\leqq j_n\leqq s_n$ which is different from all the indices t_{kn}, $k=k_0,\,k_0+1,\ldots$. Then for this j_n (5) is false for all $k\geqq k_0$, i.e. we have $a_{j_n}^{(k)}\leqq\frac{m_k d_k}{c^p}=M_k$. Thus (2) is proved, with $\mathfrak{a}^{(k_0)}=\mathfrak{e}$.

In § 28, 3. we defined local convergence. In every (F)-space, and in particular in every echelon space, local convergence is the same as topological convergence. We now determine those co-echelon spaces (of order 1) in which weak convergence is the same as local convergence.

(7) *Suppose that* $\mathfrak{a}^{(k)}=(a_i^{(k)})$, $k=1,2,\ldots$, *is a complete monotonic increasing system of steps, and that* $\lambda^\times$ *is the corresponding co-echelon space of order* 1. *Weak convergence and local convergence coincide in* $\lambda^\times$ *if and only if for each k there exists an* $N(k)$ *such that*

$$(8)\qquad \lim_{i\to\infty}\frac{a_i^{(k)}}{a_i^{(N(k))}}=0;$$

here i runs through those indices for which $a_i^{(k)}\neq 0$.

Proof. a) Suppose that (8) is satisfied and let $\mathfrak{u}^{(n)}$ be a sequence in $\lambda^\times$ which converges weakly to $\mathfrak{o}$. There then exists a k and an M such that $|u_i^{(n)}| \leqq M a_i^{(k)}$, and $\lim\limits_{n\to\infty} u_i^{(n)} = 0$, for all i and n. But then it follows from (8) that $\sup\limits_i |a_i^{(N(k))}|^{-1} |u_i^{(n)}| \to 0$ as $n \to \infty$, i.e. that $\mathfrak{u}^{(n)}$ converges strongly to $\mathfrak{o}$ in $\lambda^\times_{\mathfrak{a}(N(k))}$. But this means that $\mathfrak{u}^{(n)}$ converges locally to $\mathfrak{o}$.

b) In order to show that (8) is necessary we need only assume that the sections $\mathfrak{u}_n$ of each $\mathfrak{u}$ in $\lambda^\times$ converge locally to $\mathfrak{u}$. The sequence $u_n \mathfrak{e}_n$ must then converge locally to $\mathfrak{o}$. If in particular we take $\mathfrak{a}^{(k)}$ as $\mathfrak{u}$, $a_n^{(k)} \mathfrak{e}_n$ must converge strongly to $\mathfrak{o}$ in some suitable $\lambda^\times_{\mathfrak{a}(N(k))}$. (8) follows from this.

By (1), every co-echelon space which satisfies (8) is an (M)-space; conversely the system of steps of an (M)-space always satisfies (1), but need not always satisfy (8). If $\lambda^\times$ is a co-echelon space which is an (M)-space but which does not satisfy (8), there must always be vectors $\mathfrak{u}$ in $\lambda^\times$ whose sections converge strongly to $\mathfrak{u}$, but which do not converge strongly to $\mathfrak{u}$ in any $\lambda^\times_{\mathfrak{a}(k)}$.

In § 31, 5., we shall discuss such an example in more detail.

Every co-echelon space $\lambda^\times[\mathfrak{T}_k]$ of order 1 which is an (M)-space is bornological, by § 29, 4.(4), and so it is the topological inductive limit of the spaces $\lambda^\times_{\mathfrak{a}(k)}[\mathfrak{T}_b]$, where $\mathfrak{T}_b$ is the norm topology. It follows directly from (8) that every bounded subset of $\lambda^\times[\mathfrak{T}_k]$ is relatively compact in some suitable $\lambda^\times_{\mathfrak{a}(N)}[\mathfrak{T}_b]$. The compact subsets of $\lambda^\times[\mathfrak{T}_k]$ therefore coincide with the sets which are compact in some $\lambda^\times_{\mathfrak{a}(N)}[\mathfrak{T}_b]$.

If (1) is satisfied, but not (8), $\lambda^\times[\mathfrak{T}_k]$ contains closed and bounded, and therefore compact, sets which are not compact in any $\lambda^\times_{\mathfrak{a}(N)}[\mathfrak{T}_b]$.

10. Further investigations into sequence spaces. We have considered sequence spaces with their structure determined by the α-dual $\lambda^\times$. It is also possible to follow rather different lines. For example, one can introduce the β-dual λ^β of a sequence space λ, which consists of all those $\mathfrak{v} \in \omega$ for which the scalar product $\mathfrak{v}\mathfrak{x} = \sum\limits_{i=1}^{\infty} v_i x_i$ converges (not necessarily absolutely) for all $\mathfrak{x} \in \lambda$. $\lambda^\beta \supset \lambda^\times$, and we now obtain the dual pair $\langle \lambda^\beta, \lambda \rangle$. This extension leads to rather complicated questions, however, as was pointed out by KÖTHE and TOEPLITZ [2]. In recent times this idea has been taken up again by CHILLINGWORTH [1], MATTHEWS [1] and GARLING [2], [3].

The theory of perfect spaces has only been developed here to a level which enables us to produce some examples which are important for the general theory. For further results reference may be made to the work of TOEPLITZ and the author, mentioned above, to the works of ALLEN, COOKE and their school (for this, cf. COOKE [1], [2]), and to the further works of the author. In two works which have recently appeared (T. and Y. KŌMURA [1], PIETSCH [1]), further important developments have been made to the theory, and PIETSCH [1] has generalized it to spaces of sequences whose terms are elements of an arbitrary locally convex space.

Another general class of sequence spaces was introduced by ZELLER [1]. A sequence space λ is called an (FK)-space if a locally convex topology $\mathfrak{T}'$ is defined on λ, under which λ is an (F)-space; further it is required that each mapping $\mathfrak{a}=(a_k)\to a_k$ is a continuous linear functional on $\lambda[\mathfrak{T}']$. Thus if a sequence $\mathfrak{a}^{(n)}$ converges to $\mathfrak{a}$, it must also converge coordinatewise.

Above all, this class of sequence spaces has found applications to problems in function theory, and to the theory of summability (cf. ZELLER [1], [2]).

The theory of perfect spaces has been carried over to spaces of functions by COOPER [1] and DIEUDONNÉ [7]. Details of this will be given in the second volume.

§ 31. Counterexamples

1. The dual of l^∞. Our aim is to show that c_0 is a closed subspace of l^∞ which has no topological complement. We begin by giving a representation of the dual of l_d^∞, where d is any infinite cardinal.

Let I be the set of indices on which the elements $\mathfrak{x}=(\xi_\alpha)$, $\alpha\in I$, of l_d^∞ are defined. We shall write $l^\infty(I)$ instead of l_d^∞. If M is a subset of I, let $l^\infty(M)$ be the sectional subspace of $l^\infty(I)$ defined by M. If u is a continuous linear functional on $l^\infty(I)$, let u_M denote both the restriction of u to $l^\infty(M)$, and the continuous linear functional on $l^\infty(I)$ which is obtained from it by assigning the value 0 to the elements of $l^\infty(I\sim M)$.

(1) *If* $M_1,\ldots,M_n$ *are pairwise disjoint subsets of* I,

$$\sum_{i=1}^{n}\|u_{M_i}\|\leqq\|u\|. \tag{2}$$

For if elements $\mathfrak{x}_i$ are chosen in $l^\infty(M_i)$ in such a way that $\|\mathfrak{x}_i\|\leqq 1$ and $u_{M_i}\mathfrak{x}_i\geqq\|u_{M_i}\|-\frac{\varepsilon}{n}$, $\mathfrak{x}=\sum_{i=1}^{n}\mathfrak{x}_i$ has norm $\|\mathfrak{x}\|\leqq 1$, and $u\mathfrak{x}=\sum_{i=1}^{n}u_{M_i}\mathfrak{x}_i$ $\geqq\sum\|u_{M_i}\|-\varepsilon$, from which (2) follows.

Let $\mathfrak{J}$ be the collection of all subsets M of I. Let $\mathfrak{e}_M$ be the element of $l^\infty(I)$ whose coordinates are 1 on M and 0 on the complement of M. $\mathfrak{e}_M$ is thus the characteristic function of M. For each $u\in(l^\infty)'$, we now define a set function $\varphi(M)$ on $\mathfrak{J}$ with values in K, by setting

$$\varphi(M)=u\,\mathfrak{e}_M,\qquad M\in\mathfrak{J}. \tag{3}$$

If we define $(\alpha_1\varphi_1+\alpha_2\varphi_2)(M)$ to be $\alpha_1\varphi_1(M)+\alpha_2\varphi_2(M)$, the mapping $u\to\varphi$ is clearly linear.

The set functions $\varphi(M)$ are finitely additive, i.e. we have

$$\varphi\left(\bigcup_{i=1}^{n}M_i\right)=\sum_{i=1}^{n}\varphi(M_i),\quad\text{for pairwise disjoint } M_i. \tag{4}$$

This follows directly from (3). If we define the variation of φ to be

$$V(\varphi) = \sup \sum_{i=1}^{n} |\varphi(M_i)|, \tag{5}$$

where the supremum is taken over all systems of finitely many pairwise disjoint subsets M_i of I, it follows from (2) that $V(\varphi) \leqq \|u\| < \infty$.

We denote by $BV(I)$ the space of all finitely additive set functions (or measures) $\varphi(M)$ on I, with the norm given by (5).

(6) *The dual of* $l^\infty(I)$ *is norm isomorphic to* $BV(I)$.

Proof. The collection of linear combinations $\sum_{i=1}^{n} \alpha_i \mathfrak{e}_{M_i}$, with pairwise disjoint M_i, forms a dense linear subspace H of l^∞. From this it follows that the mapping $u \to \varphi$ is a one-one mapping of $(l^\infty)'$ onto a linear subspace of $BV(I)$.

Conversely, if φ is an element of $BV(I)$, the equation

$$u\left(\sum_{i=1}^{n} \alpha_i \mathfrak{e}_{M_i}\right) = \sum_{i=1}^{n} \alpha_i \varphi(M_i) \tag{7}$$

defines a linear functional on H. If $\sum a_i \mathfrak{e}_{M_i}$ has norm less than or equal to 1, $|\alpha_i| \leqq 1$ for all i, and so

$$\left| u\left(\sum_{i=1}^{n} \alpha_i \mathfrak{e}_{M_i}\right)\right| \leqq \sum_{i=1}^{n} |\alpha_i| \, |\varphi(M_i)| \leqq V(\varphi).$$

Thus each $\varphi \in BV(I)$ defines a continuous linear functional on H, and so it defines an element $u \in (l^\infty)'$, with $\|u\| \leqq V(\varphi)$. Since conversely we have $V(\varphi) \leqq \|u\|$, (6) is proved.

The set-function φ corresponding to u is also written as

$$u\mathfrak{x} = \int_I \mathfrak{x} \, d\varphi; \tag{8}$$

the right-hand side is defined for $\mathfrak{x} \in H$ by (7), and is defined for arbitrary $\mathfrak{x} \in l^\infty$ by taking limits $\mathfrak{x}^{(n)} \to \mathfrak{x}$, where $x^{(n)} \in H$.

A corresponding result also holds for L^∞ (cf. HILDEBRANDT [1], FICHTENHOLZ and KANTOROVITCH [1], YOSIDA and HEWITT [1]).

Let $c_0(I)$ be the space of all null-sequences in $l^\infty(I)$. Then $c_0(I)' = l^1(I)$. Thus if $u \in (l^\infty)'$, the restriction $\tilde{u}$ of u to $c_0(I)$ is given by $\tilde{u}\mathfrak{x} = \sum_\alpha c_\alpha \xi_\alpha$, where $\sum |c_\alpha| < \infty$. But this expression also defines a continuous linear functional on the whole of l^∞, which we again denote by $\tilde{u}$. In this way we obtain the canonical norm-isomorphic embedding of $l^1(I)$ in $l^1(I)''=l^\infty(I)'$. We set $u=\tilde{u}+\tilde{\tilde{u}}$, and in this way obtain an algebraically complementary decomposition

(9) $$(l^\infty)' = l^1(I) \oplus c_0(I)^\perp$$

of $(l^\infty)'$.

In order to show that this is a topological decomposition, we prove the following result of DIXMIER's [1]:

(10) *Suppose that E is a* (B)*-space. There is a continuous projection P of norm* 1 *from E''' onto E'.*

Proof. If to each $u \in E'''$ we assign its restriction $\tilde{u}=Pu$ to $E \subset E''$, P is a projection of E''' onto E'. Since the norm of $\tilde{u}$ in E' is less than or equal to the norm of u in E''', P has norm 1, and is therefore continuous.

(11) *The decomposition* (9) *is continuous with respect to the norm topology of* $(l^\infty)'$.

In (10) we put $E=c_0(I)$. The projection P of $(l^\infty)'$ onto l^1 is then continuous, and has null space $c_0^\perp$. The assertion now follows from § 15, 8.(1).

Every $\tilde{u}\mathfrak{x}$ has the form $\tilde{u}\mathfrak{x}=\sum c_\alpha \xi_\alpha$, with $\sum |c_\alpha|<\infty$; the corresponding measure is given by $\tilde{\varphi}(M)=\sum\limits_{\beta \in M} c_\beta$. If M_i are countably many pairwise disjoint subsets of I, we clearly have $\tilde{\varphi}\left(\bigcup\limits_{i=1}^\infty M_i\right)=\sum\limits_{i=1}^\infty \tilde{\varphi}(M_i)$, i.e. the measure $\tilde{\varphi}$ is σ-additive. Conversely if I is countable and if a $\varphi \in BV(I)$ is σ-additive, it follows easily on applying (2) to the one-point sets that the corresponding linearly functional lies in $l^1(I)$. Every linear functional $\tilde{\tilde{u}} \in c_0^\perp$ vanishes on c_0, and so the corresponding measure $\tilde{\tilde{\varphi}}$ vanishes on all the finite subsets of I, and is not σ-additive.

2. Subspaces of l^∞ and l^1 with no topological complements. We continue with the investigation of $(l^\infty)'$.

(1) *Let $M_1, M_2, \dots$ be countably many pairwise disjoint finite subsets of I. For each $u \in (l^\infty)'$ and each $\varepsilon>0$, there exists a subsequence M_{n_k} such that the restriction u_M of u to $l^\infty(M)$, where $M=\bigcup\limits_{k=1}^\infty M_{n_k}$, satisfies the inequality $\|u_M\| \leqq \varepsilon$.*

For $\bigcup\limits_{i=1}^\infty M_i$ can be decomposed into countably many pairwise disjoint sets $M^{(i)}$ of the form M, and it follows from 1.(2) that $\sum\limits_{i=1}^\infty \|u_{M^{(i)}}\| \leqq \|u\|$; the assertion follows from this.

(2) *If $u^{(n)}$ is a sequence in $l^\infty(I)'$, and if the M_i are defined as in* (1), *there exists a subsequence M_{j_k} of the M_i for whose union N we have $u_N^{(n)} = \tilde{u}_N^{(n)}$, for all n.*

Proof. By (1), there is a subsequence of the M_i on whose union $M^{(1)}$ we have $\|u_{M^{(1)}}^{(1)}\| \leqq 1$. Applying (1) again, there is a proper subset $M^{(2)}$ of $M^{(1)}$, consisting of certain M_i, on which $\|u_{M^{(2)}}^{(1)}\| \leqq \frac{1}{2}$ and $\|u_{M^{(2)}}^{(2)}\| \leqq \frac{1}{2}$. In the general case, there is a proper subset $M^{(n)}$ of $M^{(n-1)}$ for which $\|u_{M^{(n)}}^{(k)}\| \leqq \dfrac{1}{n}$, for $k \leqq n$. Let N be a set which contains exactly one M_{j_n} from each $M^{(n)} \sim M^{(n+1)}$. We form $l^\infty(N)$. If $\mathfrak{x}$ is an arbitrary element of $l^\infty(N)$ with $\|\mathfrak{x}\| \leqq 1$, and if $\mathfrak{x}^{(n)}$ is the section of $\mathfrak{x}$ defined by $M_{j_1} \cup \cdots \cup M_{j_n}$, then $|u_N^{(k)}(\mathfrak{x} - \mathfrak{x}^{(n)})| \leqq \dfrac{1}{n}$ for $n \geqq k$. Thus for each $\mathfrak{x} \in l^\infty$, $u_N^{(k)}(\mathfrak{x})$ is given by $\sum_{r=1}^{\infty} v_{\alpha_r}^{(k)} \xi_{\alpha_r}$, $v_{\alpha_r}^{(k)} = u_N^{(k)} \mathfrak{e}_{\alpha_r}$, where the α_r are the countably many indices in N. But then the vector $(v_{\alpha_r}^{(k)})$ which represents $u_N^{(k)}$ lies in $l^1(N)$.

(3) *If the sequence $u^{(n)} \in (l^\infty)'$ converges weakly to $\mathfrak{o}$, the sequence $\tilde{u}^{(n)}$ of restrictions to c_0 converges to $\mathfrak{o}$ with respect to the l^1-norm.*

Proof. Let us suppose that this is not the case. By choosing a subsequence of the $u^{(n)}$, if necessary, we obtain a sequence $M_1, M_2, \ldots$ of pairwise disjoint finite subsets of I for which

$$\sum_{\alpha \in M_n} |\tilde{u}^{(n)} \mathfrak{e}_\alpha| \geqq \varepsilon, \tag{4}$$

for all n, and for some $\varepsilon > 0$. By (2) there is a subsequence M_{j_n} on whose union N we have $u_N^{(n)} = \tilde{u}_N^{(n)}$, for all n. By hypothesis the $u_N^{(n)} \in l^1(N)$ converge weakly to $\mathfrak{o}$, but by (4) they do not converge strongly to $\mathfrak{o}$. But this contradicts the fact that the weakly and strongly convergent sequences in l^1 are the same.

(5) *There is no topological complement of $c_0(I)$ in $l^\infty(I)$.*

Let us suppose that there is a norm complementary decomposition

$$l^\infty = c_0 \oplus H. \tag{6}$$

Let $\mathfrak{e}_n$ be the sequence of unit vectors in $l^1 = c_0'$. We extend each $\mathfrak{e}_n$ to an element e_n of $(l^\infty)'$, by putting e_n equal to 0 on H. Because of (6), e_n is $\mathfrak{T}_s(l^\infty)$-convergent to $\mathfrak{o}$ in $(l^\infty)'$. The restrictions $\tilde{e}_n = \mathfrak{e}_n$ to c_0 are however not strongly convergent to $\mathfrak{o}$ in l^1, contradicting (3).

(3) and (5) are due to Phillips [1]. For the present account, cf. Bourbaki [6], II, p. 118. Fichtenholz and Kantorovitch [1] proved the analogous result that $C[0,1]$ has no topological complement in L^∞.

It follows from (5) that the $\mathfrak{T}_b(l^\infty)$-complementary decomposition 1.(9) is not $\mathfrak{T}_s(l^\infty)$-complementary. For if this were so, then by § 20, 5.(1) there would be a corresponding $\mathfrak{T}_s((l^\infty)')$-complementary decomposition $l^\infty = c_0 \oplus H$. c_0 and H would then be closed subspaces of l^∞, and we would have a contradiction to (5), by § 15, 12.(6).

It also follows from this that the projection P of $(l^\infty)'$ onto l^1 is $\mathfrak{T}_b(l^\infty)$-continuous, but is not $\mathfrak{T}_s(l^\infty)$-continuous.

It also follows immediately from 1.(10) and (5) that c_0 cannot be the dual of a (B)-space; this we proved in a different way in § 25, 2.(7).

It is very easy to give examples of closed linear subspaces of l^1 which have no topological complement.

Let E be a separable (B)-space in which there are weakly convergent sequences which are not strongly convergent. The spaces l^p, $p>1$, provide examples. By § 22, 4.(1), E is topologically isomorphic to a quotient space l^1/H. If H had a topological complement in l^1, this would be topologically isomorphic to E. But then weak and strong convergence would have to be the same in E, since this is the case in l^1.

3. The problem of complements in l^p and L^p. The question of whether there are closed linear subspaces without topological complements in the spaces l^p and L^p, $p>1$, $p \neq 2$, was originally raised by BANACH, and a positive answer was first given by MURRAY [1]. We shall give a rather simpler construction, due to SOBCZYK [1].

A continuous endomorphism U of a locally convex space $E[\mathfrak{T}]$ is called an involution if $U^2 = I$, the identity. The set H of all x with $Ux = x$ is called the subspace of the involution. H is always linear and closed.

Projections and involutions are closely related.

(1) *If P is a continuous projection of $E[\mathfrak{T}]$ onto H, $U = 2P - I$ is an involution with subspace H. Conversely if U is an involution with subspace H, $P = \frac{1}{2}(U+I)$ is a continuous projection of E onto H.*

Proof. Clearly $(2P-I)^2 = I$ and $[\frac{1}{2}(U+I)]^2 = \frac{1}{2}(U+I)$. If $Px = x$, $(2P-I)x = x$, and conversely if $Ux = x$, $\frac{1}{2}(U+I)x = x$.

(2) *If $P = \frac{1}{2}(I+U)$ is a continuous projection of $E[\mathfrak{T}]$ onto H, then all the projections onto H are given by the mappings of the form $\tilde{P} = \frac{1}{2}(I+U+V)$, where V is a continuous endomorphism which satisfies*

$$UV = -VU = V \tag{3}$$

If $\tilde{P}$ is a projection onto H, we must have $P\tilde{P} = \tilde{P}$ and $\tilde{P}P = P$. The first condition implies that

$$\tfrac{1}{4}(I+U)(I+U+V) = \tfrac{1}{2}(I+U) + \tfrac{1}{4}(V+UV) = \tfrac{1}{2}(I+U+V),$$

so that $UV = V$.

Similarly the second condition implies that $-VU=V$.

Conversely it can be confirmed directly that every $\tilde{P}$ whose V fulfils the conditions of (3) satisfies the equations $P\tilde{P}=\tilde{P}$ and $\tilde{P}P=P$, and so, since $\tilde{P}\tilde{P}=\tilde{P}P\tilde{P}=P\tilde{P}=\tilde{P}$, it represents a projection onto H.

If H is a closed linear subspace of the (B)-space E, let $p(H)$ denote the infimum of the norms $\|P\|$ of the projections P of E onto H. Similarly let $u(H)$ be the infimum of the norms $\|U\|$ of the involutions U with subspace H. If there is no projection or involution on H, $p(H)$ and $u(H)$ are set equal to ∞. (1) implies the following inequalities

$$(4) \qquad \tfrac{1}{2}(u(H)-1) \leqq p(H) \leqq \tfrac{1}{2}(u(H)+1).$$

We now investigate the norms of projections in the n-dimensional spaces l_n^p.

(5) *Suppose that* $1 \leqq p \leqq \infty$, $p \neq 2$, $n=2^\nu$. *There is a linear subspace* $H^{(n)}$ *of* l_n^p *with* $u(H^{(n)}) \geqq n^{\left|\frac{1}{p}-\frac{1}{2}\right|}$ *and* $p(H^{(n)}) \geqq \frac{1}{2}\left(n^{\left|\frac{1}{p}-\frac{1}{2}\right|}-1\right)$.

Proof. Because of (4), it is sufficient to prove the first inequality. We may restrict our attention to the case where $p<2$. For if P is the projection onto $H^{(n)}$ and if U is the corresponding involution, then U' corresponds to P', and P' is a projection onto $H^{(n)\perp}$. Since $\|U'\|=\|U\|$, we have $u(H^{(n)\perp}) \geqq n^{\frac{1}{p}-\frac{1}{2}}=n^{\left|\frac{1}{q}-\frac{1}{2}\right|}$ for $H^{(n)\perp} \subset l_n^q$ $\left(\frac{1}{p}+\frac{1}{q}=1\right)$.

Let $\mathfrak{A}_1=\begin{pmatrix}1 & 1\\ 1 & -1\end{pmatrix}$, and generally let $\mathfrak{A}_\nu=\begin{pmatrix}\mathfrak{A}_{\nu-1} & \mathfrak{A}_{\nu-1}\\ \mathfrak{A}_{\nu-1} & -\mathfrak{A}_{\nu-1}\end{pmatrix}$. Then the $n\times n$ matrix $\mathfrak{U}=\dfrac{1}{\sqrt{n}}\mathfrak{A}_\nu$ is orthogonal and symmetric, and so it defines an involution in l_n^p.

Let $H^{(n)}$ be the subspace of this involution. By (2), every involution with $H^{(n)}$ as subspace is given by a matrix $\mathfrak{U}+\mathfrak{V}$, which satisfies the conditions of (3). If follows from (3) that the trace of $\mathfrak{V}$ is equal to zero. Thus if $\mathfrak{U}=(u_{ik})$ and $\mathfrak{V}=(v_{ik})$, there must be at least one non-negative v_{kk}. Since $\mathfrak{U}'\mathfrak{U}=\mathfrak{E}$, we have $1=\sum_{i=1}^{n} u_{ik}^2$. From $\mathfrak{V}=\mathfrak{U}\,\mathfrak{V}$ and the symmetry of $\mathfrak{U}$ it follows that

$$1 \leqq 1+v_{kk}=\sum_{i=1}^{n} u_{ik}(u_{ik}+v_{ik}).$$

Since $\left(\sum_{i=1}^{n}|u_{ik}|^q\right)^{\frac{1}{q}} \leqq \left[n\left(n^{-\frac{1}{2}}\right)^q\right]^{\frac{1}{q}}=n^{\frac{1}{2}-\frac{1}{p}}$, Hölder's inequality gives

$$1 \leqq n^{\frac{1}{2}-\frac{1}{p}}\|(\mathfrak{U}+\mathfrak{V})\,\mathfrak{e}_k\|.$$

But from this it follows that $\|\mathfrak{U}+\mathfrak{V}\| \geqq n^{\frac{1}{p}-\frac{1}{2}}$, i.e. $u(H^{(n)}) \geqq n^{\frac{1}{p}-\frac{1}{2}}$. The case $p=1$ can be dealt with in the same way without any difficulty.

SOBCZYK showed that in fact if $n=2^\nu$, $u(H^{(n)})=n^{\left|\frac{1}{p}-\frac{1}{q}\right|}$, and that this value is the largest possible value for $u(H)$, for $H \subset l_n^p$.

If we form the space $F=\bigoplus_{\nu=1}^{\infty} l_{2^\nu}^p$, $F \subset l^p$. In the same way we can construct the space $H=\bigoplus_{\nu=1}^{\infty} H^{(n)}$ from the subspaces $H^{(n)} \subset l_{2^\nu}^p$. Then $H \subset F$, and $\bar{H}$ is a closed linear subspace of l^p.

(6) *If* $1 \leqq p \leqq \infty$, $p \neq 2$, $\bar{H}$ *has no topological complement in* l^p.

By (5) it is sufficient to show that a projection P of l^p onto $\bar{H}$ must, for each $n=2^\nu$, have a greater norm than any projection of $l_{2^\nu}^p$ onto $H^{(n)}$. Let Q_n be the projection of l^p onto $l_{2^\nu}^p$ which sends the other spaces $l_{2^{\nu'}}^p$, $\nu' \neq \nu$, to zero. Since Q_n maps $\bar{H}$ onto $H^{(n)}$, $Q_n P$ is a projection of l^p onto $H^{(n)}$. Its restriction $(Q_n P)_n$ to l_n^p is a projection of l_n^p onto $H^{(n)}$. Since $\|Q_n\| \leqq 1$ and since $\|(Q_n P)_n\| \leqq \|Q_n P\| \leqq \|P\|$, the assertion now follows.

If $p=\infty$, F is contained in c_0, and the proof of (6) also holds for c_0; thus there are also closed linear subspaces of c_0 which have no topological complement.

The next simple observation enables us to find further counterexamples:

(7) *If the locally convex space E has a closed linear subspace H with no topological complement, and if E is topologically isomorphic to a closed subspace of the locally convex space F, H does not have a topological complement in F, either.*

For the restriction to E of a continuous projection of F onto H would be a continuous projection of E onto H.

From (6), (7) and § 21, 3.(6) it follows that $C[I]$, $I=[0,1]$, has a closed linear subspace with no topological complement.

Following BANACH [3], we have

(8) *If* $1 \leqq p \leqq \infty$, l^p *is norm-isomorphic to a closed linear subspace of* $L^p[0,1]$.

Proof. a) $p<\infty$. Let us put $y_i(t)=2^{i/p}$ in $\left[\frac{1}{2^i}, \frac{1}{2^{i-1}}\right]$, and zero elsewhere. Then $\|y_i\|=\left(\int_0^1 |y_i|^p\,dt\right)^{1/p}=1$, so that $y_i \in L^p$. The mapping A which sends each $\mathfrak{x}=(\xi_i) \in l^p$ to the function $x(t)=\sum_{i=1}^{\infty} \xi_i y_i(t) \in L^p$ is a norm isomorphism of l^p onto a subspace of L^p, since $\int_0^1 |x(t)|^p\,dt = \sum_{i=1}^{\infty} |\xi_i|^p$, so that $\|A\mathfrak{x}\|=\|\mathfrak{x}\|$.

b) $p=\infty$. We follow the corresponding argument through, putting $y_i(t)=1$ in $\left[\frac{1}{2^i}, \frac{1}{2^{i-1}}\right]$, and zero elsewhere.

It follows from (6), (7) and (8) that

(9) *Every space* $L^p[0,1]$, $1 \leqq p \leqq \infty$, $p \neq 2$, *has a closed linear subspace with no topological complement.*

KAKUTANI [4] has shown that if a (B)-space of at least three dimensions has continuous projections of norm 1 onto every closed linear subspace, then it must be norm isomorphic to a space l_d^2 (cf. BOURBAKI [6], II, pp. 142/4, as well). Finally LINDENSTRAUSS and TZAFRIRI proved in Israel J. Math. 9, 263—269 (1971) that a Banach space E is isomorphic to a Hilbert space l_d^2 if every closed subspace is complemented in E.

For further examples cf. JAMES [3] and KOMATUZAKI [1], [2].

4. Complements in (F)-spaces. Not only does every closed linear subspace of l_d^2 have a topological complement, but so also does every closed linear subspace of the spaces φ_d, ω_d, $\varphi_{d_1} \oplus \omega_{d_2}$, $\varphi\omega$ and $\omega\varphi$, by § 12, 1.(5) and HAGEMANN [1]. Let us remark, without proof, that this is also true for $\varphi \oplus l^2$ and $\omega \oplus l^2$; in addition ORNSTEIN [1] has recently produced an interesting class of spaces with this property. At present these seem to be the only known examples of complete locally convex spaces with this property.

We now obtain another negative result, this time for (F)-spaces. Suppose that $E[\mathfrak{T}]$ is an (F)-space. If E has an absolutely convex neighbourhood U of $\mathfrak{o}$ which contains no straight line, the corresponding Minkowski functional is a continuous norm on E. Conversely, to every continuous norm on E there corresponds a neighbourhood of $\mathfrak{o}$ with the given property.

If $E[\mathfrak{T}]$ possesses no continuous norm, $\mathfrak{T}$ can be defined by a sequence of increasing semi-norms, no one of which is a norm.

An (F)-space $E[\mathfrak{T}]$ possesses a continuous norm if and only if there is a bounded $\mathfrak{T}_s(E)$-total subset of E'.

ω is an example of an (F)-space with no continuous norm; every space $H(\mathfrak{G})$, $\mathfrak{G}$ a domain, is an example of an (F)-space with a continuous norm (cf. § 27, 3.).

(1) *Suppose that* $E[\mathfrak{T}]$ *is an* (F)-*space which is not a* (B)-*space, but which possesses a continuous norm. E has a closed linear subspace H with* $E/H \cong \omega$, *and H has no topological complement.*

Proof. Using the terminology of § 29, 1.(8), we consider a sequence $v_n \in E'_{B_n} \sim E'_{B_{n-1}}$. The v_n are linearly independent; let G be their linear span. Since the sets B_n form a fundamental sequence of bounded subsets of E', only finitely many v_n lie in each bounded set; the bounded subsets of G are thus finite-dimensional. By § 21, 10.(5), G is therefore a weakly closed subspace of E'. We set $G^{\perp}=H$, and form the space E/H. $\mathfrak{T}=\mathfrak{T}_k(E')$ and so by § 22, 2.(3) the induced quotient topology $\mathfrak{T}$ on E/H is equal to $\mathfrak{T}_k(H^{\perp})=\mathfrak{T}_k(G)$. Now the weakly compact subsets of G are finite-dimensional, so that $\mathfrak{T}=\mathfrak{T}_k(G)=\mathfrak{T}_s(G)=\mathfrak{T}_s(H')$. As a result the space E/H, which is complete under $\mathfrak{T}$, is weakly complete, and so it is topologically isomorphic to ω.

If H had a topological complement H_0, this also would have to be topologically isomorphic to ω. But there would then have to be a continuous norm on H_0, and this is not the case.

We observe that we have established that every (F)-space which is not a (B)-space has a quotient $E/H \cong \omega$.

Suppose that E is an (F)-space with a linear subspace H which has no topological complement, and which is such that $E/H \cong \omega$. Using the dual pair $\langle E', E\rangle$, we can also introduce the linear weak topology $\mathfrak{T}_{ls}(E')$ on E. We now assert that $E[\mathfrak{T}_{ls}]$ provides an example of the circumstances described at the end of § 10, 12. As was shown in § 20, 3.(2), the lattices of $\mathfrak{T}_s(E')$-closed and $\mathfrak{T}_s(E')$-closed linear subspaces of E are the same. Since H does not have a $\mathfrak{T}_s$-complement in E, neither can it have a $\mathfrak{T}_{ls}$-complement. The $\mathfrak{T}_s$-isomorphism $E/H \cong \omega$ is also a $\mathfrak{T}_{ls}$-isomorphism, so that E/H is linearly weakly compact.

These observations are due to DIEUDONNÉ [10]. (1) is a generalisation of the following example of TOEPLITZ and the author (cf. KÖTHE [12]).

Let $a(z)$ be a holomorphic function with infinitely many simple zeros z_k in the simply connected bounded domain $\mathfrak{G}$. The operation A of multiplying the functions $x(z)\in H(\mathfrak{G})$ by $a(z)$ is a continuous endomorphism of $H(\mathfrak{G})$. Let F be the image-space. Since the uniform convergence of $a(z)x_n(z)$ implies the uniform convergence of $x_n(z)$, F is sequentially closed, and so it is closed; as a result A is a topological isomorphism of $H(\mathfrak{G})$ onto F. Let us determine $F^{\perp}\subset H(\Omega \sim \mathfrak{G})$.

We are therefore looking for all those functions $u(z)$, holomorphic in some neighbourhood of $\Omega \sim \mathfrak{G}$, and vanishing at ∞, which satisfy

$$\frac{1}{2\pi i}\oint_C u(t)\,a(t)\,x(t)\,dt=0$$

for all $x(z)\in H(\mathfrak{G})$. We may restrict our attention to the subset of all $x(z)=\dfrac{1}{z-\lambda}$, $\lambda\in\Omega\sim\mathfrak{G}$, since this is total in $H(\mathfrak{G})$. We therefore require that

$$\text{(2)} \qquad \frac{1}{2\pi i}\oint_C \frac{u(t)\,a(t)}{t-\lambda}\,dt=0 \quad \text{for all } \lambda\in\Omega\sim\mathfrak{G};$$

$u(t)a(t)$ is holomorphic in some annular domain $(\Omega\sim\mathfrak{G}_1)\cap\mathfrak{G}_2$, $\mathfrak{G}_1\subset\mathfrak{G}_2\subset\mathfrak{G}$, which contains the closed curve C, and so it is the sum of two functions $c_1(t)$, which is holomorphic in $\mathfrak{G}_2$, and $c_2(t)$, which is holomorphic in $\Omega\sim\mathfrak{G}_1$ and which vanishes at ∞. Thus we have

$$\text{(3)} \qquad \frac{1}{2\pi i}\oint_C \frac{c_1(t)}{t-\lambda}\,dt+\frac{1}{2\pi i}\oint_C \frac{c_2(t)}{t-\lambda}\,dt=0, \qquad \lambda\in\Omega\sim\mathfrak{G}.$$

The first integrand is holomorphic in $\mathfrak{G}_2$, and so the integral vanishes. Considering the second term, it follows that $c_2(z)\equiv 0$ in $\Omega\sim\mathfrak{G}$, and so also in

$\Omega\sim\mathfrak{G}_1$. Since $a(z)u(z)\equiv c_1(z)$ in $(\Omega\sim\mathfrak{G}_1)\cap\mathfrak{G}_2$, we can extend $u(z)$, which is holomorphic in $\Omega\sim\mathfrak{G}_1$, analytically to all points of $\mathfrak{G}_2$ which are not zeros of $a(z)$, by putting $u(z)=\dfrac{c_1(z)}{a(z)}$. Since $a(z)$ has only finitely many zeros $z_1,\ldots,z_n$ in $\mathfrak{G}_2$, $u(z)$ has the form $\sum_{k=1}^{n}\dfrac{\alpha_k}{z-z_k}$. It follows from this that $F^{\perp}$ consists of all finite linear combinations of the functions $u_k=\dfrac{1}{z-z_k}$, and so it has the form given in the proof of (1).

It follows from (1) that $H(\mathfrak{G})/F\cong\omega$, and that F has no topological complement.

5. An (FM)-space. We now resume the discussion of § 30, 9. We start with vectors written as double sequences:

$$\begin{aligned}\mathfrak{a}^{(k)}&=(a_{11}^{(k)},a_{12}^{(k)},\ldots;a_{21}^{(k)},a_{22}^{(k)},\ldots;\ldots)\\&=(\mathfrak{b}_1^{(k)};\ldots,\mathfrak{b}_{k-1}^{(k)};k^k\,\mathfrak{e};k^{k+1}\,\mathfrak{e};\ldots),\qquad k=1,2,\ldots,\end{aligned}$$

where

$$\mathfrak{b}_j^{(k)}=(1,2^k,3^k,\ldots),\qquad \mathfrak{e}=(1,1,\ldots).$$

Using these $\mathfrak{a}^{(k)}$, we form the corresponding echelon space λ and its dual co-echelon space $\lambda^{\times}$.

Since $\mathfrak{a}^{(1)}$ is the vector with $a_{ij}^{(1)}=1$ for all i and j, $\lambda^{\times}$ contains the space $l^{\infty}=\lambda_{\mathfrak{a}^{(1)}}^{\times}$, written as a space of double sequences.

By § 30, 9.(1), $\lambda[\mathfrak{T}]$ is an (FM)-space, and $\lambda^{\times}[\mathfrak{T}_k]$ is its strong dual (M)-space. $\mathfrak{T}_k(\lambda)=\mathfrak{T}_b(\lambda)$ on $\lambda^{\times}$.

It can be seen directly that the steps $\mathfrak{a}^k$ (once their coordinates have been arranged in a sequence) do not satisfy the conditions of § 30, 9.(8). Thus strong and local convergence in $\lambda^{\times}$ are not the same. It is easy to see that $\mathfrak{a}^{(1)}$ is not the strong limit of its sections in any of the spaces $\lambda_{\mathfrak{a}^{(k)}}^{\times}$, whereas $\mathfrak{a}^{(1)}$ is the strong limit of its sections in $\lambda^{\times}$, by § 30, 7.(8). Thus the unit ball of $\lambda_{\mathfrak{a}^{(1)}}^{\times}=l^{\infty}$ is compact in $\lambda^{\times}$, but is not compact in any of the spaces $\lambda_{\mathfrak{a}^{(k)}}^{\times}$.

Let A be the linear mapping from λ which sends each $\mathfrak{x}=(x_{ij})\in\lambda$ to the vector $A\mathfrak{x}=\left(\sum_{i=1}^{\infty}x_{i1},\sum_{i=1}^{\infty}x_{i2},\ldots\right)$. Since $\mathfrak{a}^{(1)}\in\lambda^{\times}$, it follows directly that $\sum_{j=1}^{\infty}\left|\sum_{i=1}^{\infty}x_{ij}\right|\leq\sum\sum|x_{ij}|<\infty$, so that $A\mathfrak{x}\in l^1$.

The adjoint mapping A' maps each $\mathfrak{u}\in l^{\infty}$ to the element $(\mathfrak{u};\mathfrak{u};\ldots)\in\lambda^{\times}$; this can be seen from the equations

$$\mathfrak{u}(A\mathfrak{x})=\sum_j u_j\sum_i x_{ij}=(\mathfrak{u};\mathfrak{u};\ldots)\mathfrak{x}=(A'\mathfrak{u})\mathfrak{x}.$$

The linear subspace of all $(\mathfrak{u};\mathfrak{u};\ldots)$ is weakly closed in $\lambda^{\times}$; this follows from the form of the double sequences $\mathfrak{a}^{(k)}$. A' is thus a weakly continuous one-one linear mapping from l^{∞} into $\lambda^{\times}$ with a weakly closed image space. Using a theorem about mappings between (F)-spaces which will be proved in the second volume (cf. BOURBAKI [6], Vol. 2, p. 106, Ex. 5a), it follows that A is a topological homomorphism of $\lambda[\mathfrak{T}]$ onto l^1. Thus $\lambda/N[A]$ is topologically isomorphic to l^1 in the topology induced by $\mathfrak{T}_b(\lambda^\times)$.

The (FM)-space $\lambda[\mathfrak{T}]$ therefore has a quotient space which is not an (M)-space.

At the same time this example shows that a quotient of a reflexive (F)-space need not be reflexive.

Let us now consider the canonical mapping K of λ onto $\lambda/N[A]$. Every closed bounded subset B of λ is compact, and so its image $K(B)$ is compact, too. But there exist bounded sets which are not compact in $\lambda/N[A]\cong l^1$; not every bounded subset of $\lambda/N[A]$ is contained in the closure of the image $K(B)$ of a bounded set. This means that the strong topology $\mathfrak{T}_b(\lambda/N[A])$ on $H=N[A]^\perp\subset\lambda^\times$ is strictly finer than $\mathfrak{T}_b(\lambda)$.

H is an example of a closed linear subspace of a barrelled space which is not barrelled, and is also not quasi-barrelled.

Since $\lambda^\times[\mathfrak{T}_b]$ is bornological, H is an example of a closed subspace of a bornological space which is not bornological, by § 28, 1.(1).

Finally $\lambda^\times[\mathfrak{T}_b]$ is a (DF)-space whose closed linear subspace H is not a (DF)-space. This follows for example from § 29, 3.(12), since H is separable, but not quasi-barrelled.

For this example, cf. KÖTHE [6] and GROTHENDIECK [10].

6. An (LB)-space which is not complete. Our example is constructed in a similar way to the example of 5.

Let E_0 be equal to c_0, with the elements written as double sequences,

$$\mathfrak{x}=(x_{11},x_{12},\ldots;x_{21},x_{22},\ldots;\ldots).$$

Thus $\mathfrak{x}$ belongs to E_0 if and only if $\lim_{i,k\to\infty}|x_{ik}|=0$. E_0 is a (B)-space under the norm $\|\mathfrak{x}\|_0=\sup_{i,k}|x_{ik}|$.

Let $a_{ik}^{(n)}=k$ for $i\leqq n$ and for all k, and let $a_{ik}^{(n)}=1$ for $i>n$ and for all k. Let E_n be the space of all double sequences with $\lim_{i,k\to\infty}\frac{|x_{ik}|}{a_{ik}^{(n)}}=0$, with norm $\|\mathfrak{x}\|_n=\sup_{i,k}\frac{|x_{ik}|}{a_{ik}^{(n)}}$. E_n is obtained from E_0 by making a diagonal transformation, and so it is topologically isomorphic to c_0. The embedding of E_{n-1} in E_n is continuous. The topological inductive limit $E=\lim E_n$ therefore exists, provided that the hull topology $\mathfrak{T}$ on E is Hausdorff, and E is then an (LB)-space.

If we put $a_{ik}=k$ for all i and k, the norm $\|\mathfrak{x}\|_\infty = \sup \frac{|x_{ik}|}{a_{ik}}$ is weaker than $\|\mathfrak{x}\|_n$ on E_n, and so it defines a Hausdorff topology on E which is weaker than $\mathfrak{T}$. Thus $\mathfrak{T}$ is Hausdorff as well.

It is easy to determine the dual space E'. The dual E_n' of E_n is a diagonal transform of l^1, and it consists of all those double sequences $\mathfrak{u}$ with $\sum_{i,k=1}^{\infty} |u_{ik}|\, a_{ik}^{(n)} < \infty$. Since the set of unit vectors $\mathfrak{e}_{ik}$ is total in each E_n, and is therefore total in E, E' can be identified with the space of all double sequences $\mathfrak{u}$ for which $\sum_{i,k=1}^{\infty} |u_{ik}|\, a_{ik}^{(n)} < \infty$ for $n=1,2,\ldots$.

Once again let $\mathfrak{e}$ denote the vector $(1,1,\ldots)$. Let B be the set of all vectors $\mathfrak{e}^{(n)} = (\overbrace{\mathfrak{e},\ldots,\mathfrak{e}}^{n},\mathfrak{o},\mathfrak{o},\ldots)$, $n=1,2,\ldots$. B is contained in E, and is weakly bounded, since $|\mathfrak{e}^{(n)}\mathfrak{u}| \leqq \sum_{i,k=1}^{\infty} |u_{ik}| < \infty$, for each $\mathfrak{u}\in E'$. Since $\mathfrak{e}^{(n)}$ belongs to E_n, but not to E_{n-1}, B is not contained in any one of the spaces E_n. If E were quasi-complete, the closed absolutely convex cover of B, which would be complete, would have to lie in some E_n, by § 19, 5.(5); *a fortiori* the same would be true of B.

We have therefore shown that the (LB)-space E is not quasi-complete.

Applying the final remark of § 19, 5., it follows that we have also shown that the locally convex direct sum of countable many spaces c_0, which is a complete locally convex space, has a quotient which is not complete.

7. An (F)-space which is not distinguished. We now take the sequence of vectors

$$\mathfrak{b}^{(k)} = (b_{ij}^{(k)}) = (\overbrace{\mathfrak{d};\ldots;\mathfrak{d}}^{k};\mathfrak{e};\mathfrak{e};\ldots), \qquad k=1,2,\ldots,$$

where $\mathfrak{d}=(1,2,3,\ldots)$ and $\mathfrak{e}=(1,1,1,\ldots)$, as the system of steps. By § 30, 8.(1), the corresponding echelon space λ is a separable (F)-space under the normal topology. Its strong dual is the co-echelon space $\lambda^\times$ with the strong topology $\mathfrak{T}_b(\lambda)$.

In order to show that $\lambda[\mathfrak{T}]$ is not distinguished it is sufficient, by § 29, 4.(3), to show that $\lambda^\times[\mathfrak{T}_b]$ is not bornological.

Let $\mathfrak{T}$ be the hull topology of $\lambda^\times = \bigcup_{n=1}^{\infty} \lambda_{\mathfrak{b}^{(n)}}^\times$, when each $\lambda_{\mathfrak{b}^{(n)}}^\times$ is given its norm topology. If B_n denotes the set of all $\mathfrak{u}\in\lambda^\times$ with $|u_{ij}| \leqq b_{ij}^{(n)}$, the sets $V = \Gamma_{n=1}^{\infty} c_n B_n$, $c_n>0$, form a base of $\mathfrak{T}$-neighbourhoods of $\mathfrak{o}$ in $\lambda^\times$. We must show that $\mathfrak{T}$ is strictly finer than $\mathfrak{T}_b(\lambda)$ on $\lambda^\times$.

Let $V_0 = \Gamma_{n=1}^{\infty} \frac{1}{n} B_n$. Then V_0 contains no element $\mathfrak{u} = (u_{ij})$ with the property that, for each i, there is a coordinate u_{ij} with $|u_{ij}| \geqq 2$. For if $\mathfrak{u} \in \Gamma_1^N \frac{1}{n} B_n$, then $|u_{mk}| \leqq \sum_1^N \frac{|\alpha_n|}{n}$, with $\sum_1^N |\alpha_n| \leqq 1$, for $m > N$, so that in fact $|u_{mk}| \leqq 1$.

We shall show that V_0 can contain no $\mathfrak{T}_b(\lambda)$-neighbourhood of $\mathfrak{o}$. Every bounded subset of $\lambda[\mathfrak{T}]$ is a subset of a set $\bigcap_{n=1}^{\infty} (c_n B_n)^\circ$, for suitably chosen c_n. Now $\bigcap_{n=1}^{\infty} (c_n B_n)^\circ = \left(\Gamma_{n=1}^{\infty} c_n B_n\right)^\circ = V^\circ$, so that we obtain a $\mathfrak{T}_b$-base of neighbourhoods of $\mathfrak{o}$ in $\lambda^\times$ by taking the bipolars $V^{\circ\circ}$, as V runs through the $\mathfrak{T}$-neighbourhoods of $\mathfrak{o}$. We must therefore show that V_0 can contain no $V^{\circ\circ}$. Suppose that we are given $V = \Gamma_{n=1}^{\infty} c_n B_n$. We denote by $\mathfrak{e}_{ij}$ the vector (u_{ij}) with $u_{ij} = 1$ and $u_{kl} = 0$ for $(k,l) \neq (i,j)$. The element $2^{n+1} \mathfrak{e}_{n,k_n}$ lies in $c_n B_n$ for sufficiently large k_n.

Consequently $\sum_1^N \frac{1}{2^n} 2^{n+1} \mathfrak{e}_{n,k_n}$ belongs to $\Gamma_{n=1}^{N} c_n B_n \subset V$. Thus the weak limit $2 \sum_1^\infty \mathfrak{e}_{n,k_n}$ belongs to $V^{\circ\circ}$, but does not belong to V_0, by the remark made above. Thus we have shown that $\lambda^\times[\mathfrak{T}_b]$ is not bornological, and so $\lambda[\mathfrak{T}]$ is not distinguished.

Using § 19, 9.(2), $\lambda[\mathfrak{T}]$ can be considered as a closed linear subspace H of a topological product $E[\mathfrak{T}] = \prod_{i=1}^{\infty} E_i$ of (B)-spaces. Let us consider the topologies $\mathfrak{T}_b(H)$ and $\hat{\mathfrak{T}}_b(E)$ on $E'/H^\perp$. We assert that these topologies are different. For $E'[\mathfrak{T}_b(E)]$, being $\bigoplus_{i=1}^{\infty} E_i'$, is bornological, by § 28, 4.(1), and so $E'/H^\perp$, with the topology $\hat{\mathfrak{T}}_b(E)$, is also bornological, by § 28, 4.(2). On the other hand, as we have just seen, the space $E'/H^\perp$ is not bornological under the topology $\mathfrak{T}_b(H)$, as it is isomorphic to $\lambda^\times$.

This example, which is due to KÖTHE and GROTHENDIECK (cf. GROTHENDIECK [10]), was used by AMEMIYA [1] to give an example of a bornological (DF)-space whose strong bidual is not bornological.

Bibliography

Alaoglu, L.: [1] Weak topologies of normed linear spaces. Ann. Math. **41**, 252—267 (1940).

Amemiya, I.: [1] Some examples of (F)- and (DF)-spaces. Proc. Japan Acad. **33**, 169—171 (1957).

Arens, R. F.: [1] Duality in linear spaces. Duke Math. J. **14**, 787—794 (1947).

—, and J. L. Kelley: [1] Characterizations of the space of continuous functions over a compact Hausdorff space. Trans. Amer. Math. Soc. **62**, 499—508 (1947).

Ballier, F.: [1] Über lineartopologische Algebren. J. reine angew. Math. **195**, 42—75 (1956).

Banach, S.: [1] Sur les opérations dans les ensembles abstraits et leur application aux équations intégrales. Fund. Math. **3**, 133—181 (1922).

— [2] Sur les fonctionnelles lineaires, I, II. Studia Math. **1**, 211—216, 223—239 (1929).

— [3] Théorie des opérations linéaires (Monografje Matematyczne Tom I), Warszawa 1932.

—, u. S. Mazur: [1] Zur Theorie der linearen Dimension. Studia Math. **4**, 100—112 (1933).

—, et H. Steinhaus: [1] Sur le principe de la condensation de singularités. Fund. Math. **9**, 50—61 (1927).

Bessaga, C., A. Pełczyński and S. Rolewicz: [1] Some properties of the norm in F-spaces. Studia Math. **16**, 183—192 (1957).

Birkhoff, G.: [1] A note on topological groups. Compositio Math. **3**, 427—430 (1936).

— [2] Moore-Smith convergence in general topology. Ann. Math. **38**, 39—56 (1937).

— [3] Lattice theory, 3rd Ed., Amer. Math. Soc. Colloq. Publ. Vol. XXV, New York 1967.

Bohnenblust, H. F., and A. Sobczyk: [1] Extensions of functionals on complex linear spaces. Bull. Amer. Math. Soc. **44**, 91—93 (1938).

Bourbaki, N.: [1] Sur les espaces de Banach. C. R. Acad. Sci. Paris **206**, 1701—1704 (1938).

— [2] Sur certains espaces vectoriels topologiques. Ann. Inst. Fourier **2**, 5—16 (1950).

— [3] Eléments de mathématique, Livre I, Théorie des ensembles, 4 Vols. Paris: Hermann & Cie. Act. Sci. et Ind. Nos. 1141, 1212, 1243, 1258 (1951—1958).

— [4] Eléments de mathématique, Livre II, Algèbre, 6 Vols. Act. Sci. et Ind. Nos. 1144, 1032, 1044, 1102, 1179, 1261 (1947—1958).

— [5] Eléments des mathématique, Livre III, Topologie Générale, 6 Vols. Act. Sci. et Ind. Nos. 1142, 1143, 1029, 1045, 1084, 1196 (1947—1953).

— [6] Eléments de mathématique, Livre V, Espaces vectoriels topologiques, 2 Vols. Act. Sci. et Ind. Vols. 1189, 1229 (1953, 1955).

— [7] Eléments de mathématique, Livre VI, Intégration, 2 Vols. Act. Sci. et Ind. Nos. 1175, 1244 (1952, 1956).

Bourgin, D. G.: [1] Linear topological spaces. Amer. J. Math. **65**, 637—659 (1943).

BRACONNIER, J.: [1] Spectres d'espaces et de groupes topologiques. Portugaliae Math. **7**, 93—111 (1948).

CHILLINGWORTH, H. R.: [1] Generalised "dual" sequence spaces. Nederl. Akad. Wet. Proc., Ser. A **61**, 307—315 (1958).

CIVIN, P., and B. YOOD: [1] Quasi-reflexive spaces. Proc. Amer. Math. Soc. **8**, 906—911 (1957).

CLARKSON, J. A.: Uniformly convex spaces. Trans. Amer. Math. Soc. **40**, 396—414 (1936).

COLLINS, H. S.: [1] Completeness and compactness in linear topological spaces. Trans. Amer. Math. Soc. **79**, 256—280 (1955).

COOKE, R. G.: [1] Infinite matrices and sequence spaces. London: Macmillan & Co. 1950.

— [2] Linear operators. London: Macmillan & Co. 1953.

COOPER, J. L. B.: [1] Coordinated linear spaces. Proc. London Math. Soc., III. ser. **3**, 305—327 (1953).

CUDIA, D. F.: [1] Rotundity. Proc. Symp. Pure Math. **VII**, 73—97 (1963).

— [2] The geometry of Banach spaces. Smoothness. Trans. Amer. Math. Soc. **110**, 284—314 (1964).

DAY, M. M.: [1] The spaces L^p with $0<p<1$. Bull. Amer. Math. Soc. **46**, 816—823 (1940).

— [2] Reflexive Banach spaces not isomorphic to uniformly convex spaces. Bull. Amer. Math. Soc. **47**, 313—317 (1941).

— [3] Some more uniformly convex spaces. Bull. Amer. Math. Soc. **47**, 504—507 (1941).

— [4] Uniform convexity III. Bull. Amer. Math. Soc. **49**, 745—750 (1943).

— [5] Uniform convexity in factor and conjugate spaces. Ann. Math. (2) **45**, 375—385 (1944).

— [6] Strict convexity and smoothness of normed spaces. Trans. Amer. Math. Soc. **78**, 516—528 (1955).

— [7] Every L-space is isomorphic to a strictly convex space. Proc. Amer. Math. Soc. **8**, 415—417 (1957).

— [8] Normed linear spaces. Ergebnisse d. Math., N. F., H. 21. Berlin-Göttingen-Heidelberg: Springer 1958.

DIEUDONNÉ, J.: [1] Sur le théorème de Hahn-Banach. Rev. Sci. **79**, 642—643 (1941).

— [2] La dualité dans les espaces vectoriels topologiques. Ann. École Norm. **59**, 107—139 (1942).

— [3] Sur la séparation des ensembles convexes dans un espace de Banach. Rev. Sci. **81**, 277—278 (1943).

— [4] Matrices semi-finies et espaces localement linéairement compacts. J. reine angew. Math. **188**, 162—166 (1950).

— [5] Natural homomorphisms in Banach spaces. Proc. Amer. Math. Soc. **1**, 54—59 (1950).

— [6] Linearly compact spaces and double vector spaces over sfields. Amer. J. Math. **73**, 13—19 (1951).

— [7] Sur les espaces de Köthe. J. Analyse Math. **1**, 81—115 (1951).

— [8] Complex structures on real Banach spaces. Proc. Amer. Math. Soc. **3**, 162—164 (1952).

— [9] Sur les propriétés de permanence de certains espaces vectoriels topologiques. Ann. Soc. Polon. Math. **25**, 50—55 (1952).

— [10] Sur les sousespaces linéairement compacts. Bol. Soc. Math. São Paulo **6**, 53—60 (1952).

— [11] Sur un théorème de Šmulian. Arch. d. Math. **3**, 436—440 (1953).

— [12] Recent developments in the theory of locally convex vector spaces. Bull. Amer. Math. Soc. **59**, 495—512 (1953).

— [13] Sur les espaces de Montel métrisables. C. R. Acad. Sci. Paris **238**, 194—195 (1954).

— [14] Bounded sets in (F)-spaces. Proc. Amer. Math. Soc. **6**, 729—731 (1955).

— [15] Denumerability conditions in locally convex vector spaces. Proc. Amer. Math. Soc. **8**, 367—372 (1957).

—, et A. P. Gomes: [1] Sur certains espaces vectoriels topologiques. C. R. Acad. Sci. Paris **230**, 1129—1130 (1950).

—, et L. Schwartz: [1] La dualité dans les espaces (F) et (LF). Ann. Inst. Fourier **1**, 61—101 (1950).

Dixmier, J.: [1] Sur un théorème de Banach. Duke Math. J. **15**, 1057—1071 (1948).

Donoghue, W. F., and K. T. Smith: [1] On the symmetry and bounded closure of locally convex spaces. Trans. Amer. Math. Soc. **73**, 321—344 (1952).

Dunford, N., and J. T. Schwartz: [1] Linear Operators, Part I: General Theory. New York: Interscience Publishers 1958.

Eberlein, W. F.: [1] Closure, convexity and linearity in Banach spaces. Ann. Math. (2) **47**, 688—703 (1946).

— [2] Weak compactness in Banach spaces, I. Proc. Nat. Acad. Sci. USA **33**, 51—53 (1947).

Eidelheit, M.: [1] Zur Theorie der konvexen Mengen in linearen normierten Räumen. Studia Math. **6**, 104—111 (1936).

Fan, K., and I. Glicksberg: [1] Fully convex normed linear spaces. Proc. Nat. Acad. Sci. USA **41**, 947—953 (1955).

— [2] Some geometric properties of the spheres in a normed linear space. Duke Math. J. **25**, 553—568 (1958).

Fichtenholz, G., et L. Kantorovitch: [1] Sur les opérations dans l'espace des fonctions bornées. Studia Math. **5**, 69—98 (1934).

Fischer, H. R., and H. Gross: [1] Quadratic forms and linear topologies, I. Math. Ann. **157**, 296—325 (1964).

— [2] II. Math. Ann. **159**, 285—308 (1965).

— [3] III. Math. Ann. **160**, 1—40 (1965).

Fleischer, I.: [1] Locally symmetric spaces. Diss. Univ. of Chicago 1952.

— [2] Sur les espaces normés non-archimédiens. Nederl. Akad. Wet. Proc., Ser. A **57**, 165—168 (1954).

Freundlich-Smith, A.: [1] The Pontrjagin duality theorem in linear spaces. Ann. Math. **56**, 248—253 (1952).

Garling, D. J. H.: [1] Locally convex spaces with denumerable systems of weakly compact subsets. Proc. Cambridge Phil. Soc. **60**, 813—815 (1964).

— [2] On topological sequence spaces. Proc. Cambridge Phil. Soc. **63**, 977—1019 (1967).

— [3] The β- and γ-duality of sequence spaces. Proc. Cambridge Phil. Soc. **63**, 963—981 (1967).

Goldstine, H. H.: [1] Weakly complete Banach spaces. Duke Math. J. **4**, 125—131 (1938).

Graves, L. M.: [1] On the completing of a Hausdorff space. Ann. Math. **38**, 61—64 (1937).

Grothendieck, A.: [1] Sur la complétion du dual d'un espace vectoriel localement convexe. C. R. Acad. Sci. Paris **230**, 605—606 (1950).

— [2] Quelques résultats relatifs à la dualité dans les espaces (F). C. R. Acad. Sci. Paris **230**, 1561—1563 (1950).

— [3] Critères généraux de compacité dans les espaces vectoriels localement convexes. Pathologie des espaces (LF). C. R. Acad. Sci. Paris **231**, 940—942 (1950).

— [4] Quelques résultats sur les espaces vectoriels topologiques. C. R. Acad. Sci. Paris **233**, 839—841 (1951).

— [5] Sur une notion de produit tensoriel topologique d'espaces vectoriels topologiques, et une classe remarquable d'espaces vectoriels liées a cette notion. C. R. Acad. Sci. Paris **233**, 1556—1558 (1951).

— [6] Critères de compacité dans les espaces fonctionnels généraux. Amer. J. Math. **74**, 168—186 (1952).

— [7] Sur les applications linéaires faiblement compactes d'espaces du type $C(K)$. Canad. J. Math. **5**, 129—173 (1953).

— [8] Sur certains espaces de fonctions holomorphes, I, II. J. reine angew. Math. **192**, 35—64, 77—95 (1953).

— [9] Sur les espaces de solutions d'une classe générale d'équations aux dérivées partielles. J. Analyse Math. **2**, 243—280 (1953).

— [10] Sur les espaces (F) et (DF). Summa Brasil. Math. **3**, 57—123 (1954).

— [11] Espaces vectoriels topologiques, Departamento de Matemática da Universidade de São Paulo 1954.

— [12] Résumé des résultats essentiels dans la théorie des produits tensoriels topologiques et des espaces nucléaires. Ann. Inst. Fourier **4**, 73—112 (1954).

— [13] Produits tensoriels topologiques et espaces nucléaires. Mem. Amer. Math. Soc. No. 16 (1955).

— [14] Une caractérisation vectorielle-metrique des espaces L^1. Canad. J. Math. **7**, 552—561 (1955).

Hagemann, E.: [1] Das Reziprokentheorem in beliebigen linearen Koordinatenräumen. Math. Ann. **114**, 126—143 (1937).

Hahn, H.: [1] Über Folgen linearer Operationen. Mh. Math. Phys. **32**, 3—88 (1922).

— [2] Über lineare Gleichungssysteme in linearen Räumen. J. reine angew. Math. **157**, 214—229 (1926).

Halperin, I.: [1] Convex sets in linear topological spaces. Trans. Royal Soc. Canada **47**, Ser. III, 1—6 (1953).

— [2] Uniform convexity in function spaces. Duke Math. J. **21**, 195—204 (1954).

— [3] Reflexivity in the L^λ-function spaces. Duke Math. J. **21**, 205—208 (1954).

Hammer, P. C.: [1] Maximal convex sets. Duke Math. J. **22**, 103—106 (1955).

Hausdorff, F.: [1] Zur Theorie der linearen metrischen Räume. J. reine angew. Math. **167**, 294—311 (1932).

— [2] Set theory, 2nd Ed. New York: Chelsea Publishing Co. 1962.

Hellinger, E., u. O. Toeplitz: [1] Integralgleichungen und Gleichungen mit unendlich vielen Unbekannten. Enz. Math. Wiss. II C 13, 1335—1616 (1928).

Helly, E.: [1] Über lineare Funktionaloperationen. Sitzgsber. Akad. Wiss. Wien Math.-Nat. Kl. **121** IIa, 265—297 (1912).

— [2] Über Systeme linearer Gleichungen mit unendlich vielen Unbekannten. Mh. Math. Phys. **31**, 60—91 (1921).

Hermes, H.: [1] Einführung in die Verbandstheorie. Berlin-Göttingen-Heidelberg: Springer 1955.

Hildebrandt, T. E.: [1] On bounded linear functional operations. Trans. Amer. Math. Soc. **36**, 868—875 (1934).

Hille, E., and R. S. Phillips: [1] Functional analysis and semigroups. Amer. Math. Soc. Colloq. Publ. Vol. XXXI, New York 1957.

Hyers, D. H.: [1] Locally bounded linear topological spaces. Rev. Ci. Lima **41**, 555—574 (1939).

— [2] Linear topological spaces. Bull. Amer. Math. Soc. **51**, 1—21 (1945).

Ingleton, A. W.: [1] The Hahn-Banach theorem for non-Archimedean valued fields. Proc. Cambridge Phil. Soc. **48**, 41—45 (1952).

IYER, G.: [1] On the space of integral functions, I—III. J. Indian Math. Soc. (2) **12**, 13—30 (1948). — Quart. J. Math. (2) **1**, 86—96 (1950). — Proc. Amer. Math. Soc. **3**, 874—883 (1952).

JAMES, R. C.: [1] Bases and reflexivity of Banach spaces. Ann. Math. **52**, 518—527 (1950).

— [2] A non-reflexive Banach space isometric with its second conjugate space. Proc. Nat. Acad. Sci. USA **37**, 174—177 (1951).

— [3] Projections in the space (*m*). Proc. Nat. Acad. Sci. USA **41**, 899—902 (1955).

— [4] Reflexivity and the supremum of linear functionals. Ann. Math. **66**, 159—169 (1957).

— [5] Characterizations of reflexivity. Studia Math. **23**, 205—216 (1964).

KADISON, R. V.: [1] A representation theory for commutative topological algebra. Mem. Amer. Math. Soc. No. **7** (1951).

KAKUTANI, S.: [1] Ein Beweis des Satzes von M. Eidelheit über konvexe Mengen. Proc. Imp. Acad. Tokyo **13**, 93—94 (1937).

— [2] Weak convergence in uniformly convex spaces. Tôhoku Math. J. **45**, 188—193 (1938).

— [3] Weak topology and regularity of Banach spaces. Proc. Imp. Akad. Tokyo **15**, 169—173 (1939).

— [4] Some characterisations of Euclidean spaces. Jap. J. Math. **16**, 93—97 (1939).

— [5] Weak topology, bicompact set and the principle of duality. Proc. Imp. Acad. Tokyo **16**, 63—67 (1940).

KAMKE, E.: [1] Theory of sets. New York: Dover Publications Inc. 1950.

KAPLAN, S.: [1] Cartesian products of reals. Amer. J. Math. **74**, 936—954 (1952).

KELLEY, J. L.: [1] Note on a theorem of Krein and Milman. J. Osaka Inst. Sci. Tech., Part I, **3**, 1—2 (1951).

— [2] General topology. New York: D. van Nostrand 1955.

KLEE, V. L.: [1] Some characterisations of reflexivity. Rev. Ci. Lima **52**, 15—23 (1950).

— [2] Convex sets in linear spaces (I), II, III. Duke Math. J. **18**, 443—466, 875—883 (1951); **20**, 105—111 (1953).

— [3] Invariant metrics in groups (solution of a problem of Banach). Proc. Amer. Math. Soc. **3**, 484—487 (1952).

— [4] Separation properties of convex cones. Proc. Amer. Math. Soc. **6**, 313—318 (1955).

— [5] The structure of semi-spaces. Math. Scand. **4**, 54—56 (1956).

— [6] Iteration of the "lin" Operation for convex sets. Math. Scand. **4**, 231—238 (1956).

— [7] Strict separation of convex sets. Proc. Amer. Math. Soc. **7**, 735—737 (1956).

— [8] Extremal structure of convex sets. Arch. d. Math. **8**, 234—240 (1957).

— [9] Extremal structure of convex sets, II. Math. Z. **69**, 90—104 (1958).

— [10] Some new results on smoothness and rotundity in normed linear spaces. Math. Ann. **139**, 51—63 (1959).

KÖTHE, G.: [1] Die konvergenzfreien Räume abzählbarer Stufe. Math. Ann. **111**, 229—258 (1935).

— [2] Die Teilräume eines linearen Koordinatenraumes. Math. Ann. **114**, 99—125 (1937).

— [3] Lösbarkeitsbedingungen für Gleichungen mit unendlichvielen Unbekannten. J. reine angew. Math. **178**, 193—213 (1938).

— [4] Erweiterung von Linearfunktionen in linearen Räumen. Math. Ann. **116**, 719—732 (1939).

— [5] Die Qoutientenräume eines linearen vollkommenen Raumes. Math. Z. **51**, 17—35 (1947).

— [6] Die Stufenräume, eine einfache Klasse linearer vollkommener Räume. Math. Z. **51**, 317—345 (1948).

— [7] Eine axiomatische Kennzeichnung der linearen Räume vom Typus ω. Math. Ann. **120**, 634—649 (1949).

— [8] Über die Vollständigkeit einer Klasse lokalkonvexer Räume. Math. Z. **52**, 627—630 (1950).

— [9] Über zwei Sätze von Banach. Math. Z. **53**, 203—209 (1950).

— [10] Neubegründung der Theorie der vollkommenen Räume. Math. Nachr. **4**, 70—80 (1951).

— [11] Die Randverteilungen analytischer Funktionen. Math. Z. **57**, 13—33 (1952).

— [12] Dualität in der Funktionentheorie. J. reine angew. Math. **191**, 30—49 (1953).

— [13] Lineare Räume mit linearer Topologie. Proc. Int. Math. Congr. Amsterdam I, 236—237 (1954).

— [14] Bericht über neuere Entwicklungen in der Theorie der topologischen Vektorräume. Jber. DMV **59**, 19—36 (1956).

—, u. O. Toeplitz: [1] Theorie der halbfiniten unendlichen Matrizen. J. reine angew. Math. **165**, 116—127 (1931).

— [2] Lineare Räume mit unendlich vielen Koordinaten und Ringe unendlicher Matrizen. J. reine angew. Math. **171**, 193—226 (1934).

Kolmogoroff, A.: [1] Zur Normierbarkeit eines allgemeinen topologischen Raumes. Studia Math. **5**, 29—33 (1934).

Komatuzaki, H.: [1] Sur les projections dans certains espaces du type (B). Proc. Imp. Acad. Tokyo **16**, 274—279 (1940).

— [2] Une remarque sur les projections dans certains espaces du type (B). Proc. Imp. Acad. Tokyo **17**, 238—240 (1941).

Kōmura, Y.: [1] On linear topological spaces. Kumamoto J. Science, Series A, **5**, Nr. 3, 148—157 (1962).

— [2] Some examples on linear topological spaces. Math. Ann. **153**, 150—162 (1964).

Kōmura, T., and Y.: [1] Sur les espaces parfaits de suites et leurs généralisations. J. Math. Soc. Japan **15**, 319—338 (1963).

Krein, M.: [1] Sur quelques questions de la géometrie des ensembles convexes situés dans un espace linéaire normé et complet. Dokl. Akad. Nauk SSSR, N. S. **14**, 5—7 (1937).

—, and D. Milman: [1] On extreme points of regularly convex sets. Studia Math. **9**, 133—138 (1940).

—, and V. Šmulian: [1] On regularly convex sets in the space conjugate to a Banach space. Ann. Math. **41**, 556—583 (1940).

Landsberg, M.: [1] Pseudonormen in der Theorie der linearen topologischen Räume. Math. Nachr. **14**, 29—38 (1955).

— [2] Lineare topologische Räume, die nicht lokalkonvex sind. Math. Z. **65**, 104—112 (1956).

LaSalle, J. P.: [1] Pseudo-normed linear spaces. Duke Math. J. **8**, 131—135 (1941).

Leeuw, K. de, and W. Rudin: [1] Extreme points and extremum problems in H_1. Pacific J. **8**, 467—485 (1958).

Lefschetz, S.: [1] Algebraic topology. Amer. Math. Soc. Colloq. Publ. Vol. XXVII, New York 1942.

Leptin, H.: [1] Linearkompakte Moduln, I. Math. Z. **62**, 241—267 (1955). — II. Math. Z. **66**, 289—327 (1957).

Löwig, H.: [1] Komplexe Euklidische Räume von beliebiger endlicher oder unendlicher Dimensionszahl. Acta Sci. Math. Szeged **7**, 1—13 (1934).

LOOMIS, L. H.: [1] An introduction to abstract harmonic analysis. New York: D. van Nostrand 1953.

LORENTZ, G. G.: [1] Operations in linear metric spaces. Duke Math. J. **15**, 755—761 (1948).

LOVAGLIA, A. R.: [1] Locally uniformly convex Banach spaces. Trans. Amer. Math. Soc. **78**, 225—238 (1955).

MACKEY, G. W.: [1] On infinite dimensional linear spaces. Proc. Nat. Acad. Sci. USA **29**, 216—221 (1943).

— [2] On convex topological linear spaces. Proc. Nat. Acad. Sci. USA **29**, 315—319 (1943).

— [3] Equivalence of a problem in measure theory to a problem in the theory of vector lattices. Bull. Amer. Math. Soc. **50**, 719—722 (1944).

— [4] On infinite-dimensional linear spaces. Trans. Amer. Math. Soc. **57**, 155—207 (1945).

— [5] On convex topological linear spaces. Trans. Amer. Math. Soc. **60**, 519—537 (1946).

MATTHEWS, G.: [1] Generalised rings of infinite matrices. Nederl. Akad. Wet. Proc., Ser. A **61**, 298—306 (1958).

MAZUR, S.: [1] Über die kleinste konvexe Menge, die eine gegebene kompakte Menge enthält. Studia Math. **2**, 7—9 (1930).

— [2] Über konvexe Mengen in linearen normierten Räumen. Studia Math. **4**, 70—84 (1933).

— [3] Über schwache Konvergenz in den Räumen L^p. Studia Math. **4**, 128—133 (1933).

—, u. W. ORLICZ: [1] Über Folgen linearer Operationen. Studia Math. **4**, 152—157 (1933).

— [2] Sur les espaces métriques linéaires I, II. Studia Math. **10**, 184—208 (1948); **13**, 137—179 (1953).

MCSHANE, E. J.: [1] Linear functionals on certain Banach spaces. Proc. Amer. Math. Soc. **1**, 402—408 (1950).

MICHAEL, E. A.: [1] Locally multiplicatively-convex topological algebras. Mem. Amer. Math. Soc. No. **11** (1952).

MILMAN, D. P.: [1] On some criteria for the regularity of spaces of the type (B). Dokl. Akad. Nauk SSSR, N. S. **20**, 243—246 (1938).

— [2] Characteristics of extremal points of regularly convex spaces. Dokl. Akad. Nauk SSSR, N. S. **57**, 119—122 (1947) [Russian].

—, and M. A. RUTMAN: [1] On a more precise theorem about the completeness of the system of extreme points of a regularly convex set. Dokl. Akad. Nauk SSSR, N. S. **60**, 25—27 (1948) [Russian].

MONNA, A. F.: [1] Sur les espaces linéaires normés, I—VI. Nederl. Akad. Wet. Proc. **49**, 1045—1055, 1056—1062, 1134—1141, 1142—1152 (1946); **51**, 197—210 (1948); **52**, 151—160 (1949).

— [2] Espaces linéaires a une infinité dénombrable de coordonnées. Nederl. Akad. Wet. Proc. **53**, 1548—1559 (1950).

MURRAY, F. J.: [1] On complementary manifolds and projections in spaces L_p and l_p. Trans. Amer. Math. Soc. **41**, 138—152 (1937).

— [2] Quasi-complements and closed projections in reflexive Banach spaces. Trans. Amer. Math. Soc. **58**, 77—95 (1945).

NACHBIN, L.: [1] Espaços vetoriais topologicos. Rio de Janeiro: Livraria Boffoni 1948.

— [2] A characterisation of the normed vector ordered spaces of continuous functions over a compact space. Amer. J. Math. **71**, 701—705 (1949).

— [3] A theorem of the Hahn-Banach type for linear transformations. Trans. Amer. Math. Soc. **68**, 28—46 (1950).

— [4] Topological vector spaces of continuous functions. Proc. Nat. Acad. Sci. USA **40**, 471—474 (1954).

NAIMARK, M. A.: [1] Normed rings. Groningen: P. Noordhoff N. V. 1959.

NAKANO, H.: [1] Topology and linear topological spaces. Tokyo: Maruzen Co. 1951.

NATANSON, I. P.: [1] Theory of functions of a real variable. 2 Vols. New York: Frederick Ungar Publishing Co. 1955, 1960.

NEUMANN, J. VON: [1] On complete topological linear spaces. Trans. Amer. Math. Soc. **37**, 1—20 (1935).

NIKODÝM, O. M.: [1] On transfinite iterations of the weak linear closure of convex sets in linear spaces, A. B. Rend. Circ. Mat. Palermo, II. ser. **2**, 85—105 (1953); **3**, 5—75 (1954).

ORNSTEIN, D.: [1] Dual vector spaces. Ann. Math. **69**, 520—534 (1959).

PETTIS, B. J.: [1] A note on regular Banach spaces. Bull. Amer. Math. Soc. **44**, 420—428 (1938).

— [2] A proof that every uniformly convex space is reflexive. Duke Math. J. **5**, 249—253 (1939).

PHILLIPS, R. S.: [1] On linear transformations. Trans. Amer. Math. Soc. **48**, 516—541 (1940).

— [2] On weakly compact subsets of a Banach space. Amer. J. Math. **65**, 108—136 (1943).

PIETSCH, A.: [1] Verallgemeinerte vollkommene Folgenräume. Schriftenreihe d. Institute d. Math. Deutsche Akad. Wiss. Heft 12, Akademie-Verlag Berlin (1962).

PTÁK, V.: [1] On complete topological linear spaces. Čehosl. Mat. Z. **3 (78)**, 301—364 (1953) [Russian].

— [2] Weak compactness in convex topological linear space. Čehosl. Mat. Ž. **4**, **(79)**, 175—186 (1954).

— [3] On a theorem of W. F. Eberlein. Studia Math. **14**, 276—284 (1954).

— [4] Completeness and the open mapping theorem. Bull. Soc. math. France **86**, 41—74 (1958).

— [5] A combinatorial theorem on systems of inequalities and its application to analysis. Čehosl. Mat. Ž. **9 (84)**, 629—630 (1959).

— [6] A combinatorial lemma on the existence of convex means and its application to weak compactness. Proc. Symp. Pure Math. **VII**, 437—450 (1963).

RAÍKOW, D. A.: [1] Wpolne neprerywnyje spektry lokalno wypuklych prostranstw. Trudy Mockowskogo Mat. Obschestwa **7**, 413—438 (1958).

RIESZ, F.: [1] Sur les opérations fonctionnelles linéaires. C. R. Acad. Sci. Paris **149**, 974—977 (1909).

— [2] Les systèmes d'équations linéaires à une infinité d'inconnues. Paris: Gauthier-Villars 1913.

— [3] Über lineare Funktionalgleichungen. Acta Math. **41**, 71—98 (1918).

—, and B. SZ. NAGY: Functional Analysis. London and Glasgow: Blackie & Son 1956.

RITZDORFF, F.: [1] Das Trägheitsgesetz der quadratischen Formen mit halbfiniter Koeffizientenmatrix. Math. Z. **44**, 23—54 (1938).

ROBERTS, G. T.: [1] The bounded-weak topology and completeness in vector spaces. Proc. Cambridge Phil. Soc. **49**, 183—189 (1953).

ROBERTSON, W.: [1] Contributions to the general theory of linear topological spaces. Thesis, Cambridge 1954.

— [2] Completions of topological vector spaces. Proc. London Math. Soc., III. ser. **8**, 242—257 (1958).

Rolewicz, S.: [1] On a certain class of linear metric spaces. Bull. Acad. Polon. Sci., Cl. III **5**, 471—473 (1957).

Ruston, F.: [1] A note on convexity in Banach spaces. Proc. Cambridge Phil. Soc. **45**, 157—159 (1949).

— [2] A short proof of a theorem on reflexive spaces. Proc. Cambridge Phil. Soc. **45**, 674 (1949).

— [3] Conjugate Banach spaces. Proc. Cambridge Phil. Soc. **53**, 576—580 (1957).

Schatten, R.: [1] The space of completely continuous operators on a Hilbert space. Math. Ann. **134**, 47—49 (1957).

Schauder, J.: [1] Zur Theorie stetiger Abbildungen in Funktionalräumen. Math. Z. **26**, 47—65, 417—431 (1927).

— [2] Über die Umkehrung linearer stetiger Funktionaloperationen. Studia Math. **2**, 1—6 (1930).

Schubert, H.: [1] Topology. London: MacDonald. 1968.

Schwartz, L.: [1] Théorie des distributions. New Ed. Paris: Hermann & Co. 1966.

— [2] Espaces de fonctions différentiables à valeurs vectorielles. J. Analyse Math. **4**, 88—148 (1955).

Sebastião e Silva, J.: [1] As funções análiticas e a análise funcional. Port. Math. **9**, 1—130 (1950).

— [2] Sobre a topologia dos espaços funcionais analíticos. Rev. Fac. Cienc. Lisboa, II. ser. A **1**, 23—102 (1950).

— [3] Sui fundamenti della teoria dei funzionali analitici. Port. Math. **12**, 1—46 (1953).

— [4] Su certe classi di spazi localmente convessi importanti per le applicazioni. Rend. Mat. Roma, V. ser. **14**, 388—410 (1955).

Shirota, T.: [1] On locally convex vector spaces of continuous functions. Proc. Jap. Acad. **30**, 294—298 (1954).

Sierpiński, W.: [1] Sur les ensembles complets d'un espace (D). Fund. Math. **11**, 203—205 (1928).

Silva Dias, C. L. da: [1] Espaços vectoriais topológicos e sua aplicação nos espaços funcionais analíticos. Bol. Soc. Mat. Sao Paulo **5**, 1—58 (1950).

Šmulian, V.: [1] Sur les ensembles régulièrement fermés et faiblement compacts dans l'espace du type (B). C. R. URSS **18**, 405—407 (1938).

— [2] On the principle of inclusion in the space of type (B). Mat. Sbornik, N. S. **5**, 327—328 (1939) [Russian].

— [3] On some geometrical properties of the unit sphere in the space of the type (B). Mat. Sbornik, N. S. **6**, 77—94 (1939) [Russian].

— [4] Sur la dérivabilité de la norme dans l'espace de Banach. Dokl. Akad. Nauk SSSR, N. S. **27**, 643—648 (1940).

— [5] Über lineare topologische Räume. Mat. Sbornik, N. S. **7**, 425—448 (1940).

— [6] Sur la structure de la sphère unitaire dans l'espace de Banach. Mat. Sbornik, N. S. **9**, 545—561 (1941).

— [7] Sur les espaces linéaires topologiques, II. Mat. Sbornik, N. S. **9**, 727—730 (1941).

— [8] Sur les ensembles compacts et faiblement compacts dans l'espace du type (B). Mat. Sbornik, N. S. **12**, 91—95 (1943).

Sobczyk, A.: [1] Projections in Minkowski and Banach spaces. Duke math. J. **8**, 78—106 (1941).

— [2] Projections of the space (m) on its subspace (c_0). Bull. Amer. Math. Soc. **47**, 938—947 (1941).

— [3] On the extension of linear transformations. Trans. Amer. Math. Soc. **55**, 153—169 (1944).

Straszewicz, S.: [1] Über exponierte Punkte abgeschlossener Punktmengen. Fund. Math. **24**, 139—143 (1935).

Suchomlinoff, G. A.: [1] Über Fortsetzung von linearen Funktionalen in linearen komplexen Räumen und linearen Quotientenräumen. Mat. Sbornik, N. S. **3**, 353—358 (1938) [Russian].

Takenouchi, O.: [1] Sur les espaces linéaires localement convexes. Math. J. Okayama Univ. **2**, 57—84 (1952).

Taylor, A. E.: [1] Banach spaces of functions analytic in the unit circle. Studia Math. **11**, 145—170 (1950); **12**, 25—50 (1951).

— [2] Introduction to Functional Analysis. New York. John Wiley & Sons 1958.

Tillmann, H. G.: [1] Randverteilungen analytischer Funktionen und Distributionen. Math. Z. **59**, 61—83 (1953).

— [2] Dualität in der Potentialtheorie. Port. Math. **13**, 55—86 (1954).

— [3] Dualität in der Funktionentheorie auf Riemannschen Flächen. J. reine angew. Math. **195**, 76—101 (1956).

— [4] Die Fortsetzung analytischer Funktionale. Hamb. Abh. **21**, 139—193 (1957).

Toeplitz, O.: [1] Über die Auflösung unendlich vieler linearer Gleichungen mit unendlich vielen Unbekannten. Rend. Circ. Mat. Palermo **28**, 88—96 (1909).

— [2] Die linearen vollkommenen Räume der Funktionentheorie. Comment. Math. Helv. **23**, 222—242 (1949).

Tukey, J. W.: [1] Some notes on the separation of convex sets. Port. Math. **3**, 95—102 (1942).

Tychonoff, A.: [1] Ein Fixpunktsatz. Math. Ann. **111**, 767—776 (1935).

Ulam, S.: [1] Zur Maßtheorie in der allgemeinen Mengenlehre. Fund. Math. **16**, 140—150 (1930).

Ulm, H.: [1] Elementarteilertheorie unendlicher Matrizen. Math. Ann. **114**, 493—505 (1937).

Vilenkin, N. Y.: [1] Vector spaces over topological fields. Mat. Sbornik **74**, 195—208 (1953) [Russian].

Warner, S., and A. Blair: [1] On symmetry in convex topological vector spaces. Proc. Amer. Math. Soc. **6**, 301—304 (1955).

Wehausen, J. V.: [1] Transformations in linear topological spaces. Duke math. J. **4**, 157—169 (1938).

Weil, A.: [1] Sur les espaces à structure uniforme et sur la topologie générale. Act. Sci. et Ind. No. **551** (1938).

— [2] L'intégration dans les groupes topologiques et ses applications. Act. Sci. et Ind. No. **869** (1940).

Yosida, K., and E. Hewitt: [1] Finitely additive measures. Trans. Amer. Math. Soc. **72**, 46—66 (1952).

Zaanen, A. C.: [1] Linear Analysis. Groningen: P. Noordhoff 1953.

Zelinsky, D.: [1] Linearly compact modules and rings. Amer. J. Math. **75**, 79—90 (1953).

Zeller, K.: [1] FK-Räume in der Funktionentheorie I. Math. Z. **58**, 288—305 (1953). — II. Math. Z. **58**, 414—435 (1953).

— [2] Theorie der Limitierungsverfahren. Erg. d. Math., N. F., H. 15. Berlin-Göttingen-Heidelberg: Springer 1958.

Zorn, M.: [1] A remark on method in transfinite algebra. Bull. Amer. Math. Soc. **41**, 667—670 (1935).

Author and Subject Index

GPSR Compliance
The European Union's (EU) General Product Safety Regulation (GPSR) is a set of rules that requires consumer products to be safe and our obligations to ensure this.

If you have any concerns about our products, you can contact us on

ProductSafety@springernature.com

In case Publisher is established outside the EU, the EU authorized representative is:

Springer Nature Customer Service Center GmbH
Europaplatz 3
69115 Heidelberg, Germany

www.ingramcontent.com/pod-product-compliance
Lightning Source LLC
Chambersburg PA
CBHW070708260425
25758CB00003B/184
* 9 7 8 3 6 4 2 6 4 9 9 0 5 *